# Gentzen's Centenary

Reinhard Kahle • Michael Rathjen
Editors

# Gentzen's Centenary

## The Quest for Consistency

Springer

*Editors*
Reinhard Kahle
Departamento de Matemática, FCT
Universidade Nova de Lisboa
Lisbon
Portugal

Michael Rathjen
School of Mathematics
University of Leeds
Leeds
United Kingdom

ISBN 978-3-319-10102-6 ISBN 978-3-319-10103-3 (eBook)
DOI 10.1007/978-3-319-10103-3
Springer Cham Heidelberg New York Dordrecht London

Library of Congress Control Number: 2015953429

Mathematics Subject Classification (2010): 03F03, 03F05, 03F15, 03F25, 03F30, 03F35, 03-03, 01A60, 03A05



Springer is part of Springer Science+Business Media (www.springer.com)

# Preface

With kind permission of © Eckart Menzler

This volume is a tribute by several generations of proof theorists to Gerhard Gentzen, one of the greatest logicians ever to whom we owe the most profound investigation of the nature of proofs since Aristotle and Frege. The immediate stimulus for its inception was Gentzen's 100th birthday in 2009 which was celebrated with a conference in Leeds and a workshop in Coimbra at which most of the contributors to this volume spoke.

Gentzen has been described as logic's lost genius[1] whom Gödel sometimes called a better logician than himself.[2] It could be said that Gentzen and Gödel arrived, each in their own exquisite manner, at opposing extremes of a spectrum. Gödel found a very general negative result to the effect that no system embodying a correct

[1]E. Menzler-Trott: *Logic's Lost Genius: The Life of Gerhard Gentzen* (AMS, Providence, 2007).

[2]G. Kreisel: Gödel's excursions into intuitionistic logic, in: *Gödel remembered*, (Bibliopolis, Napoli, 1987) p. 169.

amount of number theory can prove its own consistency by transferring the trick of the "Liar's Paradox" from the context of truth to that of provability. Gentzen, on the other hand, established the positive result that elementary number theory is consistent, using at some crucial point the well-orderedness of a certain ordering called $\varepsilon_0$ that sprang from Cantor's normal form (for presenting ordinals). He also gave a direct proof that the latter principle is not deducible in this theory, thereby providing an entirely new proof of a mathematical incompleteness in number theory.

Gentzen can be rightly considered to be the founding father of modern proof theory. His sequent calculus and natural deduction system beautifully explain the deep symmetries of logic. They underlie modern developments in computer science such as automated theorem proving and type theory. This volume's chapters by leading proof-theorists attest to Gentzen's enduring legacy in mathematical logic and beyond. Their contributions range from philosophical reflections and re-evaluations of Gentzen's original consistency proofs and results in proof theory to some of the most recent developments in this exciting area of modern mathematical logic.

**Acknowledgements** The plan for the present book evolved at the *Leeds Symposium on Proof Theory and Constructivism*[3] in 2009 which lasted from 4th July to 15th July. The symposium consisted of three connected events one of which was the *Gentzen Centenary Conference*. We would like to thank the Kurt Gödel Society and the Deutsche Vereinigung für Mathematische Logik und für Grundlagenforschung der Exakten Wissenschaften (DVMLG) for providing funding for the Gentzen Centenary Conference. This conference was followed by a Workshop associated with the Annual Meeting of the European Association of Computer Science Logic, *CSL 2009*, with the title *Gentzen Centenary—The Quest for Consistency*[4] which took place on 12th September in Coimbra, Portugal. This workshop also received support from the Kurt Gödel Society for which we are very grateful, as well as for the support of the Portuguese Science Foundation FCT, which funded the Workshop and the edition of this volume through the projects, *Dialogical Foundation of Semantics* (LOGICCC/0001/2007), within the ESF programme LogICCC, *Hilbert's Legacy in the Philosophy of Mathematics* (PTDC/FIL-FCI/109991/2009), and *The Notion of Mathematical Proof* (PTDC/MHC-FIL/5363/2012).

The editors are especially indebted to Matthias Baaz for his enthusiastic support of both events, for stimulating the book project and for contributing towards giving the present volume its shape by providing information, encouragement, culinary highlights and counsel. Finally, let us thank the referees of the papers collected in this volume for their valuable help.

Lisbon, Portugal — Reinhard Kahle
Leeds, UK — Michael Rathjen

[3] http://www.personal.leeds.ac.uk/~matptw/.

[4] http://www.mat.uc.pt/~kahle/gentzen/.

# In Memoriam: Grigori Mints, 1939–2014

With kind permission of his wife © Marianna Rozenfeld

When this book was about to be sent to the publisher, we received the very sad news that Grigori ("Grisha") Mints had died on 30th May 2014. He was born on 7th June 1939 in Leningrad (now again St. Petersburg).

Grisha was a driving force in proof theory and constructivism and a loyal promoter of Gentzen-style proof theory. He was the pre-eminent expert on Hilbert's epsilon calculus and the leading exponent of the substitution method approach to proof theory, expanding its range of applications to strong subsystems of arithmetic. His discovery of the method of continuous cut elimination for infinitary proofs unearthed the deeper relationship between Gentzen's reduction steps on finitary derivations and infinitary proof theory. In pursuit of his wide ranging research interests, he published three books, ten edited volumes, more than 200 scholarly papers, and thousands of reviews, with the aid of which he also maintained and fostered his world spanning network of intellectual contacts through sometimes difficult years working in the Soviet Union. Vladimir Lifschitz wrote about Grisha[1]:

> ... his true calling was to study formal proofs in the spirit of pure mathematics in the best sense of the word: the main project of Grisha's professional life was to develop a clear, complete understanding of properties of proofs, so that any possible question about them will be easy to answer.

[1] https://philosophy.stanford.edu/news/professor-grigori-grisha-mints.

In this way he can be seen as one of the leading executors of Gentzen's legacy and it seems to be more than adequate to dedicate this volume, celebrating Gerhard Gentzen's centenary, to the memory of Grisha Mints.

# Contents

## Part III Results



## Part IV Developments

# Part I
# Reflections

# Gentzen's Consistency Proof in Context

**Reinhard Kahle**

## 1 Introduction

Gentzen's celebrated consistency proof—or proofs, to distinguish the different variations he gave[1]—of Peano Arithmetic in terms of transfinite induction up to the ordinal[2] $\varepsilon_0$ can be considered as the birth of modern proof theory. After the blow which Gödel's incompleteness theorems gave the original Hilbert Programme, Gentzen's result did not just provide *a* consistency proof of formalized Arithmetic, it also opened a new way to deal "positively" with incompleteness phenomena.[3] In addition, Gentzen invented, on the way to his result, *structural proof theory*, understood as the branch of proof theory studying structural (in contrast to mathematical) properties of formal systems [79, 111]. With the introduction of sequent calculus and natural deduction and the corresponding theorems about cut elimination and normalization, respectively,[4] he revolutionized the concept of derivation calculus, fundamental for all further developments of proof theory.

Here, we focus on the aspects of his work related to the quest for consistency proofs of theories with mathematical content. We like to recall the context in which the consistency proofs—one may add: "after Gödel"—have to be put, and what might be their mathematical and/or philosophical rationale. For it, we will look

---

[1] Cf., e.g., [13, 87, 105], and [114] as well as [97] in this volume.

[2] For the ordinal $\varepsilon_0$ see, for instance, [58] in this volume.

[3] See, for instance, [90] in this volume.

[4] See, for instance, [15] and [87] in this volume.

R. Kahle (✉)
CMA, CENTRIA, and DM, FCT, Universidade Nova de Lisboa, 2829-516 Caparica, Portugal
e-mail: kahle@mat.uc.pt



R. Kahle, M. Rathjen (eds.), *Gentzen's Centenary*,
DOI 10.1007/978-3-319-10103-3_1

back to Hilbert's (original) programme and the immediate lessons one may learn from Gödel's theorems. We then consider consistency proofs for Arithmetic, whose consistency, however, is not really at issue. After discussing the interesting case of Analysis, we finish with a reflection on modern proof theory as it is guided by the quest for consistency in the investigation of stronger and stronger mathematical theories.

## 2 Hilbert's Programme

Hilbert's programme originates from his own second problem in the famous Paris problem list [45] and, in its mature form, it proposes to carry out consistency proofs of axioms systems for Arithmetic and Analysis "by finitistic methods." Hilbert didn't specify exactly what he meant by "finitistic methods" and in modern formal presentations one identifies these methods—following Tait [104]—with *primitive-recursive Arithmetic*, PRA. From an abstract point of view, the main issue is that the consistency of the base theory, in which the consistency proof should be carried out, is beyond any reasonable doubt; and this should be the case for the finitistic methods, whatever they are concretely.

The idea of Hilbert's programme was somehow already conceived with the question given in 1900, and a first sketch of how a consistency proof could be performed was given by Hilbert in 1904 in his lecture at the International Congress of Mathematicians in Heidelberg [47]. It was, however, only the appearance of Brouwer's *Intuitionism* which forced Hilbert to formulate his programme in precise formal terms.[5] Finitistic Mathematics should play, in this context, the role of the part of Mathematics which is beyond any doubt concerning consistency. It was then the aim to justify the other parts of Mathematics by formal consistency proofs carried out using only finitistic means. It is worth noting that, with the choice of finitistic Mathematics as the base, Hilbert was fully in line with the intuitionistic movement—even on philosophical grounds, and it should not come as a surprise that he himself was occasionally called an intuitionist.[6] One can even find a

[5]For the development of Hilbert's programme(s), cf. e.g., [98].

[6]See Fraenkel [28, p. 154]:

> This is the point of view of HILBERT, who, therefore, picks up himself the methodical starting point of his intuitionist opponents—but for the purpose to deny their thesis; one could almost characterize him as an intuitionist.

(German original: "Dies etwa ist der Standpunkt HILBERTS, der somit den methodischen Ausgangspunkt seiner intuitionistischen Gegner — allerdings zum Zweck der Bestreitung ihrer Thesen — selbst aufnimmt; man könnte ihn geradezu als Intuitionisten bezeichnen.") Van Dalen adds to this citation [112, p. 309]: "Although the inner circle of experts in the area (e.g. Bernays, Weyl, von Neumann, Brouwer) had reached the same conclusion some time before, it was Fraenkel who put it on record." See also footnote 18.

"intuitionistic creed" given by Gentzen in 1938, when he wrote [102, p. 235][7]:

> The most consequential form of delimitation is that represented by the '*intuitionistic*' point of view, ...

What separated Hilbert from Brouwer and Weyl was the latter's attitude to ban "the other mathematics" from the mathematical discourse. In contrast, he was proposing to justify by his *Beweistheorie* Mathematics in all its extensions on the base of finitistic Mathematics. Here, Hilbert's programme gained a new aspect: besides consistency, one could now also demand *conservativity* of "higher" Mathematics over finitary Mathematics.[8]

Without any doubt, Gödel's second incompleteness theorem put an end to Hilbert's programme in its original formulation.[9] The so-called failure of Hilbert's Programme is advocated at several places, maybe most notable by Kreisel [66, Abstract and p. 352]. But which kind of "failure" was it? Surely, it was the not the one which was feared by the critics of classical mathematics. When Hermann Weyl drew on the picture of a "house built on sand" [118, p. 1] he was afraid of possible inconsistencies which could bring classical mathematics to collapse. Of course, Gödel's theorems suggest on no account that there would be an inconsistency in classical mathematics (or even Arithmetic).[10]

As far as consistency is concerned, one may compare the situation with the classical construction problems in Euclidean Geometry. There is no way to trisect an angle by compass and ruler—but there are other means to do so (for instance, using a marked ruler). Of course, in the context of a consistency proof, using other means than finitistic ones will undermine Hilbert's original philosophical starting point. But Hilbert was, by no means, a philosophical hardliner. The only piece of written evidence which we have about Hilbert's reception of Gödel's result is the cryptic short preface in the first volume of the *Grundlagen der Mathematik* [52], saying that Gödel's result "shows only that—for more advanced consistency proofs—the finitistic standpoint has to be exploited in a manner that is sharper [...],"[11] i.e., the philosophical starting point was to change. Bernays and Ackermann provide us with two additional testimonies that Hilbert soon adapted his "meta-mathematical standpoint."

---

[7]German original [32, p. 6]: "Die folgerichtigste Art der Abgrenzung ist die durch den '*intuitionistischen*' Standpunkt [...] gegebene."

[8]We may leave it open here whether Hilbert himself was advocating such a conservativity. The issue of conservativity can be considered, of course, without reference to historic figures.

[9]It is reported in the Schütte school that this was also immediately recognized in Göttingen.

[10]But one may note the puzzling lack of understanding of Russell, expressed in a letter to Leon Henkin of 1 April 1963, cf. [18, p. 89ff].

[11]Hilbert and Bernays [55, p. VII]. German original: "Jenes Ergebnis zeigt in der Tat auch nur, daß man für die weitergehenden Widerspruchsfreiheitsbeweise den finiten Standpunkt in einer schäferen Weise ausnutzen muß, [...]."

Based on Bernays's reports, Reid writes about Hilbert's reaction to Gödel's result [92, p. 198]: "At first he was only angry and frustrated, but then he began to try to deal constructively with the problem. Bernays found himself impressed that even now, at the very end of his career, Hilbert was able to make great changes in his program."

Ackermann writes in a letter to Hilbert (August 23rd, 1933)[12]: "I was particularly interested in the new meta-mathematical standpoint which you now adopt and which was provoked by Gödel's work."

Unfortunately, we have no sources which explicate in detail Hilbert's new standpoint, but it goes without saying that Gentzen's work was in line with it.[13] In fact, Bernays starts the section heading of the presentation of Gentzen's proof of the consistency of Arithmetic in [53, Sect. 5.3] with "Transgression of the previous methodological standpoint of proof theory."[14]

Thus, with a more "liberal" philosophical position consistency proofs can still be carried out, addressing Hilbert's initial concerns. And Gentzen's consistency proof was among the first ones which provided such an argument. It was not even the only one, and Gödel gave, as early as 1938, in a talk at *Zilsel's seminar* in Vienna, an interesting overview of possible alternatives to extend Hilbert's original standpoint [38, p. 95][15]:

4. *How then shall we extend?* (Extension is necessary.) Three ways are known up to now:
    1. Higher types of functions (functions of functions of number, etc.)
    2. The modal-logical route (introduction of an absurdity applied to universal sentences and a ⟦notion of⟧ "consequence").
    3. Transfinite induction, that is, inference by induction is added for certain concretely defined ordinal numbers of the second number class.

Gödel himself preferred the first alternative, worked out in [39]; he judged the second one, which is intuitionistic logic of Brouwer and Heyting augmented by a modal-like operator $B$ (for German *beweisbar*), "the worst of the three ways" [38,

---

[12] German original [1, p.1f]: "Besonders interessiert hat mich der neue meta-mathematische Standpunkt, den Sie jetzt einnehmen und der durch die Gödelsche Arbeit veranlaßt worden ist." The letter was written after Ackermann visited Göttingen, but didn't meet Hilbert and spoke only with Arnold Schmidt, who informed him about "everything" going on in Göttingen.

[13] Detlefsen, [19] in this volume, however, points out that there are some fundamental differences between Gentzen's own philosophical view and Hilbert's view.

[14] In German: "Überschreitung des bisherigen methodischen Standpunkts der Beweistheorie".

[15] German original, [38, p. 94]:

4. *Wie also erweitern?* (Erweiterung nötig.) Drei Wege ⟦sind⟧ bisher bekannt:
    1. Höhere Typen von Funktionen (Funktionen ⟦von⟧ Funktionen von Zahlen, etc.)
    2. Modalitätslogischer Weg (Einführung einer Absurdität auf Allsätze angewendet und eines "Folgerns").
    3. Transfinite Induktion, d.h., es wird der Schluß durch Induktion für gewisse konkret definierte Ordinalzahlen der zweiten Klasse hinzugefügt.

p. 103]; the third one is, of course, Gentzen's way; for a detailed discussion of (this passage from) Gödel talk at Zilsel's seminar, see [24, p. 120f]. Of course, we don't depend on Gödel's choice; what counts is that *there are* extensions of Hilbert's original standpoint which provide a rationale for modern consistency proofs.

With respect to the second aspect of Hilbert's Programme—the supposed conservativity of higher Mathematics over finitistic Mathematics—the "failure" cannot be denied: there is no way to reduce all higher Mathematics to finitistic Mathematics; even more: higher Mathematics may prove finitistic statements which are not provable with pure finitistic methods.[16] But let's draw on a comparison here: nobody will deny that Columbus failed to find the sea route to India; but he didn't sink in the Ocean, he discovered America. In the same way, Hilbert's Programme, aiming for consistency and (maybe) conservativity, didn't sink in inconsistency, but discovered *Non-Conservativity*. Exploring this new phenomena in Mathematics is the driving force of modern proof theory.

## 3 Consistency Proofs for Arithmetic

Any consistency proof has to rely on some undisputed base. This was clearly stated by Gentzen, for instance in [31, Sect. 2.31][17]:

> Such a consistency proof is once again a *mathematical proof* in which certain inferences and derived concepts must be used. Their reliability (especially their consistency) must already be *presupposed. There can be no 'absolute consistency proof'*. A consistency proof can merely *reduce* the correctness of certain forms of inference to the correctness of other forms of inference.

---

[16]See, for instance, [75] in this volume.

[17]German original: "Ein solcher Widerspruchsfreiheitsbeweis wäre nun wieder ein *mathematischer Beweis*, in dem gewisse Schlüsse und Begriffsbildungen verwendet würden. Diese müssen als sicher (insbesondere als widerspruchsfrei) bereits *vorausgesetzt* werden. Ein '*absoluter Widerspruchsfreiheitsbeweis*' ist also *nicht möglich*. Ein Widerpruchsfreiheitsbeweis kann lediglich die Richtigkeit gewisser Schlußweisen auf die Richtigkeit anderer Schlußwiesen *zurückführen*. Man wird also verlangen müssen, daß in einem Widerspruchsfreiheitsbeweis nur solche Schlußweisen der Theorie, deren Widerspruchsfreiheit man beweist, als erheblich *sicherer* gelten können."

Similarly in [32]:

> In order to carry out a consistency proof, we naturally already require certain techniques of proof whose reliability must be *presupposed* and can no longer be justified along these lines. An absolute consistency proof, i.e., a proof which is free from presuppositions is of course impossible. [102, p. 237].

German original: "Um einen Widerspruchsfreiheitsbeweis zu führen, braucht man natürlich bereits gewisse mathematische Beweismittel, deren Unbedenklichkeit man *voraussetzen* muß und auf diesem Wege schließlich nicht weiter begründen kann. Ein absoluter, d. h. voraussetzungsloser Widerspruchsfreiheitsbeweis ist selbstverständlich unmöglich."

> It is therefore clear that in a consistency proof we can use only forms of inference that count as considerably *more secure* than the forms of inference of the theory whose consistency is to be proven. [102, p. 138]

Hilbert's original choice for such a base was finitistic Mathematics, and at that time, this was even identified—by name—with intuitionistic Mathematics in the Hilbert school.[18] Now, taking Heyting's intuitionistic formalization of Arithmetic as undisputed base, there was already a consistency proof of classical Arithmetic given by the double negation interpretation, independently found by Gödel [37] and Gentzen [33][19], and even earlier by Kolmogorov [65]. In his paper Gentzen expressed explicitly, [102, Sect. 6.1, p. 66f]:[20]

> If intuitionistic arithmetic is accepted as consistent, then the consistency of classical arithmetic is also guaranteed ...

But Gentzen was not happy with this kind of consistency proof (cf. the neat discussion in [102, p. 10f]), and went on to give his celebrated consistency proof in terms of transfinite induction up to $\varepsilon_0$. This proof starts from a different base, i.e., primitive recursive arithmetic together with transfinite induction up to $\varepsilon_0$.

Here, we dispense with a presentation of Gentzen's result which can be found, if not in Gentzen's original papers, in the standard proof-theoretic literature.[21] Hilbert, of course, was excited about the proof. But Kreisel [68, p. 121] reports also of "familiar jokes (for example, by Tarski whose confidence [in the consistency] was increased by $<\varepsilon$, or by Weyl who was astonished that one should use $\varepsilon_0$-induction to prove the consistency of ordinary, that is $\omega$-induction)."[22]

Tarski's "joke" (or a variation of it) is referred in detail in [102, p. 10]: "Gentzen's proof of the consistency of arithmetic is undoubtedly a very interesting metamathematical result, which may prove very stimulating and fruitful. I cannot say, however, that the consistency of arithmetic is now much more evident to me (at any rate, perhaps, to use the terminology of the differential calculus more evident than by an epsilon) than it was before the proof was given" [109, p. 19]. However, for a "semanticist" like Tarski there cannot be any doubt about the consistency of

[18]"Concerning the use of the word *intuitionistic* [...], it should be noted that according to Bernays [[11, p. 502]], the prevailing view in the Hilbert school at the beginning of the 1930s equated finitism with intuitionism." [24, p. 117]. See also footnote 6 above.

[19]This paper was submitted in 1933, but withdrawn by Gentzen when he became known about Gödel's paper. An English translation appeared in print in 1969, [102, #2], the German version of the Galley proofs, kept by Paul Bernays, was published only in 1974.

[20]German original [33, p. 131]: "Wenn man die intuitionistische Arithmetik als widerspruchsfrei hinnimmt, so ist [...] auch die Widerspruchsfreiheit der klassischen Arithmetik gesichert."

[21]An informal presentation of the main idea of the proof is given, for instance, by Takeuti in [120, p. 128ff].

[22]A well-known proof theorists presumably *heard* the second joke from Kreisel but confused a "y" with an "i" attributing it—with reference to Kreisel—to "un grand mathématicien français" [35, p. 520, fn. 14]; this confusion is confirmed in [36, pp. 9 and 33] where André Weil is mentioned by name (without reference to Kreisel).

Arithmetic from the very onset—otherwise, even the idea of the *structure of the natural numbers* would be pointless. We mention this, because on the assumption of the existence of a structure, any correctness lemma results in a consistency proof.[23]

Hermann Weyl's joke is equally unfair, as it suppresses the whole issue of Gentzen's proof, i.e., that the induction up to $\varepsilon_0$ is applied to quantifier-free formulas, only.[24]

Universal quantification—which was eliminated by Gentzen in the induction schemata—was at the very bottom of Hilbert's concerns, much more than, for instance, the tertium-non-datur. Hilbert's early outline of a consistency proof in the 1904 Heidelberg talk [47] was criticized by Poincaré with the argument that, for any such consistency proof, Hilbert would have to reason inductively[25]; but justifying induction by induction results in a vicious circle. Only with the separation of Metamathematics—using "weak" induction—from Mathematics proper—allowing for stronger induction—he developed a tool to respond to this critics.[26] Thus, Gentzen's use of quantifier-free inductions, though being transfinite, is fundamentally in line with Hilbert's concern to address Poincaré's objection.[27]

Ackermann gave, shortly after Gentzen, a consistency proof for Arithmetic using Hilbert's *$\varepsilon$-substitution method*, cf. [2], and its discussion in [53, Sect. 2] and [54, Supplement V].[28] From a historic point of view, it is probably more an adaptation of Gentzen's proof to a specific technique favored by Hilbert than a "new" consistency

---

[23]Smullyan [100, p. 56] illustrates very well this point in connection with Gödel's (first) incompleteness result, stressing that Gödel, by using $\omega$-consistency, makes a much weaker assumption than correctness. The pointlessness of consisteny proofs by semantic methods was well stated by Shoenfield [96, p. 214]:

> The consistency proof for $P$ by means of the standard model [...] does not even increase our understanding of $P$, since nothing goes into it which we did not put into $P$ in the first place.

[24]For sure, Weyl will have known exactly what's going on here, and probably also classified his remark only as a *joke*.

[25]See [84], cited in [98, p. 7].

[26]See, for instance, [10, p. 203]. This separation might have been suggested by Brouwer to Hilbert in 1909, cf. [112, p. 302]. Sieg [98, p. 27] writes: "Hilbert claims in [[50]], that Poincaré arrived at 'his mistaken conviction by not distinguishing these two methods of induction, which are of entirely different kinds' and feels that '[u]nder these circumstances Poincaré had to reject my theory, which, incidentally, existed at that time only in its completely inadequate early stages'."

[27]It is defensible that Hilbert took Poincaré's critics more serious than, for instance, Brouwer's, cf. [61, 62]; but since Poincaré died already in 1912, Hilbert had lost him as discussion partner at the time his programme was worked out.

[28]This supplement, added to the second edition of [53] and published in 1970, also presents a consistency proof of Kalmár, based on an unpublished manuscript of 1938.

proof.[29] However, the $\varepsilon$-substitution method was recently revived by Mints for the analysis of stronger systems, cf. [4, 76, 78] and [77] in this volume.

Gödel [39] published in 1958 a conceptually different consistency proof, a worked out version of the idea already mentioned at Zilsel's seminar in 1938 (see above) which is based on functionals of higher types, known as *Gödel's* $\mathcal{T}$ (the theory) or the *Dialectica-Interpretation* (the interpretation of Arithmetic in $\mathcal{T}$). This consistency proof is quite different from Gentzen's, and it addresses particularly the *finitistic* aspect of Hilbert's programme, as the functionals of higher types can be considered as fulfilling this aspect.

Even if somebody would not be convinced by any single consistency proof, (s)he should take into account that here conceptually different approaches—intuitionism; transfinite induction; functionals of higher type—all lead to the consistency of Peano Arithmetic. For *Church's thesis* sometimes the argument is put forward that many independent approaches to computability lead to the same class of functions. We have here a similar phenomenon, where the risk—put forward for Church's thesis—of "systematically overlooking something" is even lower, and one gains some kind of independent evidences for the consistency of Arithmetic.

In any case, as the consistency of Arithmetic is not really at issue, for modern proof theory Gentzen's consistency proof must be put in the right perspective. Macintyre writes in this respect [72, p. 2426][30]:

> Much nonsense has been pronounced about Gentzen's work, even by extremely distinguished people. Consistency is not really the main issue at all. He did reveal fine structure in the unprovability of consistency of PA, as a consequence of much deeper general methodology.

## 4 Analysis

It should be clear that for Hilbert's Programme Arithmetic could have been only an intermediate goal on the way to Analysis. It was, of course, Analysis which Hermann Weyl had in mind when speaking about a "house built on sand," it was

---

[29]Cf. Bernays in [53, p. VII]:

> Currently, W. Ackermann is developing his earlier consistency proof—by use of a sort of transfinite induction as used by Gentzen—in a way that it obtains validity for the full numbertheoretic formalism.

German original: "Gegenwärtig ist W. ACKERMANN dabei, seinen früheren (...) Widerspruchsfreiheitsbeweis durch Anwendung der transfiniten Induktion in der Art, wie sie von GENTZEN benutzt wird, so auszugestalten, daß er für den vollen zahlentheoretischen Formalismus Gültigkeit erhält."

Von Plato writes in [115, end of I.4.10]: "A second proof of Gentzen's result was given by an unwilling Wilhelm Ackermann, after repeated pleadings on the part of Bernays."

[30]In the continuation of the citation, the mentioned fine structure is illustrated by the result about provably total functions of PA which one can obtain from Gentzen's work.

Analysis which Brouwer tried to "revolutionize" (using Weyl's language) within his intuitionistic philosophy. Analysis uses at its very base the definition of the real numbers, a genuine impredicative concept. It was, first of all, Poincaré who put the use of impredicative concepts into question (though he accepted the real numbers as such).[31] But also Hilbert's own student Weyl was advocating a predicative reconstruction of Mathematics in *Das Kontinuum* [117], being willing to give up a large part of traditional Mathematics. Thus, for Hilbert, a consistency proof of classical Analysis turned now from a "simple question" of his Paris problem list into an issue of defense against an intuitionistic "Putschversuch" (as he expressed it in [49]).

It is known that Gödel started from Analysis when he was still trying to fulfill Hilbert's programme; Wang [116, p. 654] reports: "In the summer of 1930, Gödel began to study the problem of proving the consistency of analysis. [...] The problem he set for himself at that time was the relative consistency of analysis to number theory." In this context he encountered the incompleteness results which, in turn, closed this lane of argumentation.

Thus, Gentzen's consistency proof of Arithmetic is now only a first step, and the search for a consistency proof of Analysis was started immediately after. We know that Gentzen was working hard on such a consistency proof even in prison in Prague just before his premature death in 1945,[32] and some remaining notes about this work are currently in the process of publication [115]. But, it is also clear that he didn't reach a final result.

In sharp contrast to intuitionistic Arithmetic, intuitionistic Analysis can hardly be considered as a base to provide a consistency proof for (classical) Analysis which would fit Hilbert's aims. One problem are the additional principles for intuitionistic Analysis proposed by Brouwer, which are inconsistent in the classical setting. This makes it doubtful whether intuitionistic Analysis (in Brouwer's formulation) could be even considered as more reliable than classical Analysis in itself.[33]

Szabo in [102, pp. 12–16] gives a short review of other early consistency results, going beyond Arithmetic, by Fitch, Lorenzen, Takeuti, Schütte, and Ackermann. None of them are accepted as fulfilling Hilbert's requirement on a consistency

[31]See, for instance, the talk on *transfinite numbers* given by Poincaré in Göttingen in 1909 in the presence of Hilbert, included in [85] and translated by Ewald in [22, 22.G] (reprinted in [62]).

[32]Szabo [102, p. viii] refers to the memories of a friend of Gentzen in the prison: "He once confided in me that he was really quite contented since now he had at last time to think about a consistency proof for analysis. He was in fact fully convinced that he would succeed in carrying out such a proof."

[33]Here, one can turn Hilbert's programme upside down and use interpretations of new intuitionistic principles to justify them on classical grounds; see, for instance, [27, p. 340]. I also remember a proof theorist, making good use of such principles, but calling them—trained in classical Mathematics and therefore believing in the standard notion of mathematical truth—"totally wrong" (as translation of the German "grob falsch").

proof.[34] But they had, of course, some impact on the development of proof theory. The most stimulating proposal was Takeuti's *Fundamental Conjecture*, saying roughly that cut-elimination holds for second-order logic, cf. [107] and the informal presentation in [120, App. B]. There were soon some proofs of it [86, 93, 103, 106], which, however, rely on set theoretic considerations. Thus, these proofs do not provide additional reliability.[35]

Similar concerns regard other approaches, like Girard's $\mathcal{F}$ [34], where the *candidates*, used in the normalization proof, are subject to the same foundational concerns as the theory itself.[36,37]

Spector [101] introduced *bar recursion* as a concept which could be used to extend the Dialectica interpretation to Analysis.[38] To serve as a consistency proof, however, one would rely on bar recursion/bar induction as valid principle. Avigad and Feferman [8, p. 370f] write in their "Evaluation of Spector's interpretation":

> Spector was careful not to claim that the generalization of bar induction to higher types, which he used to justify bar recursion for continuous functionals, should be accepted on intuitionistic grounds. In fact, he offers the following caveat:
>
>> The author believes that the bar theorem is itself questionable, and that until the bar theorem can be given a suitable foundation, the question whether bar induction is intuitionistic is premature.
>
> The question of whether bar recursion can be justified on constructive grounds was taken up in a seminar on the foundations of analysis led by G. Kreisel at Stanford in the summer of 1963. The seminar's conclusion, summarized by Kreisel in an ensuing report [[69]], was that
>
>> ...the answer is negative by a wide margin, since not even bar recursion of type 2 can be proved consistent [by constructively accepted principles].

[34]Kreisel, in [67, p. 344], sketches also an extension of "Gödel's old translation" of a system for classical Analysis to a specific intuitionistic reformulation of Analysis, involving the general Comprehension Axiom, which "*provides an intuitionistic consistency proof of classical analysis*". He himself classifies this result as "philosophically [...] not significant at all", except for "*a reduction to intuitionistic methods of proof*"—which he judges a "technical" property. In the Discussion of this proof he reminds the reader to look for alternatives:

> Quite naively, this easy proof in no way reduces the interest of a more detailed proof theoretic reduction [...]; just as Gödel's original intuitionistic consistency proof for classical arithmetic **Z** did not make Gentzen's reduction superfluous.

[35]In a discussion of these proofs, Kreisel writes [67, p. 349, footnote 16]: "[I]n terms of consistency proofs, Tait's argument would only have proved the consistency of classical analysis in third order arithmetic!"

[36]I remember a proof-theorist classifying such a normalization proof as simply "circular."

[37]The worst-case scenario was experienced by Martin-Löf, when he realized that the normalization proof of his first (inconsistent) type theory was carried out in an *inconsistent* metatheory (see Setzer's contribution in this volume [95]).

[38]For a thorough discussion of Spector's proof see [26] in this volume. Oliva and Powell [80], also in this volume, discuss some spin-offs we can get from proof-theoretic analyses in the neighborhood of Spector's approach.

When failing to prove Takeuti's Fundamental Conjecture by more elementary means, proof theory turned naturally to *Subsystems of Analysis* where impressive results were established. Following the two traditions, called *Schütte-style* and *Takeuti-style* proof theory, we are able to give today analyses up to $\Pi^1_2$ comprehension, cf. the work of Rathjen [88, 89] and Arai [5–7], respectively.[39] These analyses of subsystems of Analysis in terms of ordinals are the natural extension of Gentzen's consistency proof for Arithmetic. It is particularly rewarding to provide the *proof-theoretic strength* of a theory; with ordinals as measure one is able to compare theories from different formal realms, like set-theoretical ones, type-theoretical ones, or others like Theories of Inductive Definitions[40] and Feferman's Explicit Mathematics. In return, these frameworks can help to carry out parts of the proof-theoretic investigations.[41]

The rationale of ordinal analyses—in comparison with the approaches mentioned above—was recently described by a colleague in the following neat characterization:

> Something that makes specifically ordinal-theoretical proof-theoretical analyses of a theory particularly convincing is that in many cases there is a big difference between the metatheory and the object theory; whereas with normalisation proofs based on Tait-style computability, or Girard-style 'candidates', the meta-theory is (more-or-less) the theory itself together with a uniform reflection principle. Something would be far wrong if one couldn't prove a normalisation theorem for Church's theory of types in such a metatheory; but the extra confidence one gets in the principles formulated therein from a normalisation theorem is tiny.

Let us close this section with the reference to some subprogrammes which grew out of Gentzen-style proof theory and which reach out for Analysis.

In [23], Feferman gives a comprehensive survey on the "viable rationale" of *reductive proof theory*, using examples of "pairs" of frameworks where the first one is reduced to the second one. Whereas Hilbert's original hope about the pair ⟨infinitary, finitary⟩ is limited by Gödel's incompleteness theorems and only exemplified by reductions to PRA [23, 5.1], one can look at other pairs like ⟨uncountable infinitary, countable infinitary⟩ [23, 5.2]; ⟨impredicative, predicative⟩ [23, 5.3]; and ⟨non-constructive, constructive⟩ [23, 5.4].[42]

---

[39]See [81, 82, 94, 108] for comprehensive presentations of the background of the respective developments.

[40]See, for instance, [14] and [57] in this volume.

[41]This was exemplified, in particular, by Kripke-Platek set theory, cf. e.g., [56, 81].

[42]In the further course of the discussion, Feferman expresses some doubts about current advances in ordinal analysis with respect to the given rationale [23, p. 80]:

> Even if one succeeds in reducing the system ($\Pi^1_2$-CA) ± BI to a constructive system (whether evidently so or not), one can hardly expect that doing so will appreciably increase one's belief in its consistency (if one has any doubts about that in the first place) in view of the difficulty of checking the extremely complicated technical work needed for its ordinal analysis.

Another successful subprogramme is *Reverse Mathematics* which looks for the weakest natural subsystem of Analysis which proves a given mathematical theorem, cf. [99].

Finally, we like to mention *Applied Proof Theory*, sometimes also promoted under the name *proof mining*, which aims to extract additional mathematical information from an in-depth analysis of proofs in formal systems, cf. [64].

For all these subprogrammes the consistency issue is clearly secondary. But they all rely on the techniques which were developed to a large extent out of Gentzen's methods used for his consistency proofs.

## 5 The Quest for Consistency

It was in an informal conversation, years ago, that two distinguished proof theorists repeatedly assured each other that, for modern proof theory, "consistency is not the question." As a matter of fact, the working mathematician considers ZFC, Zermelo–Fraenkel set theory including the axiom of choice, being beyond doubt.[43] Let's have a look at Wiles's proof of Fermat's Last Theorem. As it stands, its formalization seems to require ZFC + some Grothendieck Universes on top [74]. This is an outrageously strong system for a theorem which can be formulated in Peano Arithmetic. But no Mathematician would raise a minimal doubt about Wiles's proofs *because* it makes use of such a strong theory.

As an expert in set theory, W. Hugh Woodin makes the following "prediction" [119, p. 453][44]:

> In the next ten thousand years, there will be no discovery of an inconsistency in these theories [referring to three equiconsistent theories, including ZFC + "There exist infinitely many Woodin cardinals"].

And Gaisi Takeuti points out that we cannot even imagine any longer the original concerns of Hilbert's times, [120, p. 122]:

> In the current day, axiomatic set theory is fully accepted and it is generally acknowledged that modern mathematics can be carried out in the framework of axiomatic set theory. No contradiction has arisen in axiomatic set theory, and a sense of security that no contradiction will arise in it in the future is supported by intuitive consensus. Under the current secure circumstances one cannot imagine the sense of crises of that earlier time.

---

[43]This is, admittedly, in sharp contrast to the early times of axiomatic set theory, where Poincaré, for instance, expressed his doubts about Zermelo's axiomatization of set theory in the following words, cf. [43, p. 540]:

> But even though he has closed his sheepfold carefully, I am not sure that he has not set the wolf to mind the sheep.

[44]Of course, this prediction is embedded in a thorough discussion which gives arguments for this claim. But one may note that Woodin speaks here about the *discovery* not about the *existence* of an inconsistency.

Feferman, taking up explicitly an anti-platonist position, puts the following argument forward for the consistency of standard formal theories [23, p. 72][45]:

> I, for one, have absolutely no doubt that PA and even $PA_2$ are consistent, and no genuine doubt that ZF is consistent, and there seems to be hardly anyone who seriously entertains such doubts. Some may defend a belief in the consistency of these systems by simply pointing to the fact that no obvious inconsistencies are forthcoming in them, or that these systems have been used heavily for a long time without leading to an inconsistency. [...] My own reason for believing in the consistency of these systems is quite different. Namely, in the case of PA, we have an absolutely clear intuitive model in the natural numbers, which in the case of $PA_2$ is expanded through the notion of arbitrary subset of the natural numbers. Finally, ZF has an intuitive model in the transfinite iteration of the power set operation taken cumulatively. This has nothing to do with a belief in a platonic reality whose members include the natural numbers and arbitrary sets of natural numbers, and so on. On the contrary, I disbelieve in such entities. But I have as good a conception of what arbitrary subsets of natural numbers are *supposed* to be like as I do of the basic notions of Euclidean geometry, where I am invited to conceive of points, lines and planes as being utterly fine, utterly straight, and utterly flat, resp.

With respect to the standard formal theories, used in Mathematics, one may also cite Kreisel[46]:

> The doubts about the consistency are more doubtful than the consistency itself.

There is even an ironic corollary to Gödel's second incompleteness theorem with respect to "proof obligations": Gödel tells us that we cannot prove the (absolute) consistency of a formal mathematical theory. However, if somebody *believes* that a certain theory is inconsistent, (s)he would be committed to *prove* it, as this would be, of course, always possible. And such a person needs to be reminded of a word of Dedekind from 1887: "In science, what is provable should never be believed without proof."[47] But for one who believes in the consistency of a theory, Dedekind does not apply—thanks Gödel.

Thus, what should we think of these alleged threats of inconsistencies?

One might argue that the history of Mathematics is full of examples which one may consider as inconsistencies.[48] Mathematicians may apply a new concept in a way which results in false theorems. The simple fact that the supposed theorem

[45]The argument for the intuitive model of ZF is compared with the situation for Quine's *New Foundation* where the lack of such an intuitive model gives reason to look for a (relative) consistency proof.

[46]Conveyed by Girard in French [35, p. 525]: "Les doutes quant à la cohérence sont plus douteux que la cohérence elle-même."

[47]German original: "Was beweisbar ist, soll in der Wissenschaft nicht ohne Beweis geglaubt werden." cited and translated in [20, p. 97].

[48]See, for instance, [12]: "Historically speaking, it is of course quite untrue that mathematics is free from contradiction" and later "[Contradictions] occur in the daily work of every mathematician, beginner or master of his craft, as the result of more or less easily detected mistakes, [...]"

is false implies that a proper formalization of the argument will show a formal inconsistency. However, in most cases, the solution was never a problem: either the argumentation was dismissed with invalid, or—a little bit more interesting—some fundamental assumptions about a certain mathematical area were revised which improved our understanding of the this area.

Euler, for instance, in his famous book on Algebra [21], calculated $\sqrt{-1}\sqrt{-4} = \sqrt{4} = 2$, applying the "general law" $\sqrt{a}\sqrt{b} = \sqrt{ab}$.[49] Adding this last "law" to the axioms of the field of complex numbers, of course, leads to an inconsistent theory. Such cases are not of much interest because, typically, the wrong assumption is easy to isolate and to separate from the part which will be kept after a revision.

But there are some interesting examples of inconsistencies in the history of Mathematics which transcend such simple instances and which deserve a closer inspection:

- Cantor's naive set theory;
- Frege's *Grundgesetze der Arithmetik*, and subsequent foundational systems by Curry, Church, Kreisel, and Martin-Löf;
- Reinhardt cardinals over ZFC.

Cantor's naive set theory may be based on an unreflected comprehension principle expressed in Cantor's famous first characterization of the notion of set[50]:

> By a 'set' we understand every collection to a whole $M$ of definite, well-differentiated objects $m$ of our intuition or our thought.

It was soon discovered that this characterization allows for inconsistent set constructions like the set of all cardinals (Cantor 1897, letter to Hilbert [17, letter 156]), the set of all sets (Cantor 1899, letter to Dedekind [17, letter 163]), or the set of all ordinals (Burali-Forti 1897 [113, pp. 104ff]). It is worth noting that Cantor himself did not see any problem here, but took the "paradoxes" just as *reductio-ad-absurdum* arguments of the inexistence of the respective sets; in his correspondence with Hilbert he refines, therefore, his notion of set by distinguishing it as "consistent multiplicities."[51] Thus, for Cantor it was natural that the (in)consistency of a set construction is verified *a*

[49]This example is taken from [20, p. 59].

[50]German original: "Unter einer 'Menge' verstehen wir jede Zusammenfassung $M$ von bestimmten wohlunterschiedenen Objekten $m$ unserer Anschauung oder unseres Denkens (welche die 'Elemente' von $M$ genannt werden) zu einem Ganzen." [16, p. 282]. The translation is from [44, p. 33].

[51]In German: "consistente Vielheiten," letter to Hilbert from May 5th, 1899, [17, letter 160]; as "finished set" ("fertige Menge") already in a letter from December 2nd, 1897, [17, p. 390].

*posteriori*. Hilbert did not agree with such an approach and demanded an *a priori* justification.[52]

In practical terms, this was done by Zermelo in his axiomatization of set theory [121].[53] On the theoretical side, one finds here one of the motivations for Hilbert to propose consistency proofs *for theories* to ensure the meaningfulness of their mathematical notions.[54]

Frege's aim to give a logicist foundation of Mathematics in his *Grundgesetze der Arithmetik* [29, 30] was destroyed by Russell's Paradox. It is generally assumed that Frege's *Basic Law V* is responsible for the collapse of the system, but one may consider alternatives to resolve the problem.[55] What is of interest for us, as a lesson for the history of logic, is that Frege had some kind of *justification* of his axioms (one might as well call them *meaning explanations*). The problem was, that these were *local* justifications for the single axioms, but their combination turns out to be impossible; but it explains at the same time why we can single out *different* consistent and meaningful subsystems. The fate of Frege's system raises the question to which extent we can trust any philosophical justification programme based on local justifications (or meaning explanations).[56,57] What should provide

[52]In [47] he writes, [113, p. 131]:

> *G. Cantor* sensed the contradiction just mentioned and expressed this awareness by differentiating between "consistent" and "inconsistent" sets. But, since in my opinion he does not provide a precise criterion for this distinction, I must characterize his conception on this point as one that still leaves latitude for *subjective* judgment and therefore affords no objective certainty.

In German (cited in [17, S. 436]): "G. Cantor hat den genannten Widerspruch empfunden und diesem Empfinden dadurch Ausdruck verliehen, daß er 'konsistente' und 'nichtkonsistente' Mengen unterscheidet. Indem er aber meiner Meinung nach für diese Unterscheidung kein scharfes Kriterium aufstellt, muß ich seine Auffassung über diesen Punkt als eine solche bezeichnen, die dem *subjektiven* Ermessen noch Spielraum läßt und daher keine objektive Sicherheit gewährt." An even stronger statement against Cantor's approach can be found in a lecture note from 1917, [48], cf. [59, 60].

[53]Although this axiomatization has the flaw that its justification is *extrinsic* where philosophers would prefer to have an *intrinsic* one, cf. e.g., the discussion in [73].

[54]One may note that Cantor's criterion for a "finished set" also requires a consistency proof, but somehow locally for the particular construction only. However, as far as we know, Cantor only took note of the criterion in the negative cases, to dismiss a set construction when it was shown to be inconsistent.

[55]For instance, Aczel's *Frege Structures*, [3].

[56]The situation becomes philosophically even more doubtful when such a justification depends, in addition, on the approval of a "Master". In this respect, Lorenzen complained about Brouwer [70]:

> Unfortunately, the explanation which Brouwer himself offers for this phenomenon [that some Mathematicians consider the 'tertium non datur' as unreliable] is an esoteric issue: only one who listened the Master himself understands him.

(German original: "Unglücklicherweise ist die Erklärung, die Brouwer selbst für dieses Phänomen anbietet, eine esoterische Angelegenheit: nur, wer den Meister selber hörte, versteht ihn.")

[57]A complementary view on this issue is given by Setzer [95] in this volume.

the evidence for a consistent combination if not a *global* justification—like a *model*—which, then, could also be used directly?

After Frege, there were four more prominent examples of inconsistent foundational systems: Curry's combinatory logic, Church's original $\lambda$-calculus (both subject to the Kleene-Rosser paradox), Kreisel's theory of constructions (subject to the Kreisel-Goodman paradox), and Martin-Löf's first type theory (subject to Girard's paradox). Although these systems represent three quite different approaches, it appears to us that the problems for all arise from the *philosophical* motivation rather than from a formal (logical) inaccuracy in the formalization.[58] This suggests the conclusion that philosophical motivations are apparently more dangerous for formal systems than pure mathematical motivations (as in the case of ZFC, for instance).

With a Reinhardt cardinal in ZFC we have, however, a completely different case of inconsistency. A Reinhardt cardinal is a certain *large cardinal* which was proposed by William Nelson Reinhardt in his doctoral dissertation in 1967, and shown to be inconsistent over ZFC by Kenneth Kunen in 1971. To get a glance of the fate of this cardinal—including its role in the absence of the Axiom of Choice where no inconsistency is known—one may consult [119, Sect. 20.3]; more information can be found in [63, Sect. 23]. In a simplified way, one can say that large cardinals constitute a branch of set theory which tries to settle the Continuum's Hypothesis on the basis of "new axioms."[59] It is a fascinating area which—despite in failing so far to settle ultimately the question of the Continuum's Hypothesis—produced a large amount of interesting results. The inconsistency of the Reinhardt cardinal over ZFC simply puts a bound on what one may add.

What is important for us here is that this inconsistency should not surprise one particularly. Even less should it raise a minimal doubt about the consistency of "ordinary reasoning" in Mathematics. To the contrary, large cardinal axioms are, in some sense, designed to push our axiomatic set theories to its ultimate limit; and the Reinhardt cardinals simply show that we went beyond this limit. As Kanamori puts it [63, p. 324]: "ZFC rallies at last to force a veritable Götterdämmerung for large cardinals!"

As upshot one can say that there is simply no serious threat of inconsistencies in Mathematics, if one doesn't approach intentionally its ultimate limits—or overstretch philosophical demands.

Still, there is an issue of consistency for Analysis—and, *a forteriori*, for set theory: the *impredicative* features might have just not been explored sufficiently to find a possible contradiction. And the reason for it might be that Mathematics uses only a very limited part of the formal theories, a part which resides in an innocent, consistent subsystem; in fact, *Reverse Mathematics* gives us strong evidence for

[58]This claim can be substantiated by the fact that it was not possible for any of the systems to modify it in a way that the original aims of the authors would be preserved.

[59]A thorough discussion of this issue can be found in [25].

such a claim. It was Gentzen himself who expressed the general concern in 1938 as follows, [102, p. 235]:[60]

> Indeed, it seems not entirely unreasonable to me to suppose that *contradictions* might possibly be concealed even in classical *analysis*. The fact that, so for, none have been discovered means very little when we consider that, in practice, mathematicians always work with a comparatively limited part of the logically possible complexities of mathematical constructs.

Thus, after recalling his consistency proof for elementary number theory, he came to the conclusion that "the most important [consistency] proof of all in practice, that for *analysis*, is still outstanding" [102, p. 236].[61]

By pursuing such a consistency proof, modern proof theory developed genuine techniques not only to achieve consistency results but also to analyze the fine structure of formal theories relevant for the mathematical practice.[62] In terms of our comparison above, we may say that pursuing the quest for consistency, Gentzen provided us with the tools to explore and to map the newly discovered land of unlimited mathematical strength.

**Acknowledgements** Research supported by the Portuguese Science Foundation, FCT, through the projects *Hilbert's Legacy in the Philosophy of Mathematics*, PTCD/FIL-FCI/109991/2009, *The Notion of Mathematical Proof*, PTDC/MHC-FIL/5363/2012, and the *Centro de Matemática e Aplicações*, UID/MAT/00297/2013; and by the project *Método axiomática e teoria de categorias* of the cooperation Portugal/France in the *Programa PESSOA – 2015/2016*.

## References

1. W. Ackermann, Letter to David Hilbert, August 23rd, 1933, Niedersächsische Staats- und Universitätsbibliothek Göttingen, Cod. Ms. D. Hilbert 1.
2. W. Ackermann, Zur Widerspruchsfreiheit der Zahlentheorie. Math. Ann. **117**, 162–194 (1940)
3. P. Aczel, Frege structures and the notion of proposition, truth and set, in *The Kleene Symposium*, ed. by J. Barwise, H. Keisler, K. Kunen (North-Holland, Amsterdam, 1980), pp. 31–59
4. T. Arai, Epsilon substitution method for $ID_1(\Pi^0_1 \vee \Sigma^0_1)$. Ann. Pure Appl. Log. **121**, 163–208 (2003)
5. T. Arai, Proof theory for theories of ordinals I: recursively Mahlo ordinals. Ann. Pure Appl. Log. **122**, 1–85 (2003)

[60]German original [32, p. 6]: "So scheint es mir nicht ganz ausgeschlossen, daß auch in der klassischen *Analysis* mögliche *Widersprüche* verborgen sein können. Daß man bis jetzt keine entdeckt hat, besagt nicht viel, wenn man bedenkt, daß der Mathematiker *in praxi* immer mit einem verhältnismäßig geringen Teil der an sich logisch möglichen mannigfachen Komplizierungen der Begriffsbildung auskommt."

[61]German original [32, p. 7]: "doch steht der praktisch vor allem wichtige Beweis für die *Analysis* noch aus."

[62]Cf, e.g., [83] and [91] in this volume.

6. T. Arai, Proof theory for theories of ordinals II: $\Pi_3$-reflection. Ann. Pure Appl. Log. **129**, 39–92 (2004)
7. T. Arai, Proof theory for theories of ordinals III: $\Pi_N$-reflection, in *Gentzen's Centenary: The Quest for Consistency*, ed. by R. Kahle, M. Rathjen (Springer, Heidelberg, 2015)
8. J. Avigad, S. Feferman, Gödel's functional ("Dialectica") interpretation, in *Handbook of Proof Theory*, ed. by S.R. Buss (North-Holland, Amsterdam, 1998), pp. 337–405
9. P. Benacerraf, H. Putnam (eds.), *Philosophy of Mathematics. Selected Readings* (Prentice Hall, Englewood Cliffs, 1964) (2nd edn., Cambridge University Press, Cambridge, 1983)
10. P. Bernays, Hilberts Untersuchungen über die Grundlagen der Arithmetik, in *David Hilbert: Gesammelte Abhandlungen*, vol. 3, [51] (Springer, Berlin, 1935), pp. 196–216
11. P. Bernays, Hilbert, David, in *Encyclopedia of Philosophy*, vol. 3, ed. by P. Edwards (Macmillan, New York, 1967), pp. 496–504
12. N. Bourbaki, Foundations of mathematics for the working mathematician. J. Symb. Log. **14**(1), 1–8 (1949)
13. W. Buchholz, On Gentzen's first consistency proof for arithmetic, in *Gentzen's Centenary: The Quest for Consistency*, ed. by R. Kahle, M. Rathjen (Springer, Heidelberg, 2015)
14. W. Buchholz, S. Feferman, W. Pohlers, W. Sieg, *Iterated Inductive Definitions and Subsystems of Analysis*. Lecture Notes in Mathematics, vol. 897 (Springer, Berlin, 1981)
15. S. Buss, Cut elimination *In Situ*, in *Gentzen's Centenary: The Quest for Consistency*, ed. by R. Kahle, M. Rathjen (Springer, Heidelberg, 2015)
16. G. Cantor, *Gesammelte Abhandlungen mathematischen und philosophischen Inhalts*, ed. by E. Zermelo (Springer, Berlin, 1932)
17. G. Cantor, *Briefe*, ed. by H. Meschkowski, W. Nilson (Springer, Berlin, 1991)
18. J.W. Dawson Jr., The reception of Gödel's incompleteness theorems, in *Gödel's Theorem in Focus*, ed. by S.G. Shanker (Routledge, London, 1988), pp. 74–95
19. M. Detlefsen, Gentzen's anti-formalist views, in *Gentzen's Centenary: The Quest for Consistency*, ed. by R. Kahle, M. Rathjen (Springer, Heidelberg, 2015)
20. H.-D. Ebbinghaus et al., *Numbers*. Graduate Texts in Mathematics (Springer, Berlin, 1991)
21. L. Euler, *Vollständige Anleitung zur Algebra*. (Akademie der Wissenschaften, St. Petersburg, 1770). Two volumes.
22. W.B. Ewald (ed.), *From Kant to Hilbert*. (Oxford University Press, Oxford, 1996). Two volumes.
23. S. Feferman, Does reductive proof theory have a viable rationale? Erkenntnis **53**, 63–96 (2000)
24. S. Feferman, *Lieber Herr Bernays! Lieber Herr Gödel!* Gödel on finitism, constructivity, and Hilbert's program, in *Kurt Gödel and the Foundations of Mathematics*, ed. by M. Baaz et al. (Cambridge University Press, Cambridge, 2011), pp. 111–133
25. S. Feferman, H.M. Friedman, P. Maddy, J.R. Steel, Does mathematics need new axioms? Bull. Symb. Log. **6**(4), 401–446 (2000)
26. F. Ferreira, Spector's proof of the consistency of analysis, in *Gentzen's Centenary: The Quest for Consistency*, ed. by R. Kahle, M. Rathjen (Springer, Heidelberg, 2015)
27. F. Ferreira, A. Nunes, Bounded modified realizability. J. Symb. Log. **71**(1), 329–346 (2006)
28. A. Fraenkel, *Zehn Vorlesungen über die Grundlegung der Mengenlehre*. Wissenschaft und Hypothese, vol. XXXI (Teubner, Leipzig, 1927). Reprint (Wissenschaftliche Buchgesellschaft, Darmstadt, 1972)
29. G. Frege, *Grundgesetze der Arithmetik*, vol. 1 (Hermann Pohle, Jena, 1893). Reprinted, together with vol. 2 (Olms, Hildesheim 1966)
30. G. Frege, *Grundgesetze der Arithmetik*, vol. 2 (Hermann Pohle, Jena, 1903). Reprinted, together with vol. 1 (Olms, Hildesheim 1966)
31. G. Gentzen, Die Widerspruchsfreiheit der reinen Zahlentheorie. Math. Ann. **112**, 493–565 (1936). English translation in [102, #4]
32. G. Gentzen, Die gegenwärtige Lage in der mathematischen Grundlagenforschung. Forschungen zur Logik und zur Grundlegung der exakten Wissenschaften, Neue Folge **4**, 5–18 (1938); also in Dtsch. Math. **3**, 255–268 (1939). English translation in [102, #7]

33. G. Gentzen, Über das Verhältnis zwischen intuitionistischer und klassischer Arithmetik. Archiv für mathematische Logik und Grundlagenforschung **16**, 119–132 (1974). Written in 1933, but withdrawn from publication after the appearence of [37]. English translation in [102, #2]
34. J.-Y. Girard, Interprétation fonctionnelle et élimination des coupures de l'arithmétique d'ordre supérieur, PhD thesis, Université Paris 7, 1972
35. J.-Y. Girard, Du pourquoi au comment: la théorie de la démonstration de 1950 à nos jours, in *Development of Mathematics, 1950–2000*, ed. by J.-P. Pier (Birkhäuser, Basel, 2000), pp. 515–545
36. J.-Y. Girard, *The Blind Spot*. (European Mathematical Society, Zürich, 2011)
37. K. Gödel, Zur intuitionistischen Arithmetik und Zahlentheorie. Ergebnisse eines mathematischen Kolloquiums **4**, 34–38 (1933). Translated as *On intuitionistic arithmetic and number theory* in [113], reprinted in [40] (1933e), pp. 287–295
38. K. Gödel, Vortrag bei Zilsel/Lecture at Zilsel's, 1938. Published in [42] (*1938a), pp. 86–113
39. K. Gödel, Über eine bisher noch nicht benützte Erweiterung des finiten Standpunktes. Dialectica **12**, 280–287 (1958). Reprinted with English translation in [41], pp. 241–251
40. K. Gödel, *Collected Works*, vol. I, ed. by S. Feferman et al. (Oxford University Press, Oxford, 1986)
41. K. Gödel, *Collected Works*, vol. II, ed. by S. Feferman et al. (Oxford University Press, Oxford, 1990)
42. K. Gödel, *Collected Works: Unpublished Essays and Lectures*, vol. III, ed. by S. Feferman et al. (Oxford University Press, Oxford, 1995)
43. J. Gray, *Henri Poincaré* (Princeton University Press, Princeton, 2013)
44. M. Hallett. *Cantorian Set Theory and Limitation of Size*. Oxford Logic Guides, vol. 10 (Oxford University Press, Oxford, 1984)
45. D. Hilbert, Mathematische Probleme. Vortrag, gehalten auf dem internationalen Mathematiker-Kongreß zu Paris 1900. *Nachrichten von der königl. Gesellschaft der Wissenschaften zu Göttingen. Mathematisch-physikalische Klasse*, pages 53–297, 1900. English translation: [46]
46. D. Hilbert, Mathematical Problems. Lecture Delivered Before the International Congress of Mathematicians at Paris in 1900. Bull. Am. Math. Soc. **8**, 437–479 (1902). English translation of [45]
47. D. Hilbert, Über die Grundlagen der Logik und der Arithmetik, in *Verhandlungen des Dritten Internationalen Mathematiker-Kongresses in Heidelberg vom 8. bis 13. August 1904*, ed. by A. Krazer (Teubner, Leipzig, 1905), pp. 174–185
48. D. Hilbert, Mengenlehre. Lecture Notes Summer term 1917, (Library of the Mathematical Institute, University of Göttingen), 1917
49. D. Hilbert, Neubegründung der Mathematik: Erste Mitteilung. Abhandlungen aus dem Seminar der Hamburgischen Universität **1**, 157–177 (1922)
50. D. Hilbert, Die Grundlagen der Mathematik. Abhandlungen aus dem mathematischen Seminar der Hamburgischen Universität **6**(1/2), 65–85 (1928)
51. D. Hilbert, *Gesammelte Abhandlungen*. Analysis, Grundlagen der Mathematik, Physik, Verschiedenes, Lebensgeschichte, vol. III (Springer, Berlin, 1935). 2nd edition 1970
52. D. Hilbert, P. Bernays, *Grundlagen der Mathematik I*. Die Grundlehren der mathematischen Wissenschaften in Einzeldarstellungen, vol. 40 (Springer, Berlin, 1934). 2nd edn. 1968. Partly translated into English in the bilingual edition [55]
53. D. Hilbert, P. Bernays, *Grundlagen der Mathematik II*. Die Grundlehren der mathematischen Wissenschaften in Einzeldarstellungen, vol. 50 (Springer, Berlin, 1939). 2nd edition [54]
54. D. Hilbert, P. Bernays, *Grundlagen der Mathematik II*. Die Grundlehren der mathematischen Wissenschaften in Einzeldarstellungen, vol. 50, 2nd edn. (Springer, Berlin, 1970). 1st edition [53].
55. D. Hilbert, P. Bernays, *Grundlagen der Mathematik I* (Foundations of Mathematics I). (College Publications, London, 2011). Bilingual edition of Prefaces and §§1–2 of [52]

56. G. Jäger, *Theories for Admissible Sets: A Unifying Approach to Proof Theory*. (Bibliopolis, Naples, 1986)
57. G. Jäger, D. Probst, A proof-theoretic analysis of theories for stratified inductive definitions, in *Gentzen's Centenary: The Quest for Consistency*, ed. by R. Kahle, M. Rathjen (Springer, Heidelberg, 2015)
58. H. Jervell, Climbing mount $\varepsilon_0$, in *Gentzen's Centenary: The Quest for Consistency*, ed. by R. Kahle, M. Rathjen (Springer, Heidelberg, 2015)
59. R. Kahle, David Hilbert über Paradoxien. Departamento de Matemática, Universidade de Coimbra. Preprint Number 06-17(2006)
60. R. Kahle, The universal set—A (never fought) battle between philosophy and mathematics, in *Lógica e Filosofia da Ciência*, ed. by O. Pombo, Á. Nepomuceno. Colecção Documenta, vol. 2 (Centro de Filosofia das Ciências da Universidade de Lisboa, Lisboa, 2009), pp. 53–65
61. R. Kahle, Hilbert and Poincaré and the paradoxes, in *Poincaré's Philosophy of Mathematics: Intuition, Experience, Creativity*, ed. by H. Tahiri. Cadernos de Filosofia das Ciências, vol. 11 (Centro de Filosofia das Ciências da Universidade de Lisboa, Lisboa, 2011), pages 79–93
62. R. Kahle, Poincaré in Göttingen, in *Poincaré: Philosopher of Science, Problems and Perspectives*, ed. by M. de Paz, R. DiSalle. The Western Ontario Series in the Philosophy of Science, vol. 79 (Springer, Berlin, 2014), pp. 83–99
63. A. Kanamori, *The Higher Infinite*, 2nd edn. (Springer, Berlin, 2003). 1st edition 1994
64. U. Kohlenbach. *Applied Proof Theory: Proof Interpretations and their Use in Mathematics*. (Springer, Berlin, 2008)
65. A.N. Kolmogorov, O principe tertium non datur. Matematicheskij Sbornik **32**, 646–667 (1925). Reprinted in English translation as *On the principle of the excluded middle* in [113, p. 414–447]
66. G. Kreisel, Hilbert's programme. Dialectica **12**, 346–372 (1958). Reprinted in [9, pp. 207–238 of the second edition]
67. G. Kreisel, A survey of proof theory. J. Symb. Log. **33**(3), 321–388 (1968)
68. G. Kreisel, Formal rules and questions of justifying mathematical practice, in *Konstruktionen versus Positionen*, ed, by K. Lorenz. vol. 1: Spezielle Wissenschaftslehre (de Gruyter, Berlin, 1979), pp. 99–130
69. G. Kreisel et al., *Reports of Seminar on the Foundations of Analysis* ("Stanford report"). Technical report, Stanford University, Summer 1963
70. P. Lorenzen, Logik und Agon, in *Atti del XII Congresso Internazionale di Filosofia*, vol. 4 (Sansoni, Firenze, 1960), pp. 187–194. Reprinted in [71]
71. P. Lorenzen, K. Lorenz, *Dialogische Logik* (Wissenschaftliche Buchgesellschaft, Darmstadt, 1978)
72. A. Macintyre, The mathematical significance of proof theory. Philos. Trans. R. Soc. A **363**, 2419–2435 (2005)
73. P. Maddy, *Naturalism in Mathematics* (Clarendon Press, Oxford, 1997)
74. C. McLarty, What does it take to prove Fermat's last theorem? Grothendieck and the logic of number theory. Bull. Symb. Log. **16**(3), 359–377 (2010)
75. F. Meskens, A. Weiermann, Classifying phase transition thresholds for Goodstein sequences and Hydra games, in *Gentzen's Centenary: The Quest for Consistency*, ed. by R. Kahle, M. Rathjen (Springer, Heidelberg, 2015)
76. G. Mints, Gentzen-type systems and Hilbert's epsilon substitution method. I, in *Logic, Methodology, and Philosophy of Science IX*, ed. by D. Prawitz, B. Skyrms, D. Westerstahl (Elsevier, Amsterdam, 1994), pp. 91–122
77. G. Mints, Non-deterministic epsilon substitution method for PA and $\mathrm{ID}_1$, in *Gentzen's Centenary: The Quest for Consistency*, ed. by R. Kahle, M. Rathjen (Springer, Heidelberg, 2015)
78. G. Mints, S. Tupailo, W. Buchholz, Epsilon substitution method for elementary analysis. Arch. Math. Log. **35**, 103–130 (1996)
79. S. Negri, J. von Plato, *Structural Proof Theory* (Cambridge University Press, Cambridge, 2001)

80. P. Oliva, T. Powell, A game-theoretic computational interpretation of proofs in classical analysis, in *Gentzen's Centenary: The Quest for Consistency*, ed. by R. Kahle, M. Rathjen (Springer, Heidelberg, 2015)
81. W. Pohlers, Subsystems of set theory and second order number theory, in *Handbook of Proof Theory*, ed. by S.R. Buss. Studies in Logic and Foundations of Mathematics, vol. 137 (Elsevier, Amsterdam, 1998), pp. 209–335
82. W. Pohlers, *Proof Theory*. Universitext (Springer, Berlin, 2009)
83. W. Pohlers, Semi-formal calculi and their applications, in *Gentzen's Centenary: The Quest for Consistency*, ed. by R. Kahle, M. Rathjen (Springer, Heidelberg, 2015)
84. H. Poincaré, Les mathématiques et la logique. Revue de métaphysique et de morale **14**, 294–317 (1906). Translated in [22, vol. 2, pp. 1052–1071]
85. H. Poincaré, *Sechs Vorträge über ausgewählte Gegenstände aus der reinen Mathematik und mathematischen Physik* (Teubner, Leipzig, 1910)
86. D. Prawitz, Hauptsatz for higher order logic. J. Symb. Log. **33**, 452–457 (1968)
87. D. Prawitz, Extending Gentzen's 2nd consistency proof to normalization of natural deductions in 1st order arithmetic, in *Gentzen's Centenary: The Quest for Consistency*, ed. by R. Kahle, M. Rathjen (Springer, Heidelberg, 2015)
88. M. Rathjen, Ordinal analysis of parameter free $\Pi^1_2$-comprehension. Arch. Math. Log. **44**(3), 263–362 (2005)
89. M. Rathjen, An ordinal analysis of stability. Arch. Math. Log. **44**(1), 1–62 (2005)
90. M. Rathjen, Goodstein's theorem revisited, in *Gentzen's Centenary: The Quest for Consistency*, ed. by R. Kahle, M. Rathjen (Springer, Heidelberg, 2015)
91. M. Rathjen, P.F.V. Vizcaíno, Well ordering principles and bar induction, in *Gentzen's Centenary: The Quest for Consistency*, ed. by R. Kahle, M. Rathjen (Springer, Heidelberg, 2015)
92. C. Reid, *Hilbert* (Springer, Berlin, 1970)
93. K. Schütte, Syntactical and semantical properties of simple type theory. J. Symb. Log. **25**, 303–326 (1960)
94. K. Schütte, *Proof Theory* (Springer, Berlin, 1977)
95. A. Setzer, The use of trustworthy principles in a revised Hilbert's program, in *Gentzen's Centenary: The Quest for Consistency*, ed. by R. Kahle, M. Rathjen (Springer, Heidelberg, 2015)
96. J.R. Shoenfield, *Mathematical Logic*. (Addison-Wesley, Reading, 1967). Reprinted Association for Symbolic Logic and AK Peters, 2001
97. A. Siders, A direct Gentzen-style consistency proof for Heyting arithmetic, in *Gentzen's Centenary: The Quest for Consistency*, ed. by R. Kahle, M. Rathjen (Springer, Heidelberg, 2015)
98. W. Sieg, Hilbert's programs: 1917–1922. Bull. Symb. Log. **5**(1), 1–44 (1999)
99. S.G. Simpson, *Subsystems of Second Order Arithmetic*, 2nd edn. Perspectives in Logic (Association for Symbolic Logic and Cambridge University Press, New York, 2009)
100. R. Smullyan, *Gödel's Incompleteness Theorems*. Oxford Logic Guides, vol. 19. (Oxford University Press, New York, 1992)
101. C. Spector, Provably recursive functionals of analysis: a consistency proof of analysis by an extension of principles in current intuitionistic mathematics, in *Recursive Function Theory*, ed. by F.D.E. Dekker. Proceedings of Symposia in Pure Mathematics, vol. 5 (American Mathematical Society, Providence, 1962), pp. 1–27
102. M.E. Szabo, *The Collected Papers of Gerhard Gentzen*. Studies in Logic and the Foundations of Mathematics (North-Holland, Amsterdam, 1969)
103. W.W. Tait, A non constructive proof of Gentzen's Hauptsatz for second order predicate logic. Bull. Am. Math. Soc. **72**, 980–983 (1966)
104. W.W. Tait, Finitism. J. Philos. **78**, 524–546 (1981)
105. W.W. Tait, Gentzen's original consistency proof and the bar theorem, in *Gentzen's Centenary: The Quest for Consistency*, ed. by R. Kahle, M. Rathjen (Springer, Heidelberg, 2015)

106. M. Takahashi, A proof of cut-elimination in simple type theory. J. Math. Soc. Jpn. **19**, 399–410 (1967)
107. G. Takeuti, On a generalized logic calculus. Jpn. J. Math. **23**, 39–96 (1953)
108. G. Takeuti, *Proof Theory*, 2nd edn. (North-Holland, Amsterdam, 1987)
109. A. Tarski, Contribution to the discussion of P. Bernays '*Zur Beurteilung der Situation in der beweistheoretischen Forschung*'. Revue internationale de philosophie **8**, 16–20 (1954). Reprinted in [110, pp. 711–714]
110. A. Tarski, *Collected Papers*, vol. IV (Birkhäuser, Basel, 1986)
111. A.S. Troelstra, H. Schwichtenberg, *Basic Proof Theory*. Cambridge Tracts in Theoretical Computer Science, vol. 43, 2nd edn. (Cambridge University Press, Cambridge, 2000)
112. D. van Dalen, Brouwer and Fraenkel on intuitionism. Bull. Symb. Log. **6**(3), 284–310 (2000)
113. J. van Heijenoort (ed.) *From Frege to Gödel: A Source Book in Mathematical Logic, 1879–1931*. (Harvard University Press, Cambridge, 1967)
114. J. von Plato, From *Hauptsatz* to *Hilfssatz*, in *Gentzen's Centenary: The Quest for Consistency*, ed. by R. Kahle, M. Rathjen (Springer, Heidelberg, 2015)
115. J. von Plato (ed.) *Saved from the Cellar. Gerhard Gentzen's Shorthand Notes on Logic and Foundations of Mathematics*. to appear, 201x
116. H. Wang, Some facts about Kurt Gödel. J. Symb. Log. **46**(3), 653–659 (1981)
117. H. Weyl, *Das Kontinuum*. (Veit, Leipzig, 1918). English translation: [118]
118. H. Weyl, *The Continuum: A Critical Examination of the Foundation of Analysis*. (Thomas Jefferson University Press, 1987), (Corrected re-publication, Dover 1994). English translation of [117].
119. W.H. Woodin, The transfinite universe, in *Kurt Gödel and the Foundations of Mathematics*, ed. by M. Baaz et al. (Cambridge University Press, New York, 2011), pp. 449–471
120. M. Yasugi, N. Passell (eds.) *Memoirs of a Proof Theorist* (World Scientific, Singapore, 2003). English translation of a collection of essays written by Gaisi Takeuti
121. E. Zermelo, Untersuchungen über die Grundlagen der Mengenlehre I. Math. Ann. **65**, 261–281 (1908)

# Gentzen's Anti-Formalist Views

**Michael Detlefsen**

## 1 Introduction

In June of 1936 Gentzen gave a lecture at Heinrich Scholz' seminar in Münster. The title of the lecture was "Der Unendlichkeitsbegriff in der Mathematik."[1]

In this lecture, Gentzen presented a generally optimistic view concerning the prospects for the future development of Hilbert's proof-theoretic program to establish the consistency of classical mathematics. At the same time, curiously, he expressed sympathy with a challenge to Hilbert's formalist program that is reminiscent of some of Brouwer's criticisms.

This challenge, which I'll refer to as the *Contentualist Challenge*, was essentially this: even if the consistency of classical mathematics were ultimately to be proved by finitarily acceptable means, this would not be enough to properly found it. Also necessary, in Gentzen's view, was the provision of a way to assign contents to the so-called *ideal propositions*[2] of classical mathematics. Hilbert's so-called direct proof of the consistency of arithmetic was neither designed nor equipped to provide such an assignment. As a result, it was neither designed nor equipped to satisfy conditions the satisfaction of which Gentzen regarded as necessary for the proper foundation of classical mathematics.

[1]The lecture was published in *Semesterberichte Münster*, WS 1936/37: (65–85). It was translated into English by M. E. Szabo as "The Concept of Infinity in Mathematics" and included in [19].

[2]Or what Gentzen generally referred to as *actualist* propositions.

M. Detlefsen (✉)
Department of Philosophy, University of Notre Dame, 100 Malloy Hall, Notre Dame, IN 46556, USA
e-mail: mdetlef1@nd.edu; Michael.Detlefsen.1@nd.edu



R. Kahle, M. Rathjen (eds.), *Gentzen's Centenary*,
DOI 10.1007/978-3-319-10103-3_2

Gentzen put what he took to be the crucial point this way:

> Even if the consistency were to have been proved, the propositions of actualist mathematics (*die Aussagen der an-sich-Mathematik*)[3] would remain without sense (*sinnlos*) and would therefore, as ever, have to be repudiated (*abzulehnen*). ... The whole question of "sense" (*"Sinnes"*) does not seem ... to be ready for a final settlement. ... The objection against the sense of actualist propositions must in any case not be taken too lightly; it is not entirely without merit. [16, p. 74]

Hilbert's proposed formalist defense of classical mathematics was undertaken for the purpose of justifying the use of ideal elements in our mathematical reasoning. This included, in particular, the use of *actualist* propositions as instruments to aid the conduct of the "logical" parts of mathematical reasoning.

Hilbert did not, however, propose that these so-called ideal propositions be preserved as contentual elements of mathematical thinking. He saw them as useful, perhaps even in some sense "necessary," for the conduct of logical reasoning. He did not, however, take their usefulness to consist in their presumably contentual application in our thinking. Rather, he believed that it is due to their use as ideal elements in our logical thinking—a use which, generally speaking, is similar in both character and motive to the use of such devices as negative and complex numbers in algebra and analysis and points at infinity in projective geometry.

He put the basic point this way:

> [M]athematics contains, first, formulas to which correspond contentual (*inhaltliche*) communications of finitary propositions (mainly numerical equations or inequalities, or more complex communications composed of these) and which we may call the *real propositions* (*realen Aussagen*) of the theory, and, second, formulas that—just like the numerals of contentual number theory—in themselves mean nothing but are merely things governed by our rules and must be regarded as the *ideal material* (*idealen Gebilde*) of our theory. [27, p. 8]

By adjoining the so-called ideal propositions to the real propositions, "we obtain a system of propositions in which all the simple rules of Aristotelian logic hold and all the usual methods of mathematical inference are valid" (*op. cit.*, 9). The development of such a system of "logical" reasoning, Hilbert believed, benefits our logical reasoning in ways that are generally similar to the ways in which the use of ideal elements elsewhere in mathematics benefits other parts of mathematical reasoning. Specifically, it allows us to reason with greater facility to real or contentual conclusions, and it does so without compromising reliability.

Reasoning which makes use of ideal or actualist propositions is not, however, reasoning in the traditional sense. That is, it is not reasoning which proceeds from premises which are judgments having genuine propositional contents to conclusions (ultimate or transitory) which are likewise judgments having propositional contents via inferences that represent judgments of logical relationship between genuine propositions.

---

[3] "Actualist mathematics" was a term Gentzen commonly used for classical or traditional mathematics.

In the view of the critics of formalism (e.g., Frege and Brouwer), this meant that the reasoning which makes use of the so-called ideal propositions is not, in truth, genuine reasoning at all. Rather, it is only something which has the syntactical facade of genuine reasoning. It lacks the genuine contentful premises and the genuine logical interrelationship of contentual propositions needed for genuine reasoning.

Hilbert and those in his camp (e.g., Bernays) rejected this traditional conception of reasoning. More accurately, they rejected the view that legitimate mathematical reasoning always proceeds according to the traditional contentualist plan. Sometimes, they maintained, it proceeds in decidedly non-contentualist ways for decidedly non-contentualist motives. In their view, this reflected an identifying characteristic of modern scientific thinking generally—namely, that in addition to a descriptive component, it has as well an idealizational component.

> In science we are predominantly if not always concerned with theories that are not completely given to representing reality, but whose significance (*Bedeutung*) consists in the *simplifying idealization* (*vereinfachende Idealisierung*) they offer of reality. This idealization results from the extrapolation by which the concept formations (*Begriffsbildungen*) and basic laws (*Grundsätze*) of the theory go beyond (*überschreitet*) the realm of experiential data (*Erfahrungsdaten*) and intuitive evidence (*anschauliche Evidenz*). [29, pp. 2–3]

As Hilbert and Bernays saw it, the aim of science was not simply or only to describe, but also to idealize and to simplify. Such simplification, however, sometimes called for the use of "formal" rather than contentual methods of reasoning. Accordingly, they believed, contentual interpretation is not necessary for a proper defense of ideal reasoning.

Gentzen suggested a contrary view, giving particular attention to the case of general set theory in this connection. There he speculated in particular that proof-theoretical investigations would confirm that non-denumerable cardinalities are empty appearances (*nur leerer Schein*), that concepts and sentences concerning them are contentless, and that mathematicians ought therefore to avoid making use of them.

> I believe that, for example, in general set theory a careful proof-theoretic investigation will finally show that all powers that go beyond the countable are, in a quite definite sense, only empty appearances and that one should have the good sense to do without them.[4]

In an essay published a year later, he put the point more strongly, describing the question of the content of classical mathematics (or what he called *an-sich* mathematics) as "very important" (*sehr wichtig*) (cf. [18, p. 202]).

---

[4]The German was:

> Ich glaube, dass z. B. in der allgemeinen Mengenlehre eine sorgfältige beweistheoretische Untersuchung schliesslich die Ansicht bestätigen wird, dass alle über das Abzählbare hinausgehenden Mächtigkeiten in ganz bestimmten Sinne nur leerer Schein sind und man vernünftigerweise auf diese Begriffe wird verzichten müssen. [16, p. 74]

He thus seems to have sympathized with those critics of formalism who, like Frege and Brouwer, emphasized the question of whether formalism can adequately provide for the contentual interpretation of ideal or "actualist" propositions in mathematics.

It is this seeming affinity of Gentzen's views with the traditionalist views of Brouwer and Frege that I find noteworthy. Gentzen, after all, has generally been described, and generally described himself, as an advocate and promoter of Hilbert's ideas in the foundations of mathematics. Hilbert, however, emphasized that the interpretation of ideal reasoning is not necessary either for the conduct of mathematical reasoning or for its proper foundation. This raises the question of how significant the affinities between Gentzen's and Hilbert's views really are. This is the question I want to consider here.

Examination of Gentzen's views reveals distinct sympathy with the traditional conception of reasoning as generally consisting in a finite sequence of judgments arranged according to perceived logical relationships between their contents. He believed the formalization of mathematical reasoning to be a means of preparing it for precise metamathematical investigation, but there is little indication that he saw uninterpreted formal reasoning as playing an important role in mathematical reasoning. In fact, there are counter-indications.

His formalist sympathies thus seem to have been quite limited. He held only a version of what I will call *Representational Formalism*. This is the view that the formal representation of mathematical reasoning is a legitimate and perhaps even a valuable tool for purposes of studying certain of its properties (e.g., its consistency). Whether formal methods have a place in the actual conduct of mathematical reasoning, on the other hand, is another matter, and one which is not settled by the possible usefulness of formalization as a representational tool for metamathematical investigation.

Hilbert too was a Representational Formalist. His formalist convictions went beyond this, however. In addition to believing in the representational utility of formal methods, he believed that they have an important role to play in the actual conduct of mathematical reasoning. He believed, that is, that mathematical reasoning is partially *constituted by* the use of formal, non-contentual methods of reasoning, and he believed as well that the use of such methods has played an important role in making modern mathematics the successful science that it is.

In addition to being a Representational Formalist, then, Hilbert was what I will call a *Conductive Formalist*. Gentzen was not, or at least not so fully as Hilbert. All in all, he seems to have accepted the traditional contentualist view of mathematical reasoning that Hilbert rejected. More specifically, he held that the use of formal methods in mathematical reasoning can only be fully vindicated by providing a contentual interpretation for it.

To the extent that this is correct, Gentzen's formalism was less far-reaching than Hilbert's. This, at any rate, is what I will argue here.

## 2 The Traditional and Abstract Conceptions of Axiomatization

Gentzen and Hilbert diverged as regards their views of the basic nature of reasoning. Gentzen held a more or less traditional contentualist view of reasoning. Hilbert, on the other hand, rejected the traditional view and emphasized not only the possibility of non-contentual reasoning, but also its importance to mathematics. He did not deny that much mathematical reasoning is contentual. Nor did he deny that contentual reasoning has played an important role, perhaps even a dominant role, in the development of mathematics. He maintained only that there are also non-contentual processes of reasoning, and that these have also been important to the development and success of modern mathematics.

What I am calling the traditional conception of reasoning centered on the idea that an argument is a finite, logically ordered sequence of judgments. The term "judgment" here is used in its traditional sense—that is, to signify an attitude of affirmation taken towards a proposition.

By a "logical ordering" of judgments, I mean an arrangement of the constituent judgments of an argument according to certain perceived relations of broadly logical consequence among them. The traditional conception of proof is a specialization of this view to cases where the constituent judgments making up the proof, or at least certain of them, may have special epistemic qualifications (e.g., being self-evident) and the relations of logical consequence which are taken to relate them are perceived relations of deductive consequence.

The classical source of the traditional view was Aristotle, who presented it as part of a general account of the nature of reasoning in the *Prior Analytics*, Bk. I. What is perhaps the most widely known statement of the view was given in the *Posterior Analytics*, however.

> [D]emonstrative knowledge must proceed from premisses which are true, primary, immediate, better known than, prior to, and causative of the conclusion. On these conditions only will the first principles be properly applicable to the fact which is to be proved. Deduction will be possible without these conditions, but not demonstration; for the result will not be knowledge.
>
> *Posterior Analytics*, 71b 20–25

Similar views were expressed throughout the modern era (cf. Locke (cf. [35, Bk IV, ch. xvii, §4]) and Reid (cf. [41, Essay VII, Of Reasoning, p. 475]), and also throughout the eighteenth, nineteenth, and early twentieth centuries (cf. [46, ch.I, pt. 3]; [5, pp. 45–46]; [3, §22]; [34, p. 11]; [20, p. 15] and [33, p. 384] for statements from a variety of different types of works).

Towards the end of the nineteenth century, the traditional conception of proof gave way to a conception of proof coming from the then-emerging "abstract" conception of axiomatization. This new conception of axiomatization differed profoundly from its traditional predecessor.

On the traditional conception of axiomatization, axioms were taken to be true propositions chosen out of consideration of supposed special properties of certainty and/or immediacy and/or explanatory power. Relatedly, proofs were taken to be finite sequences of judgments the propositional contents of each element of which were either to be axioms, or to be seen to follow deductively from the contents of previous elements of the sequence. This is what I will call the *t*raditional *v*iew of *p*roof (TVP).

The abstract conception, by contrast, denied that axioms are certain, self-evident, or explanatorily basic truths. It denied, in fact, that they are truths at all, or even that they are propositions.[5] Axiomatization on the abstract plan sought to separate axioms from contents. Hilbert described the basic process he took to effect this dissociation (in the case of geometry) as follows:

> We think (denken) three different systems of things. The things of the first system we call *points* and designate them $A, B, C$ ...... The things of the second system we call *lines* and designate them $a, b, c$ ...... The things of the third system we call *planes* and designate them $\alpha, \beta, \gamma$ ......
> We think (denken) the points, lines and planes in certain mutual relations ...
> The exact (genaue) and for mathematical purposes complete (vollständige) specification of these relationships is accomplished by the *axioms of geometry*. [22, ch. 1, §1][6]

In axiomatization, in Hilbert's view, we "think." We do not observe or intuit and then express the contents of our observations or intuitions in the axioms we give. Rather, we "think," with nothing given prior to or in association with this thinking to serve as its contents.[7]

Nor was this thinking taken to have indigenous contents, at least not in any ordinary sense of the term "contents." It was not a thinking as of definite objects standing in definite relations. Rather, the objects and relations of axiomatic thinking were wholly unspecified, and could be *any* objects and relations that satisfy the abstractly thought axioms.

From the abstract point of view, then, axioms were not taken to be propositions but rather, for some, propositional functions or propositional schemata (cf. [45, p. 2]; [31, §20]), and for others (e.g., Hilbert) sentences or sentence-schemata. For

---

[5]Describing the abstract viewpoint as applied to projective geometry, Whitehead wrote: "The points mentioned in the axioms are not a special determinate class of entities ... they are in fact any entities whatever, which happen to be inter-related in such a manner, that the axioms are true when considered as referring to those entities and their inter-relations. Accordingly—since the class of points is undetermined—the axioms are not propositions at all ... An axiom (in this sense) since it is not a proposition can neither be true or false." [45, p. 1].

[6]That this represented Hilbert's general conception of axiomatization is indicated by the fact that he gave a precisely parallel characterization of the axiomatic method in arithmetic in an essay published the following year (cf. [23, p. 181]).

[7]The separation of thinking from contents represented in this view is more radical than, but still reminiscent of the separation indicated by Kant in the first critique: "I can think (denken) whatever I want, provided only that I do not contradict myself. This suffices for the possibility of the concept, even though I may not be able to answer for there being, in the sum of all possibilities, an object corresponding to it. Indeed, something more is required before I can ascribe to such a concept objective validity, that is, real possibility; the former possibility is merely logical." [32, xxvi, note a].

present purposes, the difference between these alternatives is insignificant. What is important is that axioms were viewed schematically, or hypothetically—any system of objects and relations satisfying them would also satisfy the theorems that follow from them.[8]

The attributes traditionally taken to characterize axioms (e.g., certainty, self-evidentness, explanatory depth, unprovability (in some objective or quasi-objective sense), etc.) do not of course apply to such schemata. Rather, the thinking regarding choice of axioms for abstract theories seems generally to have been that it should be driven by considerations of mutual consistency and of their usefulness as starting points for the efficient deduction of some further body of theorems.[9]

## 3 The "Decontentualization" of Proof

The core element of the abstract conception of axiomatization was thus a call for the separation—or, perhaps more accurately, call*s* for various separation*s*—of axiomatic thinking from contents. More specifically for my purposes here, it was a family of calls for various separations of the conduct of proof from contentual considerations.

In this connection, it is perhaps useful to distinguish two such separations. One of these is a separation from contents for purposes of conducting the inferential parts of proofs. For convenience, I'll call this Inferential Separation.

The other concerns a separation from contents for purposes of specifying what the constitutive axioms and rules of inference of a would-be formal proof practice are. I'll refer to this as Specificational Separation.

The mature Hilbert, I believe, supported both types of separation. I will now briefly indicate what I take to be essential to each.

### *3.1 Inferential Separation*

In 1882, Pasch had raised the importance of abstracting away from contents for purposes of ensuring that the inferential parts of proofs were genuinely deductive in character.

---

[8]Cf. [2, pp. 95–96].

[9]J.W. Young put the point this way: "[W]hat is the new point of view? The self-evident truth is entirely banished. There is no such thing. What has taken the place of it? Simply a set of assumptions concerning the science which is to be developed, in the choice of which we have considerable freedom. ... [T]hey are elected for their fitness to serve, and their fitness is very largely determined by their simplicity, by the ease with which the other propositions may be derived from them." [47, p. 52].

> [I]f geometry is to be genuinely deductive, the process of inferring (Process des Folgerns) must be everywhere independent of the *sense* (*Sinn*) of geometrical concepts just as it must be independent of figures. It is only *relations* between geometrical concepts that should be taken into account in the propositions and definitions that are dealt with. In the course of a deduction ... it should *by no means be necessary* to think of the references (Bedeutung) of the geometrical concepts involved. ... [I]f it is ..., the gappiness (Lückenhaftigkeit) of the deduction and the inadequacy of the ... proof is thereby revealed unless it is possible to remove the gaps (Lücke) by modifying the reasoning used. [40, p. 98]

There seem to be both theoretical and practical claims here. On the theoretical side there is a suggestion that an inference in a geometrical proof can properly be known to be deductively valid only if its validity can in principle be known without appealing to the contents of any non-logical term (and, more specifically, any geometrical term) that occurs, whether explicitly or implicitly, in it (i.e., in its premises or its conclusion).[10]

Pasch's practical suggestion, as I see it, ran parallel to this. It suggested as a practical criterion of deductive validity that an inference's validity be practically establishable without appealing to the sense or referent of any non-logical term (specifically, the contents of any geometrical term) occurring (explicitly or implicitly) in it. In other words, it called for the separation of geometrical proof from geometrical contents for purposes of determining the deductive validity of its inferential parts. The suggestion seems to be that persistent failure of conscientious efforts to find such a practical separation of contents from assessments of validity is indication of a failure of rigor in a proof.[11]

Hilbert too endorsed a separation of logical reasoning from contents,[12] though neither the separation he proposed nor his reasons for proposing it were identical to Pasch's.

## 3.2 Specificational Separation

Pasch's proposed separation of contents from geometrical reasoning seems in significant part to have been a call for rigor. To correctly judge the deductive validity

[10]Pasch did not of course make use of any precise demarcation of logical from non-logical terms. He did, though, have a sense of what the geometrical terms or concepts in a proof were, and he insisted that the validity of a genuinely deductive inference should be knowable without making use of appeals to the senses or referents of any of the geometrical terms that occur in it.

[11]Pasch's call for Inferential Separation of contents from proofs has led some to regard him as the (or at least a) principal founder of the abstract conception of axiomatization (cf. [39, p. 143]; [42, pp. 343–344] and [47, p. 51]). As others (cf. [15, pp. 617–618]) have pointed out, though, correctly in my view, the separation of geometrical reasoning from contents that he proposed is not nearly so radical as that proposed by Hilbert.

[12]"[I]n my theory contentual inference (inhaltliche Schließen) is replaced by manipulation of signs according to rules (äußeres Handeln nach Regeln); in this way the axiomatic method attains that reliability and perfection that it can and must reach if it is to become the basic instrument of all theoretical research." [27, p. 4].

of a geometrical inference did not, in his view, require appeal to the contents of its geometrical terms. To make use of such appeals, therefore, was either to use what one did not recognize was being used, or it was to mistake what is required for deductive validity. Pasch seems to have seen the former—the use of unrecognized information in the inferential parts of proofs—as the more insidious threat and the one protection against which thus required more careful and deliberate efforts.

The use of such information in the conduct of inference constituted a failure of rigor. Pasch's call for abstraction from the meanings of geometrical terms for purposes of conducting the inferential parts of geometrical proofs was intended to provide protection against such failure.

It is not only in the inferential parts of proofs, however, that use of unrecognized information may enter. It may also enter in the identification or specification of axioms and/or rules of inference. It may be avoidance of this type of illicit use of unrecognized information that Hilbert had in mind when he declared that the specification of axioms of an axiomatic system should provide an "exact (genaue) and for mathematical purposes complete (vollständige) specification" [22, ch. 1, §1] of the objects-as-standing-in-relations that constituted what was thought in a given axiomatic "thinking" (denken). Here, I'll focus on the part of the claim concerning exactness and leave the part concerning completeness for another occasion.

What would constitute a specification of axioms that is "exact" in this sense? There is nothing I know of in Hilbert's early writings that clarifies what he had in mind. In the fuller development of his proof theory, however, he came to the view that axioms should be syntactically rather than semantically specified. More accurately, he came to the view that proper specification of axioms consisted in their being *exhibited* (i.e., in their being given in terms of their outward appearances) rather than in their being *expressed* (i.e., in their being given in terms of semantical contents). To put it differently, Hilbert's eventual view seems to have been that only such things as can be identified by their outward appearances, without application of semantic interpretation, are exactly specifiable. Accordingly, only formulae, not propositions, can ultimately satisfy the requirements of exact specification of an axiomatic thinking (denken).

If this is how Hilbert eventually came to understand the requirement that axioms be "exactly" specified, then it represents another point at (or another way in) which at least his mature understanding of axiomatic thinking saw it as involving various types of "decontentualization."

## 4 Decontentualization and Its Discontents

Weyl described the decontentualized conception of proof of Hilbert's proof theory as representing a radical departure from the views of his predecessors.

> Before Hilbert constructed his proof theory everyone thought of mathematics as a system of contentual (*inhaltliche*), meaningful (*sinnerfüllte*), and evident (*einsichtige*) truths; this point of view was the common platform of all discussions. ... Brouwer, like everyone else,

> required of mathematics that its theorems be (in Hilbert's terminology) "real propositions," meaningful truths. [43, p. 22][13]

This may largely have been true, but, as the above remarks concerning the development of the abstract conception of axiomatization indicate, it's not entirely accurate. Pasch's view of proof, with its distinctive understanding of the requirements of inferential rigor, is not adequately captured by it.[14] Neither does it accurately convey the place that abstract views of axiomatization occupied in late nineteenth and early twentieth century understandings of axiomatic method.[15]

Be this as it may, contentualist understandings of mathematical proof were certainly common and influential during the period in which Hilbert developed his proof-theoretic ideas. Since Gentzen's understanding of the nature of proof seems to have been influenced by such views, it seems sensible to briefly survey some of the more influential contentualist views of proof of Gentzen's time.

Among these, Brouwer's are perhaps particularly important because of Gentzen's expressed sympathies with intuitionist views of proof. Brouwer stressed his opposition to non-contentual conceptions of proof in his criticisms of Hilbert's program—particularly his criticisms of Hilbert's idea that to properly found traditional mathematics is essentially to prove its consistency.[16]

In Brouwer's view, to properly found traditional mathematics (or some part of it), it was necessary to establish it not merely as consistent but as truthful or correct. What is true, however, is contentful since it is contentual items only that are capable of being true or false. Proving the syntactical consistency of a theory or inferential

---

[13]See [44, p. 640] for a similar statement. See also [7, p. 336]; [8, pp. 490–492] and [10, pp. 2–5] for related ideas and arguments.

[14]Neither are the contributions of others with ideas similar to Pasch's. These contributions were noted by various early twentieth century writers. The following statement by Young is characteristic: "The abstract formulation of mathematics seems to date back to the German mathematician Moritz Pasch. At any rate, he was the first to study in detail the axioms concerning the order of points on a straight line ... But to the Italian Giuseppe Peano belongs the credit of developing this point of view systematically. His idea, which he began to elaborate about 1889, is to put the whole of mathematics on a purely formal basis ..." [47, p. 51].

[15]Here too Young gave a more accurate description: "The point of view of 50 years ago was very largely that the foundations of mathematics were axioms; and by axioms were meant self-evident truths, that is, ideas imposed upon our minds a priori, with which we must necessarily begin any rational development of the subject. So the axioms dominated our mathematical science, as it were, by the divine right of the alleged inconceivability of the opposite. And now, what is the new point of view? The self-evident truth is entirely banished. There is no such thing. What has taken the place of it? Simply a set of assumptions concerning the science which is to be developed, in the choice of which we have considerable freedom." (*op. cit.*, 52).

[16]Strictly speaking, Hilbert required more than a proof of consistency for the proper foundation of classical mathematics. He required as well that its uses of ideal methods be "successful": "[I]f the question of the justification (Berechtigung) of a procedure (Maßnahme) means anything more than proving its consistency, it can only mean determining whether the procedure fulfills its promised purpose. Indeed, success is necessary; here, too, it is the highest tribunal, to which everyone submits." [26, p. 163].

practice could not, therefore, in Brouwer's view, properly found it since it would not establish it as correct (*richtig*).

> [T]he *formalistic critique* ... in essence comes to this: the *language accompanying the mathematical mental activity* is subjected to a mathematical examination. To such an examination the laws of theoretical logic present themselves as operators acting on primitive formulas or axioms, and one sets himself the goal of transforming these axioms in such a way that the linguistic effect of the operators mentioned (which are themselves retained unchanged) can no longer be disturbed by the appearance of the linguistic figure of a contradiction. We need by no means despair of reaching this goal,[17] but nothing of mathematical value will thus be gained: an incorrect theory (*unrichtige Theorie*), even if it cannot be inhibited by any contradiction that would refute it, is none the less incorrect, just as a criminal policy is none the less criminal even if it cannot be inhibited by any court that would curb it. [7, p. 336][18]

As Brouwer saw it, then, the fundamental mistake of the formalist was the failure to appreciate the differences between operations of genuine reasoning and operations on linguistic items. The latter might resemble the former in certain ways but, in the end, these could be only superficial similarities. To fail to recognize this was to fail to see the critical difference between genuine thinking and a mere use of language—a difference featured in what was perhaps the basic element of Brouwer's foundational outlook, the so-called *First Act of Intuitionism*.

> [T]he FIRST ACT OF INTUITIONISM completely separates mathematics from mathematical language, in particular from the phenomena of language which are described by theoretical logic, and recognizes that intuitionist mathematics is an essentially languageless activity of the mind ... [9, pp. 140–141]

Formalism flouted the First Act of Intutionism. More specifically, in the intuitionists' view, it systematically overestimated the importance of language as a vehicle for the conduct of reasoning. Similarly, as they saw it, it overestimated even the importance of mathematical language as a means of representing and studying the properties of mathematical reasoning.

The intuitionists were not the only ones to object to the decontentualizing tendencies of Hilbert's abstractionist outlook. Klein, for example, described it as representing "the death of all science" [33, p. 384].

---

[17] At this point Brouwer inserted the following remark in a note: "[T]he unjustified application of the principle of excluded middle to properties of well-constructed mathematical systems can never lead to a contradiction ..."

[18] The passage in the German original is on pp. 2–3. It is perhaps worth noting that on Hilbert's view, consistency meant consistency with real mathematics. Therefore, if incorrectness is defined as proving something that is refutable by real means, then proving consistency in Hilbert's sense *would* eliminate the possibility of incorrectness on one natural understanding of that term. Perhaps on Brouwer's understanding of "unrichtige," "richtige" was intended to imply conservativeness—so that a theory would be incorrect if it proved propositions that are not themselves provable by real means, and not only if it proved propositions that are refutable by real means. Under certain conditions, of course, the two understandings extensionally coincide.

Frege too decried it and he made the contentual nature of genuine proof the focal point of his disagreements with Hilbert and those others (e.g., Heine and Thomae) he saw as advocating non-contentualist views of proof.

> [A]n inference does not consist of signs. We can only say that in the transition from one group of signs to a new group of signs, it may look now and then as though we are presented with an inference. An inference simply does not belong to the realm of signs; rather, it is the pronouncement of a judgment made in accordance with logical laws on the basis of previously passed judgments. Each of the premises is a determinate thought recognized as true; and in the conclusion, too, a determinate thought is recognized as true .... [12, p. 387]

In Frege's view too, then, a proof was a thoroughly contentual affair—specifically, it was a sequence of judgments the propositional contents of which must be judged by the prover to stand in certain logical relationships to one another. Without such logical interrelationship there can be no genuine proof, and unless the premises and conclusions of proofs have propositional contents, there can be no genuine logical relationship between them.

## 5 Hilbert's Conductive Formalism

As mentioned, Hilbert rejected this traditional contentualist conception of proof (and, more generally, the traditional contentualist view of reasoning). This should not, however, lead us to think that he denied the importance, or even the centrality, of contentual proof to the development of mathematical knowledge. He did not. In fact, he emphasized the importance of contentual reasoning to mathematics and, particularly, its indispensability to metamathematics.

Where he thought the opponents of non-contentual reasoning had gone too far was in their view that mathematical reasoning and proof has and indeed must always be contentual, or that non-contentual reasoning has played only an insignificant role in the historical development of our mathematical knowledge. In Hilbert's view, mathematical proof has often assumed non-contentual forms, and he believed the use of such forms to have been and to continue to be invaluable in our attempts to mitigate various types of complexity and/or inefficiency that commonly limit the usefulness or even the practical applicability of contentual methods of proof.

Hilbert gave various examples intended to illustrate the usefulness of non-contentual methods of reasoning in mathematics. These included the introduction of the imaginary and complex numbers to "simplify the theorems on the existence and number of roots of an equation" [26, p. 166] and the introduction of elements at infinity in projective geometry which "make the system of laws of connection as simple and perspicuous as is possible" (*loc. cit.*) and which induce the symmetries behind the dualities of projective geometry "which are so fruitful (*fruchtbare*)" (*loc. cit.*).

What he regarded as the crowning example, though, is the use of the classical laws of logic to manage what he believed are crippling complexities of non-classical (specifically, finitary) contentual logical reasoning. Classical methods of logical reasoning may not be contentual, but this ought not blind us to the fact that they may be useful, even, in some sense, indispensable to the practical conduct of (at least parts of) our logical reasoning.

Hilbert thus urged addition of the so-called ideal propositions [26, p. 174] to real contentual propositions "in order to maintain the formally simple rules of ordinary Aristotelian logic" (*ibid.*).[19] To make such an addition was, in his view, a natural and motivated application of the method of ideal methods in mathematics, a method which had proved its efficacy and trustworthiness again and again in the history of mathematics.

Hilbert seems also to have seen the application of ideal methods as pervasive both in our scientific and in our everyday reasoning.

> In our theoretical sciences we are accustomed to the use of formal thought processes (*formaler Denkprozesse*) and abstract methods ... [But] already in everyday life (*täglichen Leben*) one uses methods and concept-constructions (*Begriffsbildungen*) which require a high degree of abstraction and which only become plain through unconscious application of the axiomatic method (*nur durch unbewußte Anwendung der axiomatischen Methoden verständlich sind*). Examples include the general process of negation and, especially, the concept of infinity. [28, p. 380]

To try to do without ideal methods in our thinking would thus, in Hilbert's view, seriously impair our effectiveness as thinkers. Opposition to their use was, in Hilbert's view, largely a result of a failure to recognize that language has valuable and legitimate non-descriptive uses. Bernays memorably urged this point in offering his Faustian summary of Hilbert's formalist viewpoint.

> Where concepts fail, a sign appears at just the right time.[20] This is the methodological principle of Hilbert's theory. [1, p. 16]

Hilbert put the point more forcefully, if perhaps less picturesquely. In his view, the use of non-contentual (or, more specifically, symbolico-algebraic) methods in the conduct of our reasoning is indispensable to the fullest practical development of our mathematical knowledge (cf. [24, pp. 162–163]; [26, p. 162]; [27, pp. 7–8]). He saw it as reflecting the importance attached to the use of non-descriptive simplifying idealizations he took to be characteristic of modern science (cf. [29, pp. 2–3]). He believed that we may legitimately take advantage of the benefits of such simplification without sacrificing security in the contentual parts of our thinking.

[19]This addition was to be controlled by consistency, of course. On this point, Hilbert thought he could satisfy even Brouwer and Kronecker. What they did not accept, however, is that controlling for consistency should be enough to establish a putative body of reasoning as genuine reasoning, much less as reliable genuine reasoning.

[20]"Thus even where concepts fail, a word appears at just the right time."

Goethe, *Faust* I (Mephistopheles to a student of theology)

The German is: "Denn eben wo Begriffe fehlen, Da stellt ein Wort zur rechten Zeit sich ein." Goethe was not endorsing but criticizing such a practice of course. He presented it as a practice employed by teachers of theology to preserve a facade of contentful thinking where in fact there were only contentually empty words.

Hilbert thus embraced what I am calling *Conductive Formalism*, the view that the use of non-contentual methods of reasoning has been and continues to be important to the effective development of our mathematical knowledge. He accepted as well of course what I call *Representational Formalism*—that is, the view that the formal representation of mathematical reasoning is a tool for facilitating the rigorous and mathematically precise investigation of mathematical reasoning.

As I read him, Gentzen only fully endorsed Representational Formalism. He seems not to have taken the use of non-contentual methods in mathematics to qualify as genuine reasoning. In addition, he seems to have taken the provision of a contentual interpretation for what Hilbert termed "ideal reasoning" (and what he, Gentzen, termed actualist or *an-sich* reasoning) as important to its proper foundation.

In these important respects, then, Gentzen's views more nearly resembled the anti-formalist views of Brouwer and Frege than the formalist views of Hilbert.

## 6 Gentzen's Conductive Contentualism

Gentzen seems in fact to have gone out of his way both to comply with traditional contentualist strictures on reasoning and to make clear his endorsement of them. §9 and §17.3 of [17] provide clear confirmation of this. They are dedicated to establishing compliance with contentualist demands as regards reasoning to actualist conclusions (more accurately, reasoning to contentual conclusions expressed by actualist sentences) in number theory.

Gentzen also took pains to show that his consistency proof for number theory meets all reasonable demands of this type. He was particularly concerned to show that his consistency proof provides for the *finitary interpretation* of the actualist sentences of number theory, and he seems to have seen this as a necessary part of justifying the use of actualist methods in arithmetic.

> The most essential component (*wesentlichste Teil*) of my consistency proof ...consists precisely in its attachment of a finitary sense to actualist propositions (*daß den an-sich Aussagen ein finiter Sinn beigelegt wird*), viz. for any given proposition, if it is proven, a reduction rule (*Reduziervorschrift*) ...can be specified, and this fact represents the finitary sense of the proposition that is obtained precisely through the consistency proof. [17, p. 564]

My point and my claim is not that Gentzen was right to have described his proof as providing finitary senses for actualist sentences. It is rather that *he* seems to have seen it as doing so, and he seems to have seen its doing so as being in some way its most essential feature.

Gentzen thus seems to have affirmed the traditional view that to be fully justified, actualist methods must be contentually interpreted.

> [E]ven if it should be demonstrated that the disputed forms of inference cannot lead to mutually contradictory results, these results would nonetheless be propositions *without sense* (*sinnlose Aussagen*) and their investigation therefore a mere recreation (*eine Spielerei*);

> genuine *knowledge* (*wirkliche Erkenntnisse*) can be gained only by means of the unobjectionable (*unbedenklichen*) intuitionist (or finitist, as the case may be) forms of inference [17, p. 564] (emphases as in text)[21]

Gentzen then went on to consider what value (*Erkenntniswert*) (*loc. cit.*) there might be in uninterpreted actualist reasoning (i.e., in actualist reasoning which, though lacking interpretation, nonetheless qualifies as actualist reasoning). He allowed as how it might have some practical value (*praktischer Wert*) (*ibid.*) and not be entirely useless (*nicht ganz zwecklos*) (*ibid.*) as an instrument of thinking. This was not, however, for him an adequate substitute for its providing a genuine contentual justification for its conclusion.

This too is similar to the things intuitionists said about the value of actualist reasoning. We already noted one such point by Brouwer in his concession that actualist reasoning might be "an efficient ... technique for memorizing mathematical constructions, and for suggesting them to others" [9, p. 140].[22] He even allowed as how it might be contentually reliable over a certain range of cases.

> *Suppose that an intuitionist mathematical construction has been carefully described by means of words, and then, the introspective character of the mathematical construction being ignored for a moment, its linguistic description is considered by itself and submitted to a linguistic application of a principle of classical logic. Is it then always possible to perform a languageless mathematical construction finding its expression in the logico-linguistic figure in question?*
> After careful examination one answers this question in the affirmative (if one allows for the inevitable inadequacy of language as a mode of description) as far as the principles of contradiction and syllogism are concerned; but in the negative (except in special cases) with regard to the principle of excluded third ... [9, p. 140]

What neither Brouwer nor Gentzen was willing to grant, though, and what in the end seems to have constituted their deepest difference with Hilbert, is that actualist reasoning might be an acceptable replacement for contentual reasoning were its syntactical consistency with finitary contentual reasoning to be finitarily proven.

Gentzen's contentualist convictions seem to have stemmed from a view that non-contentual "reasoning" is not genuine reasoning at all, that it is fundamentally a type of game and that it cannot therefore properly be a part of a genuine *science* of mathematics (cf. [17, p. 564]).

This was in fact the common attitude of the late nineteenth and early twentieth centuries. The thinking was that what essentially separates science from a game is applicability. A genuine science *is* (at least potentially) applicable. A game is not. What makes genuine sciences applicable and games not is that the former, in

---

[21]Compare this to the remark by Brouwer, quoted earlier, that "even if [actualist reasoning] cannot be inhibited by any contradiction that would refute it, it is none the less incorrect, just as a criminal policy is none the less criminal even if it cannot be inhibited by any court that would curb it," [7, p. 336] brackets added.

[22]Gentzen could not have read this text of course. There are, though, earlier texts which express similar ideas. Cf. [6] and [7] for related, though less firm endorsements of the utility of classical reasoning as a certain type of instrument to guide our thinking.

contrast to the latter, express thoughts or contents and, in doing so, they *describe* the world and so become applicable to it.

Frege expressed these ideas clearly in the second volume of the *Grundgesetze*.

> Why can one make no application (*keine Anwendung machen*) of a position (*Stellung*) of chess figures? Clearly because it expresses no thought (*es keinen Gedanken ausdrückt*). ... Why can one make applications of arithmetical equalities? Only because they express thoughts (*nur weil sie Gedanken ausdrücken*). How could we possibly apply an equation which expressed nothing, was nothing more than a group of figures (*Figurengruppe*) to be transformed (*umgewandelt*) into another group of figures by certain rules! It is applicability alone (*Anwendbarkeit allein*) that raises (*erhebt*) arithmetic from a game to the rank of science. Is it a good thing (*wohlgetan*), then, to exclude from arithmetic that which is necessary for it to be a science? [11, §91]

Hilbert saw little to justify such thinking. He accepted the idea that mathematics ought to be applicable. He did not however accept the traditional *descriptive* paradigm of application—that application essentially consists in or at least requires description (i.e., expression of a true thought or content). He adhered instead to the Berkeleyan idea that, though the application of reasoning to reality may require that its conclusion be interpretable (i.e., that it admit of interpretation by a true thought or content), the same is not true of the various steps of reasoning that lead to that conclusion.

Actualist sentences were in Hilbert's view instruments of thought and their use was essentially axiomatic in character—that is, it was completely governed by explicit (i.e., syntactically stated) rules of usage. By this he seems to have meant that actualist sentences do not function contentually, and that, accordingly, their justified use does not require semantical interpretation, be it constructive or actualist in character.

As Hilbert's saw it, Brouwer operated with substantially the same scheme of distinctions and made essentially the same mistakes that Frege did. He assumed not only that application requires interpretation but also that the rules according to which the formal operations of ideal reasoning proceed are in some sense *convened* or *chosen*.

This, to Hilbert, was a distorting oversimplification. As he saw it, the rules according to which time-tested ideal reasoning proceeds are laws according to which our reasoning most effectively proceeds. We do not *merely* choose or convene them. Rather, we experiment with various instruments of reasoning in order to test their effectiveness, and we subject them to metamathematical investigation to determine their consistency with the results of real reasoning. Those which survive such testing represent the accumulated experience and prudence of the larger community of mathematical reasoners. The discovery and metamathematical vindication of such laws, in Hilbert's view, deserved to be made the chief focus of foundational investigation.

> The formula game (*Formelspiel*) that Brouwer so dismissively judges (*wegwerfend urteilt*) has, besides its mathematical value, an important general philosophical significance. For this formula game is carried out according to certain definite rules, in which the technique of our thinking is expressed. These rules form a closed system that can be discovered

> and definitively stated. The fundamental idea of my proof theory is none other than to describe the activity of our understanding, to make a protocol of the rules according to which our thinking actually proceeds. Thinking (*Das Denken*), it so happens, parallels speaking and writing: we form statements and place them one behind another. If any totality of observations and phenomena deserves to be made the object of a serious and thorough investigation (*ernsten und grundlichen Forschung*), it is this one ... [27, pp. 15–16]

The discovery and metamathematical vindication of formal methods of reasoning was thus, in Hilbert's view, far from being a game. It was rather, in a profound sense, the investigation of the laws of human thinking and, as such, deserved to be made a chief focus of foundational research in mathematics.

## 7 Conclusion

Hilbert and Gentzen were not formalists of the same type. Specifically, Hilbert advocated a version of Conductive Formalism while Gentzen did not. More specifically, Gentzen held a fairly traditional contentualist view of the nature of proof while Hilbert rejected such a view, and, indeed emphasized the importance of the use of non-contentual methods in mathematics to the overall development of mathematical knowledge.

Hilbert was in fact emphatic on this point. His conviction reflected his observation of the fruitful uses that had been made of non-contentual methods of reasoning throughout the history of mathematics. It also reflected his general view of the place of idealization in mathematics and in modern science generally.

According to this view, scientific mathematics not only does not require interpretation, it does not generally invite it. The reason is the characteristic use it makes of simplifying idealizations.

The reasoning that stems from such idealizations is not intended to be interpreted and, generally speaking, it is neither necessary nor desirable that it should be. All that is required is that it be shown not to conflict with the results of the real or contentual (i.e., the non-idealizational) parts of the given science.

Such separation of mathematical reasoning from contentual interpretation was a central element of Hilbert's formalism.

Gentzen, by contrast, was committed both to a contentualist understanding of proof and to a view to the effect that to properly found a body of mathematical reasoning requires providing an interpretation for it. In fact, as noted above, he described as a key virtue of his consistency proof for classical first-order arithmetic that it provides finitary senses for actualist propositions (cf. [17, p. 564]).[23]

[23]Gentzen in fact raised the possibility of the need for a second type of interpretation for classical arithmetic—one which provides *actualist* interpretations of actualist sentences. Cf. [17, p. 565].

If this is right, then Gentzen cannot plausibly be described as having been a formalist of the sort Hilbert was. He was not, in particular, a conductive formalist. He did not emphasize, as Hilbert did, the importance of non-contentual methods as means of conducting mathematical reasoning. He seems not in fact to have seen the use of non-contentual methods as constituting genuine reasoning at all. Still less did he see it as the glory of modern mathematics.

**Acknowledgements** I gratefully acknowledge the generous support of the *Agence nationale de la recherche* (ANR) through its *chaires d'excellence* program. Thanks too to the HPS department at the Université de Paris 7–Diderot, the Archives Poincaré and the philosophy department at the Université de Lorraine, the chairs in epistemology and the philosophy of language at the Collège de France and the University of Notre Dame for their support. Thanks finally to audiences and individuals at the Gentzen Centenary conference in Coimbra, at Cambridge University and at the fourth French PhilMath workshop for helpful discussions of ideas in this paper.

## References

1. P. Bernays, Über Hilbert's Gedanken zur Grundlegung der Arithmetik. Jahresberichte der deutsche Mathematiker-Verienigung **30**, 10–19 (1922)
2. P. Bernays, Die Bedeutung Hilberts für die Philosophie der Mathematik. Die Naturwissenschaften **10**, 93–99 (1922)
3. B. Bolzano, *Beiträge zu einer begründeteren Darstellung der Mathematik* (Caspar Widtmann, Prague, 1810). English trans. in [4]
4. B. Bolzano, *The Mathematical Works of Bernard Bolzano*, ed. by S. Russ (Oxford University Press, Oxford, 2004)
5. C. Bossut, *Essai sur l'histoire génerale des mathématiques* (Louis, Paris, 1802)
6. L.E.J. Brouwer, Intuitionism and formalism. Bull. Am. Math. Soc. **20**, 81–96 (1913)
7. L.E.J. Brouwer, Über die Bedeutung des Satzes vom ausgescholessenen Dritten in der Mathematik, insbesondere in der Funktiontheorie. Journal für die reine und angewandte Mathematik **154**, 1–7 (1923). English translation in [21], pp. 334–345. Page references are to this translation
8. L.E.J. Brouwer, Intuitionstische Betrachtungen über den Formalismus, in *Koninklijke Akademie van wetenschappen de Amsterdam, Proceedings of the section of sciences*, vol. 31 (1928), pp. 324–329. English translation with introduction in [21], pp. 490–492. Page references are to this translation
9. L.E.J. Brouwer, Historical background, principles and methods of intuitionism. S. Afr. J. Sci. **49**, 139–146 (1952)
10. L.E.J. Brouwer, *Brouwer's Cambridge Lectures on Intuitionism*, ed. by D. van Dalen (Cambridge University Press, Cambridge, 1981)
11. G. Frege, *Grundgesetze der Arithmetik: begriffsschriftlich abgeleitet*, two volumes (H. Pohle, Jena, 1893–1903)
12. G. Frege, Über die Grundlagen der Geometrie. Jahresbericht der Deutschen Mathematiker-Vereinigung **15**, 293–309, 377–403, 423–430 (1906). English translation in [13]
13. G. Frege, *Gottlob Frege: On the Foundations of Geometry and Formal Theories of Arithmetic*, trans. and ed., with an introduction, by Eike-Henner W. Kluge, (Yale University Press, New Haven, 1971)
14. G. Frege, *Gottlob Frege: Philosophical and Mathematical Correspondence*, ed. by G. Gabriel et al. (University of Chicago Press, Chicago, 1980)

15. H. Freudenthal, The main trends in the foundations of geometry in the 19th century, in [39], pp. 613–621.
16. G. Gentzen, Der Unendlichkeitsbegriff in der Mathematik, in *Semesterberichte Münster*, WS 1936/37, pp. 65–85
17. G. Gentzen, Die Widerspruchsfreiheit der reinen Zahlentheorie. Math. Ann. **112**, 493–565 (1936)
18. G. Gentzen, Unendlichkeitsbegriff und Widerspruchsfreiheit der Mathematik, in *Travaux du IXeme Congrès internationale de philosophie*, vol. VI (Hermann, Paris, 1937)
19. G. Gentzen, *The Collected Papers of Gerhard Gentzen*, ed. and translator by M.E. Szabo (North-Holland, Amsterdam, 1969)
20. S. Hawtrey, *An Introduction to the Elements of Euclid, Pt. I*, 2nd edn. (Longmans, Green & Co., London, 1878)
21. J. van Heijenoort, *From Frege to Gödel: A Source Book in Mathematical Logic, 1879–1931* (Harvard University Press, Cambridge, 1967)
22. D. Hilbert, *Grundlagen der Geometrie* (Teubner, Leipzig, 1899)
23. D. Hilbert, Über den Zahlbegriff. Jahresbericht der Deutschen Mathematiker-Vereinigung **8**, 180–183 (1900)
24. D. Hilbert, Neubegründung der Mathematik. Erste Mitteilung. Abhandlungen aus dem Mathematischen Seminar der Hamburgischen Universität, vol. 1 (1922), pp. 157–177
25. D. Hilbert, The New Grounding of Mathematics—Part I, English trans. by P. Mancosu of [24], in [37]
26. D. Hilbert, Über das Unendliche. Math. Ann. **95**, 161–190 (1926)
27. D. Hilbert, Die Grundlagen der Mathematik, in *Abhandlungen aus dem mathematischen Seminar der Hamburgischen Universität*, vol. 6 (1928), pp. 65–85. Reprinted in *Hamburger Einzelschriften* 5, pp. 1–21. Teubner, Leipzig, 1928. Page references are to the latter
28. D. Hilbert, *Naturerkennen und Logik: Königsberg 1930* (Springer, Berlin, 1971). Printed in [30], vol. III. Page references are to this printing
29. D. Hilbert, P. Bernays, *Grundlagen der Mathematik*, vol. I (Springer, Berlin, 1934)
30. D. Hilbert, *Gesammelte Abhandlungen*, three volumes (Chelsea Publishing Company, New York, 1932–1935)
31. E.V. Huntington *The Fundamental Propositions of Algebra* (Galois Institute Press/Long Island University, Brooklyn, 1911)
32. I. Kant, *Kritik der reinen Vernunft*, 2nd edn. (J.F. Hartknoch, Riga, 1787)
33. F. Klein, *Elementarmathematik vom höheren Standpunkte aus*, Teil. 2 (Geometrie, Teubner, Leipzig, 1909)
34. D. Lardner, *The First Six Books of Euclid's Elements*, 6th edn. (Taylor & Walton, London, 1838)
35. J. Locke, *An Essay Concerning Human Understanding in Four books*. Printed for Tho. Basset, and sold by Edw. Mory, London, 1690
36. C. MacLaurin, *A Treatise of Fluxions in Two Books*, vol. I (T.W. & T. Ruddimans, Edinburgh, 1742)
37. P. Mancosu, *From Brouwer to Hilbert: The Debate on the Foundations of Mathematics in the 1920s* (Oxford University Press, Oxford, 1998)
38. E.H. Moore, On the projective axioms of geometry. Trans. Am. Math. Soc. **3**, 142–158 (1902)
39. E. Nagel, et al. (eds.), Logic, methodology and philosophy of science, in *Proceedings of the 1960 International Congress* (Stanford University Press, Stanford, 1962)
40. M. Pasch, *Vorlesungen über neuere Geometrie* (Teubner, Leipzig, 1882)
41. T. Reid, *Essays on the Intellectual Powers of Man*. Printed for John Bell, and G. G. J. & J. Robinson, London, 1785. Page references are to the 7th edn., printed in 1872
42. O. Veblen, A system of axioms for geometry. Trans. Am. Math. Soc. **5**, 343–384 (1904)
43. H. Weyl, Diskussionsbemerkungen zu dem zweiten Hilbertschen Vortrag über die Grundlagen der Mathematik. *Abhandlungen aus dem mathematischen Seminar der Hamburgischen Universität*, vol. 6 (1928), pp. 86–88. Reprinted in *Hamburger Einzelschriften* 5, pp. 22–24, Teubner, Leipzig, 1928. English translation in [21]

44. H. Weyl, David Hilbert and his mathematical work. Bull. Am. Math. Soc. **50**, 612–654 (1944)
45. A.N. Whitehead, *The Axioms of Projective Geometry* (University Press, Cambridge, 1906)
46. C. Wolff, *A Treatise of Algebra* (A. Bettesworth & C. Hitch, London, 1739)
47. J.W. Young, *Lectures on Fundamental Concepts of Algebra and Geometry* (Macmillan, New York, 1911)

# The Use of Trustworthy Principles in a Revised Hilbert's Program

**Anton Setzer**

**Abstract** After the failure of Hilbert's original program due to Gödel's second incompleteness theorem, relativized Hilbert's programs have been suggested. While most metamathematical investigations are focused on carrying out mathematical reductions, we claim that in order to give a full substitute for Hilbert's program, one should not stop with purely mathematical investigations, but give an answer to the question why one should believe that all theorems proved in certain mathematical theories are valid.

We suggest that, while it is not possible to obtain absolute certainty, it is possible to develop trustworthy core principles using which one can prove the correctness of mathematical theories. Trust can be established by both providing a direct validation of such principles, which is necessarily non-mathematical and philosophical in nature, and at the same time testing those principles using metamathematical investigations. We investigate three approaches for trustworthy principles, namely ordinal notation systems built from below, Martin-Löf type theory, and Feferman's system of explicit mathematics. We will review what is known about the strength up to which direct validation can be provided.

## 1 Reducing Theories to Trustworthy Principles

In the early 1920s Hilbert suggested a program for the foundation of mathematics, which is now called Hilbert's program. As formulated in [40], "it calls for a formalization of all of mathematics in axiomatic form, together with a proof that this axiomatization of mathematics is consistent. The consistency proof itself was to be carried out using only what Hilbert called 'finitary' methods. The special epistemological character of finitary reasoning then yields the required justification of classical mathematics." Because of Gödel's second incompleteness theorem, Hilbert's program can be carried out only for very weak theories. Because of this failure (see, e.g., [40, 44]) a relativized Hilbert's program has been suggested by

A. Setzer (✉)
Department of Computer Science, Swansea University, Singleton Park, Swansea SA2 8PP, UK
e-mail: a.g.setzer@swan.ac.uk



R. Kahle, M. Rathjen (eds.), *Gentzen's Centenary*,
DOI 10.1007/978-3-319-10103-3_3

Kreisel (Zach [44] cites [17–19]), and then further developed by Feferman [7–10]. In the approach by Feferman [7, 9], one considers two frameworks $\mathcal{F}_1$ and $\mathcal{F}_2$. $\mathcal{F}_1$ could mean infinitary, $\mathcal{F}_2$ finitary, or $\mathcal{F}_1$ mean nonconstructive, $\mathcal{F}_2$ constructive (see p. 367 of [7]). Consider for $i \in \{1, 2\}$ certain theories $T_i$ formulated in languages $\mathcal{L}_i$ corresponding to frameworks $\mathcal{F}_i$. Let $\Phi$ be a primitive recursive subset of the formulae of $\mathcal{L}_1 \cap \mathcal{L}_2$. Let $U$ be a third theory, usually a very weak theory such as PRA. Then combining [8, 10], we have $T_1 \leq T_2[\Phi]$ in $U$, if there exists a partial recursive function $f$ such that

1. if $p$ is a proof in $T_1$ of a formula $\varphi$ in $\Phi$, then $f(p)$ is a proof of $\varphi$ in $T_2$;
2. (1) can be shown in $U$.

Feferman presents many examples of such reductions.

This program of reductive proof theory gives rise to many interesting connections between various theories which provides us with a broad picture of mathematical theories and their relationship. While being very insightful and resulting in lots of metatheorems, it fails to answer the initial question by Hilbert, namely: do I know that my original theory $T_1$ is consistent? Or widening it in the sense of Kreisel and Feferman: If I have proved in theory $T_1$ a mathematical statement, do I know that it is valid? If we take say a proof of Fermat's last theorem, do we know that there is actually no counter example to this theorem? From Gödel's second incompleteness theorem it follows that there is no mathematical argument that excludes that there is at the same time a proof of Fermat's last theorem in a theory $T_1$ and a counter example (unless $T_1$ is very weak), without assuming at least the consistency of another theory of at least equal strength.

Many mathematicians evade this problem and say that all they want is to have a proof which can be formulated in, for instance, Zermelo–Fraenkel set theory. However, this is not what mathematics is intended for. Mathematics is not just a glass bead game in the sense of Hesse [15], a formal game of finding strings of symbols which follow certain decidable rules. The goal of mathematics is, as any science, to establish truth about real properties. In case of Fermat's last theorem, we want to know whether there are no numbers violating it.

What we can do, in the sense of Kreisel and Feferman, is to reduce $T_1$ to another theory $T_2$, which is essentially as strong as $T_1$, and then obtain that $T_2$ proves as well the mathematical theorems of $T_1$ we are interested in. Any mathematical argument will only reduce $T_2$ further to another theory $T_3$. So in order not to continue going in circles, we need to reduce $T_1$ to one theory $T_2$ for which we can give reasons why we believe that everything it proves (possibly restricted to a subset of statements) is valid.

At this point pure mathematical reasoning ends. No matter what we do, we cannot obtain absolute certainty. However we can establish trust. Trust does not mean blind faith. Trust is established by convincing ourselves in the best possible way that what we trust in does not break. This means that we carefully investigate the principles underlying $T_2$, examine them, and give an argument why we can trust them. However, such an analysis can never be done in a purely mathematical way—if we do this, then we just reduce $T_2$ to a third theory $T_3$, namely the theory in which the argument of the correctness of $T_2$ is formulated, and we just have added a new theory to our chain of theories.

However, what we can do, and many constructive and semi-constructive theories have been developed for this purpose, is to formulate theories $T_2$ where these principles are as pure and clean as possible. Then we can carry out two further steps:

1. We can formulate as precisely as possible an argument why we believe that we can trust in those principles. Note that this is no longer a purely mathematical argument. However, making it as precise as possible is a very valuable exercise, since it could reveal any possible flaws in those principles.
2. Since an argument as in (1) does not have the status of a mathematical theorem, it can never provide absolute certainty.[1] Therefore what is needed is to carry out additional testing. Note that mathematicians will in many cases still test their mathematical theorems even if they have proven them, however usually only in order to detect possible flaws in their proofs.

How do we test a theory?

- We can look at theorems provable in $T_2$ and check whether the theorems actually are true (e.g. in case of Fermat's last theorem that there is no counter example). However, there is one problem, namely that by the results of reverse mathematics we know that most mathematical theorems require very little proof theoretic strength. So such tests do not explore the limits of the theory.
  Peter Dybjer has in [3] suggested to develop meaning explanations for Martin-Löf type theory (MLTT) based on the principle that for each judgement of type theory a test is given. The judgement is valid if it passes all tests. Once carried out in full ([3] provides only the basic idea) one obtains for every provable judgements of type theory a test for its validity. Dybjer's article was a major inspiration for this part of the article.
- Ordinal analysis, or any other proof theoretic analysis (e.g. normalisation proofs) is a very strong test, because it tests the theory at its limits. However, this does not establish absolute certainty. When the author was pointing out to Per Martin-Löf that Michael Rathjen had told the author that he knows that $\Pi^1_2$-CA is consistent because he has proof theoretically analysed it, Martin-Löf pointed out that he had an inconsistent type theory and a normalisation proof of it. The problem was that the normalisation proof was carried out in an inconsistent theory. So even a cut elimination or normalisation argument does not guarantee the consistency of the theory.

[1] Of course even mathematical theorems can never give absolute certainty as outlined before. One can think as suggested by one of the referees that a short carefully checked mathematical proof that uses no controversial principles is the paradigm of practical certainty. However, unless one uses extremely weak principles, Gödel's incompleteness theorem applies here as well—even though it is unlikely that an inconsistency is used, we cannot exclude it.

Does this mean that we should give up proof theoretical analysis and normalisation proofs? No, not at all. If a theory is inconsistent, it is likely but not guaranteed that the inconsistency will be found when analysing it proof theoretically. A proof theoretic analysis is up to now one of the strongest ways to stretch a theory to its limits, because it requires to use principles which cannot be reduced to simpler ones. We can often reduce theories which are more expressive to less expressive ones of equal strength in such a way that the reduction shows that they are equiconsistent. However, we cannot reduce a proof theoretic stronger theory to a weaker one, unless both are inconsistent. A proof theoretic analysis needs to distinguish theories of different strength and therefore needs to make use of the principles which are responsible for its strength and which cannot be reduced to weaker ones.

One reason why a proof theoretic analysis is of big significance was pointed out by one of the referees of this article, who wrote "Something that makes specifically ordinal-theoretical proof-theoretical analyses of a theory particularly convincing is that in many cases there is a big difference between the metatheory and the object theory; whereas with normalisation proofs based on Tait-style computability, or Girard-style 'candidates', the metatheory is (more-or-less) the theory itself together with a uniform reflection principle. Something would be far wrong if one could not prove a normalisation theorem for Church's theory of types in such a metatheory; but the extra confidence one gets in the principles formulated therein from a normalisation theorem is tiny."

- In general, any metamathematical analysis of a theory is a test of it. It requires to investigate all axioms and rules of the theory in detail. And if there is an inconsistency in a theory, there is the possibility that one discovers it when carrying out this analysis.[2] If one does not discover any problem, we know at least that any derivation of an inconsistency must be increasingly complicated, since it escaped such a careful analysis. So even if a theory is eventually found to be inconsistent, it is likely that most proofs carried out in it do not make use of it, and we can replace them by proofs in a weaker theory, which does not have this inconsistency.

Therefore there is the need to define mathematical theories in which we can put our trust and describe as clearly as possible the reasons why we trust in the consistency of those theories.

### 1.1 Does the Consistency Problem Matter?

When discussing the problem about consistency, many mathematicians will wonder why there is a problem. Zermelo–Fraenkel set theory (ZF) has been in use since

[2]However, we can never be certain since the metatheory in which the analysis is carried out would be inconsistent as well.

1922. Most of mathematics can be carried out in extensions of it, and it has been analysed thoroughly by set theorists.

However, as we know from reverse mathematics, most mathematical proofs can be carried out in theories which are proof theoretically very weak compared to ZF, therefore mathematical proofs will not explore the limits of ZF. Metamathematical investigations have not really stretched theories having the strength of ZF or greater by themselves, but only investigated such theories relative to other theories of strength of at least that of ZF. Proof theory has succeeded to analyse in unpublished form (Arai [1], see as well [2]) theories of strength Kripke–Platek set theory $+\Pi_1$-Collection $+\mathrm{V} = \mathrm{L}$ (which embeds $(\Sigma^1_3 - \mathrm{DC}) + \mathrm{BI}$ and $(\Sigma^1_3 - \mathrm{AC}) + \mathrm{BI}$). In fully published form Rathjen has analysed [33] the theory of Kripke–Platek set theory plus the existence of one stable ordinal, which embeds $(\Delta^1_2 - \mathrm{CA}) + (\mathrm{BI}) + (\Pi^1_2 - \mathrm{CA})^-$, where $(\Pi^1_2 - \mathrm{CA})^-$ is parameter free $\Pi^1_2 - \mathrm{CA}$. These theories have strength well below that of ZF, and already here interesting phenomena were discovered which were very difficult to harness proof theoretically. Writing down those results has taken a long time. Most likely the reason why an analysis has been so difficult is that our technology is not evolved enough to harness that strength. However, as long as we have not analysed proof theoretically full set theory, it cannot be ruled out that there is an inconsistency lurking somewhere.[3]

Martin-Löf said in his talk at the conference "100 years of intuitionism" at Cerisy ([24], p. 254) that we are not certain that set theory is consistent. He stressed his point using a quote by Woodin.[4] He talked as well about the second failure of Hilbert's problem, which is due to technical difficulties in reaching $\Pi^1_2 - \mathrm{CA}$ and beyond.[5]

Many mathematicians have experienced that sometimes when they get stuck with proving a theorem the underlying reason is that the theorem is actually false. This psychological argument does not prove anything, especially, since when getting mathematically stuck, often all that is needed is a better idea in order to prove the theorem. However, it should provide at least for the highly sceptical scientist a strong motivation to continue with the proof theoretic project. Hilbert said "We need

[3] And even if we have, a validation argument needs to be carried out.

[4] "Just as those who study large cardinals must admit the possibility that the notions are not consistent" [43, p. 330].

[5] Martin-Löf puts $\Pi^1_2 - \mathrm{CA}$ on the other side of the "abyss", because the analysis by Rathjen only reduces it to some set theoretic ordinal notation system. Rathjen is here following a successful tradition in the Schütte school of proof theory, and the author believes that this is already the major step in constructivising this theory. The author does not see at this moment any principal reason apart from effort and time why the resulting ordinal notation system cannot be proved to be well-founded in a suitable constructive theory. However, as long as such a reduction to a fully constructive theory has not been carried out, the analysis by Rathjen remains incomplete, and one could therefore at this moment in time place $\Pi^1_2 - \mathrm{CA}$, as Martin-Löf did, on the other side of the "abyss". See however the discussion in Sect. 5 about the limits of constructivism, which indicates that it might be very difficult to carry out the necessary constructivisation.

to know, we will know".[6] The future development of proof theory will hopefully decide whether set theory is consistent or not.[7] Of course, even if one ever found an inconsistency, it most likely has no effect on everyday mathematics (which is often anyway on the surface carried out in naive set theory, which is inconsistent).

## 2 Well-Foundedness of Ordinal Notation Systems

Since the work of Gentzen, the main step in proving the consistency of reference theories in proof theory is ordinal analysis; other theories are then reduced using various techniques to these reference theories.[8] Ordinal analysis amounts to showing that the consistency of a theory can be shown in PRA + $\mathrm{TI}^{\mathrm{qf}}(\alpha)$. Here PRA is primitive recursive arithmetic, and $\mathrm{TI}^{\mathrm{qf}}(\alpha)$ is the principle of quantifier free transfinite induction up to $\alpha$ for a specific ordinal notation system. The formula $\mathrm{TI}^{\mathrm{qf}}(\alpha)$ is defined as follows: Let $\varphi(x)$ be a quantifier free formula in the language of PRA. The formula $\mathrm{Prog}(\varphi, \alpha)$, meaning $\varphi$ is progressive up to $\alpha$, is defined as $\forall\beta < \alpha.((\forall\gamma < \beta.\varphi(\gamma)) \to \varphi(\beta))$. Now $\mathrm{TI}^{\mathrm{qf}}(\alpha)$ is the statement that for all such quantifier-free formulae $\varphi$ we have that $\mathrm{Prog}(\varphi, \alpha)$ implies $\forall\beta < \alpha.\varphi(\beta)$. We will in the following sometimes replace in notions such as $\mathrm{TI}^{\mathrm{qf}}(\alpha)$ the ordinal $\alpha$ by an ordinal notation system $(A, <)$. Here, an ordinal notation system $(A, <)$ is a linearly ordered set $(A, <)$, such that $A$ is a primitive recursive subset of $\mathbb{N}$ and $< \subseteq A \times A$ is primitive recursive. So with notations such as $\mathrm{TI}^{\mathrm{qf}}(\alpha)$ we introduce as well for ordinal notation systems $(A, <)$ the notion $\mathrm{TI}^{\mathrm{qf}}(A, <)$ for which we write as well $\mathrm{TI}^{\mathrm{qf}}(A)$.

We assume that Tait's article [41], in which he argues that PRA corresponds to finitary methods, provides sufficient arguments for validating the proof principles of PRA. So in order to validate PRA + $\mathrm{TI}^{\mathrm{qf}}(\alpha)$, one needs to validate the principle of $\mathrm{TI}^{\mathrm{qf}}(\alpha)$. So assume $\varphi$ is progressive up to $\alpha$. Since $\varphi$ is quantifier free, it is decidable, and we get $\varphi(\beta) \vee \neg\varphi(\beta)$, and can argue indirectly. Assume that for $\beta_0 := \beta$ we have that $\varphi(\beta_0)$ does not hold. Then by searching through all ordinal notations and using the decidability of $\varphi$, we can find recursively an ordinal $\beta_1 < \beta$ such that $\neg\varphi(\beta_1)$ holds. Continuing we find $\beta_2$ such that $\neg\varphi(\beta_2)$ holds. By continuing his process we obtain a recursive sequence $\beta_0 > \beta_1 > \cdots$ such that $\neg\varphi(\beta_i)$ holds for all $i$. Note that this argument requires Markov's principle, however not as a principle of our theory, but as a metamathematical principle. Note as well that, if we have any

[6]German: "Wir müssen wissen. Wir werden wissen."

[7]Of course in case of a positive answer a validation argument needs then to be carried out.

[8]Of course often consistency is shown using normalisation proofs without ordinal analysis, however, as pointed out before when quoting the referee in Sect. 1.1, in a proof theoretic analysis a reduction to a quite different (very slim) theory is carried out whereas in normalisation proofs we usually reduce the consistency to a slight extension of the theory in question, and therefore do not gain such a deep understanding of the proof theoretically strong principles.

proof of a theorem which is not correct, it must contain (unless there is a problem with PRA) a concrete quantifier free $\varphi$ and a concrete $\beta < \alpha$ for which the principle of transfinite induction up to $\beta < \alpha$ is violated. From $\varphi$ and $\beta$ we will then obtain a concrete infinite descending sequence. So in order to validate our theory, we need to validate that there is no recursive infinite descending sequence of ordinals $< \alpha$, which we call NRDS($\alpha$).

We will look now at the steps towards validating that $\epsilon_0$ is well-founded. First of all, we can rule out an infinite descending recursive descending sequence of natural numbers and therefore validate NRDS($\omega$). If we assume NRDS($A, <_A$) and NRDS($B, <_B$) for linearly ordered sets ($A, <_A$) and ($B, <_B$) we can validate NRDS($A \times B, <_{\text{lex}}$) where $<_{\text{lex}}$ is the lexicographic ordering on $A \times B$ w.r.t. $<_A, <_B$. For if we had an infinite descending sequence $(a_n, b_n)_{n \in \mathbb{N}}$, we immediately see that $a_0 \geq_A a_1 \geq_A a_2 \geq \cdots$. Furthermore, for every $n$ we can find $m > n$ s.t. $a_m <_A a_n$. For as long as $a_n = a_m$ for $n < m$ we have $b_n >_B b_{n+1} >_B \cdots >_B b_m$. This descending recursive sequence of $b_i$ will eventually stop, so there must be an $m > n$ s.t. $a_m <_A a_n$, which we can find recursively. By iterating this we find an infinite descending sequence $(a_{n_k})_{k \in \omega}$ in $A$, which does not exist. Note that the purpose of this exercise is not proving in a formal theory $\mathrm{TI}^{\mathrm{qf}}(A \times B)$ but that we can get a direct insight into NRDS($A \times B$) and therefore of $\mathrm{TI}^{\mathrm{qf}}(A \times B)$.

Up to now we were working with recursive sequences, which corresponds to quantifier free induction. Using the validation of well-foundedness of $\omega$ and of the lexicographic ordering on the products, we can validate transfinite induction up to $\omega^n$ which is provable in PRA which has proof theoretic ordinal $\omega^\omega$. In order to prove transfinite induction up to an ordinal $\alpha \geq \omega^\omega$, quantifier free induction on $\omega$ is no longer sufficient. This translates into the non-existence of descending (possibly non-recursive) sequences in $\alpha$, which we call NDS($\alpha$). For instance, induction over arbitrary arithmetical formulae corresponds to non-existence of arithmetically definable descending sequences in $\omega$. Note that NDS($\alpha$) implies NRDS($\alpha$) which as stated before validates $\mathrm{TI}^{\mathrm{qf}}(\alpha)$.

So we will now, instead of validating NRDS($\alpha$), validate the stronger principle NDS($\alpha$), which means we leave a fully constructive approach.[9] Even if it is nonconstructive, we consider it still to be possible to carry out a validation argument based on this notion. We can in our opinion validate NDS($\omega$), which means we can get a direct insight that this principle is valid. Using the same argument as before we can in our opinion validate that the principle NDS is closed under forming the lexicographic ordering for the product of two orderings.

Now assume NDS($A, <_A$). Consider $A_{\text{dec}}$, the set of finite sequences (or lists) of elements $(a_1, \ldots, a_k)$ of $A$ such that $a_1 >_A \cdots >_A a_k$. Let $<_{\text{lex}}$ be the lexicographic ordering on finite sequences of elements in $A$ based on $<_A$. We vali-

[9]Constructive, if one regards Markov's principle as constructive.
In fact we will need NDS($A', <_{A'}$) only for intermediate notation systems ($A', <_{A'}$) used for validating NDS($\alpha$). For the final system, NRDS($\alpha$) is all what is required, which is implied by NDS($\alpha$).

date $\mathrm{NDS}(A_{\mathrm{dec}}, <_{\mathrm{lex}})$. Assume a descending sequence $(a_{n,0}, a_{n,1}, \ldots, a_{n,k_n-1})_{n\in\omega}$. We immediately see that $a_{n,0}$ is defined (i.e. $k_n \geq 1$) and weakly descending, i.e. $a_{0,0} \succeq_A a_{1,0} \succeq_A a_{2,0} \succeq_A \cdots$. Because there is no infinite descending sequence in $A$, this sequence must eventually become constant. Assume it is constant from $n = n_0$ onwards. Then for $n \geq n_0$ we have that $a_{n,1}$ is defined (i.e. $k_n \geq 2$) and forms a descending sequence $a_{n_0,1} \succeq_A a_{n_0+1,1} \succeq_A a_{n_0+2,1} \succeq_A \cdots$ in $A$. That sequence will eventually become constant for $n \geq n_1$ for some $n_1$. Therefore $a_{n,2}$ is descending for $n \geq n_1$ onwards and will become constant for $n \geq n_2$ for some $n_2$. By continuing this process we obtain a sequence of natural numbers $(n_i)_{i\in\omega}$ and have $a_{n_0,0} = a_{n_1,0} >_A a_{n_1,1} = a_{n_2,1} >_A a_{n_2,2} >_A \cdots$. So we obtain an infinite descending sequence $a_{n_0,0} >_A a_{n_1,1} >_A \cdots$ in $A$ which does not exist, and have therefore shown that there is no infinite descending sequence in $(A_{\mathrm{dec}}, <_{\mathrm{lex}})$. Note that we cannot determine $n_0, n_1, \ldots$, so $\mathrm{NRDS}(A)$ is not sufficient to carry out this argument.

This argument validates transfinite induction on $(A_{\mathrm{dec}}, <_{\mathrm{lex}})$. Ordering on ordinals in Cantor Normal Form (CNF) $\alpha = \omega^{\alpha_1} n_1 + \cdots + \omega^{\alpha_k} n_k$ is the same as the double lexicographic ordering on $((\alpha_1, n_1), \ldots, (\alpha_k, n_k))$. Let $(A, <)$ be an ordinal notation system. Let $\mathrm{CNF}(A)$ be the set of terms obtained by applying once CNF to elements in $A$, ordered correspondingly. $\mathrm{CNF}(A)$ is isomorphic to a subset of $((A\times(\omega\setminus 0), <_{\mathrm{lex}})_{\mathrm{dec}}, <_{\mathrm{lex}})$[10] which in turn is isomorphic to $((A\times\omega, <_{\mathrm{lex}})_{\mathrm{dec}}, <_{\mathrm{lex}})$. The order type of $\mathrm{CNF}(A)$ is $\omega^\alpha$, if the order type of $A$ is $\alpha$. This means that, if we have validated $\mathrm{NDS}(\alpha)$, we have validated $\mathrm{NDS}(\omega^\alpha)$.

Therefore we can validate $\mathrm{NDS}(\omega_n)$ and therefore at least $\mathrm{TI}^{\mathrm{qf}}(\omega_n)$ where $\omega_0 = \omega$, $\omega_{n+1} = \omega^{\omega_n}$. Since $\epsilon_0 = \sup_{n\in\omega} \omega_n$ we have validated quantifier free transfinite induction up to all ordinals less than $\epsilon_0$.

Gentzen showed that $\mathrm{PRA} + \mathrm{TI}^{\mathrm{qf}}(\epsilon_0)$ proves the consistency of PA, which was considered as a proof of the consistency of PA. The belief that this proof shows the consistency of PA (in an absolute way) must be based on some argument which validates $\mathrm{PRA} + \mathrm{TI}^{\mathrm{qf}}(\epsilon_0)$, and we have given one such argument. The above argument has shown the validity of the consistency of PA. Therefore it follows, for instance, that, if we have shown in PA Fermat's last theorem, then there can be no counter example.

In our articles [36, 37] we extended this approach to ordinal notation systems from below. Up to the strength of $(\Pi^1_1 - \mathrm{CA})_0$ we were able to give arguments, which we regard as a validation of transfinite quantifier-free induction up to those ordinals. When reaching higher ordinals, the direct insight into the well-foundedness rests necessarily upon principles of increasing proof theoretic strength. Note that according to the results of reverse mathematics, most real mathematical theorems can be shown in $(\Pi^1_1 - \mathrm{CA})_0$ , so most of mathematics can be validated by pure ordinal analysis. Beyond that strength, we could develop ordinal notation systems from below, but could only give a formal well-foundedness proof, which then needs to be carried out in another theory of at least equal strength. It is no accident that

[10]Those sequences $((a_1, n_1), \ldots, (a_k, n_k))$ s.t.$a_1 > \cdots > a_k$.

this happens when moving from $(\Pi^1_1 - \mathrm{CA})_0$ to $\Pi^1_1 - \mathrm{CA}$, since the argument is based on the concept of well-foundedness, which is a $\Pi^1_1$-concept, and one needs in some form a principle, which goes beyond $\Pi^1_1$, in order to validate $\Pi^1_1 - \mathrm{CA}$.

## 3 Martin-Löf Type Theory

With increasing strength, ordinal notation systems for describing the proof theoretic ordinal of theories become increasingly complicated. Therefore, the complexity of the well-foundedness proofs for these ordinal notation systems increases as well. Correspondingly, it becomes increasingly difficult, if possible at all, to validate the well-foundedness of the ordinal notation system directly. A solution for this problem is to make a second step and prove the well-foundedness of the ordinal notation system in a second theory for which one can carry out a validation argument more directly. Hilbert wanted originally to validate theories involving the infinite by reducing them to finitary methods. A suitable generalisation of finitary methods are constructive theories, in which the elements of sets are still finite objects, or terms. In order to deal with function spaces, we need reduction rules for terms, for instance $n + \mathrm{S}(m)$ reduces to $\mathrm{S}(n + m)$. This allows to determine elements of function types as terms which applied to elements of the argument type are elements of the result type, or reduce to such an element. So infinite objects (full functions) are replaced by finite objects (programs or terms).

The addition of recursive functions as finitary objects was the motivation of Gödel in his Dialectica paper [13], where he writes (p. 282, translation p. 245 of [11]): "It is the second requirement that must be dropped. This fact has hitherto been taken into account by our adjoining to finitary mathematics parts of intuitionistic logic and the theory of ordinals. In what follows we shall show that, for the consistency proof of number theory, we can use, instead, the notion of computable function of finite type on the natural numbers and certain elementary principles of construction for such functions."[11]

Gödel's Dialectica interpretation was still referring to classical logic, and is usually used mainly as a proof theoretical tool rather than being considered as an approach to obtaining a foundation of mathematics. A more radical approach was taken in MLTT.[12] MLTT is, as Martin-Löf phrased it once to the author (we

[11]"Es ist die zweite Forderung, welche fallen gelassen werden muss. Dieser Tatsache wurde bisher dadurch Rechnung getragen, dass man Teile der intuitionistischen Logik und Ordinalzahltheorie zur finiten Mathematik adjungierte. Im folgenden wird gezeigt, dass man statt dessen für den Widerspruchsfreiheitsbeweis der Zahlentheorie auch den Begriff der berechenbaren Funktion endlichen Types über den natürlichen Zahlen und gewisse sehr elementare Konstruktionsprinzipien für solche Funktionen verwenden kann."

[12]The standard reference is Martin-Löf's book [20]. The article [28] contains a good concise summary of the rules of MLTT (starting p. 162), however the rules for $\omega$ and $\Omega$, which make it a partial type theory, the topic of that article, need to be omitted. Another listing can be found

unfortunately do not remember the precise wording), the most serious attempt to develop a theory such that we have an insight that all judgements are valid. Those not familiar with MLTT are often perplexed by the large number of its rules. The reason for having such a large number of rules is that this theory is not defined so that it has a shortest description. Instead it is designed so that we can get an insight into the validity of all provable judgements.

In MLTT we have non-dependent judgements of the form

- $a : A$ for $a$ is of type $A$,
- $a = b : A$ for $a, b$ are equal elements of type $A$,
- $A : \mathrm{Set}$ for $A$ is a set,
- $A = B : \mathrm{Set}$ for $A, B$ are equal sets.

Dependent judgements have the form $x_1 : A_1, \ldots, x_n : A_n \Rightarrow \theta$ where $\theta$ is a non-dependent judgement, with free variables in $x_1, \ldots, x_n$.

We have as rules

- structural rules (rules for dealing with contexts, assumptions, and the definitional equalities $a = b : A$ and $A = B : \mathrm{Set}$);
- formation rules (which introduce sets, e.g. conclude $\mathbb{N} : \mathrm{Set}$);
- introduction rules (which introduce a canonical element, an element starting with a constructor, e.g. for $\mathbb{N}$ derive $0 : \mathbb{N}$ and from $a : \mathbb{N}$ derive $\mathrm{S}(a) : \mathbb{N}$);
- elimination rules, e.g. higher type primitive recursion in case of $\mathbb{N}$,
- equality rules (e.g. deriving that if $t(x)$ is defined by higher type primitive recursion into type $B(x)$, with base case $a : B(0)$, that $t(0) = a : B(0)$);
- and equality versions of the formation, introduction and elimination rules (e.g. deriving $\mathrm{S}(a) = \mathrm{S}(a') : \mathbb{N}$ from $a = a' : \mathbb{N}$).

The validation argument for MLTT is done via meaning explanations.[13] In meaning explanations, one determines the meaning of each judgement. Then one validates for each rule that, if the premises are valid w.r.t. meaning explanations, so is the conclusion. Therefore all judgements provable are valid.

Elements of sets can be canonical elements, which are formed by the introduction rules. For instance, $\mathrm{S}(2 + 2)$ is a canonical element of $\mathbb{N}$. Non-canonical elements are considered by Martin-Löf (see, e.g., [20]) as programs, which evaluate to a canonical element. Canonical elements are special cases of non-canonical elements,

in the author's article [35], where everything was made precise in order to be able to carry out a proof theoretic analysis. Arne Ranta's book [29] contains a nice introduction to MLTT. Nordström et al.'s book [26] is an excellent reference for MLTT, and there is the more recent and more concise handbook version [25].

[13] We could not find a definite and complete reference to meaning explanations. Martin-Löf's articles and book [20–23] introduce meaning explanations when discussing the rules of type theory. Tasistro's Ph.D. thesis [42] describes meaning explanations if one uses explicit substitutions (see as well a short reference in the more accessible article [12]). The author has in [39] given an account of his understanding of meaning explanation with a variation in order to accommodate coalgebraic data types defined by their elimination rules.

which as programs evaluate to themselves. Martin-Löf (private communication) considers the concept of a program, for which we have a direct insight how it operates, as crucial for understanding his meaning explanations.

The meaning of $A$ : Set is given by determining what its canonical elements are and when two canonical elements are equal. The meaning of $a : A$ is that $a$ is a non-canonical element of $A$. The meaning of the judgement $a = a' : A$ is that $a, a'$ are equal elements of $A$, which means that they evaluate to equal canonical elements of $A$.

In case of $\mathbb{N}$ we have that 0 is a canonical element, and, if $n$ is an element of $\mathbb{N}$, then $\mathrm{S}(n)$ is a canonical element of it. 0 is equal to 0, and if $n, m$ are two equal elements of $\mathbb{N}$, then $\mathrm{S}(n)$ and $\mathrm{S}(m)$ are equal canonical elements of it.

The meaning of the judgement $A = B$ : Set is that $A$ and $B$ are equal sets which means that canonical elements of $A$ are canonical elements of $B$ and vice versa, and equal canonical elements of $A$ are equal canonical elements of $B$ and vice versa.

For determining the meaning of dependent judgements, we introduce abbreviations $\vec{x}$ for $x_1, \ldots, x_n$, similar for $\vec{a}$, $\vec{a}'$ (referring to $a_i'$), and $\vec{x}_k$ for $x_1, \ldots, x_k$, similar for $\vec{a}_k, \vec{a}_k'$. A dependent judgement

$$x_1 : A_1, x_2 : A_2(x_1), \ldots, x_n : A_n(\vec{x}_{n-1}) \Rightarrow \theta(\vec{x})$$

is valid if for every choice of elements

$$a_1 : A_1, a_2 : A_2(a_1), \ldots, a_n : A_n(\vec{a}_{n-1})$$

the judgement $\theta(\vec{a})$ is valid. One needs as well that for equal elements

$$a_1 = a_1' : A_1, a_2 = a_2' : A_2(a_1), \ldots, a_n = a_n' : A_n(\vec{a}_{n-1})$$

the equality judgements in the conclusion holds: If $\theta = (A : \mathrm{Set})$ we require that $A(\vec{a}) = A(\vec{a}') : \mathrm{Set}$ holds, in case $\theta = (a : A)$ we require that $a(\vec{a}) = a(\vec{a}') : A(\vec{a})$ holds. Judgement $A = B$ : Set presupposes $A$ : Set, $B$ : Set, judgement $a : A$ presupposes $A$ : Set, judgement $a = b : A$ presuppose $a : A, b : A$. The judgement

$$x_1 : A_1, \ldots, x_n : A_n(\vec{x}_{n-1}) \Rightarrow \theta(\vec{x})$$

presupposes $A_1$ : Set, $x_1 : A_1 \Rightarrow A_2(x_1)$ : Set, etc., and as well

$$x_1 : A_1, \ldots, x_n : A_n(\vec{x}_{n-1}) \Rightarrow \theta'(\vec{x})$$

for any presupposition $\theta'(\vec{x})$ of $\theta(\vec{x})$.

Adding the meaning of the presuppositions of judgements (applied transitively) to the meaning of a judgement gives the full meaning of the judgement.

Now one can easily validate structural rules, formation rules, introduction rules, and their equality versions. Elimination rules are more difficult to validate (and that's where an increasingly high level of trust is required). In case of $\mathbb{N}$, in the simple case where we derive $x : \mathbb{N} \Rightarrow t(x) : B(x)$ by primitive recursion, we validate that $t(0) : B(0)$ and if we have $x : \mathbb{N}$ and $t(x) : B(x)$ are valid, so is $t(\mathrm{S}(x)) : B(\mathrm{S}(x))$. Now one sees that for each element of $a$ of $\mathbb{N}$ as given by the meaning explanations $t(a) : B(a)$. This holds first for canonical elements, by going through what we said constitutes a canonical element of $\mathbb{N}$, and checking for each canonical element $a$ that $t(a) : B(a)$ is validated. For non-canonical elements, the reduction of $t(a)$ is given by first reducing $a$ to canonical form 0 or $\mathrm{S}(a')$, and then applying the reductions corresponding to the base case or induction step. Therefore the rules are validated as well for non-canonical elements.

The key principle one needs to trust is the correctness of the elimination rules for the inductively defined sets $\mathbb{N}$, W-type, and universes. We cannot get around the fact that we cannot prove the consistency of MLTT, so when moving to proof theoretically stronger principles, one needs to trust the validity of the rules for proof theoretically stronger sets. We cannot avoid this, but the author believes that one can trust in the principles involved.

### *3.1 Induction-Recursion and the Mahlo Universe*

The validation of principles works well for concrete inductive-recursively defined sets, as long as we do not make use of the full logical framework, which allows to have $A$ : Set or even higher types in the context.[14] Therefore, one can validate Palmgren's superuniverse ([27], Sect. 3), but not Palmgren's higher order universes ([27], Sect. 5) or the external Mahlo universe ([4], Sect. 6.3), which reaches at least the strength of KPM ([4], Sect. 6.4). The strength of Palmgren's superuniverse is not known ([30, 31] analyse only the metapredicative version without the W-type), but substantially exceeds that of MLTT with W and one universe.[15] The latter theory was analysed by Griffor and Rathjen [14] and Setzer [34, 35], and has strength slightly bigger than Kripke–Platek set theory with one recursively inaccessible, KPI.

For the Mahlo universe we have given meaning explanations in our article [38] (not yet published). However, we cannot say that the validity of its rules are as fully

[14]When introducing his version of meaning explanations, the author usually avoids the logical framework. The reason is that he has not yet found an account of meaning explanations of the logical framework, which does not consider Set as a Russell style subuniverse of Type, and which he considers as fully satisfactory. If Set is treated as a universe, one adds considerable proof theoretic strength. Especially, with the rules for inductive-recursive definitions Set is closed under the introduction rules of (a Russell style variant of) the internal Mahlo universe. In the community of MLTT, inductive-recursive definitions is often considered as the limit of what can be at the moment justified without making use of the Mahlo universe principle. Martin-Löf has given presentations about how to treat the logical framework without adding additional strength, however we could not find yet a written account of it needed in order to judge it completely.

[15]It is easy to conjecture the precise strength, and it would not be difficult albeit time consuming to analyse the full version of it.

convincing as they are for inductive-recursive definitions without use of the full logical framework.

## 4 Feferman's System of Explicit Mathematics and the Extended Predicative Mahlo Universe

In [16] Kahle and the author have published an extended predicative version of the Mahlo universe. This version is developed in Feferman's system of explicit mathematics [5, 6]. It uses the fact that in Feferman's system one has access to the collection of all terms, and therefore can form for every term a subuniverse of the Mahlo universe which is relatively closed under this term considered as a partial function. In MLTT all objects have a type and are therefore total. Therefore in MLTT we do not have access to the collection of all terms, which in general are only partial objects.

We regard this version [16] as being predicative (in an extended sense) and believe that this theory can be validated. Feferman's theory has been developed in second order logic,[16] and optimised towards a short and concise theory. While this makes metamathematical investigations easy and makes it easily accessible to non-specialists, it causes problems when validating the provable statements.[17] It seems however that this is not a principal problem. It should be possible to present Feferman's theories in a style which is very close to that of MLTT, and develop meaning explanations. This way hopefully one could validate the extended predicative Mahlo universe in the sense of this article.

With [16] we have not reached the limit of this methodology. We have developed draft versions which reach at least the strength of Kripke–Platek set theory extended by $\Pi_3$-reflection, and it is likely that we can go far beyond with that strength.

## 5 The Limit of Constructivism

In [32, Sect. 6] Rathjen introduces assumptions (A0)–(A3) about possible extensions $\mathrm{MLTT}^+$ of Martin-Löf Type Theory, of which the most important one is assumption (A3):

(A3) Every inductive definition $\Phi : \mathrm{Pow}(\mathbb{N}) \to \mathrm{Pow}(\mathbb{N})$ for generating the elements of a type $A$ in $\mathrm{MLTT}^+$ and its pertinent decoding function are

[16]Not much of second order logic is actually used, its use is mainly for convenience rather than need.

[17]We note that this is the opinion of the second author of [16] only, who is the author of the current article.

definable by set-theoretic $\Sigma$-formulae. These formulae may contain further sets as parameters, where these sets correspond to previously defined types.

He shows (Theorem 6.1) that under these assumptions a set $M$ such that $M \prec_1$ V is a model of MLTT$^+$. Here $M \prec_1$ V means that $M$ is a $\Sigma_1$-elementary substructure of V, where V is the set theoretic universe. This determines a limit to a constructive program based on MLTT.

In his argument, Rathjen already admits that due to the acceptance of the Mahlo universe as an acceptable extension of MLTT, a more strict assumption had to be abandoned, namely that sets are introduced by monotone inductive definitions. This already indicates that it is very difficult to determine an upper bound for a constructive program. While it may be difficult to go beyond principle (A3), we believe that this is only a temporary limitation—it is likely that new constructive principles will emerge, which will be considered as acceptable but go beyond this principle. However, drawing this line is of great benefit, since it determines the requirements a new principle needs to fulfil in order to go beyond that limit.

**Acknowledgements** The author wants to thank the anonymous referees for extraordinarily detailed refereeing and many very valuable comments; Fredrik Nordvall Forsberg and Håkon Gylterud for careful proof reading; and Reinhard Kahle for his encouragement to writing such a rather philosophical article and for his patience while waiting for the completion of this article. Research for this article was supported by EPSRC grant EP/G033374/1.

## References

1. T. Arai, Proof theory of theories of ordinals III: $\Pi_1$ collection. Unpublished notes (1997)
2. T. Arai, A sneak preview of proof theory of ordinals. arXiv:1102.0596v1 [math.LO] (2011). http://arxiv.org/abs/1102.0596
3. P. Dybjer, Program testing and the meaning explanations of intuitionistic type theory, in *Epistemology Versus Ontology* (Springer, Dordrecht, 2012), pp. 215–241
4. P. Dybjer, A. Setzer, Induction-recursion and initial algebras. Ann. Pure Appl. Log. **124**, 1–47 (2003)
5. S. Feferman, A language and axioms for explicit mathematics, in *Algebra and Logic*, ed. by J. Crossley. Lecture Notes in Mathematics, vol. 450 (Springer, Berlin/Heidelberg, 1975), pp. 87–139. doi:10.1007/BFb0062852
6. S. Feferman, Constructive theories of functions and classes, in *Logic Colloquium '78 Proceedings of the Colloquium Held in Mons*, ed. by D.D. Maurice Boffa, K. McAloon. Studies in Logic and the Foundations of Mathematics, vol. 97 (North Holland, Amsterdam, 1979), pp. 159–224
7. S. Feferman, Hilbert's program relativized: proof-theoretical and foundational reductions. J. Symb. Log. **53**(2), 364–384 (1988)
8. S. Feferman. What rests on what? The proof-theoretic analysis of mathematics, in *Philosophy of Mathematics. Proceedings of the Fifteenth International Wittgenstein-Symposium, Part 1*, ed. by J. Czermak (Hölder-Pichler-Tempsky, Vienna, 1993), pp. 147–171. Reprinted in Solomon Feferman: In the light of Logic. Oxford University Press, Oxford, 1998, Ch. 10, 187–208
9. S. Feferman, Why a little bit goes a long way: logical foundations of scientifically applicable mathematics. PSA **2**, 442–455 (1993). Reprinted in Feferman: In the light of logic. Oxford University Press, Oxford, 1998, Ch. 14, pp. 284–298

10. S. Feferman, Does reductive proof theory have a viable rationale? Erkenntnis **53**(1–2), 63–96 (2000)
11. S. Feferman, J.W. Dawson Jr., S.C. Kleene, G.H. Moore, R.M. Solovay, J. van Heijenoort (eds.), *Kurt Gödel. Collected Works: Volume II: Publications 1938–1974* (Oxford University Press, Oxford, 1990)
12. D. Fridlender, A proof-irrelevant model of Martin-Löf's logical framework. Math. Struct. Comput. Sci. **12**, 771–795 (2002)
13. K. Gödel, Über eine bisher noch nicht benützte Erweiterung des finiten Standpunktes. Dialectica **12**(3–4), 280–287 (1958). Translation in [11], pp. 240–251
14. E. Griffor, M. Rathjen, The strength of some Martin-Löf type theories. Arch. Math. Log. **33**(5), 347 (1994)
15. H. Hesse, *Das Glasperlenspiel. Versuch einer Lebensbeschreibung des Magister Ludi Josef Knecht samt Knechts hinterlassenen Schriften. (English: The Glass Bead Game)* (Fretz & Wasmuth, Zürich, 1943)
16. R. Kahle, A. Setzer, An extended predicative definition of the Mahlo universe, in *Ways of Proof Theory*, ed. by R. Schindler. Ontos Series in Mathematical Logic (Ontos Verlag, Frankfurt (Main), 2010), pp. 309–334
17. G. Kreisel, A variant of Hilbert's theory of the foundations of arithmetic. Br. J. Philos. Sci. **IV**(14), 107–129 (1953)
18. G. Kreisel, A survey of proof theory. J. Symb. Log. **33**(3), 321–388 (1968)
19. G. Kreisel, Hilbert's programme, in *Philosophy of Mathematics: Selected Readings*, 2nd edn, ed. by P. Benacerraf, H. Putnam (Cambridge University Press, Cambridge, 1983), pp. 207–238. Revised version of article in Dialectica 12(3–4), pp. 346–372, 1958
20. P. Martin-Löf, *Intuitionistic Type Theory* (Bibliopolis, Naples, 1984)
21. P. Martin-Löf, Truth of a proposition, evidence of a judgement, validity of a proof. Synthese **73**, 407–420 (1987)
22. P. Martin-Löf, On the meaning of the logical constants and the justification of the logical laws. Nord. J. Philos. Log. **1**(1):11–60 (1996). Short course given at the meeting Teoria della Dimostrazione e Filosofia della Logica, organized in Siena, 6–9 April 1983, by the Scuola di Specializzazione in Logica Matematica of the Università degli Studi di Siena. Available from http://www.hf.uio.no/filosofi/njpl/vol1no1/meaning/meaning.html
23. P. Martin-Löf, An intuitionistic theory of types, in *Twenty-Five Years of Constructive Type Theory*, ed. by G. Sambin, J. Smith (Oxford University Press, Oxford, 1998), pp. 127–172. Reprinted version of an unpublished report from 1972
24. P. Martin-Löf, The Hilbert-Brouwer controversy resolved? in *One Hundred Years of Intuitionism (1907–2007)*, ed. by M. Atten, P. Boldini, M. Bourdeau, G. Heinzmann. Publications des Archives Henri Poincaré, Publications of the Henri Poincaré Archives (Birkhäuser, Basel, 2008), pp. 243–256
25. B. Nordstrom, K. Petersson, J. Smith, Martin-löf's type theory, in *Handbook of Logic in Computer Science: Logic and Algebraic Methods*, vol. 5, ed. by S. Abramsky, D.M. Gabbay, T. Maibaum (Oxford University Press, Oxford, 2001), pp. 1–38
26. B. Nordström, K. Petersson, J.M. Smith, *Programming in Martin-Löf's Type Theory. An Introduction.* (Oxford University Press, Oxford, 1990). Book out of print. Online version available via http://www.cs.chalmers.se/Cs/Research/Logic/book/
27. E. Palmgren, On universes in type theory, in *Twenty Five Years of Constructive Type Theory*, ed. by G. Sambin, J. Smith (Oxford University Press, Oxford, 1998), pp. 191–204
28. E. Palmgren, V. Stoltenberg-Hansen, Domain interpretations of Martin-Löf's partial type theory. Ann. Pure Appl. Log. **48**, 135–196 (1990)
29. A. Ranta, *Type-Theoretical Grammar* (Oxford University Press, Oxford, 1995)
30. M. Rathjen, The strength of Martin-Löf type theory with a superuniverse, part I. Arch. Math. Log. **39**(1), 1–39 (2000)
31. M. Rathjen, The strength of Martin-Löf type theory with a superuniverse, part II. Arch. Math. Log. **40**(3), 207–233 (2001)

32. M. Rathjen, The constructive Hilbert program and the limits of Martin-Löf type theory. Synthese **147**, 81–120 (2005)
33. M. Rathjen, An ordinal analysis of parameter free $\Pi^1_2$-comprehension. Arch. Math. Log. **44**(3), 263–362 (2005)
34. A. Setzer, Proof theoretical strength of Martin-Löf type theory with W-type and one universe, Ph.D. thesis, Universität München, 1993. http://www.cs.swan.ac.uk/~csetzer
35. A. Setzer, Well-ordering proofs for Martin-Löf type theory. Ann. Pure Appl. Log. **92**, 113–159 (1998)
36. A. Setzer, Ordinal systems, in *Sets and Proofs*, ed. by S. Barry Cooper, J. Truss (Cambridge University Press, Cambridge, 1999), pp. 301–331
37. A. Setzer, Ordinal systems part 2: one inaccessible, in *Logic Colloquium '98*, ed. by S. Buss, P. Hajek, P. Pudlak. ASL Lecture Notes in Logic 13 (Peters Ltd., Natick, 2000), pp. 426–448
38. A. Setzer, Universes in type theory Part II – autonomous Mahlo. Ann. Pure Appl. Log. (2011, submitted). http://www.cs.swan.ac.uk/~csetzer/articles/modelautomahlomain.pdf
39. A. Setzer, Coalgebras as types determined by their elimination rules, in *Epistemology Versus Ontology*, ed. by P. Dybjer, S. Lindström, E. Palmgren, G. Sundholm. Logic, Epistemology, and the Unity of Science, vol. 27 (Springer, Dordrecht, 2012), pp. 351–369. doi:10.1007/978-94-007-4435-6_16
40. Stanford Encyclopedia of Philosophy, Hilbert's program (2003). http://plato.stanford.edu/entries/hilbert-program/#1
41. W.W. Tait, Finitism. J. Philos. **78**(9), 524–546 (1981)
42. A. Tasistro, Substitution, record types and subtyping in type theory, with applications to the theory of programming, Ph.D. thesis, Department of Computing Science, University of Gothenburg, Gothenburg, 1997
43. W.H. Woodin, The tower of Hanoi, in *Truth in Mathematics*, ed. by H.H.G. Dales, G. Oliveri (Oxford University Press, Oxford, 1998), pp. 329–351
44. R. Zach, Hilbert's program then and now, in *Philosophy of Logic*, ed. by D. Jacquette (North-Holland, Amsterdam, 2007), pp. 411–447

# Part II
# Gentzen's Consistency Proofs

# On Gentzen's First Consistency Proof for Arithmetic

**Wilfried Buchholz**

## 1 Introduction

If nowadays "Gentzen's consistency proof for arithmetic" is mentioned, one usually refers to [3] while Gentzen's first (published) consistency proof, i.e. [2], is widely unknown or ignored. The present paper is intended to change this unsatisfactory situation by presenting [2, IV. Abschnitt] in a slightly modified and modernized form.

The method from [2] can be roughly summarized as follows: By recursion on the build-up of $d$, for each derivation $d$ in a suitably designed finitary proof system $\mathcal{Z}$ of first order arithmetic a family $(d[n])_{n\in|\mathsf{tp}(d)|}$ of *reduced* $\mathcal{Z}$-derivations is defined such that

$$\frac{\ldots\mathsf{End}(d[n])\ldots(n\in|\mathsf{tp}(d)|)}{\mathsf{End}(d)}\quad\text{(where }\mathsf{End}(d)\text{ denotes the endsequent of }d)$$

forms an inference $\mathsf{tp}(d)$ in cutfree $\omega$-arithmetic with repetition rule $\mathsf{Rep}$. Obviously, if $d$ is a derivation of falsum $\perp$, i.e. if $\mathsf{End}(d)=\perp$, then $\mathsf{tp}(d)$ can only be an instance of $\mathsf{Rep}$, so that $d[0]$ is again a derivation of $\perp$. In a second step, to each $d$ an ordinal $o(d)<\varepsilon_0$ is assigned such that $o(d[n])<o(d)$ for all $n\in|\mathsf{tp}(d)|$. Then the consistency of $\mathcal{Z}$ follows by (quantifierfree) transfinite induction up to $\varepsilon_0$.

Actually Gentzen's terminology is somewhat different. First (in Sect. 13 of [2]) Gentzen defines *reduction steps on sequents*. Such a reduction step $\mathcal{I}$ may involve a certain 'option' (Wahlfreiheit), so that the result of applying $\mathcal{I}$ to a sequent $\Pi$ actually is a family of sequents $\big(\mathcal{I}(\Pi,n)\big)_{n\in|\mathcal{I}|}$. Then (in Sect. 14 of [2]) for each

---

W. Buchholz (✉)
Ludwig-Maximilians Universität München, Munich, Germany
e-mail: buchholz@mathematik.uni-muenchen.de



R. Kahle, M. Rathjen (eds.), *Gentzen's Centenary*,
DOI 10.1007/978-3-319-10103-3_4

$\mathcal{Z}$-derivation $d$ (whose endsequent is not an axiom) a *reduction step on derivations*, $d \hookrightarrow (d[n])_{n\in|\mathcal{I}|}$, is defined such that $\forall n \in |\mathcal{I}|(\mathsf{End}(d[n]) = \mathcal{I}(\mathsf{End}(d), n))$, where $\mathcal{I}$ is a reduction step on sequents, uniquely determined by $d$. Here, in contrast to Gentzen, we also regard $\mathsf{Rep}$ as a reduction step on sequents—with $|\mathsf{Rep}| = \{0\}$ and $\mathsf{Rep}(\Pi, 0) = \Pi$.

The outline of the paper is as follows. In Sects. 2 and 3 we repeat relevant parts of [2] using to a great extent Gentzen's own words (in the translation by Szabo [5]). In the course of this we do not hesitate to deviate from the original text (in content or form) whenever we think it is appropriate or facilitates understanding. The main point where we deviate from [2] (besides omitting conjunction &) is the following: In the reduction steps on sequents concerning an antecedent formula $\forall xF$ or $\neg A$ (13.51, 13.53) we always require that this formula is retained in the reduced sequent while Gentzen allows to omit it. As a consequence we also have to modify the reduction steps on atomic $\mathcal{Z}$-derivations (which will be deferred till Sect. 6). In Sect. 4 we present the main definitions and proofs of Sect. 3 in a more condensed style (and with some further modifications). This facilitates the work in Sect. 5 where we assign to each $\mathcal{Z}$-derivation $d$ an ordinal $o(d) < \varepsilon_0$ and prove that each reduction step on a derivation $d$ lowers its ordinal, i.e. we prove that $o(d[n]) < o(d)$ for all $n \in |\mathsf{tp}(d)|$. Our ordinal assignment is essentially that of [4] which on first sight looks very different from Gentzen's original assignment in [2], where certain finite decimal fractions were used as notations for ordinals $< \varepsilon_0$. But in the appendix we will show that actually both ordinal assignments are rather closely related. In Sect. 7 we give an interpretation of $\mathcal{Z}$ in an infinitary system $\mathcal{Z}^\infty$. This way we obtain a semantic explanation for Gentzen's reduction steps on $\mathcal{Z}$-derivations and for the ordinal assignment of Sect. 5. Finally, in Sect. 8 we indicate how the approach of Sects. 4, 5 can easily be adapted to calculi with multisuccedent sequents.

## 2 Formal Language, Reduction Steps on Sequents

The following symbols will serve for the formation of formulae: *Variables* (for natural numbers) which are divided into *free* and *bound* variables; the constant 0 and the unary function symbol $\mathsf{S}$ (successor); *predicate symbols* (each of a fixed arity); the logical connectives $\neg$, $\forall$.[1]

*Terms* are generated from the constant 0 and free variables by iterated application of $\mathsf{S}$. The terms $0, \mathsf{S}0, \mathsf{SS}0, \ldots$ are called *numerals*. In the following we identify numerals and natural numbers.

*Formulas:*

1. If $P$ is an $n$-ary predicate symbol and $t_1, \ldots, t_n$ are terms, then $Pt_1 \ldots t_n$ is a (prime) formula. If $t_1, \ldots, t_n$ are numerals, then $Pt_1 \ldots t_n$ is called a *minimal formula*.

[1] We omit conjunction '&' in order to keep the focus on the essential things.

2. If $A$ is a formula, then so is $\neg A$.
3. From a given formula we obtain another formula by replacing a free variable by a bound variable $x$ not yet occurring in the formula and prefixing $\forall x$.

We assume that to each minimal formula a *truth value* "true" or "false" is assigned.

We use $\perp$ as abbreviation for some fixed *false minimal formula* (e.g. $0 = \mathsf{S}0$).

*Abbreviation.* $A \approx B \;:\Leftrightarrow$ either $A = B$ or $A, B$ are both false minimal formulas.

*Remark* $A \approx \perp \;\Leftrightarrow\; A$ is a false minimal formula.

A *sequent* is an expression of the form $\Gamma \rightarrow B$ where $\Gamma$ is a finite sequence of formulae.

The formulae in $\Gamma$ are called the *antecedent formulae* and $B$ the *succedent formula* of the sequent. We also call $\Gamma$ the *antecedent* of $\Gamma \rightarrow B$.

A formula (sequent) is called *closed* if no free variable occurs in it.

*Abbreviations.*

$A \in \Gamma \;:\Leftrightarrow\; A$ occurs in the sequence $\Gamma$.

$\Gamma \subseteq \Gamma' \;:\Leftrightarrow$ for all formulas $A$, if $A \in \Gamma$ then $A \in \Gamma'$ (e.g. $A, B, A, A \subseteq B, B, A, C$).

**Definition (Reduction Steps on Sequents)**

On a closed sequent $\Pi$ an individual *reduction step* can be carried out in the following way.

13.21. Suppose that the succedent formula of the sequent $\Pi$ has the form $\forall x F(x)$. In that case we replace it by a formula $F(n)$, i.e. by a formula which results from $F(x)$ by the substitution of an *arbitrarily chosen numeral* $n$ for the variable $x$.

13.23. Suppose that the succedent formula of the sequent $\Pi$ has the form $\neg A$. In that case we replace it by the formula $\perp$ and, at the same time, adjoin the formula $A$ to the antecedent of the sequent.

13.4. Suppose that the succedent formula of the sequent $\Pi$ is a true minimal formula; or: that the succedent formula is a false minimal formula and that one of the antecedent formulae of $\Pi$ is also a false minimal formula. Then we say that the sequent $\Pi$ has (or, is in) *endform*, and no reduction step is defined.

13.5. Suppose that the succedent formula of $\Pi$ is a false minimal formula, and that none of the antecedent formulae of $\Pi$ is a false minimal formula. In that case the following two different kinds of reduction step are permissible (counterpart of 13.2):

13.51. Suppose that an antecedent formula has the form $\forall x F(x)$. We adjoin a formula $F(k)$ ($k$ an arbitrary numeral) to the antecedent.

13.53. Suppose that an antecedent formula has the form $\neg A$. We replace the succedent formula by $A$.

In condensed form these reduction steps are described by the following schemata (reading them bottom-up):

$$(\mathsf{R}_{\forall x F(x)})\frac{\ldots\Gamma\!\to\! F(n)\ldots(n\!\in\!\mathbb{N})}{\Gamma\!\to\!\forall x F(x)}; \qquad (\mathsf{R}_{\neg A})\frac{A,\Gamma\!\to\!\bot}{\Gamma\!\to\!\neg A};$$

$$(\mathsf{L}^k_{\forall x F(x)})\frac{F(k),\Gamma\!\to\! C}{\Gamma\!\to\! C} \text{ with } C \approx \bot \text{ and } \forall x F(x) \in \Gamma;$$

$$(\mathsf{L}^0_{\neg A})\frac{\Gamma\!\to\! A}{\Gamma\!\to\! C} \text{ with } C \approx \bot \text{ and } \neg A \in \Gamma.$$

In the sequel, each of the symbols $\mathsf{R}_{\forall x F}$, $\mathsf{R}_{\neg A}$, $\mathsf{L}^k_{\forall x F}$, $\mathsf{L}^0_{\neg A}$ is used as the name of the respective reduction step (as shown above). But the above schemata can also be read as inferences in $\omega$-arithmetic; therefore, these symbols will also be called *inference symbols*. Another reason is that this term has already been used in several previous publications (e.g., in [1])—and "reduction step symbol" would sound too clumsy.

## 3 Reduction Steps on Derivations

**Definition (The System $\mathcal{Z}$ of Pure Number Theory)**
Derivable objects of $\mathcal{Z}$ are sequents.

The *axioms* (or *initial sequents*) of $\mathcal{Z}$ will be specified in Sect. 6.
*Inference Rules*

$\forall$-introduction: $\dfrac{\Gamma\!\to\! F(a)}{\Gamma\!\to\!\forall x F(x)}$ , if the free variable $a$ does not occur in the conclusion.

$\neg$-introduction: $\dfrac{A,\Gamma\!\to\!\bot}{\Gamma\!\to\!\neg A}$

complete induction: $\dfrac{\Gamma\!\to\! F(0) \qquad F(a),\Gamma\!\to\! F(\mathsf{S}a)}{\Gamma\!\to\! F(t)}$ , if $a$ does not occur in the conclusion.

chain rule: $\dfrac{\Gamma_0\!\to\! A_0 \;\ldots\ldots\; \Gamma_l\!\to\! A_l}{\Gamma\!\to\! C}$ ,

if there exists $j \le l$ such that $C \approx A_j$ and $\forall i \le j(\Gamma_i \subseteq \Gamma, A_0, \ldots, A_{i-1})$. In addition we require that no free variable is vanishing, i.e., that every free variable occurring in one of the premises $\Gamma_i\!\to\! A_i$ also occurs in the conclusion $\Gamma\!\to\! C$.

*Abbreviation.*
$d \vdash \Gamma\!\to\! C \;:\Leftrightarrow\; d$ is a $\mathcal{Z}$-derivation (i.e., a derivation in $\mathcal{Z}$) and the endsequent of $d$ is $\Gamma\!\to\! C$.

A derivation is called *closed* if its endsequent is closed.

For each closed derivation $d$, whose endsequent is not in endform (13.4) we shall now define *the reduction step on d* and at the same time prove the following: by such a step the derivation is transformed into another closed derivation and its endsequent is thereby modified in the following way: At most one reduction step according to 13.2 or 13.5 is carried out on the sequent. (It may thus happen that an endsequent remains entirely unchanged.) The reduction step on a derivation is unambiguous, except in the case in which the endsequent undergoes a transformation according to a reduction step on sequents involving a *choice* (13.21); here, the choice may be made arbitrarily; if this has been done, the reduction step is then also unambiguous. If the endsequent of $d$ has endform according to 13.4, *no* reduction step is defined for this derivation.

**Definitions**

1. The result of carrying out the reduction step on $d$ is denoted by $d[n]$ where in case 13.21, $n$ is the *'arbitrarily choosen numeral'*, and $n = 0$ otherwise.
2. If the reduction step on $d$ causes a reduction step on the endsequent $\Pi$ of $d$, then $\mathsf{tp}(d)$ denotes the name of this latter reduction step[2] and $\mathsf{tp}(d)(\Pi, n)$ denotes the result of applying $\mathsf{tp}(d)$ to $\Pi$, where $n$ plays the same role as in 1.
3. If the reduction step on $d$ does not change the endsequent of $d$, we set $\mathsf{tp}(d) :=$ $\mathsf{Rep}$ and $\mathsf{Rep}(\Pi, n) := \Pi$.
4. The *arity* of $d$ is defined by $\mathsf{arity}(d) := \begin{cases} \mathbb{N} & \text{if } \mathsf{tp}(d) = \mathsf{R}_{\forall xF}, \\ \emptyset & \text{if the endsequent of } d \text{ has endform}, \\ \{0\} & \text{otherwise.} \end{cases}$

Summing up, by recursion on the build-up of $d$ we will define $\mathsf{tp}(d)$ and $d[n]$ and simultaneously prove

**Theorem 3.1** *If $d$ is a closed $\mathcal{Z}$-derivation of $\Pi$, then $d[n] \vdash \mathsf{tp}(d)(\Pi, n)$ for all $n \in \mathsf{arity}(d)$.*

In the following we assume that $d$ is a closed $\mathcal{Z}$-derivation whose endsequent is *not* in endform.

14.21. The axioms of $\mathcal{Z}$ are treated later (in Sect. 6).

14.22. We now consider the case where the endsequent is the result of the application of a *rule of inference* and we presuppose that for the derivations of the premises the reduction step has already been defined and the validity of the associated assertion (i.e. Theorem 3.1) demonstrated.

14.23. Suppose that the endsequent of $d$ is the result of a $\forall$-*introduction* or a $\neg$-*introduction*. It (i.e. the endsequent) is then eliminated and its premise taken for the new endsequent, where, in the case of a $\forall$-introduction, every occurrence of the free variable $a$ must be replaced throughout the derivation $d_0$ of this premise by an arbitrarily chosen numeral $n$.

[2] Cf. end of Sect. 2.

The derivation has obviously remained correct, and the endsequent has become a reduced endsequent in the sense of 13.21 or 13.23.
In other words:

$$\text{If } d = \left\{ \begin{array}{c} d_0(a) \\ | \\ \Gamma \to F(a) \\ \hline \Gamma \to \forall x F(x) \end{array} \right. \text{, then } d[n] := d_0(n) \quad \text{and } \mathsf{tp}(d) := \mathsf{R}_{\forall x F(x)}.$$

$$\text{If } d = \left\{ \begin{array}{c} d_0 \\ | \\ A, \Gamma \to \bot \\ \hline \Gamma \to \neg A \end{array} \right. \text{, then } d[0] := d_0 \text{ and } \mathsf{tp}(d) := \mathsf{R}_{\neg A}.$$

14.24. Suppose that the endsequent of $d$ is the result of a *'complete induction'*.

$$d = \left\{ \begin{array}{cc} d_0 & d_1(a) \\ | & | \\ \Gamma \to F(0) & F(a), \Gamma \to F(\mathsf{S}a) \\ \hline \multicolumn{2}{c}{\Gamma \to F(k)} \end{array} \right.$$

(Since $d$ is closed, the induction term is a numeral $k$.)

Then we set

$$d[0] := \left\{ \begin{array}{ccccc} d_0 & d_1(0) & d_1(1) & & d_1(k-1) \\ | & | & | & & | \\ \Gamma \to F(0) & F(0), \Gamma \to F(1) & F(1), \Gamma \to F(2) & \ldots\ldots & F(k-1), \Gamma \to F(k) \\ \hline \multicolumn{5}{c}{\Gamma \to F(k)} \end{array} \right.$$

and $\mathsf{tp}(d) := \mathsf{Rep}$.

14.25. The last case to be considered is that in which the endsequent is the conclusion of a *'chain-rule'* inference: $d = \left\{ \begin{array}{ccc} d_0 & & d_l \\ | & & | \\ \Gamma_0 \to A_0 & \ldots\ldots & \Gamma_l \to A_l \\ \hline \multicolumn{3}{c}{\Theta \to D} \end{array} \right.$

The premise whose succedent formula provides the succedent formula of the endsequent, I shall call the *'major premise'*. If the succedent of the endsequent is a false minimal formula, we choose as major premise the *first* premise (in the given order) whose succedent formula is *also* a false minimal formula. This does not change the correctness of the 'chain-rule' inference.

So there is a $j \leq l$ such that $A_j \approx D, \forall i \leq j(\Gamma_i \subseteq \Theta, A_0, \ldots, A_{i-1})$ and, if $A_j$ is a false minimal formula then none of $A_0, \ldots, A_{j-1}$ is a false minimal formula.

From these preliminaries it follows that the major premise $\Gamma_j \to A_j$ can in no case be in endform (13.4), for otherwise the endsequent $\Theta \to D$ would obviously also have to be in endform, and this was assumed not to be the case. Hence a *reduction step* can be carried out on the derivation of the major premise. In respect of this reduction step, i.e. in respect of $\mathsf{tp}(d_j)$, I distinguish four cases (14.251–14.254).

14.251. Suppose that the major premise undergoes a change according to 13.2 in the reduction step on its derivation $d_j$, i.e. $\mathsf{tp}(d_j) = \mathsf{R}_{A_j}$ and $A_j = D$. In that case the endsequent is subjected to the appropriate reduction step for sequents according to 13.2; any *choice* that arises is to be made arbitrarily. The reduction step for derivations is then carried out on the derivation $d_j$ of the major premise and, whenever a choice exists, the *same* choice is to be made as before. The succedent formulae of both sequents are now the same once again and the 'chain-rule' inference is once again correct. Thus, the reduction step for the entire derivation $d$ is completed.

In other words, $\mathsf{tp}(d) := \mathsf{tp}(d_j)$ and

$$d[n] := \left\{ \frac{\begin{matrix} d_0 & & d_j[n] & & d_l \\ | & & | & & | \\ \Gamma_0 \to A_0 \ldots & & \Gamma_j \to F(n) & \ldots & \Gamma_l \to A_l \end{matrix}}{\Theta \to F(n)} \right. \quad \text{if } A_j = D = \forall x F(x);$$

$$d[0] := \left\{ \frac{\begin{matrix} d_0 & & d_j[0] & & d_l \\ | & & | & & | \\ \Gamma_0 \to A_0 \ldots & & A, \Gamma_j \to \bot & \ldots & \Gamma_l \to A_l \end{matrix}}{A, \Theta \to \bot} \right. \quad \text{if } A_j = D = \neg A.$$

14.252. Suppose that the major premise undergoes a change according to 13.5 in the reduction step on its derivation, and the affected antecedent formula $B$ also occurs in the antecedent of the endsequent, i.e. $\mathsf{tp}(d_j) = \mathsf{L}_B^k$ with $B \in \Theta$. In that case the reduction step is carried out on the derivation of the major premise and the endsequent is modified according to the *corresponding* reduction step on sequents (13.5), so that the 'chain-rule' inference becomes again correct.

In other words, $\mathsf{tp}(d) := \mathsf{tp}(d_j)$ and

$$d[0] := \left\{ \frac{\begin{matrix} d_0 & & d_j[0] & & d_l \\ | & & | & & | \\ \Gamma_0 \to A_0 \ldots & & F(k), \Gamma_j \to A_j & \ldots & \Gamma_l \to A_l \end{matrix}}{F(k), \Theta \to D} \right. \quad \text{if } B = \forall x F(x);$$

$$d[0] := \left\{ \frac{\begin{matrix} d_0 & & d_j[0] & & d_l \\ | & & | & & | \\ \Gamma_0 \to A_0 \ldots & & \Gamma_j \to A & \ldots & \Gamma_l \to A_l \end{matrix}}{\Theta \to A} \right. \quad \text{if } B = \neg A.$$

14.253. (Principal case) Suppose that the major premise, say $\Delta \to C$, undergoes a change according to 13.5 in the reduction step on its derivation and that the affected antecedent formula $(V)$ is a formula that does not occur among the antecedent formulae of the endsequent, since it agrees with the succedent formula of an earlier premise; suppose further that *this* premise, call it $\Gamma \to V$, undergoes a *change* during the reduction step on its derivation which, in that case, must necessarily be a change according to 13.2. (Since $V$ cannot be a minimal formula.)—Remember that the endsequent of the whole derivation has the form $\Theta \to D$. I shall distinguish *two subcases* depending on whether $V$ has the form $\forall x F(x)$ or $\neg A$.

Suppose first that $V$ has the form $\forall x F(x)$. In that case an antecedent formula $F(k)$ is adjoined in the reduction step according to 13.51 on $\Delta \to C$; in the reduction step on $\Gamma \to \forall x F(x)$ which must be carried out according to 13.21, the *same* symbol $k$ may be chosen for the numeral to be substituted, so that $\Gamma \to F(k)$ results. We now form three 'chain-rule' inferences: the premises of the first are those of the original 'chain-rule' inference, but with $\Gamma \to F(k)$ in place of $\Gamma \to \forall x F(x)$; its conclusion: $\Theta \to F(k)$. A correct result. The premises of the second are those of the original 'chain-rule' inference, except that $\Delta \to C$ is replaced by the sequent that was reduced according to 13.51; its conclusion: $F(k), \Theta \to D$. This is also a correct 'chain-rule' inference. The third 'chain-rule' inference again yields the endsequent $\Theta \to D$ from $\Theta \to F(k)$ and $F(k), \Theta \to D$. Together with each one of the sequents used we must of course write down the complete derivation of each sequent so that altogether we now have another correct derivation.

If $V$ has the form $\neg A$, then $\Delta \to C$ is reduced to $\Delta \to A$, and $\Gamma \to \neg A$ to $A, \Gamma \to \bot$. We now form, as before, two 'chain-rule' inferences with the conclusions $A, \Theta \to \bot$ and $\Theta \to A$. With their order interchanged, these two yield by a third 'chain-rule' inference again $\Theta \to D$. (Note that $D$, like $C$ and $\bot$, is a false minimal formula.)

In other words,

$$\text{if } d = \left\{ \begin{array}{c} d_i \qquad\quad d_j \\ \mid \qquad\quad\;\; \mid \\ \ldots \Gamma \to V \ldots \Delta \to C \ldots \\ \hline \Theta \to D \end{array} \right. \quad \text{with major premise } \Delta \to C,\ \mathsf{tp}(d_j) = \mathsf{L}_V^k \text{ and } V \notin \Theta,$$

we set $\mathsf{tp}(d) := \mathsf{Rep}$, while the reduced derivation $d[0]$ depends on the form of $V$.

$$\text{If } V = \forall x F(x), \text{ then } d[0] := \left\{ \begin{array}{c} d\{0\} \qquad\qquad d\{1\} \\ \mid \qquad\qquad\quad \mid \\ \Theta \to F(k) \quad F(k), \Theta \to D \\ \hline \Theta \to D \end{array} \right.$$

$$\text{where } d\{0\} := \left\{ \begin{array}{c} d_i[k] \qquad\qquad d_j \\ \mid \qquad\qquad\quad \mid \\ \ldots\ \Gamma \to F(k) \ldots\ \Delta \to C \ldots \\ \hline \Theta \to F(k) \end{array} \right.$$

$$\text{and } d\{1\} := \left\{ \begin{array}{c} d_i \qquad\qquad d_j[0] \\ \mid \qquad\qquad\quad \mid \\ \ldots \Gamma \to V \ldots\ F(k), \Delta \to C \ldots \\ \hline F(k), \Theta \to D \end{array} \right.$$

$$\text{If } V = \neg A \text{, then } d[0] := \left\{ \begin{array}{cc} d\{0\} & d\{1\} \\ | & | \\ \Theta \to A & A, \Theta \to \bot \\ \hline & \Theta \to D \end{array} \right.$$

$$\text{where} \quad d\{0\} := \left\{ \begin{array}{cc} d_i & d_j[0] \\ | & | \\ \ldots \Gamma \to V \ldots & \Delta \to A \ldots \\ \hline & \Theta \to A \end{array} \right.$$

$$\text{and} \quad d\{1\} := \left\{ \begin{array}{cc} d_i[0] & d_j \\ | & | \\ \ldots \ A, \Gamma \to \bot \ldots & \Delta \to C \ldots \\ \hline A, \Theta \to \bot & \end{array} \right.$$

14.254. We are still left with the following possibilities: the major premise remains *unchanged* in the reduction step on its derivation; or: its change is of the kind assumed at 14.253, and the premise $\Gamma \to V$ remains *unchanged* in the reduction step on its derivation.—In both cases we carry out the reduction step on the derivation of the unchanged remaining premise, and this completes the reduction.

However, if this reduction step on the derivation of the unchanged remaining premise is according to 14.253, we proceed somewhat *differently*, namely: we carry out this reduction step, but *without* completing the prescribed third 'chain-rule' inference; instead, we take the two premises of this 'chain-rule' inference and insert them in place of its conclusion in the sequence of premises of that 'chain-rule' inference which concludes the derivation as a whole. This obviously leaves that 'chain-rule' inference correct. The endsequent is not changed.

Let us have a closer look on one of these cases; namely, the case where the premise $\Delta \to C$ $(= \Gamma_j \to A_j)$ remains unchanged in the reduction step on its derivation $d_j$, and where this reduction step is according to 14.253.

$$\text{Then } d_j[0] = \left\{ \begin{array}{cc} d_j\{0\} & d_j\{1\} \\ | & | \\ \Gamma_j \to B & B, \Gamma_j \to A'_j \\ \hline & \Gamma_j \to A_j \end{array} \right. \text{ for some } B \text{ and } A'_j \approx A_j \approx D.$$

We set $\mathsf{tp}(d) := \mathsf{Rep}$ and

$$d[0] := \left\{ \begin{array}{cccccc} d_0 & d_{j-1} & d_j\{0\} & d_j\{1\} & d_{j+1} & d_l \\ | & | & | & | & | & | \\ \Gamma_0 \to A_0 \ldots & \Gamma_{j-1} \to A_{j-1} & \Gamma_j \to B & B, \Gamma_j \to A'_j & \Gamma_{j+1} \to A_{j+1} \ldots & \Gamma_l \to A_l \\ \hline & & & \Theta \to D & & \end{array} \right.$$

The definition of the *reduction step on a derivation* and the proof of Theorem 3.1 are now complete. As an immediate consequence from Theorem 3.1 one obtains

**Corollary 3.1** *If* $d \vdash \to \bot$, *then* $d[0] \vdash \to \bot$.

*Proof* By Theorem 3.1 we get $d[0] \vdash \mathsf{tp}(d)(\to \bot, 0)$. Since no reduction step is applicable to $\to \bot$, it cannot be that $\mathsf{tp}(d)$ is $\mathsf{R}_A$ or $\mathsf{L}^k_A$. Hence $\mathsf{tp}(d) = \mathsf{Rep}$ and thus $\mathsf{tp}(d)(\to \bot, 0) = \to \bot$. □

*Remark (Consistency of $\mathcal{Z}$)* In Sect. 5 we will assign to each $\mathcal{Z}$-derivation $d$ an ordinal $o(d) < \varepsilon_0$ and prove that $o(d[n]) < o(d)$ whenever $d[n]$ is defined (Theorem 5.2). Together with Corollary 3.1 this implies the consistency of $\mathcal{Z}$ via (quantifierfree) induction up to $\varepsilon_0$.

## 4 Reduction Steps on Derivations Revisited

In this section we present the contents of Sects. 2, 3 in a more condensed style. In the course of this we also carry out some minor modifications on Gentzen's original approach, namely

- In the reduction steps $\mathsf{L}^k_{\forall xF}$ and $\mathsf{L}^0_{\neg A}$ it is no longer required that the succedent $C$ is a false minimal formula. Accordingly the notion "endform" will be modified, and the condition "$A_j \approx C$" in the chain rule will be replaced by "$A_j \in \{C, \bot\}$".
- Each chain-rule inference will now have an explicitly shown *rank* which is an upper bound on the ranks of all its cut formulas.

*Some preliminary definitions and abbreviations*

1. $A \approx \top \;:\Leftrightarrow\; A$ is a true minimal formula.
2. $\Gamma{\to}C$ has (or, is in) *endform* $:\Leftrightarrow\; C \approx \top$ or $\Gamma$ contains a false minimal formula.
3. $\mathrm{rk}(C) := \begin{cases} 0 & \text{if } C \text{ is prime,} \\ \mathrm{rk}(A)+1 & \text{if } C = \forall x A \text{ or } C = \neg A. \end{cases}$
4. If $X$ is a formula or sequent, then $\mathrm{FV}(X)$ denotes the set of all free variables occurring in $X$.
5. $\Pi$ ranges over sequents; for $\Pi = \Gamma{\to}C$ we set $A,\Pi := A,\Gamma{\to}C$ and $\Pi{\cdot}A := \Gamma{\to}A$.
6. An *inference symbol* is an expression of one of the following three kinds: $\mathsf{R}_A$ with $\mathrm{rk}(A) > 0$ or $A \approx \top$, $\mathsf{L}^k_A$ with $\mathrm{rk}(A) > 0$ or $A \approx \bot$, $\mathsf{Rep}$.
7. For each inference symbol $\mathcal{I}$ we define
   - its *arity* $|\mathcal{I}| := \begin{cases} \mathbb{N} & \text{if } \mathcal{I} = \mathsf{R}_{\forall xF}, \\ \emptyset & \text{if } \mathcal{I} = \mathsf{R}_A \text{ or } \mathcal{I} = \mathsf{L}^k_A \text{ with } \mathrm{rk}(A) = 0, \\ \{0\} & \text{otherwise,} \end{cases}$
   - the result of applying (the reduction step denoted by) $\mathcal{I}$ to $\Pi$ with choice $n$:

$$\mathcal{I}(\Pi, n) := \begin{cases} \Pi{\cdot}F(n) & \text{if } \mathcal{I} = \mathsf{R}_{\forall xF}, \\ F(k), \Pi & \text{if } \mathcal{I} = \mathsf{L}^k_{\forall xF}, \\ A, \Pi{\cdot}\bot & \text{if } \mathcal{I} = \mathsf{R}_{\neg A}, \\ \Pi{\cdot}A & \text{if } \mathcal{I} = \mathsf{L}^k_{\neg A}, \\ \Pi & \text{otherwise,} \end{cases}$$

- the relation $\mathcal{I} \lhd \Pi$ ($\mathcal{I}$ is *permissible* for $\Pi$):

$$\mathcal{I} \lhd \Gamma{\to}C \quad :\Leftrightarrow \quad \mathcal{I} = \mathsf{R}_C \text{ or } \mathcal{I} = \mathsf{L}^k_A \text{ with } A \in \Gamma \text{ or } \mathcal{I} = \mathsf{Rep}.$$

**Definition** The figure $\dfrac{\Gamma_0{\to}A_0 \ \ \ldots \ \ \Gamma_l{\to}A_l}{\Gamma{\to}C}$ is called a *chain-rule inference of rank r* if there exists a $j \le l$ such that $A_j \in \{C, \bot\}$ and $\forall i \le j(\Gamma_i \subseteq \Gamma, A_0, \ldots, A_{i-1})$ and $\forall i < j(\mathrm{rk}(A_i) \le r)$.

**Inductive Definition of** $d \vdash \Pi$ ($d$ is a $\mathcal{Z}$-derivation with endsequent $\Pi$)

1. Atomic derivations (axioms): cf. Sect. 6.
2. If $d_0 \vdash \Gamma{\to}F(a)$ and $a \notin \mathrm{FV}(\Gamma{\to}\forall x F(x))$, then $\mathrm{I}^a_{\forall x F(x)} d_0 \vdash \Gamma{\to}\forall x F(x)$.
3. If $d_0 \vdash A, \Gamma{\to}\bot$, then $\mathrm{I}_{\neg A} d_0 \vdash \Gamma{\to}\neg A$.
4. If $d_0 \vdash \Gamma{\to}F(0)$ and $d_1 \vdash F(a), \Gamma{\to}F(\mathsf{S}a)$ and $a \notin \mathrm{FV}(\Gamma{\to}F(t))$, then $\mathrm{Ind}^{a,t}_F d_0 d_1 \vdash \Gamma{\to}F(t)$.
5. If $d_i \vdash \Pi_i$ with $\mathrm{FV}(\Pi_i) \subseteq \mathrm{FV}(\Pi)$ for $i = 0, \ldots, l$, and if $\dfrac{\Pi_0 \ \ \ldots \ \ \Pi_l}{\Pi}$ is a *chain-rule inference of rank r*, then $\mathsf{K}^r_\Pi d_0 \ldots d_l \vdash \Pi$.

A derivation is called *closed* iff its endsequent is closed.

**Lemma 4.1** *Assume* $\Pi_i = \Gamma_i{\to}A_i$ $(i = 0, \ldots, j_0)$ *and* $\Pi = \Gamma{\to}C$ *with* $A_{j_0} \in \{C, \bot\}$ *and* $\forall i \le j_0(\Gamma_i \subseteq \Gamma, A_0, \ldots, A_{i-1})$.

Further, let $\mathcal{I}_0, \ldots, \mathcal{I}_{j_0}$ be inference symbols such that
$\forall i \le j_0(\mathcal{I}_i \lhd \Pi_i \ \&\ \mathcal{I}_i \ntriangleleft \Pi)$.
Then $\exists i, j, k(i < j \le j_0 \ \&\ \mathcal{I}_i = \mathsf{R}_{A_i} \ \&\ \mathcal{I}_j = \mathsf{L}^k_{A_i} \ \&\ 0 < \mathrm{rk}(A_i))$.

*Proof* From $\mathcal{I}_{j_0} \lhd \Pi_{j_0} \ \&\ \mathcal{I}_{j_0} \ntriangleleft \Pi \ \&\ A_{j_0} \in \{C, \bot\}$ it follows that $\mathcal{I}_{j_0} \in \mathsf{L}$ (i.e. $\mathcal{I}_{j_0} = \mathsf{L}^k_B$ for some $B$ and $k$). Hence there exists the least $j \le j_0$ such that $\mathcal{I}_j \in \mathsf{L}$. Assume $\mathcal{I}_j = \mathsf{L}^k_B$. Then $\mathsf{L}^k_B \lhd \Pi_j \ \&\ \mathsf{L}^k_B \ntriangleleft \Pi$ which implies $B \in \Gamma_j \setminus \Gamma \subseteq \{A_0, \ldots, A_{j-1}\}$. So we have $\mathcal{I}_j = \mathsf{L}^k_{A_i}$ for some $i < j$. By minimality of $j$ and since $i < j \le j_0$, we have $\mathcal{I}_i \notin \mathsf{L}$ and $\mathcal{I}_i \lhd \Pi_i \ \&\ \mathcal{I}_i \ntriangleleft \Pi$, which implies $\mathcal{I}_i = \mathsf{R}_{A_i}$. Finally, from $\mathsf{R}_{A_i} \lhd \Pi_i$ and $\mathsf{L}^k_{A_i} \lhd \Pi_j$ we conclude $(\mathrm{rk}(A_i) = 0 \Rightarrow A_i \approx \top)$ and $(\mathrm{rk}(A_i) = 0 \Rightarrow A_i \approx \bot)$, hence $\mathrm{rk}(A_i) > 0$. □

**Definition 4.2** ($\mathsf{tp}(d)$ **and** $d[n]$) For each closed $\mathcal{Z}$-derivation $d$ we now define an inference symbol $\mathsf{tp}(d)$ and, for each $n \in |\mathsf{tp}(d)|$, a closed $\mathcal{Z}$-derivation $d[n]$. In the main case 5.1. where $d$ is 'critical' we also define the auxiliary derivations $d\{0\}, d\{1\}$ and the formula $\mathsf{A}(d)$. The whole definition proceeds by recursion on the build-up of $d$. In parallel we observe that $\mathsf{tp}(d)$ is permissible for $\Pi$ (i.e., $\mathsf{tp}(d) \lhd \Pi$) whenever $d \vdash \Pi$.

1. $d$ atomic: cf. Sect. 6.
2. $d = \mathrm{I}^a_{\forall x F} d_0$: Then $\mathsf{tp}(d) := \mathsf{R}_{\forall x F}$ and $d[n] := d_0(a/n)$.
3. $d = \mathrm{I}_{\neg A} d_0$: Then $\mathsf{tp}(d) := \mathsf{R}_{\neg A}$ and $d[0] := d_0$.
4. $d = \mathrm{Ind}^{a,k}_F d_0 d_1$ with $d_0 \vdash \Gamma{\to}F(0)$ and $d_1 \vdash F(a), \Gamma{\to}F(\mathsf{S}a)$:

Then $\mathsf{tp}(d) := \mathsf{Rep}$ and $d[0] := \mathsf{K}^r_{\Gamma\to F(k)} d_0\, d_1(a/0)\dots d_1(a/k-1)$, where $r := \mathrm{rk}(F)$.

5. $d = \mathsf{K}^r_\Pi d_0 \dots d_l$ with $\Pi = \Gamma\to C$ and $d_i \vdash \Pi_i = \Gamma_i \to A_i$ $(i \le l)$:
   Abbreviation: $\mathsf{K}^r_{\Pi'}(i/d'_1\dots d'_m) := \mathsf{K}^r_{\Pi'} d_0 \dots d_{i-1} d'_1 \dots d'_m d_{i+1}\dots d_l$.
   Let $j_0$ be minimal s.t. $A_{j_0} \in \{C, \bot\}$ & $\forall i \le j_0(\Gamma_i \subseteq \Gamma, A_0, \dots, A_{i-1})$.
   We say that $d$ is *critical* if $\forall i \le j_0(\mathsf{tp}(d_i) \ntriangleleft \Pi)$.

   5.1. $d$ critical:
   Then due to Lemma 4.1, and since $\forall i \le l(\mathsf{tp}(d_i) \triangleleft \Pi_i)$ there exists a pair $(i, j)$ such that
   $i < j \le j_0$, $\mathsf{tp}(d_i) = \mathsf{R}_{A_i}$, $\mathsf{tp}(d_j) = \mathsf{L}^k_{A_i}$ (for some $k$) and $0 < \mathrm{rk}(A_i)$.
   We take the least such pair and set $\mathsf{tp}(d) := \mathsf{Rep}$ and
   $d[0] := \mathsf{K}^{r-1}_\Pi d\{0\} d\{1\}$ where

$$d\{0\} := \mathsf{K}^r_{\Pi\cdot\mathsf{A}(d)} \begin{cases} (i/d_i[k]) & \text{if } A_i = \forall x F, \\ (j/d_j[0]) & \text{if } A_i = \neg A, \end{cases}$$

$$d\{1\} := \mathsf{K}^r_{\mathsf{A}(d),\Pi} \begin{cases} (j/d_j[0]) & \text{if } A_i = \forall x F, \\ (i/d_i[0]) & \text{if } A_i = \neg A, \end{cases}$$

$$\text{and}\quad \mathsf{A}(d) := \begin{cases} F(k) & \text{if } A_i = \forall x F, \\ A & \text{if } A_i = \neg A. \end{cases}$$

   5.2. $d$ not critical: Let $i \le j_0$ be minimal such that $\mathsf{tp}(d_i) \triangleleft \Pi$.

   5.2.1. $d_i$ critical:
   Then $\mathsf{tp}(d) := \mathsf{Rep}$ and $d[0] := \mathsf{K}^{r'}_\Pi(i/d_i\{0\}, d_i\{1\})$ with $r' := \max\{\mathrm{rk}(\mathsf{A}(d_i)), r\}$.

   5.2.2. $d_i$ not critical: Then $\mathsf{tp}(d) := \mathsf{tp}(d_i)$ and $d[n] := \mathsf{K}^r_{\mathsf{tp}(d)(\Pi,n)}(i/d_i[n])$.

**Lemma 4.3** *If $d \vdash \Pi$, then* $\mathsf{tp}(d) \triangleleft \Pi$.

**Theorem 4.4** *For $d \vdash \Pi$ the following holds:*

*(a) If $d = \mathsf{K}^r_\Pi d_0 \dots d_l$ is critical, then $d\{0\} \vdash \Pi\cdot\mathsf{A}(d)$, $d\{1\} \vdash \mathsf{A}(d), \Pi$, and* $\mathrm{rk}(\mathsf{A}(d)) < r$.

*(b)* $\forall n \in |\mathsf{tp}(d)|\big(d[n] \vdash \mathsf{tp}(d)(\Pi, n)\big)$.

Proof by simultaneous induction on the build-up of $d$:

(a) The premise "$d$ critical" yields that we are in Case 5.1 of Definition 4.2.
Subcase $A_i = \forall x F$:
By assumption we have $d_\nu \vdash \Pi_\nu$ for all $\nu \le l$. From $d_i \vdash \Pi_i$ and $d_j \vdash \Pi_j$ together with $\mathsf{tp}(d_i) = \mathsf{R}_{A_i}$ and $\mathsf{tp}(d_j) = \mathsf{L}^k_{A_i}$ we get $d_i[k] \vdash \Pi_i\cdot F(k)$ and $d_j[0] \vdash F(k), \Pi_j$ by IH(b).
Since $\dfrac{\Pi_0 \dots \Pi_{i-1}\, \Pi_i\cdot F(k) \dots}{\Pi\cdot F(k)}$ and $\dfrac{\Pi_0 \dots \Pi_{j-1}\ F(k), \Pi_j \dots \Pi_{j_0}\ \dots}{F(k), \Pi}$ are chain inferences of degree $r$,
we conclude $d\{0\} \vdash \Pi\cdot F(k)$ and $d\{1\} \vdash F(k), \Pi$. Further $\mathrm{rk}(\mathsf{A}(d)) = \mathrm{rk}(F(k)) < \mathrm{rk}(A_i) \le r$.

Subcase $A_i = \neg A$:
Similar to the previous case, only that now $d_j[0] \vdash \Pi_j{\cdot}A$ and $d_i[0] \vdash A, \Pi_i{\cdot}\bot$, and we apply the chain inferences $\dfrac{\Pi_0 \ \ldots \ \Pi_{j-1} \ \ \Pi_j{\cdot}A \ \ldots}{\Pi{\cdot}A}$ and $\dfrac{\Pi_0 \ \ldots \ \Pi_{i-1} \ \ A, \Pi_i{\cdot}\bot \ \ldots}{A, \Pi}$ to obtain $d\{0\} \vdash \Pi{\cdot}A$ and $d\{1\} \vdash A, \Pi$.

(b) We follow the case distinction of Definition 4.2.:

1. $d$ atomic: cf. Sect. 6. 2.–4.: Left to the reader.
5. $d = \mathsf{K}^r_\Pi d_0 \ldots d_l$:
   - 5.1. $d$ critical: Then $\mathsf{tp}(d) = \mathsf{Rep}$ and $d[0] = \mathsf{K}^{r-1}_\Pi d\{0\}d\{1\}$. By (a) we have $d\{0\} \vdash \Pi{\cdot}\mathsf{A}(d)$, $d\{1\} \vdash \mathsf{A}(d), \Pi$, and $\mathrm{rk}(\mathsf{A}(d)) < r$. Hence $d[0] \vdash \Pi$, i.e. $\forall n \in |\mathsf{tp}(d)|(d[n] \vdash \mathsf{tp}(d)(\Pi, n))$.
   - 5.2. $d$ not critical, and $i$ is minimal s.t. $\mathsf{tp}(d_i) \lhd \Pi$:
     - 5.2.1. $d_i$ critical: By IH(a) we have, $d_i\{0\} \vdash \Pi_i{\cdot}\mathsf{A}(d_i)$ and $d_i\{1\} \vdash \mathsf{A}(d_i), \Pi_i$. Further,
$$\frac{\Pi_0 \ \ldots \ \Pi_{i-1} \ \ \Pi{\cdot}\mathsf{A}(d_i) \ \ \mathsf{A}(d_i), \Pi_i \ \ \Pi_{i+1} \ \ldots \ \Pi_l}{\Pi}$$
is a chain inference of degree $r' := \max\{\mathrm{rk}(\mathsf{A}(d_i)), r\}$.
Hence $d[0] = \mathsf{K}^{r'}_\Pi(i/d_i\{0\}d_i\{1\}) \vdash \Pi$, which yields the claim, since $\mathsf{tp}(d) = \mathsf{Rep}$.
     - 5.2.2. $d_i$ not critical: Then $\mathsf{tp}(d) = \mathsf{tp}(d_i)$, and by IH(b) we have $d_i[n] \vdash \mathsf{tp}(d_i)(\Pi_i, n)$ for all $n \in |\mathsf{tp}(d_i)|$.
Further, $\dfrac{\Pi_0 \ \ldots \ \Pi_{i-1} \ \ \mathsf{tp}(d_i)(\Pi_i, n) \ \ \Pi_{i+1} \ \ldots \ \Pi_l}{\mathsf{tp}(d_i)(\Pi, n)}$ is a chain inference of rank $r$.
Since $\mathsf{tp}(d) = \mathsf{tp}(d_i)$, we conclude $d[n] = \mathsf{K}^r_{\mathsf{tp}(d)(\Pi,n)}(i/d_i[n]) \vdash \mathsf{tp}(d)(\Pi, n)$ for all $n \in |\mathsf{tp}(d)|$.

**Corollary** *If* $d \vdash \to\bot$, *then* $d[0] \vdash \to\bot$.

*Proof* From $d \vdash \to\bot$ by Lemma 4.3 we get $\mathsf{tp}(d) \lhd \to\bot$, which implies $\mathsf{tp}(d) = \mathsf{Rep}$. Now by Theorem 4.4b we conclude $d[0] \vdash \mathsf{Rep}(\to\bot, 0)$, i.e. $d[0] \vdash \to\bot$. □

## 5 Ordinal Assignment and Termination Proof

In this section we will assign to each $\mathcal{Z}$-derivation $d$ an ordinal $o(d) < \varepsilon_0$ and prove that if $d$ is a closed $\mathcal{Z}$-derivation then $o(d[n]) < o(d)$ for all $n \in |\mathsf{tp}(d)|$. The ordinal $o(d)$ will be defined via the auxiliary notions $\mathrm{dg}(d)$ (degree of $d$) and $\tilde{\mathsf{o}}(d)$ (pre-ordinal of $d$).[3]

[3]This ordinal assignment is essentially that of [4].

**Definition of** $\mathrm{dg}(d) < \omega$ **and** $\tilde{o}(d), o(d) < \varepsilon_0$
For atomic $d$ cf. Sect. 6.
Otherwise

$$\mathrm{dg}(d) := \begin{cases} \mathrm{dg}(d_0) & \text{if } d = \mathrm{I}^a_{\forall xF} d_0 \text{ or } d = \mathrm{I}_{\neg A} d_0, \\ \max\{\mathrm{dg}(d_0){-}1, \mathrm{dg}(d_1){-}1, r\} & \text{if } d = \mathrm{Ind}^{a,t}_F d_0 d_1 \text{ with } r := \mathrm{rk}(F), \\ \max\{\mathrm{dg}(d_0){-}1, \dots, \mathrm{dg}(d_l){-}1, r\} & \text{if } d = \mathsf{K}^r_\Pi d_0 \dots d_l, \end{cases}$$

$$\tilde{o}(d) := \begin{cases} \tilde{o}(d_0) + 1 & \text{if } d = \mathrm{I}^a_{\forall xF} d_0 \text{ or } d = \mathrm{I}_{\neg A} d_0, \\ \omega^{\tilde{o}(d_0)} \# \omega^{\tilde{o}(d_1)+1} & \text{if } d = \mathrm{Ind}^{a,t}_F d_0 d_1, \\ \omega^{\tilde{o}(d_0)} \# \dots \# \omega^{\tilde{o}(d_l)} & \text{if } d = \mathsf{K}^r_\Pi d_0 \dots d_l, \end{cases}$$

$o(d) := \omega_{\mathrm{dg}(d)}(\tilde{o}(d))$, where $\omega_0(\alpha) := \alpha, \omega_{n+1}(\alpha) := \omega^{\omega_n(\alpha)}$.

*Remark* $\tilde{o}(d(a/t)) = \tilde{o}(d)$ and $\mathrm{dg}(d(a/t)) = \mathrm{dg}(d)$.

**Lemma 5.1** *For each closed $\mathcal{Z}$-derivation $d$ the following holds:*

*(a) If $d$ is not critical, then* $\mathrm{dg}(d[n]) \le \mathrm{dg}(d)$ & $\tilde{o}(d[n]) < \tilde{o}(d)$, *for all* $n \in |\mathsf{tp}(d)|$.
*(b) If $d$ is critical, then:*

*(i)* $\mathrm{dg}(d\{\nu\}) \le \mathrm{dg}(d)$ & $\tilde{o}(d\{\nu\}) < \tilde{o}(d)$, *for* $\nu = 0, 1$.
*(ii)* $\mathrm{dg}(d[0]) < \mathrm{dg}(d)$ & $\tilde{o}(d[0]) < \omega^{\tilde{o}(d)}$ & $\mathrm{rk}(\mathsf{A}(d)) < \mathrm{dg}(d)$.

Proof by induction on the build-up of $d$:
*Notation:* In the following we omit the subscript of $\mathsf{K}^r_\Pi$.
Assume $d \vdash \Pi$. As before we follow the case distinction of Definition 4.2.

1. $d$ atomic: cf. Sect. 6.
2. $d = \mathrm{I}^a_{\forall xF} d_0$: Then $\mathsf{tp}(d) = \mathsf{R}_{\forall xF}$ and $d[n] = d_0(a/n)$.
So we have $\mathrm{dg}(d[n]) = \mathrm{dg}(d_0(a/n)) = \mathrm{dg}(d_0) = \mathrm{dg}(d)$ and
$\tilde{o}(d[n]) = \tilde{o}(d_0(a/n)) = \tilde{o}(d_0) < \tilde{o}(d)$.
3. $d = \mathrm{I}_{\neg A} d_0$: similar to 2.
4. $d = \mathrm{Ind}^{a,k}_F d_0 d_1$:
Then $\mathsf{tp}(d) = \mathsf{Rep}$ and $d[0] = \mathsf{K}^r d_0\, d_1(a/0) \dots d_1(a/k{-}1)$, where $r = \mathrm{rk}(F)$.
So we have $\mathrm{dg}(d[0]) \le \max\{\mathrm{dg}(d_0){-}1, \mathrm{dg}(d_1){-}1, r\} = \mathrm{dg}(d)$ and
$\tilde{o}(d[0]) = \omega^{\tilde{o}(d_0)} \# \omega^{\tilde{o}(d_1)} \cdot k < \omega^{\tilde{o}(d_0)} \# \omega^{\tilde{o}(d_1)+1} = \tilde{o}(d)$.
5. $d = \mathsf{K}^r d_0 \dots d_l$:

   5.1. $d$ critical: Then $\mathsf{tp}(d) = \mathsf{Rep}$ and $d[0] = \mathsf{K}^{r-1} d\{0\} d\{1\}$ where either $d\{0\} = \mathsf{K}^r(i/d_i[k])$ & $d\{1\} = \mathsf{K}^r(j/d_j[0])$ or $d\{0\} = \mathsf{K}^r(j/d_j[0])$ & $d\{1\} = \mathsf{K}^r(i/d_i[0])$.
   By IH(a), $\mathrm{dg}(d_i[k]) \le \mathrm{dg}(d_i)$ & $\tilde{o}(d_i[k]) < \tilde{o}(d_i)$ and $\mathrm{dg}(d_j[0]) \le \mathrm{dg}(d_j)$ & $\tilde{o}(d_j[0]) < \tilde{o}(d_j)$.
   This yields $\mathrm{dg}(d\{\nu\}) \le \mathrm{dg}(d)$ & $\tilde{o}(d\{\nu\}) < \tilde{o}(d)$ for $\nu = 0, 1$.

Hence $\mathrm{dg}(d[0]) = \max\{\mathrm{dg}(d\{0\}){-}1, \mathrm{dg}(d\{1\}){-}1, r{-}1\} < \mathrm{dg}(d)$ and $\tilde{\mathrm{o}}(d[0]) = \omega^{\tilde{\mathrm{o}}(d\{0\})} \# \omega^{\tilde{\mathrm{o}}(d\{1\})} < \omega^{\tilde{\mathrm{o}}(d)}$.
By Theorem 4.4a we have $\mathrm{rk}(\mathsf{A}(d)) < r$, thence $\mathrm{rk}(\mathsf{A}(d)) < \mathrm{dg}(d)$.

5.2. $d$ not critical, and $i$ is minimal s.t. $\mathsf{tp}(d_i) \lhd \Pi$:

5.2.1. $d_i$ critical: Then $\mathsf{tp}(d) = \mathsf{Rep}$ and $d[0] = \mathsf{K}^{r'}(i/d_i\{0\}d_i\{1\})$ with $r' = \max\{\mathrm{rk}(\mathsf{A}(d_i)), r\}$.
By IH(b) we have $\mathrm{dg}(d_i\{\nu\}) \le \mathrm{dg}(d_i)$ & $\tilde{\mathrm{o}}(d_i\{\nu\}) < \tilde{\mathrm{o}}(d_i)$ for $\nu = 0, 1$,
and also $\mathrm{rk}(\mathsf{A}(d_i)) < \mathrm{dg}(d_i)$.
The latter yields $r' \le \max\{\mathrm{dg}(d_i){-}1, r\} \le \mathrm{dg}(d)$. Hence
$\mathrm{dg}(d[0]) = \max\{\mathrm{dg}(d_0){-}1, \dots, \mathrm{dg}(d_i\{0\}){-}1, \mathrm{dg}(d_i\{1\}){-}1, \dots, \mathrm{dg}(d_l){-}1, r'\} \le$
$\le \max\{\mathrm{dg}(d_0){-}1, \dots, \mathrm{dg}(d_i){-}1, \dots, \mathrm{dg}(d_l){-}1, r'\} \le \mathrm{dg}(d)$ and
$\tilde{\mathrm{o}}(d[0]) = \omega^{\tilde{\mathrm{o}}(d_0)} \# \dots \# \omega^{\tilde{\mathrm{o}}(d_i\{0\})} \# \omega^{\tilde{\mathrm{o}}(d_i\{1\})} \# \dots \# \omega^{\tilde{\mathrm{o}}(d_l)} <$
$\omega^{\tilde{\mathrm{o}}(d_0)} \# \dots \# \omega^{\tilde{\mathrm{o}}(d_i)} \# \dots \# \omega^{\tilde{\mathrm{o}}(d_l)} = \tilde{\mathrm{o}}(d)$.

5.2.2. $d_i$ not critical: Then $\mathsf{tp}(d) = \mathsf{tp}(d_i)$ and $d[n] = \mathsf{K}^r(i/d_i[n])$.
By IH(a), $\mathrm{dg}(d_i[n]) \le \mathrm{dg}(d_i)$ and $\tilde{\mathrm{o}}(d_i[n]) < \tilde{\mathrm{o}}(d_i)$.
Hence $\mathrm{dg}(d[n]) = \max\{\mathrm{dg}(d_0){-}1, \dots, \mathrm{dg}(d_i[n]){-}1, \dots, \mathrm{dg}(d_l){-}1, r\} \le \mathrm{dg}(d)$ and
$\tilde{\mathrm{o}}(d[n]) = \omega^{\tilde{\mathrm{o}}(d_0)} \# \dots \# \omega^{\tilde{\mathrm{o}}(d_i[n])} \# \dots \# \omega^{\tilde{\mathrm{o}}(d_l)} < \omega^{\tilde{\mathrm{o}}(d_0)} \# \dots \# \omega^{\tilde{\mathrm{o}}(d_i)} \# \dots \# \omega^{\tilde{\mathrm{o}}(d_l)} = \tilde{\mathrm{o}}(d)$.

**Theorem 5.2** *If $d$ is a closed $\mathcal{Z}$-derivation, then $o(d[n]) < o(d)$ for all $n \in |\mathsf{tp}(d)|$.*

*Proof* By Lemma 5.1 we have $\tilde{\mathrm{o}}(d[n]) < \omega_{\mathrm{dg}(d)-\mathrm{dg}(d[n])}(\tilde{\mathrm{o}}(d))$ and thus $o(d[n]) = \omega_{\mathrm{dg}(d[n])}(\tilde{\mathrm{o}}(d[n])) < \omega_{\mathrm{dg}(d)}(\tilde{\mathrm{o}}(d)) = o(d)$. □

## 6 Treatment of Atomic Derivations

At several places in the preceding sections we had postponed the treatment of atomic derivations. This will now be caught up.

The *logical axioms of $\mathcal{Z}$* are all sequents of the following kinds:

- $\Gamma \to A$ with $A \in \Gamma$.
- $\Gamma \to F(t)$ with $\forall x F(x) \in \Gamma$.
- $\Gamma \to \bot$ with $A, \neg A \in \Gamma$.
- $\Gamma \to A$ with $A$ atomic and $\neg\neg A \in \Gamma$.

The *mathematical axioms of $\mathcal{Z}$* are given by a set of sequents $\mathrm{Ax}(\mathcal{Z})$ satisfying the following conditions:

- $\Pi \in \mathrm{Ax}(\mathcal{Z}) \Rightarrow \Pi(a/t) \in \mathrm{Ax}(\mathcal{Z})$ and $A, \Pi \in \mathrm{Ax}(\mathcal{Z})$.
- $\mathrm{FV}(\Pi) = \emptyset \Rightarrow (\Pi \in \mathrm{Ax}(\mathcal{Z}) \Leftrightarrow \Pi$ has endform$)$.

**Definition of the atomic $\mathcal{Z}$-derivations**

0. If $\Pi \in \mathrm{Ax}(\mathcal{Z})$, then $\mathsf{Ax}^0_\Pi \vdash \Pi$.
1. If $\Pi = \Gamma \to C$ with $C \in \Gamma$, then $\mathsf{Ax}^1_\Pi \vdash \Pi$.

2.1. If $\Pi = \Gamma \to F(t)$ with $\forall x F \in \Gamma$, then $\mathsf{Ax}^{\forall x F,t}_\Pi \vdash \Pi$.
2.2. If $\Pi = \Gamma \to \bot$ with $\neg A, A \in \Gamma$, then $\mathsf{Ax}^{\neg A,0}_\Pi \vdash \Pi$.
3. If $\Pi = \Gamma \to A$ with $\mathrm{rk}(A) = 0$ & $\neg\neg A \in \Gamma$, then $\mathsf{Ax}^{\neg\neg}_\Pi \vdash \Pi$.

**Definition of $\mathsf{tp}(d)$ and $d[n]$ for closed atomic $\mathcal{Z}$-derivations $d$**

0. $d = \mathsf{Ax}^0_{\Gamma\to C}$: Then $\Gamma \to C$ has endform, and we set
$$\mathsf{tp}(d) := \begin{cases} \mathsf{R}_C \text{ if } C \approx \top, \\ \mathsf{L}^0_A \text{ if } C \not\approx \top \text{ and } A \text{ is the first formula in } \Gamma \text{ s.t. } A \approx \bot. \end{cases}$$
1. $d = \mathsf{Ax}^1_{\Gamma\to C}$ with $C \in \Gamma$:

1.1. $\mathrm{rk}(C) = 0$: Then $\mathsf{tp}(d) := \begin{cases} \mathsf{R}_C \text{ if } C \approx \top, \\ \mathsf{L}^0_C \text{ if } C \approx \bot. \end{cases}$

1.2. $\mathrm{rk}(C) > 0$: Then $\mathsf{tp}(d) := \mathsf{R}_C$ and $d[n] := \mathsf{Ax}^{C,n}_{\mathsf{tp}(d)(\Pi,n)}$.

2. $d = \mathsf{Ax}^{C,k}_\Pi$: Then $\mathsf{tp}(d) := \mathsf{L}^k_C$ and $d[0] := \mathsf{Ax}^1_{\mathsf{tp}(d)(\Pi,0)}$.
3. $d = \mathsf{Ax}^{\neg\neg}_{\Gamma\to A}$:

3.1. $A \approx \top$: Then $\mathsf{tp}(d) := \mathsf{R}_A$.

3.2. $A \approx \bot$: Then $\mathsf{tp}(d) := \mathsf{L}^0_{\neg\neg A}$ and $d[0] := \mathrm{I}_{\neg A}\mathsf{Ax}^0_{A,\Gamma\to\bot}$.

**Lemma 6.1** *If $d \vdash \Pi$ with $\mathrm{FV}(\Pi) = \emptyset$ and $d$ atomic, then:*
*(a)* $\mathsf{tp}(d) \lhd \Pi$.
*(b)* $d[n] \vdash \mathsf{tp}(d)(\Pi, n)$ *for all* $n \in |\mathsf{tp}(d)|$.

*Proof*

(a) Left to the reader.
(b) Abbreviation: $\Pi' := \mathsf{tp}(d)(\Pi, n)$.

1.2. $d = \mathsf{Ax}^1_\Pi$ with $\Pi = \Gamma \to C$ and $C \in \Gamma$ & $\mathrm{rk}(C) > 0$:
Then $\mathsf{tp}(d) = \mathsf{R}_C$ and $\Pi' = \begin{cases} \Gamma \to F(n) \text{ if } C = \forall x F(x), \\ A, \Gamma \to \bot \text{ if } C = \neg A. \end{cases}$
Hence $d[n] = \mathsf{Ax}^{C,n}_{\Pi'} \vdash \Pi'$.

2.1. $d = \mathsf{Ax}^{\forall x F,k}_\Pi$ with $\Pi = \Gamma \to F(k)$: Then $\mathsf{tp}(d) = \mathsf{L}^k_{\forall x F}$ and $\Pi' = F(k), \Gamma \to F(k)$. Hence $d[0] = \mathsf{Ax}^1_{\Pi'} \vdash \Pi'$.

2.2. $d = \mathsf{Ax}^{\neg A,0}_\Pi$ with $\Pi = \Gamma \to \bot$ and $A, \neg A \in \Gamma$: Then $\mathsf{tp}(d) = \mathsf{L}^0_{\neg A}$ and $\Pi' = \Gamma \to A$. Hence $d[0] = \mathsf{Ax}^1_{\Pi'} \vdash \Pi'$.

3.2. $d = \mathsf{Ax}^{\neg\neg}_\Pi$ with $\Pi = \Gamma \to A$, $A \approx \bot$, and $\neg\neg A \in \Gamma$:
Then $d' := \mathsf{Ax}^0_{A,\Gamma\to\bot} \vdash A, \Gamma \to \bot$ and thus $d[0] = \mathrm{I}_{\neg A} d' \vdash \Gamma \to \neg A$.
Further, $\Pi' = \mathsf{L}^0_{\neg\neg A}(\Pi, n) = \Gamma \to \neg A$.

**Definition of** $\mathrm{dg}(d), \tilde{o}(d), o(d)$ **for atomic** $\mathcal{Z}$**-derivations** $d$
$\mathrm{dg}(d) := 0$ and $o(d) := \omega_{\mathrm{dg}(d)}(\tilde{o}(d)) = \tilde{o}(d)$, where
$\tilde{o}(\mathsf{Ax}^0_\Pi) := 0,\ \tilde{o}(\mathsf{Ax}^1_{\Gamma\to C}) := 2\mathrm{rk}(C),\ \tilde{o}(\mathsf{Ax}^{C,t}_\Pi) := 2\mathrm{rk}(C) - 1,\ \tilde{o}(\mathsf{Ax}^{\neg\neg}_\Pi) := 2.$

**Lemma 6.2** *If $d$ is a closed atomic $\mathcal{Z}$-derivation, then $o(d[n]) < o(d)$ for all $n \in |\mathsf{tp}(d)|$.*

*Proof* Left to the reader. □

## 7 Embedding of $\mathcal{Z}$ into an Infinitary System $\mathcal{Z}^\infty$

In this section we give an interpretation of the finitary system $\mathcal{Z}$ in an infinitary system $\mathcal{Z}^\infty$ of $\omega$-arithmetic. This way we obtain an explanation of the reduction steps on $\mathcal{Z}$-derivations and the assignment of ordinals to $\mathcal{Z}$-derivations introduced in Sects. 4–6.

Derivable objects of $\mathcal{Z}^\infty$ are closed sequents $\Pi = \Gamma \to C$.

The inference symbols of $\mathcal{Z}^\infty$ are:

$\mathsf{R}_A$ with $\mathrm{rk}(A) > 0$ or $A \approx \top$, $\mathsf{L}^k_A$ with $\mathrm{rk}(A) > 0$ or $A \approx \bot$, and $\mathsf{Cut}_D$ for arbitrary sentences $D$.

We set $\mathsf{Cut}_D \lhd \Pi$ for each $\Pi$, $|\mathsf{Cut}_D| := \{0,1\}$, $\mathsf{Cut}_D(\Pi,0) := \Pi{\cdot}D$ and $\mathsf{Cut}_D(\Pi,1) := D,\Pi$.

$$\mathrm{rk}(\mathcal{I}) := \begin{cases} \mathrm{rk}(D) \text{ if } \mathcal{I} = \mathsf{Cut}_D, \\ -1 \quad \text{otherwise.} \end{cases}$$

The following definition introduces the relation $\mathfrak{d} \vdash^\alpha_m \Pi$ which is short for "*$\mathfrak{d}$ is a $\mathcal{Z}^\infty$-derivation of $\Pi$ with ordinal height $\leq \alpha$ and cutrank $\leq m$*".

**Inductive Definition of** $\mathfrak{d} \vdash^\alpha_m \Pi$
If $\mathcal{I}$ is an inference symbol of $\mathcal{Z}^\infty$ with $\mathrm{rk}(\mathcal{I}) < m$, and if
$\mathcal{I} \lhd \Pi \ \&\ \forall n \in |\mathcal{I}| \exists \alpha_n < \alpha(\mathfrak{d}_n \vdash^{\alpha_n}_m \mathcal{I}(\Pi,n))$, then $\mathcal{I}(\mathfrak{d}_n)_{n\in|\mathcal{I}|} \vdash^\alpha_m \Pi$.

**Definition of** $\mathsf{last}(\mathfrak{d})$**:** If $\mathfrak{d} = \mathcal{I}(\mathfrak{d}_n)_{n\in|\mathcal{I}|}$, then $\mathsf{last}(\mathfrak{d}) := \mathcal{I}$.

*Remark* If $\mathfrak{d} \vdash^\alpha_m \Pi$, then $\mathsf{last}(\mathfrak{d}) \lhd \Pi$.

**Theorem and Definition 7.1** *If* $\dfrac{\Pi_0 \ \ldots\ \Pi_l}{\Pi}$ *is a chain inference of rank $r \leq m$, and if $\mathfrak{d}_i \vdash^{\alpha_i}_{m+1} \Pi_i$ for $i = 0,\ldots,l$, then there exists a $\mathcal{Z}^\infty$-derivation $\mathfrak{d} = \mathcal{K}^r_\Pi(\mathfrak{d}_0,\ldots,\mathfrak{d}_l) \vdash^\alpha_m \Pi$ with $\alpha := \omega^{\alpha_0}\#\ldots\#\omega^{\alpha_l}$.*

Proof by induction on $\alpha$:
Assume $\Pi = \Gamma \to C$ and $\Pi_i = \Gamma_i \to A_i$, and let $j_0$ be minimal such that
$A_{j_0} \in \{C,\bot\}\ \&\ \forall i \leq j_0(\Gamma_i \subseteq \Gamma, A_0,\ldots,A_{i-1})$.

1. $\forall i \le j_0(\mathsf{last}(\mathfrak{d}_i) \not\triangleleft \Pi)$. By Lemma 4.1 there is the least pair $(i,j)$ such that $i < j \le j_0$, $\mathsf{last}(\mathfrak{d}_j) = \mathsf{L}^k_{A_i}$ (for some $k$), $\mathsf{last}(\mathfrak{d}_i) = \mathsf{R}_{A_i}$, and $0 < \mathrm{rk}(A_i) \le r$. Then $\mathfrak{d}_i = \mathsf{R}_{A_i}(\mathfrak{d}_{in})_n$ and $\mathfrak{d}_j = \mathsf{L}^k_{A_i}\mathfrak{d}_{j0}$.
   Let $\mathfrak{d} := \mathsf{Cut}_D(\mathfrak{e}_0, \mathfrak{e}_1)$ with $D := \begin{cases} F(k) & \text{if } A_i = \forall x F, \\ A & \text{if } A_i = \neg A, \end{cases}$ and
   $\mathfrak{e}_0 := \mathcal{K}^r_{\Pi \cdot D}\begin{cases} (i/\mathfrak{d}_{ik}) & \text{if } A_i = \forall x F, \\ (j/\mathfrak{d}_{j0}) & \text{if } A_i = \neg A \end{cases}$ and $\mathfrak{e}_1 := \mathcal{K}^r_{D,\Pi}\begin{cases} (j/\mathfrak{d}_{j0}) & \text{if } A_i = \forall x F, \\ (i/\mathfrak{d}_{i0}) & \text{if } A_i = \neg A. \end{cases}$
   The IH yields $\mathfrak{e}_0 \vdash^{\alpha'}_m \Pi \cdot D$ and $\mathfrak{e}_1 \vdash^{\alpha''}_m D, \Pi$ with $\alpha', \alpha'' < \alpha$.
   Since $\mathrm{rk}(D) < \mathrm{rk}(A_i) \le r \le m$, it follows that $\mathfrak{d} \vdash^{\alpha}_m \Pi$.
2. Otherwise:
   Let $i \le j_0$ be minimal such that $\mathsf{last}(\mathfrak{d}_i) \triangleleft \Pi$, and let $\mathcal{I} := \mathsf{last}(d_i)$.

   2.1. $\mathcal{I} = \mathsf{Cut}_D$: Then $\mathfrak{d}_i = \mathsf{Cut}_D(\mathfrak{d}_{i0}, \mathfrak{d}_{i1})$ with
   $\mathfrak{d}_{i0} \vdash^{\alpha_{i0}}_{m+1} \Pi_i \cdot D \ \& \ \mathfrak{d}_{i1} \vdash^{\alpha_{i1}}_{m+1} D, \Pi_i \ \& \ \alpha_{i0}, \alpha_{i1} < \alpha \ \& \ \mathrm{rk}(D) \le m$.
   We set $\mathfrak{d} := \mathcal{K}^{r'}_{\Pi}(\mathfrak{d}_0, \ldots, \mathfrak{d}_{i-1}, \mathfrak{d}_{i0}, \mathfrak{d}_{i1}, \mathfrak{d}_{i+1}, \ldots, \mathfrak{d}_l)$ with
   $r' := \max\{\mathrm{rk}(D), r\} \le m$.
   From $\mathfrak{d}_{i0} \vdash^{\alpha_{i0}}_{m+1} \Pi_i \cdot D$ & $\mathfrak{d}_{i1} \vdash^{\alpha_{i1}}_{m+1} D, \Pi_i$ & $\alpha_{i0}, \alpha_{i1} < \alpha_i$ and $\mathfrak{d}_\nu \vdash^{\alpha_\nu}_{m+1} \Pi_\nu$ for $\nu \in \{0, \ldots, l\} \setminus \{i\}$ by IH we obtain $\mathfrak{d} \vdash^{\beta}_m \Pi$ with
   $\beta := \omega^{\alpha_0} \# \ldots \# \omega^{\alpha_{i-1}} \# \omega^{\alpha_{i0}} \# \omega^{\alpha_{i1}} \# \omega^{\alpha_{i+1}} \# \ldots \# \omega^{\alpha_l} < \alpha$.

   2.2. $\mathcal{I} \notin \mathsf{Cut}$: Then $\mathfrak{d} := \mathcal{I}\big(\mathcal{K}^r_{\mathcal{I}(\Pi,n)}(i/\mathfrak{d}_{in})\big)_{n \in |\mathcal{I}|}$, where $\mathfrak{d}_i = \mathcal{I}(\mathfrak{d}_{in})_{n \in |\mathcal{I}|}$.

**Abbreviation** $\mathcal{Z}^\infty \vdash^{\alpha}_m \Pi \ :\Leftrightarrow\ \exists \mathfrak{d}$ such that $\mathfrak{d} \vdash^{\alpha}_m \Pi$.

**Corollary 7.2** $\mathcal{Z}^\infty \vdash^{\alpha}_{m+1} \Pi \ \Rightarrow\ \mathcal{Z}^\infty \vdash^{\omega^\alpha}_m \Pi$.
*(Follows from Theorem 7.1 for $l = 0$.)*

Having the operations $\mathcal{K}^r_\Pi$ at hand it is now easy to embed $\mathcal{Z}$ into the infinitary system $\mathcal{Z}^\infty$.

**Definition** of a $\mathcal{Z}^\infty$-derivation $d^\infty$ for each closed $\mathcal{Z}$-derivation $d$

1. For atomic $d$ we define $d^\infty := \mathsf{tp}(d)\big(d[n]^\infty\big)_{n \in |\mathsf{tp}(d)|}$ by recursion on $o(d) < \omega$. Especially, in case $d = \mathsf{Ax}^{\neg\neg}_{\Gamma \to A}$ with $A \approx \bot$ we have $d^\infty = \mathsf{L}^0_{\neg\neg A} d[0]^\infty = \mathsf{L}^0_{\neg\neg A}(\mathsf{I}_{\neg A}\mathsf{Ax}^0_{A,\Gamma\to\bot})^\infty = \mathsf{L}^0_{\neg\neg A}\mathsf{R}_{\neg A}\mathsf{tp}(\mathsf{Ax}^0_{A,\Gamma\to\bot}) = \mathsf{L}^0_{\neg\neg A}\mathsf{R}_{\neg A}\mathsf{L}^0_{A}$.
2. $(\mathsf{I}^a_{\forall x F} d_0)^\infty := \mathsf{R}_{\forall x F}\big(d_0(a/n)^\infty\big)_{n \in \mathbb{N}}$.
3. $(\mathsf{I}_{\neg A} d_0)^\infty := \mathsf{R}_{\neg A} d_0^\infty$.
4. $(\mathrm{Ind}^{a,k}_F d_0 d_1)^\infty := \mathcal{K}^r_{\Gamma \to F(k)}(d_0^\infty, d_1(a/0)^\infty, \ldots, d_1(a/k-1)^\infty)$.
5. $(\mathsf{K}^r_\Pi d_0 \ldots d_l)^\infty := \mathcal{K}^r_\Pi(d_0^\infty, \ldots, d_l^\infty)$.

**Theorem 7.3** *If $d \vdash \Pi$ and $\mathrm{FV}(\Pi) = \emptyset$, then $d^\infty \vdash^{\tilde{o}(d)}_{\mathrm{dg}(d)} \Pi$.*

Proof by induction on the build-up of $d$ using Theorem 7.1:
Assume $\Pi = \Gamma \to C$.

1. $d$ atomic: Left to the reader.
2. $d = \mathrm{I}^a_{\forall xF} d_0$: Then $C = \forall x F$ and $d_0(n) \vdash \Gamma \to F(n)$.
   By IH, $d_0(n)^\infty \vdash^{\tilde{o}(d_0)}_{\mathrm{dg}(d_0)} \Gamma \to F(n)\ (\forall n)$.
   Hence $d^\infty = \mathsf{R}_{\forall xF}(d_0(n)^\infty)_{n \in \mathbb{N}} \vdash^{\tilde{o}(d)}_{\mathrm{dg}(d)} \Pi$.
3. $d = \mathrm{I}_{\neg A} d_0$: Similar to 2.
4. $d = \mathrm{Ind}^{a,k}_F d_0 d_1$ with $d_0 \vdash \Gamma \to F(0)$, $d_1 \vdash F(a), \Gamma \to F(\mathsf{S}a)$, and $\Pi = \Gamma \to F(k)$:
   By IH, $d_0^\infty \vdash^{\tilde{o}(d_0)}_{\mathrm{dg}(d_0)} \Gamma \to F(0)$ and $d_1(a/n)^\infty \vdash^{\tilde{o}(d_1)}_{\mathrm{dg}(d_1)} F(n), \Gamma \to F(\mathsf{S}n)\ (\forall n)$.
   From this by Theorem 7.1 we obtain $d^\infty = \mathcal{K}^r_{\Gamma \to F(k)}(d_0^\infty, d_1(a/0)^\infty, \ldots, d_1(a/k{-}1)^\infty) \vdash^{\tilde{o}(d)}_{\mathrm{dg}(d)} \Gamma \to F(k)$, since $r \le \mathrm{dg}(d)$ and $\mathrm{dg}(d_0), \mathrm{dg}(d_1) \le \mathrm{dg}(d){+}1$ and $\omega^{\tilde{o}(d_0)} \# \omega^{\tilde{o}(d_1)} \# \ldots \# \omega^{\tilde{o}(d_1)} < \omega^{\tilde{o}(d)}$.
5. $d = \mathsf{K}^r_\Pi d_0 \ldots d_l$ with $d_i \vdash \Pi_i\ (i = 0, \ldots, l)$:
   Note that $\mathrm{dg}(d) = \max\{\mathrm{dg}(d_0){-}1, \ldots, \mathrm{dg}(d_l){-}1, r\}$ and therefore
   (1) $\mathrm{dg}(d_i) \le \mathrm{dg}(d){+}1$, (2) $r \le \mathrm{dg}(d)$.
   By IH we have $d_i^\infty \vdash^{\tilde{o}(d_i)}_{\mathrm{dg}(d_i)} \Pi_i$ and therefore, by (1), $d_i^\infty \vdash^{\tilde{o}(d_i)}_{\mathrm{dg}(d)+1} \Pi_i\ (i = 0, \ldots, l)$. From this by (2) and Theorem 6.1 we conclude

$$d^\infty = \mathcal{K}^r_\Pi(d_0^\infty, \ldots, d_l^\infty) \vdash^{\alpha}_{\mathrm{dg}(d)} \Pi \text{ with } \alpha = \omega^{\tilde{o}(d_0)} \# \ldots \# \omega^{\tilde{o}(d_l)} = \tilde{o}(d).$$

**Corollary 7.4** *If* $d \vdash \Pi$ *and* $\mathrm{FV}(\Pi) = \emptyset$, *then* $\mathcal{Z}^\infty \vdash^{o(d)}_0 \Pi$.

**Theorem 7.5**

*(i) If* $\mathsf{tp}(d) = \mathsf{Rep}$, *then* $d^\infty = \begin{cases} \mathsf{Cut}_{\mathsf{A}(d)}(d\{0\}^\infty, d\{1\}^\infty) & \textit{if } d \textit{ critical}, \\ d[0]^\infty & \textit{otherwise}. \end{cases}$

*(ii) If* $\mathcal{I} := \mathsf{tp}(d) \neq \mathsf{Rep}$, *then* $d^\infty = \mathcal{I}\big(d[n]^\infty\big)_{n \in |\mathcal{I}|}$.

*Proof by induction over the build-up of d, comparing definitions 4.2 and 7.1.*

## 8 Multisuccedent Sequents

The approach of Sects. 4,5 can easily be adapted to calculi with multisuccedent sequents by generalizing the chain-rule as follows[4]:

(GCR) The figure $\dfrac{\Pi_0 \ \ldots \ \Pi_l}{\Pi}$ is called a *(generalized) chain-rule inference of rank r* if $\Pi$ can be derived from (weakenings of) the sequents $\Pi_0, \ldots, \Pi_l$ by a finite number of cuts of rank $\le r$.

By adding this rule to the proof system of [3] and taking the ordinal assignment from Sect. 5 of the present paper a certain simplification of [3] can be achieved, especially the somewhat unpleasant concept of "Höhenlinie" can be avoided.

[4] A similar rule is used in [4].

In the following we review the essential concepts of Sects. 4,5 in a kind of axiomatic presentation, thereby adjusting everything to the multisuccedent context. The main ingredient here is Lemma 8.1 which replaces Lemma 4.1. The above rule (GCR) will be captured by the inductively defined relation "$(\Pi_0, \ldots, \Pi_l) \Vdash_r \Pi$".

**Definitions** A *sequent* is an expression $\Gamma{\to}\Delta$ where $\Gamma$ and $\Delta$ are finite (possibly empty) sequences of formulas.

For $\Pi = \Gamma{\to}\Delta$ we set
$\mathsf{L}(\Pi) := \Gamma$ and $\mathsf{R}(\Pi) := \Delta$;
$A, \Pi := A, \Gamma \to \Delta$ and $\Pi, A := \Gamma \to \Delta, A$.

*Inference symbols* $\mathsf{R}_A$, $\mathsf{L}^k_A$, $\mathsf{Rep}$ and their *arities* are the same as in Sect. 4.

For each inference symbol $\mathcal{I}$, sequent $\Pi$, and $n \in |\mathcal{I}|$ the sequent $\mathcal{I}(\Pi, n)$ is defined by

$$\mathcal{I}(\Pi, n) := \begin{cases} \Pi, F(n) \text{ if } \mathcal{I} = \mathsf{R}_{\forall x F}, \\ F(k), \Pi \text{ if } \mathcal{I} = \mathsf{L}^k_{\forall x F}, \\ A, \Pi \quad\; \text{if } \mathcal{I} = \mathsf{R}_{\neg A}, \\ \Pi, A \quad\; \text{if } \mathcal{I} = \mathsf{L}^0_{\neg A}, \\ \Pi \qquad\;\; \text{otherwise.} \end{cases}$$

The relation $\mathcal{I} \lhd \Pi$ is defined by:
$\mathsf{R}_A \lhd \Pi \;:\Leftrightarrow\; A \in \mathsf{R}(\Pi)$ ,
$\mathsf{L}^k_A \lhd \Pi \;:\Leftrightarrow\; A \in \mathsf{L}(\Pi)$ ,
$\mathsf{Rep} \lhd \Pi \;:\Leftrightarrow\; 0 = 0$.
*Abbreviation.* $\Pi \subseteq \Pi' \;:\Leftrightarrow\; \mathsf{L}(\Pi) \subseteq \mathsf{L}(\Pi')$ & $\mathsf{R}(\Pi) \subseteq \mathsf{R}(\Pi')$.

**Inductive Definition of** $(\Pi_0, \ldots, \Pi_l) \Vdash_r \Pi$ Let $\vec{\Pi} := (\Pi_0, \ldots, \Pi_l)$.

1. If $\Pi_i \subseteq \Pi$ for some $i \leq l$, then $\vec{\Pi} \Vdash_r \Pi$.
2. If $\vec{\Pi} \Vdash_r \Pi, C$ and $\vec{\Pi} \Vdash_r C, \Pi$ with $\text{rk}(C) \leq r$, then $\vec{\Pi} \Vdash_r \Pi$.

**Lemma 8.1 ("Existence of a Suitable Cut")** *If* $\vec{\Pi} = (\Pi_0, \ldots, \Pi_l) \Vdash_r \Pi$ *and* $\forall i \leq l(\mathcal{I}_i \lhd \Pi_i \;\&\; \mathcal{I}_i \not\lhd \Pi)$, *then there are* $i, j \leq l$ *such that* $\mathcal{I}_i = \mathsf{R}_B \;\&\; \mathcal{I}_j = \mathsf{L}^k_B \;\&\; \text{rk}(B) \leq r$ *for some* $B, k$.

Proof by induction over the definition of $\vec{\Pi} \Vdash_r \Pi$:
From the second premise we conclude $\forall i \leq l(\Pi_i \not\subseteq \Pi)$. Together with $\vec{\Pi} \Vdash_r \Pi$ this implies that there exists a $C$ of rank $\leq r$ such that $\vec{\Pi} \Vdash_r \Pi, C$ and $\vec{\Pi} \Vdash_r C, \Pi$.

Case 1: $\forall i \leq l(\mathcal{I}_i \not\lhd \Pi, C)$ or $\forall i \leq l(\mathcal{I}_i \not\lhd C, \Pi)$.
Then the claim follows immediately from the IH.
Case 2: Otherwise. Then there exist $i, j \leq l$ such that $\mathcal{I}_i \lhd \Pi, C$ and $\mathcal{I}_j \lhd C, \Pi$.
From $\mathcal{I}_i \lhd \Pi, C \;\&\; \mathcal{I}_i \not\lhd \Pi$ it follows that $\mathcal{I}_i = \mathsf{R}_C$.
From $\mathcal{I}_j \lhd C, \Pi \;\&\; \mathcal{I}_j \not\lhd \Pi$ it follows that $\mathcal{I}_j = \mathsf{L}^k_C$ for some $k$.

**Assumption 0** $\mathcal{D}$ is a set of (derivation) terms, and to each $d \in \mathcal{D}$ there is assigned a sequent $\mathsf{End}(d)$, an inference symbol $\mathsf{tp}(d)$, and, for each $n \in |\mathsf{tp}(d)|$, a term $d[n] \in \mathcal{D}$.

**Abbreviation** $d \vdash \Pi \;:\Leftrightarrow\; d \in \mathcal{D} \;\&\; \mathsf{End}(d) = \Pi$.

**Assumption 1** If $(\Pi_0, \ldots, \Pi_l) \;\Vdash_r\; \Pi$ and $d_0 \vdash \Pi_0, \ldots, d_l \vdash \Pi_l$, then $\mathsf{K}^r_\Pi d_0 \ldots d_l \vdash \Pi$.

**Definitions** Assume $d = \mathsf{K}^r_\Pi d_0 \ldots d_l \vdash \Pi$ with $d_i \vdash \Pi_i$ and $\mathsf{tp}(d_i) \lhd \Pi_i$ for all $i \leq l$.

- $d$ is critical $:\Leftrightarrow\; \forall i \leq l(\mathsf{tp}(d_i) \ntriangleleft \Pi)$.
- If $d$ is critical we take the least pair $(i, j)$ such that $i, j \leq l \;\&\; \mathsf{tp}(d_i) = \mathsf{R}_B \;\&\; \mathsf{tp}(d_j) = \mathsf{L}^k_B \;\&\; \mathrm{rk}(B) \leq r$ for some $B, k$ (which exists according to Lemma 8.1), and define

$$\mathsf{A}(d) := \begin{cases} F(k) \text{ if } B = \forall x F(x), \\ A \quad\;\; \text{if } B = \neg A, \end{cases}$$

$$d\{0\} := \mathsf{K}^r_{\Pi,\mathsf{A}(d)} \begin{cases} (i/d_i[k]) \text{ if } B = \forall x F, \\ (j/d_j[0]) \text{ if } B = \neg A, \end{cases}$$

$$d\{1\} := \mathsf{K}^r_{\mathsf{A}(d),\Pi} \begin{cases} (j/d_j[0]) \text{ if } B = \forall x F, \\ (i/d_i[0]) \;\text{ if } B = \neg A. \end{cases}$$

**Assumption 2** If $d = \mathsf{K}^r_\Pi d_0 \ldots d_l \vdash \Pi$ with $d_i \vdash \Pi_i$ and $\mathsf{tp}(d_i) \lhd \Pi_i$ for all $i \leq l$, then the following holds

(a) If $d$ is critical, then $\mathsf{tp}(d) = \mathsf{Rep}$ and $d[0] = \mathcal{K}^{r-1}_\Pi d\{0\} d\{1\}$.
(b) If $d$ is not critical and $i \leq l$ is the least number s.t. $\mathsf{tp}(d_i) \lhd \Pi$, then

$$\mathsf{tp}(d) = \begin{cases} \mathsf{Rep} \quad \text{if } d_i \text{ critical}, \\ \mathsf{tp}(d_i) \text{ otherwise}. \end{cases}$$

$$d[n] = \begin{cases} \mathsf{K}^{r'}_\Pi(i/d_i\{0\}d_i\{1\}) \text{ with } r' := \max\{\mathrm{rk}(\mathsf{A}(d_i)), r\} \text{ if } d_i \text{ critical}, \\ \mathsf{K}^r_{\mathsf{tp}(d)(\Pi,n)}(i/d_i[n]) \qquad\qquad\qquad\qquad\qquad\qquad\qquad \text{otherwise}. \end{cases}$$

**Assumption 3** There are mappings $\mathrm{dg} : \mathcal{D} \to \omega$ and $\tilde{\mathsf{o}} : \mathcal{D} \to On$ such that such that for each $d = \mathsf{K}^r_\Pi d_0 \ldots d_l$ we have $\mathrm{dg}(d) = \max\{\mathrm{dg}(d_0){-}1, \ldots, \mathrm{dg}(d_l){-}1, r\}$, and $\tilde{\mathsf{o}}(d) = \omega^{\tilde{\mathsf{o}}(d_0)} \# \ldots \# \omega^{\tilde{\mathsf{o}}(d_l)}$.

**Abbreviations** For $d \in \mathcal{D}$ and $\Pi := \mathsf{End}(d)$ we set:

$$d \in \mathcal{D}_1 \;:\Leftrightarrow\; \mathsf{tp}(d) \lhd \Pi \;\&\; \forall n \in |\mathsf{tp}(d)| \big(d[n] \vdash \mathsf{tp}(d)(\Pi, n)\big),$$

$$d \in \mathcal{D}_2 \;:\Leftrightarrow\; \begin{cases} \mathrm{dg}(d[0]) < \mathrm{dg}(d) & \text{if } d \text{ critical}, \\ \forall n \in |\mathsf{tp}(d)|(\mathrm{dg}(d[n]) \leq \mathrm{dg}(d)) & \text{otherwise}, \end{cases}$$

$$d \in \mathcal{D}_3 \;:\Leftrightarrow\; \begin{cases} \tilde{\mathsf{o}}(d[0]) < \omega^{\tilde{\mathsf{o}}(d)} & \text{if } d \text{ critical}, \\ \forall n \in |\mathsf{tp}(d)|(\tilde{\mathsf{o}}(d[n]) < \tilde{\mathsf{o}}(d)) & \text{otherwise}. \end{cases}$$

**Theorem 8.2** *For* $\nu = 1, 2, 3$ *the following holds:*
*If* $d = \mathsf{K}^r_\Pi d_0 \ldots d_l \in \mathcal{D}$ *with* $d_0, \ldots, d_l \in \mathcal{D}_\nu$, *then* $d \in \mathcal{D}_\nu$.

*Proof* Cf. the proofs of Theorem 4.4 and Lemma 5.1. □

# Appendix

In this appendix we will show how Gentzen's original ordinal assignment [2, Sect. 15] can be transformed into the assignment which we have used in Sect. 5. This transformation consists in essentially four steps.

Step 1: We do not use exactly the same set of decimal fractions as Gentzen did. Gentzen defined his set of *Ordnungszahlen* (let's call it $\mathcal{O}_G$) by: $\mathcal{O}_G := \{n.u : n \in \mathbb{N} \ \& \ u \in \mathcal{M}_n\}$ where $\mathcal{M}_0 := \{1, 11, 111, \ldots, 2\}$, $\mathcal{M}_{n+1} := \{u_0 0^{n+1} u_1 0^{n+1} \ldots 0^{n+1} u_l : l \geq 0 \ \& \ u_0, \ldots, u_l \in \mathcal{M}_n \ \& \ 0.u_l <_{\mathbb{R}} \cdots <_{\mathbb{R}} 0.u_0\}$. This corresponds to representing ordinals in base 2 Cantor normal form, while here we shall use base $\omega$. Instead of $\mathcal{O}_G$ we define the set $\mathcal{O} := \{n.u : n \in \mathbb{N} \ \& \ u \in M_n\}$, where $M_0 := \{1\}$, $M_{n+1} := \{u_0 0^{n+1} u_1 0^{n+1} \ldots 0^{n+1} u_l : l \geq 0 \ \& \ u_0, \ldots, u_l \in M_n \ \& \ 0.u_l \leq_{\mathbb{R}} \cdots \leq_{\mathbb{R}} 0.u_0\}$.

Step 2: We define an embedding of $(\mathcal{O}, <_{\mathbb{R}})$ into the set theoretic ordinals, namely for each 'Ordnungszahl' $n.u \in \mathcal{O}$ we define an ordinal $|n.u| \in On$ such that $\forall n.u, m.v \in \mathcal{O}(n.u <_{\mathbb{R}} m.v \Rightarrow |n.u| < |m.v|)$ (Lemma 3).

Step 3: We modify Gentzen's assignment of 'Ordnungszahlen' to derivations [2, Sect. 15.2] according to the alterations made in step 1. For each derivation $d$ we define its *numerus* $\rho(d) \in \mathbb{N}$, *mantissa* $\mu(d) \in \bigcup_{n\in\mathbb{N}} M_n$, and *'Ordnungszahl'* $\mathrm{Ord}(d) := \rho(d).\mu(d) \in \mathcal{O}$. Actually we only consider the crucial case where $d$ ends with a chain-rule inference.

Step 4: We show how the ordinal $|\mathrm{Ord}(d)|$ can be defined directly by recursion on the build-up of $d$, without referring to the decimal fraction $\mathrm{Ord}(d)$. Then we compare the involved recursion equations with the corresponding equations in the definition of $\tilde{\mathrm{o}}(d)$, $o(d)$ in Sect. 5.

**Step 1.**
Let $\{0, 1\}^+$ denote the set of all finite nonempty words $u$ over the alphabet $\{0, 1\}$, and let
$\{0, 1\}^{(+)} := \{u \in \{0, 1\}^+ :$ the first and the last letter of $u$ is $1\}$.
Further, let $0^n$ denote the word consisting of $n$ zeros. Each expression $n.u$ (with $n \in \mathbb{N}$ and $u \in \{0, 1\}^{(+)}$) will be identified with the real number denoted by it in the usual way.

**Definition of** $M_n \subseteq \{0, 1\}^{(+)}$

1. $M_0 := \{1\}$;
2. $M_{n+1} := \{u_0 0^{n+1} u_1 0^{n+1} \ldots 0^{n+1} u_l : l \geq 0 \ \& \ u_0, \ldots, u_l \in M_n \ \& \ 0.u_l \leq_{\mathbb{R}} \cdots \leq_{\mathbb{R}} 0.u_0\}$.

Further we set $M := \bigcup_{n\in\mathbb{N}} M_n$. The elements of $M$ are called *mantissas*.

**Definition** $\mathrm{h} : M \to \mathbb{N}$, $\mathrm{h}(u) := \min\{n : u \in M_n\}$.

*Remark* $M_n \subseteq M_{n+1}$, and $\mathrm{h}(u)$ is the maximal number of consecutive zeros in $u$.

**Lemma 1** *If* $u = u_0 0^{n+1} \dots 0^{n+1} u_l \in M_{n+1}$ *and* $v = v_0 0^{n+1} \dots 0^{n+1} v_k \in M_{n+1}$ *with* $u_0, \dots, u_l, v_0, \dots, v_k \in M_n$, *then* $0.u <_{\mathbb{R}} 0.v$ *if, and only if,* $l < k \;\&\; \forall i \leq l(u_i = v_i)$ *or* $\exists j \leq \min\{l,k\}\big(\forall i < j(u_i = v_i) \;\&\; 0.u_j <_{\mathbb{R}} 0.v_j\big)$.

*Proof* Straightforward. □

**Definition** $\mathcal{O} := \{n.u : n < \omega \;\&\; u \in M_n\}$ *(Ordnungszahlen)*

**Step 2.**

**Definition of** $|u|_n \in On$ **for** $u \in M_n$

1. $|1|_0 := 0$.
2. If $u = u_0 0^{n+1} \dots 0^{n+1} u_l \in M_{n+1}$, then $|u|_{n+1} := \omega^{|u_0|_n} + \dots + \omega^{|u_l|_n}$.

As usual we set $\omega_0(\alpha) := \alpha$, $\omega_{n+1}(\alpha) := \omega^{\omega_n(\alpha)}$.

**Lemma 2** *For* $u \in M_n$ *the following holds:*

*(a)* $|u|_{n+k} = \omega_k(|u|_n)$,
*(b)* $\omega_n(0) \leq |u|_n < \omega_{n+1}(0)$.

**Definition** For $n.u \in \mathcal{O}$ let $|n.u| := |u|_n \in On$.

**Lemma 3** $n.u \in \mathcal{O} \;\&\; m.v \in \mathcal{O} \;\&\; n.u <_{\mathbb{R}} m.v \;\Rightarrow\; |n.u| < |m.v|$.

Proof by induction on the length of $u$:
Case $n < m$: Then $|n.u| = |u|_n < \omega_{n+1}(0) \leq \omega_m(0) \leq |v|_m = |m.v|$.
Case $n = m$: Then $0.u <_{\mathbb{R}} 0.v$ and $u, v \in M_n$ with $n > 0$. Hence $u = u_0 0^n \dots 0^n u_l \in M_n$ and $v = v_0 0^n \dots 0^n v_k \in M_n$ with $u_0, \dots, u_l, v_0, \dots, v_k \in M_{n-1}$. By Lemma 1 it follows that one of the following two cases applies:

(i) $l < k \;\&\; \forall i \leq l(u_i = v_i)$: Then trivially $|u|_n < |v|_n$.
(ii) $\forall i < j(u_i = v_i) \;\&\; 0.u_j <_{\mathbb{R}} 0.v_j$ for some $j \leq \min\{l,k\}$:Then $\forall i \in \{j, \dots, l\}(0.u_i <_{\mathbb{R}} 0.v_j)$ and thus, by IH, $\forall i \in \{j, \dots, l\}(|u_i|_{n-1} < |v_j|_{n-1})$. Hence $|u|_n = \omega^{|v_0|_{n-1}} + \dots + \omega^{|v_{j-1}|_{n-1}} + \omega^{|u_j|_{n-1}} + \dots + \omega^{|u_l|_{n-1}} < \omega^{|v_0|_{n-1}} + \dots + \omega^{|v_j|_{n-1}} \leq |v|_n$.

**Step 3.**

The following are more or less Gentzen's own words (in [2, 15.2])—of course with some alterations enforced by the modifications made in step 1.
*To each given derivation* $d$ *we assign an 'Ordnungszahl'* $\mathrm{Ord}(d) := \rho(d).\mu(d) \in \mathcal{O}$ *according to the following recursive rule: (...) If the endsequent of* $d$ *is the conclusion of a 'chain-rule' inference (i.e., if* $d = \mathsf{K}^r_\Pi d_0 \dots d_l$*) we consider the mantissas* $u_i = \mu(d_i)$ *of the 'Ordnungszahlen' of the derivations* $d_i$*; suppose that* $\nu$ *is the maximum number of consecutive zeros in all of these mantissas (i.e.,* $\nu = \max_{i \leq l} \mathrm{h}(u_i)$*). The mantissas are written down from left to right according to their size (the largest one first) and any two successive mantissas are seperated by* $\nu+1$ *zeros. (It may well be that several successive mantissas are equal.)*

*The result is the mantissa $\mu(d)$ of the ordinal number for the whole derivation; i.e., $\mu(d) := u_{\sigma(0)}0^{\nu+1}u_{\sigma(1)}0^{\nu+1}\dots 0^{\nu+1}u_{\sigma(l)}$ where $\sigma$ is an appropriate permutation of $\{0,\dots,l\}$, and $u_i = \mu(d_i)$. As the numerus $\rho(d)$ we take the least natural number $\rho$ whose excess over the maximum number of consecutive zeros in the mantissa is $\geq 0$ and, firstly, is not more than 1 less than the corresponding excess in any of the ordinal numbers for the derivations of the premises and, secondly, is not less than the rank of the succedent formula of any one of the premises preceding the major premise (14.25).* W.l.o.g. we may assume here that $l \geq 1$ and therefore $\mathrm{h}(\mu(d)) = \nu{+}1$. So $\rho(d)$ is the least number $\rho$ such that (i) $\rho - (\nu{+}1) \geq \rho(d_i) - \mathrm{h}(u_i) - 1$ for $i = 0,\dots,l$, and (ii) $\rho - (\nu{+}1) \geq r$, which amounts to: $\rho(d) - \mathrm{h}(\mu(d)) = \max(\{\rho(d_i) - \mathrm{h}(\mu(d_i)) - 1 : i \leq l\} \cup \{r\})$.

**Step 4.**

Let $\mathrm{h}(d) := \mathrm{h}(\mu(d))$, $\mathrm{exc}(d) := \rho(d) - \mathrm{h}(d)$, and $\widehat{\mathrm{o}}(d) := |\mu(d)|_{\mathrm{h}(d)}$
Then

(1) $|\mathrm{Ord}(d)| = \omega_{\mathrm{exc}(d)}(\widehat{\mathrm{o}}(d))$,
and for $d = \mathsf{K}^r_\Pi d_0 \dots d_l$ we have the recursion equations
(2) $\mathrm{h}(d) = \max_{i\leq l}\mathrm{h}(d_i) + 1$, and
(3) $\mathrm{exc}(d) = \max(\{\mathrm{exc}(d_i) - 1 : i \leq l\} \cup \{r\})$.
(4) $\widehat{\mathrm{o}}(d) = \omega^{\alpha_0} \# \cdots \# \omega^{\alpha_l}$ with $\alpha_i := \omega_{\nu-\mathrm{h}(d_i)}(\widehat{\mathrm{o}}(d_i))$ and $\nu := \max_{i\leq l}\mathrm{h}(d_i)$.

Proof of (1) and (4):

(1) $|\mathrm{Ord}(d)| = |\rho(d).\mu(d)| = |\mu(d)|_{\rho(d)} = \omega_{\rho(d)-\mathrm{h}(d)}(\widehat{\mathrm{o}}(d)) = \omega_{\mathrm{exc}(d)}(\widehat{\mathrm{o}}(d))$.
(4) By definition, $\mu(d) = u_{\sigma(0)}0^{\nu+1}\dots 0^{\nu+1}u_{\sigma(l)}$ with $u_i = \mu(d_i)$ and $\nu = \max_{i\leq l}\mathrm{h}(d_i)$. Hence $\nu{+}1 = \mathrm{h}(\mu(d)) = \mathrm{h}(d)$, $\widehat{\mathrm{o}}(d) = |\mu(d)|_{\nu+1} = \omega^{|u_0|_\nu} \# \cdots \# \omega^{|u_l|_\nu}$, and $|u_i|_\nu = |\mu(d_i)|_\nu \overset{L.2a}{=} \omega_{\nu-\mathrm{h}(d_i)}(|\mu(d_i)|_{\mathrm{h}(d_i)})$.

*Observation:* In case that $\mathrm{h}(\mu(d_0)) = \cdots = \mathrm{h}(\mu(d_l))$ we have
(5) $\widehat{\mathrm{o}}(d) = \omega^{\widehat{\mathrm{o}}(d_0)} \# \cdots \# \omega^{\widehat{\mathrm{o}}(d_l)}$.

Now compare (1), (3), (5) with the corresponding clauses in the definitions of $o(d)$, $\mathrm{dg}(d)$, $\tilde{\mathrm{o}}(d)$ in Sect. 5:

(1)' $o(d) = \omega_{\mathrm{dg}(d)}(\tilde{\mathrm{o}}(d))$
(3)' $\mathrm{dg}(d) = \max(\{\mathrm{dg}(d_i){-}1 : i \leq l\} \cup \{r\})$
(5)' $\tilde{\mathrm{o}}(d) = \omega^{\tilde{\mathrm{o}}(d_0)}\#\cdots\#\omega^{\tilde{\mathrm{o}}(d_l)}$

**Acknowledgements** I would like to thank the anonymous referee for careful reading and valuable comments which helped to improve the paper.

## References

1. W. Buchholz, Explaining Gentzen's Consistency Proof within infinitary Proof Theory, in *Computational Logic and Proof Theory. 5th Kurt Gödel Colloquium, KGC'97*, ed. by G. Gottlob, A. Leitsch, D. Mundici. Lecture Notes in Computer Science, vol. 1298 (Springer, New York, 1997)
2. G. Gentzen, Die Widerspruchsfreiheit der reinen Zahlentheorie. Math. Ann. **112**, 493–565 (1936)
3. G. Gentzen, Neue Fassung des Widerspruchsfreiheitsbeweises für die reine Zahlentheorie. Forschungen zur Logik und zur Grundlegung der exakten Wissenschaften. Neue Folge **4**, 19–44 (1938)
4. L.M. Kogan-Bernstein, Simplification of Gentzen's reductions in the classical arithmetic. Zap. Nauchn. Sem. LOMI **105**, 45–52 (1981), English translation: J. Math. Sci. **22**(3) (1983)
5. M.E. Szabo, *The collected Papers of Gerhard Gentzen* (Amsterdam, North-Holland 1969)

# From *Hauptsatz* to *Hilfssatz*

**Jan von Plato**

**Abstract** Gentzen found his original consistency proof of arithmetic late in 1934. His work in pure logic was a preliminary to the result. Archival sources show that the consistency proof was based on an explicit semantic notion of correctness as *reducibility* of sequents and a proof that steps of derivation maintain reducibility. A crucial point in the latter was Gentzen's *Hilfssatz* that stated, in analogy to his famous *Hauptsatz*, that composition of sequents maintains reducibility. The *Hilfssatz* was needed essentially for the case of the rule of complete induction. It was the point at which Gentzen's proof superseded standard arithmetic methods in favour of an induction on well-founded trees, i.e., what came later to be called bar induction. After criticisms by Bernays and Gödel, the first proof evolved into one based on transfinite induction. Traces of the *Hilfssatz* that was founded on intuitionistic ideas disappeared, and Gentzen developed instead transfinite induction further into a general ordinal proof theory.

## 1 The Situation in 1932

Gerhard Gentzen, a student of Paul Bernays, set as his goal in early 1932 "to clear the consistency problem of mathematics, at least for arithmetic," as he wrote in a letter (see Menzler-Trott 2007, p. 31). A perplexing situation regarding consistency had arisen with the arrival of Gödel's incompleteness theorem, a result that had become known during the fall of 1930. It was at once well received, especially through the forceful endorsement on the part of Johann von Neumann. Bernays had been in contact with Gödel, to clarify the consequences of the result for Hilbert's enterprise of "securing the foundations of mathematics" through a consistency proof. In fact, the preface Bernays wrote to the first volume of the *Grundlagen der Mathematik*, dated March 1934 and published in 1934, tells the following: The manuscript was in practice finished in 1930, but the whole project had to be thought through again when Gödel's result became known: A finitary, "absolutely reliable" consistency proof of the kind envisaged by Hilbert would not be possible.

J. von Plato (✉)
Department of Philosophy, University of Helsinki, Helsinki, Finland
e-mail: jan.vonplato@helsinki.fi



R. Kahle, M. Rathjen (eds.), *Gentzen's Centenary*,
DOI 10.1007/978-3-319-10103-3_5

Some, von Neumann as foremost, declared the foundational enterprise dead: The consistency of mathematics would remain forever unprovable in some absolute sense. Bernays, instead, sought a way out through intuitionism that he took to go beyond the Hilbertian finitism. There is no unclarity as to Bernays' assessment: Brouwer had it right on all essential points and, especially, the law of excluded middle has so far (about 1931–1932) no justification beyond finitary situations. As an aside, the reader of his book may wonder at the appearance of Hilbert's name as a co-author, but there were other reasons for that, especially for the second volume that the expelled Jewish professor Bernays could have never published otherwise in Nazi-Germany in 1939.

Bernays describes finitism as a categorical build-up of mathematics, in the sense that nothing is assumed, but everything is built up finitistically from decidable concepts. Brouwer's intuitionism brings to this picture the new element that also hypothetical proofs are considered, and mathematical constructions made on top of such assumed proofs. Bernays wrote (ibid., p. 43):

> The methodological point of "intuitionism" that is at the basis of Brouwer, is formed by a certain *extension of the finitary position* [Erweiterung der finiten Einstellung], namely, an extension in so far as Brouwer allows the introduction of an assumption about the presence of a consequence, resp. of a proof, even if such a consequence, resp. proof, is not determined in respect of its visualizable constitution [nicht...nach anschaulicher Beschaffenheit bestimmt]. For example, from Brouwer's point of view, propositions of the following forms are allowed: "If proposition $B$ holds under assumption $A$, also $C$ holds," and also: "The assumption that $A$ is refutable leads to a contradiction," or in Brouwer's mode of expression, "the absurdity of $A$ is absurd."

The essence of intuitionism as given here is that it is permitted to assume conditionals, and even more simply, the presence of a hypothetical proof. One would think that this was no novelty in principle, for what are mathematical axioms if not conditionals that are assumed? Bernays thinks instead that there is no hypothetical element in the practice of logicism or formalism. In this light, Gentzen's departure from these traditions in his setting up of natural deduction in 1932 is the more remarkable, because the most central idea in natural deduction is to consider hypothetical inferences.

Bernays proceeds with the discussion in very general terms, the problem being always how to extend the finitary standpoint, and ends with the conclusion that we are still far away from even a solution to the consistency problem of arithmetic (p. 44). The solution was instead much closer than he could imagine, for Gentzen had it by the end of the year 1934.

## 2 Groundwork for the Consistency Proof

Gentzen, who was just 22 years old in 1932, would take nothing of the defeatism of von Neumann. Where his confidence came from is not known, but it got confirmed in less than a year, by the interpretation of classical Peano arithmetic in intuitionistic

Heyting arithmetic. It was a result that proved Bernays' admission by which the help of Brouwer's intuitionistic mathematics would be needed to overcome the dead-end of Hilbert's *Beweistheorie*. The consistency proof itself was finished late in 1934. The steps of events will be described in this section in five installments: (1) The logical analysis of "actual proofs in mathematics." (2) The semantical explanation of the logical forms of propositions used in mathematics, with the subformula property and normalization as crucial elements. (3) The elimination of indirect proofs through the Gödel–Gentzen translation. (4) The surfacing of transfinite ordinals. (5) The first consistency proof, end of 1934.

## 2.1 Actual Proofs in Mathematics

Gentzen began by a study of how "one actually carries through proofs in mathematics." He observed that the prevailing method of formally presenting proofs did not match the practice: Mathematical statements were formalized in the language of logic, and especially the starting points of proofs, namely the mathematical axioms. Logic itself was also axiomatized, with axioms such as $(A \supset (B \supset C)) \supset (B \supset (A \supset C))$. The "horseshoe" implication symbol was invented by Giuseppe Peano, just an inverted capital letter C that got later stylized into $\supset$. It reveals what the above axiom does: If you read $\supset$ as "consequence" (for the C inverted) or "follows" or "if..., then ...," whatever is handiest, you get

If from $A$ it follows that $C$ follows from $B$, then from $B$ it follows that $C$ follows from $A$.

Think of $A$ and $B$ as assumptions, and the axiom prescribes that $C$ follows, whichever the order is in which you take the assumptions $A$ and $B$.

The rest of the logical axioms have similar intuitive meanings. They were clear to Frege who mainly invented the axioms. Later his identification of the principles of proof turned into "symbolic logic," interpreted as a formal game, and the meaning of the axioms was by and large forgotten.

There were just two rules of inference in axiomatic logic: From $A \supset B$ and $A$ to infer $B$ was the propositional one, and universal generalization the other. In the latter, a universal quantifier could be introduced if a statement was proved for an arbitrary object, as denoted by an **eigenvariable**. The precise statement of conditions for universal generalization was a great achievement of Frege's.

The application of Frege's logic to mathematical proofs, as in the work of Peano and Russell, proceeds through expressing the mathematical axioms with the language of logic, and in the application of the two principles of proof. Here is a simple example, the axiomatic theory of equality. The axioms are reflexivity, symmetry, and transitivity:

$$a = a, \quad a = b \supset b = a, \quad a = b \;\&\; b = c \supset a = c.$$

It is next to impossible to put the logical codification of mathematical proofs in terms of axiomatic logic into actual use. Say, the expression of transitivity of equality in the Euclidean style, $a = c \,\&\, b = c \supset a = b$, already has an axiomatic proof that cannot be shown here because it is too broad to be printed. The proofs are often so wicked that the only feasible way to construct them would be to do them first in a calculus of natural deduction, then to apply a translation algorithm into proofs in axiomatic logic.

When Gentzen started his program in early 1932, he had no difficulty in putting the ruling axiomatic logical tradition aside. The aim of axiomatic logic had been dictated by Frege's and Russell's doctrine of **logicism**, by which logical axioms express the most basic logical truths and logical proofs just add more truths to the basic stock. The whole notion is empty for Gentzen because, as emphasized by Franks (2010), pure logic has no subject matter for him. Logical principles, Gentzen's rules of proof, show how to move from given assumptions to a conclusion. Gentzen would grant, at most, that if the assumptions are **correct** (*richtig*), also the conclusion should be.

The conceptual order in Gentzen is different from that of logicism. In the latter, there cannot be any doubt that logical proofs preserve correctness, because, if we take the doctrine seriously, such proofs are based on the ultimate notion of logical truth in a simplest possible manner. The axioms are such truths, and if $A \supset B$ and $A$ are, also $B$ is. There are no hypotheses, so this inductive argument is strictly local in character.

In logicism, mathematical truth is subordinate, and perhaps even reducible, to logical truth. If the reduction succeeds, the foundational problems of mathematics are solved for good. In Gentzen, instead, the very problem is to find a notion of correctness, in the first place for arithmetic, that is supported by logical inferences.

By September 1932, Gentzen had finalized his set of logical principles of proof, what is known as natural deduction (*natürliches Schliessen*, perhaps more properly rendered as natural inference, or even natural reasoning). His analysis of "actual proofs" in mathematics led to intuitionistic logic, a topic well-defined after Arend Heyting's axiomatization of 1930 that had the axiomatization of *Principia Mathematica* as a basis.

A year later, Heyting explained the logical connectives in terms of proof, or perhaps better, sufficient conditions for proof: $A \,\&\, B$ is proved whenever $A$ and $B$ have been proved separately, $A \vee B$ is proved whenever one of $A$ and $B$ has been proved, $A \supset B$ is proved whenever any proof of $A$ turns into some proof of $B$. For the quantifiers, $\forall x A(x)$ is proved whenever $A(y)$ is proved for an arbitrary $y$, and $\exists x A(x)$ is proved whenever $A(a)$ is proved for some object $a$. It was realized soon that the explanation of implication need not reduce a proof of $A \supset B$ into something simpler, for $A$ could have been obtained by any proof.

There is in the collection of stenographic notes that Gentzen wrote a set from the fall of 1932, some 25 big stenographic pages, with a few pages added in the next spring and ten more in October 1934. The title is "Formal conception of the notion of contentful correctness in pure number theory, relation to proof of consistency" (*Die formale Erfassung des Begriffs der inhaltlichen Richtigkeit in der reinen Zahlentheorie, Verhältnis zum Widerspuchsfreiheitsbeweis*).[1] Most of it was written within a month in October–November, and it was meant to be a groundwork for systematic formal studies, after the basic structure of mathematical reasoning had been cleared in September. I abbreviate the manuscript in the same way he did, as **INH**. The first task in it is to explain the notion of **correctness** for intuitionistic logic, quite similarly to Heyting's explanations. In the case of $A \& B$ and $A \vee B$, a reduction is achieved, but $A \supset B$ remained problematic.

Bernays was well aware of the problem, namely that in a case of iterated implications such as $(A \supset B) \supset C$, the correctness of $C$ depends on the correctness of another conditional statement $A \supset B$. This is a problem of well-foundedness. A related problem is circularity: If, as in Heyting's explanation, a proof of $A \supset B$ takes any proof of $A$ and gives as a result some proof of $B$, the notion to be explained, namely proof, is already assumed.

Once correctness for statements has been explained, it can be applied to statements in proofs. Here is the lesson from Gentzen's analysis:

**Reduction to Components** *If $A \supset B$ is provable, it should have a proof that is somehow made up from the components of $A \supset B$.*

The correctness of a notion of proof with this property would not be circular.

What is the notion Gentzen was searching after? Looking at his rules of natural deduction, a specific feature of most of the rules strikes the eye:

$$\frac{A \quad B}{A \& B}\,\&I \qquad \frac{A \& B}{A}\,\&E \qquad \frac{A \& B}{B}\,\&E \qquad \frac{A}{A \vee B}\,\vee I_1 \qquad \frac{B}{A \vee B}\,\vee I_2 \qquad \frac{A \supset B \quad A}{B}\,\supset E$$

In the introduction rules, the premisses are **subformulas** of the conclusion, in the elimination rules, it is the other way around. There remain the introduction rule for implication and elimination rule for disjunction that have a schematic character different from the above:

$$\frac{\begin{matrix}[A]\\ \vdots \\ B\end{matrix}}{A \supset B}\,\supset I \qquad\qquad \frac{A \vee B \quad \begin{matrix}[A]\\ \vdots \\ C\end{matrix} \quad \begin{matrix}[B]\\ \vdots \\ C\end{matrix}}{C}\,\vee E$$

**Intuitionistic propositional logic** results when the rule of **falsity elimination** is added to these rules: There is a constant proposition called falsity and denoted $\bot$, with negation defined by $\neg A \equiv A \supset \bot$, and with the rule $\bot E$ by which any formula can be concluded from $\bot$. Intuitionistic predicate logic is obtained by adding the quantifier rules:

[1] I translate *inhaltlich* as contentful. Gödel suggested in the 1960s "contentual," but my translation is at least an English word. Georg Kreisel dislikes it: He told me in July 2010 that one should just use the word *meaning*. *Inhaltlich*, then, would be *meaningfully*, or perhaps *in terms of meaning*. I regret not having asked what he thinks of Gödel's invented word.

$$\frac{A(y)}{\forall x A(x)}\,\forall I \qquad \frac{\forall x A(x)}{A(t)}\,\forall E \qquad \frac{A(t)}{\exists x A(x)}\,\exists I \qquad \frac{\exists x A(x) \quad \begin{matrix}[A(y)]\\ \vdots \\ C\end{matrix}}{C}\,\exists E$$

In rules $\forall I, \exists E$, $y$ is an eigenvariable.

The introduction rules of Gentzen's natural deduction are formal versions of Heyting's explanations. For the elimination rules, different motivations and criteria have been presented, as discussed in von Plato (2012).

## 2.2 Normalization

At this point, in October 1932, the task is to establish a subformula property for formal proofs, or **derivations** (*Herleitungen*), by the new rules of natural deduction. Going through the combinatorial possibilities, one notices cases such as

$$\begin{matrix}\dfrac{\dfrac{\begin{matrix}\vdots\\ A\end{matrix} \quad \begin{matrix}\vdots\\ B\end{matrix}}{A \,\&\, B}\,\&I}{A}\,\&E \\ \vdots\end{matrix} \qquad\qquad \begin{matrix}\dfrac{\dfrac{\begin{matrix}[A]\\ \vdots\\ B\end{matrix}}{A \supset B}\,\supset I \quad \begin{matrix}\vdots\\ A\end{matrix}}{B}\,\supset E \\ \vdots\end{matrix}$$

There is a local "peak"(*Gipfel*) in a derivation, $A \,\&\, B$ or $A \supset B$, that need not belong as a part to the conclusion of the whole derivation or some open assumption the conclusion depends on. These peaks can be eliminated:

$$\begin{matrix}\dfrac{\dfrac{\begin{matrix}\vdots\\ A\end{matrix} \quad \begin{matrix}\vdots\\ B\end{matrix}}{A \,\&\, B}\,\&I}{A}\,\&E \\ \vdots\end{matrix} \quad \text{becomes} \quad \begin{matrix}\vdots\\ A\\ \vdots\end{matrix} \qquad\qquad \begin{matrix}\dfrac{\dfrac{\begin{matrix}[A]\\ \vdots\\ B\end{matrix}}{A \supset B}\,\supset I \quad \begin{matrix}\vdots\\ A\end{matrix}}{B}\,\supset E \\ \vdots\end{matrix} \quad \text{becomes} \quad \begin{matrix}\vdots\\ A\\ \vdots\\ B\\ \vdots\end{matrix}$$

There is a subtlety with the second proof transformation: Rule $\supset I$ is displayed schematically, with an arbitrary number of copies of the open assumption $A$ closed by the introduction of $A \supset B$. If $A$ was used $n$ times in the derivation, the transformed derivation can be presented by the scheme

$$\begin{array}{ccc} \vdots & & \vdots \\ A & \overset{n\times}{\ldots} & A \\ & \vdots & \\ & B & \\ & \vdots & \end{array}$$

The derivation of $A$ and what it depends on gets multiplied any number of times.

Things are not so obvious with disjunction (neither with existence, universality is easy). There are the transformations for $\lor I$ followed by $\lor E$, as in the first of the $I$-rules:

$$\dfrac{\dfrac{\begin{array}{c}\vdots\\ A\end{array}}{A \lor B}{\scriptstyle \lor I} \qquad \begin{array}{c}[A]\\ \vdots\\ C\end{array} \qquad \begin{array}{c}[B]\\ \vdots\\ C\end{array}}{\begin{array}{c}C\\ \vdots\end{array}}{\scriptstyle \lor E} \qquad \text{becomes} \qquad \begin{array}{c}\vdots\\ A\\ \vdots\\ C\\ \vdots\end{array}$$

There is in addition the possibility that a disjunction or existence elimination separates an introduction from an elimination, say, if $C$ is of the form $D \,\&\, E$ and has been derived in a minor premiss by rule $\&I$, then to be eliminated by $\&E$ applied to the conclusion. The hidden non-normality is made explicit by a **permutative** conversion:

$$\dfrac{\dfrac{\begin{array}{c}\vdots\\ A \lor B\end{array} \quad \dfrac{\begin{array}{c}\overset{1}{A}\\ \vdots\\ D\end{array} \quad \begin{array}{c}\vdots\\ E\end{array}}{D \,\&\, E}{\scriptstyle \&I} \quad \begin{array}{c}\overset{1}{B}\\ \vdots\\ D \,\&\, E\end{array}}{D \,\&\, E}{\scriptstyle \lor E,1}}{\begin{array}{c}D\\ \vdots\end{array}}{\scriptstyle \&E_1} \qquad \text{becomes} \qquad \dfrac{\begin{array}{c}\vdots\\ A \lor B\end{array} \quad \dfrac{\dfrac{\begin{array}{c}\overset{1}{A}\\ \vdots\\ D\end{array} \quad \begin{array}{c}\vdots\\ E\end{array}}{D \,\&\, E}{\scriptstyle \&I}}{D}{\scriptstyle \&E_1} \quad \dfrac{\begin{array}{c}\overset{1}{B}\\ \vdots\\ D \,\&\, E\end{array}}{D}{\scriptstyle \&E_1}}{\begin{array}{c}D\\ \vdots\end{array}}{\scriptstyle \lor E,1}$$

Now the $I$-$E$ pair in the derivation of the first minor premiss can be eliminated.

Gentzen left first out $\lor$ and $\exists$, by translating $A \lor B$ into $\neg(\neg A \,\&\, \neg B)$ and $\exists x A(x)$ into $\neg \forall x \neg A(x)$. Now he got the **normalization theorem** for the $\lor, \exists$-free fragment of predicate logic:

**Normalization Theorem** *All derivations can be so transformed that no I-rule is followed by the corresponding E-rule.*

The main difficulty in the proof is to give a measure or weight to derivations such that the elimination of a local peak such as $A \supset B$ (a non-normality) reduces the weight more than the multiplication of the derivation of $A$ by any number $n$. It is known since 2005 that Gentzen solved the problem some time late in 1932 and included at some stage even a treatment of the rules for disjunction and existence with the permutative conversions as in the above example. Namely, I found in February 2005 a handwritten version of a plan and partial execution of his thesis that contained as the greatest surprise a detailed proof of normalization for intuitionistic natural deduction, otherwise attributed to Prawitz (1965) (to which Raggio's proof of the same year can be added). An English translation of Gentzens proof, 13 journal pages, together with my introduction, was published in von Plato (2008).

The thesis manuscript contains a stenographic addition by which the **subformula property** of normal derivations is an immediate corollary to normalization:

**Subformula Property** *All formulas in a normal derivation are subformulas of the conclusion or some open assumption.*

**Consistency** is an immediate consequence of these results: If $A \,\&\, \neg A$ were derivable, also $\bot$ would be derivable, and therefore any formula, but $\bot$ has no normal derivation by the subformula property, therefore no derivation at all.

How to extend all of the above to arithmetic, that was the new formulation of the consistency problem.

## 2.3 *Elimination of Indirect Proofs*

The proof of the normalization theorem in two stages, first without $\vee, \exists$, then for the full language, bore an unexpected fruit: Gentzen noticed that the **principle of indirect proof** could be dispensed with if $\vee$ and $\exists$ were absent, subject to a little adjustment.

Two treatments of negation were given in the thesis manuscript, either as a primitive notion with separate rules, or as defined by $\neg A \equiv A \supset \bot$. Even the printed thesis lists both notions and their respective rules. They are, for the defined notion, special cases of the implication rules:

$$\frac{\begin{matrix}[A]\\ \vdots \\ \bot\end{matrix}}{\neg A}\supset I \qquad\qquad \frac{\neg A \quad A}{\bot}\supset E$$

The rules of primitive negation are:

$$\frac{\begin{matrix}[A]\\ \vdots \\ B\end{matrix} \quad \begin{matrix}[A]\\ \vdots \\ \neg B\end{matrix}}{\neg A}\neg I \qquad\qquad \frac{\neg A \quad A}{C}\neg E$$

Both rules are derivable if the defined notion of negation is used. The introduction rule is not pure, in the sense that it contains already a negation in a premiss. All other rules are such that no other connective than the one introduced or eliminated appears in the rule schemes.

Gentzen's thesis manuscript gives transformations to repeated applications of the primitive rules of negation, but these transformations do not follow any general pattern for the simplification of derivations. If the transformations are reproduced with the use of the rules for defined negation, they turn out to be instances of standard conversion patterns of natural deduction (see von Plato 2012 for a detailed presentation). In conclusion, the defined notion of negation is the well-behaving one.

**Classical natural deduction** results if the rule of **indirect proof** is added to intuitionistic logic:

$$\dfrac{\begin{matrix}[\neg A]\\ \vdots \\ \bot\end{matrix}}{A}\,DN$$

The nomenclature *DN* stands for double negation, which is explained as follows: If instead of *DN* rule $\supset I$ is applied, the conclusion is $\neg\neg A$, and double negation elimination gives the conclusion $A$.

If the conclusion of rule *DN* is a premiss in an elimination rule, there is no direct guarantee for the subformula property. This problem is clear from a text fragment Gentzen later dated as being from "about January 1933." It is titled *Decision in classical predicate calculus reducible to decision in intuitionistic calculus with only* $\supset$ *and* ( )*?* (There is written ▷∀◁ above the notation for the universal quantifier ( ), where the triangles indicate a later addition.) The object of the paper is to translate derivations in classical natural deduction to derivations by the rules for implication and universal quantification and with an added constant proposition $\mathcal{F}$ that stands for the false formula (i.e., a fragment of what is called today minimal logic). To this purpose, Gentzen first transforms the formulas of classical predicate logic into equivalent ones that contain only implication, universal quantification, and $\mathcal{F}$. The rules for negation are:

$$RA:\quad \dfrac{\dfrac{\overset{1}{[\mathfrak{A}]}\atop\mathfrak{B}\qquad \overset{1}{[\mathfrak{A}]}\atop\mathfrak{B}\supset\mathcal{F}}{\mathcal{F}}\,FE}{\mathfrak{A}\supset\mathcal{F}}\,FI1 \qquad REND:\quad \dfrac{\mathfrak{A}\supset\mathcal{F}.\supset\mathcal{F}}{\mathfrak{A}}\,\text{“}DN\text{”}\ \text{(law of double negation)}$$

This is directly from the manuscript. Numerical labels identify occurrences of closed assumptions, *RA* stands for *reductio* and *REND* for something like "reduction of negation doubled." The order of premisses in rule *FE* ("follows-elimination") was changed later in the winter of 1932–1933.

The last point is to change every atomic formula into its double negation. Now derivations can be so transformed that rule *DN* is applied to the components of its conclusion. If *DN* has been applied to conclude an implication $\mathfrak{B}\supset\mathfrak{C}$, the transformation, again from the manuscript, is:

$$\cfrac{\cfrac{\cfrac{\cfrac{\cfrac{\cfrac{\cfrac{\overset{1}{\mathfrak{B}} \quad \overset{3}{\mathfrak{B}\supset\mathfrak{C}}}{\mathfrak{C}}\,{\scriptstyle FE} \qquad \overset{2}{\mathfrak{C}\supset\mathcal{F}}}{\mathcal{F}}\,{\scriptstyle FE}}{\mathfrak{B}\supset\mathfrak{C}.\supset\mathcal{F}}\,{\scriptstyle FI\,3} \qquad \mathfrak{B}\supset\mathfrak{C}.\supset\mathcal{F}:\supset F}{\mathcal{F}}\,{\scriptstyle FE}}{\mathfrak{C}\supset\mathcal{F}.\supset\mathcal{F}}\,{\scriptstyle FI\,2}}{\mathfrak{C}}\,{\scriptstyle DN}\text{ for }\mathfrak{C}}{\mathfrak{B}\supset\mathfrak{C}}\,{\scriptstyle FI\,1}$$

A similar transformation is made if *DN* is applied to a universally quantified formula. In the end, *DN* is applied to double negations of what were atomic formulas before the transformation added two negations. With four negations at the head of each atomic formula, rule *DN* just eliminates two of them, but this can be done without the classical rule. Therefore, as Gentzen concludes: "It is obvious that the inference *DN* can be completely eliminated by these steps."

The atomic formulas of arithmetic are equations. If they don't contain free variables, they are decidable, as Gentzen well understood. Rule *DN* applied to the double-negation of a numerical equation $n = m$ has the same force as the law of excluded middle, $n = m \vee \neg n = m$, and which of the disjuncts is the case can be decided. Therefore *DN* need not be applied to atomic formulas without free variables. In particular, Gentzen could conclude in January 1933:

**Relative Consistency** *If a contradiction is derivable in classical Peano arithmetic, it is already derivable in a fragment of intuitionistic Heyting arithmetic.*

This was, of course, not Gentzen's terminology, but the result was clear: As mentioned, one of the central aims of the Hilbert school had been to "secure the transfinite arguments of arithmetic." These contain in particular the indirect existence proofs, with $\exists x A(x)$ concluded if $\forall x \neg A(x)$ led to a contradiction. Gentzen's result showed that such steps were not a "dubious" component in arithmetic proofs.

The general conclusion from Gentzen's result, obtained at the same time by Gödel, was:

**Intuitionistic Consistency** *The consistency problem of arithmetic has an intuitionistic sense and, therefore, possibly an intuitionistic solution.*

A further conclusion was that intuitionism does indeed go, as described by Bernays in general terms, beyond Hilbert's "strictly finitistic methods."

Gödel seems not to have pursued the idea of an intuitionistic solution to the consistency problem, even if he reflected on his incompleteness theorems in a talk of 1933 given in Boston. It is titled "The present situation in the foundations of mathematics" and got published from a handwritten English manuscript in the third volume of his *Collected Works* (1995). He notes (pp. 50–51) that consistency is a purely syntactic notion, so that "the whole matter becomes merely a combinatorial question about the handling of symbols according to given rules." Further, "the chief point in the desired proof of freedom from contradiction is that it must be conducted by perfectly unobjectionable methods." These methods are codified in what he calls "system A," and of which he lists some principles. He then notes that such—finitistic—methods cannot lead to a proof, so that the hope for a consistency proof by "Hilbert and his disciples ... has vanished entirely in view of some recently discovered facts." (ibid., p. 52). Next Gödel notes that intuitionism goes clearly beyond what is finitistic. In particular, he refers (p. 53) to the interpretation of classical arithmetic in intuitionistic arithmetic as one that gives an intuitionistic proof of consistency, but adds later that this foundation "is of doubtful value." Gödel ends his talk by the remark that "there remains the hope that in future one may find other and more satisfactory methods of construction beyond the limits of system A, which may enable us to found classical arithmetic and analysis upon them. This question promises to be a fruitful field for further investigations." It seems that only the appearance of Gentzen's proof in 1935 made him take this possibility seriously.

## 2.4 *The Surfacing of Transfinite Ordinals*

Gentzen found out, probably in early 1933, that his proof idea for the consistency of intuitionistic arithmetic, therefore also for Peano arithmetic, would not be realizable. He had added, right at the start when he developed intuitionistic natural deduction, a **rule of induction**:

$$\frac{\begin{matrix} & A(y) \\ \vdots & \vdots \\ A(1) & A(y+1) \end{matrix}}{A(t)}\,CI$$

The conclusion of *CI* (for Complete Induction, *vollständige Induktion*) gives by rule $\forall I$, when a fresh variable $x$ is chosen for the arbitrary term $t$, $\forall x A(x)$. As with indirect proof, there need not remain any trace of the conclusion of *CI* in any of the open assumptions or in the endformula of a finished derivation in arithmetic, so that the subformula property is not guaranteed to hold.[2] Neither can one restrict the induction formula to some specific class of formulas to get a sufficient control over the structure of derivations.

The first occurrence of transfinite induction in Gentzen is already in 1932, in **INH** (date 9.X.):

> A new idea: Is it possible to perform appropriate reductions so that one takes the longest proposition, or a proposition that is of the highest value according to some other assignment

[2] A notion of normal derivability can be applied even in the absence of the subformula property, with easy proofs of the disjunction and existence properties, cf. von Plato (2006).

> that is invariant under reductions, and eliminates it in all its places of occurrence, without multiplying propositions of the same value?
> The assignment of values will go into the transfinite with *CI*'s.

It is not difficult to see where the last comes from: If instead of the universal generalization of the conclusion of rule *CI*, a numerical instance $A(n)$ is concluded by *CI*, the derivation should have a lower value than the derivation of $\forall x A(x)$. (This is mentioned explicitly in the popular article Gentzen 1936a.) The only way out is that an uppermost *CI* with a fresh variable in the conclusion has the value $\omega$. The next thing to determine is what happens when there are several nested *CI*'s. There are some remarks about the possible ordinal assignments made during the spring of 1933, but nothing definitive: It seems to be a line abandoned for the time being.

With the original aim of Gentzen's study temporarily lost, he concentrated on pure classical logic, found his sequent calculus, and proved the famous *Hauptsatz*, cut elimination theorem, during the rest of the spring of 1933 (as detailed in von Plato 2012). Among the sporadic remarks about arithmetic added to the manuscript **INH**, March to June 1933, one dated IV.33 states that "the need to use transfinite induction in the consistency proof seems certain to me." More statements are found in the next section.

As to the use of transfinite numbers in a metamathematical context, the precedents contain at least: Hertz (1923), Ackermann (1924), Hilbert (1926), and Brouwer (1926).

## 2.5 *Consistency, End of 1934*

With the thesis finished in May 1933, Gentzen had other things to worry about than the consistency of arithmetic and analysis. The mathematics department of Göttingen was in ruins after the Nazi takeover and his professor Bernays fired as a "non-Aryan." Gentzen took up his research in 1934, helped by a little scholarship. One thing he tried was to use type theory as the language of mathematics. A result from the spring of 1934 is a consistency proof of Hermann Weyl's system of predicative analysis. Very little is known about the proof: One letter from Bernays to Weyl tells that Gentzen was not able to reproduce it without his notes in 1937, when he met Bernays in Paris (in Menzler-Trott 2007, p. 82). The result seems to have been a by-product of the attempts at producing a proof of the consistency of arithmetic, thus, not a strong result. Jean Cavaillès mentions in his book *Méthode axiomatique et formalisme* that the method of the consistency proof for arithmetic "extends without modification to mathematical theories in which the predicates and functions are decidable or calculable in finitary terms: so for the constructive part of analysis" (1938, p. 162). Gentzen wrote on 11 December 1935 to Bernays about the discussions he had with Cavaillès who was visiting Göttingen at the time (see Menzler-Trott 2007, p. 64).

By the end of 1934, Gentzen had found a proof of consistency of arithmetic. A letter to Bernays of 12 May 1938 tells about a much later proof, the one that became standard through Gentzen (1938b): "How I have obtained the consistency proof from the methods of proof in my dissertation is, I believe, now somewhat easy to see in the new version" (Menzler-Trott 2007, p. 95). As we shall see in the end of Chapter IV, the very first proof used a sequent calculus, instead of the natural calculus of the 1935 proof submitted for publication in August of that year. There Gentzen (1936, p. 512) notes that the proof would be simpler, though "less natural," if a sequent calculus were used. The analogy to cut elimination that he mentions is the *Hilfssatz* to be treated in detail in Chapter IV.

There is even a letter of 11 April 1934 to Bernays by which a consistency proof by transfinite induction existed already at that time (Menzler-Trott, p. 54). First Gentzen writes that "the consistency of mathematics is equivalent to the carrying over of the *Hauptsatz* of my dissertation from predicate logic to type theory" (*Stufenlogik*, second-order logic with an axiom of infinity). He hopes to achieve such a consistency proof soon "by force," after which he adds: "It remains to modify the proof so that only permitted forms of inference are used. I hope to achieve this, in analogy to arithmetic only, through transfinite numbers."

It is known also through discussions that Kreisel has had with Bernays that the use of transfinite induction in the published 1936 proof was, in contrast to the proof submitted for publication in 1935, a return to "an earlier idea" (as in Kreisel 1987, p. 174), discarded in favor of the 1935 proof for reasons that are at least to some extent explained in **INH**.

## 3 The Meaning of Consistency

### *3.1 "Where Is the Gödel-Point Hiding?"*

There was, obviously, no easy way to a consistency proof of arithmetic by transfinite induction. Within a week from the surfacing of the "new idea" of using such induction, Gentzen in his characteristic manner set already out to determine what he was actually trying to do: He asked in **INH** (date 16.X.32) what meaning a consistency proof can have:

> Why is a consistency proof through a coarse contentful explanation,
> $\mathfrak{A}$ & $\mathfrak{B}$ correct when $\mathfrak{A}$ correct and $\mathfrak{B}$ correct, $\mathfrak{A} \rightarrow \mathfrak{B}$ correct when from the correctness of $\mathfrak{A}$ the one of $\mathfrak{B}$ follows, $x\, \mathcal{A}x$ when $\mathcal{A}\nu$ correct for all numbers, $\neg\mathfrak{A}$ correct when $\mathfrak{A}$ not correct,
> after Gödel not formal? Does it contain a circularity? One infers: The logical axioms are correct, the mathematical axioms are correct, inference scheme and substitution give correct from correct, therefore all things provable are correct.

He asks at one place: "*Where is the Gödel-point hiding?*" It took him just a few days more to come to the conclusion that the notion of correctness in arithmetic transcends what can be expressed and proved in arithmetic (**INH**, date 21.X.), an insight usually associated with Tarski:

> I believe I can now see clearly why a consistency proof cannot be formalized through the giving of a coarse contentful meaning. To wit, because the meaning is not formalizable, and this naturally always: in the usual formalism, e.g., of Gödel.

Within two days, the proof strategy was clear (**INH**, date 23.X.):

> One shows now through ordinary inferences, i.e., without *CI*: There is to each proof with a numerical result a proof with a lower value and the same result. Namely, one shows existence of a peak, this peak can be reduced. The assignment of values follows according to 88.3 bottom ff.[3] So, the value of a proof is a system of transfinite numbers of the form: a polynomial in $\omega$ with natural coefficients. (To be replaced by $\omega^{a_1} + [..] + \omega^{a_\nu}$.)
>
> The main inference can be seen as a transfinite induction over a decidable proposition, namely the proposition: The numerical result is correct.
>
> ⋮
>
> There must obtain, in my opinion, some kind of a connection between the informal element in the non-formalizable definition of "correctness" and the non-formalizable (?) transfinite induction. For both of them seem to make possible a non-formalizable proof of consistency.

The attempts do not lead to any definitive result, and by early November, they peter out.

The manuscript **INH** continues by remarks that stem from February, April, and June 1933. In April, there is a clear division of proofs of consistency into three types:

1. A "purely-formal" proof.
2. A "semi-contentful" proof.
3. A proof through reducibility.

The ordinal that is needed in a purely formal proof is estimated to be $\omega^{\omega^\omega}$. The published proof of 1936 contains remarks about such a proof (§10.7). The third type of proof should proceed through the "peak theorem," i.e., through normalization.

There follow what Gentzen by a later addition indicated as *General thoughts about the proof of consistency:*

> The idea as a whole: Each proof has a (transfinite) *value*. Consistency of a system of proofs can be shown only through a proof that has a higher value than all of these. Therefore the theorem of Gödel.

The idea became the central one in **ordinal proof theory** that arose as a generalization of the proof theory of arithmetic. After the quoted passage, there are the added words "taken over to **WTZ**." That signum stands for something like *Widerspruchsfreiheit transfinite Zahlen* (consistency transfinite numbers) and fits well with the published 1936 consistency proof, but no pages of such a series of notes are left. They have probably finished in the garbage dump in the 1960s and 1970s in Göttingen, where Gentzen's manuscripts for his published papers had been kept.

[3] The numbers refer to the stenographic series **D** in which each sheet such as 88 contains four pages, from 88.1. to 88.4. This series became by sheet 92 renamed **INH**.

By June 1933, the consistency problem is formulated in terms of sequent calculus:

> (VI 33) *The possibility of a transfinite assignment of values* seems almost sure, more or less on the basis of the reducibility theorem. Let us take the following consideration: Proofs that become continuously smaller with reduction are assigned values according to the number of their sequents. ... One should just be able to classify each proof directly in a correct way. The best should be to begin with simple calculi.

Now there is a leap to October 1934 when the consistency proof seems already finished. We read (date X.34):

> One must distinguish between the semi-contentful proof that associates to each formula resp. sequent a semi-contentful concept of correctness, and the proof by the concept of reducibility that works with reductions of a derivation. This one leads over to the purely-formal proof that considers only the reductions of a derivation of a contradiction.

It is the semi-contentful proof, or, in Kreisel's terms, the proof that is partly in terms of meaning, that would give a true insight into the significance of consistency, and that Gentzen sets out to write down towards the end of 1934. The passages of **INH** from October 1934 contain already references to a series with the signum **WAV** that stands for *Widerspruchsfreiheit Arithmetik Veröffentlichung* (consistency arithmetic publication) and of which some pages have been preserved. They deal mainly with the production of sequents with formulas in prenex normal form and with a variant of Gentzen's reduction procedure for the classical sequent calculus *LK* of the doctoral thesis, to be discussed below. The writing proceeded chapter by chapter in the spring of 1935, each chapter sent to Bernays as it got ready. The latter made comments concerning which only Gentzen's replies have been preserved: These comments provoked some changes after which Gentzen submitted his long manuscript, some hundred typewritten pages, to the *Mathematische Annalen* where it was received on 11 August 1935. A copy was sent to Weyl.

The quote from X.34 above refers to "the concept of reducibility that works with reductions of a derivation." There are two distinct notions that are called reducibility. One is the **syntactic** notion of conversion of non-normalities in derivations, and the analogous situation with the induction rule: The rule has as a conclusion a numerical instance, and the step is resolved into a number of instances of logical rules. This notion can be applied also to derivations in sequent calculus, because of the correspondence between natural and sequent derivations. On the other hand, Gentzen's search for a meaning to a consistency proof had led him to a general **semantic** notion of reducibility that applies to formulas and sequents. In the above list of three suggested consistency proofs of April 1933, the second, "semi-contentful" type uses the semantic notion of *reducibility of sequents*, the third instead the syntactic notion of *reducibility of derivations*. Confusion can be created when the reducibility of sequents in the semantic sense is applied to the sequents of a derivation. The aim with the notion of reducibility of sequents was to give a finitary interpretation to arithmetic. "Finitary" here has to be taken in broad terms, not in the way of the strict finitism of Hilbert. It turns out that by the end of 1935,

Gentzen's variant of finitism encompassed the whole of the second number class, i.e., the constructive transfinite ordinals.

## *3.2 Brouwerian Insights*

Gentzen's **reduction procedure** for sequents is intended as a semantic explanation of arithmetic. The reduction rules are modeled upon "The mathematics of finite domains," the title of Gentzen's §7, in which the quantifiers can be replaced by conjunctions and disjunctions, and classical propositional logic dictates what the conditions of correctness for the formulas are: $A \& B$ is correct when both $A$ and $B$ are correct, $\neg A$ is correct when $A$ is false, etc. The correctness of the rules of inference of propositional logic is almost immediate.

Let us note that Gentzen's view of classical logic is exactly the same as Brouwer's: It is the logic of finite domains. This is the second of the "four insights" in Brouwer's paper *Intuitionistische Betrachtungen über den Formalismus* (Intuitionistic considerations on formalism). It was printed in the *Sitzungsberichte der Preussischen Akademie der Wissenschaften* in 1928 and I have more than one reason to believe that Gentzen had studied it carefully.

Brouwer's first insight was that "the formalists" have to differentiate between the generation of theorems in formal systems and the contentful theory of these systems, and that the latter is based on "the intuitionistic theory of the set of natural numbers." The second insight was cited above. The third insight was that excluded middle equals the assumption of the solvability of every mathematical problem. The fourth insight is most relevant for Gentzen: *"The recognition that a contentful justification of formalistic mathematics by a proof of its consistency contains a vicious circle."* This is directly the terminology of Gentzen's initial ponderings in **INH**. Brouwer's insights are also seen in action in Gentzen (1936a), among others, in: "I believe that, for example, in the general theory of sets a careful proof theoretic investigation will finally confirm the opinion that all powers exceeding the countable are, in a quite definite sense, only empty appearances and one will have to have the good sense to do without these concepts." All in all, a trusted disciple, from among "the formalists" to boot, had emerged as if by itself, to whom the typically Brouwerian exclamation in the beginning of the *Betrachtungen* applies:

> The acceptance of these insights is only a question of time, because they are the results of pure reflection and hence contain no disputable element, so that anyone who has once understood them must accept them.

Gentzen refers to Brouwer's paper at the very end of his long article. It is, in addition to the reference to Brouwer's 1924 paper on the continuity of real functions in Gentzen (1938a), his only reference to a work of Brouwer's. He would, instead, refer freely to Heyting's formalization of intuitionistic logic. I think these facts just tell us what Gentzen thought proper to refer to as a Göttingen logician whose future depends on the opinion of Hilbert, rather than what he was indebted to in his work.

How could he, who in the spring of 1935 had applied for an assistantship with Hilbert, have stated the simple truth: The consistency proof of 1935 resolved, in the words of Bernays, the "contemporary fiasco" of Hilbert's *Beweistheorie*, by the methods of Brouwer's intuitionistic mathematics. This, namely, is what I am going to suggest below.

# 4 The Plan and Circumstances of the Original Proof

## *4.1 Outline*

The bearing idea of Gentzen seems to have been: The consistency of arithmetic is proved by giving a special semantic explanation of correctness in arithmetic, either of formulas or of sequents. Next, this notion is applied to formal derivations. Finally, it is shown that there is no derivation of a contradiction that would be correct in the semantic sense.

By the above, consistency was a by-product of the more ambitious idea of giving a constructive semantics to intuitionistic arithmetic. Syntax and semantics have to match each other, and it has to be laid down what is achieved by a consistency proof, in particular, that it does not somehow assume what it sets out to prove. In his discussion of these topics in **INH**, Gentzen carefully avoids talking about the traditionally central notion of semantics, namely truth. He talks, like Brouwer, about **correctness** (*Richtigkeit*) and says that a statement **holds** (*gilt*).

Bernays had the submitted proof with him when he sailed to New York in September 1935. On board was Gödel; His position as the king of logicians was reflected in his status on board, in the first class. I have seen a postcard in the Bernays collection of the ETH-Zurich in which Gödel requests a meeting with Bernays, for the fired professor had to travel in a tourist class and could not just like that go and meet Gödel. During the fall term, the two commented on Gentzen's proof, but only the answers of the latter have been preserved. They contain some information, though in a form that is often bound to frustrate the reader, such as the following passage from a letter of 11 December 1935 (Menzler-Trott 2007, p. 64):

> The possible changes indicated by Gödel were known to me, but are in fact inapplicable from the finite standpoint because of their impredicative character.

Gentzen answered to the criticisms by changing the semantically based consistency proof into one that uses the now generally known transfinite induction principle, with essential changes of large parts of the manuscript sent to the journal in February 1936. They contained, as mentioned, a turn into an older idea, and various passages from **INH** make evident this remark of Bernays, transmitted through Kreisel's recollections in (1987, pp. 173–175). By good luck, the proof originally submitted for publication was preserved by Bernays in the form of galleys. They were published in English translation in the Szabo edition of Gentzen's papers in 1969, and in the German original in 1974. Even if Bernays kept the proofs for forty years, they have been lost in connection with the 1974 publication in the *Archiv für mathematische Logik und Grundlagenforschung* (later *Archive for Mathematical Logic*).

The net effect of the criticisms was a proof that mixed elements from the purely formal and semi-contentful approaches, instead of arriving at the former through the third proof idea, that of a proof through the syntactic notion of reducibility. The presentation suffered from these changes, but Gentzen was happy with the overall result he had found during the fall of 1935, namely, that a clear-cut transfinite induction can replace his original proof, with a precise "Gödel-point," the transfinite ordinal $\varepsilon_0$ that characterized Peano arithmetic.

## 4.2 The Setting of the Original Proof

The version submitted in August 1935, referred to here as Gentzen (1935), got mutilated by the changes Gentzen made. Gentzen (1935) gives a semantics for the derivability relation in arithmetic, expressed as a sequent $A_1, \ldots, A_n \rightarrow C$. There is just a single conclusion $C$ from the assumptions in the list $A_1, \ldots, A_n$, instead of a finite number of possible cases as in the classical sequent calculus *LK* of the doctoral thesis Gentzen (1934–35).

When the sequent notation is used, there is a double sense to derivability: The arrow is like the "vertical dots" in the inference schemes of natural deduction. On the other hand, there is the notion of derivability of a sequent by the rules of sequent calculus. Thus, these rules relate derivabilities in the first sense to each other, in the way exemplified by the left sequent calculus rule for disjunction, say. Disjunction elimination becomes the sequent rule: If $C$ is derivable from $A$ and from $B$, it is derivable from $A \vee B$. With assumptions added, we have the correspondence:

$$\frac{A \vee B \quad \begin{matrix}[A], \Gamma \\ \vdots \\ C\end{matrix} \quad \begin{matrix}[B], \Delta \\ \vdots \\ C\end{matrix}}{C} \vee E \qquad \rightsquigarrow \qquad \frac{A, \Gamma \rightarrow C \quad B, \Delta \rightarrow C}{A \vee B, \Gamma, \Delta \rightarrow C} L\vee$$

There are above the inference line of rule $\vee E$ two schematic derivations that are given as two corresponding sequents above the inference line of rule $L\vee$. Its conclusion gives the final situation of derivability of rule $\vee E$.

The correspondence goes in the same way for the other rules. For simplicity, I have taken the situation in which the major premiss of rule $\vee E$ in natural deduction is an assumption. The correspondence between natural deduction and sequent calculus was understood rather well by Gentzen, though not in full (see my 2012 for an exhaustive treatment).

In Gentzen (1935), a semantics of derivability in the first sense, as represented by the dots or arrows, is given. Then it is applied to derivability in the second sense.

The rest of this section is structured as follows: (3) The reduction of sequents. (4) The calculus *NLK*. (5) The reduction of derivations in *NLK*. (6) The consistency theorem. (7) Consistency: the three first proofs.

## 4.3 The Reduction of Sequents

The atomic formulas of arithmetic are decidable equalities between numerical terms. It follows that the whole propositional part of arithmetic is decidable. Gentzen's **reduction procedure** is carried over from the classical propositional logic of formulas to sequents, as exemplified by the following: If $A \,\&\, B$ in the antecedent of a sequent $A \,\&\, B, \Gamma \rightarrow C$ is false, one of $A$ and $B$ is false, and each can be tried in turn in the place of $A \,\&\, B$. If $\neg A$ in $\neg A, \Gamma \rightarrow C$ is false, it is deleted and the sequent changed into $\Gamma \rightarrow A$.

Gentzen's essential idea is to extend the procedure from the finitary domain to quantified formulas, i.e., to apply the "transfinite sense" of $\forall x A(x)$ in a certain way. Gentzen calls it "the in-itself sense" (*der an-sich Sinn*).

A way to think of the reduction procedure is that the correctness of a sequent $\Gamma \rightarrow C$ is guaranteed if, in whatever way $C$ may have as a consequence a false claim, it can be shown that some assumption in $\Gamma$ likewise presupposes a falsity. Then, whenever the assumptions $\Gamma$ hold, also $C$ holds. Say, to put it in figurative terms, we have a sequent of the form $\Gamma \rightarrow \forall x A(x) \,\&\, \forall x B(x)$ and an omniscient opponent who can **reason classically** by the in-itself sense of things and to whom the infinity of the natural numbers is not an obstacle. Such a creature can decide when $\forall x A(x) \,\&\, \forall x B(x)$ is false in its eyes, with, say, $\forall x A(x)$ a false conjunct, next to take a falsifying instance $A(n)$ out of the infinitely many possibilities. Our task is to show that, even if we don't have the opponent's classical and transfinite capacities, we can make finitarily choices *after* the opponent's choices so that some assumption in $\Gamma$ turns out false. It is this "finitary sense" that Gentzen is after in his semantical explanations.

The reduction of sequents is effected by suitable moves in what I, continuing to speak in suggestive terms of Gentzen's procedure, call a "falsification game" in which first certain "S-moves" are taken in the succedents of sequents, followed by "A-moves" in the antecedent.

**S-moves:**

**SVar.** *The sequent* $\Gamma \rightarrow C$ *has free variables. Numbers are chosen at will to instantiate these until there are no free variables left.*

**S&.** *The sequent is* $\Gamma \rightarrow A \,\&\, B$ *and either* $\Gamma \rightarrow A$ *or* $\Gamma \rightarrow B$ *is chosen at will.*

**S¬.** *The sequent is* $\Gamma \rightarrow \neg A$ *and the reduced sequent is* $A, \Gamma \rightarrow 0 = 1$.

**S∀.** *The sequent is* $\Gamma \rightarrow \forall x A(x)$ *and some instance* $\Gamma \rightarrow A(n)$ *is chosen at will.*

**Order of Precedence:** *Move* **SVar** *comes before the other* **S**-*moves.*

The **S**-moves are classical, for the falsifier knows how to end up with the worst possible case, here, a false equation as a conclusion. Each **S**-step simplifies the succedent of the sequent to be reduced until an equation $m = n$ remains. If the equation is true, the attempt at falsifying the sequent failed. Otherwise, when no **S**-move is applicable and $m = n$ is false, the task is to show that some of the assumptions must contain a falsity, too. To do this, the following steps can be taken in the antecedent:

**A-moves:**

**A&.** *The sequent is* $A \,\&\, B, \Gamma \to m = n$ *with* $m = n$ *false. The reduced sequent is* $A, A \,\&\, B, \Gamma \to m = n$ *or* $B, A \,\&\, B, \Gamma \to m = n$.

**A¬.** *The sequent is* $\neg A, \Gamma \to m = n$ *with* $m = n$ *false. The reduced sequent is* $\neg A, \Gamma \to A$.

**A∀.** *The sequent is* $\forall x A(x), \Gamma \to m = n$ *with* $m = n$ *false. The reduced sequent is* $A(k), \forall x A(x), \Gamma \to m = n$ *for some* $k$.

**Order of Precedence:** **S**-*moves come always before* **A**-*moves.*

In the first of the **A**-steps, the conjunction is repeated, for it can happen that one needs at some later stage also the other conjunct. It would be possible to have a single move with $A, B$ that replaces $A \,\&\, B$ with no repetition. The negation step seems to be classical, in that $\neg A$ in the antecedent and a falsity in the succedent does not lead to the intuitionistically derivable $\neg\neg A$ in the succedent, but to $A$; However, as said, the reasoning in the succedent part is classical.

The aim of the reduction procedure is to ensure that a false formula in the antecedent part of a sequent can be produced, whenever a false numerical equation has appeared in the succedent. Note that if a negation at left is reduced, there will be an **S**-step, unless it was a negation of an equality. Given a sequent $\Gamma \to C$, the result of reduction is, provided the process terminates, a sequent to which no reduction step applies.

Let us now check that, indeed, the reasoning in the succedent side is classical even if the domain is infinite:

Let the sequent be $\Gamma \to A \,\&\, B$. The **S**-steps should turn out something false out of the succedent, by which the succedent itself is also false. If that is so, then $\neg(A \,\&\, B)$ is true, i.e., classically $\neg A \vee \neg B$ is true. The worst case is produced by a choice, say $\neg A$, that gives a sequent $\Gamma \to A$ with a false succedent. Note that the conjuncts in $A \,\&\, B$ may very well be "transfinite," universally quantified formulas, and that it need not be decidable which of them is false. This does not matter, because **A**-steps have to be such that they apply to any choice that may have been taken in the succedent.

Let the sequent be $\Gamma \to \forall x A(x)$. As above, if the succedent is false, then $\neg\forall x A(x)$ is true, so classically $\exists x \neg A(x)$ is true. There is an instance $\neg A(k)$ true "in itself," and the sequent $\Gamma \to A(k)$ with a false succedent has to be dealt with.

Finally, if the sequent is $\Gamma \to \neg A$, there are no choices and the reduction goes on with $A, \Gamma \to 0 = 1$.

**Definition 1 (Irreducibility, Endform, Correctness)**

(i) *A sequent is* **irreducible** *if no reduction move applies to it.*
(ii) *A sequent* $\Gamma \rightarrow m = n$ *is in* **endform** *if either* $m = n$ *is true or there is some false equality in* $\Gamma$.
(iii) *A sequent* $\Gamma \rightarrow C$ *is* **correct** *if for each choice of* **S***-moves there are* **A***-moves such that* $\Gamma \rightarrow C$ *reduces to endform.*

We say often simply that a sequent is reducible if it is reducible to endform. The aim of Gentzen's consistency proof is to show that all derivable sequents are reducible. It follows that the sequent $\rightarrow 0 = 1$ is not derivable, because it is irreducible but not in endform: No atom in the antecedent is false, because there are none.

## 4.4 *The Calculus NLK*

As can be seen, the reduction of sequents is an idea independent of a particular logical calculus. To emphasize this important aspect, I reversed the order of presentation of the calculus and the reduction procedure from that in Gentzen (1935). In fact, it is this aspect that made it possible for Gentzen to change the calculus into another one in the published proof, instead of rewriting the whole paper (as he perhaps should have done).

The calculus in Gentzen (1935) is what is today called "natural deduction in sequent calculus style." It can be found already in the handwritten thesis manuscript, with the nomenclature *NLK* where the letters stand for "natürlich-logistisch klassisch." *NLK* is an obvious intermediate stage in the translation from natural deduction proper into sequent calculus: The idea is simply to display for each formula occurrence in a natural derivation all the open assumptions the formula depends on. There is a fundamental difference to sequent calculus proper, because there are no left rules for conjunction, implication, and universal quantification. To finish the translation to sequent calculus, Gentzen inserts cuts (see von Plato 2012 for details).

A further aspect of *NLK* is its classical character: Gentzen knew that classical logic would not be necessary but used it anyway. My guess is that he did it mainly for expository purposes, so that his intended general reader of the *Mathematische Annalen* would not be put off by a reliance on such esoteric things as intuitionistic logic. I shall now present the rules of *NLK*, as they are given in Gentzen's paper, except for the fraktur type: These rules are direct translations into the notation of sequent calculus of the rules of classical natural deduction and the induction rule. In rules that have more than one premiss, the contexts $\Gamma, \Delta, \ldots$ are accordingly added up in the antecedent of the conclusion. To the assumptions of natural deduction correspond "logical groundsequents" of the form $A \rightarrow A$. *Refutation* is from the German *Widerlegung*, abbreviated *Wid* below (Table 1).

The rules of inference are given in a linear form of sentences. There is, in fact, not a single inference line printed in the whole work. Here again, Gentzen perhaps

**Table 1** The rules of Gentzen's calculus NLK

| |
|---|
| *&-introduction*: The sequents $\Gamma \rightarrow A$ and $\Delta \rightarrow B$ give the sequent $\Gamma, \Delta \rightarrow A \,\&\, B$ |
| *&-elimination*: $\Gamma \rightarrow A \,\&\, B$ gives $\Gamma \rightarrow A$ resp. $\Gamma \rightarrow B$ |
| *$\vee$-introduction*: $\Gamma \rightarrow A$ gives $\Gamma \rightarrow A \vee B$ resp. $\Gamma \rightarrow B \vee A$ |
| *$\vee$-elimination*: $\Gamma \rightarrow A \vee B$ and $A, \Delta \rightarrow C$ and $B, \Theta \rightarrow C$ give $\Gamma, \Delta, \Theta \rightarrow C$ |
| *$\forall$-introduction*: $\Gamma \rightarrow A(a)$ gives $\Gamma \rightarrow \forall x A(x)$ on the condition that the free variable *a does not occur* in $\Gamma$ nor in $\forall x A(x)$ |
| *$\forall$-elimination*: $\Gamma \rightarrow \forall x A(x)$ gives $\Gamma \rightarrow A(t)$ |
| *$\exists$-introduction*: $\Gamma \rightarrow A(t)$ gives $\Gamma \rightarrow \exists x A(x)$ |
| *$\exists$-elimination*: $\Gamma \rightarrow \exists x A(x)$ and $A(a), \Delta \rightarrow C$ give $\Gamma, \Delta \rightarrow C$ on the condition that the free variable *a does not occur* in $\Gamma, \Delta, C$ nor in $\exists x A(x)$ |
| *$\supset$-introduction*: $A, \Gamma \rightarrow B$ gives $\Gamma \rightarrow A \supset B$ |
| *$\supset$-elimination*: $\Gamma \rightarrow A$ and $\Delta \rightarrow A \supset B$ give $\Gamma, \Delta \rightarrow B$ |
| Rule of "*refutation*": $A, \Gamma \rightarrow B$ and $A, \Delta \rightarrow \neg B$ give $\Gamma, \Delta \rightarrow \neg A$ |
| "*Elimination of double negation*": $\Gamma \rightarrow \neg\neg A$ gives $\Gamma \rightarrow A$ |
| Rule of "*complete induction*": $\Gamma \rightarrow A(0)$ and $A(x), \Delta \rightarrow A(x+1)$ give $\Gamma, \Delta \rightarrow A(t)$ |

wanted to appeal to a general readership, to whom the notation of two-dimensional proof trees with their inference lines was completely unknown at the time. Those few specialists who had read his doctoral thesis were an exception.

Formal derivations within Gentzen's calculus consist of series of sequents, with the following definition (p. 513):

> A derivation consists of a number of sequents in succession, such that each of these is either a "groundsequent" or results from some previous sequents through a "structural modification" or a "rule of inference."

To deal with the explicit listing of the assumptions in the antecedent parts of sequents, Gentzen adds the following "structural modifications:"

1. **Exchange** of the order of assumptions in the list.
2. **Contraction** of two occurrences of an assumption into one.
3. **Weakening** of an antecedent by the addition of an assumption.
4. **Change** of a bound variable by a **fresh** one.

Gentzen writes (pp. 513–514) that these rules are "purely formal in nature and inconsequential in their content; they have to be mentioned explicitly because of the peculiarities of the formalism."

The calculus is completed by adding what Gentzen calls "mathematical groundsequents." They have the form $\rightarrow A$, with $A$ a mathematical axiom. The right axioms are not listed, instead, Gentzen writes that for the consistency proof, it is not so essential what the mathematical axioms are. He gives as examples the following:

$$\forall x\, x = x, \quad \forall x \forall y (x = y \supset y = x), \quad \forall x \forall y \forall z (x = y \,\&\, y = z \supset x = z),$$

$$\forall x \neg x + 1 = x, \quad \forall x \forall y\, x + y = y + x, \quad \forall x \forall y \forall z\, (x + y) + z = x + (y + z).$$

Gentzen was convinced that the rule of induction was the only one that created real problems for the consistency proof. The rest of the arithmetic principles could be dealt with in whatever way was easiest. One such was given in the doctoral thesis. It contains "as an application of the sharpened *Hauptsatz*" a consistency proof for induction-free arithmetic (IV §3). Axioms are allowed to appear in the antecedent parts of sequents in a classical calculus and consistency is proved by the midsequent theorem. An alternative method was to formulate the axioms as groundsequents with free parameters, in the form

$$\rightarrow a = a, \quad a = b \rightarrow b = a, \quad a = b, b = c \rightarrow a = c,$$

$$a + 1 = a \rightarrow \ , \quad \rightarrow a + b = b + a, \quad \rightarrow (a + b) + c = a + (b + c).$$

In the consistency proof of 1938, such groundsequents contain after some transformations only numerical terms, and it can be decided whether they are correct, i.e., whether an equation in the succedent is a true numerical equation or an equation in the antecedent a false one.[4]

For this presentation, we grant to Gentzen what he presumes, namely, that the arithmetical principles except that of complete induction will not cause problems. It will be sufficient to prove the consistency of the system of classical natural deduction augmented by the rule of complete induction.

The rules of *NLK* exhibit some strange features: Why does the classical calculus *NLK* contain a full set of connectives and quantifiers? Further, there was no normalization theorem for the classical calculus. How could the ideas about a meaning explanation through normalization be carried over to a consistency proof in terms of *NLK*?

The essential difference of *NLK* with respect to a proper sequent calculus is that the elimination rules for conjunction, implication, and universal quantification operate on the right part of sequents. The corresponding sequent calculus rules operate on the left, antecedent parts of sequents. Looking at rules $\vee E$ and $\exists E$, we notice the following: If the first premiss is a logical groundsequent, $A \vee B \rightarrow A \vee B$ resp. $\exists x A(x) \rightarrow \exists x A(x)$, and if it is left unwritten, the rules turn out identical to the left rules of sequent calculus. I have followed this way in my 2009 article, with an intuitionistic sequent calculus for Heyting arithmetic, and given a proof of its consistency directly along the lines of Gentzen's proof.

When Gentzen comes to the proof of consistency in his paragraph 14, he has already removed the connectives $\vee, \supset$, and $\exists$ by the obvious translations into the fragment with just $\&, \neg$, and $\forall$. Even the inferences by the rules for the former group are transformed in the obvious way. Gentzen notes, at the end of paragraph 12, that a transformed derivation "is an essentially intuitionistically acceptable number-theoretic derivation: namely, the 'elimination of a double negation' could, where it

[4]After Gentzen's times, the axioms have been put aside by various degrees of hand-waving, in the style of: "It's all primitive recursively decidable, so why bother?" For a proper proof-theoretical treatment of the arithmetical axioms, see Siders (2015).

is used, be replaced by other rules of inference." We have seen above how this goes through, in the note of January 1933, and there is even more reason to ask why the classical rule is kept.

## 4.5 The Reduction of Derivations

The main part of Gentzen's original consistency proof consists of a few lemmas that I state as follows, with some typical cases of the proofs covered:

**Lemma 2** *Initial sequents $A \to A$ are correct.*

The proof is by induction on the length of $A$. Assume **SVar**-moves to have been taken, so that there are no free variables. There are four cases of which we show two:

**1.** $A$ is an equality $m = n$, and we have $m = n \to m = n$. By the decidability of numerical equality, if $m = n$ is true, $A \to A$ is in endform, and the same if $m = n$ is false.

**2.** $A$ is $B \& C$. Then $B \& C \to B \& C$ reduces by **S&** to $B \& C \to B$ or to $B \& C \to C$. Case 2.1. Consider the first time when the reduction of $B \& C \to B$ by arbitrary **S**-moves gives a sequent of the form $B \& C, \Gamma \to m = n$, i.e., the first time for an **A**-move. The sequent $B \to B$ is reducible by the inductive hypothesis, so the same sequence of **S**-moves as for $B \& C \supset B$ gives the reducible sequent $B, \Gamma \to m = n$. Application of **A&** to $B \& C, \Gamma \to m = n$ gives $B, B \& C, \Gamma \to m = n$. When formula $B \& C$ in the antecedent is left intact, the sequent reduces exactly as $B, \Gamma \to m = n$. Case 2.2. If $B \& C \to B \& C$ is reduced by **S&** to $B \& C \to C$, the proof is as above, with $C$ in place of $B$.

We see here in action the method of simulating in **A**-moves the choices made in the preceding **S**-moves. The remaining two cases of $A \equiv \forall x B(x)$ and $A \equiv \neg B$ are treated similarly. QED.

I give the **rule of composition** as the inference scheme:

**Composition of two sequents**

$$\frac{\Gamma \to D \qquad D, \Delta \to C}{\Gamma, \Delta \to C}\,\textit{Comp}$$

Gentzen takes it for granted that derivations can be composed in his calculus *NLK*.

**Lemma 3 (Closure of Derivability Under Composition)** *If the sequents $\Gamma \to D$ and $D, \Delta \to C$ are derivable in NLK and possible eigenvariables distinct, also the sequent $\Gamma, \Delta \to C$ obtained by composition is derivable in NLK.*

The proof would be straightforward were the calculus intuitionistic, as in von Plato (2009, lemma 5.2). I have not tried to determine how a proof with Gentzen's rules would go through, but let's assume it does.

Next in Gentzen's article comes the crucial property of the whole proof of consistency, one that he named the *Hilfssatz* in obvious analogy to his famous *Hauptsatz*, or cut elimination theorem for predicate logic. It states that composition preserves the correctness of sequents in the sense of the above definition:

**Hilfssatz 4 (Closure of Reducibility Under Composition)** *If the sequents $\Gamma \to D$ and $D, \Delta \to C$ are reducible to endform and possible eigenvariables distinct, their composition into $\Gamma, \Delta \to C$ is reducible to endform.*

*Proof* The proof is by induction on the length of the composition formula $D$. We can assume possible free variables to have been removed by **SVar**.

1. $D \equiv m = n$. Then the first premiss of *Comp* reduces to $\Gamma^* \to 0 = 1$, or $\Gamma^* \to m = n$ if move **S**$\neg$ was never applied. Assume **S**-moves to have been applied to the conclusion $\Gamma, \Delta \to C$ until $\Gamma, \Delta, \Delta^* \to k = l$ is produced, in which $k = l$ can be assumed false and $\Delta^*$ consists of those formulas, possibly none, that applications of **S**$\neg$ have brought to the antecedent. Leaving $\Delta, \Delta^*$ intact, the sequence of **A**-moves that reduces $\Gamma \to m = n$ to the endform $\Gamma^* \to 0 = 1$ (or $\Gamma \to m = n$), reduces $\Gamma, \Delta, \Delta^* \to k = l$ to an endform.
We note that if $\Gamma \to m = n$ is reducible and $m = n$ false, the equation $0 = 1$ can replace $m = n$: Compose $\Gamma \to m = n$ with the sequent in endform $m = n \to 0 = 1$ to get $\Gamma \to 0 = 1$.
2. $D \equiv A \& B$. The composition is

$$\frac{\Gamma \to A \& B \quad A \& B, \Delta \to C}{\Gamma, \Delta \to C}\,Comp$$

By assumption, $\Gamma \to A \& B$ is reducible, so both of $\Gamma \to A$ and $\Gamma \to B$ are. Consider the second premiss $A \& B, \Delta \to C$. Either there is no application of **A**& to $A \& B$ in its reduction and $A \& B$ can be removed. Then $\Delta \to C$ is reducible, and therefore also $\Gamma, \Delta \to C$. Else **A**& is applied at some stage to a reducible sequent $A \& B, \Delta^* \to 0 = 1$ with, say, the reducible sequent $A, A \& B, \Delta^* \to 0 = 1$ as result. We now apply *Comp*:

$$\frac{\Gamma \to A \quad A, A \& B, \Delta^* \to 0 = 1}{A \& B, \Gamma, \Delta^* \to 0 = 1}\,Comp$$

By the inductive hypothesis, *Comp* applied to shorter formulas maintains reducibility, so $A \& B, \Gamma, \Delta^* \to 0 = 1$ is reducible. The reduction of $\Gamma, \Delta \to C$ by the arbitrarily chosen **S**-moves that reduce the premiss $A \& B, \Delta \to C$ to $A \& B, \Delta^* \to 0 = 1$, gives the sequent $\Gamma, \Delta^* \to 0 = 1$ that is reducible to endform by the same **A**-moves as $A \& B, \Gamma, \Delta^* \to 0 = 1$.

**3.** $D \equiv \forall x A(x)$. The composition is

$$\frac{\Gamma \to \forall x A(x) \quad \forall x A(x), \Delta \to C}{\Gamma, \Delta \to C}\, Comp$$

By assumption, $\Gamma \to \forall x A(x)$ is reducible, so $\Gamma \to A(n)$ is reducible for any choice of $n$. As in **2**, either there is no application of **A**$\forall$ to $\forall x A(x)$ in the reduction of the second premiss and $\forall x A$ can be removed. Then $\Delta \to C$ is reducible, and therefore also $\Gamma, \Delta \to C$. Else **A**$\forall$ is applied at some stage to a reducible sequent $\forall x A(x), \Delta^* \to 0 = 1$, with the reducible sequent $A(k), \forall x A(x), \Delta^* \to 0 = 1$ as result. With the instance $k$ also in the first premiss, application of *Comp* to the shorter formula $A(k)$ gives

$$\frac{\Gamma \to A(k) \quad A(k), \forall x A(x), \Delta^* \to 0 = 1}{\forall x A(x), \Gamma, \Delta^* \to 0 = 1}\, Comp$$

The conclusion is reducible by the inductive hypothesis. The reduction of $\Gamma, \Delta \to C$ by the arbitrarily chosen **S**-moves that reduce the premiss $\forall x A(x), \Delta \to C$ to $\forall x A(x), \Delta^* \to 0 = 1$, gives the sequent $\Gamma, \Delta^* \to 0 = 1$ that is reducible to endform by the same **A**-moves as $\forall x A(x), \Gamma, \Delta^* \to 0 = 1$.

**4.** $D \equiv \neg A$. The composition is

$$\frac{\Gamma \to \neg A \quad \neg A, \Delta \to C}{\Gamma, \Delta \to C}\, Comp$$

In the reduction of the second premiss of *Comp*, if **A**$\neg$ is never applied to $\neg A$, it can be deleted and what remains, the sequent $\Delta \to C$, is reducible. Then also $\Gamma, \Delta \to C$ is reducible. Otherwise there is a reducible sequent $\neg A, \Delta^* \to 0 = 1$, to which in turn **A**$\neg$ is applied to give the reducible sequent $\neg A, \Delta^* \to A$.

The first premiss of *Comp* reduces by **S**$\neg$ to $A, \Gamma \to 0 = 1$. Application of *Comp* to the shorter formula $A$ gives

$$\frac{\neg A, \Delta^* \to A \quad A, \Gamma \to 0 = 1}{\neg A, \Gamma, \Delta^* \to 0 = 1}\, Comp$$

The conclusion is reducible by the inductive hypothesis.

As above, if in the reduction of $\neg A, \Gamma, \Delta^* \to 0 = 1$ move **A**$\neg$ is never applied to $\neg A$, it can be deleted and the remaining sequent $\Gamma, \Delta^* \to 0 = 1$ is reducible. This is the sequent produced from $\Gamma, \Delta \to C$ by the arbitrary initial **S**-moves that gave $\neg A, \Delta^* \to 0 = 1$, so $\Gamma, \Delta \to C$ is reducible.

If instead in the reduction of $\neg A, \Gamma, \Delta^* \to 0 = 1$ move **A**$\neg$ is applied at some stage to $\neg A$ in a reducible sequent $\neg A, \Gamma^*, \Delta^{**} \to 0 = 1$, the reducible sequent $\neg A, \Gamma^*, \Delta^{**} \to A$ is obtained. Composition with $A, \Gamma \to 0 = 1$ gives $\neg A, \Gamma, \Gamma^*, \Delta^{**} \to 0 = 1$ that is reducible. Therefore, continuing this analysis, at some stage the formula $\neg A$ in the antecedent of the result of composition must remain unreduced and can be deleted. The resulting sequent is then reducible. QED.

The proof seems innocent enough, even if the very last steps are a bit tedious. They bring perhaps to mind methods in proofs of underivability through failed proof search.

Gentzen tries to persuade the reader of the constructive character of the reduction procedure by reformulating the *Hilfssatz* in the following terms (cf. Gentzen 1935, sec. 14.44): "If reduction procedures for $\Gamma \rightarrow \mathfrak{D}$ and $\mathfrak{D}, \Delta \rightarrow \mathfrak{C}$ are known, a reduction procedure for $\Gamma, \Delta \rightarrow \mathfrak{C}$ can also be given." These, however, are just words; There is no difference of substance to the formulation above.

## 4.6 *The Consistency Theorem*

The final component in Gentzen's consistency proof is to show that the rules of inference preserve correctness of sequents:

**Theorem 5** *If the sequent* $\Gamma \rightarrow C$ *is derivable, it reduces to endform.*

The proof is by induction on the last step of a derivation. If $\Gamma \rightarrow C$ is a logical groundsequent, it is correct as shown above by the lemma. Otherwise consider the last rule of the derivation and show that if the premisses reduce to endform, also the conclusion reduces. The cases are the structural modifications, seven logical rules, and *CI*.

Gentzen goes through the two cases for $\forall$. Then he notes that the three conjunction rules go through similarly. The cases for $\forall$ are:

1. The last rule is $\forall I$. The conclusion is $\Gamma \rightarrow \forall x A(x)$, and it reduces by **S**$\forall$ to $\Gamma \rightarrow A(m)$. The premiss $\Gamma \rightarrow A(y)$ is by assumption reducible, with $y$ the eigenvariable. Rule **SVar** produces a sequent $\Gamma \rightarrow A(n)$ that is reducible for any choice of $n$, in particular, the choice $m$. Therefore the conclusion of $\forall I$ is reducible.
2. The last rule is $\forall E$. The premiss is $\Gamma \rightarrow \forall x A(x)$. **S**-moves applied to the conclusion $\Gamma \rightarrow A(t)$ produce the sequent $\Gamma \rightarrow A(m)$. The premiss is reducible for any choice of value for $x$, therefore $\Gamma \rightarrow A(t)$ is reducible by the same **A**-moves as for $\Gamma \rightarrow \forall x A(x)$.

Next comes a peculiar turn, when Gentzen writes (14.44) that for the two negation rules and *CI*, the *Hilfssatz* is put into use. Namely, the question is: If we leave out rule *CI*, should we not get a standard proof of the consistency of classical first-order logic as a result? Moreover, the classical rule is dispensable in *NLK*. What has a principle such as the *Hilfssatz* to do in this connection?

The situation is clarified in my 2009 paper that uses an intuitionistic calculus and a normalization theorem. The overall result is contained in the:

**Observation** *With the intuitionistic calculus NLI, the Hilfssatz is needed only for showing that rule CI preserves the correctness of derivations.*

Gentzen naturally knew the above by the result of his thesis, namely, that if induction is left out, the proof of consistency can be carried through finitistically in the tradition of Hilbert's program. Some logical groundwork had simply remained undone; Maybe there was some haste for a poor scholarship holder who tried to secure an academic position in extremely difficult times: Gentzen had in fact applied for scholarships in Germany, inquired Weyl about the Rockefeller foundation financing and about a stay in Princeton, and obtained a position as a teacher in a lyceum in Stralsund in case nothing else worked. Moreover, he had no one to talk to, with people expelled from Göttingen.

Finally, we look at the crucial step of the consistency proof, namely the case of rule *CI*.

**3.** The last rule is *CI*. The premisses are $\Gamma \to A(1)$ and $A(y), \Delta \to A(y+1)$, the conclusion $\Gamma, \Delta \to A(t)$. In its reduction, if $t$ has free variables, application of **Svar** gives some numerical term $n$ in place of $t$. In the second premiss, any application of rule **SVar** gives a reducible sequent, so that $A(m), \Delta \to A(m+1)$ is derivable and reducible for any $m$. An $n-1$-fold composition of $\Gamma \to A(1)$ with $A(1), \Delta \to A(2), \ldots, A(n-1), \Delta \to A(n)$ gives

$$\dfrac{\dfrac{\Gamma \to A(1) \quad A(1), \Delta \to A(2)}{\Gamma, \Delta \to A(2)}\,Comp \qquad A(2), \Delta \to A(3)}{\Gamma, \Delta^2 \to A(3)}\,Comp$$

$$\vdots$$

$$\dfrac{\Gamma, \Delta^{n-2} \to A(n-1) \qquad A(n-1), \Delta \to A(n)}{\Gamma, \Delta^{n-1} \to A(n)}\,Comp$$

Thus, the sequent $\Gamma, \Delta^{n-1} \to A(n)$ is derivable by the admissibility of composition and reducible by the *Hilfssatz*. For the conclusion $\Gamma, \Delta \to A(t)$ of *CI*, an **S**-move reduces it into $\Gamma, \Delta \to A(n)$ and $\Gamma, \Delta \to A(n)$ is reducible because $\Gamma, \Delta^{n-1} \to A(n)$ is.

By hindsight, we have one more aspect of later calculi of proof search present in the reduction procedure. Namely, it has to be shown that the rule of contraction preserves reducibility, and this is secured because there is a possible repetition of a formula for rules that are not invertible.

With the above lemmas and preparations, consistency can be easily concluded: As noted above, the sequent $\to 0 = 1$ is irreducible but not in endform, therefore it is not derivable.

**Corollary 6** *The system NLK+CI+arithmetic axioms is consistent.*

More is achieved than the unprovability of $0 = 1$, namely, it follows that from derivability follows correctness, or soundness in more recent logical terminology.

## 4.7 The Earliest Preserved Consistency Proof

The above account of the consistency proof is essentially based on the preserved galley proofs of Gentzen's article in its original 1935 form. We have now a background against which it is possible to understand recently transcribed stenographic manuscripts from the fall of 1934. These are, first, the last ten pages of **INH**, written in October of that year. Secondly, there is the manuscript **BZ**, for *Beweistheorie der Zahlentheorie* (Proof theory of arithmetic), written between August 1934 and March 1935, with pages 1–6 and 9–12 preserved. The third one is **WAV**, mentioned already above, and written around October 1934, but without dates and with the pages 55–56, 77–80, and 83–86 preserved. It consists of preliminary notes for the preparation of the final manuscript, judging from the pages that have been preserved as well as from occasional references to it in the other manuscripts. These notes have direct connections to the article that Gentzen prepared in the spring of 1935.

There are parts in **BZ** and **WAV** that treat the same topic, the preparation of sequent derivations in which all formulas are in prenex normal form. The propositional part of arithmetic is decidable, and Gentzen wanted to delimit propositional steps in derivations to a "finitary" part, above a "transfinite part" that contains steps of inference with the quantifiers, a separation that follows from the midsequent theorem for derivations in the classical sequent calculus *LK* that he used at this stage. The aim was to have a consistency proof that is "more concentrated on what is essential" ( **WAV**, p. 78). One idea in **WAV** is to minimize the number of proper rules of inference, through the use of groundsequents, such as $A \,\&\, B \rightarrow A$, $A \rightarrow A \vee B$, and $\forall x A(x) \rightarrow A(t)$. The reducibility of such sequents follows easily from the reducibility of initial sequents; Say, when an **A**-move is met with the first one, $A \,\&\, B$ is replaced by $A$, and then reduction steps can be applied in the antecedent as in the reduction of $A \rightarrow A$.

**WAV** contains the earliest preserved proof of consistency of arithmetic, detailed out in three pages and based on the reduction procedure (pp. 78–80). It is titled "the second proof of correctness (*LK* consistency proof)," and by this proof, it becomes further clear that the first proof was also based on the reduction procedure, but with the intuitionistic sequent calculus *LI* augmented by the classical sequent $\neg\neg A \rightarrow A$. When the classical "symmetric calculus" is used, as Gentzen calls it, the reduction procedure has to be defined also for disjunction and existence. (He prefers to leave implication out, because it breaks the symmetry of *LK*.) The details of the reduction procedure for symmetric sequents are not spelled out, but it is clear how they are to be taken: The arbitrary choices (moves by the opponent in my terminology above) extend now to the antecedent part, with the aim of producing a true numerical equation at left. Thus, the opponent is able to make a best possible choice in the case of an antecedent formula $\exists x A(x)$, for a true instance $A(t)$. Afterwards the respondent can reply to such a choice in the succedent by choosing the same instance $A(t)$. Analogously, the opponent chooses one of the disjuncts in $A \vee B$ in the antecedent, and the respondent in the succedent. Whenever the opponent has produced a false equation in the succedent or a true equation in the antecedent, the respondent is in turn, with the aim of producing a false equation in the antecedent or a true one in the succedent. Whenever this is the case, a sequent is in endform, a notion that coincides with the one above for single-succedent sequents.

As before, the proof of consistency proceeds by showing that derivable sequents reduce to endform. Propositional connectives are handled by logical groundsequents, as above, and for conjunction in the succedent by $A, B \to A \& B$ and disjunction in the antecedent by the dual $A \vee B \to A, B$. The quantifier rules are straightforward. There remain *CI* and the crux of the proof, namely that the composition of sequents in the form of a **mix rule** (*Mischung*), or **multicut** in more recent terminology, maintains reducibility (**WAV**, p. 79):

> Let the reducibility of both upper sequents be already shown. That for the lower sequent to be shown. We do a complete induction after the grade of the mix. That is now: The number of $\forall$ and $\exists$ at the head of the mix formula $\mathfrak{M}$.
>
> $$\frac{\Gamma \to \Delta\,(\mathfrak{M}) \quad \Theta\,(\mathfrak{M}) \to \Lambda}{\Gamma\,\Theta^* \to \Delta^*\,\Lambda}$$

The notion of grade indicates that the formulas are in prenex normal form. The proof that the grade of the mix formula can be lowered ends with the words (**WAV**, p. 80): "This somewhat peculiar inference is subjected to detailed criticism in Section IV," clearly a reference to the paper Gentzen was writing. In that paper, the proof through a reduction procedure obtained a third form, through the classical natural calculus *NLK* that uses the sequent notation. Thus, what I have called the original proof was by Gentzen's count in **WAV** actually the third one. Moreover, **INH** and **BZ** contain references to a lost series **WTZ**, clearly for "consistency transfinite numbers," but the few indications of ordinals in that attempted proof do not yet contain the Gentzen ordinal $\varepsilon_0$ of 1936.

## 5 Nature and Reception of the Original Proof

Gentzen was obviously happy and content with his original proof. A lot of work had gone into it, both formal and conceptual: The detailed discussions in **INH**, especially, give an indication of the importance of the latter for Gentzen. Others, however, felt that something was missing: Gentzen (1935) contains general discussions about the significance of consistency proofs and it even singles out the *Hilfssatz* as central, but it does not indicate clearly what the crucial points in the proof of the latter are. Specifically, the termination of the reduction process is not treated in precise terms.

### *5.1 The Problem of Termination of the Reduction Procedure*

There has been an extensive correspondence between Gentzen and Bernays about the consistency proof, as well as some letters between Gentzen and Weyl, and Gentzen and Van der Waerden. Only the letters of Gentzen to Bernays have been preserved. The first of these letters, dated June 23, 1935, was sent from Gentzen's hometown Stralsund by the Baltic Sea and included the "final part" of the consistency paper. It went then on to discuss the suggestions made by Bernays and notes, among others, that the existence property of arithmetic follows for formulas $\exists x A(x)$, "in case $A(x)$ is not transfinite." Towards the end Gentzen writes that he wanted to discuss in the final chapter transfinite ordinal numbers and their relation to reduction procedures and construction procedures, and then continues: "In the end, these things did not seem ripe for a presentation yet but could perhaps find place in a later separate publication."

A second letter written three weeks later, 14 July, contains:

> I have written in fact nonsense on pp. 75–76; I held my eye on an older form of the notion of reduction, in which the reduction steps are uniquely determined. The passages could be corrected more or less as follows: At 15.21, reducibility should be replaced by: 'There is a number $\nu$ so that for each series $\mathcal{R}_\nu$ of $\nu$ numbers, a series of at most $\nu$ sequents can be given such that the first one is $\mathfrak{S}$q, and each of these is formed from the preceding one through a reduction step, and the last one has endform, and further, *the possible choices are determined through the associated numbers from the series* $\mathcal{R}_\nu$.' Correspondingly under 15.23: "For each *infinite* series $\mathcal{R}$ of numbers, a finite series of sequents can be given, the first of which ..." as before. – I have, however, cancelled these passages completely, because they are not fully necessary; perhaps I could give sometime later complete proofs to both theorems in a special publication.

The uniquely determined sequence of reduction steps should refer to a reduction procedure for derivations of the false formula $0 = 1$.

The above passage is reminiscent of Brouwer's explanation of bar induction in his (1924), where the connection to transfinite induction is also made—a pity we don't have Gentzen's proof of his 15.23 preserved. He states quite clearly that the choice sequences in steps of reduction, represented as sequences of natural numbers, lead to endform in a finite number of steps. Gentzen's use of natural numbers in the description of the reduction procedure brings him very close to Brouwer who in 1924 formulated the bar theorem as follows:

> If to each element of a set $M$ a natural number $\beta$ is associated, $M$ is decomposed by this association into a well-ordered species $S$ of subsets $M_\alpha$, such that each of these is determined by a finite initial segment of choices. To each element of the same $M_\alpha$ is associated the same natural number $\beta_\alpha$.

In Gentzen, $M$ consists of the collection of reduction sequences of sequents and the choices to single reduction steps.

Now there is a big gap in the correspondence, until 4 November, with a four-page tightly and very orderly written letter sent to Bernays in Princeton, where also Gödel was. The former, possibly with the help of the latter, had taken up the central problem of the proof, as can be gathered from Gentzen's answer:

> I have considered all these aspects already myself, including the geometrical image of branching line segments. You are quite right that the finiteness of even a single reduction path for the sequent $\Gamma, \Delta \rightarrow \mathfrak{C}$ can get grounded on the finiteness of a whole series of different reduction paths for $\mathfrak{D}, \Delta \rightarrow \mathfrak{C}$. But this does nothing for my proof idea!

He then writes that he had added an explanation to the article—it would be in the end of Sect. 14 (p. 112 of the German version of 1974)—the proofs of which he had sent back a few days earlier:

> Let the following be remarked to avoid misunderstandings: The type of reduction of $\mathfrak{D}, \Delta \rightarrow \mathfrak{C}$ into $\mathfrak{D}, \Delta^* \rightarrow \mathfrak{C}^*$ can eventually depend on a *choice* (14.6 2 1) that takes place in the reduction of the mix-sequent $\Gamma, \Delta \rightarrow \mathfrak{C}$.[5] The same holds of each further step of reducing back, and, it can be added, the new *mix*-sequent $\Gamma, \Delta^* \rightarrow \mathfrak{C}^*$ etc. need in no way always be the reduced one of the preceding sequent (14.6 2 3). So, the number of steps of proof can be very different, according to the result of the individual choices; the only thing that is certain is that it is in every case *finite*. To prove a claim for *every* possible choice, it is sufficient to prove it for *one* specific, arbitrary choice. Therefore *it is* sufficient in the entire proof to keep an eye on just one *single specific* sequence of reductions of the sequent $\mathfrak{D}, \Delta \rightarrow \mathfrak{C}$, and thereby on just one *single specific* finite series of steps of proof.

No second round of proofs is known that would contain this passage. The terminology of mix-sequents is that of the doctoral thesis, where cut formulas were called mix-formulas. This terminology is used also in the consistency proof of October 1934, mentioned above. As to why the termination is not addressed in the paper, Gentzen writes that "since you don't seem so far to have said anything concerning the recognition of the finiteness of the forms of inference, I have left them out of the consistency proof; also because there would be still one thing and another to clarify." Gentzen had obviously a great desire to publish what he had to offer so far.

## 5.2 *The Essence of Gentzen's Hilfssatz*

Gentzen's letter of 4 November contains a description of what he calls "the essence of the somewhat peculiar inductive inference" in the *Hilfssatz*, namely, why the reduction procedure should terminate:

> A proposition $\forall x F(x)$ is proved if each of the infinitely many special cases $F(\nu)$ is proved. Let each of these again be equivalent to a proposition $\forall x F_\nu(x)$, each special case $F_\nu(\mu)$ of these propositions again equivalent to a proposition $\forall x F_{\nu,\mu}(x)$, etc. Let the following be known: Each arbitrary series of specializations $\forall x F(x)$, $F(\nu) \supset\subset \forall x F_\nu(x)$, $F_\nu(\mu) \supset\subset \forall x F_{\nu,\mu}(x), \ldots$ ends after a finite number of components in a formula $F_{\nu\mu\ldots}(\varrho)$, the *correctness of which is known*. To be proved now: $\forall x F(x)$ is correct. To this end, I infer as follows: The correctness of $\forall x F(x)$ is secured if $F(\nu)$ holds for whichever arbitrarily chosen $\nu$. So let us assume that we had chosen a specific number $\nu$, and it remains just to prove $F(\nu)$. This is $\supset\subset \forall x F_\nu(x)$. Now I infer just as before, namely, that to show that this proposition holds, it suffices to take whichever arbitrarily determined special case, say $F_\nu(\mu)$, etc. This chain of inferences must end after a finite number of steps, because each arbitrary sequence $\forall x F(x), F(\nu), F_\nu(\mu), \ldots$ had to be finite. Thereby $\forall x F(x)$ is proved.

He says that this is "an analogy" that should be compared to "the image of the branching sequence of line segments." The latter can be depicted as follows, with Gentzen's example:

[5] We saw this situation in the above consistency proof, in the case of rule *CI* in which $A(t)$ in the succedent was reduced to $A(n)$.

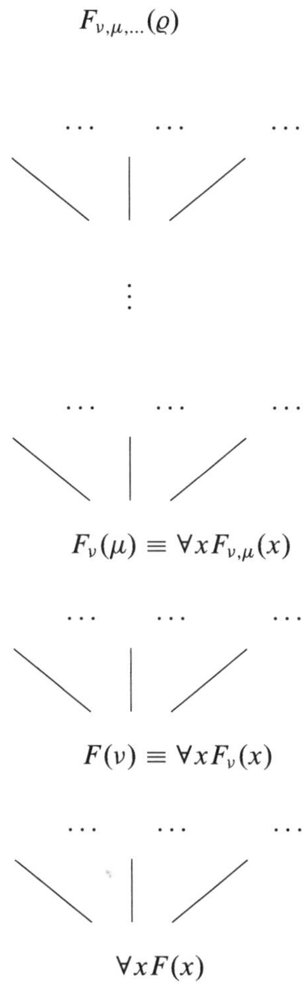


There is no bound on how many universal quantifiers can occur in a formula, and therefore a denumerably branching tree of any finite height can occur. Moreover, the "analogy" begins with a peculiar requirement that each of the infinitely many instances of $\forall x F(x)$ be proved. Is the analogy an appeal to infinitary proof theory, or to the infinite capacities of the classical reasoner? "What do you think, now, about this way of inference? Shouldn't it be finite?" These are his questions to Bernays, but he adds at once the parenthetical remark:

> If one turns the proof into an indirect one, i.e., begins like this: Assume that $\forall x F(x)$ does not hold, then there is a counterexample $\nu$ so that $F(\nu)$ does not hold, so neither ⊃⊂ $\forall x F_\nu(x)$, etc, then the *tertium non datur* enters.

Now we can read the suggestion as the choice of a path in a reduction tree that has a denumerable branching at each node. If there is at least one sequence of choices such that the topformula $F_{\nu\mu\ldots}(\varrho)$ gets falsified, we have established $\neg F_{\nu\mu\ldots}(\varrho)$. If not, i.e., if no counterexample was found, proceeding all the way down to the root of the reduction tree we get that the assumption that $\forall x F(x)$ does not hold is false, and a classical step of double negation elimination (the *tertium non datur*) gives $\forall x F(x)$.

Gentzen is well aware that a new type of proof is about to surface here. One reason for expecting something new is, naturally, that the proof must go beyond those that can be justified in arithmetic. Turning now to the reduction rules, we notice that sequences of moves in the succedent and antecedent can alternate any number of times, and each block of succedent moves can produce an initial segment in the Baire space of a denumerably branching tree.

## 5.3 A Lost Connection: Consistency Proofs and Bar Induction

Gentzen (1935) was received, in a literal sense, by Bernays and Weyl. Parts of the paper were changed in February 1936, by which the galleys of the original version had been prepared before that date, and the paper must have gone to print clearly earlier. In fact, Menzler-Trott (2007, p. 61) reproduces a letter from Gentzen to Hellmuth Kneser, written 27 October 1935, in which it is stated that the first galley proofs have already arrived. He also wrote there that Van der Waerden, then a professor at Leipzig, had commented very positively on the proof.

As mentioned, Cavaillès was in Göttingen in the fall of 1935. His book contains a discussion of Gentzen's proof, with a description of the reduction procedure and the problem of its termination, but along the treatment by transfinite induction of the published version (1938, pp. 165–170). A letter from Cavaillès to Albert Lautmann indicates that Gentzen had read the text and "repaired passages where to him I had oversimplified" (cf. Menzler-Trott, p. 82).

Weyl gave his copy to Stephen Kleene who, by his own telling, got a job from Wisconsin and gave the copy back after only two days. That was very unfortunate for the development of proof theory and foundational study in general. It took another fifteen years before Kleene took up Gentzen's work, in an article about sequent calculus (Kleene 1952a), and in the *Introduction to Metamathematics*. In the latter, the *Hauptsatz* is presented in detail and applied to a consistency proof of arithmetic without the induction rule (p. 463). For the full consistency proof, there is just a "brief heuristic account of the method used by Gentzen" (p. 476). It is all based on the published proof. Richard Vesley worked with Kleene on the constructive theory of ordinals and together they studied Brouwer's work. He has told me (in an e-mail of 3 March 2011) that he is sure that they never discussed the extent to which Gentzen had been influenced by Brouwer's intuitionistic theory of ordinals.

Bernays (1970) recalled that the main point of criticism was Gentzen's implicit use of the **fan theorem**, a principle of Brouwer's intuitionistic mathematics, by which, if all branches of a finitely branching tree are finite, the tree consists of a finite number of nodes. The same is explained in his prefatory words to the publication of Gentzen (1935) in 1974 (p. 97):

> A methodical objection was made against the original proof, namely that it used implicitly a principle usually described today as the "fan theorem," by which each branching figure that branches only finitely at each point and in which each thread ends after a finite number of component parts, can on the whole have only a finite extension.

The fan theorem is a special case of the **bar theorem** in which latter the branchings are denumerably infinite. These terminologies are much later than the results, but it is still a bit strange that Bernays explicitly describes the finite branching, when Gentzen's proof clearly has denumerable branching. A detailed proof of Gentzen's *Hilfssatz* can indeed be given by the use of bar induction; It makes Gentzen's "peculiar inductive inference" of termination of reduction crystal clear (Siders and von Plato, this volume). As we saw, Bernays writes that Gentzen's use of bar induction was "implicit," and if so, then he had come to use that principle independently of Brouwer, which would be remarkable.

In Brouwer (1924) to which Gentzen refers in his (1938a), the bar theorem is called "the main theorem on well-ordered sets." The additional remarks in Brouwer (1924a) make quite explicit the associated principle of transfinite induction on well-founded trees the bar theorem rests on (p. 645). The theorem was known to Gödel, and also to von Neumann who also was in Princeton at that time and must have heard discussions about Gentzen's result.[6] Kreisel had extensive discussions with Bernays about Gentzen's original proof, and he writes (1987, p. 173) that "Gödel and von Neumann criticized the original—posthumously published—version." There is a more general principle behind the fan and the bar theorem; In Kreisel (1976, p. 201) we find stated that both Gödel and von Neumann "naturally knew the theory of choice sequences that Brouwer had developed systematically, and especially the problematic assumption (of which Brouwer was particularly proud), namely that all functions $F$ with arbitrary choice sequences of natural numbers as arguments and natural numbers as values... can be produced inductively. The best-known corollary is the *fan theorem.*"

During and after the criticisms by Bernays and Gödel, seconded by von Neumann and possibly even Weyl (as suggested by a letter of Weyl's for which see Menzler-Trott 2007, p. 58), Gentzen laid the foundation of today's **ordinal proof theory**: It can be seen clearly from his letters how this topic emerged in a few months' time, with the consequence that the semantical explanation of sequents through a notion of reducibility and the consistency proof by induction on well-founded trees receded in the background. By his (1938b), after having closed his new proof of consistency by a presentation of transfinite induction, he writes that he puts no specific weight on the notion of reducibility of derivable sequents and ends up with what seems almost a contradiction in terms: "I resorted to it at the time as *one* argument against radical intuitionism." This paper was the second part of an issue of Heinrich Scholz'

---

[6] I owe the information about von Neumann's knowledge of Brouwer's "fundamental theorem on finite sets" to Dirk van Dalen: He kindly sent me a copy of a letter of von Neumann's to Brouwer, from April 1929, that contains a constructive proof of the existence of a winning strategy in chess by the fan theorem. The letter is found in van Dalen (2011).

publication series on logic and foundations. The first part was Gentzen's essay on *The present situation in mathematical foundational research.* There he contrasts the lessons from intuitionism, presumably those of Brouwer's four insights, against "radical intuitionism, that rejects as senseless everything in mathematics that does not correspond to the constructive point of view." Gentzen became a Brouwerian intuitionist in 1932 but then found by 1936 that Brouwer's constructive ordinals codify intuitionistic principles in more conventional terms, it seems.

The fate of the original proof was that it was simply put aside, just like Gentzen had put aside his detailed proof of normalization for natural deduction, the former saved only because Bernays had kept the galley proofs, the latter only because he had kept Gentzen's handwritten notes. Gentzen's use of induction on well-founded trees had been saved also in another sense, the extent of which is yet to be fully determined: Namely, as shown by the titles of topics in Gödel's stenographic notes in his *Arbeitshefte*, there are at least 150 pages of work of his on Gentzen's proof, with such suggestive titles as *Principal lemma of Gentzen's consistency proof with choice sequences* (*Arbeitsheft* 11, p. 28). In the earlier *Arbeitsheft* 4 (p. 39), there is the title *Gentzen with choice sequences*. The proof ends on p. 50 with: "**Theorem.** Induction Principle. $[(n)\mathfrak{A}(\Phi_n)] \supset \mathfrak{A}(\Phi)$ ). $\mathfrak{A}$(const.) $\supset (\Phi)\mathfrak{A}(\Phi)$." The meaning is that if from the assumption that every one-step continuation $\Phi_n$ of a reduction sequence $\Phi$ has the property $\mathfrak{A}$ it follows that $\Phi$ has the property $\mathfrak{A}$, then from the base case $\mathfrak{A}$(const.) follows that all reduction sequences have the property $\mathfrak{A}$.

A picture starts emerging from a study of Gentzen's original proof, the letters he wrote to Bernays, Gödel's titles in the *Arbeitshefte,* his "Zilsel" lecture of 1938 and the Yale lecture of 1941, and Kreisel's recollections: Namely, Gödel's **no-counterexample** interpretation of the Zilsel lecture derives from Gentzen's original proof (cf. also Tait 2005). Secondly, concerning the **Dialectica-interpretation**, Kreisel (1987, p. 175) writes: "At first Gödel, like von Neumann, was ill at ease with Gentzen's use of functionals, albeit of *lowest* type. But when Gödel returned to the subject, about 5 years later, he used *all finite* types." The connections between Gentzen's proof, Gödel, and bar induction are suggestive enough, but the source materials are at present not sufficiently known for these matters to be discussed in any conclusive way—so here is where we must rest for now.

**Sources and Acknowledgement** Prof. Christian Thiel of Erlangen University received in 1984 two folders of stenographic notes from the sister of Gentzen, Waltraut Student. They had been left in the Gentzen family's summer place on the Baltic island of Rügen in 1944. During a visit to Erlangen in February 2005, I was able to study the parts Thiel had transcribed, about half of the material including pages 1–4 of the series **BZ**. The short manuscripts about natural deduction from September 1932 and January 1933 caught my interest, and Thiel transcribed them soon. My numerous visits to Erlangen led to the complete transcription of **INH** for Gentzen's centenary year in 2009. For the correctness of the rest of the transcriptions, from **BZ** and **WAV** as cited in this paper, I am responsible. I also wish to thank here Bill Howard, Bill Tait, and Thierry Coquand for extensive exchanges over several years on the theme of Gentzen, Gödel, and bar induction.

# References

Ackermann, W. (1924) Begründung des "tertium non datur" mittels der Hilbertschen Theorie der Widerspruchsfreiheit. *Mathematische Annalen*, vol. 93, pp. 1–36.

Bernays, P. (1970) On the original Gentzen consistency proof for number theory. In J. Myhill et al., eds, *Intuitionism and Proof Theory*, pp. 409–417, North-Holland.

Brouwer, L. (1924) Beweis, dass jede volle Funktion gleichmässig stetig ist. *Koninklijke Akademie van Wetenschappen te Amsterdam, Proceedings of the Section of Sciences*, vol. 27, pp. 189–193.

Brouwer, L. (1924a) Bemerkungen zum Beweise der gleichmässigen Stetigkeit voller Funktionen. Ibid., pp. 644–646.

Brouwer, L. (1926) Zur Begründung de intuitionistischen Mathematik. III. *Mathematische Annalen*, vol. 96, pp. 451–488.

Brouwer, L. (1928) Intuitionistische Betrachtungen über den Formalismus. *Sitzungsberichte der Preussischen Akademie der Wissenschaften*, pp. 48–52.

Cavaillès, J. (1938) *Méthode axiomatique et formalisme*. Republished in *Oeuvres completés de Philosophie des sciences*, Hermann, Paris 1994.

van Dalen, D. (2005) *Mystic, Geometer, and Intuitionist: The Life of L.E.J. Brouwer*. Volume II, Oxford U.P.

van Dalen, D. (2011) *The Selected Correspondence of L.E.J. Brouwer*. Springer, Berlin.

Dawson, J. and C. Dawson (2005) Future tasks for Gödel scholars. *The Bulletin of Symbolic Logic*, vol. 11, pp. 150–171.

Franks, C. (2010) Cut as consequence. *History and Philosophy of Logic*, vol. 31, pp. 349–379.

Gentzen, G. (1934–35) Untersuchungen über das logische Schliessen. *Mathematische Zeitschrift*, vol. 39, pp. 176–210 and 405–431.

Gentzen, G. (1935) Der erste Widerspruchsfreiheitsbeweis für die klassische Zahlentheorie. First printed in *Archiv für mathematische Logik*, vol. 16 (1974), pp. 97–118.

Gentzen, G. (1936) Die Widerspruchsfreiheit der reinen Zahlentheorie. *Mathematische Annalen*, vol. 112, pp. 493–565.

Gentzen, G. (1936a) Der Unendlichkeitsbegriff in der Mathematik. *Semester-Berichte Münster, WS 1936/37*, pp. 65–80. English translation in Menzler-Trott (2007), pp. 343–350.

Gentzen, G. (1938a) Die gegenwärtige Lage in der mathematischen Grundlagenforschung. *Forschungen zur Logik und zur Grundlegung der exakten Wissenschaften*, vol. 4, pp. 1–18.

Gentzen, G. (1938b) Neue Fassung des Widerspruchsfreiheitsbeweises für die reine Zahlentheorie. *Forschungen zur Logik und zur Grundlegung der exakten Wissenschaften*, vol. 4, pp. 19–44.

Gentzen, G. (1969) *The Collected Papers of Gerhard Gentzen*, ed. M. Szabo.

Gentzen, G. (2008) The normalization of derivations. *The Bulletin of Symbolic Logic*, vol. 14, pp. 245–257.

Gödel, K. (1933) The present situation in the foundations of mathematics. In Gödel (1995), pp. 45–53.

Gödel, K. (1938) Lecture at Zilsel's. In Gödel (1995), pp. 86–113.

Gödel, K. (1941) In what sense is intuitionistic logic constructive? In Gödel (1995), pp. 189–200.

Gödel, K. (1995) *Collected Works*, vol. 3. Oxford U. P.

Hertz, P. (1923) Über Axiomensysteme für beliebige Satzsysteme. Teil II. *Mathematische Annalen*, vol. 89, pp. 76–102.

Heyting, A. (1930) Die formalen Regeln der intuitionistischen Logik. *Sitzungsberichte der Preussischen Akademie von Wissenschaften, Physikalisch-mathematische Klasse*, pp. 42–56.

Hilbert, D. (1926) Über das Unendliche. *Mathematische Annalen*, vol. 95, pp. 161–190.

Hilbert, D. and P. Bernays (1934, 1939) *Grundlagen der Mathematik I–II*. Springer.

Kanckos, A. (2011) *A Possible and Necessary Consistency Proof*. PhD Thesis, University of Helsinki.

Kleene, S. (1952) *Introduction to Metamathematics*, North-Holland.

Kleene, S. (1952a) Permutability of inferences in Gentzen's caluli LK and LJ. *Memoirs of the American Mathematical Society*, No. 10, pp. 1–26.

Kreisel, G. (1976) Wie die Beweistheorie zu ihren Ordinalzahlen kam und kommt? *Jahresbericht der Deutschen Mathematiker-Vereinung*, vol. 78, pp. 177–223.

Kreisel, G. (1987) Gödel's excursions into intuitionistic logic. In P. Weingartner and L. Schmetterer, eds, *Gödel Remembered*, pp. 65–186, Bibliopolis, Naples.

Menzler-Trott, E. (2007) *Logic's Lost Genius: The Life of Gerhard Gentzen*. American Mathematical Society.

von Plato, J. (2006) Normal derivability and existence property in Heyting arithmetic. *Acta Philosophica Fennica*, vol 78 (2006), pp. 159–163.

von Plato, J. (2008) Gentzen's proof of normalization for natural deduction. *The Bulletin of Symbolic Logic*, vol. 14, pp. 240–244.

von Plato, J. (2009) Gentzen's original proof of the consistency of arithmetic revisited. In G. Primiero and S. Rahman (eds.) *Acts of Knowledge - History, Philosophy and Logic*, pp. 151–171, College Publications.

von Plato, J. (2012) Gentzen's proof systems: byproducts in a work of genius. *The Bulletin of Symbolic Logic*, vol. 18, pp. 313–367.

Prawitz, D. (1965) *Natural Deduction: A Proof-Theoretical Study*. Almqvist & Wicksell.

Raggio, A. (1965) Gentzen's Hauptsatz for the systems NI and NK, *Logique et analyse*, vol. 8, pp. 91–100.

Siders, A. (2015) A direct Gentzen-style consistency proof for Heyting arithmetic. This volume.

Siders, A. and J. von Plato (2015) Bar induction in the proof of termination of Gentzen's reduction procedure. This volume.

Tait, B. (2005) Gödel's reformulation of Gentzen's first consistency proof of arithmetic for arithmetic: the no-counterexample interpretation. *The Bulletin of Symbolic Logic*, vol. 11, pp. 225–238.

# Appendix: Bar Induction in the Proof of Termination of Gentzens Reduction Procedure

**Annika Siders and Jan von Plato**

## 1 Introduction

We shall give an explicit formulation to the use of bar induction in Gentzen's original proof of consistency, as a continuation of the analysis in the preceding essay about the *Hilfssatz*, referred to here as *HH*.

The article Bernays (1970) was the first one to explain in print the ideas in Gentzen's original proof of consistency, and it also made clear that the proof was in the end based on bar induction. There is a review of Bernays' article by Joseph Shoenfield in which the latter writes that "the progress made in formalizing intuitionistic systems in recent years should make it possible to formalize this proof and thus see exactly what intuitionistic principles are needed to carry it out" (*Mathematical Reviews*, MR0276062).

## 2 Bar Induction in the 1935 Proof

We prove that derivable sequents reduce to endform. As the basic predicate $B$ in the induction, the property is used that the succedent of a derivable sequent is an atomic formula, here an equation. For the inductive predicate $I$, we use the property that a derivable sequent with an atomic formula as a succedent reduces to endform. For the proof, we show first that reduction steps in the succedent preserve the derivability of a sequent:

**Lemma** *If* $\Gamma \rightarrow C$ *is a derivable sequent and an* **S**-*move is applied to it, a derivable sequent is obtained.*

A. Siders • J. von Plato
Department of Philosophy, University of Helsinki, Helsinki, Finland
e-mail: annika.siders@helsinki.fi; jan.vonplato@helsinki.fi

We go through the possible **S**-moves in turn:

**SVar.** If $\Gamma \to C$ has free variables, numbers are chosen at will to instantiate these until there are no free variables left. Derivability is maintained under substitution so that the reduced sequent is derivable.

**S&.** The sequent is $\Gamma \to A \,\&\, B$, and both of the reduced sequents $\Gamma \to A$ and $\Gamma \to B$ are derivable by rule $\&E$.

**S¬.** The sequent is $\Gamma \to \neg A$. The following derivation by the rules of the calculus *NLK* shows that $A, \Gamma \to 0 = 1$ is derivable, with *Wk*, *Ref*, and *DN* standing for the rules of weakening, refutation, and elimination of double negation, respectively:

$$\dfrac{\dfrac{\dfrac{A \to A}{\neg 0 = 1, A \to A}\,Wk \qquad \dfrac{\Gamma \to \neg A}{\neg 0 = 1, \Gamma \to \neg A}\,Wk}{A, \Gamma \to \neg\neg 0 = 1}\,Ref}{A, \Gamma \to 0 = 1}\,DN$$

**S∀.** The sequent is $\Gamma \to \forall x A(x)$, and any instance $\Gamma \to A(n)$ is derivable by rule $\forall E$. QED.

**Theorem** *Derivable sequents reduce to endform.*

For a proof, we go through the four conditions for bar induction:

1. $B$ has to be decidable. This is the case.
2. For any given derivable sequent $\Gamma \to C$ and any sequence of reduction steps, there is a step in the sequence by which the succedent formula has turned into an equality. To show this, consider the reductions steps: If there are free variables in $\Gamma \to C$, move **Svar** must be applied first, to substitute them by constants. Thereafter the other **S**-moves must be applied, each producing a shorter formula in the succedent until it is an equation.
3. Given a derivable sequent such that each applicable reduction step produces a sequent that reduces to endform, to show that the sequent before the reduction reduces to endform. This is immediate.
4. Finally, it has to be shown that if a derivable sequent has been reduced so that it has the property $B$, i.e., is of the form $\Gamma \to m = n$, it is a derivable sequent that reduces to endform. The derivability part follows by the lemma. The rest is an induction on the last rule in the derivation of $\Gamma \to m = n$. If $m = n$ is true, the sequent is in endform. Therefore we may assume $m = n$ to be false.

The possible cases are:

4.1. $\Gamma \to m = n$ is an initial sequent. Then the antecedent is the false equation $m = n$ and the sequent in endform.

4.2. $\Gamma \to m = n$ is a "mathematical groundsequent," for which we take the formulation with free parameters, as in *HH*, Section IV.4, with all free variables removed by steps of **Svar**:

$$\begin{array}{l}\to m = m, \quad n = m \to m = n, \quad m = k, k = n \to m = n,\\ k + 1 = k \to 0 = 1, \quad \to h + k = k + h, \quad \to (h + k) + l = h + (k + l).\end{array}$$

The reflexivity groundsequent is in endform and symmetry has a false antecedent $n = m$ whenever the succedent $m = n$ is false. With transitivity, if $m = n$ is false, if $m = k$ in the antecedent is true, then $k = n$ in the antecedent is false and similarly if $k = n$ is true. With $k + 1 = k \rightarrow 0 = 1$, the antecedent is false, and for the rest, the succedent is true.

4.3. The last rule is a logical one. There are the cases $\&E, \forall E$, and *DN*.

4.3.1. The last rule is $\&E$:

$$\frac{\Gamma \rightarrow A \,\&\, m = n}{\Gamma \rightarrow m = n}\,\&E$$

The premiss reduces to endform by assumption, and therefore also the conclusion. The reduction is similar if the second form of rule $\&E$ is applied.

4.3.2. The last rule is $\forall E$:

$$\frac{\Gamma \rightarrow \forall x.\, x = n}{\Gamma \rightarrow m = n}\,\forall E$$

The premiss reduces to endform by assumption, and therefore also the conclusion. The reduction is similar if the right member of the equation was quantified.

4.3.3. The last rule is *DN:*

$$\frac{\Gamma \rightarrow \neg\neg m = n}{\Gamma \rightarrow m = n}\,DN$$

The first step of reduction for the premiss gives $\neg m = n, \Gamma \rightarrow 0 = 1$. If step $\mathbf{A}\neg$ is applied to $\neg m = n$, the reduced sequent is $\neg m = n, \Gamma \rightarrow m = n$ with a false equation in the succedent. Therefore some other reduction step must be applied, and if $\mathbf{A}\neg$ is applied at some later stage to $\neg m = n$, a similar useless loop is produced. Therefore $\neg m = n$ in the antecedent can be left intact and $\Gamma \rightarrow m = n$ reduces to endform by the same steps as the sequent $\neg m = n, \Gamma \rightarrow 0 = 1$.

4.4. The last rule is *CI* with $\Gamma \equiv \Gamma', \Gamma''$ and the conclusion $\Gamma', \Gamma'' \rightarrow m = n$:

$$\frac{\Gamma' \rightarrow m = 0 \quad m = x, \Gamma'' \rightarrow m = x + 1}{\Gamma', \Gamma'' \rightarrow m = n}\,CI$$

If $m = 0$ is false, the conclusion reduces to endform by the same steps as $\Gamma' \rightarrow m = 0$. If $m = 0$ is true, **Svar** gives in particular for the second premiss the reducible sequent $m = 0, \Gamma'' \rightarrow m = 0 + 1$ with a false succedent. The steps of reduction leave the true equation $m = 0$ intact and apply as well for the reduction of $\Gamma', \Gamma'' \rightarrow m = n$.

By 1–4, the conditions for bar induction are satisfied and all derivable sequents have the property $I$, i.e., reduce to endform. QED.

## References

Bernays, P. (1970) On the original Gentzen consistency proof for number theory. In J. Myhill et al., eds, *Intuitionism and Proof Theory*, pp. 409–417, North-Holland.

Shoenfield, J. (1972) Review of Bernays (1970). *Mathematical Reviews*, MR0276062.

# A Note on How to Extend Gentzen's Second Consistency Proof to a Proof of Normalization for First Order Arithmetic

**Dag Prawitz**

**Abstract** The purpose of this note is to show that the normalization theorem can be proved for first order Peano arithmetic by adapting to natural deduction the method used in Gentzen's second consistency proof. Gentzen explained the intuitive idea behind his proof by informally arguing for the possibility of a normalization theorem of natural deduction, but what he actually proved was a special case of the Hauptsatz for a sequent calculus formalization of arithmetic.

To transfer Gentzen's method to natural deduction, I shall assign his ordinals to notations for natural deductions that use an explicit operation of substitution. The idea is first worked out for predicate logic. The main problems reside there and consist in finding a normalization strategy that harmonizes with the ordinal assignment. The result for predicate logic is then extended to arithmetic without effort, and thereby full normalization of natural deductions in first order arithmetic is achieved.

## 1 Introduction

Gentzen's two most important results, his Hauptsatz (cut elimination theorem) and his consistency proof for arithmetic, were both clearly inspired by insights that he got by reflecting on his system of natural deduction. This becomes especially clear when Gentzen [6][1] explains the basic idea behind his second[2] published consistency proof.

[1]In the sequel, I shall refer to pages in the original German paper "Neue Fassung des Widerspruchsfreiheitsbeweises für die reine Zahlentheorie" by writing Gentzen [6] and to pages in the English translation of the paper in *The Collected Papers of Gerhard Gentzen* [7] by writing Gentzen [7].

[2]Counting a first proof of the consistency of elementary number theory that Gentzen withdrew from publication after its planned publication had advanced as far as to galley proofs (see [7] or [2]), this is really his third consistency proof.

D. Prawitz (✉)
Filosfiska institutionen, Stockholms universitet, Universitetsvaegen 10D, 10691 Stockholm, Sweden
e-mail: dag.prawitz@philosophy.su.se



R. Kahle, M. Rathjen (eds.), *Gentzen's Centenary*,
DOI 10.1007/978-3-319-10103-3_6

In this proof, Gentzen shows that there can be no derivation of a contradiction in the sequent calculus that he had set up for arithmetic (or elementary number theory as he called it), because any such derivation would be reducible to a simpler derivation of the same thing; more precisely, he assigns (transfinite) ordinal numbers to the derivations and shows that as long as the derivations have any logical complexity, the ordinals are lowered by suitably chosen reductions of them. Before going into the technical details of the proof, Gentzen gives a lucid account of why it should always be possible to simplify such a derivation. The account is given in the form of an argument that refers to some crucial features of his system of natural deduction, and runs more or less as follows.[3]

A contradiction can certainly not arise as long as one only proceeds according to the rules set up for arithmetical identities or other atomic sentences. A derivation that ends in a contradiction must therefore contain logically compound sentences. Somewhere in the derivation there must then appear a sentence of maximal complexity. In general, the only way in which such a "complexity extremum" can arise is by a sentence that enters into the derivation by the application of an introduction inference and is then used in a subsequent elimination inference. But it is reasonable to assume that one could then as well go directly from the premisses of the introduction to the conclusion of the elimination. One would thereby remove the intermediate sentence, standing between the introduction and the elimination, which is of higher complexity than the surrounding ones. This would lower the peak of the derivation.

In a footnote, Gentzen remarks at this point: "precisely the same line of thought, incidentally, underlies the proof of the 'Hauptsatz' of my dissertation." One could remark, even more to the point, that precisely this line of thought is the idea behind the normalization theorem of natural deduction, which says precisely that *maximum formulas*, that is, formula occurrences that stand as the conclusion of an introduction inference and as the major premiss of an elimination inference, can be removed from the deduction.[4]

Gentzen then goes on saying that in fact, the situation is not as simple as in the sketched argument, because, in the case of number theory, logically compound sentences can be inferred not only by the application of logical rules but also by the use of mathematical induction. Although they can be reduced in an obvious way when the term $t$ in the inferred sentence $A(t)$ is a numeral ($t$ standing at the relevant argument place for the induction), no reduction can be made if $t$ is a variable. This means, he says, that it may not be possible to perform a reduction at the very peak of the derivation, but he reassures: "It is nonetheless possible in each case to locate a formula in the derivation which represents a 'relative extremum', viz., a formula

[3] Gentzen [6, pp. 26–28], [7, pp. 261–263].

[4] It has recently been revealed that Gentzen was not only aware of the possibility of such a theorem, but that, in an early draft of his dissertation (found in Bernays' archive, see [27]), he also stated and proved the theorem for intuitionistic logic, essentially in the way it was later proved (Prawitz [20]—the independent proof by Raggio [21] is a little different).

which is introduced by the introduction of its terminal connective and whose further use in the derivation then consists in the elimination of that connective, and which is therefore reducible."

After this beautiful exposition of the underlying idea of his proof, Gentzen says somewhat disappointingly that the basic idea has been presented against the background of natural deduction and cannot be more than a superficial indication of the actual proof, which will be carried out, not for natural deduction, but for the sequent calculus. Earlier in the paper[5] he had motivated his choosing this formalism instead of natural deduction by giving two reasons. One was the problem in natural deduction caused by the special position of a classical law of negation that has to be added to the intuitionistic system, which is, he says, "completely removed in a seemingly magical way"[6] by going over to sequent calculus. The second reason was that the natural succession of sentences in an informal proof, which is by and large retained in natural deduction, is replaced in sequent calculus by an artificial arrangement that can be made with respect to certain aims, which proves to have technical advantages in the consistency proof. Having made these remarks, Gentzen proceeds to the precise consistency proof, which turns out to involve a lot of technical complications.

Gentzen's proof can be described as establishing the Hauptsatz for the special case of derivations that end in a contradiction (technically the same as ending in the empty sequent): It is shown that such derivations, if there were any, would be reducible to cut free form, which demonstrates that there really are not any. When Gentzen after having finished his first published consistency proof continued to work on another version of the proof, his hope was most likely to obtain the consistency as a corollary of the general Hauptsatz for arithmetic.[7]

Today we know that the addition of a classical rule to intuitionistic natural deduction does not need to cause any real problem with respect to normalization,[8] and we have a lot of experiences of how to prove normalization theorems. It is therefore natural to ask if the idea behind Gentzen's proof could not after all be

---

[5]Gentzen [6, pp. 24–25]; Gentzen [7, pp. 259–260].

[6]Gentzen [7, p. 259]. The German text reads: "... [die Sonderstellung der Negation ist] ... auf eine fast wie Zauberei anmutende Weise vollständig behoben", Gentzen [6, p. 25].

[7]This presumption is supported by what has been found in Gentzen's Nachlass by Jan von Plato. It turns out that at an early stage of the work on his dissertation, Gentzen had expected to obtain the consistency of arithmetic directly from a normalization theorem for natural deduction. When that failed, he restricted his dissertation essentially to predicate logic, and then proved the consistency of arithmetic along other lines. But it is clear that he did not abandon his original idea. A witness to this is even found in his plans for a book on the foundations of mathematics. In a notebook concerned with these plans, he writes: "to assimilate the proof of the hillock theorem to the proof of consistency" (translation by von Plato); "the hillock theorem" is here Gentzen's name for the normalization theorem of natural deduction ("der Hügelsatz" in German—but in other contexts often "der Gipfelsatz").

[8]If we choose $\bot$, &, $\supset$, and $\forall$ as logical constants, the normalization theorem for classical logic even takes a simpler form than for intuitionistic logic [20].

applied directly to natural deductions so as to yield the full normalization theorem for arithmetic.[9] The purpose of this note is to show that the question can be answered positively.

Proofs of the normalization theorem for natural deductions have already been given by Jervell [9] and Martin-Löf [13] (and later by Leivant [12]). Their proofs used quite different means than those of Gentzen, and do not answer the question whether a normalization theorem for arithmetic can be obtained by combinatorial means of the kind employed by Gentzen. In other later works, Gentzen's result has been reworked and extended, usually by making excursions into infinite derivations (see Sect. 2.1). In this note, I show how Gentzen's method can be transferred directly to natural deduction and how they can then be used to obtain a stronger result, the full normalization theorem.

This also throws some additional light upon the relation between natural deduction and sequent calculus. Results for sequent calculus often have analogues for systems of natural deduction that are more easily established there. It has therefore been puzzling why Gentzen's second consistency proof has been so difficult to carry over to natural deduction. At this point it should be said that Gentzen was quite right about the second reason that he invoked for preferring the sequent calculus: it allows a greater flexibility as to how the inferences of a proof may be ordered. To get the desired result for natural deduction, I have found it necessary to bring in an explicit operation of substituting one deduction for an assumption in another deduction, or, what is the same, composing two natural deductions.[10] When such an operation is made explicit in a system of natural deduction, one gets essentially the same flexibility with respect to the ordering of inferences as one has in the sequent calculus.

The rest of the paper is organized as follows. In Sect. 2, I explain the problems that one meets when trying to prove either the normalization theorem or the Hauptsatz for arithmetic. Gentzen's assignment of ordinals is shown to be a natural attempt to deal with one of the major problems (besides allowing transfinite induction up to a sufficiently high ordinal). Its significance in this respect is most easily seen if one restricts oneself to predicate logic. The finite ordinal assigned to a deduction $\mathcal{D}$ can then be seen as an estimation of an upper bound on the length of the normal deduction to which $\mathcal{D}$ reduces. The remaining main problems appear already when one stays within predicate logic and consist essentially in problems

[9]This question has been raised by several people, but has remained unanswered. Recently, Kanckos [10, 11] showed that a closed natural deduction of $\perp$ in a system for Heyting arithmetic would reduce to normal form by using vectors from Howard [8], instead of Gentzen's ordinal assignment. For my own part, I outlined an approach to a positive answer in lectures at Stockholm University 1979, a conference at Oxford 1980, and one at Siena 1984, and, in more detail, in (professor Ettore Casari's Saturday) seminars at Università degli Studi di Firenze 1991. The approach was not brought to a conclusion at these times, but agreed with the present solution in being built on the idea of adding an operation of explicit substitution.

[10]Von Plato [28] has also drawn attention to the key position of this operation (under the name composition) when comparing natural deductions and sequent calculus derivations.

about how to harmonize the ordinal assignment with a normalization strategy. In Sect. 3, it is described how one of these problems can be tackled by bringing in the explicit substitution operation mentioned above as a technical device. After the normalization theorem for predicate logic is being proved in this manner in Sect. 4, the result is extended to arithmetic in Sect. 5 in an effortless way.[11] The analogue of the obstacle that prevented Gentzen's consistency proof from being a proof of the general Hauptsatz is overcome here, and full normalization is achieved. In a final Sect. 6, explicit substitution is considered as an operation of independent interest, and a normalization theorem is proved for an enriched system of natural deduction for arithmetic containing explicit substitution as an inference rule.

# 2 The Problems and How to Overcome Them

## *2.1 To Find a Suitable Induction Measure*

If first order arithmetic is embedded in a suitable infinitary system, for instance, replacing the rule of induction by the $\omega$-rule, the normalization theorem or the Hauptsatz may be proved with the same ease as for first order predicate logic: Using reductions that replace a maximum formula or cut formula by ones of lower degree, one can make a primary induction over the maximal degree of the maximum or cut formulas and a secondary induction over the length of the deduction; in the infinitary case, the length will be a transfinite ordinal.[12]

When staying within a finitary system of first order arithmetic with the usual rule of mathematical induction, such a strategy is not possible, because an $\forall$-reduction may have the effect that an inference by mathematical induction becomes reducible, and its reduction may in turn create maximum or cut formulas of higher degree than the removed one. This is the kind of problem that Gentzen referred to in the account of the basic idea behind his proof quoted in Sect. 1. One must thus find another measure to make induction over, and this is what was created by Gentzen's assignment of ordinals.[13]

---

[11]After finishing this paper, I have found that Gentzen planned to organize the book mentioned in footnote 7 in the same way, first treating predicate logic using a finite version of his ordinal assignment and then extending the result to arithmetic using transfinite ordinals.

[12]Examples of such systems, essentially like classical sequent calculi, are found in, for instance, Schütte [25] and Tait [26]. The latter considers not only inference rules with infinitely many premisses but also infinitely long sentences. Martin-Löf [14] develops an intuitionistic system of natural deduction of that kind.

[13]An alternative is to enrich the infinitary system with information allowing one to extract a finitary normal derivation from the normalized infinitary one, as was first outlined by Mints [15]. This general idea has later been worked out in more detail and in different ways by Buchholz [3] and Mints [18]. Another alternative is presented by Mints [16], who defines other reductions and another ordinal assignment.

The transfinite ordinals needed to prove the normalization theorem for arithmetical systems of different strengths come out very naturally in the case of infinitary systems. In contrast, Gentzen's ordinal assignment is sometimes deemed to be somewhat artificial or ad hoc. However, his assignment when adapted to predicate logic can be understood as a straightforward estimation of the length of the normal deduction to which a deduction reduces. As will be explained (Sect. 2.3), the assignment appears as an easily obtained improvement of a better known bound on how much the length of a deduction increases when normalized.[14] In a subsequent subsection (Sect. 2.4.1), I explain how such an assignment can be a useful measure to make induction over when one wants to prove that reduction sequences terminate. The next subsection is only a background to the explanation of Gentzen's assignment that follows afterwards.

## 2.2 An Upper Bound on the Lengths of the Normalized Deductions

### 2.2.1 A Well-known Fact

It is well known that the size of a deduction may grow exponentially when normalized and that there is the following upper bound, where length is used as a measure of size:

*A natural deduction $\mathcal{D}$ of length n reduces to a normal deduction of length less than*

$$2^{2^{\cdot^{\cdot^{\cdot^{2^n}}}}} \qquad \textit{the base } 2 \textit{ is to occur } d \textit{ times in the tower}$$

*where $d > 0$ is the highest degree of a maximum formula in $\mathcal{D}$.*

Usually I shall write an iterated exponentiation of this kind in the more linear form $2_d(n)$.

The systems of natural deduction that I consider here are the usual ones but confined to introduction and elimination rules for &, $\supset$, and $\forall$ and arbitrary rules for atomic formulas, among which may be the classical rule for falsehood, $\bot E_C$, restricted by the requirement that the conclusion be atomic, which is sufficient for classical logic (a system called C' in [20], but $\bot E_C$ is now restricted). By an inference rule for atomic formulas, I mean a rule such that the premisses and the conclusion of an application of the rule are atomic.

[14]The better known bound is credited to Schütte [25], who noted that the analogue to the fact stated in Sect. 2.2.1 holds for his infinitary system mentioned in footnote 12; the length $n$ is then in general a transfinite ordinal. The statement in Sect. 2.2.1 was proved for Gentzen's intuitionistic system of natural deduction by Pereira [19]. Cellucci [4] establishes several results concerning how the length of natural deduction increases by normalization, including negative results on how much the upper bound can be improved.

A deduction is *normal* when it has no maximum formula. The *degree of a formula* $A$, written degr($A$), can be defined as the number of constants &, $\supset$, and $\forall$ in $A$ within the scope of each other (although it is only $\supset$ that matters at the moment); the definition will be modified in Sect. 6.4. The *degree of a deduction* is defined as the highest degree of a maximum formula in the deduction; the degree is 0 if there are no such occurrences. By the *length* of a deduction, I understand its number of nodes or, in other words, the number of formula occurrences.

The normal deduction to which a deduction $\mathcal{D}$ reduces will be denoted by $|\mathcal{D}|$. In order to see how Gentzen's assignment improves the bound given above slightly, it is instructive first to see why $2_d(n)$ is an upper bound for the length of $|\mathcal{D}|$, when $\mathcal{D}$ is a deduction of length $n$ and degree $d$. To this end, I first make the following easy observation.

### 2.2.2 Observation Concerning ⊃-Reductions

Given a deduction $\mathcal{D}$ of the form shown to the left below, where $\Pi$ is a deduction of $A$ with length $k$ and $\Sigma$ is a deduction of $B$ with length $m$, the deduction $\mathcal{D}^*$ shown to the right, obtained from $\mathcal{D}$ by implication reduction, has length less or equal to $k \cdot m$

$$\mathcal{D} = \quad \dfrac{\dfrac{\Pi}{A} \qquad \dfrac{\begin{matrix}[A]\\ \Sigma\\ B\end{matrix}}{A \supset B}}{B} \qquad\qquad \mathcal{D}^* = \quad \begin{matrix}\Pi\\ [A]\\ \Sigma\\ B\end{matrix}$$

The correctness of the observation follows immediately from the fact that there are less or equal to $m$ occurrences of $A$ in $\Sigma$.

The notation $[A]$ in the left figure is used to indicate the formula occurrences of the form $A$ that stand as free assumptions in $\Sigma$ but are bound (discharged) in $\mathcal{D}$ by the $\supset$-introduction exhibited. The right figure is to be understood as denoting the deduction $\mathcal{D}^*$ obtained from $\Sigma$ by substituting $\Pi$ for each of the assumptions indicated by $[A]$; in other words, for each such assumption $A$ in $\Sigma$, the deduction $\Pi$ is put on the top of $\Sigma$ in place of $A$.[15]

[15]As long as there is no line under $\Pi$ that separates it from $A$, the deduction $\Pi$ is taken to include $A$. We can choose either to make explicit that $\Pi$ has $A$ as its end-formula by writing

$$\begin{matrix}\Pi\\ A\end{matrix}$$

—sometimes written $\Pi/A$ to keep it on one line—or leave that out, writing just $\Pi$ (in the same way as we may indicate the free occurrences of $x$ in a formula $A$ by writing $A(x)$ but may also leave that implicit, writing just $A$). A linear notation for the result of substituting $\Pi/A$ for the assumptions $[A]$ in $\Sigma$ is: $\Pi/[A]/\Sigma$.

### 2.2.3 Lowering the Degree of a Deduction Stepwise

As we go downwards in a deduction and successively remove by reductions the maximum formulas of highest degree, the lengths of the obtained deductions may thus be estimated by successively squaring the previous lengths, which means that the estimated length grows exponentially. More precisely, the following holds.

A deduction $\mathcal{D}$ of degree $d > 0$ and length $n$ can be reduced to a deduction $\mathcal{D}'$ of degree $d' < d$ and length $n' < 2^n$.

From this follows of course the fact quoted at the beginning of Sect. 2.2.1. The assertion is easily proved by induction over the number of maximum formulas of degree $d$ in $\mathcal{D}$. It will be of interest to strengthen the conclusion a little to get: A deduction of degree $d > 0$ and length $n + 1$ ($n > 0$) can be reduced to a deduction of degree $d' < d$ and length $n' < 2^n + 1$.

Let $\mathcal{D}$ be a deduction of degree $d$ and length $n + 1$ with a lowermost maximum formula $F$ of degree $d$, which is to say that there is no other maximum formula of degree $d$ below $F$ in $\mathcal{D}$. Let $\mathcal{D}_1$ be the part of $\mathcal{D}$ (which may coincide with $\mathcal{D}$) whose end-formula is the formula immediately below F. The crucial case is when $F$ has the form of an implication $A \supset B$. Then $\mathcal{D}_1$ is of the form shown to the left in the figure displayed above with the length $n_1 + 1 = k + m + 2$, where $k$ and $m$ are the lengths of the parts $\Pi$ and $\Sigma$ of $\mathcal{D}_1$. By the induction assumption (to be applied only when $\Pi$ or $\Sigma$ is of degree $d$, and then using the first, weaker statement), they reduce to deductions $\Pi'$ and $\Sigma'$ of degree less than $d$ and lengths $k' < 2^k$ and $m' < 2^m$.

Let $\mathcal{D}'$ be the deduction to which $\mathcal{D}$ reduces by carrying out the same reductions as those by which $\Pi$ and $\Sigma$ reduce to $\Pi'$ and $\Sigma'$. Then $\mathcal{D}'$ has the form shown to the left below and reduces to the deduction $\mathcal{D}'^*$ of the form shown to the right below:

$$\mathcal{D}' = \begin{matrix} \mathcal{D}_1' \\ B \\ \mathcal{D}_0 \end{matrix} = \dfrac{\begin{matrix} \Pi' \\ A \end{matrix} \quad \dfrac{\begin{matrix} [A] \\ \Sigma' \\ B \end{matrix}}{A \supset B}}{\begin{matrix} B \\ \mathcal{D}_0 \end{matrix}} \quad \text{reduces to} \quad \mathcal{D}'^* = \begin{matrix} \mathcal{D}_1'^* \\ B \\ \mathcal{D}_0 \end{matrix} = \begin{matrix} \Pi' \\ [A] \\ \Sigma' \\ B \\ \mathcal{D}_0 \end{matrix}$$

Applying the observation concerning $\supset$-reductions made above, we get the following result concerning the length $n_1'^*$ of $\mathcal{D}'^*$:

$$n_1'^* < k' \cdot m' < 2^k \cdot 2^m = 2^{k+m} < 2^{k+m+1} + 1 = 2^{n_1} + 1.$$

The analogue result when $F$ is a conjunction or a universal quantification is obtained more trivially, since in that case the length decreases when $\mathcal{D}'$ is reduced to $\mathcal{D}'^*$, as seen below for $\forall$:

$$\mathcal{D} = \begin{array}{c} \Pi(a) \\ A(a) \\ \hline \forall x A(x) \\ \hline A(t) \\ \mathcal{D}_0 \end{array} \text{ reduces to } \mathcal{D}' = \begin{array}{c} \Pi'(a) \\ A(a) \\ \hline \forall x A(x) \\ \hline A(t) \\ \mathcal{D}_0 \end{array},$$

$$\text{which reduces to } \mathcal{D}'^{*} = \begin{array}{c} \Pi'(t) \\ A(t) \\ \mathcal{D}_0 \end{array}$$

If $F$ is the only lowermost maximum formula in $\mathcal{D}$ of degree $d$ and $\mathcal{D}_1$ coincides with $\mathcal{D}$, we have proved what we want. If $\mathcal{D}_1$ is a proper part of $\mathcal{D}$, then the length of $\mathcal{D}$ is $n + 1 = n_1 + 1 + p$ with $p > 0$, and it holds for the length $n'^{*}$ of $\mathcal{D}'^{*}$ that $n'^{*} = n_1'^{*} + p < 2^{n_1} + 1 + p < 2^{n_1+p} + 1 = 2^n + 1$.

If there are other lowermost maximum formulas $F_i$ of degree $d$ in $\mathcal{D}$ besides $F$, we repeat the procedure for them. Let there be altogether $q > 1$ such formulas whose deductions $\mathcal{D}_i$ are of the length $n_i + 1$. The length of $\mathcal{D}$ is then $n + 1 = n_1 + 1 + n_2 + 1 + \cdots + n_q + 1 + p$, and the deduction we finally arrive at has a degree less than $d$ and a length less than $2^{n_1} + 1 + 2^{n_2} + 1 + \cdots + 2^{n_q} + 1 + p$, which is less than $2^n + 1$, as was to be shown.

## 2.3 The Gentzen Measure: A Lower Upper Bound—Definition of Critical Inferences

The bound on the length of the normal deduction given in Sect. 2.2.1 can be improved as is easily seen by inspecting the proof above. I first consider an example of this.

Let $\mathcal{D}$ be a natural deduction of the following form

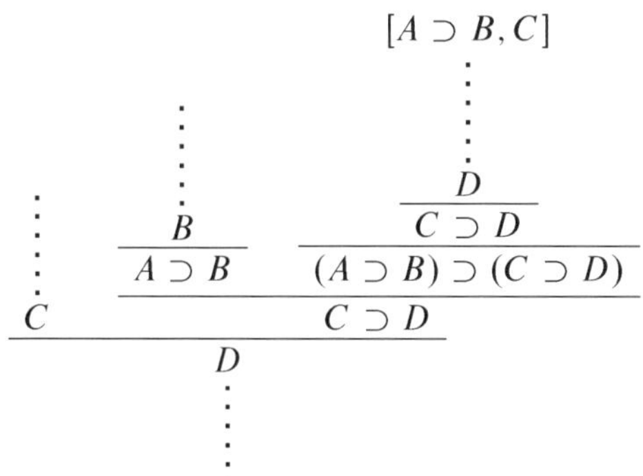

where $A$, $B$, $C$, and $D$ are atomic formulas and where there are arbitrarily many maximum formulas of degree 1 or 2 above the exhibited maximum formula $(A \supset B) \supset (C \supset D)$ and above $A \supset B$ but no other maximum formulas. Let $\alpha$, $\beta$, and $\gamma$ be the number of nodes of the upper parts of $\mathcal{D}$ that constitute deductions of C, $(A \supset B)$, and $(A \supset B) \supset (C \supset D)$, respectively, and let $\delta$ be the number of other nodes of $\mathcal{D}$.

By applying the proof given in Sect. 2.2.3 (rather than the result in Sect. 2.2.1), it is seen that $\mathcal{D}$ reduces to a deduction $\mathcal{E}$ of degree 1 and length less than $2^{\beta+\gamma} + \alpha + \delta$. In $\mathcal{E}$ the implication $(C \supset D)$ is a maximum formula and its deduction has length less than $2^{\beta+\gamma} + 1$. A second application of the proof gives that $\mathcal{E}$ reduces to a normal deduction $\mathcal{F}$ of length less than $2^{\alpha+2^{\beta+\gamma}+1} + \delta$-1, which is to be compared to the bound $2^{2^{\alpha+\beta+\gamma+\delta}}$ given in Sect. 2.2.1.

This better estimation of the length of $\mathcal{F}$ will agree with what I shall call the *Gentzen measure* of the deduction $\mathcal{D}$, which will now be defined. It will amount to a generalization of the example above, and to that end we need Gentzen's notion of *level* ("Höhe") and *level line* ("Höhelinie").

Following Gentzen, we associate to each inference an *inference line* drawn between the premiss(es) of the inference and its conclusion. Some inference lines are level lines. To begin with, in a deduction $\mathcal{D}$ of degree $d$, the inference lines immediately below the lowermost maximum formulas of degree $d$ are to be level lines. Thus, in the example above the two formulas $A \supset B$ and $(A \supset B) \supset (C \supset D)$ stand on a level line.

The idea is that the Gentzen measure to be assigned to a formula occurrence $F$ standing immediately below one of these level lines is to be an estimation, according to the proof above, of the length of the deduction of $F$ obtained after having removed by reductions all maximum formulas of degree $d$ above the level line. Thus, the measure to be assigned to $F$ should be $2^{\alpha} + 1$, where $\alpha$ is the sum of the Gentzen measures of the premisses (which in this case coincide with the lengths of their deductions), or, rather, $2_j(\alpha) + 1$, where $d - j$ is the highest degree of a maximum formula below $F$, in order to estimate the length of the deduction when all maximum formulas of degree higher than $d - j$ above the level line have been eliminated. When determining the next level line below $F$, we must take into account that there may be formulas of degrees less than $d$ that will become maximum formulas as the result of the reductions that remove the maximum formulas of higher degrees. We want the Gentzen measure to be an estimation of how much the deduction expands by reductions that remove not only the original maximum formulas but also the additional maximum formulas that can arise after those reductions, and so on. This means that we must pay attention not only to maximum formulas but also to *potential* maximum formulas, in other words, formulas that may become maximum formulas as the result of reductions. In the example above, the occurrence of $C \supset D$ under the level line is such a potential maximum formula.

Ideally we should determine which formulas of a deduction $\mathcal{D}$ could become maximum formulas after reductions by only referring to structural properties of $\mathcal{D}$. I shall not try to do this here,[16] but shall instead consider all major premisses of elimination inferences as potential maximum formulas, only excepting those whose

[16]Such a definition of potential maximum formula could be given by using ideas presented by Sanz [22].

degree is greater than the degree $d$ of $\mathcal{D}$; they are certainly not potential maximum formulas. In other words, all elimination inferences whose maximum formula is of degree $<d$ are placed on a par with the ones whose major premiss is a maximum formula, although this may generate unnecessarily many level lines.[17]

To have a common term for these inferences, I shall say that an elimination inference in a deduction $\mathcal{D}$ of degree $d$ is *critical* if its major premiss is either a maximum formula or has a degree less than $d$. We shall later enlarge the category of critical inference, but the notion will also be somewhat narrowed down in Sect. 6.4. A critical inference is said to be of *degree d* if its major premiss is of the degree $d$ ($\bot\text{E}_\text{C}$-inferences, having degree 0, are inessential).

The *level* of a formula occurrence $A$ is now defined as the greatest degree of a critical inference whose conclusion stands below $A$; if there is no such inference, the level of $A$ is 0. An inference line is defined as a *level line* if the level $h$ of the formula(s) immediately above the line is (are) higher than the level $h'$ of the formula immediately below the line. The difference $j = h - h'$ will be called the *jump* at the level line, and I shall say that the inference in question *contains* a level line with jump $j$. This agrees with the previous explanation, since clearly an inference line is a level line if and only if it is the inference line of a critical inference of some degree $d$ such that the degree of each critical inference that stands below (if any) is less than $d$. If there are two formulas standing on a level line, they have the same level $h$, and $h$ is identical to the degree of the major premiss of the inference in question.

What I am calling the *Gentzen measure* of a formula occurrence $A$ in a natural deduction $\mathcal{D}$, written $G_\mathcal{D}(\text{A})$, can now be defined. Note that it depends on what counts as level lines, which in turn depends on what counts as critical inferences. Note also that it differs from Gentzen's assignment of ordinals in using 2 instead of $\omega$ as the base for the exponentiation; thus, the values assigned will be finite ordinals, instead of transfinite ordinals less than $\varepsilon_0$. The definition runs as follows:

1. If $A$ is a top-formula of $\mathcal{D}$, $G_\mathcal{D}(A) = 1$.
2. If $A$ is immediately below exactly one formula $B$, let $n$ be $G_\mathcal{D}(B)$. If $A$ is immediately below two formulas $B$ and $C$, let $n$ be $G_\mathcal{D}(B) + G_\mathcal{D}(C)$. Then,
   (a) $G_\mathcal{D}(A) = n + 1$, provided $A$ is not immediately below a level line, and
   (b) $G_\mathcal{D}(A) = 2_j(n) + 1$, in case $A$ is immediately below a level line with jump $j$.

The Gentzen measure of the deduction $\mathcal{D}$, written $G\ (\mathcal{D})$, is the same as $G_\mathcal{D}(A)$, where $A$ is the end-formula of the deduction $\mathcal{D}$. A normal deduction $\mathcal{D}$ has clearly no level lines, and its length is therefore identical to $G\ (\mathcal{D})$.

[17]In the sequent calculus for predicate logic, the level lines are determined by the actual cuts. An inference line is a level line if and only if it is the inference line of a cut such that all cuts further down have lower degree than that cut; the degree of a cut being defined as the degree of its cut formula. Since maximum formulas are what correspond to cut formulas, one could expect that the level lines in natural deduction should be similarly determined by the actual maximum formulas, but this would lead to an entirely wrong notion.

The already mentioned fact that the Gentzen measure of a deduction $\mathcal{D}$ is an (improved) upper bound on the length of the normal deduction to which it reduces, in short *for any non-normal deduction* $\mathcal{D}$, *the length of* $|\mathcal{D}| < G(\mathcal{D})$, can now easily be proved by the strategy used in the proof in Sect. 2.2.3, as was illustrated in the example.[18]

## 2.4 The Two Main Problems

### 2.4.1 How the Gentzen Measure Can Be Used to Prove that Reduction Sequences Terminate

The fact just stated above does not mean per se that the Gentzen measure can be used to prove by induction that reduction sequences terminate. When a reduction increases the length of the deduction, the Gentzen measure of the deduction may also go up. This surely happens if a maximum formula that stands above a level line, but not immediately above it, is removed by a length-increasing reduction in the usual way: the part of the deduction above the level line will expand but the level line will remain the same, and hence the Gentzen measure of the formula immediately below the line will increase.

However, Gentzen's idea was that when a reduction is made with respect to a cut, the new cuts and their derivations are to be put under the closest level line below the old cut. Then the part of the derivation under that level line will instead expand, but that is compensated for by a certain contraction of the part above the level line.

The idea can be illustrated more precisely using the ordinals that Gentzen assigned to sequents in a derivation, which are like the Gentzen measure defined above with some differences, the main of which is explained in footnote 17. Say that $S_l$ is the sequent immediately below a level line with jump 1 in a derivation $\mathcal{D}$ and that $\alpha$ is the sum of the ordinals assigned to the sequents immediately above the line. Then Gentzen assigns $\omega^{\alpha}$ to $S_1$ and an ordinal $\omega^{\alpha} + \beta$ to the sequent $S_2$ that stands immediately above the next level line further down. Let $\mathcal{D}^*$ be a reduction of $\mathcal{D}$ which leaves these level lines unchanged but removes a cut standing higher up above the first level line, replacing it with cuts of lower degree. If now, following Gentzen, the new cuts are placed under this first level line, the sequent corresponding to $S_1$ in $\mathcal{D}^*$ is assigned a value $\omega^{\alpha^*}$ where $\alpha^* < \alpha$, and the sequent $S_2$ is assigned a value $\omega^{\alpha^*} + \gamma + \beta$ where $\gamma < \omega^{\alpha^*}$, because the new cuts have lower degrees and their premisses have shorter derivations relative to the cuts that they replace. Thus, we achieve that the value assigned to $S_2$ in $\mathcal{D}^*$ is lower than that assigned to $S_2$ in $\mathcal{D}$, since

$$\omega^{\alpha^*} + \gamma + \beta < \omega^{\alpha^*} + \omega^{\alpha^*} + \beta < \omega^{\alpha^*+1} + \beta \leq \omega^{\alpha} + \beta.$$

[18] A similar result was proved for the sequent calculus for predicate logic by Pereira [19].

In other words, the value assigned to a derivation $\mathcal{D}$ is an approximation of the length of $|\mathcal{D}|$ that gets better and better at each reduction—the upper bound gets successively lower. In this way, it is accomplished that one can prove by induction over the Gentzen measure that suitably chosen reduction sequences must terminate.

The problem is, however, that not even in sequent calculus can inferences always be moved around as one may like. What Gentzen showed was that if the end-sequent is empty, a suitable cut can be found which can be removed by putting the new cuts replacing it under a level line situated below the old cut. This can be generalized quite easily to the case when the formulas of the end-sequent contain no quantifiers,[19] but is not easily generalized to the case when quantifiers are involved. Gentzen seems not to have overcome this problem, and therefore never proved the general Hauptsatz for arithmetic.

### 2.4.2 The Problem of ⊃-Reductions

The project to use the Gentzen measure in a proof of the normalization theorem of natural deduction meets an additional problem. Even when a deduction $\mathcal{D}$ does not expand at an ⊃-reduction, the Gentzen measure may anyway increase. The reduction $\mathcal{D}^*$ of $\mathcal{D}$ is formed, as we recall, by putting the deduction of the minor premiss $A$ of an ⊃-elimination (⊃E) of $\mathcal{D}$ at the top of the deduction of the major premiss $A \supset B$, while the conclusion $B$ of the ⊃E and the premiss $B$ of the ⊃-introduction (⊃I) standing above merge into one occurrence of $B$ in $\mathcal{D}^*$. What happens when one is to calculate the Gentzen measure of this occurrence of $B$ in $\mathcal{D}^*$ is therefore that the measure of $A$, which was added to the measure of the premiss $B$ to get the measure of the conclusion $B$ in $\mathcal{D}$, is instead being put as an exponent in the expression that was used to calculate the Gentzen measure of the premiss $B$ in $\mathcal{D}$. This may of course give $B$ in $\mathcal{D}^*$ a value greater than that given to the conclusion $B$ in $\mathcal{D}$. Clearly, the restructuring of the deduction that takes place at an ⊃-reduction tallies badly with the idea to prove by induction over the Gentzen measure that reductions terminate. This is in contrast to the situation in the sequent calculus where the elimination of cuts do not involve any restructuring of the derivation of this kind.

To see how this problem can be overcome, note that describing an ⊃-reduction requires the use of a sign for the substitution that takes place at such a reduction. A description of a deduction that uses a suitable notation for the operation of substitution may not need to be restructured at reductions in the same way as the deductions themselves, and can be more flexible as to the order in which inferences are presented. This opens for the possibility of assigning the Gentzen measure to such descriptions of deductions rather than to the deductions themselves, and to achieve in this way that the measure goes down at reductions.

This idea will be described in more detail in Sect. 3. We will have to define new reductions for the descriptions (Sect. 4.2), and some complications seem then

[19] A fact noted and used by Scarpellini [23].

unavoidable. On the other hand, within the framework of natural deduction, it turns out to be easy to define a strategy for normalization that overcomes what corresponds to the obstacle to proving a general Hauptsatz mentioned in the preceding subsection, thereby allowing us to prove a full normalization theorem.

Before developing these ideas further, I shall take up some questions concerning the relation between natural deduction and the sequent calculus.

## 2.5 On the Relation Between Natural Deduction and the Sequent Calculus

One may suggest that the problem to transfer Gentzen's work on the sequent calculus for arithmetic to natural deduction could easily be solved by asking what corresponds in natural deduction to Gentzen's ordinal assignment. Besides ignoring the problems discussed above, the suggestion overlooks the fact that there is no unambiguous answer to the raised question.

Derivations in the sequent calculus (SQ) can be translated to natural deductions in a straightforward way: an initial sequent $A \rightarrow A$ is translated to a deduction consisting of just $A$, when a succedent rule has been applied in SQ one applies the corresponding introduction rule to the end-formula in natural deduction, and when an antecedent rule has been applied one instead enlarges the natural deduction upwards by applying the corresponding elimination rule. If the cut rule is applied to two sequents, one joins together the corresponding natural deductions $\Pi/A$ and $\Sigma$, where $A$ corresponds to the cut formula, by using the substitution operation $\Pi/A/\Sigma$ (see footnote 15). This may result in $A$ becoming a maximum formula, but if the derivation in SQ is cut-free, it is translated in this way into a normal natural deduction.

A derivation in the sequent calculus may accordingly be seen as an instruction for how to build a corresponding natural deduction by working in two directions, downwards and upwards, and joining two deductions by the operation of substitutions in case cuts have been used in SQ. The instructions given by two different derivations in SQ may result in the same natural deduction; the order in which the deduction is to be constructed according to the instructions differs, but the result becomes the same [20, pp. 90–91].

If one asks about a natural deduction what ordinal Gentzen assigns to the corresponding derivation in the sequent calculus, the problem is therefore that there are many derivations with different measures that correspond to a given natural deduction. One may of course define a particular translation of natural deductions to derivations in SQ. There are at least two such translation procedures proposed in the literature. Gentzen [5] defined one that proceeds inductively on how a natural deduction is constructed downwards: an assumption $A$ is translated to the initial sequent $A \rightarrow A$, when an introduction rule has been applied in natural deduction the corresponding succedent rule is applied to the translation into SQ of the deduction(s) of the premiss(es), and when an elimination rule is applied, one proceeds from the translation of the deduction(s) of the premiss(es) by performing a cut with a suitable sequent that is derived by the use of the corresponding antecedent rule. For instance, in the case of ⊃-elimination the left figure below is translated to the right one:

$$\dfrac{\begin{matrix}\Gamma \\ \Pi \\ A\end{matrix} \qquad \begin{matrix}\Delta \\ \Sigma \\ A \supset B\end{matrix}}{B} \qquad\qquad \dfrac{\begin{matrix}\Sigma^* \\ \Delta \to A \supset B\end{matrix} \qquad \dfrac{\begin{matrix}\Pi^* \\ \Gamma \to A\end{matrix} \quad B \to B}{\Gamma, A \supset B \to B}}{\Delta, \Gamma \to B}$$

The obtained derivation in the sequent calculus will in this way use a cut each time an elimination rule is used in natural deduction, although the natural deduction may be normal.

The Gentzen measure defined above for natural deduction turns out to be fairly close to the ordinal that Gentzen assigns to the derivation in SQ obtained by this translation procedure (after having replaced ω by 2 as the base for exponentiation): the exponentiations that determine the ordinal that Gentzen assigns to a derivation depend on the actual cuts (footnote 17), while the exponentiations that determine the Gentzen measure defined for a natural deduction depend on the elimination inferences occurring in the deduction, which correspond to cuts according to the translation procedure. The main difference is caused by eliminations having been excepted from being counted as critical when their degrees are equal to or higher than that of the deduction and the major premiss is not a maximum formula.

The structure of the derivation $\mathcal{D}^*$ obtained by Gentzen's translation of a natural deduction $\mathcal{D}$ is however quite different from the structure of $\mathcal{D}$ since it contains cuts even when $\mathcal{D}$ is normal. Matters are quite different when we come to the other translation, which is the one that I have used [20, pp. 91–93] in order to get Gentzen's Hauptsatz as a corollary of the normal form theorem for natural deduction. It translates in the same way applications of introduction rules to applications of succedent rules, but if an elimination rule has been applied at the end of a natural deduction $\mathcal{D}$, one does not consider its immediate sub-deduction(s), but goes upwards in the main branch of the deduction. If the deduction is normal, one reaches a top formula that stands as the major premiss of an elimination inference. One can then apply the corresponding antecedent rule to the translation(s) of the sub-deduction(s) of $\mathcal{D}$ from which $\mathcal{D}$ was obtained by applying the elimination rule. In this way one obtains a cut free derivation in the sequent calculus.

When generalized to non-normal deductions (as described by [19]), we may hit upon a maximum formula before reaching the top when going upwards in the main branch. We then divide the deduction into two shorter parts at this point, letting the maximum formula belong to both parts, and to the corresponding derivations in the sequent calculus we apply a cut. The only cuts that will occur in the derivation obtained by this translation are ones that correspond to maximum formulas in the natural deduction.

However, if one would take as measure of a natural deduction the ordinal Gentzen assigned to the derivation in the sequent calculus obtained by this translation, one may get a quite different measure from the one defined above. It is not easy to see how this measure could be defined directly for natural deduction, that is, without going via the sequent calculus, and it is therefore difficult to say whether it would be a useful measure.

# 3 The Substitution Schema

## *3.1 Two Ways to Understand the Schema*

I shall refer to the schema below as the *schema* or *rule of substitution*:

$$\dfrac{\begin{matrix} \Pi \\ A \end{matrix} \qquad \begin{matrix} [A] \\ \Sigma \\ B \end{matrix}}{B}$$

There are two alternative ways in which this schema may be taken.

### 3.1.1 The Schema as an Inference Rule

It may be understood as stating in the usual schematic way an inference rule that allows one to infer a conclusion $B$ from the two premisses $A$ and $B$ and to bind (discharge) at the same time a number of assumptions designated by $[A]$. Given a deduction $\Pi$ of $A$ from $\Gamma$ and a deduction $\Sigma$ of $B$ from $\Delta$, we get a new deduction of $B$ from $\Gamma \cup \Delta\text{-}\{A\}$ if the set $[A]$ contains all assumptions of the form $A$ that $B$ depends on in $\Sigma$, and from $\Gamma \cup \Delta$ otherwise.

The rule is trivially derivable in systems of natural deduction: given a deduction $\Pi$ of $A$ from $\Gamma$ and a deduction $\Sigma$ of $B$ from $\Delta$ we can form a new deduction of $B$ from $\Gamma \cup \Delta\text{-}\{A\}$ without using the substitution rule by simply substituting $\Pi$ for the free (undischarged) assumptions $A$ in $\Sigma$, which I have denoted in the above by writing

$$\begin{matrix} \Pi \\ [A] \\ \Sigma \\ B \end{matrix}$$

or in linear form $\Pi/[A]/\Sigma$, where $[A]$ refers to the free assumptions $A$ in $\Sigma$.

Although the rule is derivable, one may add it as a new primitive inference rule for several reasons. One point in doing so is simply that it allows a more compact way of writing; instead of having to replace each occurrence of $A$ in $[A]$ in $\Sigma$ by a copy of $\Pi$, one writes $\Pi$ only once. It is like counting an expression like $\mathbf{S}x/tA$ as a formula of the object language, interpreted in the same way as the formula obtained by substituting the term $t$ for the variable $x$ in $A$, and it amounts exactly to this, if natural deductions are translated to terms in a $\lambda$-calculus.[20] In this form, the schema has attracted attention within computer science under the name *explicit substitution* after a paper by Abadi et al. [1].

Taking the schema as an inference rule and adding it to a system of natural deduction, one can give implication reduction the following form

$$\dfrac{\begin{matrix}\Pi\\ A\end{matrix} \qquad \dfrac{\begin{matrix}[A]\\ \Sigma\\ B\end{matrix}}{A \supset B}}{B} \qquad \text{reduces to} \qquad \dfrac{\begin{matrix}\Pi\\ A\end{matrix} \qquad \begin{matrix}[A]\\ \Sigma\\ B\end{matrix}}{B}$$

When implication reduction is given this form, it cuts down the length of the deduction, just as &- and $\forall$-reductions do. This must not deceive us into thinking that thereby we have got rid of the problem of deductions growing in size when normalized. As we shall see soon, that an $\supset$-reduction not any longer expands the deduction is only a temporary gain. The great advantage of giving $\supset$-reduction this new form in this connection is that it involves no restructuring of the deductions; the mutual relations between the formula occurrences are kept intact—thereby the second major problem noted above (Sect. 2.4.2) is overcome.

### 3.1.2 The Schema Taken as Meta-notation for Substitution

Alternatively, instead of seeing it as a new inference rule, the substitution schema may be taken as just another notation for substitution, on a par with the other two notations for substitution that I have been using. If we are only interested in proving the normalization theorem for the usual systems of natural deduction, the substitution schema is still of interest, because the syntactical objects that we get when adding it to the usual notations for applying inference rules are of help as intermediate objects to show that the usual reductions of natural deductions terminate. The idea is to assign Gentzen measures to these syntactical objects and to define reductions for them that lower the measure. The general idea will be explained more precisely in rest of this section. It will then be used in Sects. 4 and 5 to prove the normalization theorem for the ordinary systems of natural deduction for predicate logic and Peano arithmetic, respectively.

In the final Sect. 6, I will prove a normalization theorem for an enriched system of natural deduction that arises when the substitution rule is added as a primitive inference rule, taken to be of interest in its own right.

[20] I first learned about adding such a substitution notation to the language of predicate logic in 1960 from Ettore Casari, who studied it in his doctoral thesis at the University of Münster.

### 3.2 The Objects to Which the Gentzen Measure Is Assigned

We shall consequently be dealing with syntactical objects that are like usual, tree-formed natural deductions except for possibly also containing applications of the schema of substitution. If confusions could arise otherwise, I shall call them *deduction notations*, and shall refer to ordinary natural deductions as *standard* natural deductions. I use the same syntactical variables to refer to objects in one of the two domains, and shall speak of just deductions when no ambiguity can arise.

Interpreting the substitution schema in the intended way, a deduction notation denotes a standard natural deduction. When $\mathcal{D}$ stands for a deduction notation, I shall write $\downarrow\mathcal{D}$ to refer to the standard natural deduction that it denotes, in other words, the natural deduction one obtains by carrying out the substitutions indicated by the instances of the substitution schema that occur in $\mathcal{D}$. The operation $\downarrow$ is defined in the obvious way; the formal details depend on how a standard natural deduction is defined precisely. This will not be entered into here, but the following remarks are appropriate. When using the schema of substitution as exhibited in Sect. 3.1, there must be something indicating for which occurrences of $A$ in $\Sigma$, one is to substitute $\Pi$, just as the schema for $\supset$E presupposes that it is told which assumptions get bound by the inference. Gentzen used numerals to this end, and in examples below I follow him in this. By not using variables as in the linear $\lambda$-notation for indicating assumptions and the places at which substitutions are to be made, certain problems are avoided. In particular, although the tree-form used for presenting deduction notations does not specify the order in which the substitutions indicated by different applications of the substitution schema are to be carried out, it is clear that the same result is obtained regardless of what order is chosen.

The terminology commonly employed when speaking about natural deductions will be used also for deduction notations. I find it convenient to speak as if the substitution schema stated an inference rule, even if we do not need to adopt that perspective. Thus, in applications of the substitution schema as displayed Sect. 3.1, the occurrence of $B$ that appears under the line will be referred to as the *conclusion* of the substitution (inference), and $A$ and $B$ above the line will be referred to as the *major and minor premiss*, respectively. The top formulas in the deduction $\Sigma$ indicated by $[A]$ are said to be *assumptions* bound by the substitution, although in the natural deduction that is denoted by the deduction notation, they do not stand as assumptions. The degree of the major premiss of a substitution counts as the *degree* of the substitution.

### 3.3 The Need of New Reductions

In Sect. 4, new reductions will be defined for the deduction notations, but it is appropriate to say something about the need of them already now. It must be noted that use of the substitution schema may conceal maximum formulas. Consider as an example an $\supset$-reduction where the minor premiss of the removed application of $\supset$E is a universal formula, say $\forall x A(x)$, inferred by an $\forall$-introduction ($\forall$I), and where the assumptions of the form $\forall x A(x)$ that are bound by the likewise removed $\supset$I occur as premisses in $\forall$-eliminations ($\forall$E). In the result of the usual form of $\supset$-reduction, there will be maximum formulas of the form $\forall x A(x)$. When using the new form of $\supset$-reduction, $\forall x A(x)$ will instead occur as the conclusion of $\forall$I when standing as the major premiss of the substitution that replaces the $\supset$E, and will occur as the (major) premisses of $\forall$E when standing as assumptions bound by the substitution.

To normalize the deduction, we must of course remove such hidden maximum formulas by some form of reductions. When doing this, the single application of the substitution schema, which may have been designating a number of substitutions of a deduction $\Pi / A(a) / \forall x A(x)$ for different assumptions $\forall x A(x)$ from which $A(t)$ is inferred, will have to be replaced by applications of the substitution schema for substituting $\Pi / A(t)$ for different assumptions $A(t)$. Since the term $t$ may vary with different occurrences of $\forall x A(x)$ as assumptions, we shall need as many new applications of the substitution schema as there are different terms $t$ of this kind. This will cause the same exponential growth as before.

Making these reductions stepwise, we reduce the left deduction below to the one at the right:

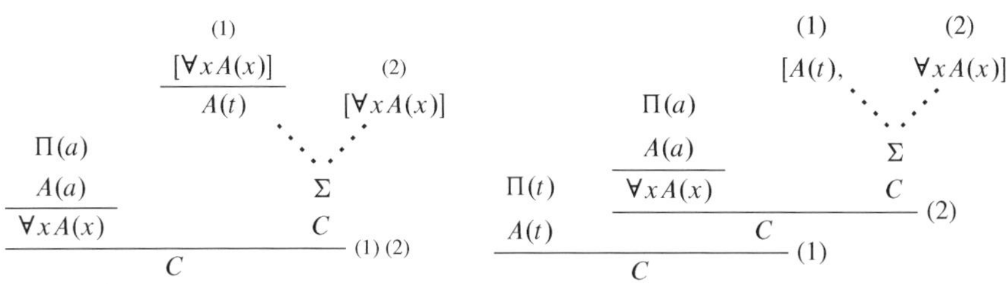

The left tree shows a deduction where a substitution with $\forall x A(x)$ as the major premiss binds not only one assumption of the form $\forall x A(x)$ marked (1), which stands as premiss of an $\forall$E with $A(t)$ as conclusion, but also a number of other assumptions referred to by $[\forall x A(x)]$, marked (2). In the right tree, the application of $\forall$E-inference exhibited in the left tree is removed. As an effect $A(t)$ occurs now as a new assumption, marked (1). It is bound by a second substitution, placed under the first one, as indicated by the attached numeral (1), while the assumptions $[\forall x A(x)]$ marked (2) are bound as before by the first one.

Two important points should now be noted. Firstly, the Gentzen measure (to be defined for deduction notations in the next subsection) decreases at this reduction, only if the first substitution contains a level line. If it does not, the new substitution must be placed further down, under the first level line that can be found further down, just as in the sequent calculus, following Gentzen's idea (Sect. 2.4.1).[21]

[21] When using induction over the degree of the maximum formulas (Sect. 2.1) instead of the Gentzen measure, there is of course no need to move the new substitution in this way.

Secondly, in the right tree, the part represented by Σ may not be a correct deduction if taken in isolation, because there may be an ∀I in Σ that has a formula $F(t)$ as premiss and $\forall xF(x)$ as conclusion; in other words, $t$ may be a variable bound by this ∀I, and it being free in $A(t)$ violates the condition for applying ∀I. If we have to put the new substitution under the first level line that can be found below the original one, the same problem applies to the part of Θ that stands above this level line and below the original substitution. This is the exact analogue of the problem mentioned above that Gentzen met and constituted an obstacle for generalizing his consistency proof to a proof of the full Hauptsatz (Sect. 2.4.1). This problem will now be overcome by choosing a normalization strategy that picks out for reduction the first possible one when going from below in the main thread (as defined below) whose last inference is an elimination. In that way it will be guaranteed that there are no introduction inferences at all below $A(t)$.

## 3.4 The Gentzen Measure Applied to Deduction Notations

In addition to the inferences that have already been designated as critical (Sect. 2.3), all applications of the substitution schema are counted as *critical*, regardless of their degrees. (This means that there is no point any longer in excepting eliminations of degree greater than the degree of the deduction from being counted as critical; however, in Sect. 6.4 the notion of critical inference will be narrowed down in another way.)

Relative to the now extended notion of critical inference, the definitions of *level* and *level line* are exactly as before. Even the definition of the *Gentzen measure* runs in the same way with the one difference that in the clause assigning a measure to the conclusion of a substitution there is no addition of 1; accordingly:

If the sum of the Gentzen measures of the premisses of a substitution is $n$, the Gentzen measure of the conclusion is equal to $n$ if the substitution does not contain a level line, and is equal to $2_j(n)$ when the substitution contains a level line with jump $j$.[22]

## 3.5 Branches, Threads, and Segments

A *branch* of a deduction notation is a sequence of formula occurrences that one gets by first picking out a top-formula and then adding successively the formula occurrences immediately below until the first formula occurrence is reached that stands as the minor premiss of an ⊃E or as the major premiss of a substitution; if no

[22]However, there would be no harm in adding 1 to $n$ or to $2_j(n)$ in this case too. There is also an option in the original definition of the Gentzen measure (Sect. 2.3): when assigning a value to a formula immediately below a level line, it is possible to avoid the addition of 1 to $2_j(n)$.

such occurrence is reached, the last formula of the branch is the end-formula of the tree, and we then have a *main branch*.

A *thread* of a deduction notation is defined similarly except that the first element of the sequence is to be a top-formula not bound by a substitution, and that when we reach the major premiss of a substitution, the next element in the thread is a formula occurrence standing as an assumption bound by that substitution (if any). The thread ends when one reaches the minor premiss of an $\supset$E or the major premiss of a vacuous substitution, or arrives at the end-formula if no such premiss appears. In the latter case we have a *main thread* of the deduction. Going in the opposite direction we find the main thread by going from the end-formula upwards, always choosing the major premiss at an elimination and the minor premiss at a substitution. A main thread in a deduction notation $\mathcal{D}$ thus corresponds to a main branch in $\downarrow\mathcal{D}$.

There can be consecutive parts of a thread where the elements are occurrences of the same formula $A$. It will sometimes be convenient to distinguish between the different occurrences of $A$ by adding superscripts, $A^1$, $A^2$, and so on. By a *segment* of a thread, I understand such a part $A^1, A^2, \ldots, A^n$ of a thread that extends as far as possible, in other words, where for each $i < n$, $A^i$ is either the major premiss of a substitution and $A^{i+1}$ is an assumption bound by the substitution or $A^i$ is the minor premiss of a substitution and $A^{i+1}$ is the conclusion of the substitution, and where $A^1$ is not the conclusion of a substitution or a premiss bound by a substitution while $A^n$ is not the minor premiss of a substitution, nor the major premiss of a substitution that binds an assumption. For instance, the sequences of occurrences of $C$ exhibited in the last tree displayed above constitute a segment or at least the last part of a segment that may have a beginning higher up in $\Sigma$. I shall say that a segment is *straight* if it is a part of a branch, in which case all of its elements except the last one are minor premisses of substitutions, and that it is *curved* if one of its elements is bound by a substitution.

A *segment* will be said to be the *conclusion* of the inference that its first element is the conclusion of, if any, and to be a *top-segment* otherwise. It will be said to be the (major or minor) *premiss* of the inference that its last element is the (major or minor) *premiss* of, if any, and to be an *end-segment* otherwise. A *maximum segment* is a segment that is the conclusion of an introduction and major premiss of an elimination.

For a deduction notation $\mathcal{D}$ without any maximum formula or maximum segment, it holds obviously that $\downarrow\mathcal{D}$ is a normal natural deduction in the standard sense.

# 4 Proof of the Normalization Theorem for Predicate Logic Using the Gentzen Measure

## *4.1 Main Structure of the Proof*

By induction over the Gentzen measure for deduction notations, I shall now prove: *for every deduction notation $\mathcal{D}$, the standard deduction $\downarrow\mathcal{D}$ reduces to normal form.* Since for a standard natural deduction $\mathcal{D}$, it holds that $\downarrow\mathcal{D} = \mathcal{D}$, this proves the usual normalization theorem for standard deductions. I presuppose the usual definition of what it is for a standard deduction $\mathcal{D}$ to *reduce immediately* to $\mathcal{D}^*$; reducibility is the transitive closure of this relation.

As in the previous sections, I am considering systems of natural deduction for predicate logic with the usual introduction and elimination rules for $\perp$, &, $\supset$, and $\forall$ and arbitrary inference rules for atomic formulas (see Sect. 2.2.1).[23]

The proof has two main cases depending on the form of the deduction (notation) $\mathcal{D}$.[24]

*Case 1* applies when

(i) the end-formula or end-segment of $\mathcal{D}$ is the conclusion of an elimination, and
(ii) the main thread of $\mathcal{D}$ contains a maximum formula or a maximum segment.

This is the crucial case, and for this case two things will be shown:

(a) that a reduction $\mathcal{D}^*$ of $\mathcal{D}$ can be defined for the last one of the maximum formulas and maximum segments in the main thread such that $\downarrow\mathcal{D}$ reduces to $\downarrow\mathcal{D}^*$,
(b) that for a reduction $\mathcal{D}^*$ so defined, $G(\mathcal{D}^*) < G(\mathcal{D})$.

Thus, according to the induction assumption, $\downarrow\mathcal{D}^*$ reduces to normal form. By the transitivity of the reduction relation and the fact that $\downarrow\mathcal{D}$ reduces to $\downarrow\mathcal{D}^*$, it follows that $\downarrow\mathcal{D}$ reduces to normal form. The substance of the proof consists in defining these reductions (Sect. 4.2), and showing that they lower the Gentzen measure (Sect. 4.3).

*Case 2* is the negation of case 1, which means that either (i) the end-formula or end-segment of $\mathcal{D}$ is the conclusion of an introduction inference or of an instance of an inference rule for atomic formulas, or (ii) the end-formula or end-segment of $\mathcal{D}$ is the conclusion of an elimination inference, but there is no maximum formula or maximum segment in the main thread, or (iii) the end-formula or the end-segment is the conclusion of no inference.

---

[23]In Gentzen's consistency proof, the logical constants are $\neg$, &, $\vee$, $\forall$, and $\exists$. The possibility that several formulas occur in the succedent is essential there, why the proof does not easily extend to intuitionistic logic. To include $\supset$ is of course essential in natural deduction.

[24]Martin-Löf's [13] proof of the normalization theorem has two similar main cases, and in this respect, the main strategy of my proof is the same as his.

In case (iii), the deduction $\downarrow\mathcal{D}$ is already normal. In cases (i) and (ii), I consider the immediate sub-deduction(s) of $\mathcal{D}$ or the immediate sub-deduction(s) of the result $\mathcal{D}'$ of letting the last inference R that is not a substitution commute with the substitutions below R. They (assuming that they are two) have lower Gentzen measures than $\mathcal{D}$ and hence the standard natural deductions that they denote reduce to normal form. Letting $\mathcal{D}^*$ be the standard natural deduction formed from them by attaching R, it is seen that $\mathcal{D}^*$ is in normal form and that $\downarrow\mathcal{D}$ reduces to $\mathcal{D}^*$. I return to some details below after having carried through the tasks of case 1.

## *4.2 Immediate Reductions Defined for Deduction Notations*

### 4.2.1 Maximum Formulas

For *maximum formulas*, &- and ∀-reductions are as usual, while the usual ⊃-reductions are replaced with ones that use substitution, as already shown above (Sect. 3.1.1).

### 4.2.2 Straight Maximum Segments

For *straight maximum segments* the reductions are the obvious ones, indicated by the deductions below where in each case the deduction (notation) to the right is an immediate reduction of the one to the left ($i = 1$ or 2 in the uppermost two figures).

$$\dfrac{\dfrac{\begin{matrix}\Sigma_n\\ C_n\end{matrix}\quad \dfrac{\begin{matrix}\Sigma_2\\ C_2\end{matrix}\quad \dfrac{\begin{matrix}\Sigma_1\\ C_1\end{matrix}\quad \dfrac{\begin{matrix}\Pi_1\\ A_1\end{matrix}\quad \begin{matrix}\Pi_2\\ A_2\end{matrix}}{A_1\&A_2}}{A_1\&A_2}}{\overline{A_1\&A_2}}}{A_1\&A_2}}{\begin{matrix}A_i\\ \Sigma_0\end{matrix}} \qquad\qquad \dfrac{\begin{matrix}\Sigma_n\\ C_n\end{matrix}\quad \dfrac{\begin{matrix}\Sigma_2\\ C_2\end{matrix}\quad \dfrac{\begin{matrix}\Sigma_1\\ C_1\end{matrix}\quad \begin{matrix}\Pi_i\\ A_i\end{matrix}}{A_i}}{\overline{A_i}}}{\begin{matrix}A_i\\ \Sigma_0\end{matrix}}$$

$$\dfrac{\begin{matrix}\Pi_1\\ A\end{matrix}\quad \dfrac{\begin{matrix}\Sigma_n\\ C_n\end{matrix}\quad \dfrac{\begin{matrix}\Sigma_2\\ C_2\end{matrix}\quad \dfrac{\begin{matrix}\Sigma_1\\ C_1\end{matrix}\quad \dfrac{\begin{matrix}[A]\\ \Pi_2\\ B\end{matrix}}{A\supset B}}{A\supset B}}{\overline{A\supset B}}}{A\supset B}}{\begin{matrix}B\\ \Sigma_0\end{matrix}} \qquad\qquad \dfrac{\begin{matrix}\Pi_1\\ A\end{matrix}\quad \dfrac{\begin{matrix}\Sigma_n\\ C_n\end{matrix}\quad \dfrac{\begin{matrix}\Sigma_2\\ C_2\end{matrix}\quad \dfrac{\begin{matrix}\Sigma_1\\ C_1\end{matrix}\quad \begin{matrix}[A]\\ \Pi_2\\ B\end{matrix}}{B}}{\overline{B}}}{B}}{\begin{matrix}B\\ \Sigma_0\end{matrix}}$$

$$\cfrac{\begin{matrix}\Sigma_n\\ C_n\end{matrix}\quad \cfrac{\begin{matrix}\Sigma_2\\ C_2\end{matrix}\quad \cfrac{\begin{matrix}\Sigma_1\\ C_1\end{matrix}\quad \cfrac{\begin{matrix}\Pi(a)\\ A(a)\end{matrix}}{\forall xA(x)}}{\forall xA(x)}}{\forall xA(x)}}{\cfrac{\forall xA(x)}{\begin{matrix}A(t)\\ \Sigma_0\end{matrix}}} \qquad\qquad \cfrac{\begin{matrix}\Sigma_n\\ C_n\end{matrix}\quad \cfrac{\begin{matrix}\Sigma_2\\ C_2\end{matrix}\quad \cfrac{\begin{matrix}\Sigma_1\\ C_1\end{matrix}\quad \begin{matrix}\Pi(t)\\ A(t)\end{matrix}}{A(t)}}{A(t)}}{\begin{matrix}A(t)\\ \Sigma_0\end{matrix}}$$

Clearly, if $\mathcal{D}^*$ is one of these reductions of $\mathcal{D}$, then $\downarrow\mathcal{D}^*$ is an immediate reduction of $\downarrow\mathcal{D}$.

### 4.2.3 Simplifications

It will be convenient to introduce reductions that remove all vacuous substitutions and clearly superfluous substitutions where one of the premisses stands as a top-formula. The three possible cases of deduction notations that contain such substitutions are shown to the left below, where in the first case no assumption is bound by the substitution. Deduction notations of one of these forms are said to reduce immediately to the respective deduction notation shown to the right.

$$\cfrac{\begin{matrix}\Pi\\ A\end{matrix}\quad \begin{matrix}\Sigma\\ B\end{matrix}}{\begin{matrix}B\\ \Theta\end{matrix}} \qquad \cfrac{\begin{matrix}\Pi\\ A\end{matrix}\quad [A]}{\begin{matrix}A\\ \Theta\end{matrix}} \qquad \cfrac{A\quad \begin{matrix}[A]\\ \Sigma\\ B\end{matrix}}{\begin{matrix}B\\ \Theta\end{matrix}} \qquad \begin{matrix}\Sigma\\ B\\ \Theta\end{matrix} \qquad \begin{matrix}\Pi\\ A\\ \Theta\end{matrix} \qquad \begin{matrix}\Sigma\\ B\\ \Theta\end{matrix}$$

I call these reduction *simplifications*. If $\mathcal{D}^*$ is a simplification of $\mathcal{D}$, then clearly $\downarrow\mathcal{D} = \downarrow\mathcal{D}^*$.

### 4.2.4 Immediate Reductions Defined for Maximum Curved Segments

Let $\mathcal{D}$ be a deduction with a maximum curved segments $A^1, A^2, \ldots, A^n$ that has been simplified as much as possible according to the above. Then $A^n$ is the only element of the segment that stands as a top-formula. I consider first the general case when $A$ has the form $\forall xA(x)$, which was exemplified in Sect. 3.3. $\mathcal{D}$ has then the form exhibited in the figure below.

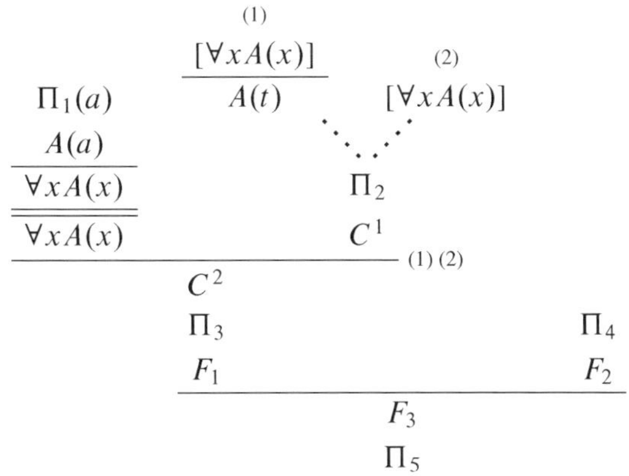

The double lines above the major premiss $\forall x A(x)$ of the exhibited substitution in $\mathcal{D}$ indicate the possibility of a number of other substitutions in which $\forall x A(x)$ is the minor premiss. To distinguish the different occurrences of the formula $C$ (standing as the minor premiss and as the conclusion of the substitution), they are equipped with different superscripts.

As was explained above (Sects. 2.4.1 and 3.3), the new substitution that is added when forming the immediate reduction $\mathcal{D}^*$ of $\mathcal{D}$ must be put under a level line. The inference from $F_1$ and $F_2$ to $F_3$ is to be the first inference in $\mathcal{D}$ below $C^2$ that contains a level line, in case the substitution does not do so; if it does, $\Pi_3/F_1$, $\Pi_4/F_2$, and $F_3$ all fall away. The first such inference may be one with only one premiss, in which case $\Pi_4/F_2$ falls away. It is to be noted that the inference may be another substitution with $F_2$ as major or minor premiss (if $F_2$ is the major premiss it should really stand to the left of $F_1$ to follow the convention that I have used). Otherwise $\mathcal{D}$ is as already described above (Sect. 3.3).

As was also explained above, leaving out the assumption of $\forall x A(x)$ marked (1) when forming $\mathcal{D}^*$ gives rise to a new assumption $A(t)$, and in the general case, $t$ may be a variable bound by an application of $\forall$I occurring further below, which would break the restriction on $\forall$I. Therefore, an immediate reduction of $\mathcal{D}$ at this maximum segment is defined only when the following condition is satisfied: if the term $t$ is or contains a parameter, no $\forall$I-inference in $\Pi_2$ below the formula $A(t)$ and no $\forall$I-inference in $\Pi_3$ below $C^2$ binds it. This condition is obviously satisfied because of the normalization strategy of the proof: when we are to make a reduction of the maximum segment in question, that is when case 1 applies, there are no introductions in the main thread after this maximum segment.

Given that this condition is satisfied and provided furthermore that there is no open assumption of $\Pi_1(a)$ that is bound in $\Pi_3$, the immediate reduction $\mathcal{D}^*$ of $\mathcal{D}$ at this maximum segment takes the following form.

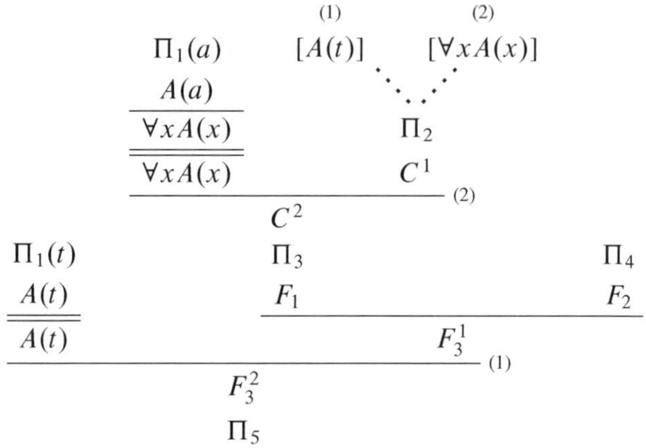

When we are to make a reduction of this kind, the normalization strategy guarantees that the proviso is satisfied to the extent that no $\supset$-introduction in $\Pi_3$ can bind an assumption occurring in $\Pi_1(a)$, and this can be added to the condition for the $\forall$-reduction to be defined. But there is nothing to prevent there being a substitution in $\Pi_3$ below $C^2$ that binds an assumption occurring in $\Pi_1(a)$. If there is such a substitution, it has to be repeated in the left sub-deduction above so that also $A(t)$ and hence $F_3^2$ become independent of this assumption.

Therefore, in case there is an assumption of $\Pi_1(a)$ that is bound in $\Pi_3$ by a substitution the immediate reduction $\mathcal{D}^*$ of $\mathcal{D}$ becomes more involved and is defined as follows. Let $\Theta_1/H_1, \Theta_2/H_2, \ldots,$ and $\Theta_m/H_m$ be all the deductions that appear in $\Pi_3$ as the deduction of the major premiss of a substitution whose minor premiss is a formula occurrence in $\Pi_3$ below $C^2$ or is $C^2$, taken in the order in which they appear in the deduction $\mathcal{D}$. Then, $\mathcal{D}^*$ is to have the following form.

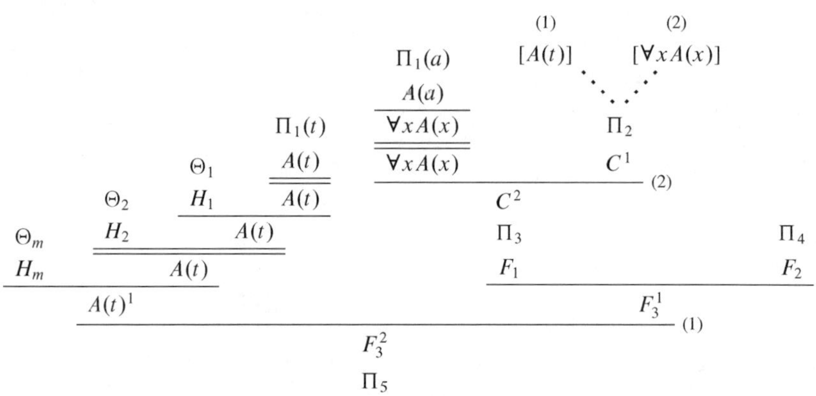

Each deduction $\Theta_i/H_i$ $(i < m)$ in $\mathcal{D}^*$ exhibited above is to be the deduction of the major premiss of a substitution that binds the same assumptions in $\Pi_1(t)$ or in some $\Theta_j$ for $j < i$ that the corresponding substitution in $\Pi_3$ binds. Because of these added substitutions it is guaranteed that $F_3^2$ depends on the same assumptions as $F_3^1$. Some of these substitutions may be vacuous and may then be left out.

Note that if the deduction notation $\mathcal{D}$ immediately reduces to $\mathcal{D}^*$ according to the now defined $\forall$-reduction for curved segments, then $\downarrow\mathcal{D}^*$ is a standard $\forall$-reduction of $\downarrow\mathcal{D}$.

The &-reductions for curved segments are defined analogously and are not exhibited.

When the formulas of the maximum curved segment to be reduced are occurrences of an implication $A \supset B$, the deduction $\mathcal{D}$ has the following form

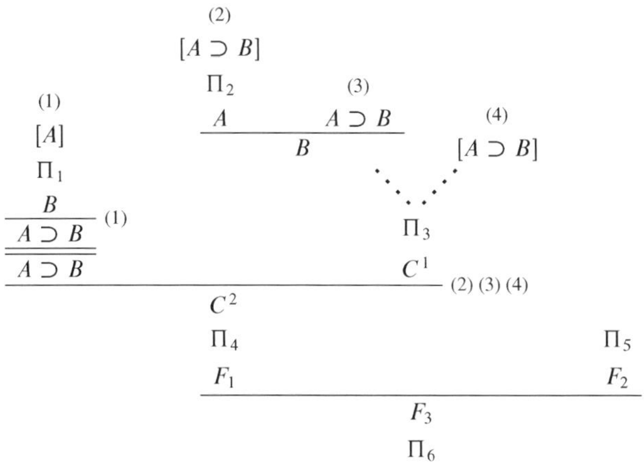

where as in the case of $\forall$-reductions the inference line below $F_1$ and $F_2$ is the first level line below the substitution in question (or is the inference line of that substitution if it is a level line, in which case $\Pi_3/F_1$, $\Pi_4/F_2$, and $F_3$ all fall away as before).

To help the understanding of the general idea behind $\supset$-reductions for curved segments, I first consider the case when there is no assumption on which the last formula $A$ of $\Pi_2$ in $\mathcal{D}$ depends that is bound by an inference in $\Pi_3$ or $\Pi_4$ and no assumption in $\Pi_1$ that is bound in $\Pi_4$. The immediate reduction $\mathcal{D}^*$ of $\mathcal{D}$ at this curved segment then assumes the form:

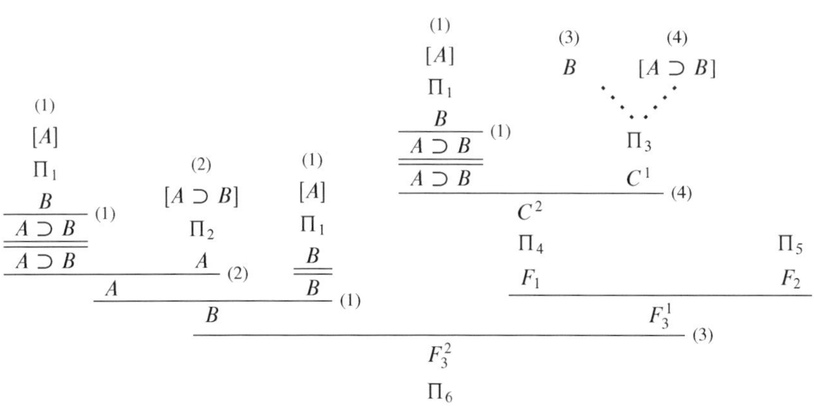

As indicated in the figure, the exhibited substitution in $\mathcal{D}$ binds assumptions of the form $A \supset B$ marked (2), (3), and (4). The one marked (3) is supposed to be the last formula in the curved maximum segment in question, and the reduction $\mathcal{D}^*$ of $\mathcal{D}$ is formed by leaving out the $\supset$E-inference in which it was the major premiss, which means that in $\mathcal{D}^*$ the former conclusion $B$ of the cancelled $\supset$E becomes an assumption instead, now marked (3). It becomes bound by a new substitution, placed under the level line. The original substitution remains as before (except, of course, for not binding the assumptions marked (2) or (3) in $\mathcal{D}$). The deduction of the major premiss of the new substitution is obtained by making use of the deduction of the major premiss of the original substitution, substituting $B$ for $A \supset B$ in its end-segment and leaving out the last inference of $\Pi_1$, that is, the exhibited $\supset$I-inference, which was binding assumptions $A$ in $\mathcal{D}$. In order to bind these assumptions $A$ in $\mathcal{D}^*$, we add another substitution in which we use the deduction $\Pi_2$ of the minor premiss of the $\supset$E in $\mathcal{D}$; in other words, we move this deduction from its original position in $\mathcal{D}$ to the position as a deduction of the minor premiss of this substitution. But since the end-formula $A$ of this deduction $\Pi_2$ may depend on assumptions of the form $A \supset B$, marked (2), we have to add yet another substitution in which a copy of the given deduction of $A \supset B$ is used as deduction of the major premiss.

The proviso stated above can to some extent be taken as a condition for an $\supset$-reduction to be defined: no assumption on which the last formula $A$ of $\Pi_2$ in $\mathcal{D}$ depends is to be bound by an $\supset$-introduction in $\Pi_3$ or $\Pi_4$, and no assumption in $\Pi_1$ is to be bound by an $\supset$-introduction in $\Pi_4$. That this condition is satisfied is guaranteed by the normalization strategy as in the case of $\forall$-reductions, but again nothing prevents there being substitutions in $\Pi_3$ or $\Pi_4$ that bind such assumptions. If there are such substitutions, all substitutions that occur along the main thread from $B$ to $F_1$ except the one where $C^1$ is the minor premiss have to be added to the deduction of the major premiss $B$ of the substitution in $\mathcal{D}^*$ exhibited above.

The general form of an $\supset$-reduction has accordingly to be more involved and is defined as follows. Let $\Sigma_1/G_1, \Sigma_2/G_2, \ldots,$ and $\Sigma_n/G_n$ be all the deductions that appear in $\Pi_3$ as the deduction of the major premiss of a substitution whose minor premiss is a formula occurrence in $\Pi_3$ that is different from $C^1$ and stands below $B$ or is $B$, taken in the order in which they appear in $\Pi_3$, and let $\Theta_1/H_1, \Theta_2/H_2, \ldots,$ and $\Theta_m/H_m$ be all the deductions that appear in $\Pi_4$ as the deduction of the major premiss of a substitution whose minor premiss is a formula occurrence in $\Pi_4$ below $C^2$ or is $C^2$, taken in the order in which they appear in $\Pi_4$. Then the immediate reduction $\mathcal{D}^*$ of $\mathcal{D}$ at the curved maximum segment in question is to have the form exhibited on the next page.

Each deduction $\Sigma_i/G_i$ $(i < n)$ in $\mathcal{D}^*$ exhibited in the figure is to be the deduction of the major premiss of a substitution that binds the same assumptions in $\Pi_2$ or in some $\Sigma_j, j < i$, that the corresponding substitution in $\mathcal{D}$ was binding. Similarly, each deduction $\Theta_i/H_i$ $(i < m)$ exhibited in $\mathcal{D}^*$ is to be the deduction of the major premiss of a substitution. If the minor premiss is $B$, the substitution is to bind the same assumptions in $\Pi_1$ or in some $\Theta_j, j < i$, that the corresponding substitution in the part $\Pi_3$ of $\mathcal{D}$ was binding. If the minor premiss is $A$, the substitution is to bind the

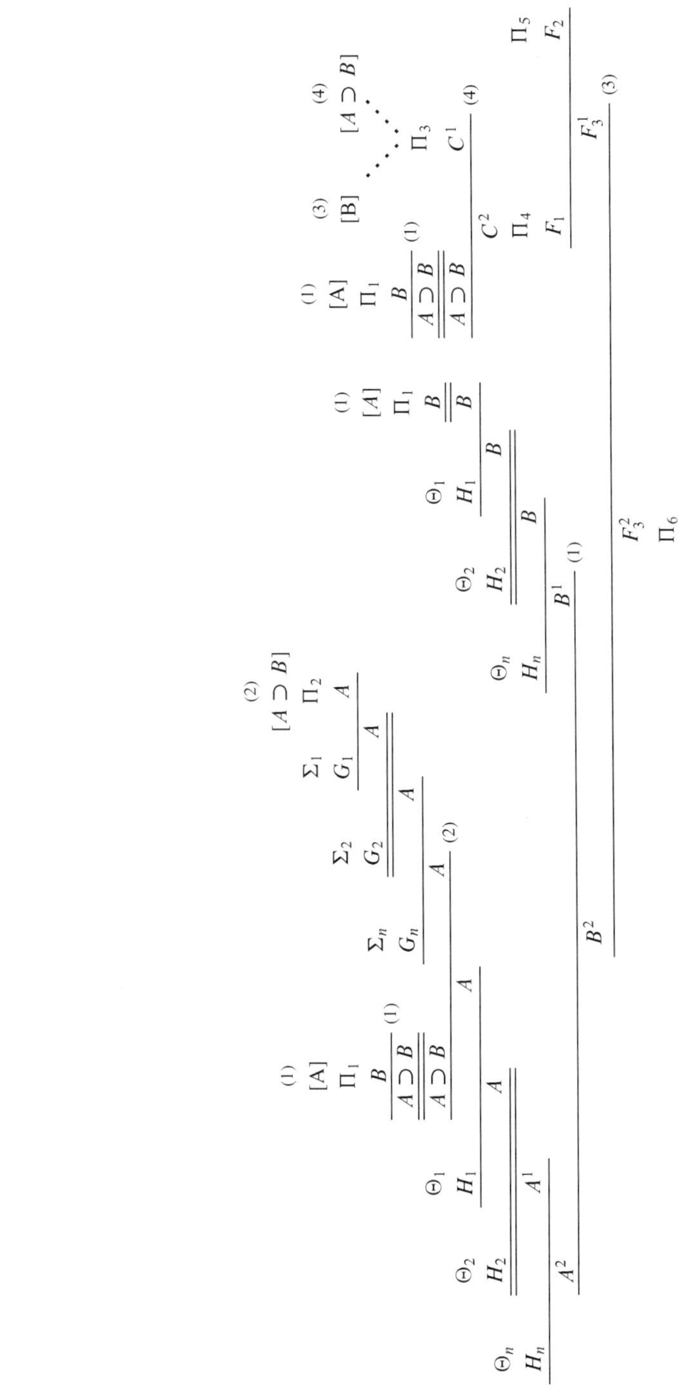

same assumptions in $\Pi_2$, in some $\Sigma_j$, $j \le n$, in $\Pi_1$, or in some $\Theta_j$, $j < i$, that the corresponding substitution in the part $\Pi_4$ of $\mathcal{D}$ was binding. Because of these added substitutions it is again guaranteed that $F_3^2$ depends on the same assumptions as $F_3^1$. Some of these substitutions may be vacuous and may then be left out.

It is to be noted again that even in the case of this more involved reduction, $\downarrow\mathcal{D}^*$ is an $\supset$-reduction of $\downarrow\mathcal{D}$ as defined for standard natural deduction.

## *4.3 Verifying That the Reductions Lower the Gentzen Measure*

Immediate reductions of deduction notations lower the Gentzen measure regardless of where they are carried out, and hence any sequence of immediate reductions will terminate in a deduction notation $\mathcal{D}$ for which there is no immediate reduction defined. This fact does not give us a strong normalization theorem, however, because $\mathcal{D}$ may contain a maximum segment that cannot be removed due to the conditions stated in the definition of $\supset$- and $\forall$-reductions for curved maximum segments not being satisfied, and $\downarrow\mathcal{D}$ will then not be normal. As already explained above, the maximum formulas and segments must be removed in a certain order laid down when the main structure of the proof was described. But in the verifications below of the fact that the reductions lower the Gentzen measure, this order need not be presupposed.

The verifications are to generalize to the case of arithmetic, and therefore only arguments that hold also for transfinite ordinals will be used. The verifications are essentially as in Gentzen's proof (see also [24] for properties of ordinals that are essential), but since the reductions are different I shall refer to the facts needed to see that the measure goes down.

### 4.3.1 Eliminations of Maximum Formulas

It is easily seen that reductions that eliminate maximum formulas lower the Gentzen measure since now even $\supset$-reductions shorten the length. As a model for how the verification goes, I consider in some detail the case when $\mathcal{D}^*$ is an $\supset$-reduction of $\mathcal{D}$ that eliminates a maximum formula $A \supset B$. Adding superscripts to the variables used in the definition of the reduction (Sect. 3.1.1), let $B^1$ and $B^2$ in $\mathcal{D}$ be the premiss of the $\supset$I and the conclusion of the $\supset$E, respectively, and let $B^1$ and $B^2$ in $\mathcal{D}^*$ be the minor premiss and the conclusion, respectively, of the substitution that replaces $\supset$E.

Let us first assume that $A \supset B$ in $\mathcal{D}$ does not stand on a level line. We note the following equalities and inequalities:

$$G_{\mathcal{D}^*}(B^2) = G_{\mathcal{D}^*}(A) + G_{\mathcal{D}^*}(B^1) =$$
$$= G_{\mathcal{D}}(A) + G_{\mathcal{D}}(B^1) < G_{\mathcal{D}}(A) + G_{\mathcal{D}}(B^1) + 2 = G_{\mathcal{D}}(B^2).$$

The first equality holds because the substitution that replaces the $\supset$E-inference cannot involve a level line when the $\supset$E-inference did not; that the inference does not contain a level line means that there is a critical inference of degree $d \geq \text{degr}(A \supset B)$ whose conclusion stands below $B^2$ in $\mathcal{D}$, which implies that $A$ and $B^2$ in $\mathcal{D}$ have the same level $d$—note that since this inference stands below $B^2$ in $\mathcal{D}^*$ too and $d > \deg(A)$, both $A$ and $B^2$ in $\mathcal{D}^*$ have still the same level $d$. The second equality holds because nothing is changed in the deduction above any of $A$ and $B^1$; in particular, the levels remain the same for the reasons just stated. This lowering of the Gentzen measure of $B^2$ in $\mathcal{D}^*$ propagates down in the deduction to the end-formula since neither is there a change in the deduction below $B^2$; in particular, the levels do not change. Hence, $G(\mathcal{D}^*) < G(\mathcal{D})$.

Let us now assume instead that $A \supset B$ in $\mathcal{D}$ stands on a level line with jump $j$ from level $h_1$ to $h_2$, where $h_1 = \text{degr}(A \supset B)$. The substitution that in $\mathcal{D}^*$ replaces the $\supset$E then contains a level line if $\text{degr}(A) > h_2$, and in any case, $G_{\mathcal{D}^*}(B^2) = 2_{j^*}\left(G_{\mathcal{D}^*}(A) + G_{\mathcal{D}^*}(B^1)\right)$, if we set $j^* = \max(\text{degr}(A), h_2) - h_2$ (note that $2_0(n) = n$). Unlike the previous sub-case, we cannot assert that $G_{\mathcal{D}^*}(A) = G_{\mathcal{D}}(A)$ and $G_{\mathcal{D}^*}(B^1) = G_{\mathcal{D}}(B^1)$, because now the level of $A$ and $B^1$ is changed, namely from $h_1$ in $\mathcal{D}$ to $\max(\text{degr}(A), h_2)$ in $\mathcal{D}^*$, and the level of formulas above $A$ and $B^1$ may drop equally much. The difference of level $h_1 - \max(\text{degr}(A), h_2)$, which may also be written $j - j^*$, can have the effect that new level lines with jumps $\leq j - j^*$ appear in $\mathcal{D}^*$ above $A$ or $B^1$ or that the jump of a level line that occurred already in $\mathcal{D}$ above $A$ or $B^1$ increases when passing to $\mathcal{D}^*$ with at most $j - j^*$. If so, the Gentzen measure of the formulas immediately below these level lines will increase, but there is a bound on how much the measure can increase, because there is a bound on the jumps of the new level lines and on how much the jumps of old level lines may go up. I satisfy myself here with asserting that for the formula occurrences $F$ concerned, which stand above $A$ or $B^1$ or are identical to $A$ or $B^1$, it holds that $G_{\mathcal{D}^*}(F) \leq 2_{j-j^*}(G_{\mathcal{D}}(F))$, in particular $G_{\mathcal{D}^*}(A) \leq 2_{j-j^*}(G_{\mathcal{D}}(A))$ and $G_{\mathcal{D}^*}(B^1) \leq 2_{j-j^*}\left(G_{\mathcal{D}}(B^1)\right)$.

This possible increase of the Gentzen measure of formulas above the original level line that we find when comparing $\mathcal{D}$ and $\mathcal{D}^*$ is more than compensated for by the fact that the level line in $\mathcal{D}$ is replaced by one with the lower jump $j^*$ in $\mathcal{D}^*$ or disappears if $j^* = 0$, as is now seen by noting the following facts[25]:

$$G_{\mathcal{D}^*}(B^2) = 2_{j^*}(G_{\mathcal{D}^*}(A) + G_{\mathcal{D}^*}(B^1)) \leq$$
$$\leq 2_{j^*}[2_{j-j^*}(G_{\mathcal{D}}(A)) + 2_{j-j^*}(G_{\mathcal{D}}(B^1))] \leq$$

[25]This is a recurring theme in the verifications. An inference containing a level line may also disappear without being replaced by any new inference, as is the case with the simplifications (section 4.2.3), which is a special case of the one above. Gentzen [6, p. 41], [7, pp. 281–282] deals at length with the same phenomenon that occurs when eliminating a cut. It may also happen that an inference line that was a level line in a deduction $\mathcal{D}$ ceases to be a level line in a reduction $\mathcal{D}^*$ of $\mathcal{D}$, although the inference remains the same. This may, for instance, be the case when some elimination inferences are excepted from being counted as critical, as in Sects. 2.3 and 6.4. If the jump of the level line was $j$, then for the formulas $A$ immediately above the line it holds again that $G_{\mathcal{D}^*}(A) \leq 2_j(G_{\mathcal{D}}(A))$. But since the line is not a level line in $\mathcal{D}^*$, it holds for the formula $B$ immediately below the line that $G_{\mathcal{D}^*}(B) \leq (G_{\mathcal{D}}(B))$.

$$\leq 2_{j^*}[2_{j-j^*}(G_{\mathcal{D}}(A) + G_{\mathcal{D}}(B^1))] =$$
$$= 2_{j^*+j-j^*}(G_{\mathcal{D}}(A) + G_{\mathcal{D}}(B^1)) = 2_j(G_{\mathcal{D}}(A) + G_{\mathcal{D}}(B^1)) <$$
$$< 2_j(G_{\mathcal{D}}(A) + G_{\mathcal{D}}(B^1) + 1) + 1 = G_{\mathcal{D}}(B^2).$$

As before, this is sufficient to see that $G\,(\mathcal{D}^*) < G\,(\mathcal{D})$.

#### 4.3.2 Eliminations of Straight Maximum Segments

The reasoning for straight maximum segments is essentially the same. Adding superscripts $1, 2, \ldots, n, n+1$, to the letters used in the definition of $\supset$-reductions for straight maximum segments, and assuming that $j$ and $j^*$ are as above or that $j = j^* = 0$ if no level lines are involved, we get that $G_{\mathcal{D}^*}\left(B^{n+1}\right) = 2_{j^*}\left(G_{\mathcal{D}^*}(A) + G_{\mathcal{D}^*}\left(B^n\right)\right)$ and that $G_{\mathcal{D}^*}(A) \leq 2_{j-j^*}G_{\mathcal{D}}(A)$, and find now that $G_{\mathcal{D}^*}\left(B^n\right) < 2_{j-j^*}\left(G_{\mathcal{D}}\left((A \supset B)^n\right)\right)$. As above, it then follows that $G_{\mathcal{D}^*}\left(B^{n+1}\right) < 2_j\left(G_{\mathcal{D}}(A) + G_{\mathcal{D}}\left((A \supset B)^n\right)\right) < G_{\mathcal{D}}\left(B^2\right)$ where $B^2$ in $\mathcal{D}$ refers to the conclusion of the $\supset$E-inference.

#### 4.3.3 Simplifications

That simplifications lower the Gentzen measure is also easy to see. It is trivial if the eliminated substitution does not contain a level line. If it contains a level line with jump $j$, we get in the second of the three cases that $G_{\mathcal{D}^*}\,(A) \leq 2_j\left(G_{\mathcal{D}}\left(A^1\right)\right) < 2_j\left(G_{\mathcal{D}}\left(A^1\right) + 1\right) = G_{\mathcal{D}}\left(A^3\right)$, where $\mathcal{D}^*$ is the reduction of $\mathcal{D}$ and $A^1$ is the major premiss while $A^3$ is conclusion of the substitution in $\mathcal{D}$ that becomes eliminated in $\mathcal{D}^*$; for the first of these equalities or inequalities, see what was said above about the general phenomenon that occur when a level line disappears (Sect. 4.3.1 and footnote 25). In the first and third case, we get similarly that $G_{\mathcal{D}^*}\,(B) \leq 2_j\left(G_{\mathcal{D}}\left(B^1\right)\right) < 2_j\left(G_{\mathcal{D}}(A) + G_{\mathcal{D}}(B^1)\right) = G_{\mathcal{D}}(B^2)$, where $B^1$ is the minor premiss and $B^2$ is the conclusion of the substitution in $\mathcal{D}$.

#### 4.3.4 Eliminations of Curved Maximum Segments

The crucial cases are the $\supset$-, &-, and $\forall$-reductions for maximum curved segments, since the length of the deduction now increases because of new substitutions inserted under the previous one. Gentzen's idea to put the derivations that are added below a level line (Sect. 2.4.1) comes now into play. As an illustration, consider a case of $\forall$-reduction where the Gentzen measure of the conclusion $F_3$ under the level line in the original deduction $\mathcal{D}$ is $2^n$; $n$ being the sum of the Gentzen measures of the premisses above. In the deduction $\mathcal{D}^*$ resulting from the reduction, the Gentzen measure of the corresponding formula $F_3^1$ will be $2^{n^*}$ where $n^* < n$, because something has disappeared from the original deduction above the level line. The Gentzen measure of the conclusion of the added substitution, $F_3^2$, will be $2^{n^*}+m$ where $m<2^{n^*}$, assuming that the added substitution does not involve a level line. Hence,

$$G_{\mathcal{D}^*}\left(F_3^2\right) = 2^{n^*} + m < 2^{n^*} + 2^{n^*} \leq 2^{n^*+1} \leq 2^n = G_{\mathcal{D}}\left(F_3\right).$$

In the case of $\supset$-reduction, the reasoning is similar but a little more complicated. The verifications below are obtained along this line, there are only some more details, essentially because we have to pay attention to other possible level lines that may occur.

In all the cases of reductions that remove curved maximal segments, we want to see that $G_{\mathcal{D}^*}\left(F_3^2\right) < G_{\mathcal{D}}\left(F_3\right)$; letters as in the definitions of the reductions. Let the exhibited level line in $\mathcal{D}$ have a jump $j$ from level $h_1$ to level $h_2$. Then $G_{\mathcal{D}}\left(F_3\right) = 2_j\left(G_{\mathcal{D}}\left(F_1\right) + G_{\mathcal{D}}\left(F_2\right)\right)$, assuming that the inference from $F_1$ and $F_2$ to $F_3$ is a substitution. Otherwise we have to add 1, which case is exactly parallel. As we recall, $F_2$ may fall away, and $F_3$ may be identical with $C^2$, which only simplify the verifications.

Consider first the general case of $\forall$-reduction and the sub-case when the new substitution inference does not contain a level line. In line with the schematic argument given above, we shall make use of the two facts (1) and (2) stated below:

(1) $G_{\mathcal{D}^*}\left(F_1\right) + 1 \leq G_{\mathcal{D}}\left(F_1\right)$ and
(2) $G_{\mathcal{D}^*}\left(A(t)^1\right) < 2_j\left(G_{\mathcal{D}^*}\left(F_1\right) + G_{\mathcal{D}^*}\left(F_2\right)\right).$

(1) is consequence of the fact that the sub-deduction of $F_1$ in $\mathcal{D}^*$ has one node less than the sub-deduction of $F_1$ in $\mathcal{D}$. (2), or more precisely that $G_{\mathcal{D}^*}\left(A(t)^1\right) < 2_j\left(G_{\mathcal{D}^*}\left(F_1\right)\right)$ is a consequence of the fact that the deduction of the major premiss $A(t)^1$ of the last exhibited substitution inference in $\mathcal{D}^*$ contains at least one node less than the deduction of $F_1$ in $\mathcal{D}^*$ and that the levels of its formulas may increase with at most $j$—that the levels increase is due to the fact that the formulas do not stand over the exhibited level line, as the corresponding ones in the deduction of $F_1$ do; the jump of the latter has been assumed to be $j = h_1 - h_2$.

From the two facts (1) and (2) and the trivial fact that $G_{\mathcal{D}^*}\left(F_2\right) = G_{\mathcal{D}}\left(F_2\right)$, we get the following equalities and inequalities:

$$\begin{aligned} G_{\mathcal{D}^*}\left(F_3^2\right) &= G_{\mathcal{D}^*}(A(t)^1) + 2_j\left(G_{\mathcal{D}^*}\left(F_1\right) + G_{\mathcal{D}^*}\left(F_2\right)\right) < \\ &< 2_j\left(G_{\mathcal{D}^*}\left(F_1\right) + G_{\mathcal{D}^*}\left(F_2\right)\right) + 2_j\left(G_{\mathcal{D}^*}\left(F_1\right) + G_{\mathcal{D}^*}\left(F_2\right)\right) \leq \\ &\leq 2_j\left(G_{\mathcal{D}^*}\left(F_1\right) + G_{\mathcal{D}^*}\left(F_2\right) + 1\right) \leq 2_j\left(G_{\mathcal{D}}\left(F_1\right) + G_{\mathcal{D}}\left(F_2\right)\right) = \\ &= G_{\mathcal{D}}\left(F_3\right). \end{aligned}$$

To verify the other sub-case when the new substitution contains a level line, we note that its jump must be $j' = \deg(A(t)) - h_2$. The jump of the original substitution will then go down to $j^* = j - j' = h_1 - \deg(A(t))$. Note that $j^*$ must still be greater than 0 and that $j' + j^* = j$. The fact (2) invoked above is now replaced by

$G_{\mathcal{D}^*}\left(A(t)^1\right) < 2_{j^*}\left(G_{\mathcal{D}^*}(F_1) + G_{\mathcal{D}^*}(F_2)\right)$, since the levels decrease with at most $h_1 - \deg(A(t)) = j^*$. We then get instead:

$$G_{\mathcal{D}^*}(F_3^2) = 2_{j'}\left[G_{\mathcal{D}^*}(A(t)^1) + 2_{j^*}\left(G_{\mathcal{D}^*}(F_1) + G_{\mathcal{D}^*}(F_2)\right)\right] <$$

$$< 2_{j'}\left[2_{j^*}\left(G_{\mathcal{D}^*}(F_1) + G_{\mathcal{D}^*}(F_2)\right) + 2_{j^*}\left(G_{\mathcal{D}^*}(F_1) + G_{\mathcal{D}^*}(F_2)\right)\right] \leq$$

$$\leq 2_{j'+j^*}\left(G_{\mathcal{D}^*}(F_1) + G_{\mathcal{D}^*}(F_2) + 1\right) \leq 2_j\left(G_{\mathcal{D}}(F_1) + G_{\mathcal{D}}(F_2)\right) =$$

$$= G_{\mathcal{D}}(F_3).$$

Consider finally the $\supset$-reductions (the case of &-reduction being analogous to $\forall$-reduction). The deduction $\mathcal{D}^*$ resulting from the reduction is now considerably much longer than $\mathcal{D}$, containing not only two new copies of the deduction $\Pi_1$ of $B$ in $\mathcal{D}$, but also a deduction of $A$, which may be not much shorter than the given deduction of $F_3$ in $\mathcal{D}$. It is now crucial that in $\mathcal{D}^*$ the deduction of the minor premiss $C^1$ contains at least two nodes less than the corresponding deduction in $\mathcal{D}$.

Given again that the jump of the displayed level line in $\mathcal{D}$ is $j = h_1 - h_2$, and that thus $G_{\mathcal{D}}(F_3) = 2_j\left(G_{\mathcal{D}}(F_1) + G_{\mathcal{D}}(F_2)\right)$, we define $f$ to be max(degr($B$), $h_2$) $- h_2$. If $f > 0$, then $F_3^2$ stands in $\mathcal{D}^*$ immediately under a level line, and the jump of the level line under which $F_3^1$ stands in $\mathcal{D}^*$ will be $h_1 -$ degr($B$). In any case, $G_{\mathcal{D}^*}\left(F_3^2\right) = 2_f\left(G_{\mathcal{D}^*}(B^2) + G_{\mathcal{D}^*}\left(F_3^1\right)\right)$, and $G_{\mathcal{D}^*}\left(F_3^1\right) = 2_g\left(G_{\mathcal{D}^*}(F_1) + G_{\mathcal{D}^*}(F_2)\right)$, setting $g = h_1 -$ max(degr($B$), $h_2$). Note that $g > 0$ and that $f + g = j$.

Let $k =$ max(degr($A$), max(degr($B$), $h_2$)) $-$ max(degr($B$), $h_2$). If $k > 0$, then $B^2$ stands immediately under a level line with jump $k$, and in any case $G_{\mathcal{D}^*}(B^2) = 2_k\left(G_{\mathcal{D}^*}\left(A^2\right) + G_{\mathcal{D}^*}\left(B^1\right)\right)$.

We find that $G_{\mathcal{D}^*}\left(F_3^2\right) = 2_f\left[2_k\left(G_{\mathcal{D}^*}\left(A^2\right) + G_{\mathcal{D}^*}\left(B^1\right)\right) + 2_g\left(G_{\mathcal{D}^*}(F_1) + G_{\mathcal{D}^*}(F_2)\right)\right]$. We want to compare the values $G_{\mathcal{D}^*}(A^2)$ and $G_{\mathcal{D}^*}\left(B^1\right)$ with $G_{\mathcal{D}}(F_3)$ and the latter value with $G_{\mathcal{D}^*}(F_3)$.

To this end, consider the value we get if we change the calculation of the Gentzen measure of $F_1$ in $\mathcal{D}$ in the one respect that we leave out from the calculation the inference from $A$ and $A \supset B$ to $B$; more precisely, we let the value of the conclusion $B$ be the same as that of the minor premiss $A$ (instead of the sum of the values of $A$ and $A \supset B$ plus 1, which is the Gentzen measure of $B$ since the inference cannot contain a level line). Call this value $\alpha$.

Clearly, $G_{\mathcal{D}^*}(F_1) \leq \alpha$ and $\alpha + 2 \leq G_{\mathcal{D}}(F_1)$. Trivially, $G_{\mathcal{D}^*}(F_2) = G_{\mathcal{D}}(F_2)$. Define $m = h_1 -$ max(degr($A$), max(degr($B$), $h_2$)), and note for later use that $k + m = g$. We can now see that $G_{\mathcal{D}^*}\left(A^2\right) \leq 2_m\left(\alpha + G_{\mathcal{D}}(F_2)\right)$ by comparing the deduction of $A^2$ in $\mathcal{D}^*$ with the deduction of $F_3$ in $\mathcal{D}$. We first note that all the inferences that occur in the former occur also in the latter, and that this holds even if we leave out the inference from $A$ and $A \supset B$ to $B$ in the deduction of $F_3$ in $\mathcal{D}$. This would imply that $G_{\mathcal{D}^*}\left(A^2\right) \leq \left(\alpha + G_{\mathcal{D}}(F_2)\right)$, if we did not have to take into account that there may be a change with respect to level lines when making these two computations. We should now recall that the inference from $F_1$ and $F_2$

to $F_3$ could be a substitution with $F_2$ as the major premiss, in which case $\Theta_{\mathrm{m}}/\mathrm{H}_{\mathrm{m}}$ is identical to $\Pi_5/F_2$. If so, $A^2$ stands immediately under a level line with jump $m$, and $G_{\mathcal{D}^*}\left(A^2\right) \le 2_m\left(\alpha + G_{\mathcal{D}}\left(F_2\right)\right)$. If the inference is not a substitution of this kind, the level of $A^2$ in $\mathcal{D}^*$ is anyway max(d($A$), max(d($B$), $h_2$)), while the level of $F_1$ in $\mathcal{D}$ had the higher level $h_1$. Thus, the difference between the level of $A^2$ in $\mathcal{D}^*$ and the level of $F_1$ in $\mathcal{D}$ is $h_1 - m$. As we saw above (when verifying the case of $\supset$-reduction for maximum formulas in Sect. 4.3.1), this lowering of level with $m$ can cause an increase of the Gentzen measure from $n$ to at most $2_m(n)$, which gives again the consequence that $G_{\mathcal{D}^*}\left(A^2\right) \le 2_m\left(\alpha + G_{\mathcal{D}}\left(F_2\right)\right)$. For the same kind of reasons, $G_{\mathcal{D}^*}\left(B^1\right) \le 2_m\left(\alpha + G_{\mathcal{D}}\left(F_2\right)\right)$.

Given these facts, we get

$$
\begin{aligned}
&G_{\mathcal{D}^*}\left(F_3^2\right) = \\
&= 2_f\left[2_k\left(G_{\mathcal{D}^*}\left(A^2\right) + G_{\mathcal{D}^*}\left(B^1\right)\right) + 2_g\left(G_{\mathcal{D}^*}\left(F_1\right) + G_{\mathcal{D}^*}\left(F_2\right)\right)\right] \le \\
&\le 2_f\left[2_k\left(2_m\left(\alpha + G_{\mathcal{D}}\left(F_2\right)\right) + 2_m\left(\alpha + G_{\mathcal{D}}\left(F_2\right)\right)\right) + 2_g\left(\alpha + G_{\mathcal{D}}\left(F_2\right)\right)\right] \le \\
&\le 2_f\left[2_k\left(2_m\left(\alpha + G_{\mathcal{D}}\left(F_2\right) + 1\right)\right) + 2_g\left(\alpha + G_{\mathcal{D}}\left(F_2\right)\right)\right] = \\
&= 2_f\left[2_{k+m}\left(\alpha + G_{\mathcal{D}}\left(F_2\right) + 1\right) + 2_g\left(\alpha + G_{\mathcal{D}}\left(F_2\right)\right)\right] = \\
&= 2_f\left[2_g\left(\alpha + G_{\mathcal{D}}\left(F_2\right) + 1\right) + 2_g\left(\alpha + G_{\mathcal{D}}\left(F_2\right)\right)\right] < 2_f\left[2_g\left(\alpha + G_{\mathcal{D}}\left(F_2\right) + 2\right)\right] = \\
&= 2_{f+g}\left(\alpha + G_{\mathcal{D}}\left(F_2\right) + 2\right) = 2_j\left(\alpha + G_{\mathcal{D}}\left(F_2\right) + 2\right) \le \\
&\le 2_j\left(G_{\mathcal{D}}\left(F_1\right) + G_{\mathcal{D}}\left(F_2\right)\right) = G_{\mathcal{D}}\left(F_3\right).
\end{aligned}
$$

### 4.4 *Details of Case 2*

In case 2, something more should be said concerning the two sub-cases (i) and (ii). Most of what is said applies in the same way to these two sub-cases. Consider first their common sub-case when the end formula $A$ of $\mathcal{D}$ does not stand as the conclusion of a substitution. Let $\mathcal{D}_1$ and $\mathcal{D}_2$ be the immediate sub-deductions of $\mathcal{D}$ obtained by leaving out the last inference R; if there is only one immediate sub-deduction of $\mathcal{D}$, disregard what is said about $\mathcal{D}_2$. Since $G\left(\mathcal{D}_1\right) < G\left(\mathcal{D}\right)$ and $G\left(\mathcal{D}_2\right) < G\left(\mathcal{D}\right)$, the standard deductions $\downarrow\mathcal{D}_1$ and $\downarrow\mathcal{D}_2$ reduce to normal deductions $\mathcal{D}_1^*$ and $\mathcal{D}_2^*$ according to the induction assumption. Let $\mathcal{D}^*$ be the deduction $\left(\mathcal{D}_1^*, \mathcal{D}_2^*/A\right)$ obtained by attaching R to $\mathcal{D}_1^*$ and $\mathcal{D}_2^*$. The standard deduction $\downarrow\mathcal{D}$ is reduced to $\mathcal{D}^*$ by the reductions that bring $\downarrow\mathcal{D}_1$ to $\mathcal{D}_1^*$ followed by reductions that bring $\downarrow\mathcal{D}_2$ to $\mathcal{D}_2^*$. It remains to see that $\mathcal{D}^*$ is normal, too. This is obvious in case (i), when R is an application of an introduction rule or a rule for atomic formulas. In case (ii), when R is an elimination inference, we note that there was no maximum formula or maximum segment in the main thread of $\mathcal{D}$, and hence there is no maximum formula or maximum segment in the main thread of $\mathcal{D}_2$, assumed to be the deduction of the major premiss of R, neither any maximum formula in the main branch of $\downarrow\mathcal{D}_2$. The reduction sequence that brings $\downarrow\mathcal{D}_2$ to $\mathcal{D}_2^*$ then does not change anything in the main branch, and hence the last inference of $\mathcal{D}_2^*$ cannot be an introduction.

Consider now the other common sub-case when the end formula of $\mathcal{D}$ is the conclusion of a substitution. Then $\mathcal{D}$ has the form of the first deduction shown below; again, disregard $\Sigma_2/A_2$ if the end-segment is the conclusion of an inference rule with only one premiss. Let $\mathcal{D}'$ be the second deduction shown below, obtained by letting the inference R, which the end-segment is the conclusion of, permute with the substitutions connected with the end-segment.

$$\cfrac{\begin{matrix}\Pi_n\\ C_n\end{matrix} \qquad \cfrac{\begin{matrix}\Pi_2\\ C_2\end{matrix} \qquad \cfrac{\begin{matrix}\Pi_1\\ C_1\end{matrix} \qquad \cfrac{\begin{matrix}\Sigma_1\\ A_1\end{matrix} \qquad \begin{matrix}\Sigma_2\\ A_2\end{matrix}}{A}}{A}}{A}}{A}$$

$$\cfrac{\cfrac{\begin{matrix}\Pi_n\\ C_n\end{matrix} \qquad \cfrac{\begin{matrix}\Pi_2\\ C_2\end{matrix} \qquad \cfrac{\begin{matrix}\Pi_1\\ C_1\end{matrix} \qquad \begin{matrix}\Sigma_1\\ A_1\end{matrix}}{A_1}}{A_1}}{A_1} \qquad \cfrac{\begin{matrix}\Pi_n\\ C_n\end{matrix} \qquad \cfrac{\begin{matrix}\Pi_2\\ C_2\end{matrix} \qquad \cfrac{\begin{matrix}\Pi_1\\ C_1\end{matrix} \qquad \begin{matrix}\Sigma_2\\ A_2\end{matrix}}{A_2}}{A_2}}{A_2}}{A}$$

It is possible that $G(\mathcal{D}') > G(\mathcal{D})$, but for the immediate sub-deductions $\mathcal{D}_1$ and $\mathcal{D}_2$ of $\mathcal{D}'$ it holds that $G(\mathcal{D}_1) > G(\mathcal{D})$ and $G(\mathcal{D}_2) > G(\mathcal{D})$. Hence, $\downarrow\mathcal{D}_1$ and $\downarrow\mathcal{D}_2$ reduce to normal deductions $\mathcal{D}_1^*$ and $\mathcal{D}_2^*$. Again, let $\mathcal{D}^*$ be $(\mathcal{D}_1^*, \mathcal{D}_2^*/A)$. As above, $\downarrow\mathcal{D}'$ reduces to $\mathcal{D}^*$ and $\mathcal{D}^*$ is normal. Since $\downarrow\mathcal{D} = \downarrow\mathcal{D}'$ we have shown that $\downarrow\mathcal{D}$ reduces to normal form.

## 5 Extension to First Order Arithmetic

### *5.1 A System for First Order Arithmetic*

To extend the normalization result to classical first order Peano arithmetic, I consider a system of natural deduction obtained by adding the rule of mathematical induction to a system of natural deduction for predicate logic of the kind considered in previous section that contains appropriate rules for atomic formulas. I leave open the exact choice of inference rules for the atomic formulas since it does not matter for the normalization result, but the language should contain the individual constant 0 and a symbol s for successor. I am following Gentzen in restricting the closed terms to be the numerals (formed from 0 and s), which means that addition and multiplication have to be covered by ternary predicates.

The inference rule of mathematical induction is taken in the following form

$$\frac{A(0) \qquad \begin{matrix}[A(a)] \\ A(sa)\end{matrix}}{A(t)}$$

where $t$ is a (closed or open) individual term, and where $a$ is a parameter closed by the inference (in other words, $a$ should satisfy the conditions of eigenvariables).

## 5.2 *Induction as an Elimination Rule for N*

It appears from the work by Martin-Löf [13] that the induction rule could be understood as the elimination rule for the predicate of being a natural number, say $N(a)$, which has as introduction rules the axiom $N(0)$ and the rule to infer $N(sa)$ from $N(a)$. Then the rule of mathematical induction has $N(t)$ as a third premiss, which is here omitted since I have not required that the language should contain the predicate of being a natural number. If one has such a third premiss, it counts as a major premiss. Thus, if such a major premiss stands as the conclusion of an introduction, it is a maximum formula. Its occurrence in a deduction $\mathcal{D}$ is removed by a reduction of $\mathcal{D}$ that takes the following form: if the major premiss is $N(0)$, then $\mathcal{D}$ reduces to the deduction of the first minor premiss $A(0)$; if the major premiss is $N(st)$, above which stands a sub-deduction $\Pi/N(t)$, then $\mathcal{D}$ reduces to a deduction obtained from the deduction of the second minor premiss, the induction step, by substituting $t$ for $a$, and by substituting for the assumption $A(t)$ another application of the induction rule taking $\Pi/N(t)$ as the deduction of the major premiss.

What is to count as an appropriate extension of the normalization theorem of predicate logic to arithmetic is then obvious. Its statement is literally the same as for predicate logic: every deduction reduces to normal form, i.e. to a deduction that contains no maximum formula, and hence, cannot be reduced further.

## 5.3 *Reducible Induction Inferences*

In the light of Martin-Löf's work, the induction rule as stated above is to be seen as an abridged elimination rule. Since its major premiss is left tacit here, we have to state separately what it is for an induction inference to be reducible. For convenience, I shall follow Gentzen, and say that an application of the induction rule is *reducible* when the term $t$ that is substituted for the parameter $a$ in the induction formula to get the conclusion $A(t)$ is a numeral and that a deduction of the form shown to the left below *immediately reduces* to $\Pi_1/A(0)$, if $n = 0$, and to the deduction shown to the right when $n$ is a numeral distinct from 0.

$$
\begin{array}{c}
\dfrac{\begin{array}{c}\Pi_1\\ A(0)\end{array} \qquad \begin{array}{c}[A(a)]\\ \Pi_2(a)\\ A(sa)\end{array}}{A(t)}\\
\Sigma
\end{array}
\qquad\qquad
\begin{array}{c}
\dfrac{\begin{array}{c}
\dfrac{\dfrac{\begin{array}{c}\Pi_1\\ A(0)\end{array} \quad \begin{array}{c}[A(0)]\\ \Pi_2(0)\\ A(s0)\end{array}}{A(s0)} \quad \begin{array}{c}[A(s0)]\\ \Pi_2(s0)\\ A(ss0)\end{array}}{A(ss0)}\\
\hline\hline
A(n-1)
\end{array} \quad \begin{array}{c}[A(n-1)]\\ \Pi_2(n-1)\\ A(n)\end{array}}{A(n)}\\
\Sigma
\end{array}
$$

The deduction to the right is obtained by a series of $n$ substitutions ($n$ being used ambiguously both for numerals and numbers); $n-1$ stands of course for the $n$:th numeral, counting 0 as the first numeral. A reduction proceeds in this way in one sweep like in Gentzen's proof, instead of stepwise as described in Sect. 5.2.

The definition of what counts as an immediate reduction is to be understood in two ways. It states that a standard natural deduction of the form shown to the left reduces immediately to the standard natural deduction shown to the right above, obtained by an iterated operation of substitution. But it also states that a deduction notation of the form shown to the left reduces immediately to the deduction notation shown to the right above. If $\mathcal{D}$ is a deduction notation of that form, and $\mathcal{D}^*$ is the deduction notation that is an immediate reduction of $\mathcal{D}$, then obviously $\downarrow\mathcal{D}^*$ is an immediate reduction of $\downarrow\mathcal{D}$.

A standard natural deduction in arithmetic is defined to be *normal* when it contains no maximum formula and no reducible induction inference.

## 5.4 *The Gentzen Measure of Arithmetical Deductions*

The *degree* of an induction inference is the degree of its induction formula $A(a)$. All induction inferences count as *critical*. The definitions of *level* and *level line* are then as before. The definition of the *Gentzen measure* also runs as before with the difference that $\omega$ instead of 2 is the base of the iterated exponentiations in connection with level lines, and that the sum of two Gentzen measures is always to be understood as the natural sum of ordinals.

Furthermore, there is a new clause for the case that $A(t)$ stands in a deduction $\mathcal{D}$ as the conclusion of an induction inference, which in the same manner as before depends on whether its inference line is a level line or not: given that $G_{\mathcal{D}}(A(0)) = \alpha$ and $G_{\mathcal{D}}(A(sa)) = \beta$, where $A(0)$ and $A(sa)$ are the two premisses of the induction inference, we define $G_{\mathcal{D}}(A(t)) = \alpha + \beta \cdot \omega$, if the inference line is not a level line, and $G_{\mathcal{D}}(A(t)) = \omega_j(\alpha + \beta \cdot \omega)$, if the inference line is a level line with the jump $j$.

## 5.5 Branches and Threads

In the definitions of *branch* and *thread*, we now stipulate that they may also begin at a conclusion of an induction inference and that they end when a premiss of an induction inference is reached. Thus, going from below, the first formula of a main branch or thread is found if one has reached the conclusion of an induction.

## 5.6 Proof of the Normalization Theorem

It is now very easy to extend the proof of Sect. 4 to arithmetic. Let us first verify that the Gentzen measure is lowered by a reduction of an induction inference when the numeral $n$ in the conclusion $A(n)$ is different from 0; the case when $n$ is 0 being trivial.

Note first of all that if the induction inference in the deduction $\mathcal{D}$ does not contain a level line, then none of the new substitutions in the reduction $\mathcal{D}^*$ of $\mathcal{D}$ contains a level line, while if the inference contains a level line, then the last, and only the last, of the new substitutions in $\mathcal{D}^*$ contains a level line, the jump being the same. Let $j$ be the jump at the level line that the induction inference contains, if there is such a level line, and let $j$ be 0 otherwise.

We then find that for the conclusion $A(n)$ of the last exhibited substitution in the reduction $\mathcal{D}^*$

$$\begin{aligned} G_{\mathcal{D}^*}(A(n)) &= \omega_j\left(G_{\mathcal{D}^*}(A(0)) + G_{\mathcal{D}^*}(A(1)) + \cdots + G_{\mathcal{D}^*}(A(n))\right) = \\ &= \omega_j\left(G_{\mathcal{D}}(A(0)) + G_{\mathcal{D}}(A(sa)) + \cdots + G_{\mathcal{D}}(A(sa))\right) < \\ &< \omega_j\left(G_{\mathcal{D}}(A(0)) + G_{\mathcal{D}}(A(sa)) \times \omega\right) = \omega_f\left(G_{\mathcal{D}}(A(n))\right). \end{aligned}$$

Here, the iterated natural sum has $n$ terms, and $A(n)$ in the last term of the identity refers to the conclusion of the induction inference in $\mathcal{D}$.

It remains to say how the two main cases of the proof are to be defined now. Case 1 becomes: The end formula or end segment of the deduction (notation) $\mathcal{D}$ is the conclusion of an elimination or of an induction, and the main thread of $\mathcal{D}$ contains a maximum formula, a maximum segment or the conclusion of a reducible induction inference. As before we locate the last maximum formula or segment in the main thread, if any, and let $\mathcal{D}^*$ be the reduction of $\mathcal{D}$ performed at that place. If there is no maximum formula or maximum segment in the main thread, the thread starts at the conclusion of a reducible induction inference (given how case 1 is now defined), and we let $\mathcal{D}^*$ be the reduction of $\mathcal{D}$ for this induction inference. As we have seen $G(\mathcal{D}^*) < G(\mathcal{D})$. By the induction assumption, $\downarrow\mathcal{D}^*$ reduces to normal form and hence so does $\downarrow\mathcal{D}$, since $\downarrow\mathcal{D}$ reduces to $\downarrow\mathcal{D}^*$.

Case 2 is still the negation of case 1 as now defined, and the proof in this case remains the same with the sole addition that in case (ii) is now to be included the possibility that the end-formula or end-sequent of the main thread of $\mathcal{D}$ is the conclusion of an induction. The proof goes in the same way.

# 6 A Normalization Theorem for a System of Natural Deduction with Explicit Substitution

## 6.1 *Adding the Substitution Schema as an Inference Rule*

In the above, the schema of substitution has been used in deduction notations as a means to prove the normalization theorem for standard systems of natural deduction. But as noted, the schema may be understood as stating an inference rule in its own right, which may be added for good reasons to standard systems of natural deduction (Sect. 3.1.1). When taking this perspective, one naturally asks for a normalization theorem for such an enriched system of natural deduction, which so far has not been obtained. This is the theme of this section.

I shall consider the system obtained by adding the substitution rule to the system for classical first order Peano arithmetic defined in Sect. 5. The first thing to do is to define additional immediate reductions for this system. To prove the normalization theorem for it, I shall then make some small changes of previous definitions.

## 6.2 *Additional Immediate Reductions*

To the immediate reductions that were defined for maximum formulas, maximum segments, and induction inferences occurring in deduction notations and the ones called simplifications (Sects. 4.2 and 5.3), I add certain permutations of substitutions with other inferences as immediate reductions. Distinguishing two cases depending on whether the minor premiss of the substitution is the conclusion of an inference with one or two premises, they are as indicated below; the last exhibited inference in the left figures is the substitution inference:

$$\dfrac{\begin{matrix}\Pi\\ A\end{matrix}\qquad \dfrac{\begin{matrix}[A]\\ \Sigma\\ B\end{matrix}}{C}}{\begin{matrix}C\\ \Theta\end{matrix}} \quad \text{reduces immediately to} \quad \begin{matrix}\dfrac{\dfrac{\begin{matrix}\Pi\\ A\end{matrix}\qquad \begin{matrix}[A]\\ \Sigma\\ B\end{matrix}}{B}}{C}\\ \Theta\end{matrix}$$

$$\begin{matrix}\dfrac{\begin{matrix}\Pi\\ A\end{matrix}\qquad \dfrac{\begin{matrix}[A]\\ \Sigma_1\\ B_1\end{matrix}\qquad \begin{matrix}[A]\\ \Sigma_2\\ B_2\end{matrix}}{C}}{C}\\ \Theta\end{matrix} \quad \text{reduces immediately to} \quad \begin{matrix}\dfrac{\dfrac{\begin{matrix}\Pi\\ A\end{matrix}\qquad \begin{matrix}[A]\\ \Sigma_1\\ B_1\end{matrix}}{B_1}\qquad \dfrac{\begin{matrix}\Pi\\ A\end{matrix}\qquad \begin{matrix}[A]\\ \Sigma_2\\ B_2\end{matrix}}{B_2}}{C}\\ \Theta\end{matrix}$$

### 6.3 Normal Forms

A deduction that has no immediate reduction cannot contain any substitution inference, since a substitution inference can always be permuted upwards or be removed by a simplification. However, an irreducible deduction may still contain maximum segments because of the restrictions which the reductions defined for curved maximum segments are provided with. Accordingly, the deductions that we are now considering are defined as *normal* when containing no maximum formula or segment, no reducible induction inference, and no substitution inference, which is not the same as saying that it has no immediate reduction.

The normalization theorem will be proved by defining a complete normalization strategy, in contrast to the partial ones defined in Sects. 4 and 5. It generates a reduction sequence that will be shown to terminate in a normal deduction. The theorem obviously contains the standard normalization theorem as a special case: if a sequence of immediate reductions starts from a standard deduction and terminates in a normal deduction, it also ends in a standard deduction, although the intermediate deductions may contain applications of the substitution rule.

### 6.4 Critical Inferences and the Degree of a Formula

The notion of critical inference was introduced as a rough approximation of the class of inferences whose major premiss is a potential maximum formula (Sect. 2.3). Still without aiming at an optimal notion of critical inference, I shall narrow down the previously defined notion, which will be essential for the proof of the normalization theorem.

To this end, I need the notion of *order* of threads in a deduction defined as follows: A main thread has order 0. A thread has order $n + 1$, if it ends with a minor premiss of an $\supset$-elimination whose major premiss belongs to a thread of order $n$, a major premiss of a vacuous substitution whose minor premiss belongs to a thread of order $n$, or a premiss of an induction whose conclusion belongs to a thread of order $n$.

The new definition of *critical inference* in a deduction is then defined by saying that every substitution inference is critical and that, in addition, an elimination or induction inference is critical, except when the thread of order $n$ to which its conclusion belongs satisfies the condition that there is no maximum formula or maximum segment in this thread or in a thread of order lower than $n$ and no reducible induction whose conclusion belongs to this thread or to a thread of order lower than $n$. A normal deduction will consequently have no critical inference.

One can safely make these exceptions in the definition of critical, because when the stated condition is satisfied for a thread of a deduction $\mathcal{D}$, no reduction of $\mathcal{D}$ can have the effect that the major premiss of an elimination in the thread becomes a maximum formula or that an induction whose conclusion stands in the thread becomes reducible. An effect of this definition is that a critical inference in a deduction may cease to be critical after a reduction has been performed, but as we have seen this is not a problem (Sect. 4.3.2 and footnote 25). What is essential in order that the Gentzen measure is to decrease at reductions is that the inverse cannot happen: if an inference is critical in a reduction $\mathcal{D}^*$ of $\mathcal{D}$, then the corresponding inference in $\mathcal{D}$ must also have been classified as critical.

We also need to redefine the degree of a formula so that atomic formulas get the degree 1; thus, $\text{degr}(A) = 1$, if $A$ is atomic; $\text{degr}(\forall x A) = \text{degr}(A) + 1$; $\text{degr}(A \& B) = \text{degr}(A \supset B) = \max(\text{degr}(A), \text{degr}(B)) + 1$. The point of this redefinition is that we want to show that applications of the substitution rule can be removed by permutations upwards and that, to this end, we have to take care of the fact that the deduction may expand exponentially at such permutations even when the formulas involved are atomic.

## 6.5 The Normalization Strategy

Deductions will be normalized by first considering a main thread, making reductions according to the first one of the cases specified below that is applicable. When none of the cases is applicable to the main threads, one continues with the threads of order 1, and so on. When several choices are open for how to proceed, one can specify as one pleases which one is to be chosen, so as to get a unique order of reductions.

The precise instruction for how to form a *reduction sequence*, that is, a sequence of deductions such that each one reduces to the next one, is as follows: Given that the last element in the sequence obtained so far is a deduction $\mathcal{D}$, the next deduction $\mathcal{D}^*$ in the sequence is obtained by finding a thread in $\mathcal{D}$ of some order $n$ such that one of the instructions (1)–(4) is applicable but none of them is applicable to threads of order less than $n$, and then proceed as the first one of the applicable cases says:

1. There is a simplification to make. $\mathcal{D}^*$ is obtained by making one of them.
2. There is a segment in the thread that stands as the conclusion of an introduction and after this segment all formula occurrences in the thread, if any, are conclusions of introductions. $\mathcal{D}^*$ is obtained by making a series of commutative reductions so that the introduction comes to stand under the segment in question; see Sect. 4.4.

3. There is a maximum formula, a maximum segment, or the conclusion of a reducible induction in the thread. $\mathcal{D}^*$ is got by removing the last maximum formula or segment in the thread by a reduction or, if there is none, by removing the induction by a reduction.
4. There is a segment in the thread that stands as the conclusion of an elimination or an induction and after this segment all formula occurrences in the thread, if any, are conclusions of eliminations or introductions. $\mathcal{D}^*$ is got by making a series of commutative reductions so that the elimination or the induction comes to stand under the segment in question; the result is again as described graphically in Sect. 4.4.

We have to show that in case (3) there is a reduction to be made. Given this, it is clear that when the reduction sequence terminates because none of the cases are applicable to any thread, the deduction is normal.

It is sufficient to show that when instruction (3) is to be followed by carrying out an $\supset$- or $\forall$-reduction for a curved maximum segment, the condition stipulated for such a reduction (Sect. 4.2) is satisfied. Using the same letters as in the definitions of these reductions, and referring to the assumption $A \supset B$ or $\forall x A$ that is bound by the substitution in question as the assumption $D$, we have to show that there is no conclusion $E$ of an introduction inference standing below $D$ and above the exhibited level line. To this end, note to begin with that such an $E$ would have to belong either to the same thread of order $n$ as $D$ or to a thread of a lower order. Since cases (1)–(4) are not to be applicable to threads of order lower than $n$, there can be no premiss of a substitution in such threads. The minor premiss $C^1$ and the conclusion $C^2$ of the substitution at which $D$ is bound must therefore belong to the same thread as $D$.

Furthermore, from the fact that $C^1$ and $C^2$ belong to the same thread as $D$ and the facts that cases (1) and (2) are not to be applicable to this thread and that there is to be no maximum formula or segment in the thread after $D$, it follows that the conclusion $E$ cannot stand above $C^2$. Hence, $E$ would have to stand under $C^1$ and above the level line. For the same reasons as already invoked, the premiss $F_1$ of the inference that contains the level line and that $E$ stands above or is identical with cannot be the major premiss of an elimination nor the premiss of a substitution, if it belongs to the same thread as $D$. Hence, if $F_1$ belongs to the same thread as $D$, it must be the minor premiss of an elimination or of the premiss of an induction in view of the fact the inference contains a level line. It follows that the conclusion $F_3$ of this inference that contains the level line must belong to a thread of order lower than $n$. But this contradicts that the inference contains a level line: since its conclusion belongs to a thread whose order is less than $n$, and which therefore contains no maximum formula or segment and no reducible induction, the inference cannot be critical according to the new definition of this notion.

As seen, the instruction of how to form the reduction sequence is such that when (1) or (3) is applied, the obtained deduction is an immediate reduction of the preceding one, while when (2) or (4) is applied, the obtained deduction can be obtained by a sequence of immediate reductions from the preceding one.

## 6.6 Verifying That the Gentzen Measure Goes Down in a Reduction Sequence

The normalization theorem is now obtained by verifying the fact that the Gentzen measure goes down in the reductions sequences, which must therefore terminate.[26]

That $G(\mathcal{D}^*) < G(\mathcal{D})$ when $\mathcal{D}^*$ is obtained from a deduction $\mathcal{D}$ according to instruction (1) or (3) has already been verified in the preceding sections. Consider now the case when $\mathcal{D}^*$ is obtained according to instruction (2). I use the symbols occurring in the graphic description of the case in Sect. 4.4, differentiating between the different occurrences of the same formula in the displayed segments by superscripts $1, 2, \ldots, n$, and $n+1$. We first note that the inference line contained in the lowest exhibited substitution, where $C_n$ is the major premiss, must contain a level line and that there is no level line further down in $\mathcal{D}$, because all the inferences below this line are either introductions or their conclusions belong to threads of lower order than the order of the thread to which $A^{n+1}$ belongs. Accordingly, its jump is $j = \text{degr}(C_n)$. For the same reasons, the two corresponding substitutions in $\mathcal{D}^*$ where $C_n$ is the major premiss are also level lines with the same jump $j$. Recall that $j \geq 1$ (Sect. 6.4). Let $\alpha = \max\left(G_{\mathcal{D}^*}(C_n) + G_{\mathcal{D}^*}\left(A_1^n\right), G_{\mathcal{D}^*}(C_n) + G_{\mathcal{D}^*}\left(A_2^n\right)\right)$. Clearly, $\alpha + 2 \leq G_{\mathcal{D}}(C_n) + G_{\mathcal{D}}(A^n)$, because the part of $\mathcal{D}$ that ends with $A^n$ has at least two nodes more than the part of $\mathcal{D}^*$ that ends with $A_i^n (i = 1 \text{ or } 2)$. Consequently, we find that

$$\begin{aligned}
G_{\mathcal{D}^*}(A) &= G_{\mathcal{D}^*}\left(A_1^{n+1}\right) + G_{\mathcal{D}^*}\left(A_2^{n+1}\right) + 1 = \\
&= \omega_j\left(G_{\mathcal{D}^*}(C_n) + G_{\mathcal{D}^*}\left(A_1^n\right)\right) + \omega_j\left(G_{\mathcal{D}^*}(C_n) + G_{\mathcal{D}^*}\left(A_2^n\right)\right) + 1 \leq \\
&\leq \omega_j(\alpha) + \omega_j(\alpha) + 1 < \omega_j(\alpha + 1) + 1 < \omega_j(\alpha + 2) \leq \\
&\leq \omega_j\left(G_{\mathcal{D}}(C_n) + G_{\mathcal{D}}(A^n)\right) = G_{\mathcal{D}}\left(A^{n+1}\right).
\end{aligned}$$

Finally, let $\mathcal{D}^*$ be the reduction obtained from $\mathcal{D}$ according to instruction (4). Since this instruction is applied when the thread in question and threads of lower order do not contain any maximum formula or segment or the conclusion of a reducible induction, the elimination or induction inference that is permuted with the segment in question to get $\mathcal{D}^*$ is not a critical inference, nor is there a critical inference below the conclusion $C_n$. The same reasoning that showed in the preceding case that $G_{\mathcal{D}^*}(A) < G_{\mathcal{D}}\left(A^{n+1}\right)$ therefore applies again.

**Acknowledgements** Per Martin-Löf encouraged me to make investigations that led to the addition of Sect. 6 to a first version of this paper. Luiz Carlos Pereira and Michael Hahn made valuable

[26] In the proofs of the normalization theorems, I have not paid attention to their status as consistency proofs, but now the statement that Gentzen measure goes down in a reduction sequence, from which it follows by transfinite induction that it must terminate, has the simple form that Gentzen required of a statement proved by induction in a consistency proof.

comments after reading manuscripts at different stages and pointed out a deficiency in an earlier version of the proof. Sama Agahi made the figures that are inserted in the text. I am grateful to all of them.

## References


1. M. Abadi, L. Cardelli, P.-L. Curien, J.-J. Levy, Explicit Substitutions. J. Funct. Program. **1**, 375–416 (1991)
2. P. Bernays, On the original Gentzen consistency proof for number theory, in *Intuitionism and Proof Theory*, ed. by J. Myhill et al. (North-Holland, Amsterdam, 1980), pp. 409–417
3. W. Buchholz, Explaining Gentzen's consistency proof within an infinitary proof theory, in *Computational logic and proof theory, KGC 1997*, ed. by G. Gottlob et al., Lecture Notes in Computer Science, vol. 1289 (Springer, Berlin, 1997)
4. C. Cellucci, Proof theory and complexity. Synthese **62**, 173–189 (1985)
5. G. Gentzen, Untersuchungen über das logische Schließen, Math. Z. **39**, 176–210, 405–431 (1934–1935)
6. G. Gentzen, Neue Fassung des Widerspruchsfreiheitsbeweises für die reine Zahlentheorie, *Forschungen zur Logik und zur der exakten Wissenschaften. Neue Folge, Heft 4* (Hirzel, Leipzig, 1938), pp. 19–44
7. G. Gentzen, in *The Collected Papers of Gerhard Gentzen*, ed. by M.E. Szabo (North-Holland, Amsterdam, 1969)
8. W.A. Howard, Assignments of ordinals to terms for primitive recursive functionals of finite type, in *Intuitionism and Proof Theory*, ed. by J. Myhill et al. (North-Holland, Amsterdam, 1970), pp. 443–458
9. H. Jervell, A normal form in first order arithmetic, in *Proceedings of the Second Scandinavian Logic Symposium*, ed. by J.E. Fenstad (North Holland, Amsterdam, 1971), pp. 93–108
10. A. Kanckos, Consistency of Heyting arithmetic in natural deduction. Math. Log. Q. **56**, 611–624 (2010)
11. A. Kanckos, A possible and necessary consistency proof, in *Philosophical Studies from the University of Helsinki* (University of Helsinki, Helsinki, 2011)
12. D. Leivant, Strong normalization for arithmetic (variations on a theme of Prawitz), in *Proof Theory Symposium Kiel 1974*, ed. by A. Dold et al., Lecture Notes in Mathematics, vol. 500 (Springer, Berlin, 1975), pp. 182–197
13. P. Martin-Löf, Hauptsatz for the intuitionistic theory of iterated inductive definitions, in *Proceedings of the Second Scandinavian Logic Symposium*, ed. by J.E. Fenstad (North Holland, Amsterdam, 1971), pp. 179–216
14. P. Martin-Löf, Infinite terms and a system of natural deduction. Compos. Math. **2**, 92–103 (1972)
15. G. Mints, Normalization of finite terms and derivations via infinite ones, in *V. International Congress for LMPS, Abstracts of Papers,* 1975, pp. 215–219 (reprinted in [17, pp. 73–76])
16. G. Mints, A new reduction sequence for arithmetic (in Russian), in *Zapiski nauchnyh seminarov LOMI*, vol. 88 (1979), pp. 106–130 (translated in [17, pp. 77–96])
17. G. Mints, *Selected Papers in Proof Theory* (Bibliopilis and North-Holland, Napoli, 1992)
18. G. Mints, Reduction of finite and infinite derivations. Ann. Pure Appl. Log. **104**, 167–188 (2000)
19. L.C. Pereira, On the estimation of the length of normal derivations, in *Philosophical Studies*, vol. 4 (Akademilitteratur, Stockholm, 1982)
20. D. Prawitz, *Natural Deduction. A Proof-Theoretical Study* (Almqvist & Wicksell, Uppsala, 1965) (Reprinted: Dover Publications, Mineola, 2006)
21. A. Raggio, Gentzen's Hauptsatz for the systems NI and NK. Log. Anal. **30**, 91–100 (1965)

22. W.C. Sanz, Big bound, in *Natural Deduction, Rio 2001*, mimeographed abstracts of invited talks, PUC, Rio de Janeiro, 2001
23. B. Scarpellini, Applications of Gentzen's second consistency proof. Math. Ann. **181**, 325–354 (1969)
24. K. Schütte, *Beweistheorie* (Springer, Berlin, 1960)
25. K. Schütte, Beweistheoretische Erfassung der unendlichen Induktion in der Zahlentheorie. Math. Ann. **122**, 369–389 (1951)
26. W. Tait, Normal derivability in classical logic, in *The Syntax and Semantics of Infinitary Languages*, ed. by J. Barwise, Lecture Notes of Mathematics, vol. 72 (Springer, Berlin, 1968), pp. 204–236
27. J. von Plato, Gentzen's proof of normalization for intuitionistic natural deduction. Bull. Symb. Log. **14**, 240–257 (2008)
28. J. von Plato, Talk at Workshop on Proof Theory and Constructivism, 2012, Helsinki, 23 March 2012

# A Direct Gentzen-Style Consistency Proof for Heyting Arithmetic

**Annika Siders**

## 1 Introduction

Gerhard Gentzen was the first to give a proof of the consistency of Peano Arithmetic and in all he worked out four different proofs between 1934 and 1939. The second proof was published as [1], the third as [2], and the fourth as [3]. The first proof was published posthumously in English translation in [4] and in the German original as [5].

A study of the papers Gentzen left behind shows that he worked on yet another fifth proof between 1939 and 1943. The aim was to rework the 1938 proof with an intuitionistic sequent calculus, to get a direct proof of the consistency of intuitionistic Heyting Arithmetic. Gentzen's attempts are preserved in the form of close to a hundred large pages of stenographic notes, with the signum BTJZ that stands for "Proof theory of intuitionistic number theory".

The aim of this paper is to give a direct Gentzen-style proof of the consistency of intuitionistic arithmetic. It is based on Gentzen's classical proof from 1938 formulated by Gaisi Takeuti in [10, ch. 2, § 12]. Takeuti's proof can be considered the standard proof today. The proof is carried out by giving a reduction procedure (as in our Lemma 5.5.1) for every derivation of the empty sequent that represents a contradiction in the system. By giving every sequent an ordinal it is shown that the reduction procedure terminates.

Gentzen and Takeuti used semantical arguments to prove a lemma (our Lemma 5.4.8) stating that there is no so-called simple derivation of the empty sequent. Their proof is short, but we shall instead show that the lemma can be proved purely proof-theoretically by formulating the arithmetical axioms as rules

A. Siders (✉)
Department of Philosophy, University of Helsinki, Unionkatu 38 A, 00014, Helsinki, Finland
e-mail: annika.siders@helsinki.fi

 
R. Kahle, M. Rathjen (eds.), *Gentzen's Centenary*,
DOI 10.1007/978-3-319-10103-3_7

instead of initial sequents and by considering all possible combinations of these rules, as in our Lemma 5.4.6.

We shall assume that the reader has basic knowledge of ordinals and refer to [10] for a more detailed treatment of the subject. For further reading and description of Gentzen's manuscripts we also recommend the thorough discussion of Gentzen's work found in [9].

## 2 The Sequent Calculus G0i

A *sequent* is an expression of the form $\Gamma \rightarrow A$, where the *antecedent* $\Gamma$ is a (possibly empty) multiset. A multiset is a finite list of formulas where the order of the formulas does not matter but the multiplicity of the formulas does, in contrast to ordinary sets. The *succedent* $A$ is a formula, but can also be empty. The rules for the intuitionistic sequent calculus G0i, from [7] except that we have no rule of weakening, are as follows.

**Initial sequent:**

$$A, \Gamma \rightarrow A$$

**Logical rules:**

$$\frac{A, B, \Gamma \rightarrow C}{A\&B, \Gamma \rightarrow C}\ L\& \qquad \frac{\Gamma \rightarrow A \quad \Gamma' \rightarrow B}{\Gamma, \Gamma' \rightarrow A\&B}\ R\&$$

$$\frac{A, \Gamma \rightarrow C \quad B, \Gamma' \rightarrow C}{A \vee B, \Gamma, \Gamma' \rightarrow C}\ L\vee \qquad \frac{\Gamma \rightarrow A}{\Gamma \rightarrow A \vee B}\ R\vee \qquad \frac{\Gamma \rightarrow B}{\Gamma \rightarrow A \vee B}\ R\vee$$

$$\frac{\Gamma \rightarrow A}{\sim A, \Gamma \rightarrow}\ L\sim \qquad \frac{A, \Gamma \rightarrow}{\Gamma \rightarrow \sim A}\ R\sim$$

$$\frac{\Gamma \rightarrow A \quad B, \Gamma' \rightarrow C}{A \supset B, \Gamma, \Gamma' \rightarrow C}\ L\supset \qquad \frac{A, \Gamma \rightarrow B}{\Gamma \rightarrow A \supset B}\ R\supset$$

$$\frac{A(t/x), \Gamma \rightarrow C}{\forall x A, \Gamma \rightarrow C}\ L\forall \qquad \frac{\Gamma \rightarrow A(y/x)}{\Gamma \rightarrow \forall x A}\ R\forall$$

$$\frac{A(y/x), \Gamma \rightarrow C}{\exists x A, \Gamma \rightarrow C}\ L\exists \qquad \frac{\Gamma \rightarrow A(t/x)}{\Gamma \rightarrow \exists x A}\ R\exists$$

**Structural rules:**

$$\frac{A, A, \Gamma \rightarrow C}{A, \Gamma \rightarrow C}\ LC$$

$$\frac{\Gamma \rightarrow A \quad A, \Gamma' \rightarrow C}{\Gamma, \Gamma' \rightarrow C}\ Cut$$

In the quantifier rules the expression $A(t/x)$ means that every free occurrence of $x$ in $A$ is substituted with the term $t$. In the rules $L\exists$ and $R\forall$ the standard variable restriction holds that $y$, also called the eigenvariable of the rule, must not be free in the conclusion of the rule. The formula that is introduced in the conclusion of a logical rule, for example $A\&B$ in the conjunction rules, is the *principal formula* of the rule. The formulas that the rule is applied on are the *auxiliary* formulas. In the structural rules the principal formula is the formula that the rules are applied on, in this case $A$. The formula is also called contraction or cut formula. The multiset $\Gamma$ in the sequents is called the *context* of the rule. We use a calculus with arbitrary contexts in all initial sequents and hence no rule of weakening is needed.

## 3 Heyting Arithmetic

**Definition 3.0.1** A *term* is the constant 0 or a variable and if $t$ and $t'$ are terms then also $s(t), t + t'$ and $t \cdot t'$ are terms. We say that a term is *closed* if it does not contain any variable.

We also define *numerals* inductively in the following way: 0 is a numeral and if $\overline{n}$ is a numeral, then also $s(\overline{n})$ is a numeral. Numerals are formal expressions for the natural numbers and $\overline{n}$ is $n$ copies of $s$ followed by a 0.

The axioms of Heyting Arithmetic can be formulated as rules of natural deduction, expanding the logical calculus. Together with an induction rule the logical and arithmetical rules constitute the system of Heyting Arithmetic (HA). Negri and von Plato [6] developed the general method for converting mathematical axioms into rules for the primary purpose of proving cut elimination in systems of sequent calculus. The specific system for arithmetic was first used by von Plato [8] to prove the disjunction and existential properties. These rules act on the succedent part of the sequents and have arbitrary contexts. As a special case we get rules without premises.

**Rules for the equality relation:**

$$\frac{}{\Gamma \rightarrow t = t}\ Ref \qquad \frac{\Gamma \rightarrow t = t'}{\Gamma \rightarrow t' = t}\ Sym$$

$$\frac{\Gamma_1 \rightarrow t = t' \quad \Gamma_2 \rightarrow t' = t''}{\Gamma_{1-2} \rightarrow t = t''}\ Tr$$

**Recursion rules:**

$$\frac{}{\Gamma \to t + 0 = t} +Rec0 \qquad \frac{}{\Gamma \to t + s(t') = s(t + t')} +Recs$$

$$\frac{}{\Gamma \to t \cdot 0 = 0} \cdot Rec0 \qquad \frac{}{\Gamma \to t \cdot s(t') = t \cdot t' + t} \cdot Recs$$

**Replacement rules:**

$$\frac{\Gamma \to t = t'}{\Gamma \to s(t) = s(t')} sRep$$

$$\frac{\Gamma \to t = t'}{\Gamma \to t + t'' = t' + t''} +Rep_1 \qquad \frac{\Gamma \to t' = t''}{\Gamma \to t + t' = t + t''} +Rep_2$$

$$\frac{\Gamma \to t = t'}{\Gamma \to t \cdot t'' = t' \cdot t''} \cdot Rep_1 \qquad \frac{\Gamma \to t' = t''}{\Gamma \to t \cdot t' = t \cdot t''} \cdot Rep_2$$

**Infinity rules:**

$$\frac{\Gamma \to s(t) = 0}{\Gamma \to} Inf_1 \qquad \frac{\Gamma \to s(t) = s(t')}{\Gamma \to t = t'} Inf_2$$

**Induction rule:**

$$\frac{\Gamma_1 \to A(0/x) \quad A(y/x), \Gamma_2 \to A(sy/x) \quad A(t/x), \Gamma_3 \to D}{\Gamma_{1-3} \to D} Ind$$

In the arithmetical rules $t, t'$ and $t''$ are terms. In the induction rule $y$ is the eigenvariable of the rule and it should not occur free in the conclusion. The induction formula $A$ is arbitrary.

**Definition 3.0.2** A *valid derivation* in HA is an initial sequent or an arithmetical rule without premises or is obtained by applying a rule on valid derivations of the premises of the rule.

The end-piece is inductively defined for any given derivation. The end-sequent is included in the derivation. If a sequent concluded by a structural rule or Ind is included in the end-piece, then all the premises of the rule are also included in the end-piece. An arithmetical or logical rule borders on the end-piece if the conclusion of the rule is included in the end-piece.

A formula $A$ is a *descendant* of a formula $B$ if $A$ is in the context of the conclusion of a rule and $B$ is an identical formula in the context of a premise or if $A$ is the principal formula of the rule and $B$ is an auxiliary formula in a premise. Furthermore, if $A$ a descendant of $B$ and $B$ is a descendant of $C$, then $A$ is a descendant of $C$. If $A$ is a descendant of $B$, then $B$ is a *predecessor* of $A$.

## 4 The Ordinal of a Derivation

We define the height of a sequent in a derivation.

**Definition 4.0.3** (i) The *grade* of a formula is the number of logical symbols in the formula. The grade of a Cut or an *Ind* is the grade of the cut or the induction formula.
(ii) The *height* of a sequent $S$ in a derivation $P$, denoted $h(S; P)$ or $h(S)$, is the maximum of the grades of the cuts and inductions below $S$ in $P$.

Note that the height of the end-sequent is 0 and that the premises of a rule all have the same height. If $S_1$ is a sequent under another sequent $S_2$, then $h(S_1) \leqslant h(S_2)$.

We denote the natural sum of two ordinals $\mu$ and $\nu$ by $\mu\#\nu$. We shall also use the following notation: for an ordinal $\alpha$ and a natural number $n, \omega_n(\alpha)$ is inductively defined as $\omega_0(\alpha) \equiv \alpha$ and $\omega_{n+1}(\alpha) \equiv \omega^{\omega_n(\alpha)}$. Thus, we have

$$\omega_n(\alpha) \equiv \left.\omega^{\cdot^{\cdot^{\omega^{\alpha}}}}\right\} n \text{ powers of } \omega.$$

The limit of $\omega_n(0)$ when $n$ approaches infinity is $\epsilon_0$, an ordinal which in some ways is characteristic for the strength of derivability in Heyting Arithmetic.

We can now give every derivation in HA an ordinal.

**Definition 4.0.4** The *ordinal of a sequent* $S$ in a derivation $P$, denoted $o(S; P)$ or $o(S)$, is defined inductively as follows:

1. An initial sequent has the ordinal 1.
2. The conclusion of an arithmetical rule without premises has the ordinal 1.
3. If $S$ is the conclusion of a contraction, then the ordinal is the same as the ordinal of the premise.
4. If $S$ is the conclusion of a one-premise arithmetical or logical rule, where the ordinal of the premise is $\mu$, then $o(S) = \mu + 1$.
5. If $S$ is the conclusion of a two-premise arithmetical or logical rule, where the ordinals of the premises are $\mu$ and $\nu$, respectively, then $o(S) = \mu\#\nu$.
6. If $S$ is the conclusion of a cut where the premises have the ordinals $\mu$ and $\nu$, then $o(S) = \omega_{k-l}(\mu\#\nu)$, or

$$\left.\omega^{\cdot^{\cdot^{\omega^{\mu\#\nu}}}}\right\} k - l \text{ powers of } \omega,$$

   where $k$ is the height of the premises and $l$ is the height of the conclusion.
7. If $S$ is the conclusion of an induction and the premises have the ordinals $\mu_1, \mu_2$ and $\mu_3$ and the height $k$ and the conclusion has the height $l$, then the ordinal of the conclusion is $o(S) = \omega_{k-l+1}(\mu_1\#\mu_2\#\mu_3)$.

*The ordinal of a derivation* $P$, denoted $o(P)$, is the ordinal of the end-sequent. Thus, every derivation has an ordinal less than $\epsilon_0$.

If the height remains unchanged in a cut, then the ordinal of the conclusion in case 6 is $\mu\#\nu$, whereas the ordinal of the corresponding case 7 is $\omega^{\mu_1\#\mu_2\#\mu_3}$.

# 5 The Consistency of Heyting Arithmetic

## *5.1 The Theorem of Consistency*

**Definition 5.1.1** A system is said to be *inconsistent* if the empty sequent $\rightarrow$ is derivable. If the system is not inconsistent, it is *consistent*.

**Theorem 5.1.2 (The Consistency of Heyting Arithmetic)** *The empty sequent* $\rightarrow$ *is not derivable in HA, that is, HA is consistent.*

To prove this theorem we give a reduction procedure for derivations. Assume that there is a derivation of the empty sequent. We can assume that the arithmetical rules are applied before logical and structural rules in the derivation. If needed, it is possible to change the order of the rules according to Lemma 5.2.3, even though this may increase the ordinal of the derivation. The permutation only has to be performed once before the reduction procedure. By the reduction procedure we conclude that if there is a derivation of the empty sequent, then there is a reduced derivation with a lower ordinal and another reduced derivation and so on. Then we would have an infinite succession of decreasing ordinals all less than $\epsilon_0$, but this is impossible and the reduction procedure must terminate. Therefore we cannot have a derivation of the empty sequent. Thus, the system of Heyting Arithmetic, HA, is consistent.

The reduction procedure for derivations is described in Lemma 5.5.1, but before we give the proof we need some additional results.

## *5.2 Properties of Derivations*

**Definition 5.2.1** A *thread* in a derivation is a sequence of sequents in a derivation, for which the following holds:

1. It begins with an initial sequent or the conclusion of an arithmetical rule without premises.
2. Every sequent but the last one is a premise of a rule and the sequent is followed by the conclusion of that rule.

**Lemma 5.2.2** *Assume that* $S_1$ *is a sequent in a derivation* $P$. *Let* $P_1$ *be the subderivation ending with* $S_1$ *and let* $P_1'$ *be another derivation ending with* $S_1$. *Now let* $P'$ *be the derivation that we get when* $P_1'$ *is substituted for* $P_1$ *in* $P$.

*If* $o(S_1; P') < o(S_1; P)$, *then* $o(P') < o(P)$.

*Proof* For every thread in $P$, passing through $S_1$, we show that the following holds: If $S$ is a sequent in a thread at or below $S_1$ and if $S'$ is the corresponding sequent to $S$ in $P'$, then $o(S'; P') < o(S; P)$. According to the assumption the proposition holds if $S = S_1$. The heights of the sequents below $S$ in $P$ and $S'$ in $P'$ are the same and for every ordinal $\alpha, \beta$ and $\gamma$ that satisfy $\alpha < \beta$, we have $\alpha\#\gamma < \beta\#\gamma$. Thus, the inequality is retained for every rule applied. If we then let $S$ be the end-sequent of the derivation, we obtain the inequality for the derivations. □

**Lemma 5.2.3** *In a derivation we can permute the order of the rules and first apply the arithmetical rules and then Ind and the logical and structural rules.*

*Proof* If we have a logical rule followed by an arithmetical rule, then the arithmetical rule is not applied on the principal formula of the logical rule, since this formula is compound. Hence we can permute the order of the rules and apply the arithmetical rule first.

Assume that we have an instance of contraction followed by an arithmetical rule. If the arithmetical rule is not applied on the contraction formula, then we can permute the order of the rules. We now consider the case that the arithmetical rule is applied on the contraction formula. If the arithmetical rule is a one-premise rule, then we can apply the arithmetical rule on each copy of the formula followed by an instance of contraction. If, on the other hand, the arithmetical rule has two premises, that is if the rule is an instance of transitivity, then we can apply transitivity on each copy of the formula, multiplying the derivation of the other transitivity premise, and then apply contraction on the principal formula of the transitivity and also on possible formulas in the context of the multiplied premise.

If we have an instance of Cut followed by an arithmetical rule, then we can permute the order of the rules and the same holds for an instance of *Ind* followed by an arithmetical rule. □

Note that this change in the order of the rules can increase the ordinal of the derivation.

**Lemma 5.2.4** *(i) For an arbitrary closed term $t$ there exists a numeral $\overline{n}$ such that $\rightarrow t = \overline{n}$ can be derived without Ind or Cut.*

*(ii) Let $t$ and $t'$ be closed terms for which $\rightarrow t = t'$ can be derived without Ind or Cut and let $q$ be an arbitrary term. Now the sequent $\rightarrow q(t/x) = q(t'/x)$ is derivable without Ind or Cut.*

*(iii) Let $t$ and $t'$ be closed terms for which $\rightarrow t = t'$ can be derived without Ind or Cut and let $q$ and $r$ be terms. Then the sequent $q(t/x) = r(t/x) \rightarrow q(t'/x) = r(t'/x)$ can be derived without Ind or Cut.*

*(iv) Let $t$ and $t'$ be closed terms for which $\rightarrow t = t'$ can be derived without Ind or Cut. Then for an arbitrary formula $A$ the sequent $A(t/x) \rightarrow A(t'/x)$ can be derived without Ind or Cut.*

*Proof* (i) For the constant 0 the proposition holds. Assume that the proposition holds for the closed terms $t$ and $t'$, that is there are $n$ and $m$, such that $\rightarrow$

$t = \overline{n}$ and $\rightarrow t' = \overline{m}$ can be derived without *Ind* or Cut. Then the sequent $\rightarrow s(t) = s(\overline{n})$ is derivable with *sRep* where $s(\overline{n}) \equiv \overline{n+1}$.
The sequent $\rightarrow t + t' = \overline{n+m}$ can be derived as follows. First we get a derivation of $\rightarrow t + t' = \overline{n} + \overline{m}$.

$$\dfrac{\dfrac{\rightarrow t = \overline{n}}{\rightarrow t + t' = \overline{n} + t'}\,+Rep_1 \qquad \dfrac{\rightarrow t' = \overline{m}}{\rightarrow \overline{n} + t' = \overline{n} + \overline{m}}\,+Rep_2}{\rightarrow t + t' = \overline{n} + \overline{m}}\,Tr$$

Furthermore, if $m = 0$ we have $\rightarrow \overline{n} + \overline{0} = \overline{n+0}$ with $+Rec0$ since $\overline{n+0} \equiv \overline{n}$. If $m > 0$, that is $m = sm'$ for some $m'$, then we have as induction hypothesis a derivation of $\rightarrow \overline{n} + \overline{m'} = \overline{n+m'}$.

$$\dfrac{\dfrac{}{\rightarrow \overline{n} + \overline{sm'} = s(\overline{n} + \overline{m'})}\,+Recs \qquad \dfrac{\rightarrow \overline{n} + \overline{m'} = \overline{n+m'}}{\rightarrow s(\overline{n} + \overline{m'}) = s(\overline{n+m'})}\,sRep}{\rightarrow \overline{n} + \overline{sm'} = s(\overline{n+m'})}\,Tr$$

We now have $\rightarrow \overline{n} + \overline{m} = \overline{n+m}$ for every $m$. With transitivity on the conclusions of these derivations we get the result $\rightarrow t + t' = \overline{n+m}$.
The sequent $\rightarrow t \cdot t' = \overline{n \cdot m}$ is derivable in a similar manner.

(ii) If $q$ is the constant 0 or a variable different from $x$, then the sequent is derivable with *Ref*. If $q$ is the variable $x$, then we already have the derivation according to the assumption. Now assume that $q \equiv s(q')$ and as induction hypothesis we have a derivation of $\rightarrow q'(t/x) = q'(t'/x)$ that fulfils the requirements. Then we get $\rightarrow s(q'(t/x)) = s(q'(t'/x))$ with *sRep*. If $q \equiv q' + q''$ we get the following derivation where we write $q'(t)$ and $q''(t)$ instead of $q'(t/x)$ and $q''(t/x)$.

$$\dfrac{\dfrac{\rightarrow q'(t) = q'(t')}{\rightarrow q'(t) + q''(t) = q'(t') + q''(t)}\,+Rep_1 \qquad \dfrac{\rightarrow q''(t) = q''(t')}{\rightarrow q'(t') + q''(t) = q'(t') + q''(t')}\,+Rep_2}{\rightarrow q'(t) + q''(t) = q'(t') + q''(t')}\,Tr$$

If $q \equiv q' \cdot q''$ the derivation is similar.

(iii) According to (ii) we have derivations of $\rightarrow q(t) = q(t')$ and $\rightarrow r(t) = r(t')$ that fulfil the requirements. We can now construct the derivation:

$$\dfrac{\dfrac{\dfrac{\rightarrow q(t) = q(t')}{\rightarrow q(t') = q(t)}\,Sym \qquad q(t) = r(t) \rightarrow q(t) = r(t)}{q(t) = r(t) \rightarrow q(t') = r(t)}\,Tr \qquad \rightarrow r(t) = r(t')}{q(t) = r(t) \rightarrow q(t') = r(t')}\,Tr$$

(iv) The proof is carried out by induction on the complexity of the formula. If $A$ is an atomic formula, then the proposition is proved in (iii).

If $A \equiv B\&C$ and we as induction hypothesis have that $B(t/x) \rightarrow B(t'/x)$ and $C(t/x) \rightarrow C(t'/x)$ are derivable without *Ind* or Cut, then we get the derivation:

$$\dfrac{\dfrac{B(t/x) \rightarrow B(t'/x) \quad C(t/x) \rightarrow C(t'/x)}{B(t/x), C(t/x) \rightarrow B(t'/x)\&C(t'/x)}\, R\&}{B(t/x)\&C(t/x) \rightarrow B(t'/x)\&C(t'/x)}\, L\&$$

Assume that $A \equiv \forall y B$. If $x \equiv y$, then $x$ is not free in $A$ and $A(t/x) \rightarrow A(t'/x)$ is an initial sequent. On the other hand if $x$ is not $y$, then we have by induction hypothesis that $(B(z/y))(t/x) \rightarrow (B(z/y))(t'/x)$, where $x \neq z$, can be derived without *Ind* or Cut. Because $t$ and $t'$ are closed terms they do not contain $y$ and we may change the order of the substitutions, that is $(B(z/y))(t/x) = (B(t/x))(z/y)$ and $(B(z/y))(t'/x) = (B(t'/x))(z/y)$. We now get the derivation:

$$\dfrac{\dfrac{(B(t/x))(z/y) \rightarrow (B(t'/x))(z/y)}{\forall y B(t/x) \rightarrow (B(t'/x))(z/y)}\, L\forall}{\forall y B(t/x) \rightarrow \forall y B(t'/x)}\, R\forall$$

The other cases are similar. □

In (i) of the lemma, we only state the existence of a numeral that equals the closed term, not that this numeral is unique. The uniqueness of the numeral is equivalent to the consistency of simple derivations proved in Lemma 5.4.8.

## 5.3 Cut Elimination in HA

We shall give a direct proof of cut elimination in the system HA. Note that the Cut rule is a special case of our induction rule, if the induction formula has no occurrence of the variable $x$. In this case the second premise of the induction is an initial sequent and we have a form of vacuous induction. Thus, cuts can be eliminated by replacing them with inductions. But as the cut elimination Theorem 5.3.2 shows, we can also properly eliminate cut.

**Definition 5.3.1** The *length* of a derivation in HA is defined inductively.

An initial sequent has the length 1.

The length of the conclusion of an arithmetical rule without premises is 1.

The length of the conclusion of the rule *Sym* is the same as the length of the premise.

The length of the conclusion of a one-premise rule (except *Sym*), where the premise has the length $\alpha$ is $\alpha + 1$.

The length of the conclusion of a two-premise rule, where the premises have the lengths $\alpha$ and $\beta$ is $\alpha + \beta$.

The length of the conclusion of *Ind*, where the premises have the lengths $\alpha$, $\beta$ and $\gamma$ is $\alpha + \beta + \gamma$.

**Theorem 5.3.2 (Cut Elimination in HA)** *If there is a derivation of the sequent* $\Gamma \to D$ *in HA, such that the derivation contains no induction, then we can transform the derivation into a derivation of the sequent without Cut, without introducing additional vacuous inductions.*

*Proof* The proof is by induction on the grade of the cut formula with a subinduction on the length of the derivation. We assume that there are no instances of Cut above the cut we consider.

We assume that the right cut premise has been derived with $n - 1$ instances of contraction on the cut formula, where $n \geqslant 1$. We consider the premise of the first contraction.

1. Firstly we consider the case that the premise is an initial sequent.

$$\dfrac{\Gamma_1 \to A \qquad \dfrac{A^n, \Gamma_2 \to A}{A, \Gamma_2 \to A}\,LC^{n-1}}{\Gamma_{1-2} \to A}\,Cut$$

In this case we can add the missing context $\Gamma_2$ in the derivation of the left cut premise and get the sought derivation without Cut.
We now assume that the premise of the contraction has been derived by a rule $R$.

$$\dfrac{\Gamma_1 \to A \qquad \dfrac{\overline{A^n, \Gamma_2 \to D}\,R}{A, \Gamma_2 \to D}\,LC^{n-1}}{\Gamma_{1-2} \to D}\,Cut$$

If rule $R$ is an instance of *Sym* we can permute the contractions and the cut above the *Sym*. The length of the cut remains unchanged. Thus, we may assume that $R$ is not *Sym*.
2. If rule $R$ is an arithmetical rule without premises, then also the conclusion of the cut is an instance of the same rule.
3. If rule $R$ is an arithmetical one-premise rule, then $A$ is not principal in the rule. We can then permute the contractions and the cut above the arithmetical rule, decreasing the length of the cut.
4. Suppose rule $R$ is *Tr*.

$$\dfrac{\Gamma_1 \to A \qquad \dfrac{\dfrac{A^k, \Gamma_1' \to t = t' \quad A^l, \Gamma_2' \to t' = t''}{A^n, \Gamma_2 \to t = t''}\,Tr}{A, \Gamma_2 \to t = t''}\,LC^{n-1}}{\Gamma_{1-2} \to t = t''}\,Cut$$

where $\Gamma_2 = \Gamma'_{1-2}$ and $n = k + l$. We then transform the derivation decreasing the length of the cuts on $A$.

$$\dfrac{\dfrac{\Gamma_1 \to A \quad \dfrac{A^k, \Gamma'_1 \to t = t'}{A, \Gamma'_1 \to t = t'}\,LC^{k-1}}{\Gamma_1, \Gamma'_1 \to t = t'}\,Cut \qquad \dfrac{\Gamma_1 \to A \quad \dfrac{A^l, \Gamma'_2 \to t' = t''}{A, \Gamma'_2 \to t' = t''}\,LC^{l-1}}{\Gamma_1, \Gamma'_2 \to t' = t''}\,Cut}{\dfrac{\Gamma_1^2, \Gamma_2 \to t = t''}{\Gamma_{1-2} \to t = t''}\,contractions}\,Tr$$

5. If rule $R$ is a logical one-premise rule where $A$ is not principal, then we can permute the contractions and the cut above the rule, decreasing the length of the cut.
6. If rule $R$ is a logical two-premise rule where $A$ is not principal, then we transform the derivation as in case 4, decreasing the length of the cuts.
7. Suppose rule $R$ is a logical rule where $A$ is principal. We consider the rule with which the left premise of the cut has been derived.

**7.1** If the left cut premise is an initial sequent, then the formula $A$ is in $\Gamma_1$. Thus, we can get the conclusion of the cut by adding the missing context $\Gamma_1$ without $A$ in the derivation of the right cut premise.

**7.2** The left cut premise has not been derived by an arithmetical rule, since the formula $A$ has logical structure.

**7.3** If the left cut premise has been derived by a logical one-premise rule where $A$ is not principal, then we can permute the cut above the rule.

**7.4** If the left cut premise has been derived by a logical two-premise rule where $A$ is not principal, that is $L\supset$ or $L\vee$, then we can in the case of $L\vee$ apply Cut twice, once on each premise of the logical rule and then apply the logical and in the case of $L\supset$ apply Cut before the rule.

**7.5** If the left cut premise has been derived by a logical rule where $A$ is principal, then we consider the derivation according to the form of the formula. We consider the case where $A$ is a conjunction $B\&C$.

$$\dfrac{\dfrac{\Gamma'_1 \to B \quad \Gamma''_1 \to C}{\Gamma_1 \to B\&C}\,R\& \qquad \dfrac{\dfrac{B, C, (B\&C)^{n-1}, \Gamma_2 \to D}{(B\&C)^n, \Gamma_2 \to D}\,L\&}{B\&C, \Gamma_2 \to D}\,LC^{n-1}}{\Gamma_{1-2} \to D}\,Cut$$

In the derivation

$$\dfrac{\dfrac{\Gamma'_1 \to B \quad \Gamma''_1 \to C}{\Gamma_1 \to B\&C}\,R\& \qquad \dfrac{B, C, (B\&C)^{n-1}, \Gamma_2 \to D}{B, C, B\&C, \Gamma_2 \to D}\,LC^{n-2}}{B, C, \Gamma_{1-2} \to D}\,Cut$$

the cut length is shorter. Thus, we have by the induction hypothesis a derivation of the sequent $B, C, \Gamma_{1-2} \rightarrow D$ without Cut. We now construct the following derivation, where the grades of the cut formulas are less.

$$\dfrac{\Gamma_1' \rightarrow B \qquad \dfrac{\Gamma_1'' \rightarrow C \quad B, C, \Gamma_{1-2} \rightarrow D}{B, \Gamma_1'', \Gamma_{1-2} \rightarrow D}\,Cut}{\dfrac{\Gamma_1^2, \Gamma_2 \rightarrow D}{\Gamma_{1-2} \rightarrow D}\,\vdots\ contractions}\,Cut$$

The other cases of cut formula are treated in a similar manner.

**7.6** If the left cut premise has been derived by a contraction, then we can permute the cut above the rule.

**7.7** If the left cut premise has been derived by *Ind*, then we can permute the cut above the rule.

**8.** If rule $R$ is an instance of contraction, where $A$ is not principal, then we can permute the contractions and the cut above the rule, decreasing the length of the cut.

**9.** Suppose rule $R$ is an instance of *Ind*.

$$\dfrac{\Gamma_1 \rightarrow A \qquad \dfrac{\dfrac{A^k, \Gamma_1' \rightarrow B(0/x) \quad A^l, B(y/x), \Gamma_2' \rightarrow B(sy/x) \quad A^m, B(t/x), \Gamma_3' \rightarrow D}{A^n, \Gamma_2 \rightarrow D}\,Ind}{A, \Gamma_2 \rightarrow D}\,LC^{n-1}}{\Gamma_{1-2} \rightarrow D}\,Cut$$

Here we have $\Gamma_2 = \Gamma_{1-3}'$ and $n = k + l + m$. We transform the derivation as in case 4, decreasing the length of the cuts on $A$. □

This direct proof of cut elimination in Heyting Arithmetic is an extension of the proof given in [7]. Note that contrary to Gentzen's original proof of cut elimination for sequent calculus in his thesis of 1933, our proof is carried out without introducing any rule of multicut.

## 5.4 Consistency Proof for Simple Derivations

**Definition 5.4.1** A *simple derivation* is a derivation without free variables and without *Ind* that contains only atomic formulas.

Thus, in a simple derivation we have only initial sequents, arithmetical and structural rules, and in addition there are no compound formulas in the contexts.

Our aim is now to show that there is no simple derivation of the empty sequent, but first we consider only the case that the derivation does not contain rule $Inf_2$.

**Definition 5.4.2** We define inductively if the *value* of a closed term is 0 or 1. The constant 0 has value 0. A term of the form $s(t)$ has value 1. A term of the form $t+t'$ has value 0 if both $t$ and $t'$ have value 0 and otherwise it has value 1. A term of the form $t \cdot t'$ has value 0 if $t$ or $t'$ has value 0 and otherwise it has value 1.

According to the definition a closed term has value 0 if it equals 0 and value 1 if it is greater than 0.

**Lemma 5.4.3** *There is no simple derivation of the empty sequent without rule* $Inf_2$.

*Proof* Assume that there is a derivation of the empty sequent without rule $Inf_2$. According to Theorem 5.3.2 there is then a derivation of the empty sequent without Cut (and this new derivation without Cut is also without $Inf_2$ and *Ind*). Furthermore, we note that in a cut-free simple derivation of the empty sequent all sequents have an empty antecedent, since formulas in the antecedent can only disappear through cut. Therefore there are no initial sequents or instances of contraction in the derivation, but only arithmetical rules.

Now, the last rule of the derivation must be $Inf_1$, because all other rules give as a conclusion a sequent with a formula in the succedent. Thus, we have a derivation of the sequent $\rightarrow s(t) = 0$ for some term $t$.

In a simple derivation there are only closed terms and therefore every term has a value. We now prove by induction on the length of the derivation that every sequent in the derivation of $\rightarrow s(t) = 0$ has the property that the succedent is a formula $t = t'$ where $t$ and $t'$ have the same value.

**Base Case of the Induction** As stated we have no initial sequents in the derivation and thus we only consider the conclusions of the arithmetical rules without premises as the base case. We want to prove that the terms of the principal formula in the succedent have the same value.

In *Ref* both terms of the principal formula, $t = t$, have the same value. In $+Rec0$ the terms $t + 0$ and $t$ of the principal formula, $t + 0 = t$, have the same value. In $+Recs$ the principal formula is of the form $t + s(t') = s(t + t')$. Both $t + s(t')$ and $s(t + t')$ in $+Recs$ have the value 1. In $\cdot Rec0$ the principal formula is of the form $t \cdot 0 = 0$. The constant 0 has the value 0 and therefore the term $t \cdot 0$ also has the same value. In $\cdot Recs$ the principal formula is of the form $t \cdot s(t') = t \cdot t' + t$. If the term $t$ has the value 1, then both terms $t \cdot s(t')$ and $t \cdot t' + t$ have the value 1. If $t$, on the other hand, has the value 0, then both terms have the value 0.

**Induction Step** Assume as induction hypothesis that the proposition holds for the premises of an arithmetical rule, that is the terms of the formulas in the succedents of the premises have the same value.

In *Sym* we can conclude that if the terms $t$ and $t'$ in the formula $t = t'$ have the same value, then the same applies for the formula $t' = t$ in the conclusion. In *Tr* we can see that if the terms of the formula $t = t'$ and $t' = t''$ have the same values, then the terms of the formula $t = t''$ have the same value. In *sRep* both terms of the formula $s(t) = s(t')$ in the conclusion have the value 1. In $+Rep_1$, if the terms of the formula $t = t'$ in the premise have the same value, then also the terms of the formula $t + t'' = t' + t''$ in the conclusion have the same value. The same holds for rule $+Rep_2$ and the $\cdot Rep$-rules.

Because all sequents in the derivation have an empty antecedent, rule $Inf_1$ gives the empty sequent as the conclusion and thus it can occur only as the last rule in the derivation.

Thus, we have completed the induction and have proved that in a simple derivation of the sequent $\rightarrow s(t) = 0$, all sequents have in the succedent an equation, where the terms have the same value. On the other hand, the terms $s(t)$ and 0 have different values. This is a contradiction and therefore there cannot exist any simple derivation of the empty sequent. □

**Lemma 5.4.4** *If we have a derivation of a sequent* $\Gamma \rightarrow D$*, then there is a derivation of the same length of the sequent where all instances of Sym come directly after arithmetical rules without premises or after initial sequents.*

*Proof* Suppose that we have a premise of *Sym* derived by a rule that is not an arithmetical rule without premises. If the rule is a one-premise arithmetical rule, that is *sRep*, *+Rep*, *·Rep*, or $Inf_2$, we can permute the instance of *Sym* above the other rule. If we have two instances of *Sym* we have a loop and can delete both rules. If the rule is logical (except $L\vee$), structural, or an instance of *Inf*, we can also permute the *Sym* above the other rule.

If the rule is an instance of *Tr*, then the derivation is:

$$\dfrac{\dfrac{\Gamma_1 \rightarrow t = t' \quad \Gamma_2 \rightarrow t' = t''}{\Gamma_{1-2} \rightarrow t = t''}\,Tr}{\Gamma_{1-2} \rightarrow t'' = t}\,Sym$$

We can then instead apply *Sym* on each premise followed by *Tr*.

$$\dfrac{\dfrac{\Gamma_2 \rightarrow t' = t''}{\Gamma_2 \rightarrow t'' = t'}\,Sym \quad \dfrac{\Gamma_1 \rightarrow t = t'}{\Gamma_1 \rightarrow t' = t}\,Sym}{\Gamma_{1-2} \rightarrow t'' = t}\,Tr$$

This does not alter the length of the derivation. The case of $L\vee$ is similar. □

**Lemma 5.4.5** *There is a derivation of the sequent* $\rightarrow 0 \cdot c = 0$ *(without* $Inf_2$*) for every closed term* $c$*.*

*Proof* Firstly we show by induction that for every numeral $\overline{m}$ we have a derivation of the sequent $\rightarrow 0 \cdot \overline{m} = 0$. We can derive $\rightarrow 0 \cdot 0 = 0$ with *·Recs*. Now assume that $\overline{m}$ is $s\overline{n}$ for some numeral $\overline{n}$ and we have a derivation of $\rightarrow 0 \cdot \overline{n} = 0$. We then get the derivation

$$\dfrac{\dfrac{}{\rightarrow 0 \cdot s(\overline{n}) = 0 \cdot \overline{n} + 0}\,\cdot Recs \quad \dfrac{\dfrac{}{\rightarrow 0 \cdot \overline{n} + 0 = 0 \cdot \overline{n} + 0}\,+Rec0 \quad \rightarrow 0 \cdot \overline{n} = 0}{\rightarrow 0 \cdot \overline{n} + 0 = 0}\,Tr}{\rightarrow 0 \cdot s(\overline{n}) = 0}\,Tr$$

Thus, the proposition holds for every numeral.

For every closed term $c$ there is a numeral $\overline{m}$ for which the sequent $\rightarrow c = \overline{m}$ is derivable (without $Inf_2$), this according to Lemma 5.2.3(i). We then get the sought derivation

$$\dfrac{\dfrac{\rightarrow c = \overline{m}}{\rightarrow 0 \cdot c = 0 \cdot \overline{m}}\cdot Rep_2 \qquad \rightarrow 0 \cdot \overline{m} = 0}{\rightarrow 0 \cdot c = 0}\,Tr$$

□

**Lemma 5.4.6** *If there is a simple derivation of the sequent* $\rightarrow s(t) = s(t')$ *without the rule* $Inf_2$*, then there is a simple derivation of the sequent* $\rightarrow t = t'$ *without* $Inf_2$.

*Proof* The proof is by induction on the length of the derivation. We assume that if there is a shorter derivation of some sequent $\rightarrow s(a) = s(b)$, then we have a derivation of $\rightarrow a = b$ without rule $Inf_2$.

Assume that we have a simple derivation of a sequent $\rightarrow s(t) = s(t')$ without $Inf_2$. We can by Theorem 5.3.2 assume that the derivation is cut free. Thus, every sequent in the derivation has an empty antecedent. By Lemma 5.4.4 we can assume that all instances of *Sym* come directly after arithmetical rules without premises (note that there are no initial sequents in the derivation because the antecedents are empty).

We consider the form of the derivation. The last rule can be *sRep*, *Ref*, *Sym*, or *Tr*.

1. Assume that the last rule of the derivation is *sRep*. The premise of the rule is $\rightarrow t = t'$ and we can remove the rule and get the sought derivation.
2. Assume that the last rule is *Ref*. Then $t \equiv t'$ and the sequent $\rightarrow t = t'$ is also derivable with *Ref*.
3. Assume that the last rule is *Sym*. Since the premise of *Sym* is derived by an arithmetical rule without premises the only possibility is that this rule is *Ref*. The case is treated as in case 2.
4. The remaining possibility is that the last rule is derived by *Tr*. We trace up in the derivation along the left premise until we reach a sequent not derived by *Tr*. The derivation is of the form

$$\dfrac{\begin{array}{c}\dfrac{\rightarrow s(t) = a_1 \qquad \rightarrow a_1 = a_2}{\rightarrow s(t) = a_2}\,Tr \\ \vdots\; Tr - rules \\ \rightarrow s(t) = a_n\end{array} \qquad \rightarrow a_n = s(t')}{\rightarrow s(t) = s(t')}\,Tr \tag{5.4.1}$$

where $n \geqslant 1$ and the sequent $\rightarrow s(t) = a_1$ is not derived by *Tr*.
If one of the other *Tr*-premises $\rightarrow a_i = a_{i+1}$ is derived by *Tr*

$$\dfrac{\to s(t)=a_i \qquad \dfrac{\to a_i=a \qquad \to a=a_{i+1}}{\to a_i=a_{i+1}}}{\to s(t)=a_{i+1}}$$

we can change the order of the *Tr*-rules without altering the length of the derivation.

$$\dfrac{\dfrac{\to s(t)=a_i \qquad \to a_i=a}{\to s(t)=a} \qquad \to a=a_{i+1}}{\to s(t)=a_{i+1}}$$

Hence we can assume that the derivation is of the form (5.4.1) and that none of the premises $\to a_i = a_{i+1}$ have been derived by *Tr*.
If some term $a_i$ is of the form $s(t'')$, then the sequent $\to s(t) = a_i$ is the sequent $\to s(t) = s(t'')$. We can then alter the order of the *Tr*-rules and get a derivation of the same length.

$$\dfrac{\to s(t)=s(t'') \qquad \dfrac{\begin{array}{c}\to s(t'')=a_{i+1}\\ \vdots \\ \to s(t'')=a_n\end{array} \qquad \to a_n=s(t')}{\to s(t'')=s(t')}}{\to s(t)=s(t')}$$

The derivations of the sequents $\to s(t) = s(t'')$ and $\to s(t'') = s(t')$ are shorter and therefore we have derivations of the sequents $\to t = t''$ and $\to t'' = t'$. By *Tr* we get the sought derivation of $\to t = t'$.
We can now assume that the derivation has the form (5.4.1) and that no term $a_i$ has the form $s(t'')$. We consider the different possibilities to derive the *Tr*-premises.

**4.1** Assume that one of the premises has been derived by *Ref*. We now have a loop in the derivation since the conclusion of the following *Tr* is the same as the other premise. We can delete the rule *Tr* and get a shorter derivation. Thus, we may assume that no premise has been derived by *Ref*.

**4.2** Assume that two adjacent *Tr*-premises have been derived by the same replacement rule $+Rep_1$, $+Rep_2$, $\cdot Rep_1$, or $\cdot Rep_2$ or that three adjacent *Tr*-premises have been derived by two instances of the same replacement rule with one instance of the other replacement rule in between. As an example we consider the following derivation.

$$\dfrac{\dfrac{\to s(t)=a+b \qquad \dfrac{\to b=c}{\to a+b=a+c}\,{\scriptstyle +Rep_2}}{\to s(t)=a+c}\,{\scriptstyle Tr} \qquad \dfrac{\to c=d}{\to a+c=a+d}\,{\scriptstyle +Rep_2}}{\to s(t)=a+d}\,{\scriptstyle Tr}$$

We can then apply *Tr* on the premises of the replacement rules and get a shorter derivation.

$$\dfrac{\to s(t) = a+b \qquad \dfrac{\dfrac{\to b=c \quad \to c=d}{\to b=d}\,Tr}{\to a+b=a+d}\,{+Rep_2}}{\to s(t)=a+d}\,Tr$$

Thus, we can assume that we at most have two adjacent *Tr*-premises derived by $+Rep$ or $\cdot Rep$ and that these rules have different indexes.

**4.3** Assume that some of the *Tr*-premises have been derived by *Sym* and $+Rec0$. We consider the rightmost premise derived in this way. It cannot be the last *Tr*-premise $\to a_n = s(t)$ since the sequent is of the form $\to a_i = a_i + 0$. Thus, the derivation is of the form

$$\dfrac{\dfrac{\dfrac{\dfrac{}{\to a_i+0=a_i}\,{+Rec0}}{\to a_i=a_i+0}\,Sym}{\to s(t)=a_i+0}\,Tr? \qquad \dfrac{}{\to a_i+0=b}\,R}{\to s(t)=b}\,Tr \tag{5.4.2}$$

where *Tr*? indicates that if $a_i \equiv s(t)$ we have no rule there, but if $a_i \not\equiv s(t)$ we have a *Tr*-rule there.

Rule $R$ can according to the form of the term be *Sym*, $+Rec$, $+Rep_1$, or $+Rep_2$ and if the rule is *Sym*, then the premise can be derived by $\cdot Recs$. We consider the different alternatives.

**4.3.1** Assume that $R$ is $+Rec0$. Then $b \equiv a_i$. If $a_i \equiv s(t)$, then we have derived an instance of *Ref* and if $a_i \not\equiv s(t)$, then we have a loop in the derivation with the sequent $\to s(t) = a_i$ two times. By eliminating the loop we get a shorter derivation.

**4.3.2** Assume that $R$ is $+Rep_1$. Now $b \equiv c + 0$ and the derivation (5.4.2) is

$$\dfrac{\dfrac{\dfrac{\dfrac{}{\to a_i+0=a_i}\,{+Rec0}}{\to a_i=a_i+0}\,Sym}{\to s(t)=a_i+0}\,Tr? \qquad \dfrac{\to a_i=c}{\to a_i+0=c+0}\,{+Rep_1}}{\to s(t)=c+0}\,Tr$$

We can transform the derivation into a shorter derivation.

$$\dfrac{\dfrac{\to a_i=c}{\to s(t)=c}\,Tr? \qquad \dfrac{\dfrac{}{\to c+0=c}\,{+Rec0}}{\to c=c+0}\,Sym}{\to s(t)=c+0}\,Tr$$

**4.3.3** Assume that $R$ is *Sym* and that the premise of this rule is derived by $\cdot Recs$. Now $a_i \equiv 0 \cdot c$, $b \equiv 0 \cdot s(c)$ and the derivation (5.4.2) is

$$\dfrac{\dfrac{\to s(t)=0\cdot c \qquad \dfrac{\dfrac{}{\to 0\cdot c+0=0\cdot c}\;\mathit{+Rec0}}{\to 0\cdot c=0\cdot c+0}\;\mathit{Sym}}{\to s(t)=0\cdot c+0}\;\mathit{Tr} \qquad \dfrac{\dfrac{}{\to 0\cdot s(c)=0\cdot c+0}\;\mathit{\cdot Recs}}{\to 0\cdot c+0=0\cdot s(c)}\;\mathit{Sym}}{\to s(t)=0\cdot s(c)}\;\mathit{Tr}$$

According to Lemma 5.4.5 there is a derivation of the sequent $\to 0\cdot s(c)=0$ (without rule $\mathit{Inf}_2$). With *Tr* we get a derivation of the sequent $\to s(t) = 0$ without $\mathit{Inf}_2$. Thus, applying $\mathit{Inf}_1$ we get a derivation of the empty sequent without $\mathit{Inf}_2$. This is a contradiction according to Lemma 5.4.3.

**4.3.4** Assume that $R$ is $+\mathit{Rep}_2$. Then $b \equiv a_i + c$ and we have another *Tr*-premise to the right derived by a rule $R'$. The derivation (5.4.2) is

$$\dfrac{\dfrac{\dfrac{\dfrac{\dfrac{}{\to a_i+0=a_i}\;\mathit{+Rec0}}{\to a_i=a_i+0}\;\mathit{Sym}}{\to s(t)=a_i+0}\;\mathit{Tr?} \qquad \dfrac{\to 0=c}{\to a_i+0=a_i+c}\;\mathit{+Rep_2}}{\to s(t)=a_i+c}\;\mathit{Tr} \qquad \dfrac{}{\to a_i+c=d}\;R'}{\to s(t)=d}\;\mathit{Tr} \tag{5.4.3}$$

Considering the form of the formula $a_i + c = d$ the rule $R'$ can be *Sym*, *+Rec*0, *+Recs*, or $+\mathit{Rep}_1$ (note that according to 4.2 the rule cannot be $+\mathit{Rep}_2$) and if it is *Sym*, then the *Sym*-premise can only be derived by *·Recs*. We consider the different possibilities.

**4.3.4.1** Assume that $R'$ is *+Rec*0. The derivation is treated as in case 4.3.1.

**4.3.4.2** Assume that $R'$ is *+Recs*. Now $c \equiv s(e)$ and $d \equiv s(a_i + e)$. The sequent $\to 0 = c$ is then $\to 0 = s(e)$. This gives a contradiction as in case 4.3.3.

**4.3.4.3** Assume that $R'$ is $+\mathit{Rep}_1$. Now $d \equiv a_i + e$ and the derivation (5.4.3) is

$$\dfrac{\dfrac{\dfrac{\dfrac{\dfrac{}{\to a_i+0=a_i}\;\mathit{+Rec0}}{\to a_i=a_i+0}\;\mathit{Sym}}{\to s(t)=a_i+0}\;\mathit{Tr?} \qquad \dfrac{\to 0=c}{\to a_i+0=a_i+c}\;\mathit{+Rep_2}}{\to s(t)=a_i+c}\;\mathit{Tr} \qquad \dfrac{\to a_i=e}{\to a_i+c=e+c}\;\mathit{+Rep_1}}{\to s(t)=e+c}\;\mathit{Tr}$$

We can transform the derivation into a shorter derivation.

$$\dfrac{\dfrac{\dfrac{\to a_i=e \qquad \dfrac{\dfrac{}{\to e+0=e}\;\mathit{+Rec0}}{\to e=e+0}\;\mathit{Sym}}{\to a_i=e+0}\;\mathit{Tr} \qquad \dfrac{\to 0=c}{\to e+0=e+c}\;\mathit{+Rep_2}}{\to a_i=e+c}\;\mathit{Tr}}{\to s(t)=e+c}\;\mathit{Tr?}$$

**4.3.4.4** Assume that $R'$ is *Sym* and that the *Sym*-premise has been derived by $\cdot$*Recs*. Now $a_i \equiv c \cdot e$, $d \equiv c \cdot s(e)$ and the conclusion of derivation (5.4.3) is $\rightarrow s(t) = c \cdot s(e)$. We get a simple derivation of the sequent $\rightarrow s(t) = 0$ without $Inf_2$, since we according to Lemma 5.4.5 have a simple derivation of the sequent $\rightarrow 0 \cdot s(e) = 0$.

$$\dfrac{\dfrac{\rightarrow s(t) = c \cdot s(e) \quad \dfrac{\dfrac{\rightarrow 0 = c}{\rightarrow c = 0}\,Sym}{\rightarrow c \cdot s(e) = 0 \cdot s(e)}\,\cdot Rep_1}{\rightarrow s(t) = 0 \cdot s(e)}\,Tr \qquad \rightarrow 0 \cdot s(e) = 0}{\rightarrow s(t) = 0}\,Tr$$

This is a contradiction as in case 4.3.3.

We have now treated all the possibilities of rule $R'$ and case 4.3.4 is finished. We have also treated all cases in 4.3 and thus we can assume that no *Tr*-premise in derivation (5.4.1) has been derived by *Sym* and $+Rec0$.

**4.4** We consider derivation (5.4.1). The leftmost *Tr*-premise $\rightarrow s(t) = a_1$ can only be derived by *Sym* and the premise of *Sym* by $+Recs$. The following *Tr*-premise can be derived by $+Rep_1, +Rep_2, Sym,$ or $+Recs$ and if it is derived by *Sym*, then the *Sym*-premise is derived by $\cdot Recs$. We treat the different cases simultaneously, since the derivation will ultimately have the same form disregarding some *Rep*-rules and possible instances of $\cdot Recs$. According to case 4.2 we can only have two adjacent *Tr*-premises derived by the $+Rep$-rules. We assume that we have one premise derived by $+Rep_1$ and one by $+Rep_2$. The following *Tr*-premise can be derived by $+Recs$, $+Rec0$, or *Sym* and $\cdot Recs$. If it is derived by $+Rec0$ we get a contradiction as in case 4.3.3. We assume that the premise is derived by *Sym* and $\cdot Recs$. The following two premises can be derived by $\cdot Rep_1$ and $\cdot Rep_2$ and the next only by $\cdot Recs$, because if it is derived by $\cdot Rec0$ we have a contradiction as in case 4.3.3. Again we can have two $+Rep$-rules and a number of repetition of the rules $\cdot Recs, \cdot Rep_1, \cdot Rep_2, \cdot Recs, +Rep_1$ and $+Rep_2$. The last *Tr*-premise is derived by $+Recs$.

Hence the derivation has the following form (where we have left out the sequent arrow and unnecessary parentheses):

$$\dfrac{\dfrac{\dfrac{\dfrac{}{a + sb = s(a+b)}\,+Recs}{s(a+b) = a + sb}\,Sym \quad \dfrac{a = c}{a + sb = c + sb}\,+Rep_1}{s(a+b) = c + sb}\,Tr \qquad \dfrac{sb = d}{c + sb = c + d}\,+Rep_2}{s(a+b) = c + d}\,Tr$$

From the rule $\cdot Recs$ we have $c \equiv d \cdot e$.

$$\dfrac{\dfrac{\dfrac{\vdots \atop s(a+b)=c+d \qquad \dfrac{\dfrac{}{d\cdot se=c+d}\,{\cdot Recs}}{c+d=d\cdot se}\,Sym}{s(a+b)=d\cdot se}\,Tr \qquad \dfrac{d=f}{d\cdot se=f\cdot se}\,{\cdot Rep_1}}{s(a+b)=f\cdot se}\,Tr \qquad \dfrac{se=g}{f\cdot se=f\cdot g}\,{\cdot Rep_2}}{s(a+b)=f\cdot g}\,Tr$$

From the rule $\cdot Recs$ we have $g \equiv sh$.

$$\dfrac{\dfrac{\dfrac{\vdots \atop s(a+b)=f\cdot g \qquad \dfrac{}{f\cdot g=f\cdot h+f}\,{\cdot Recs}}{s(a+b)=f\cdot h+f}\,Tr \qquad \dfrac{f\cdot h=c_2}{f\cdot h+f=c_2+f}\,{+Rep_1}}{s(a+b)=c_2+f}\,Tr \qquad \dfrac{f=d_2}{c_2+f=c_2+d_2}\,{+Rep_2}}{s(a+b)=c_2+d_2}\,Tr$$

From the formula $s(a+b) = c_2 + d_2$ we can have a repetition of $\cdot$*Recs* and *Rep*-rules. If we have $n-1$ repetitions, where $n \geqslant 1$, then the end of the derivation is

$$\dfrac{\vdots \atop s(a+b)=c_n+d_n \qquad \dfrac{}{c_n+d_n=s(a_2+b_2)}\,{+Recs}}{s(a+b)=s(a_2+b_2)}\,Tr \tag{5.4.4}$$

Here we have $c_n \equiv a_2$ and $d_n \equiv sb_2$ and also $a_2 + b_2 \equiv t'$.

If we in the derivation have at least one row of the specified rules, that is if $n > 1$, then we show that we can derive $c_i = c_{i+1}$ and $d_i = d_{i+1}$. If we don't have all *Rep*-rules in the derivation, then we have identities instead of equations and the derivation is shorter.

In the derivation we have subderivations of the formulas $d_i = f_i$ and $f_i \cdot h_i = c_{i+1}$ and we also have the identity $c_i \equiv d_i \cdot e_i$. Since $g_i \equiv sh_i$ and we have a subderivation of $se_i = g_i$, that is $se_i = sh_i$, we have by the induction hypothesis a derivation of $e_i = h_i$. Thus, we can construct a derivation of $c_i = c_{i+1}$.

$$\dfrac{\dfrac{\dfrac{d_i=f_i}{d_i\cdot e_i=f_i\cdot e_i}\,{\cdot Rep_1} \qquad \dfrac{e_i=h_i}{f_i\cdot e_i=f_i\cdot h_i}\,{\cdot Rep_2}}{d_i\cdot e_i=f_i\cdot h_i}\,Tr \qquad f_i\cdot h_i=c_{i+1}}{d_i\cdot e_i=c_{i+1}}\,Tr$$

On the other hand, we get $d_i = d_{i+1}$ with *Tr* from $d_i = f_i$ and $f_i = d_{i+1}$.

With *Tr* we get derivations of $c = c_n$ and $d = d_n$. We now construct a derivation of $t = t'$, that is $a + b = a_2 + b_2$. From the subderivation of $a = c$ and the derivation of $c = c_n$, we get with *Tr* a derivation of $a = c_n$. Since $c_n \equiv a_2$ we now have a derivation of $a = a_2$.

From the subderivation of $sb = d$ and the derivation of $d = d_n$ we get with *Tr* a derivation of $sb = d_n$. Since $d_n \equiv sb_2$ we have a derivation of $sb = sb_n$ and this derivation is shorter. According to the induction hypothesis we have a derivation of

$b = b_n$. We now get the sought derivation

$$\dfrac{\dfrac{a = a_2}{a + b = a_2 + b}\,{+Rep_1} \qquad \dfrac{b = b_2}{a_2 + b = a_2 + b_2}\,{+Rep_2}}{a + b = a_2 + b_2}\,Tr$$

Hence we have treated case 4.4 and also case 4 is finished. □

**Lemma 5.4.7** *If there is a simple derivation of the sequent* $\rightarrow t = t'$*, then there is a derivation of the same sequent without rule* $Inf_2$

*Proof* Assume that the sequent $\rightarrow t = t'$ is derivable with at least one instance of $Inf_2$ in the derivation. Then take an uppermost instance of $Inf_2$. The premise of this rule is $\rightarrow s(u) = s(v)$. According to Lemma 5.4.6 the conclusion of the rule $\rightarrow u = v$ is derivable without $Inf_2$. Thus, we can replace the subderivation with this derivation without $Inf_2$. In this way we can remove every instance of $Inf_2$ in the derivation. □

**Lemma 5.4.8** *There is no simple derivation of the empty sequent.*

*Proof* Assume that we have a simple derivation of the empty sequent. According to Theorem 5.3.2 there is a cut-free derivation of the sequent. The last rule of this derivation must be $Inf_1$ with a premise $\rightarrow s(t) = 0$ because all other rules give as the conclusion a sequent with a formula in the succedent. According to Lemma 5.4.7 the premise is derivable without $Inf_2$. Therefore we also have a derivation of the empty sequent without $Inf_2$. This is a contradiction according to Lemma 5.4.3 and thus there cannot be any simple derivation of the empty sequent. □

Gentzen and Takeuti use semantical arguments in their proofs of this lemma, while we managed to complete the proof using purely proof-theoretical means. Takeuti proves that there is either a false formula in the antecedent of a sequent in a simple proof or a true formula in the succedent. He needs these semantical arguments because he has arbitrary initial sequents in his system only specified by the requirement that they have a true atomic formula with closed terms in the succedent or a false formula in the antecedent. We managed to remove the semantical arguments from the lemma through our formulation of the system HA.

## 5.5 The Reduction Procedure for Derivations

We can now begin to describe the actual reduction procedure for derivations of the empty sequent. The main idea of the proof is that we first substitute free variables in the proof. Then according to the form of the derivation we convert inductions or cuts on compound formulas with predecessors in arithmetical rules or initial sequents. If this is not possible, then we have a so-called suitable cut. If we have a suitable cut, then we can introduce cuts on formulas of a lower grade. The problematic case is that if there are contractions on the cut formula then we cannot directly convert the suitable cut into cuts on formulas of lower grade. The problem is solved by the so-called height lines that are permuted up in the derivation by introducing additional cuts on formulas of lower grade, lowering the ordinal of the derivation.

**Lemma 5.5.1 (Reduction Procedure)** *If $P$ is a derivation of the empty sequent $\rightarrow$ in which the arithmetical rules are applied before the logical and structural rules, then there exists a derivation, $P'$, of the empty sequent, such that the arithmetical rules are applied before the logical and structural rules and $o(P') < o(P)$.*

*Proof* The proof describes a reduction procedure where a derivation $P$ is transformed into a derivation $P'$ with a lower ordinal. The reduction consists of several steps which are performed as many times as possible before proceeding to the next step and the reduction ends when a derivation with a lower ordinal is obtained.

Let $P$ be a derivation of the empty sequent $\rightarrow$. We may assume that the eigenvariables of the rules are different and that an eigenvariable occurs only above the rule in the derivation.

**Step 1.** If there are any free variables in the derivation that are not eigenvariables, then we substitute them with the constant 0. The derivation that we get is also a valid derivation of the empty sequent and it has the same ordinal as $P$.

**Step 2.** If the end-piece of $P$ contains an induction, then we perform the following reduction. Assume $I$ to be the last induction of the derivation.

$$\cfrac{\begin{array}{ccc} & P_0(x) & \\ \vdots & \vdots & \vdots \\ \Gamma_1 \overset{\mu_1}{\rightarrow} A(0) & A(x), \Gamma_2 \overset{\mu_2}{\rightarrow} A(sx) & A(t), \Gamma_3 \overset{\mu_3}{\rightarrow} D \quad (l) \end{array}}{\begin{array}{c}\Gamma_{1-3} \rightarrow D \quad (k) \\ \vdots \\ \rightarrow \end{array}} \; I$$

Here $P_0(x)$ is the subderivation ending with $A(x), \Gamma_2 \rightarrow A(sx)$ and $S$ is the sequent $\Gamma_{1-3} \rightarrow D$. The premises of $I$ all have the same height, $l$. Let $k$ be the height of the conclusion of the rule and let $\mu_i$, where $i = 1, 2, 3$, be the ordinals of the premises. Now the conclusion has the ordinal $o(\Gamma_{1-3} \rightarrow D; P) = \omega_{l-k+1}(\mu_1 \# \mu_2 \# \mu_3)$.

The term $t$ in the third premise of the rule does not contain any free variable since they were substituted in step 1. Neither does $t$ contain any eigenvariables because $I$ is the last rule with an eigenvariable in the derivation. Thus, $t$ is a closed term and there exists a number $n$, for which the sequent $\rightarrow t = \overline{n}$ is derivable without inductions or cuts [this according to Lemma 5.2.4(i)]. Therefore we have a derivation $Q$ of the sequent $A(\overline{n}) \rightarrow A(t)$ also without inductions or cuts. This according to Lemma 5.2.4(iv).

The derivation $P$ can now be reduced to $P'$ according to the following principle if $n > 0$ (if $n$ equals 0 the corresponding reduction is used but no contractions are needed. Instead the missing context $\Gamma_2$ is added in the derivation.) Let $P_0(\overline{m})$ be the derivation that we get from $P_0(x)$ when every occurrence of $x$ is substituted with $\overline{m}$ and let $\Pi$ be the derivation:

$$
\dfrac{\dfrac{\begin{array}{c}\\ \vdots \\ \Gamma_1 \to A(0)\end{array} \quad \begin{array}{c} P_0(\overline{0}) \\ \vdots \\ A(0), \Gamma_2 \to A(s0)\end{array}}{\Gamma_1, \Gamma_2 \to A(s0)}\,Cut \qquad \begin{array}{c} P_0(\overline{1}) \\ \vdots \\ A(s0), \Gamma_2 \to A(ss0)\end{array}}{\begin{array}{c}\Gamma_1, \Gamma_2^2 \to A(ss0) \\ \vdots \\ \Gamma_1, \Gamma_2^n \to A(\overline{n})\end{array}}\,Cut
$$

We reduce $P$ to the following derivation $P'$ where $\Pi$ is a subderivation:

$$
\dfrac{\dfrac{\begin{array}{c}\Pi \\ \vdots \\ \Gamma_1, \Gamma_2^n \to A(\overline{n})\end{array} \quad \begin{array}{c} Q \\ \vdots \\ A(\overline{n}) \to A(t)\end{array}}{\Gamma_1, \Gamma_2^n \to A(t)}\,Cut \qquad \begin{array}{c}\vdots \\ A(t), \Gamma_3 \to D\end{array}}{\begin{array}{c}\Gamma_1, \Gamma_2^n, \Gamma_3 \to D \\ \vdots \ \text{Contractions} \\ \Gamma_{1-3} \to D \\ \vdots \\ \to \end{array}}\,Cut
$$

All cuts shown in $\Pi$ and $P'$ are on formulas of the same grade, so all cut premises have the same height $l$. Therefore the ordinals of the premises of the first cut in $\Pi$ are $o(\Gamma_1 \to A(0); P') = \mu_1$ and $o(A(0), \Gamma_2 \to A(s0); P') = \mu_2$. The ordinal of the conclusion, $S_1'$, is then $o(S_1') = \omega_{l-l}(\mu_1 \# \mu_2) = \mu_1 \# \mu_2$. The conclusion of the second cut, $S_2'$, then has the ordinal $o(S_3') = \mu_1 \# \mu_2 \# \mu_2$ and so on. If we write $\mu * m = \mu \# \mu \# \ldots \# \mu (m$ times), we get $o(S_m') = \mu_1 \# (\mu_2 * m)$ for every $m = 1, \ldots, n$. If we denote the ordinal of $Q$ by $q$, we have $o(A(\overline{n}) \to A(t)) = q < \omega$ because $Q$ does not contain any inductions or cuts. Because each of the ordinals $\mu_1, \mu_2 * n, q$, and $\mu_3$ is less than $\omega^{\mu_1 \# \mu_2 \# \mu_3}$, the sum is also less, that is we have the inequality $\mu_1 \# (\mu_2 * n) \# q \# \mu_3 < \omega^{\mu_1 \# \mu_2 \# \mu_3}$. From this follows that $o(S; P') = \omega_{k-l}(\mu_1 \# (\mu_2 * n) \# q \# \mu_3) < \omega_{l-k+1}(\mu_1 \# \mu_2 \# \mu_3) = o(S; P)$, that is $o(S; P') < o(S; P)$. According to Lemma 5.2.2 we then have $o(P') < o(P)$.

Thus, if there is an induction in the end-piece we have reduced the derivation. Otherwise we can assume that the end-piece is free from inductions.

**Step 3.** Assume that there is a compound formula $E$ in the end-piece of the derivation. Let $I$ be the cut in the end-piece where the formula disappears. No predecessor of the formula in the left cut premise can be derived by an arithmetical rule that borders on the end-piece since the formula $E$ has logical structure. Now assume that a predecessor of the formula in the right cut premise has been derived by an arithmetical rule that borders on the end-piece.

$$
\dfrac{\dfrac{\vdots \qquad \dfrac{}{E, \Gamma_2' \to D'}\, Arithm. \atop \vdots}{\Gamma_1 \to E \quad E, \Gamma_2 \to D \quad (k)}}{\Gamma_{1-2} \to D \quad (l)}\, I
\\
\vdots
\\
\to
$$

Above the arithmetical rule that borders on the end-piece we have only other arithmetical rules and initial sequents (this according to the assumption made in the beginning of the proof.) The formula $E$ is therefore not principal in any rule above the arithmetical rule and it cannot be introduced in an initial sequent as the formula on both sides either, since no succedent of a sequent above the arithmetical rule can be compound.

Hence the formula $E$ has been introduced in the context of an arithmetical rule without premises or in an initial sequent and we can eliminate the formula and trace down in the derivation deleting the formula in the context of every arithmetical rule. Thus, we get a derivation of the sequent $\Gamma_2' \to D'$ that is otherwise similar to the derivation of $E, \Gamma_2' \to D'$.

We now divide the reduction into two cases depending on whether we have any contractions on the formula $E$ between the arithmetical rule that borders on the end-piece and the cut $I$ where the formula disappears.

**Case 1.** Assume that there are no contractions on the formula $E$ between the arithmetical rule and $I$. We now continue deleting every occurrence of $E$ and also the cut $I$, instead adding the missing context $\Gamma_1$ in the antecedent. Thus, we have a valid derivation of the sequent $\Gamma_{1-2} \to D$ and the derivation $P'$ is as follows:

$$
\dfrac{}{\Gamma_1, \Gamma_2' \to D'}\, Arithm.
\\
\vdots
\\
\Gamma_{1-2} \to D
\\
\vdots
\\
\to
$$

Now in order to calculate the ordinal of the new derivation let $S$ be a sequent in $P$ above $E, \Gamma_2 \to D$ and let $S'$ be the corresponding sequent in $P'$. We then show by induction on the number of inferences up to $E, \Gamma_2 \to D$ that the following inequality holds

$$\omega_{k_1 - k_2}(o(S; P)) \geqslant o(S'; P'), \tag{5.5.1}$$

where $k_1 = h(S; P)$ and $k_2 = h(S'; P')$ and thus $k_1 \geqslant k_2$.

If $S$ is an initial sequent or the conclusion of an arithmetical rule without premises, then $o(S; P) = o(S'; P') = 1$ and the proposition holds. Now assume that the sequent $S$ has been derived with a rule and that the claim holds for its premises. If $S$ has been derived with contraction, the heights and the ordinals of the conclusions $S$ and $S'$ are the same as for the premises and the proposition holds.

If $S$ has been derived with an arithmetical or logical one-premise rule, then the heights of the conclusions are the same as for the premises. If we let the ordinals of the premises be $\alpha$ and $\alpha'$, then we get $\omega_{k_1-k_2}(o(S; P))=\omega_{k_1-k_2}(\alpha+1) > \omega_{k_1-k_2}(\alpha)$. Since the claim holds for the premises, that is $\omega_{k_1-k_2}(\alpha) \geqslant \alpha'$, we get $\omega_{k_1-k_2}(\alpha+1) > \alpha'$ and furthermore $\omega_{k_1-k_2}(\alpha + 1) \geqslant \alpha' + 1$ and the proposition holds.

If $S$ has been derived with an arithmetical or logical two-premise rule, then again the heights of the conclusions are the same as for the premises. If we let the ordinals of the premises be $\alpha, \beta$ and $\alpha', \beta'$ we have the following inequalities for the premises of the rules $\omega_{k_1-k_2}(\alpha) \geqslant \alpha'$ and $\omega_{k_1-k_2}(\beta) \geqslant \beta'$. If $k_1 = k_2$, then we get from the inequalities of the premises $\alpha \geqslant \alpha'$ and $\beta \geqslant \beta'$ the inequality $\omega_{k_1-k_2}(o(S; P)) = o(S; P) = \alpha\#\beta \geqslant \alpha'\#\beta'$. On the other hand, if $k_1 > k_2$, we get $\omega_{k_1-k_2}(\alpha\#\beta) > \omega_{k_1-k_2}(\alpha) \geqslant \alpha'$ and $\omega_{k_1-k_2}(\alpha\#\beta) > \omega_{k_1-k_2}(\beta) \geqslant \beta'$. This gives $\omega_{k_1-k_2}(\alpha\#\beta) > \alpha'\#\beta'$ and the proposition holds.

If $S$ has been derived with a cut the premises of which have the height $m_1$ and the ordinals $\alpha$ and $\beta$ and $S'$ has been derived with a cut the premises of which have the height $m_2$ and the ordinals $\alpha'$ and $\beta'$, then we have the following inequalities for the premises $\omega_{m_1-m_2}(\alpha) \geqslant \alpha'$ and $\omega_{m_1-m_2}(\beta) \geqslant \beta'$. We then get $\omega_{k_1-k_2}(o(S; P)) = \omega_{k_1-k_2}(\omega_{m_1-k_1}(\alpha\#\beta)) = \omega_{m_1-k_2}(\alpha\#\beta) = \omega_{m_2-k_2}(\omega_{m_1-m_2}(\alpha\#\beta))$. If $m_1 = m_2$, then from the inequalities of the premises $\alpha \geqslant \alpha'$ and $\beta \geqslant \beta'$ we get the inequality $\omega_{m_2-k_2}(\omega_{m_1-m_2}(\alpha\#\beta)) = \omega_{m_2-k_2}(\alpha\#\beta) \geqslant \omega_{m_2-k_2}(\alpha'\#\beta')$. If $m_1 > m_2$, then we get $\omega_{m_1-m_2}(\alpha\#\beta) > \omega_{m_1-m_2}(\alpha) \geqslant \alpha'$ and $\omega_{m_1-m_2}(\alpha\#\beta) > \omega_{m_1-m_2}(\beta) \geqslant \beta'$. Thus, we get $\omega_{m_1-m_2}(\alpha\#\beta) > \alpha'\#\beta'$ and from this follows that $\omega_{m_2-k_2}(\omega_{m_1-m_2}(\alpha\#\beta)) > \omega_{m_2-k_2}(\alpha'\#\beta')$, that is the proposition holds.

If $S$ has been derived with an *Ind* the premises of which have the height $m_1$ and the ordinals $\alpha, \beta$ and $\gamma$ and $S'$ has been derived with an *Ind* the premises of which have the height $m_2$ and the ordinals $\alpha', \beta'$ and $\gamma'$ then we have the following inequalities for the premises $\omega_{m_1-m_2}(\alpha) \geqslant \alpha'$, $\omega_{m_1-m_2}(\beta) \geqslant \beta'$ and $\omega_{m_1-m_2}(\gamma) \geqslant \gamma'$. We then have

$$\begin{aligned}\omega_{k_1-k_2}(o(S; P)) &= \omega_{k_1-k_2}(\omega_{m_1-k_1+1}(\alpha\#\beta\#\gamma)) \\ &= \omega_{m_1-k_2+1}(\alpha\#\beta\#\gamma) = \omega_{m_2-k_2+1}(\omega_{m_1-m_2}(\alpha\#\beta\#\gamma)) \\ &\geqslant \omega_{m_2-k_2+1}(\alpha'\#\beta'\#\gamma') = o(S'; P')\end{aligned}$$

Thus, it has been proved that inequality (5.5.1) holds.

Now let $S$ be the sequent $E, \Gamma_2 \to D$ and $S'$ the corresponding sequent $\Gamma_{1-2} \to D$. If we let $o(\Gamma_1 \to E; P) = \mu_1, o(E, \Gamma_2 \to D; P) = \mu_2, o(\Gamma_{1-2} \to D; P) = \nu$ and $o(\Gamma_{1-2} \to D; P') = \nu'$ and also let $h(\Gamma_{1-2} \to D; P) = l$ and

$h(E, \Gamma_2 \rightarrow D; P) = k$, then we have $l \leqslant k$ and $h(\Gamma_{1-2} \rightarrow D; P') = l$. From the inequality we get

$$\omega_{k-l}(\mu_2) \geqslant \nu'$$

and from this follows the inequality

$$\nu = \omega_{k-l}(\mu_1 \# \mu_2) > \omega_{k-l}(\mu_2) \geqslant \nu'.$$

according to Lemma 5.2.2 we can conclude that $o(P) > o(P')$.

**Case 2.** Assume that there is at least one contraction on the formula $E$ between the arithmetical rule and $I$. Let the uppermost contraction be $I'$. Recall that we have a derivation of the sequent $\Gamma_2' \rightarrow D'$ that is otherwise similar to the derivation of $E, \Gamma_2' \rightarrow D'$. We can now reduce the derivation to the left into the one on the right by eliminating the contraction.

$$\begin{array}{c} \overline{E, \Gamma_2' \rightarrow D'}\ \textit{Arithm.} \\ \vdots \\ E, E, \Gamma_2'' \rightarrow D'' \\ \hline E, \Gamma_2'' \rightarrow D''\ I' \\ \vdots \\ E, \Gamma_2 \rightarrow D \\ \vdots \end{array} \quad \rightsquigarrow \quad \begin{array}{c} \overline{\Gamma_2' \rightarrow D'}\ \textit{Arithm.} \\ \vdots \\ E, \Gamma_2'' \rightarrow D'' \\ \vdots \\ E, \Gamma_2 \rightarrow D \\ \vdots \end{array}$$

In this reduction the ordinal is preserved and $o(P) = o(P')$. We now repeat step 3 if we can or continue with step 4 and assume that compound formulas in the end-piece of $P$ do not have predecessors in arithmetical rules that border on the end-piece. Therefore these formulas must have predecessors in initial sequents or logical rules that border on the end-piece.

**Step 4.** Assume that the end-piece contains an initial sequent $D, \Gamma \rightarrow D$. Since the end-sequent is empty both formulas $D$ (or rather descendants of both formulas) must disappear through cuts. Assume that the $D$ in the antecedent is the first formula to disappear in a cut (the other case is similar). The derivation $P$ now has the form

$$\begin{array}{c} \qquad\qquad D, \Gamma \rightarrow D \\ \vdots \qquad\qquad \vdots \\ \Gamma_1 \rightarrow D \quad D, \Gamma_2 \rightarrow D \\ \hline \Gamma_{1-2} \rightarrow D \end{array}\ \textit{Cut}$$
$$\begin{array}{c} \vdots \\ \rightarrow \end{array}$$

We can reduce $P$ into a derivation $P'$ where the cut has been eliminated by adding the missing context $\Gamma_2$ in the antecedent of the derivation of the left premise.
Since both $D$'s from the sequent $D, \Gamma \rightarrow D$ disappear through cuts, we have a cut on the other $D$ in the succedent below the sequent $\Gamma_{1-2} \rightarrow D$. Therefore the heights of the sequents remain unchanged, while the ordinal of the subderivation ending with $\Gamma_{1-2} \rightarrow D$ decreases. Thus, we get $o(P') < o(P)$ by Lemma 5.2.2. We can now proceed to step 5 and can assume that the end-piece does not contain any initial sequents but only cuts and contractions.

**Step 5.** To continue the reduction procedure we consider the compound cut formulas of the end-piece. We want to diminish the ordinal of the derivation by introducing cuts on shorter formulas. For this we need a suitable cut in the end-piece.

**Definition 5.5.2** A cut in the end-piece of a derivation is a *suitable cut* if both copies of the cut formula have predecessors that are principal in logical rules that border on the end-piece.

**Sublemma 5.5.3** *Assume that a derivation $P$ fulfils the following requirements:*

1. *The end-piece of $P$ contains at least one cut on a compound formula.*
2. *In every cut on a compound formula in the end-piece each copy of the cut formula has a predecessor in the conclusion of a logical rule that borders on the end-piece.*
3. *The principal formula of the logical rule mentioned in (2) has a descendant that disappears through a cut in the end-piece.*

*Then $P$ has a suitable cut.*

*Proof* The proof is an induction on the number of cuts on compound cut formulas in the end-piece.

In the end-piece of $P$ there is at least one cut on a compound formula according to (1). If there is only one cut, then the cut formulas of both premises have a predecessor in a logical rule bordering on the end-piece according to (2). If the principal formula of the rule was not the predecessor of the cut formula, then it would according to (3) have to disappear through another cut in the end-piece. Thus, the principal formula has to be the predecessor of the only cut and we have a suitable cut.

Now assume that $P$ has $n$ cuts on compound formulas in the end-piece. As induction hypothesis we have that any derivation with fewer such cuts has a suitable cut, provided that the derivation fulfils the stipulated requirements. Let $I$ be the last of the cuts on some compound formula, $D$.

$$\dfrac{\begin{matrix} P_1 \\ \vdots \\ \Gamma_1 \rightarrow D \end{matrix} \quad \begin{matrix} P_2 \\ \vdots \\ D, \Gamma_2 \rightarrow E \end{matrix}}{\Gamma_{1-2} \rightarrow E} \; I$$

If $I$ is a suitable cut, the proposition is proved. Therefore we assume that $I$ is not a suitable cut. Both cut formulas of the premises have, according to (2) a predecessor in the conclusion of a logical rule bordering on the end-piece. Since the cut is not a suitable cut a predecessor of one $D$ is not principal in one of the logical rules. We may assume that this is the case for the $D$ in the premise $\Gamma_1 \rightarrow D$. According to (3) a descendant of the principal formula in the logical rule disappears through a cut. If this cut was $I$, then the principal formula would be $D$, but then $I$ would be a suitable cut. Therefore there must be another cut on a compound formula and this cut is above $I$ in $P_1$ since $I$ was the last cut. Thus, $P_1$ satisfies (1). $P_1$ also inherits property (2) from $P$. None of the principal formulas in the logical rules bordering on the end-piece can disappear through the cut $I$, since that would make $I$ a suitable cut, therefore the cuts must be in $P_1$ and $P_1$ fulfils criterion (3). Therefore the subderivation $P_1$ fulfils all three requirements and according to the induction hypothesis has a suitable cut. This is also a suitable cut of the derivation $P$. □

We now continue to consider the derivation $P$ of the empty sequent. If the derivation $P$ contained only atomic formulas, then any instances of *Ind* would be in the end-piece, but this is not possible since these were reduced in step 2. Hence the derivation $P$ contains a compound formula, for otherwise the derivation would be simple which is impossible according to Lemma 5.4.8. Since the end-sequent is empty and the end-piece does not contain any instances of *Ind* all formulas in the end-piece must disappear through cuts. At least one of these formulas has logical structure. The derivation $P$ therefore satisfies the first criterion in Sublemma 5.5.3. Assume that $D$ is a compound formula that disappears though a cut in the end-piece. The formula $D$ cannot have a predecessor in an arithmetical rule that borders on the end-piece, since these were treated in step 3. Neither can a predecessor of $D$ have been introduced in an initial sequent in the end-piece, since these were treated in step 4. The only remaining possibility is that the formula has a predecessor in the conclusion of a logical rule bordering on the end-piece. This means that $P$ satisfies the second criterion in Sublemma 5.5.3. From the fact that the end-sequent is empty and that there are no inductions in the end-piece we draw the conclusion that $P$ satisfies the third criterion in Sublemma 5.5.3. Therefore $P$ fulfils all the requirements of the sublemma and $P$ contains a suitable cut.

Now consider the lowermost suitable cut $I$ and perform the following reduction according to the form of the cut formula.

**Case 1.** Assume that the cut formula of the last suitable cut is a conjunction $B\&C$. Now $P$ has the form

$$
\begin{array}{c}
\dfrac{\dfrac{\begin{array}{c}\vdots\\ \Gamma_1'' \to B\end{array} \quad \begin{array}{c}\vdots\\ \Gamma_1''' \to C\end{array}}{\begin{array}{c}\Gamma_1' \to B\&C\\ \vdots\\ \Gamma_1 \overset{\mu}{\to} B\&C\end{array}}\,R\& \qquad \dfrac{\begin{array}{c}\vdots\\ B, C, \Gamma_2' \to D'\end{array}}{\begin{array}{c}B\&C, \Gamma_2' \to D'\\ \vdots\\ B\&C, \Gamma_2 \overset{\nu}{\to} D \quad (l)\end{array}}\,L\&}{\Gamma_{1-2} \to D}\,I\\
\vdots\\
\Theta \overset{\lambda}{\to} E \quad (k)\\
\vdots\\
\to
\end{array}
$$

where $\Gamma_1' = \Gamma_1'', \Gamma_1'''$ and $\Theta \to E$ is the first sequent below $I$ that has a lower height than the premises of the cut. Such a sequent exists because the height of the end-sequent is 0 while the cut premises have a height of at least 1. Let $l$ be the height of the premises of the cut $I$ and let $h(\Theta \to E; P) = k$. Then we have $k < l$. The sequent $\Theta \to E$ must be the conclusion of a cut since the end-piece only contains contractions and cuts and the conclusion of a contraction has the same height as the premise. Furthermore, we let $o(\Gamma_1 \to B\&C) = \mu, o(B\&C, \Gamma_2 \to D) = \nu$ and $o(\Theta \to E) = \lambda$.

In the derivation of $B, C, \Gamma_2' \to D'$ we can add the formula $B\&C$ in the context and get a derivation of the sequent $B\&C, B, C, \Gamma_2' \to D'$. Now let $P_3$ be the following derivation:

$$
\begin{array}{c}
\dfrac{\begin{array}{c}\vdots\\ \Gamma_1 \overset{\mu_3}{\to} B\&C\end{array} \quad \begin{array}{c}\vdots\\ B\&C, B, C, \Gamma_2' \to D'\\ \vdots\\ B\&C, B, C, \Gamma_2 \overset{\nu_3}{\to} D\end{array}}{B, C, \Gamma_{1-2} \to D}\,J_3\\
\vdots\\
B, C, \Theta \to E
\end{array}
$$

We take the derivation of $\Gamma_1'' \to B$ and instead of applying a right conjunction rule we add the missing formulas $\Gamma_1'''$ in the context and get a derivation of the sequent $\Gamma_1' \to B$. Then we apply the cuts and contractions above the left premise of the cut $J_3$ shown in $P_3$ (this is possible because the descendant of the conjunction in the succedent disappears through the cut $J_3$ and therefore cannot be principal in another rule above the cut.) Hence we have constructed a derivation of the sequent $\Gamma_1 \to B$. We again instead of applying the cut $J_3$ add the missing context $\Gamma_2$ and get a derivation of $\Gamma_{1-2} \to B$. After this we continue with the same rules as below $P_3$ applying the same rules on the same formulas if we have a contraction or a cut on formulas in the antecedent. If we, on the other hand, in $P_3$ have a cut on the formula in the succedent (that is a cut on the formula in $P_3$ that has been replaced by the formula $B$ in the constructed derivation) we instead of applying the cut add the missing context in the antecedent of the sequent. Thus, we get a valid derivation of the sequent $\Theta \to B$ and we call this derivation $P_1$. Correspondingly we construct a derivation of the sequent $\Theta \to C$ from the derivation $P_3$ and call this derivation $P_2$.

We now compose the three derivations into the derivation $P'$:

$$
\dfrac{\begin{array}{c} P_1 \\ \vdots \\ \Theta \overset{\lambda_1}{\to} B \end{array} \qquad \dfrac{\begin{array}{c} P_2 \\ \vdots \\ \Theta \overset{\lambda_2}{\to} C \end{array} \quad \begin{array}{c} P_3 \\ \vdots \\ B, C, \Theta \overset{\lambda_3}{\to} E \end{array} \quad (m_2)}{B, \Theta^2 \to E \quad (m_1)}\ \mathit{Cut}}{\begin{array}{c} \Theta^3 \overset{\lambda_0}{\to} E \quad (k) \\ \vdots \ \text{contractions} \\ \Theta \to E \\ \vdots \\ \to \end{array}}\ \mathit{Cut}
$$

Let $m_1$ be the height of the premises of the cut on the formula $B$ and let $m_2$ be the height of the premises of the cut on the formula $C$. The premises of the cut $J_3$ in $P'$ have the height $l$ because all cuts below the premises of the cut $I$ also occur below $J_3$. And both added cuts have a lower grade than the cut formula $B \& C$. Furthermore, we have that $h(\Theta^3 \to E; P') = k$.

Assume that the grade of $B$ is higher than or equal to the grade of $C$ (otherwise we may exchange the order of the two cuts). Now we have $m_1 = m_2$. If $k$ is higher than the grade of $B$ (and the grade of $C$), then we have that $k = m_1 = m_2$ and if not $m_1$ equals the grade of $B$. In both cases we have $k \leqslant m_1$.

Let

$$
\begin{aligned}
\lambda_0 &= o(\Theta^3 \to E; P') \\
\lambda_1 &= o(\Theta \to B; P') \\
\lambda_2 &= o(\Theta \to C; P') \\
\lambda_3 &= o(B, C, \Theta \to E; P') \\
\mu_3 &= o(\Gamma_1 \to B\&C; P') \\
\nu_3 &= o(B\&C, B, C, \Gamma_2 \to D; P')
\end{aligned}
$$

Then we have that $\nu_3 < \nu$ since the heights of the sequents above remain unchanged and a logical rule has been removed. Furthermore, we have that $\mu_3 = \mu$.

Now let

$$
\dfrac{S_1' \quad S_2'}{S'}\ J'
$$

be an arbitrary rule between $J_3$ and the sequent $B, C, \Theta \rightarrow E$ in the subderivation $P_3$ of $P'$ and let

$$\frac{S_1 \quad S_2}{S} \; J$$

be the corresponding rule between $I$ and $\Theta \rightarrow E$ in $P$. Let

$$\begin{array}{ll} \alpha_1' = o(S_1'; P') & \alpha_2' = o(S_2'; P') \; \alpha' = o(S'; P') \\ \alpha_1 = o(S_1; P) & \alpha_2 = o(S_2; P) \;\; \alpha = o(S; P) \\ k_1 = h(S_1'; P) = h(S_2'; P') \; k_2 = h(S'; P') & \end{array}$$

Then we have that $\alpha = \alpha_1 \# \alpha_2$ if $S'$ is not the sequent $B, C, \Theta \rightarrow E$ and $\alpha = \omega_{l-k}(\alpha_1 \# \alpha_2)$ if $S'$ is the sequent $B, C, \Theta \rightarrow E$. On the other hand, we have that $\alpha' = \omega_{k_1-k_2}(\alpha_1' \# \alpha_2')$.

We show by induction on the number of inferences between $J_3$ and $S'$ that

$$\alpha' < \omega_{l-k_2}(\alpha) \tag{5.5.2}$$

if $S'$ is not the sequent $B, C, \Theta \rightarrow E$.

If $J'$ is $J_3$, then we have that

$$\alpha' = \omega_{l-k_2}(\mu_3 \# \nu_3) < \omega_{l-k_2}(\mu \# \nu) = \omega_{l-k_2}(\alpha)$$

because $\mu_3 = \mu$ and $\nu_3 < \nu$.

If we assume that the inequality holds for the premises of $J'$, that is $\alpha_1' < \omega_{l-k_1}(\alpha_1)$ and $\alpha_2' < \omega_{l-k_1}(\alpha_2)$ then we get that $\alpha_1' \# \alpha_2'$ is less than $\omega_{l-k_1}(\alpha_1) \# \omega_{l-k_1}(\alpha_2)$, this implies that $\alpha_1' \# \alpha_2' < \omega_{l-k_1}(\alpha_1 \# \alpha_2)$. From this follows that the inequality holds for the conclusion, because we have

$$\alpha' = \omega_{k_1-k_2}(\alpha_1' \# \alpha_2') < \omega_{k_1-k_2}(\omega_{l-k_1}(\alpha_1 \# \alpha_2)) = \omega_{l-k_2}(\alpha_1 \# \alpha_2) = \omega_{l-k_2}(\alpha).$$

Thus, it is proved that the inequality (5.5.2) holds.

The inequality (5.5.2) holds for the premises of the cut that gives the sequent $B, C, \Theta \rightarrow E$. The premises have the height $l = k_2$ and if we denote the ordinals of the premises $\alpha_1'$ and $\alpha_2'$ and for the corresponding premises in $P$ $\alpha_1$ and $\alpha_2$, we get from the inequalities of the premises that $\alpha_1' < \omega_{l-l}(\alpha_1) = \alpha_1$ and $\alpha_2' < \omega_{l-l}(\alpha_2) = \alpha_2$ hold. From this follows that $\lambda_3 = \omega_{l-m_2}(\alpha_1' \# \alpha_2') < \omega_{l-m_2}(\alpha_1 \# \alpha_2) = \omega_{l-m_2}(\kappa)$, if we let $\lambda = \omega_{l-k}(\kappa)$.

Then remains to calculate corresponding inequalities for the ordinals of the other subderivations $P_1$ and $P_2$. We consider the derivation $P_1$. There are two possibilities to consider, namely, that the last cut above the sequent $\Theta \rightarrow E$ in $P_3$ has been eliminated in the construction of $P_1$ and the possibility that there is a corresponding cut above the sequent $\Theta \rightarrow B$ in $P_1$. We show that in both cases $\lambda_1 \leqslant \lambda_3$.

Assume that there is a corresponding cut in $P_1$. The conclusion of the cut in $P_3$ has the height $m_2$, the premises have the height $l > m_2$ and the cut formula has the grade $l$. The cut formula of the cuts between $J_3$ and the cut in question have a grade lower or equal to $l$. Thus, all heights remain unchanged when the cuts are removed in $P_1$. And we conclude that $\lambda_1 \leqslant \lambda_3$.

Now assume for the other case that the last cut above the sequent $\Theta \rightarrow B$ has been eliminated. This means that the heights of the corresponding sequents in $P_1$ and $P_3$ are no longer equal. We define the notion height difference to be able to inductively prove the inequality we want.

**Definition 5.5.4** Let the premises of a cut or an induction have the height $g$ and the conclusion the height $h$. The *height difference* of the cut or the induction is $g-h$ for the cut and $g-h+1$ for the induction. The height difference between two sequents in a derivation is the sum of the height differences for all cuts and inductions between the two sequents.

The height difference between two sequents is equal to the height of the uppermost sequent, minus the height of the lowermost sequent, plus the number of inductions between the sequents.

Let $S$ be a sequent in $P_3$ with the ordinal $\alpha$ and $S'$ the corresponding sequent in $P_1$ with the ordinal $\alpha'$. We show by induction that

$$\alpha' \leqslant \omega_{h-h'}(\alpha) \tag{5.5.3}$$

where $h$ is the height difference between $S$ and the conclusion of the subderivation $P_3$, that is $B, C, \Theta \rightarrow E$ and $h'$ is the height difference between $S'$ and the conclusion of the subderivation $P_1$, that is $\Theta \rightarrow B$.

The expression is well defined if $h \geqslant h'$. The sequents $B, C, \Theta \rightarrow E$ and $\Theta \rightarrow B$ have the same height $m_1 = m_2$ and the number of inductions between $S$ and $B, C, \Theta \rightarrow E$ and between $S'$ and $\Theta \rightarrow B$ is the same. Since the cut formulas below $S'$ also occur below $S$ we have that the height of $S$ is greater or equal to the height of $S'$. This means that $h \geqslant h'$ and the expression is well defined. We can now proceed to proving the inequality (5.5.3).

If $S$ is an initial sequent or the conclusion of an arithmetical rule without premises, then $\alpha' = \alpha = 1$ and the inequality holds regardless of the size of $h - h'$.

Assume that the inequality holds for the premise of a one-premise rule. Let the height difference under the premise in $P_3$ be $h$ and in $P_1$ $h'$ and let the ordinals of the premises be $\alpha$ and $\alpha'$, respectively. The height differences under the conclusions are the same. If the rule is a contraction, the inequality of the premises is preserved. If the rule is logical or arithmetical, then we get $\alpha' \leqslant \omega_{h-h'}(\alpha) < \omega_{h-h'}(\alpha + 1)$ and from this $\alpha' + 1 \leqslant \omega_{h-h'}(\alpha + 1)$.

Assume that the inequality holds for the premises of a two-premise arithmetical or logical rule, that is $\alpha_1' \leqslant \omega_{h-h'}(\alpha_1)$ and $\alpha_2' \leqslant \omega_{h-h'}(\alpha_2)$ hold. Here $\alpha_1$ and $\alpha_2$ are the ordinals of the premises in $P_3$ and $\alpha_1'$ and $\alpha_2'$ are the ordinals of the premises in $P_1$. The height differences under the premises, $h$ and $h'$, are the same as under the conclusion. We then get $\alpha' = \alpha_1'\#\alpha_2' \leqslant \omega_{h-h'}(\alpha_1)\#\omega_{h-h'}(\alpha_2) \leqslant \omega_{h-h'}(\alpha_1\#\alpha_2) = \omega_{h-h'}(\alpha)$.

Assume that the inequality holds for the premises of a cut in $P_3$, that is $\alpha_1' \leqslant \omega_{h-h'}(\alpha_1)$ and $\alpha_2' \leqslant \omega_{h-h'}(\alpha_2)$ hold. If the cut has been eliminated in $P_1$, then $S'$ has the ordinal $\alpha_1'$. Let the height difference of the cut be $g$ in $P_3$. Now the height difference under $S$ is $h - g$ and we get the inequality $\alpha' = \alpha_1' \leqslant \omega_{h-h'}(\alpha_1) < \omega_{h-h'}(\alpha_1\#\alpha_2) = \omega_{(h-g)-h'}(\omega_g(\alpha_1\#\alpha_2)) = \omega_{(h-g)-h'}(\alpha)$. On the other hand, if the cut also occurs in $P_1$, in other words if it has not been eliminated, we let the height difference in $P_1$ be $g'$. Now the height difference under $S$ is $h - g$ and under $S'$ $h' - g'$ and we get the inequality for the conclusion $\alpha' = \omega_{g'}(\alpha_1'\#\alpha_2') \leqslant \omega_{g'}(\omega_{h-h'}(\alpha_1)\#\omega_{h-h'}(\alpha_2)) \leqslant \omega_{g'}(\omega_{h-h'}(\alpha_1\#\alpha_2)) = \omega_{g'+h-h'-g}(\omega_g(\alpha_1\#\alpha_2)) = \omega_{(h-g)-(h'-g')}(\alpha)$.

Lastly assume that the inequality holds for the premises of an instance of *Ind*, that is $\alpha_1' \leqslant \omega_{h-h'}(\alpha_1)$, $\alpha_2' \leqslant \omega_{h-h'}(\alpha_2)$ and $\alpha_3' \leqslant \omega_{h-h'}(\alpha_3)$. Let the height difference for the induction in $P_1$ be $g'$ and in $P_3$ $g$. Now the height difference under $S$ is $h - g$ and under $S'$ $h' - g'$ and we get the inequality $\alpha' = \omega_{g'}(\alpha_1'\#\alpha_2'\#\alpha_3') \leqslant \omega_{g'}(\omega_{h-h'}(\alpha_1)\#\omega_{h-h'}(\alpha_2)\#\omega_{h-h'}(\alpha_3)) \leqslant \omega_{g'}(\omega_{h-h'}(\alpha_1\#\alpha_2\#\alpha_3)) = \omega_{g'+h-h'-g}(\omega_g(\alpha_1\#\alpha_2\#\alpha_3)) = \omega_{(h-g)-(h'-g')}(\alpha)$.

Thus, it has been proved that the inequality holds. Now let $S$ be the sequent $B, C, \Theta \to E$ and $S'$ the sequent $\Theta \to B$. Then the height differences $h$ and $h'$ are 0 and we get $\lambda_1 \leqslant \omega_{h-h'}(\lambda_3) = \lambda_3$.

Regardless of if the last cut has been eliminated we thus have $\lambda_1 \leqslant \lambda_3$. Correspondingly we get $\lambda_2 \leqslant \lambda_3$. Using the inequality $\lambda_3 < \omega_{l-m_2}(\kappa)$ and the fact that $m_1 = m_2$ we then get $\lambda_1\#\lambda_2\#\lambda_3 < \omega_{l-m_1}(\kappa)$, since $l > m_1$. Furthermore, we get that $\lambda_0 = \omega_{m_1-k}(\lambda_1\#(\omega_{m_2-m_1}(\lambda_2\#\lambda_3))) = \omega_{m_1-k}(\lambda_1\#\lambda_2\#\lambda_3) < \omega_{m_1-k}(\omega_{l-m_1}(\kappa)) = \omega_{l-k}(\kappa) = \lambda$.

From the inequality $\lambda_0 < \lambda$ we get according to Lemma 5.2.2 that $o(P) > o(P')$.

**Case 2.** Assume that the cut formula of the last suitable cut is $\forall x B(x)$. The derivation $P$ then has the form

$$
\dfrac{\dfrac{\dfrac{\vdots}{\Gamma_1' \to B(y/x)}}{\Gamma_1' \to \forall x B(x)}\,R\forall \quad\vdots\quad \Gamma_1 \to \forall x B(x) \qquad \dfrac{\dfrac{\vdots}{B(t/x), \Gamma_2' \to D'}}{\forall x B(x), \Gamma_2' \to D'}\,L\forall \quad\vdots\quad \forall x B(x), \Gamma_2 \to D}{\Gamma_{1-2} \to D}\,I
$$

$$
\begin{array}{c} \Gamma_{1-2} \to D \\ \vdots \\ \Theta \to E \\ \vdots \\ \to \end{array}
$$

where the sequent $\Theta \to E$ is defined in the same way as in case 1.

From the derivation of the sequent $B(t/x), \Gamma_2' \to D_2'$ we get a derivation of the sequent $\forall x B(x), B(t/x), \Gamma_2' \to D_2'$ by adding a formula in the context. Let $P_2$ be the following derivation:

$$\dfrac{\begin{array}{c}\vdots\\ \Gamma_1 \to \forall x B(x)\end{array} \quad \begin{array}{c}\vdots\\ \forall x B(x), B(t/x), \Gamma_2' \to D' \\ \vdots \\ \forall x B(x), B(t/x), \Gamma_2 \to D\end{array}}{\begin{array}{c}B(t/x), \Gamma_{1-2} \to D \\ \vdots \\ B(t/x), \Theta \to E\end{array}}\, J_2$$

We can get a derivation of the sequent $\Gamma_1' \to B(t/x)$ from the derivation of $\Gamma_1' \to B(y/x)$ by substituting $y$ with $t$. We then apply the rules between the logical rule and $J_2$ in $P_2$ to the sequent $\Gamma_1' \to B(t/x)$ (this is possible because the quantified formula in the succedent of the sequents in $P_2$ is not principal in any rule above the cut $J_2$). We now have a derivation of the sequent $\Gamma_1 \to B(t/x)$ and can instead of applying the cut add the missing context in the antecedent and get a derivation of $\Gamma_{1-2} \to B(t/x)$. Then we apply the cuts and contractions below the cut $J_2$ on formulas in the antecedent. If we have a cut on the succedent, that is on the formula that has been replaced with $B(t/x)$, we just add the missing context in the antecedent and eliminate the cut. Thus, we obtain a valid derivation of the sequent $\Theta \to B(t/x)$ and we call this derivation $P_1$.

Now we can join the two derivations together into one derivation $P'$

$$\dfrac{\begin{array}{c}P_1\\ \vdots\\ \Theta \to B(t/x)\end{array} \quad \begin{array}{c}P_2\\ \vdots\\ B(t/x), \Theta \to E\end{array}}{\begin{array}{c}\Theta^2 \to E \\ \vdots \ \text{contractions} \\ \Theta \to E \\ \vdots \\ \to\end{array}}\, Cut$$

The ordinal calculations are similar to the ones in case 1 and for the other cases of cut formulas the proofs are also similar.

Thus, we have reduced the derivation $P$ into a derivation $P'$ with a lower ordinal and the proof of Lemma 5.5.1 is finished. We can conclude that the derivation $P'$ also fulfils the requirement that all arithmetical rules are applied before the logical and structural rules. This makes it possible to repeat the reduction and get a sequence of decreasing ordinals. □

With the proof of the reduction procedure finished we also have a proof of the consistency Theorem 5.1.2.

Some of the essential features of our proof are: Cut elimination is proved directly, without Gentzen's rule of multicut; the arithmetical axioms are treated purely syntactically; all rules with several premises have independent contexts and no rule of weakening is used. It is hoped that a comparison of our proof with Gentzen's notes in his series BTJZ will eventually show at what point his attempts failed.

## References


1. G. Gentzen, Die Widerspruchsfreiheit der reinen Zahlentheorie. Math. Ann. **112**, 493–565 (1936)
2. G. Gentzen, Neue Fassung des Widerspruchsfreiheitsbeweises für die reine Zahlentheorie. Forschungen zur Logik und zur Grundlegung der exakten Wissenschaften. **4**, 19–44 (1938)
3. G. Gentzen, Beweisbarkeit und Unbeweisbarkeit von Anfangsfällen der transfiniten Induktion in der reinen Zahlentheorie. Math. Ann. **119**, 140–161 (1943)
4. G. Gentzen, *The Collected Papers of Gerhard Gentzen*, ed. by M.E. Szabo (North-Holland, Amsterdam, 1969)
5. G. Gentzen, Der erste Widerspruchsfreiheitsbeweis für die klassische Zahlentheorie. Archiv für mathematische Logik **16**, 97–118 (1974)
6. S. Negri, J. von Plato, Cut elimination in the presence of axioms. Bull. Symb. Log. **4**, 418–435 (1998)
7. J. von Plato, A proof of Gentzen's Hauptsatz without multicut. Arch. Math. Log. **40**, 9–18 (2001)
8. J. von Plato, Normal form and existence property for derivations in Heyting arithmetic. Acta Philos. Fenn. **78**, 159–163 (2006)
9. J. von Plato, Gentzen's logic, in *Handbook of the History of Logic*, vol. 5 (North-Holland, Amsterdam, 2009), pp. 667–721
10. G. Takeuti, *Proof Theory* (North-Holland, Amsterdam, 1987)

# Gentzen's Original Consistency Proof and the Bar Theorem

W.W. Tait

The story of Gentzen's original consistency proof for first-order number theory [9],[1] as told by Paul Bernays [1, 9], [11, Letter 69, pp. 76–79], is now familiar: Gentzen sent it off to *Mathematische Annalen* in August of 1935 and then withdrew it in December after receiving criticism and, in particular, the criticism that the proof used the Fan Theorem, a criticism that, as the references just cited seem to indicate, Bernays endorsed or initiated at the time but later rejected. That particular criticism is transparently false, but the argument of the paper remains nevertheless invalid from a constructive standpoint. In a letter to Bernays dated November 4, 1935, Gentzen protested this evaluation; but then, in another letter to him dated December 11, 1935, he admits that the 'critical inference in my consistency proof is defective'. The defect in question involves the application of proof by induction to certain trees, the "reduction trees" for sequents (see below and § 1), of which it is only given that they are well-founded. No doubt because of his desire to reason "finitistically," Gentzen nowhere in his paper explicitly speaks of reduction trees, only of reduction rules that would generate such trees; but the requirement of well-foundedness, that every path taken in accordance with the rule terminates, of course makes implicit reference to the tree. Gentzen attempted to avoid the induction; but as he ultimately conceded, the attempt was unsatisfactory.

[1]The paper first appeared in print via an appendix to the translation of [5] in [8]. A somewhat revised version of it is presented in [1] and the full text, together with an introduction by Bernays, in [9].

W.W. Tait (✉)
Department of Philosophy, University of Chicago, S. Everett Avenue, 5522, Chicago, IL 60637, USA
e-mail: williamtait@mac.com



R. Kahle, M. Rathjen (eds.), *Gentzen's Centenary*,
DOI 10.1007/978-3-319-10103-3_8

Brouwer's Bar Theorem has generally been cited as what is needed to repair the argument.[2] The Bar Theorem does indeed suffice to close the gap between the well-foundedness of a reduction tree and proof by induction on it, but we will see that Brouwer's argument for the Bar Theorem in the context in question involves an argument for the claim that a proof of the well-foundedness of a reduction tree for a sequent can only be based on having the corresponding *deduction tree* of the sequent. The deduction tree in question is obtained by reading the reduction tree, which is constructed "bottom-up," as "top-down." Deduction trees are built up inductively and so proof by induction on them is valid. Moreover, given a deduction tree for a sequent, the corresponding reduction tree can be constructed; but the converse is constructively invalid. So—and this is the main point of this paper—the gap in Gentzen's argument is filled, not by the Bar Theorem, but by taking as the basic notion that of a deduction tree in the first place rather than that of a reduction tree. These deduction trees are well-known objects, namely *cut-free deductions* in a formalization of first-order number theory in the sequent calculus with the $\omega$-rule.

The formalization of number theory in the original paper as well as in the 1936 paper ultimately takes as the logical constants $\neg, \wedge$ and $\forall$. Deductions are of sequents of the form $\Gamma \vdash A$, where $A$ is a formula and $\Gamma$ a possibly null sequence of formulas. The rules of inference are the natural deduction rules: the introduction and elimination rules for the logical constants are only for the succedent formula (so that a deduction of the sequent $\Gamma \vdash A$ corresponds to a deduction of $A$ in natural deduction whose assumption formulas are all in $\Gamma$). I will refer to this system as *the formal system of first-order number theory in natural deduction*. The precise details don't really concern us, since the non-trivial parts of Gentzen's argument do not really concern these natural deductions.

His consistency argument in the original version aims at showing that a natural deduction of a sequent $\Gamma \vdash A$ is a code for constructing a reduction tree for the sequent. Since there is no reduction tree for the sequent $\vdash 1 = 2$, for example, consistency is implied. As we noted and will see, reduction trees can be replaced everywhere in the argument by the corresponding deduction trees.

In the 1936 version of the consistency proof, the notion of a reduction tree plays no role in the proof of consistency: that proof is obtained by means of the notion of a reduction procedure for *deductions* of sequents in the formal system of first-order number theory in natural deduction and an assignment of ordinals to these deductions such that each reduction step results in a decrease in ordinal. A reduction tree for a deduced sequent, along with an ordinal measure of the height of the subtrees, simply falls out of the proof. Of course, the assignment of ordinals to the subtrees allows the reduction tree to be identified with the corresponding deduction tree (since induction on the tree can be expressed by induction on the ordinals).

[2] Could Bernays' references to the "Fan Theorem," all over 30 years later, have been the result of a confusion of the Fan Theorem with the Bar Theorem?

In the original paper, on the contrary, the notion of a reduction rule for a sequent plays an essential role: the non-trivial part of the argument—and the source of difficulty—is Gentzen's argument for his Lemma. As we noted, the corresponding deduction trees are cut-free deductions in the formal system of first-order number theory in the sequent calculus with the $\omega$-rule; and the Lemma states (in terms of deduction trees) that cuts in that system can be eliminated.

So if one takes Gentzen's Lemma to be the focal point of the original paper on consistency, it is [6] rather than [5] that is the real sequel to the original paper. It is there that he formalizes first-order number theory in the sequent calculus; and, although he does not prove cut-elimination for the system obtained by adding the $\omega$-rule, the argument leads directly to that result. Indeed—and this is the secondary point of my paper—one may think of contemporary infinitary proof theory as the product of Gentzen's Lemma (understood as being about deduction trees rather than reduction tree) and his earlier *Hauptsatz* for propositional logic in the framework of the sequent calculus [4], where the former shows the way to extend the latter to infinite conjunctions and disjunctions.

Cut-elimination with the $\omega$-rule was finally explicitly proved by Lorenzen [14] and Schütte [15]. Lorenzen's proof applied to ramified analysis of finite order but does not supply ordinal bounds. Schütte's proof applies to a variant formalization of the sequent calculus and supplies the ordinal bounds. [18] contains a unified treatment of Schüttes' result and his later papers on cut-elimination for ramified analysis [16, 17] with the $\omega$-rule, using a simplified form of the sequent calculus.[3]

Although the notion of a reduction rule played no part in the consistency proof in [5], it should be noted that it retained a conceptual/philosophical role. Gentzen not only wanted a proof of consistency, he wanted a way to understand the truth of a sentence of number theory that is in some sense "finitary" but at the same time supported classical reasoning in number theory. In this respect, the original paper and the 1936 paper go beyond the original Hilbert program of finding finitary consistency proofs for formal systems. Indeed, Gödel's incompleteness theorems would seem to demand such an extension. Consistency of a particular formal system is of less interest when we know that the system, if consistent, is also incomplete. A "finitist" interpretation of classical mathematical propositions that guarantee their consistency transcends any particular formal system. Gentzen's candidate for such an interpretation was this: $A$ is true precisely if we can state a "reduction rule" for $\vdash A$, i.e. a rule for constructing a reduction tree.[4]

---

[3]See [2] for a detailed description of the relation between Gentzen's 1938 consistency proof and Schütte's 1951 result.

[4]The rule is, in itself, just a rule for constructing a certain tree—I call it a "pre-reduction rule" below. It is a reduction rule in virtue of the tree being well-founded. So, to know that $A$ is true would mean, not simply to possess the rule, but also to know that the tree is well-founded. One might have some difficulty in labelling knowledge of this kind as "finitary."

Like Gödel, Gentzen had discovered the double negation interpretation of classical first-order number theory in the corresponding intuitionistic version. If from a "finitist" point of view one were satisfied with the intuitionistic system, i.e. Heyting arithmetic, this result would provide the desired interpretation—and would certainly diminish the significance of a consistency proof for a finitist. But, also like Gödel in [10], Gentzen rejected the intuitionistic conception of logic as presented by Heyting as non-finitist. The difficulties they had with it centred on the intuitionistic meaning of implication. The "circularity" that Gentzen found in Heyting's account of the meaning of $\rightarrow$ in propositions $A \rightarrow B$, where $A$ contains $\rightarrow$ [5, Sect. 11.1], disappears when one adopts the type-theoretic approach of Curry-Howard. But, of course, when $A$ contains $\rightarrow$, its proofs are no longer to be understood as concrete finitary objects on the type-theoretic conception; rather they themselves are already objects of higher type and so it would be a stretch to regard proofs of $A \rightarrow B$, i.e. operations transforming proofs of $A$ into proofs of $B$, as in any sense "finitist."

Gentzen's interpretation of the sentences of arithmetic in terms of reduction rules has a somewhat alien flavour. But it evolved into two different and more homely interpretations of classical reasoning in number theory: the *no-counterexample interpretation* in the hands of Gödel [10][5] and Kreisel [12, 13] and a game-theoretic interpretation by Coquand [3], according to which a reduction rule is a winning strategy in a certain two-person game. These two interpretations are discussed in [19, 20] and I will not discuss them here.

**1. Reduction Rules.** A sequent is of the form $\Gamma \vdash A$, where $\Gamma$ is a set of formulas and $A$ is a formula. Gentzen defines the notion of a reduction rule for sequents of arbitrary formulas. If free variables occur in formulas in the sequent, the reductions of the sequent begin with replacing one of them throughout the sequent by an arbitrarily chosen numeral. If the sequent consists just of sentences and some formula in it contains a closed term $f(t_1, \ldots, t_n)$, then a reduction consists in replacing such a term by the numeral $\bar{k}$ for its value. We can eliminate these reduction steps by considering only sequents of sentences and by identifying the sentence $A(f(t_1, \ldots, t_n))$ with $A(\bar{k})$. Since the reductions of sequents of sentences yield only sequents of sentences, we can also cut down on the number of forms of sentences that must be treated separately by treating $A \wedge B$ and $\forall x A(x)$ as special cases of *conjunctive* sentences $\bigwedge_i A_i$: namely

$$A_0 \wedge A_1 = \bigwedge_{i<2} A_i \qquad \forall x A(x) = \bigwedge_{i<\omega} A(\bar{i}).$$

[5] That Gödel had anticipated the no-counterexample interpretation in these notes was first noticed by C. Parsons and W. Sieg in their introductory note.

Concerning the atomic sentences, Gentzen included only decidable sentences, i.e. built up from $\bar{0}$, the successor function constant, constants for other computable functions and decidable relation constants.[6] We will write

$$\perp$$

to denote any false atomic sentence, such as $1 = 2$ (Gentzen's favourite): they are interchangeable.

A sequent $\Gamma \vdash A$ of sentences is called an *axiom sequent* just in case either $A$ is a true atomic sentence or it is a false atomic sentence and $\Gamma$ contains a false atomic sentence. The rules of inference we will consider are $\bigwedge -R$

$$\frac{\cdots \qquad \Gamma \vdash A_j \qquad \cdots}{\Gamma \vdash \bigwedge_i A_i} \quad \text{(all } j\text{)}$$

and $\bigwedge -L$

$$\frac{\Gamma, \bigwedge_i A_i, A_j \vdash \perp}{\Gamma, \bigwedge_i A_i \vdash \perp}$$

for $\bigwedge$; and $\neg - R$

$$\frac{\Gamma, A \vdash \perp}{\Gamma \vdash \neg A}$$

and $\neg - L$

$$\frac{\Gamma, \neg A \vdash A}{\Gamma, \neg A \vdash \perp}$$

for $\neg$. The explicitly listed composite sentence in the conclusion of an inference is called its *principal sentence*.

**Definition** A *pre-reduction rule* $R$ for a sequent $\Gamma \vdash A$ of sentences effectively determines, for each $n$, the $R$-admissible sequences $\langle \Gamma_0 \vdash A_0, \ldots, \Gamma_n \vdash A_n \rangle$ of sequents of sentences as follows:

- $\langle \Gamma \vdash A \rangle$ is the only $R$-admissible sequence of length 1.

[6]When in [7] he comes to the problem of determining the bound on the provable ordinals, he needs to essentially redo the argument for the case that the atomic formulas also include $t \in V$, where $V$ stands for an indeterminate set of numbers and $t$ a numerical term. But it is obvious how to treat this extension. The atomic sentences must be extended to include the expressions $n \in V$ and the *axiom sets*, defined below, has to be extended to include sequents of the form $\Gamma, n \in V \vdash n \in V$.

- All $R$-admissible sequences of length $n+2$ are one-element extensions of $R$-admissible sequences of length $n+1$. Let $\langle \Gamma \vdash A, \ldots, \Delta \vdash B \rangle$ be $R$-admissible. We specify its $R$-admissible one-element extensions.
  - If $\Delta \vdash B$ is an axiom sequent, then there are no $R$-admissible extensions.
  - If $\Delta$ consists only of true atomic sentences and $B$ is a false atomic sentence, then $\langle \Gamma \vdash A, \ldots, \Delta \vdash B, \Delta \vdash B \rangle$ is its only $R$-admissible one-element extension.
  - Otherwise $R$ determines an inference with conclusion $\Delta \vdash B$ and $\langle \Gamma \vdash A, \ldots, \Delta \vdash B, \Theta \vdash C \rangle$ is $R$-admissible for every premise $\Theta \vdash C$ of that inference.

A pre-reduction rule $R$ for $\Gamma \vdash A$ is a *reduction rule* for $\Gamma \vdash A$ iff every $\omega$-sequence $\langle \Gamma \vdash A, \Delta \vdash B, \Theta \vdash C, \ldots \rangle$ of sequents of sentences contains a finite initial segment that is not $R$-admissible. □

Our definition of a sequent $\Gamma \vdash A$ differs from Gentzen's in that, for him, $\Gamma$ is a *sequence* of sentences rather than a set. But that makes no difference in the definition of a reduction rule. Notice that the inferences specified above have the property that the antecedent of the conclusion is a subset of the antecedent of each premise. Gentzen also allows another form of both $\forall - L$ and $\neg - L$, namely

$$\frac{\Gamma, A_j \vdash \bot}{\Gamma, \bigwedge_i A_i \vdash \bot}$$

and

$$\frac{\Gamma, \vdash A}{\Gamma, \neg A \vdash \bot}$$

But, since adding a sentence to the antecedent of each premise and the conclusion of any inference in our sense is again an inference, it is clear that a reduction rule for $\Gamma \vdash A$ in the wider sense of Gentzen can be transformed into one in our sense.

Let $R$ be a reduction rule for $\Gamma \vdash A$, where $\Gamma \vdash A$ is not an axiom sequent. Then there is a unique inference

$$\frac{\ldots \qquad \Gamma_i \vdash A_i \qquad \ldots}{\Gamma \vdash A}$$

such that the $R$-admissible sequences of length 2 are precisely $\langle \Gamma \vdash A, \Gamma_i \vdash A_i \rangle$ of length 2. The principal sentence of this inference is called the *principal sentence of* $R$ and the sequents $\Gamma_i \vdash A_i$ are called the *reducts* of $\Gamma \vdash A$ determined by $R$.

If $R$ is a pre-reduction rule for $\Gamma \vdash A$, the *reduction tree* $\mathcal{T}_R$ for $\Gamma$ determined by $R$ has as its nodes the $R$-admissible sequences, where, for nodes $\mu$ and $\nu$, $\mu <_{\mathcal{T}_R} \nu$ means that $\mu$ is a proper initial segment of the $R$-admissible sequence $\nu$. The root of the tree is of course the one-element sequence $\langle \Gamma \vdash A \rangle$. The condition that the pre-reduction rule $R$ be a reduction rule is precisely that $\mathcal{T}_R$ be a well-founded tree.

Let $\langle \Gamma \vdash A, \ldots, \Delta \vdash B \rangle$ be an $R$-admissible sequence. $R$ determines a reduction rule $R'$ for $\Delta \vdash B$: the $R'$-admissible sequences are just those $\langle \Delta \vdash B \ldots, \Theta \vdash C \rangle$ such that $\langle \Gamma \vdash A, \ldots, \Delta \vdash B \ldots, \Theta \vdash C \rangle$ is $R$-admissible. We say in this case that $R'$ is a *reduction sub-rule* of $R$. Note that $\mathcal{T}_{R'}$ is isomorphic to the subtree $\mathcal{T}_{R,\mu} = \{\nu | \mu \leq_{\mathcal{T}_R} \nu\}$, where $\mu = \langle \Gamma \vdash A, \ldots, \Delta \vdash B \rangle$. We will sometimes confuse the two.

Given the reduction rule $R$ for $\Gamma \vdash A$ and a set $\Theta$ of sentences, we may obtain a reduction rule $R'$ for $\Gamma \cup \Theta \vdash A$: the $R'$ admissible sequences are the sequences $\langle \Gamma \cup \Theta \vdash A, \ldots, \Delta \cup \Theta \vdash B \rangle$ such that $\langle \Gamma \vdash A, \ldots, \Delta \vdash B \rangle$ is $R$-admissible. $\mathcal{T}_{R'}$ is of course isomorphic to $\mathcal{T}_R$ and we will sometimes confuse them.

Let $R$ be a reduction rule for the sequent $\Gamma \vdash A$ of sentences. To each node $\langle \Gamma \vdash A, \ldots, \Delta \vdash B \rangle$ of $\mathcal{T}_R$ assign the sequent $\Delta \vdash B$. Then the connection between the assignments to successive nodes is given by the rules of inference. Note that when $\bigwedge_i A_i = \forall x A(x)$ the rule for conjunctions is the $\omega$-rule. Since $\mathcal{T}_R$ is well-founded, every path upward through it terminates in an axiom set. Moreover, the root of $\mathcal{T}_R$, the bottom-most node $\langle \Gamma \vdash A \rangle$, is assigned $\Gamma \vdash A$. Thus, from a *classical* point of view, $\mathcal{T}_R$ is a cut-free deduction of $\Gamma \vdash A$ in a sequent calculus formalization of first-order number theory with the $\omega$-rule.

**2. Induction on Trees.** From a constructive standpoint the situation is more complicated. Deductions are top-down, starting with axiom sequents and passing from premises to conclusion and finally to the sequent deduced. In this form, proof by induction on the deduction tree (or on its height) is fully justified. But reduction trees for $\Gamma \vdash A$ are built bottom-up, starting with $\Gamma \vdash A$ and passing up from conclusion to premises and finally to axiom sets. They are to be well-founded, but that does not constructively justify the principle of induction applied to them. That is exactly the problem that Gentzen failed to avoid in the original consistency proof.

We say that a property $P$ defined on the nodes of a connected tree $\mathcal{T}$ is $\mathcal{T}$-*inductive* if $P(\nu)$ for every $\nu$ immediately above $\mu$ implies $P(\mu)$. The *principle of induction* on $\mathcal{T}$ states that every $\mathcal{T}$-inductive property defined on its nodes holds for all of its nodes.

If a tree $\mathcal{T}$ satisfies the principle of induction, then it is well-founded. (For the property $P(\mu)$ of a node $\mu$ of $\mathcal{T}$ that the subtree $\mathcal{T}_\mu = \{\nu \in \mathcal{T} \mid \mu \leq_{\mathcal{T}} \nu\}$ is well-founded is an inductive property.) Classically, we can easily infer from the well-foundedness of a tree $\mathcal{T}$ that it satisfies the principle of induction. If not, choose an inductive property $P$ which is not possessed by every node of $\mathcal{T}$ and, having defined $\mu_0 < \ldots < \mu_n$ where the $\mu_i$ do not have the property $P$, the inductiveness of $P$ implies that there is a $\mu_{n+1} > \mu_n$ which also fails to have $P$. Iterating this construction, we obtain an infinite path of nodes up through the tree that do not have the property $P$. Constructively, however, this argument of course fails: from the fact that $\mu_n$ does not have $P$ it follows from the inductiveness of $P$ that it is not the case that all successor nodes $\mu$ of $\mu_n$ have the property $P$; but that does not imply that there exists such a successor node which does not have $P$.

Call a tree *inductive* if it is in the least class $\mathcal{I}$ of connected trees such that whenever $0 \leq \alpha \leq \omega$ and $\mathcal{T}_i \in \mathcal{I}$ for all $i < \alpha$, then the connected tree with whose immediate subtrees are precisely the $\mathcal{T}_i$ is inductive. Inductive trees obviously satisfy the principle of induction.

One example is the constructive ordinals of the second number class. Let $\mathcal{T}_0$ and $\mathcal{T}_1$ be connected trees. $\mathcal{T}_0 \subseteq \mathcal{T}_1$ means that $\mathcal{T}_0$ is a substructure of $\mathcal{T}_1$, i.e. that each node of $\mathcal{T}_0$ is a node of $\mathcal{T}_1$ and that $\mu <_{\mathcal{T}_0} \nu$ implies $\mu <_{\mathcal{T}_1} \nu$. $\mathcal{T}_0 \prec \mathcal{T}_1$ means that $\mathcal{T}_0 \subseteq (\mathcal{T}_1)_\mu$ for some node $\mu$ in $\mathcal{T}_1$ other than its root. 0 is the one-node tree with node $\emptyset$, $\alpha + 1$ is the connected tree with root $\{\alpha\}$ and whose only immediate subtree is $\alpha$, and when $\alpha_n \prec \alpha_{n+1}$ for each $n$, $\lim_n \alpha_n$ is the connected tree with root $\{\alpha_n \mid n < \omega\}$ and whose immediate subtrees are the distinct $\alpha_n$. $\alpha + \beta$ is defined as usual by induction on $\beta$:

$$\alpha + 0 = \alpha \qquad \alpha + (\beta + 1) = (\alpha + \beta) + 1 \qquad \alpha + \lim_n \beta_n = \lim_n (\alpha + \beta_n).$$

We can always assign ordinal bounds $|\mathcal{T}|$ on the height of an inductive tree $\mathcal{T}$, with $|\mathcal{T}'| < |\mathcal{T}|$ when $\mathcal{T}'$ is a proper subtree of $\mathcal{T}$: let $\mathcal{T}_i$ for $i < \alpha$ be its immediate subtrees and assume $|\mathcal{T}_i|$ is defined for each $i < \alpha$. If $\alpha < \omega$ set $|\mathcal{T}| = |\mathcal{T}_0| + \cdots + |\mathcal{T}_{\alpha-1}| + 1$. Otherwise, $|\mathcal{T}| = \lim_n(|\mathcal{T}_0| + \cdots + |\mathcal{T}_n| + 1)$. Conversely, let $\mathcal{T}$ be any tree and suppose that we have assigned an ordinal $|\mathcal{T}_\mu|$ to each subtree $\mathcal{T}_\mu$ of $\mathcal{T}$ so that $|\mathcal{T}_\mu| < |\mathcal{T}_\nu|$ when $\mu <_\mathcal{T} \nu$. Then $\mathcal{T}$ satisfies the principle of induction, since it can be reduced to the principle of induction on the ordinals.

The immediately relevant example of inductive trees is given by what we are calling the deduction trees. If $\Gamma \vdash A$ is an axiom sequent, then a one-node tree with node $\Gamma \vdash A$ is a deduction tree. If $\Gamma \vdash A$ is the conclusion of an inference and $\mathcal{D}_i$ is a deduction of the $i$th premise, $i < \alpha$, then the tree with root $\{\mathcal{D}_i \mid i < \alpha\}$ and immediate subtrees $\mathcal{D}_i$ $(i < \alpha)$ is a deduction of $\Gamma \vdash A$.[7]

Corresponding to the notion of a reduction rule, we also have the notion of a *deduction rule* for $\Gamma \vdash A$. Such a rule $R$ determines a deduction tree $\mathcal{D}_R$ for $\Gamma \vdash A$ as follows: if $\Gamma \vdash A$ is an axiom sequent then $\mathcal{D}_R$ is the one-node tree with node $\Gamma \vdash A$. Otherwise, $R$ determines an inference with the conclusion $\Gamma \vdash A$ and a deduction rule $R_i$ for each premise $\Gamma_i \vdash A_i$ of the inference. $\mathcal{D}_R$ is a tree with root $\{\mathcal{D}_{R_i} \mid i < \alpha\}$ and immediate subtrees $\mathcal{D}_{R_i}$.

A deduction rule $R$ for $\Gamma \vdash A$ determines a reduction rule $R'$ for $\Gamma \vdash A$ as follows. The construction is by induction on $\mathcal{D}_R$. If $\Gamma \vdash A$ is an axiom set, $\langle \Gamma \vdash A \rangle$ is $R'$-admissible. Otherwise, $R'$ determines an inference with conclusion $\Gamma \vdash A$ and a deduction rule $R'$ for each premise $\Gamma_i \vdash A_i$. The $R'$-admissible sequences

[7]It is not excluded that some premise of an inference is identical with the conclusion. Therefore, we have to distinguish the node of a deduction tree from the sequent attached to it. For simplicity, I have defined the nodes of deduction trees to be in general infinitary objects. The finitist will want to replace these with suitable codes.

are of the form $\langle \Gamma \vdash A, \Gamma_i \vdash A_i, \ldots, \Delta \vdash B \rangle$ such that $\langle \Gamma_i \vdash A_i, \ldots, \Delta \vdash B \rangle$ is $R'_i$-admissible. Notice that the converse construction, of a deduction rule $R$ from a reduction rule $R'$, is not constructively valid, since induction on $\mathcal{T}_{R'}$ is not valid.

While we are on the subject of induction on trees, let me mention a construction to which we will refer below. Suppose $\mathcal{T}_0$ and $\mathcal{T}_1$ are trees. We define $\mathcal{T} = \mathcal{T}_0 \times \mathcal{T}_1$ as follows: its nodes are $\mu = (\mu_0, \mu_1)$, where $\mu_e$ is a node of $\mathcal{T}_e$.

$$\mu <_{\mathcal{T}} \nu \iff [\mu_0 <_{\mathcal{T}_0} \nu_0 \;\&\; \mu_1 \leq_{\mathcal{T}_1} \nu_1] \; or \; [\mu_0 \leq_{\mathcal{T}_0} \nu_0 \;\&\; \mu_1 <_{\mathcal{T}_1} \nu_1.]$$

Observe that, if both $\mathcal{T}_0$ and $\mathcal{T}_1$ satisfy the principle of induction, then so does $\mathcal{T}_0 \times \mathcal{T}_1$.

**3. Gentzen's Lemma.** The only non-trivial step in Gentzen's demonstration that a reduction rule for a sequent can be extracted from a natural deduction of it is the proof of the following, which we formulate in terms of sets rather than sequents:

**Lemma** [9, Sects. 14.44 and 14.6]. If there are reduction rules $R_0$ for $\Gamma, D \vdash C$ and $R_1$ for $\Gamma \vdash D$, then there is a reduction rule for $\Gamma \vdash C$.

The proof of the Lemma is familiar if we think of deductions rather than reductions. It proceeds by induction on the *rank* $|D|$ of $D$, where $|A| = 0$ when $A$ is atomic and

$$|\neg A| = |A| + 1 \qquad\qquad |\bigwedge_i A_i| = \sup_i(|A_i| + 1).$$

The inductive assumption is that the Lemma holds for all $B$ with $|B| < |D|$, and then we want to conclude from this that it holds also for $D$. The proof of this involves an induction within the induction on the rank of $A$; namely an induction on the tree $\mathcal{T}_{R_0}$. In the original paper, Gentzen tried to avoid this induction, and so I will put each case of its application below in square brackets.

**Case 1.** $\Gamma, D \vdash C$ is an axiom sequent. Then $C$ is atomic. If it is true, then $\Gamma \vdash C$ is an axiom sequent. So assume $C$ is false. If $D$ is also a false atomic sentence, then the reduction rule $R'_1$ for $\Gamma$ is obtained replacing $D$ by $C$ in $R_1$.

**Case 2.** $\Gamma, D \vdash C$ is not an axiom sequent.

**Case 2a.** $D$ is not the principal sentence of $R_0$. Then the reducts of $\Gamma, D \vdash C$ determined by $R_0$ have the form $\Gamma_i, D \vdash C_i$ with the proper sub-reduction rule $R_0^i$. [By induction on $\mathcal{T}_{R_0}$, we may assume that there is a reduction rule $R^i$ for $\Gamma_i, \Gamma \vdash C_i$. The $R$-reduction tree for $\Gamma \vdash C$ has as its immediate sub-trees the $\mathcal{T}_{R^i}$.]

**Case 2b.** $D$ is the principal sentence of $R_0$. Thus, $C = \bot$.

**Case 2bi.** $D = \neg E$. Then the unique reduct is $\Gamma, D \vdash E$. [By the induction hypothesis, we may assume that there is a reduction rule $R'$ for $\Gamma \vdash E$.] $R_1$ reduces $\Gamma \vdash D$ to $\Gamma, E \vdash C$ with immediate sub-rule $R'_1$. $|E| < |D|$ and so by induction on $|D|$ applied to $R'$ and $R'_1$, there is a reduction rule $R$ for $\Gamma \vdash C$.

**Case 2bii.** $D = \bigwedge_i D_i$ and $R_0$ reduces $\Gamma, D \vdash C$ to $\Gamma, D, D_j \vdash C$ with corresponding sub-reduction rule $R'_0$. [By induction on $\mathcal{T}_{R_0}$, we may assume that there is a reduction rule $R'$ for $\Gamma, D_j \vdash C$.] The reducts of $\Gamma \vdash D$ determined by $R_1$ are the $\Gamma \vdash D_i$ for each $i$, with sub-reduction rules $R_1^i$. Since $|D_j| < |D|$, the induction hypothesis on rank applied to $R'$ and $R_1^j$ yields $R$. □

As we indicated, the square-bracketed parts of the argument, explicitly invoking induction on $\mathcal{T}_{R_0}$, do not appear in Gentzen's original paper. His argument rather is as follows: with the rule $R_1$ for $\Gamma \vdash D$ fixed, we reduce the problem of finding a reduction rule for $\Gamma \vdash C$ to that of finding one for a reduct $\Gamma' \vdash C'$ of $\Gamma, D \vdash C$ as determined by $R_0$. If $\Gamma', D \vdash C'$ is not an axiom sequent, then we reduce this problem of finding a reduction rule for $\Gamma'' \vdash C''$, where $\Gamma'', D \vdash C''$ is a reduct of $\Gamma', D \vdash C'$ as determined by $R_0$—and so on:

> Continuing in this way, we must *reach the end in finitely many steps*, i.e. the completion of the proof. [8, Sect. 14.63]

This may sound convincing as a constructive argument until we ask: how many steps? Of course there is no answer to this because it depends upon which path we take. In particular, if we reach stage $n$ and $C^{(n)}$ in $\Gamma^{(n)}, D \vdash C^{(n)}$ is of the form $\bigwedge_i C_i$, then the $n+1$st stage $\Gamma^{(n)}, D \vdash C_j$ depends on the "free choice" of $j$. In the November 4, 1935 letter to Bernays, Gentzen seems to have been arguing that, because the choice of $j$ is *free*, we are really thinking about a generic path $\langle \Gamma, D \vdash C, \Gamma', D \vdash C', \ldots \rangle$ through $\mathcal{T}_{R_0}$ which therefore presumably has a generic finite length $x$. $\Gamma^{(x)} \vdash C^{(x)}$ has a reduction rule and so, working backward, has $\Gamma \vdash C$. But as we noted in the introductory remarks, Gentzen soon gave up on this argument.

**4. The Bar Theorem.** Gentzen's reference to "free choices" seems to point to Brouwer's function theory; but Gentzen seems to have ignored the one feature of Brouwer's theory that would ground his argument: the Bar Theorem. Whether or not he explicitly rejected it, he certainly did not employ it in his argument. The setting for the Bar Theorem is the notion of a *spread law*.

**Definition** A *spread law* $S$ effectively determines, for each $n$, the $S$-admissible sequences $\langle a_0, \ldots, a_n \rangle$ of elements of a decidable set $M$ as follows:

- $S$ determines for which $a \in M$ the one-element sequence $\langle a \rangle$ is $S$-admissible.
- All $S$-admissible sequences of length $n+2$ are one-element extensions of $S$-admissible sequences of length $n+1$. Given the $S$-admissible sequence $\langle a_0, \ldots, a_n \rangle$, $S$ determines for which $a \in M$ $\langle a_0, \ldots, a_n, a \rangle$ is $S$-admissible.

Moreover, it is required that every $S$-admissible sequence have a proper extension. We will consider only *connected* spread laws $S$, i.e. such that there is exactly one one-element $S$-admissible sequence $\langle a_0 \rangle$, called its *root*. □

If $S$ satisfies all the conditions of being a spread law except the condition that every $S$-admissible sequence have an $S$-admissible extension, we can turn it into a spread law $S^{\#}$ by the condition that every $S$-admissible sequence is $S^{\#}$-admissible and, if $\langle a, \dots, b\rangle$ is either maximal $S$-admissible or contains #, then $\langle a, \dots, b, \#\rangle$ is $S^{\#}$-admissible.

Let $S$ be a spread law. An $S$-*sequence* is an $\omega$-sequence $\langle a, b, \dots\rangle$ such that each finite initial segment is $S$-admissible. $[S]$ denotes the set of all $S$-sequences. A *bar* on $S$ is a set $B$ of $S$-admissible sequences such each $\theta \in [S]$ has an initial segment in $B$. When $B$ is decidable, we can assume that the initial segment is unique. $\mathcal{T}_S$ is the tree of $S$-admissible sequences.

**Bar Theorem.** If

(i) $B$ is a decidable bar on the connected spread $S$ with root $\langle a_0\rangle$,
(ii) Every element of $B$ has the property $P$,
(iii) $P$ is inductive on $\mathcal{T}_S$,

then $P(\langle a_0\rangle)$. □

We can apply the Bar Theorem to validate the induction on $\mathcal{T}_{R_0}$ in the proof of Gentzen's Lemma. The spread law $S$ in this case is $R_0^{\#}$. So we have a connected spread with root $\Gamma, D \vdash C$. The assertion that $R_0$ is a reduction rule for $\Gamma, D \vdash C$ and in particular is well-founded implies that the set $[R_0]$ of maximal $R_0$-admissible sequences is a decidable bar on $S$. (Here we are using the fact that a reduction rule is to be effective.) Let $P$ be the property of $S$-admissible sequences $\sigma$ that, if $\sigma$ is the $R_0$-admissible sequence $\langle \Gamma, D \vdash C, \dots, \Gamma', D \vdash C'\rangle$, then there is a reduction rule for $\Gamma', \Gamma \vdash C'$. Every element $\langle \Gamma \vdash C, \dots, \Gamma' \vdash C'\rangle$ of the bar $[R_0]$ has the property $P$, since $\Gamma' D \vdash C'$ is an axiom sequent. (If $C'$ is a true atomic sentence or $\Gamma'$ contains a false atomic sentence other than $D$, $\Gamma', \Gamma \vdash C'$ is an axiom set. If $C'$ and $D$ are both false atomic, a reduction rule for $\Gamma', \Gamma \vdash C'$ is easily obtained from the reduction rule $R_1$ for $\Gamma \vdash D$.) Gentzen proved that $P$ is inductive on $\mathcal{T}_S$. So by the Bar Theorem, $\langle \Gamma, D \vdash C\rangle$ has the property $P$, i.e. $\Gamma \vdash C$ has a reduction rule.

But a cynic might wonder at Brouwer's magic: simply by calling $[R_0^{\#}]$ a "spread" he could conclude from the fact that $\mathcal{T}_{R_0}$ is well-founded that it satisfies the induction principle. But let's look at Brouwer's argument for the Bar Theorem. It begins with the doctrine that a proof of such an implication consists in a method of transforming a proof of the antecedent [conditions (i)–(iii)] into a proof of the conclusion. (This simply reflects the intuitionistic meaning of implication.) Now consider the condition $(i)$. We say that $B$ *bars* the $S$-admissible sequence $\sigma$ if each $\theta \in [S]$ of which $\sigma$ is an initial segment has an initial segment in $B$. So condition $(i)$ states that $B$ bars $\langle a_0\rangle$. Brouwer argues—and this is the crucial step—that ultimately the only way to prove that $B$ bars $\langle a_0\rangle$ is by using the axioms

$$B \text{ bars } \sigma$$

for each $\sigma \in B$ and the inferences:

$$\frac{\cdots \qquad B \text{ bars } \sigma_i \qquad \cdots}{B \text{ bars } \sigma} \quad (\text{all } i)$$

where the $\sigma_i$ are all the $S$-admissible one-element extensions of $\sigma$. Thus, to have a proof that $B$ is a bar on $S$ is to have an (in general infinitary) deduction tree $\mathcal{D}$ using just these axioms and inferences. So then, using the conditions (ii) and (iii) of the Bar Theorem, the proof of its conclusion is obtained by replacing the property "$B$ bars $x$" by the property $P(x)$ in the deduction. This is proved by induction on $\mathcal{D}$. This is permissible because deductions, unlike reduction trees, satisfy the principle of induction.

Of course, in our case we concluded (i), i.e. that $[R_0]$ is a bar on $S = R_0^\#$, not by such a deduction of the statement that $[R_0]$ bars $\langle \Gamma, D \vdash C \rangle$, but from the fact that $\mathcal{T}_{R_0}$ is well-founded. But, whether or not one finds the crucial step in Brouwer's argument for the Bar Theorem convincing in general, *its application in this instance requires that the proof of well-foundedness of* $\mathcal{T}_{R_0}$ *ultimately be a deduction* $\mathcal{D}$ *in the above sense that* $[R_0]$ *is a bar on* $R_0^\#$. This is what is presupposed by an application of the Bar Theorem to the proof of Gentzen's Lemma.

But now let $\mathcal{D}'$ be obtained by replacing

$$[R_0] \text{ bars } \langle \Gamma, D \vdash C, \ldots, \Gamma', D \vdash C' \rangle$$

throughout $\mathcal{D}$ by

$$\Gamma', D \vdash C'$$

Then $\mathcal{D}'$ is a deduction tree for $\Gamma, D \vdash C$. Indeed, it is just the $R_0$-reduction tree $\mathcal{T}_{R_0}$ read top-down. So *the application of the Bar Theorem already presupposes a deduction tree for* $\Gamma, D \vdash C$. Thus:
*It was not the Bar Theorem that Gentzen needed; it was the switch from the basic notion of a reduction tree to that of a deduction tree. Reading the above proof of the Lemma with* $R_0$ *and* $R_1$ *understood as rules for deduction rather than of reduction, the square bracketed inductions are valid.*

**5. Cut-elimination.** Staying within the domain of deduction trees then, rather than reduction trees, Gentzen's Lemma is constructively valid. If we add to the rules of inference given in § 1 the *cut rule*

$$\frac{\Delta, D \vdash C \qquad \Delta \vdash D}{\Delta \vdash C}$$

then we have one formalization of the rules of inference for first-order number theory with the $\omega$-rule. $D$ is called the *cut-formula* of this cut. Call deductions in this system *deduction trees with cuts*. By Gentzen's Lemma, every deduction tree with cuts can be reduced to one without cuts. Simply iterate the operation of eliminating the top-most cuts.

In terms of ordinal bounds on the height of the trees, this is not the most efficient way to eliminate cuts. The more efficient method is essentially the transfinite version of Gentzen's *Hauptsatz* for first-order logic in the sequent calculus. The *cut-degree* of a deduction is the least ordinal greater than the rank $|D|$ of $D$ for all cut-formulas $D$ in it [where $|A| = 0$ for atomic $A$, $|\neg A| = |A|+1$, and $|\bigwedge_i A_i| = lub_i(|A_i|+1)$]. If $\alpha$ is the height of the given deduction of $\Gamma \vdash A$ and its cut-degree is some $m < \omega$, the bound we get on the height of the cut-free deduction is $2^\alpha_m$, where $2^\alpha_0 = \alpha$ and $2^\alpha_{n+1} = 2^{2^\alpha_n}$. The efficient proof proceeds by eliminating cuts of maximum rank $m+1$, replacing them with cuts of maximum rank $m$ at the cost of increasing the height of the deduction from $\alpha$ to $2^\alpha$. (If the cut-degree is $\omega$, then the bound is $\epsilon_\alpha$.) The bound we get from Gentzen's Lemma is higher.

**6. The Sequent Calculus and the Set Calculus.** If we were to admit an arbitrary sentence $B$ in the rule of $\bigwedge -L$

$$\frac{\Gamma, \bigwedge_i A_i, A_j \vdash B}{\Gamma, \bigwedge_i A_i \vdash B} \quad \text{(some } j\text{)}$$

instead of restricting it to $B = \bot$, the rules of inference would be the natural cut-free rules for first-order number theory with the $\omega$-rule with the logical constants $\forall, \wedge, \neg$. This would of course imply a greater freedom for a reduction rule $R$ for $\Gamma \vdash A$. Namely the reducts of $\Delta \vdash B$ could be of the form $\Delta' \vdash B$ even when $B$ is composite, and not just of the form $\Delta \vdash B'$. So, faced with an admissible $\langle \Gamma \vdash A, \ldots, \Delta \vdash B\rangle$, where $B$ is composite, $R$ could choose the one-element extension $\langle \Gamma \vdash A, \ldots, \Delta \vdash B, \Delta' \vdash B\rangle$ rather than being restricted to one of the form $\langle \Gamma \vdash A, \ldots, \Delta \vdash B, \Delta \vdash B'\rangle$. By induction on the sentence $B$, it is easy to see that for all $B$, if there is a reduction rule for $\Gamma, \bigwedge_i A_i, A_j \vdash B$, then there is one for $\Gamma, \bigwedge_i, A_i \vdash B$. So the general case of the inference is derivable from the special case.

But why did Gentzen restrict reductions to the case $B = \bot$? The answer is that it leads to a simpler proof of his main lemma—or more accurately, in view of the gap in his argument, it seems fair to answer rather than it made it easier for him to convince himself that there was no gap. Let $D = \bigwedge_i D_i$, $R_0$ be a reduction rule for $\Gamma, D \vdash C$ and $R_1$ a reduction rule for $\Delta \vdash D$ and suppose that $R_0$ chooses the reduct $\Gamma, D, D_j \vdash C$. By induction (which we have now justified by replacing the reduction trees by the corresponding deduction trees) we obtain a reduction rule for $\Gamma, D_j, \Delta \vdash C$. Now, on Gentzen's more restricted notion of reduction, the only possible reductions of $\Delta \vdash D$ are all the sequents $\Delta \vdash D_i$, including the case of $\Delta \vdash D_j$. So now the cut is reduced to the simpler cut-formula $D_j$. But with the more liberal notion of a reduction, the reduct of $\Delta \vdash D$ that $R_1$ chooses might be of the form $\Delta' \vdash D$. Clearly, in this case, the proof that cuts can be eliminated involves more symmetry between $R_0$, the deduction of $\Gamma, D \vdash C$ and $R_1$, the deduction of $\Delta \vdash D$. In fact it requires induction, not on $\mathcal{T}_{R_0}$, but on $\mathcal{T}_{R_0} \times \mathcal{T}_{R_1}$. Gentzen's attempt to avoid/disguise the induction would certainly have been even less plausible in this case.

The symmetry that is revealed by taking the more general form of $\bigwedge$-elimination is made even more evident with two changes in the formalization. One is a change in the logical constants, replacing negation $\neg$ by disjunction $\vee$ and existential quantifier $\exists$, except we will admit negation of atomic formulas. Classically at least, if $A$ is atomic, the choice between $A$ and $\neg A$ as more basic is arbitrary. We will refer to them both as *prime sentences*. Similarly, there is no ground for treating $\neg \bigwedge_i A_i$ as more logically complex than $\bigwedge_i A_i$: for it is expressed by $\bigvee_i \neg A_i$, where

$$A_0 \vee A_1 := \bigvee_{i<2} A_i \qquad \exists x A(x) := \bigvee_{i<\omega} A(\bar{i}).$$

So henceforth, sentences will be built up from prime sentences by means of $\bigvee$ and $\bigwedge$. The *complement* $\bar{A}$ of a sentence $A$ is defined by

$$\bar{A} := \neg A \qquad \overline{\neg A} := A$$

if $A$ is atomic and

$$\overline{\bigvee_i A_i} := \bigwedge_i \overline{A_i} \qquad \overline{\bigwedge_i A_i} := \bigvee_i \overline{A_i}$$

So negation is no longer a logical constant: it is only used to arbitrarily mark one of two complementary prime sentences.

The other change is this: there is no reason in the classical sequence calculus to restrict sequents to one succedent; but moreover, there is no reason to retain sequents at all. The sequent $A_1, \ldots, A_m \vdash B_1, \ldots, B_n$ has the same classical meaning as $\vdash \overline{A_1}, \ldots, \overline{A_m}, B_1, \ldots, B_n$, and so we may as well just take as the units of deduction the corresponding sets $\{\overline{A_1}, \ldots, \overline{A_m}, B_1, \ldots, B_n\}$ the set $\{\overline{A_1}, \ldots, \overline{A_m}, B_1, \ldots, B_n\}$ understood as expressing the disjunction of its elements. In place of axiom sequents, we now have *axiom sets*, i.e. sets of sentences containing a true prime sentence.

The rules of inference now take the simple form:

$$\frac{\Gamma_0, \bigvee_i A_i, A_j}{\Gamma, \bigvee_i A_i} \quad \text{(some } j\text{)}$$

and

$$\frac{\cdots \qquad \Gamma_j, \bigwedge_i A_i, A_j \qquad \cdots}{\Gamma, \bigwedge_i A_i} \quad \text{(all } j\text{)}$$

and the cut-rule takes the form

$$\frac{\Gamma, A \qquad \Gamma, \bar{A}}{\Gamma}$$

where $A$ is called the *cut-formula*. Gentzen's Lemma now takes the form that, given cut-free deduction $\mathcal{D}_0$ of $\Gamma, A$ and $\mathcal{D}_1$ of $\Gamma, \bar{A}$, there is a deduction of $\Gamma$ involving only cuts with cut-formulas of rank $< |A|$. The argument is again essentially Gentzen's, except that symmetry demands again that the proof must be by induction on $\mathcal{D}_0 \times \mathcal{D}_1$ rather than on just one of the trees $\mathcal{D}_e$. Indeed, the proof is just an extension of Gentzen's proof of his *Hauptsatz* for propositional logic in the framework of the sequent calculus to the case of infinite disjunctions and conjunctions.

**Acknowledgements** My understanding of the philosophical background of Gentzen's work on consistency was enhanced by reading the unpublished manuscript "On the Intuitionistic Background of Gentzen's 1936 Consistency Proof and Its Philosophical Aspects" by Yuta Takahashi.

## References

1. P. Bernays, On the original Gentzen consistency proof for number theory, *Intuitionism and Proof Theory*, ed. by A. Kino, J. Myhill, R. Vesley (North-Holland, Amsterdam, 1970), pp. 409–418
2. W. Buchholz, Explaining Gentzen's consistency proof within infinitary proof theory, in *Computational Logic and Proof Theory. Fifth Kurt Gödel Colloquium, KGC '97*, ed. by G. Gottlob, A. Leitsch, D. Mundici. Lecture Notes in Computer Science, vol. 1298 Springer, Berlin (1997)
3. T. Coquand, A semantics of evidence for classical arithmetic. J. Symb. Log. **60**, 325–337 (1995)
4. G. Gentzen, Untersuchungen über das logische Schliessen I, II. Math. Z. **39**, 176–210, 405–431 (1935). English translation in [8, pp. 68–131]
5. G. Gentzen, Die Widerspruchfreiheit der reinen Zahlentheorie. Math. Ann. **112**, 493–565 (1936) English translation in [8, pp. 132–213]
6. G. Gentzen, Nue Fassung des Widerspruchsfreiheitsbeweises für die reine Zahlentheorie, in *Forschungen zur Logik und der Grundlagen der exakten Wissenschaften, New Series, No. 4* (1938), pp. 5–18. English translation in [8, pp. 252–286]
7. G. Gentzen, Beweisbarkeit und Unbeweisbarkeit von Anfangsfällen der transfiniten Induktion in der reinen Zahlentheorie. Math. Ann. **119**, 140–161 (1943). English translation in [8, pp. 287–308].
8. G. Gentzen, *The Collected Papers of Gerhard Gentzen* (North-Holland, Amsterdam, 1969). Translated and edited by M. E. Szabo
9. G. Gentzen, Der erste Widerspruchsfreiheitsbeweis für die klassische Zahlentheorie. Archiv für Mathematische Logik und Grundlagenforschung **16**, 97–118 (1974). Edited with a brief introduction by P. Bernays
10. K. Gödel, Lecture at Zilsel's, in *Collected Works*, vol. III (Oxford University Press, Oxford, 1938), pp. 87–113.
11. K. Gödel, *Collected Works*, vol. IV (Oxford University Press, Oxford, 2003)
12. G. Kreisel, On the interpretation of non-finitist proofs—part i. J. Symb. Log. **16**, 241–267 (1951)
13. G. Kreisel, On the interpretation of non-finitist proofs—part ii. interpretations of number theory, applications. J. Symb. Log. **17**, 43–58 (1952)
14. P. Lorenzen, Algebraische und logistische Untersuchungen über freie Verbände. J. Symb. Log. **16**, 81–106 (1951)

15. K. Schütte, Beweisetheoretische Erfassung der unendlichen Induktion in der Zahlentheorie. Math. Ann. **122**, 369–389 (1951)
16. K. Schütte, Beweistheoretische Untersuchung der verzweigten Analysis. Math. Ann. **124**, 123–147 (1952)
17. K. Schütte, Eine Grenze für die Beweisbarkeit der transfinitien Induction in der verzweigten Typenlogik. Archiv für mathematische Logik und Grundlagenforschung **7**, 45–60 (1964)
18. W. Tait, Normal derivability in classical logic, in *The Syntax and Semantics of Infinitary Languages*, ed. by J. Barwise (Springer, Berlin, 1968), pp. 204–236
19. W. Tait, Gödel's unpublished papers on foundations of mathematics. Philos. Math. **9**, 87–126 (2001). Reprinted in [21], pp. 276–313
20. W. Tait, Gödel's reformulation of Gentzen's first consistency proof for arithmetic. Bull. Symb. Log. **11**(2), 225–238 (2005)
21. W. Tait, *The Provenance of Pure Reason: Essays in the Philosophy of Mathematics and Its History* (Oxford University Press, Oxford, 2005)

# Goodstein's Theorem Revisited

**Michael Rathjen**

**Abstract** Prompted by Gentzen's 1936 consistency proof, Goodstein found a close fit between descending sequences of ordinals $< \varepsilon_0$ and sequences of integers, now known as Goodstein sequences. This chapter revisits Goodstein's 1944 paper. In light of new historical details found in a correspondence between Bernays and Goodstein, we address the question of how close Goodstein came to proving an independence result for **PA**. We also present an elementary proof of the fact that already the termination of all special Goodstein sequences, i.e. those induced by the shift function, is not provable in **PA**. This was first proved by Kirby and Paris in 1982, using techniques from the model theory of arithmetic. The proof presented here arguably only uses tools that would have been available in the 1940s or 1950s. Thus we ponder the question whether striking independence results could have been proved much earlier? In the same vein we also wonder whether the search for strictly mathematical examples of an incompleteness in **PA** really attained its "holy grail" status before the late 1970s. Almost no direct moral is ever given; rather, the paper strives to lay out evidence for the reader to consider and have the reader form their own conclusions. However, in relation to independence results, we think that both Gentzen and Goodstein are deserving of more credit.

## 1 History

This paper grew out of a Goodstein lecture that I gave at the Logic Colloquium 2012 in Manchester. The lecture touched on many of Goodstein's papers, though, this chapter will just be concerned with his best known result [7] from 1944. Whilst reading [7], I formed the overwhelming impression that Goodstein came very close to proving an independence result. Recent archival searches by Jan von Plato have brought to light a remarkable correspondence between Goodstein and Bernays. These letters confirm that this impression was not unfounded. Goodstein's paper [7] originally bore the title "A note on Gentzen's theorem", thereby referring to Gentzen's 1936 paper [5] which proved the consistency of first order number

M. Rathjen (✉)
School of Mathematics, University of Leeds, Leeds LS2 JT, England
e-mail: M.Rathjen@leeds.ac.uk



R. Kahle, M. Rathjen (eds.), *Gentzen's Centenary*,
DOI 10.1007/978-3-319-10103-3_9

theory[1] by transfinite induction up to the ordinal $\varepsilon_0$. He sent it to Church in 1942 for publication in the JSL. Church sent the paper to Bernays for refereeing. Bernays then contacted Goodstein directly and included a long list of remarks and suggestions in his letter [2] dated 1 September 1942. As a result of these comments, Goodstein altered his paper considerably and also changed the title to "On the restricted ordinal theorem". By the latter he meant the proposition that every strictly descending sequence of ordinals below $\varepsilon_0$ is necessarily finite. As Gentzen [5] showed, this implies the consistency of first order number theory. Crucially in his paper Goodstein proved that this statement is equivalent to a statement $P$ about integers, now known as the termination of Goodstein sequences. From Bernays' letter it is clear that the original version of Goodstein's paper contained a claim about the unprovability of $P$ in number theory. Bernays in his letter correctly pointed out that $P$, being of $\Pi^1_1$ form, is not a statement that can be formalized in Gentzen's system of first order number theory as it talks about all descending sequences.

> The system $\underline{A}$ cannot be exactly the system denoted by Gentzen as "reine Zahlentheorie", since this one contains no function variables and so your theorem $P$ is not expressible in it. However the Gentzen proof surely can be extended to the case that free function variables are added to the considered formal system. [2]

Unfortunately Goodstein then removed the entire passage about the unprovability of $P$. He could have followed Bernays' suggestion or he could have found an independence result for **PA** proper by scrutinizing Gentzen's proof which only utilizes the termination of primitive recursive sequences of ordinals.[2] The latter principle is expressible in the language of **PA** and (in light of Goodstein's own work) can be shown to be equivalent to the termination of primitive recursive Goodstein sequences (see Theorem 2.8).

Barwise [1] in the Handbook of Mathematical Logic added an editor's note to the famous paper by Paris and Harrington [11]:

> Since 1931, the year Gödel's Incompleteness Theorems were published, mathematicians have been looking for a strictly mathematical example of an incompleteness in first-order Peano arithmetic, one which is mathematically simple and interesting and does not require the numerical coding of notions from logic. The first such examples were found early in 1977, when this Handbook was almost finished.

Barwise describes the problem of finding a natural mathematical incompleteness in Peano arithmetic almost as a "holy grail problem" of mathematical logic. As Goodstein almost found a mathematical example in the 1940s one wonders whether this problem was perceived as so important back then.[3] In his paper Goodstein identifies as his main objective to determine which initial cases of Gentzen's

[1]First order number theory or *reine Zahlentheorie* as it was called by Gentzen is essentially the same system as what is nowadays called Peano arithmetic, **PA**.

[2]This statement may be a bit too strong since it assumes that Goodstein had penetrated the details of Gentzen's rather difficult paper [5].

[3]In view of the impact Hilbert's problem list had on mathematics and of how Hilbert's work and ideas furnished the young Gödel with problem to solve, one might guess that if one of the

restricted ordinal theorem can be proven by finitist means. Here initial cases refer to the ordinals (in Goodstein's notation) $\vartheta_n$ ($\vartheta_0 = \omega$, $\vartheta_{k+1} = \omega^{\vartheta_k}$) and the pertaining assertions $P(n)$ that all descending sequences below $\vartheta_n$ are necessarily finite. Interestingly enough, Goodstein claimed that *"$P(n)$ is capable of a finite constructive proof for any assigned n"* [7, p. 39]. Bernays referred to this claim in his letter from 29 September 1943 [3] when he wrote *"I think, that the methodological difficulties appearing already in the case of $\omega^{\omega^{\omega}}$, ..., will induce you to speak in a more reserved form about it."* Bernays' criticism was very justified, indeed, but this time Goodstein did not heed his advice.

In the next section we give an account of Goodman's theorem and illuminate its origins in Gentzen's work. We also give two independence results from **PA** that by and large can be credited to Gentzen (Theorem 2.8) and Goodstein (Theorem 2.9), respectively. We leave it to the reader to assay whether they meet Barwise's criteria of being mathematically simple and interesting and not requiring the numerical coding of notions from logic.

In the last section we give an elementary proof of Kirby's and Paris' 1982 result [9] that the termination of special Goodstein sequences induced by the shift function is not provable in **PA**. Yet another proof was presented by Cichon [4] in 1983. Our main technique consists in making descending sequences of ordinals $< \varepsilon_0$ slow. We like to think that this elementary proof could have been found in the 1940s or at least 1950s. Such a thought could be considered to be unfair to the logicians who, after a lot of hard technical work, established this independence result. This is not our intention and we like to stress that their techniques were certainly not available before the 1970s. On the other hand, we definitely think that Goodstein and Gentzen deserve at least some credit for "their" independence results. Another question that seems to be relevant in this context is the following: could it be that the problem of finding statements independent from **PA** did not occupy centre stage in mathematical logic before the 1970s, thereby accounting for their late arrival?[4]

## 2 Cantor Normal Forms

Let $\varepsilon_0$ be the least ordinal $\beta$ such that $\omega^\beta = \beta$. Every ordinal $0 < \alpha < \varepsilon_0$ can be written in a unique way as

$$\alpha = \omega^{\alpha_1} \cdot k_1 + \cdots + \omega^{\alpha_n} \cdot k_n \tag{1}$$

luminaries of mathematical logic had declared the importance of this problem in the 1940s, the young ones would have leapt at this chance and followed Gentzen's and Goodstein's lead.

[4]For what it's worth, here is some anecdotal evidence. Around 1979, Diana Schmidt proved that Kruskal's theorem elementarily implies that the ordinal representation system for $\Gamma_0$ is well-founded [12]. She even wrote (p. 61) that she didn't know of any applications of her result to proof theory. This is quite surprising since in conjunction with proof-theoretic work of Feferman and Schütte from the 1960s it immediately implies the nowadays celebrated result that Kruskal's theorem is unprovable in predicative mathematics.

where $\alpha > \alpha_1 > \cdots > \alpha_n$ and $0 < k_1, \ldots, k_n < \omega$. This we call the **Cantor normal form** of $\alpha$. By writing $\alpha =_{NF} \omega^{\alpha_1} \cdot k_1 + \cdots + \omega^{\alpha_n} \cdot k_n$ we shall convey that (1) obtains.

The ordinals $\alpha_i$ with $\alpha_i \neq 0$ can also be written in Cantor normal form with yet smaller exponents. As this process terminates after finitely many steps every ordinal $< \varepsilon_0$ can be represented in a unique way as a term over the alphabet $\omega, +, \cdot, 0, 1, 2, 3, \ldots$ which we call its **complete Cantor normal form**.

In what follows we identify ordinals $< \varepsilon_0$ with their representation in complete Cantor normal form. Henceforth, unless indicated otherwise, ordinals are assumed to be smaller than $\varepsilon_0$ and will be denoted by lowercase Greek letters. By $|\alpha|$ we denote the length of $\alpha$ in complete Cantor normal form (viewed as a string of symbols). More precisely, if $\alpha =_{NF} \omega^{\alpha_1} \cdot k_1 + \cdots + \omega^{\alpha_n} \cdot k_n$ we define

$$|\alpha| = \max\{|\alpha_1|, \ldots, |\alpha_n|, k_1, \ldots, k_n\} + 1.$$

By $\mathrm{C}(\alpha)$ we denote the highest integer coefficient that appears in $\alpha$, i.e. inductively this can be defined by letting $\mathrm{C}(0) = 0$ and

$$\mathrm{C}(\alpha) = \max\{\mathrm{C}(\alpha_1), \ldots, \mathrm{C}(\alpha_n), k_1, \ldots, k_n\}$$

where $\alpha =_{NF} \omega^{\alpha_1} \cdot k_1 + \cdots + \omega^{\alpha_n} \cdot k_n$.

There is a similar Cantor normal form for positive integers $m$ to any base $b$ with $b \geq 2$, namely we can express $m$ uniquely in the form

$$m = b^{n_1} \cdot k_1 + \cdots + b^{n_r} \cdot k_r \tag{2}$$

where $m > n_1 > \cdots > n_r \geq 0$ and $0 < k_1, \ldots, k_r < b$. As each $n_i > 0$ is itself of this form we can repeat this procedure, arriving at what is called the **complete** $b$**-representation** of $m$. In this way we get a unique representation of $m$ over the alphabet $0, 1, \ldots, b, +, \cdot$.

For example $7\,625\,597\,485\,157 = 3^{27} \cdot 1 + 3^4 \cdot 2 + 3^1 \cdot 2 + 3^0 \cdot 2 = 3^{3^3} + 3^{3+1} \cdot 2 + 3^1 \cdot 2 + 2$.

**Definition 2.1** Goodstein [7] defined operations mediating between ordinals $< \varepsilon_0$ and natural numbers.

For naturals $m > 0$ and $c \geq b \geq 2$ let $\mathrm{S}^b_c(m)$ be the integer resulting from $m$ by replacing the base $b$ in the complete $b$-representation of $m$ everywhere by $c$. For example $\mathrm{S}^3_4(34) = 265$, since $34 = 3^3 + 3 \cdot 2 + 1$ and $4^4 + 4 \cdot 2 + 1 = 265$.

For any ordinal $\alpha$ and natural $b \geq 2$ with $b > \mathrm{C}(\alpha)$ let $\hat{\mathrm{T}}^\omega_b(\alpha)$ be the integer resulting from $\alpha$ by replacing $\omega$ in the complete Cantor normal form of $\alpha$ everywhere by $b$. For example

$$\hat{\mathrm{T}}^\omega_3(\omega^{\omega+1} + \omega^2 \cdot 2 + \omega \cdot 2 + 1) = 3^{3+1} + 3^2 \cdot 2 + 3 \cdot 2 + 1 = 106.$$

Conversely, for naturals $m \geq 1$ and $b \geq 2$ let $\mathrm{T}^b_\omega(m)$ be the ordinal obtained from the complete $b$-representation of $m$ by replacing the base $b$ everywhere with $\omega$. Thus $\mathrm{T}^3_\omega(106) = \omega^{\omega+1} + \omega^2 \cdot 2 + \omega \cdot 2 + 1$ and

$$\mathrm{T}^3_\omega(34) = \mathrm{T}^4_\omega(\mathrm{S}^3_4(34)) = \mathrm{T}^4_\omega(265) = \omega^\omega + \omega \cdot 2 + 1.$$

We also set $\mathrm{T}^b_\omega(0) = 0$ and $\hat{\mathrm{T}}^\omega_b(0) := 0$.

Goodstein's main insight was that given two ordinals $\alpha, \beta < \varepsilon_0$ one could replace the base $\omega$ in their complete Cantor normal forms by a sufficiently large number $b$ and the resulting natural numbers $\hat{\mathrm{T}}^\omega_b(\alpha)$ and $\hat{\mathrm{T}}^\omega_b(\beta)$ would stand in the same ordering as $\alpha$ and $\beta$. This is simply a consequence of the fact that the criteria for comparing ordinals in Cantor normal form are the same as for natural numbers in base $b$-representation, as spelled out in the next lemma.

**Lemma 2.2**

*(i) Let $\alpha =_{NF} \omega^{\alpha_1} \cdot k_1 + \cdots + \omega^{\alpha_r} \cdot k_r$ and $\beta =_{NF} \omega^{\beta_1} \cdot l_1 + \cdots + \omega^{\beta_s} \cdot k_s$. Then $\alpha < \beta$ if and only if either of the following obtains:*

1. *There exists $0 < j \leq \min(r, s)$ such that $\alpha_i = \beta_i$ and $k_i = l_i$ for $i = 1, \ldots, j-1$ and $\alpha_j < \beta_j$, or $\alpha_j = \beta_j$ and $k_j < l_j$.*
2. *$r < s$ and $\alpha_i = \beta_i$ and $k_i = l_i$ hold for all $1 \leq i \leq r$.*

*(ii) Let $b \geq 2$, $n = b^{a_1} \cdot k_1 + \cdots + b^{a_r} \cdot k_r$ and $m = b^{a'_1} \cdot l_1 + \cdots + b^{a'_s} \cdot l_s$ be $b$-representations of integers $n$ and $m$, respectively. Then $n < m$ if and only if either of the following obtains:*

1. *There exists $0 < j \leq \min(r, s)$ such that $a_i = a'_i$ and $k_i = l_i$ for $i = 1, \ldots, j-1$ and $a_j < a'_j$, or $a_j = a'_j$ and $k_j < l_j$.*
2. *$r < s$ and $a_i = a'_i$ and $k_i = l_i$ hold for all $1 \leq i \leq r$.*

**Lemma 2.3** *Let $m, n, b$ be naturals, $b \geq 2$, and $\alpha, \beta$ be ordinals with $\mathrm{C}(\alpha), \mathrm{C}(\beta) < b$.*

*(i) $\hat{\mathrm{T}}^\omega_b(\mathrm{T}^b_\omega(m)) = m$.*
*(ii) $\mathrm{T}^b_\omega(\hat{\mathrm{T}}^\omega_b(\alpha)) = \alpha$.*
*(iii) $\alpha < \beta \Leftrightarrow \hat{\mathrm{T}}^\omega_b(\alpha) < \hat{\mathrm{T}}^\omega_b(\beta)$.*
*(iv) $m < n \Leftrightarrow \mathrm{T}^b_\omega(m) < \mathrm{T}^b_\omega(n)$.*

*Proof* (i) and (ii) are obvious. (iii) and (iv) follow from Lemma 2.2. □

**Definition 2.4** Given any natural number $m$ and non-decreasing function

$$f : \mathbb{N} \to \mathbb{N}$$

with $f(0) \geq 2$ define

$$m^f_0 = m, \quad \ldots \quad , m^f_{i+1} = \mathrm{S}^{f(i)}_{f(i+1)}(m^f_i) \dot{-} 1$$

where $k \dot{-} 1$ is the predecessor of $k$ if $k > 0$, and $k \dot{-} 1 = 0$ if $k = 0$.

We shall call $(m_i^f)_{i\in\mathbb{N}}$ a **Goodstein sequence**. Note that a sequence $(m_i^f)_{i\in\mathbb{N}}$ is uniquely determined by $f$ once we fix its starting point $m = m_0^f$.

The case when $f$ is just a shift function has received special attention. Given any $m$ we define $m_0 = m$ and $m_{i+1} := \mathrm{S}_{i+3}^{i+2}(m_i) \dot{-} 1$ and call $(m_i)_{i\in\mathbb{N}}$ a **special Goodstein sequence**. Thus $(m_i)_{i\in\mathbb{N}} = (m_i^{\mathrm{id}_2})_{i\in\mathbb{N}}$, where $\mathrm{id}_2(x) = x + 2$. Special Goodstein sequences can differ only with respect to their starting points. They give rise to a recursive function $f_{good}$ defined as follows: $f_{good}(m)$ is the least $i$ such that $m_i = 0$ where $(m_i)_{i\in\mathbb{N}}$ is the special Goodstein sequence starting with $m_0 = m$.

**Theorem 2.5** *(Goodstein 1944) Every Goodstein sequence terminates, i.e. there exists k such that $m_i^f = 0$ for all $i \geq k$.*

*Proof* If $m_i^f \neq 0$ one has

$$\mathrm{T}_\omega^{f(i)}(m_i^f) = \mathrm{T}_\omega^{f(i+1)}(\mathrm{S}_{f(i+1)}^{f(i)}(m_i^f)) > \mathrm{T}_\omega^{f(i+1)}(m_{i+1}^f)$$

by Lemma 2.3(iv) since $\mathrm{S}_{f(i+1)}^{f(i)}(m_i^f) = m_{i+1}^f + 1$. Hence, as there are no infinitely descending ordinal sequences, there must exist a $k$ such that $m_k^f = 0$. □

The statement of the previous theorem is not formalizable in **PA**. However, the corresponding statement about termination of special Goodstein sequences is expressible in the language of **PA** as a $\Pi_2$ statement. It was shown to be unprovable in **PA** by Kirby and Paris in 1982 [9] using model-theoretic tools. The latter article prompted Cichon [4] to find a different (short) proof that harked back to older proof-theoretic work of Kreisel's [10] from 1952 which identified the so-called $< \varepsilon_0$-recursive functions as the provably recursive functions of **PA**. Other results pivotal to [4] were ordinal-recursion-theoretic classifications of Schwichtenberg [13] and Wainer [16] from around 1970 which showed that the latter class of recursive functions consists exactly of those elementary in one of the fast growing functions $F_\alpha$ with $\alpha < \varepsilon_0$. As $F_{\varepsilon_0}$ eventually dominates any of these functions it is not provably total in **PA**. Cichon verified that $F_{\varepsilon_0}$ is elementary in the function $f_{good}$ of Definition 2.4. Thus termination of special Goodstein sequences is not provable in **PA**.

Returning to Goodstein, he established a connection between sequences of natural numbers and descending sequences of ordinals. Inspection of his proof shows that, using the standard scale of reverse mathematics, it can be carried out in the weakest system, $\mathbf{RCA}_0$, based on recursive comprehension (see [15]).

**Theorem 2.6** *(Goodstein 1944) Over $\mathbf{RCA}_0$ the following are equivalent:*

*(i) Every Goodstein sequence terminates.*
*(ii) There are no infinitely descending sequences of ordinals*

$$\varepsilon_0 > \alpha_0 > \alpha_1 > \alpha_2 > \cdots .$$

*Of course, when we speak about ordinals $< \varepsilon_0$ in* $\mathbf{RCA}_0$ *we mean Cantor normal forms.*

*Proof* "(ii)⇒(i)" follows from Theorem 2.5. For the converse, assume (i) and, aiming at a contradiction, suppose we have a strictly descending sequence of ordinals $\varepsilon_0 > \alpha_0 > \alpha_1 > \alpha_2 > \cdots$. Define a function $f : \mathbb{N} \to \mathbb{N}$ by letting $f(i) = \max\{\mathrm{C}(\alpha_0), \ldots, \mathrm{C}(\alpha_i)\} + 1$. $f$ is non-decreasing. Let $m := \hat{\mathrm{T}}^{\omega}_{f(0)}(\alpha_0)$. We claim that

$$\hat{\mathrm{T}}^{\omega}_{f(i)}(\alpha_i) \leq m_i^f \tag{3}$$

and to this end use induction on $i$. It's true for $i = 0$ by definition. Inductively assume $m_i^f \geq \hat{\mathrm{T}}^{\omega}_{f(i)}(\alpha_i)$. Then

$$\mathrm{S}^{f(i)}_{f(i+1)}(m_i^f) \geq \mathrm{S}^{f(i)}_{f(i+1)}(\hat{\mathrm{T}}^{\omega}_{f(i)}(\alpha_i))$$

and hence

$$\mathrm{S}^{f(i)}_{f(i+1)}(m_i^f) \geq \mathrm{S}^{f(i)}_{f(i+1)}(\hat{\mathrm{T}}^{\omega}_{f(i)}(\alpha_i)) = \hat{\mathrm{T}}^{\omega}_{f(i+1)}(\alpha_i) > \hat{\mathrm{T}}^{\omega}_{f(i+1)}(\alpha_{i+1}) \tag{4}$$

where the last inequality holds by Lemma 2.3(iii) since $\alpha_{i+1} < \alpha_i$ and

$$\mathrm{C}(\alpha_{i+1}), \mathrm{C}(\alpha_i) < f(i+1).$$

From (4) we conclude that $m_{i+1}^f = \mathrm{S}^{f(i)}_{f(i+1)}(m_i^f) \dot{-} 1 \geq \hat{\mathrm{T}}^{\omega}_{f(i+1)}(\alpha_{i+1})$, furnishing the induction step.

Since $m_k^f = 0$ for a sufficiently large $k$, (3) yields that $\hat{\mathrm{T}}^{\omega}_{f(k)}(\alpha_k) = 0$, and hence $\alpha_k = 0$, contradicting $\alpha_k > \alpha_{k+1}$. □

Whereas it's not possible to speak about arbitrary Goodstein sequences in **PA**, one can certainly formalize the notion of a primitive recursive sequence of naturals in this theory. As a result of the proof of the previous Theorem we have:

**Corollary 2.7** *Over* **PA** *the following are equivalent:*

*(i) Every primitive recursive Goodstein sequence terminates.*
*(ii) There are no infinitely descending primitive recursive sequences of ordinals*

$$\varepsilon_0 > \alpha_0 > \alpha_1 > \alpha_2 > \cdots.$$

A very coarse description of Gentzen's result [5] one often finds is that he showed that transfinite induction up to $\varepsilon_0$ suffices to prove the consistency of first order number theory (also known as Peano arithmetic, **PA**). What Gentzen actually did is much more subtle. He defined a reduction procedure on derivations (proofs) and showed that if successive application of a reduction step on a given derivation always leads to a non-reducible derivation in finitely many steps, then the consistency of

**PA** follows. The latter he ensured by assigning ordinals to derivations in such a way that a reduction step applied to a reducible derivation results in a derivation with a smaller ordinal. Let us explain in more detail how this is done in the later [6] which uses the sequent calculus. Firstly, he defined an assignment ord of ordinals to derivations of **PA** such for every derivation $D$ of **PA** in his sequent calculus, $\mathrm{ord}(D)$ is an ordinal $< \varepsilon_0$. He then defined a reduction procedure $\mathcal{R}$ such that whenever $D$ is a derivation of the empty sequent in **PA** then $\mathcal{R}(D)$ is another derivation of the empty sequent in **PA** but with a smaller ordinal assigned to it, i.e.

$$\mathrm{ord}(\mathcal{R}(D)) < \mathrm{ord}(D). \tag{5}$$

Moreover, both ord and $\mathcal{R}$ are primitive recursive functions and only finitist means are used in showing (5). As a result, if $\mathrm{PRWO}(\varepsilon_0)$ is the statement that there are no infinitely descending primitive recursive sequences of ordinals below $\varepsilon_0$, then the following are immediate consequences of Gentzen's work.

**Theorem 2.8** *(Gentzen 1936, 1938)*

*(i) The theory of primitive recursive arithmetic,* **PRA**, *proves that* $\mathrm{PRWO}(\varepsilon_0)$ *implies the consistency of* **PA**.
*(ii) Assuming that* **PA** *is consistent,* **PA** *does not prove* $\mathrm{PRWO}(\varepsilon_0)$.

*Proof* For (ii), of course, one invokes Gödel's second incompleteness theorem. □

So it appears that an attentive reader could have inferred the following from [5–7] in 1944:

**Theorem 2.9** *Termination of primitive recursive Goodstein sequences is not provable in* **PA**.

*Proof* Use Theorem 2.8(ii) and Corollary 2.7. □

# 3 Slowing Down

The key to establishing that already termination of special Goodstein sequences is beyond **PA** is to draw on Theorem 2.8 and to show that infinite descending sequences can be made slow. This technology was used in a paper by Simpson [14, Lemma 3.6] where it is credited to Harvey Friedman. It would be good to know where this padding technique was used for the first time.

**Definition 3.1** Addition of ordinals $\alpha + \beta$ is usually defined by transfinite recursion on $\beta$. For ordinals given in complete Cantor normal form addition can be defined explicitly. We set $\alpha + 0 := \alpha$ and $0 + \alpha := \alpha$. Now let $\alpha, \beta$ be non-zero ordinals, where $\alpha =_{NF} \omega^{\alpha_1} \cdot k_1 + \cdots + \omega^{\alpha_r} \cdot k_r$ and $\beta =_{NF} \omega^{\beta_1} \cdot l_1 + \cdots + \omega^{\beta_s} \cdot l_s$. If $\alpha_1 < \beta_1$, then $\alpha + \beta := \beta$. Otherwise there is a largest $1 \leq i \leq r$ such that $\alpha_i \geq \beta_1$. If $\alpha_i = \beta_1$, then

$$\alpha + \beta := \omega^{\alpha_1} \cdot k_1 + \cdots + \omega^{\alpha_{i-1}} \cdot k_{i-1} + \omega^{\beta_1} \cdot (k_i + l_1) + \omega^{\beta_2} \cdot l_2 + \cdots + \omega^{\beta_{l_s}} \cdot l_s.$$

If $\alpha_i > \beta_1$, then

$$\alpha + \beta := \omega^{\alpha_1} \cdot k_1 + \cdots + \omega^{\alpha_i} \cdot k_i + \omega^{\beta_1} \cdot l_1 + \cdots + \omega^{\beta_{l_s}} \cdot l_s.$$

With the help of addition we can also explicitly define multiplication $\omega^\alpha \cdot \beta$ as follows: $\omega^\alpha \cdot 0 := 0$. If $\beta =_{NF} \omega^{\beta_1} \cdot l_1 + \cdots + \omega^{\beta_s} \cdot l_s$ then

$$\omega^\alpha \cdot \beta := \omega^{\alpha+\beta_1} \cdot l_1 + \cdots + \omega^{\alpha+\beta_s} \cdot l_s.$$

We shall use $\omega \cdot \beta$ to stand for $\omega^1 \cdot \beta$.

Next we recall an elementary result that was known in the 1950s (e.g. [8]).

**Lemma 3.2** *For a function $\ell : \mathbb{N} \to \mathbb{N}$ define $\ell^0(l) = l$ and $\ell^{k+1}(l) = \ell(\ell^k(l))$. The* Grzegorczyk hierarchy *$(f_l)_{l \in \mathbb{N}}$ is generated by the functions $f_0(n) = n+1$ and $f_{l+1}(n) = (f_l)^n(n)$.*

*For every primitive recursive function $h$ of arity $r$ there is an $n$ such $h(\vec{x}) \leq f_n(\max(2, \vec{x}))$ holds for all $\vec{x} = x_1, \ldots, x_r$.*

*Proof* The proof proceeds by induction on the generation of primitive recursive function, using properties of the hierarchy $(f_l)_{l \in \mathbb{N}}$. It is straightforward but a bit tedious. We shall give it in the appendix. □

**Lemma 3.3 (PA)** *Let $f : \mathbb{N} \to \mathbb{N}$ be primitive recursive. Then there exists a primitive recursive function $g : \mathbb{N}^2 \to \omega^\omega$ such that*

*(1) $g(n, m) > g(n, m+1)$ whenever $m < f(n)$.*
*(2) There exists a constant $K$ such $|g(n, m)| \leq K \cdot (n + m + 1)$ holds for all $n, m$.*

*Proof* By Lemma 3.2 is suffices to show this for any $f = f_l$ in the hierarchy $(f_l)_{l \in \mathbb{N}}$. We will actually obtain a $0 < k < \omega$ such that $g : \mathbb{N}^2 \to \omega^k$. To find $g$ we proceed by induction on $l$.

**Base Case**: $f(n) = n + 1$. Define $g$ by

$$g(n, m) = (n + 2) \dot{-} m.$$

**Induction Step**: Let $g : \mathbb{N}^2 \to \omega^k$ satisfy the conditions (1) and (2) for $f$, and let $f'$ be defined by diagonalizing over $f$, i.e. $f'(k) = f^k(k)$, where $f^0(l) = l$ and $f^{k+1}(l) = f(f^k(l))$. If $m < f'(n)$ define $g'(n, m)$ by letting

$$g'(n, m) = \omega^k \cdot (n - i) + g(f^i(n), j),$$

where $i$ and $j$ are the unique integers such that

$$m = f(n) + f^2(n) + \cdots + f^i(n) + j,$$

$i < n$ and $j < f^{i+1}(n)$. If $m \geq f'(n)$ set $g'(n, m) = 0$.

We first show that $g'$ satisfies requirement (1) for $f'$. So suppose $m < f'(n)$. Let $m = f(n) + f^2(n) + \cdots + f^i(n) + j$ with $j < f^{i+1}(n)$. We distinguish two cases. If also $j + 1 < f^{i+1}(n)$, then

$$g'(n,m+1) = \omega^k \cdot (n-i) + g(f^i(n), j+1) < \omega^k \cdot (n-i) + g(f^i(n), j) = g'(n,m)$$

holds by the inductive assumption on $g$ and $f$ since $j < f(f^i(n))$. The other possible case is that $j + 1 = f^{i+1}(n)$ and then we have

$$\begin{aligned} g'(n,m+1) &= \omega^k \cdot (n-(i+1)) + g(f^{i+1}(n), 0) < \omega^k \cdot (n-i) + g(f^i(n), j) \\ &= g'(n,m) \end{aligned}$$

since $\omega^k \cdot (n-(i+1)) + \omega \leq \omega^k \cdot (n-i)$ as $k > 0$.

$g'$ also satisfies requirement (2) for $f'$ since

$$\begin{aligned} |g'(n,m)| &\leq \text{constant} \cdot n + \text{constant} \cdot (f^i(n) + m + 1) \\ &\leq \text{constant} \cdot (n + m + 1). \end{aligned}$$

□

**Corollary 3.4 (PA)** *From a given primitive recursive strictly descending sequence $\varepsilon_0 > \beta_0 > \beta_1 > \beta_2 > \cdots$ one can construct a slow primitive recursive strictly descending sequence $\varepsilon_0 > \alpha_0 > \alpha_1 > \alpha_2 > \cdots$, where slow means that there is a constant $K$ such that*

$$|\alpha_i| \leq K \cdot (i+1)$$

*holds for all $i$.*

*Proof* By the previous Lemma let $g : \mathbb{N}^2 \to \omega^\omega$ be chosen such that $g(n,m) > g(n,m+1)$ for every $m < |\beta_{n+1}|$ and $|g(n,m)| \leq K \cdot (n+m+1)$ holds for all $n, m$. Now set

$$\alpha_j = \omega^\omega \cdot \beta_n + g(n,m)$$

where $j = |\beta_0| + |\beta_1| + \cdots + |\beta_n| + m$ for $m < |\beta_{n+1}|$. For such $j$ one computes that

$$\begin{aligned} |\alpha_j| &\leq \text{constant} \cdot |\beta_n| + \text{constant} \cdot (n + m + 1) \\ &\leq \text{constant} \cdot (j + 1). \end{aligned}$$

We also need to determine $\alpha_i$ for $i < |\beta_0|$. For instance let

$$\alpha_i = \omega^\omega \cdot \beta_0 + |\beta_0| + 1 - i$$

for $i < |\beta_0|$. Clearly we can choose a constant $K_0$ such that $|\alpha_i| \leq K_0 \cdot (i+1)$ for $i < |\beta_0|$. □

**Theorem 3.5 (PA)** *Let $\varepsilon_0 > \alpha_0 > \alpha_1 > \alpha_2 > \cdots$ be a slow primitive recursive descending sequence of ordinals, i.e. there is a constant $K$ such that $|\alpha_i| \leq K \cdot (i+1)$ for all $i$. Then there exists a primitive recursive descending sequence $\varepsilon_0 > \beta_0 > \beta_1 > \beta_2 > \cdots$ such that $\mathrm{C}(\beta_r) \leq r+1$ for all $r$.*

*Proof* Obviously $K > 0$. Let $\omega_0 = 1$ and $\omega_{n+1} = \omega^{\omega_n}$. As $\alpha_0 < \varepsilon_0$ we find $s < \omega$ such that $\omega \cdot \alpha_0 < \omega_s$ and $K < s$. Now put

$$\beta_j := \sum_{i=0}^{K-1-j} \omega_{s-i}$$

for $j = 0, \ldots, K-1$, and

$$\beta_{K\cdot(n+1)+i} := \omega \cdot \alpha_n + (K-i)$$

for $n < \omega$ and $0 \leq i < K$. By construction, $\beta_r > \beta_{r+1}$ for all $r$. As $\mathrm{C}(\omega_r) = 1$ for all $r$, one has $\mathrm{C}(\beta_j) = 1$ for all $j = 0, \ldots, K-1$. Moreover, as $\mathrm{C}(\alpha_n) \leq |\alpha_n| \leq K \cdot (n+1)$, it follows that

$$\mathrm{C}(\beta_{K\cdot(n+1)+i}) = \mathrm{C}(\omega \cdot \alpha_n + (K-i)) \leq K \cdot (n+1) + 1,$$

since multiplying by $\omega$ increases the coefficients by at most 1. As a result, $\mathrm{C}(\beta_r) \leq r+1$ for all $r$. □

**Lemma 3.6 (PA)** *Let $\varepsilon_0 > \beta_0 > \beta_1 > \beta_2 > \cdots$ be a primitive recursive descending sequence of ordinals such that $\mathrm{C}(\beta_n) \leq n+1$. Then the special Goodstein sequence $(m_i)_{i\in\mathbb{N}}$ with $m_0 = \hat{\mathrm{T}}_2^{\omega}(\beta_0)$ and $m_{i+1} = \mathrm{S}_{i+3}^{i+2}(m_i) \dot{-} 1$ does not terminate.*

*Proof* We claim that

$$m_k \geq \hat{\mathrm{T}}_{k+2}^{\omega}(\beta_k) \tag{6}$$

holds for all $k$.

For $k = 0$ this holds by definition. Assume this to be true for $i$, i.e. $m_i \geq \hat{\mathrm{T}}_{i+2}^{\omega}(\beta_i)$. Let $\delta = \mathrm{T}_{\omega}^{i+2}(m_i)$. Since $\mathrm{C}(\beta_i) < i+2$ it follows from Lemma 2.3(iii) that $\delta \geq \beta_i$, and hence $\delta > \beta_{i+1}$. As $\mathrm{C}(\delta), \mathrm{C}(\beta_{i+1}) < i+3$ it follows from Lemma 2.3(iii) that $\hat{\mathrm{T}}_{i+3}^{\omega}(\delta) > \hat{\mathrm{T}}_{i+3}^{\omega}(\beta_{i+1})$. Thus, since

$$m_{i+1} = \mathrm{S}_{i+3}^{i+2}(m_i) \dot{-} 1 = \hat{\mathrm{T}}_{i+3}^{\omega}(\delta) \dot{-} 1,$$

we arrive at $m_{i+1} \geq \hat{\mathrm{T}}_{i+3}^{\omega}(\beta_{i+1})$ as desired.

Equation (6) entails $m_k \neq 0$ for all $k$. □

In sum, what we have done amounts to an elementary proof of the following result due to Kirby and Paris [9, Theorem 1(ii)]:

**Corollary 3.7** *The statement that any special Goodstein sequence terminates is not provable in* **PA**.

*Proof* Let $GS$ be the statement that every special Goodstein sequence terminates. Arguing in **PA** and assuming $GS$, we obtain from Lemma 3.6, Theorem 3.5 and Corollary 3.4 that there is no infinite primitive recursive descending sequence of ordinals below $\varepsilon_0$, i.e. PRWO($\varepsilon_0$). However, by Theorem 2.8 the latter is not provable in **PA**. □

**Acknowledgements** The author acknowledges support by the EPSRC of the UK through Grant No. EP/G029520/1.

He is also indebted to Jan von Plato for showing him letters from the Goodstein-Bernays correspondence.

## Appendix

It remains to prove Lemma 3.2. To this end the following is useful.

**Lemma A.1** *Recall that for a function $h : \mathbb{N} \to \mathbb{N}$ we defined $h^0(l) = l$ and $h^{k+1}(l) = h(h^k(l))$. Also recall that the hierarchy $(f_l)_{l \in \mathbb{N}}$ is generated by the functions $f_0(n) = n + 1$ and $f_{l+1}(n) = (f_l)^n(n)$. We shall write $f_l^n$ rather than $(f_l)^n$.*

*Let $f$ be any of the functions $f_l$ in this hierarchy. Then $f$ satisfies the following properties:*

*(i) $f(x) \geq x + 1$ if $x > 0$.*
*(ii) $f^z(x) \geq x$ for all $x, z$.*
*(iii) If $x < y$ then $f(x) < f(y)$ and $f^z(x) < f^z(y)$.*
*(iv) $f_{l+1}(x) \geq f_l(x)$ whenever $x > 0$.*

*Proof* (i)–(iii) will be proved simultaneously by induction on $l$. (i) and (iii) are obvious for $f = f_0$ and (ii) follows via a trivial induction on $z$. Now assume that (i)–(iii) hold for $f_k$ and $l = k + 1$. For $x > 0$ one then computes

$$f_l(x) = f_k^x(x) = f_k(f_k^{x-1}(x)) \geq f_k(x) \geq x + 1$$

using the properties for $f_k$. (ii) follows from this by induction on $z$. As to (iii), note that

$$f_l(x+1) = f_k^{x+1}(x+1) = f_k(f_k^x(x+1)) > f_k(f_k^x(x)) \geq f_k^x(x) = f_l(x),$$

using the properties for $f_k$, and thus (iii) follows by straightforward inductions on $y$ and $z$.

If $x > 0$, then $f_{l+1}(x) = f_l^x(x) = f_l(f_l^{x-1}(x)) \geq f_l(x)$ by (ii) and (iii). □

*Proof of Lemma 3.2* We want to prove that for or every primitive recursive function $h$ of arity $r$ there is an $n$ such $h(\vec{x}) \leq f_n(\max(2, \vec{x}))$ holds for all $\vec{x} = x_1, \ldots, x_r$.

We show this by induction on the generation of the primitive recursive functions. Clearly for all $n$ we have $h(\vec{x}) \leq f_n(\max(2, \vec{x}))$ by Lemma A.1(i) if $h$ is any of the initial functions $x \mapsto 0$, $\vec{x} \mapsto x_i$, and $x \mapsto x + 1$.

Now let $h$ be defined by $h(\vec{x}) = g(\varphi_1(\vec{x}), \ldots, \varphi_s(\vec{x}))$ and assume that the assertion holds for $g, \varphi_1, \ldots, \varphi_s$. By Lemma 3.2(iv) we can then pick an $n$ such that $g(\vec{y}) \leq f_n(\max(2, \vec{y}))$ and $\varphi_i(\vec{x}) \leq f_n(\max(2, \vec{y}))$ hold for all $\vec{y}, \vec{x}$ and $1 \leq i \leq s$. As a result,

$$\begin{aligned} h(\vec{x}) \leq f_n(\max(2, f_n(\max(2, \vec{x})))) = f_n(f_n(\max(2, \vec{x}))) = \\ f_n^2(\max(2, \vec{x})) \leq f_n^{\max(2, \vec{x})}(\max(2, \vec{x})) = f_{n+1}(\max(2, \vec{x})), \end{aligned}$$

showing that $f_{n+1}$ is a majorant for $h$.

Now suppose $h$ is defined by primitive recursion from $g$ and $\varphi$ via $h(\vec{x}, 0) = g(\vec{x})$ and $h(\vec{x}, y+1) = \varphi(\vec{x}, y, h(\vec{x}, y))$ and that $f_n$ majorizes $g$ and $\varphi$, i.e. $g(\vec{x}) \leq f_n(\max(2, \vec{x}))$ and $\varphi(\vec{x}, y, z) \leq f_n(\max(2, \vec{x}))$. We claim that

$$h(\vec{x}, y) \leq f_n(\max(2, \vec{x}, y)). \tag{7}$$

We prove this by induction on $y$. For $y = 0$ we have $h(\vec{x}, y) = g(\vec{x}) \leq f_n(\max(2, \vec{x})) = f_n^1(\max(2, \vec{x}))$. For the induction step we compute

$$\begin{aligned} h(\vec{x}, y+1) &= \varphi(\vec{x}, y, h(\vec{x}, y)) \leq f_n(\max(2, \vec{x}, y, h(\vec{x}, y))) \\ &\leq f_n(\max(2, \vec{x}, y, f_n^{y+1}(\max(2, \vec{x}, y)))) = f_n(f_n^{y+1}(\max(2, \vec{x}, y))) \\ &= f_n^{y+2}(\max(2, \vec{x}, y)) \end{aligned}$$

where the second "≤" uses the inductive assumption and the penultimate "=" uses Lemma A.1.

From the claim (7) we get with Lemma A.1, letting $w = \max(2, \vec{x}, y)$, that

$$\begin{aligned} h(\vec{x}, y) &\leq f_n^{y+1}(\max(2, \vec{x}, y)) \leq f_n^{w+1}(w) = f_n(f_n^w(w)) = f_n(f_{n+1}(w)) \\ &\leq f_{n+1}(f_{n+1}(w)) = f_{n+1}^2(w) \leq f_{n+1}^w(w) = f_{n+2}(w). \end{aligned}$$

As a result, $h(\vec{x}, y) \leq f_{n+2}(\max(2, \vec{x}, y))$. □

## References

1. J. Barwise (ed.), *Handbook of Mathematical Logic* (North-Holland, Amsterdam, 1977)
2. P. Bernays, *Letter to Goodstein, dated September 1st, 1942*, Bernays collection of the ETH Zürich

3. P. Bernays, *Letter to Goodstein, dated September 29th, 1943*, Bernays collection of the ETH Zürich
4. A. Cichon, A short proof of two recently discovered independence results using recursion theoretic methods. Proc. Am. Math. Soc. **87**, 704–706 (1983)
5. G. Gentzen, Die Widerspruchsfreiheit der reinen Zahlentheorie. Math. Ann. **112**, 493–565 (1936)
6. G. Gentzen, *Neue Fassung des Widerspruchsfreiheitsbeweises für die reine Zahlentheorie*, Forschungen zur Logik und zur Grundlegung der exacten Wissenschaften, Neue Folge 4 (Hirzel, Leipzig, 1938), pp. 19–44
7. R.L. Goodstein, On the restricted ordinal theorem. J. Symb. Log. **9**, 33–41 (1944)
8. A. Grzegorczyk, *Some Classes of Recursive Functions*. Rozprawy Mate No. IV (Warsaw, 1953)
9. L. Kirby, J. Paris, Accessible independence results for Peano arithmetic. Bull. Lond. Math. Soc. **14**, 285–293 (1982)
10. G. Kreisel, On the interpretation of non-finitist proofs II. J. Symb. Log. **17**, 43–58 (1952)
11. J. Paris, L. Harrington, A mathematical incompleteness in Peano arithmetic, in *Handbook of Mathematical Logic*, ed. by J. Barwise (North-Holland, Amsterdam, 1977), pp. 1133–1142
12. D. Schmidt, *Well-Partial Orderings and Their Maximal Order Types* (Habilitationsschrift, Universität Heidelberg, 1979)
13. H. Schwichtenberg, Eine Klassifikation der $\varepsilon_0$-rekursiven Funktionen. Zeitschrift für mathematische Logik und Grundlagen der Mathematik **17**, 61–74 (1971)
14. S.G. Simpson, Nichtbeweisbarkeit von gewissen kombinatorischen Eigenschaften endlicher Bäume. Archiv für mathematische Logik **25**, 45–65 (1985)
15. S.G. Simpson, *Subsystems of Second Order Arithmetic*, 2nd edn. (Cambridge University Press, Cambridge, 2009)
16. S.S. Wainer, A classification of the ordinal recursive functions. Archiv für Mathematische Logik und Grundlagenforschung **13**, 136–153 (1970)

# Part III
# Results

# Cut Elimination In Situ

**Sam Buss**

**Abstract** We present methods for removing top-level cuts from a sequent calculus or Tait-style proof without significantly increasing the space used for storing the proof. For propositional logic, this requires converting a proof from tree-like to dag-like form, but at most doubles the number of lines in the proof. For first-order logic, the proof size can grow exponentially, but the proof has a succinct description and is polynomial time uniform. We use direct, global constructions that give polynomial time methods for removing all top-level cuts from proofs. By exploiting prenex representations, this extends to removing all cuts, with final proof size near-optimally bounded superexponentially in the alternation of quantifiers in cut formulas.

## 1 Introduction

Gentzen's technique of cut elimination, together with the closely related normalization, is arguably the most important construction of proof theory. The importance of cut elimination lies partly in its connections to constructivity, and indeed cut elimination is algorithmic and can be carried out effectively. The present paper focuses on algorithms for cut elimination in the setting of pure propositional logic and pure first-order logic. We introduce methods for removing top-level cuts from a proof without significantly increasing the space used for generating the proof. Of course, it is well known that eliminating top-level cuts can make proof size grow exponentially, so it requires some special care to describe the resulting proof without any significant increase in space. For propositional logic, our methods require converting a proof from tree-like to dag-like form, but at most double the number of lines in the proof. For first-order logic, the proof size can grow exponentially; in fact, both the number of lines in the proof and the size of the terms can grow exponentially. However, our constructions give polynomial size dag representations for the terms, and succinct descriptions of the proof that give a polynomial time uniform description of the proof and its terms.

S. Buss (✉)
Department of Mathematics, University of California, San Diego, La Jolla, CA 92093-0112, USA
e-mail: sbuss@ucsd.edu

 
R. Kahle, M. Rathjen (eds.), *Gentzen's Centenary*,
DOI 10.1007/978-3-319-10103-3_10

Along with the small space usage, our cut elimination methods give direct, global constructions. We define direct, concrete descriptions of the proof that results from eliminating the top-level cuts. Our construction is "global" in that it operates on the entire proof and eliminates all top-level cuts at once.

Our constructions synthesize and generalize a number of prior results from proof complexity and continuous cut elimination. Our immediate motivation arose from the desire to find global versions of the polynomial time algorithms for the continuous cut elimination used by Aehlig-Beckmann [1] and Beckmann-Buss [4]. Continuous cut elimination was developed by Mints [11, 12] for the analysis of higher order logics, and [1] introduced its use for the analysis of bounded arithmetic. In particular, [1, 4] required polynomial time constructions of proofs. Like Mints, they create proofs step-by-step and use a special REP (for "repetition" or "repeat") inference to slow down the construction of proofs. In contrast, we shall give direct (not step-by-step) constructions, and avoid the use of a REP inference.

There is extensive prior work giving upper bounds on the complexity of cut elimination in propositional and first-order logic, including [2, 5–8, 13, 14, 16–18]. Some of the best such bounds measure the complexity of proofs in terms of the height of proofs [5, 7, 13, 16–18]. Loosely speaking, these results work by removing top-level connectives from cut formulas, at the cost of exponentiating the height of the proof, and repeating this to remove all cuts from a proof. Zhang [17] and Gerhardy [7] bound the height of cut free proofs in terms of the nesting of quantifiers in cut formulas; namely, if quantifiers are nested to depth $d$ without any intervening propositional connectives, then cut elimination requires a height increase of only an exponential stack of 2's of height $d + 2$. They further show that cut-elimination can remove a top-level block of $\exists$ and $\vee$ (respectively, $\forall$ and $\wedge$) connectives at the cost of a single exponential increase in proof height.

In contrast, the present paper works with proof size rather than proof height. Somewhat counterintuitively, blocks of arbitrarily nested $\exists$ and $\wedge$ connectives (respectively, $\forall$ and $\vee$ connectives) can be removed all at once, with a single exponential increase in proof size.

Krajíček [9, 10], Razborov [15], and Beckmann-Buss [3] have given complexity bounds for reducing the depth (alternation of $\vee$'s and $\wedge$'s) of formulas in constant depth propositional Frege or Tait-style proofs. Reducing the depth of formulas in a proof is essentially equivalent to removing the outermost blocks of like (propositional) connectives from cut formulas. Krajíček [9] and later Beckmann and Buss [3] show that the depth of formulas in a constant depth proof can be reduced from $d + 1$ to $d$ at the cost of converting the proof from tree-like format to dag-like format with only a polynomial increase in proof size. Our Theorem 3 below is similar to Lemma 6 of [3] in this regard, but gives a more explicitly uniform construction, and works even if there are multiple nested outermost like quantifiers that need to be eliminated.

This paper deals with cut elimination for a Tait-style calculus instead of a Gentzen sequent calculus. In the setting of classical logic, our results all apply immediately to cut elimination in a Gentzen sequent calculus.[1] We assume the reader has some familiarity with sequent calculi or Tait calculi, but Sect. 2 begins with formal definitions of our Tait-style proof system, including definitions of proof size and cut formula complexity. It also describes the basic ideas behind the later constructions. Section 3 shows that, for tree-like propositional proofs, outermost like connectives in cut formulas can be removed at the cost of converting the proof to dag-like form, while at most doubling the number of lines in the proof. Sections 4 and 5 extend this to first-order logic, but now, instead of forming a dag-like proof of the same size, the number of lines in the proof can become exponentially larger. However, the exponentially long proof still has a direct, global, polynomial-time specification. For expository purposes, Sect. 4 first shows how to eliminate all top-level like quantifiers from cut formulas. Section 5 then combines the earlier constructions to show how to eliminate all outermost $\forall$ and $\vee$ connectives. In light of the duality of the Tait calculus, this is the same as removing all top-level $\exists$ and $\wedge$ connectives. Our constructions use direct methods that reduce the cut-formula complexity for multiple cuts simultaneously.

So far, we have discussed only the problem of removing the top-level connectives from cut formulas. Obviously, the process could be iterated to remove all cuts. Define the *alternating quantifier depth* of a formula as the maximum number of alternating blocks of existential and universal quantifiers along any branch in the tree representation of the formula (with negations pushed to the atoms, but allowing $\wedge$ and $\vee$ connectives to appear arbitrarily along the branch). Let $\mathsf{aqd}(P)$ be the maximum alternating quantifier depth of any cut formula in the proof $P$. Section 6 proves that it is possible to convert $P$ into a cut free proof of the same end cedent, with the size of $P$ bounded by $2_d^{|P|}$ for $d = \mathsf{aqd}(P) + O(1)$. Here $|P|$ is the number of lines in $P$, and the superexponential function $2_d^a$ is defined by $2_0^a = a$ and $2_{i+1}^a = 2^{2_i^a}$. This improves on what can be obtained straightforwardly using the constructions of Sects. 3–5 or from the prior bounds obtained by Zhang [17], Gerhardy [7], and Beckmann and Buss [5], since we bound the height of the stack of two's in terms of the number of alternations of quantifiers without regard to intervening $\wedge$'s or $\vee$'s. The basic idea for the proof in Sect. 6 is to first modify $P$ so that all cut formulas are in prenex form, and then apply the results of Sect. 4. The results of Sect. 6 do not depend on either Sects. 3 or 5; but we do appeal to constructions of [5, 7, 17] to handle removing cuts on quantifier free formulas.

[1] Tait systems do not work as well as the Gentzen sequent calculus for non-classical systems such as intuitionistic logic. Thus our results would need to be modified to apply to intuitionistic logic, for instance.

# 2 Preliminaries

## 2.1 Tait Calculus

Our first-order Tait system uses logical connectives $\wedge$, $\vee$, $\exists$ and $\forall$, and a language of function symbols, constant symbols, and predicate symbols. Terms and atomic formulas are defined as usual. A *literal* is either an atomic formula $P(\vec{s})$ or a negated atomic formula $\overline{P(\vec{s})}$. Formulas are formed using connectives $\wedge$, $\vee$, $\forall$ and $\exists$. The negation of complex formulas is inductively defined by defining $\overline{(\overline{p})}$, $\overline{B \wedge C}$, $\overline{B \vee C}$, $\overline{(\exists x)A}$, and $\overline{(\forall x)A}$ to be the formulas $p$, $\overline{B} \vee \overline{C}$, $\overline{B} \wedge \overline{C}$, $(\forall x)\overline{A}$, and $(\exists x)\overline{A}$, respectively.

We adopt a convention from the Gentzen sequent calculus and assume that first-order variables come in two sorts: free variables (denoted with letters $a, b, c, \ldots$) and bound variables (denoted with letters $x, y, \ldots$). Free variables cannot be quantified and must appear only freely. A bound variable $x$ may occur in formulas only within the scope of a quantifier $(\forall x)$ or $(\exists x)$ that binds it.

A line of a Tait calculus proof, called a *cedent*, consists of a set of formulas. The intended meaning of a cedent is the disjunction of its members. The allowable rules of inference are shown in Fig. 1. It should be noted that an initial cedent $A, \overline{A}$ must have $A$ atomic. We allow Tait proofs to be either tree-like or dag-like. The usual conditions for eigenvariables apply to $\forall$ inferences. The formulas introduced in the lower cedents of inferences are called the *principal* formula of the inference: these are the formulas $A \wedge B$, $A \vee B$, $(\exists x)A(x)$, and $(\forall x)A(x)$ in Fig. 1. The formulas eliminated from the upper cedent are called *auxiliary* formulas: these are the formulas $A$, $B$, $A(s)$, $A(b)$, $A$, and $\overline{A}$ in the figure. The auxiliary formulas of a cut inference are called *cut formulas*. Formulas that appear in the sets $\Gamma$ and $\Gamma_i$ are called *side* formulas.

The $\wedge$ and cut inferences have two cedents as hypotheses, which are designated the *left* and *right* upper cedents. For a cut inference, we require that the outermost connective of the left cut formula $A$ not be an $\wedge$ or $\exists$ connective; equivalently, the outermost connective of the right cut formula $\overline{A}$ is not $\vee$ or $\forall$. This restriction on $A$'s outermost connective causes no loss of generality, since the order of the upper cedents can always be reversed. (We sometimes display cuts with upper cedents out of order, however.) For an $\wedge$ inference, the left–right order of the upper cedents is

$$\text{Axiom: } A, \overline{A} \qquad\qquad \text{Weakening: } \frac{\Gamma}{\Gamma, \Delta}$$

$$\wedge{:}\ \frac{A, \Gamma_1 \qquad B, \Gamma_2}{A \wedge B, \Gamma_1, \Gamma_2} \qquad\qquad \vee{:}\ \frac{A, B, \Gamma}{A \vee B, \Gamma}$$

$$\exists{:}\ \frac{A(s), \Gamma}{(\exists x)A(x), \Gamma} \qquad\qquad \forall{:}\ \frac{A(b), \Gamma}{(\forall x)A(x), \Gamma}$$

$$\text{Cut: } \frac{A, \Gamma_1 \qquad \overline{A}, \Gamma_2}{\Gamma_1, \Gamma_2}$$

**Fig. 1** The rules of inference for a Tait system. The lines of the proof are to be interpreted as sets of formulas. The formula $A$ of the *axiom* rule must be atomic. The free variable $b$ of the $\forall$ inference is called an *eigenvariable* and may not occur in the lower cedent

dictated by the order of the conjunction; except in the case where $A$ and $B$ are the same formula, and then the upper cedents are put in some arbitrary left–right order.

The left–right ordering of upper cedents allows us to define the postordering of the cedents of a tree-like proofs. The *postordering* of the nodes of a tree $T$ is the order of the nodes output by the following recursive traversal algorithm: Starting at the root of $T$, the traversal algorithm first recursively traverses the child nodes in left-to-right order, and then outputs the root node. The postorder traversal of the underlying proof tree induces an ordering of the cedents in the proof.

Axioms (initial cedents) and weakening inferences are ignored when measuring the size or height of $P$. Thus, the *size*, $|P|$, of a Tait proof $P$ is defined as the number of $\vee$, $\wedge$, $\forall$, $\exists$, and cut inferences in $P$. The *height*, $h(P)$, of $P$ is the maximum number of these kinds of inferences along any branch of $P$.

The fact that cedents are *sets* rather than multisets or sequences means that if a formula is written twice on a line, it appears only once in the cedent. For instance, in the $\vee$ inference, it is possible that $A \vee B$ is a member of $\Gamma$. It is also possible that $A$ (say) appears in $\Gamma$, in which case both $A$ and $A \vee B$ appear in the conclusion of the inference. This latter possibility, however, makes our analysis of cut elimination more awkward, since we will track occurrences of formulas along paths in the proof tree. The problem is that there will be an ambiguity about how to track the formula $A$ in the case where it "splits into two," for example in an $\vee$ inference by both being a member of $\Gamma$ and being used to introduce $A \vee B$. The ambiguity can be avoided by considering proofs that satisfy the following "auxiliary condition":

**Definition** A Tait proof $P$ satisfies the *auxiliary condition* provided that no inference has an auxiliary formula also appearing as a side formula. Specifically, referring to Fig. 1, the auxiliary condition requires the following to hold:

(a) In an $\vee$ inference, neither $A$ nor $B$ may occur in $\Gamma$.
(b) In an $\wedge$ inference, neither $A$ nor $B$ may occur in $\Gamma_1$ or $\Gamma_2$.
(c) In an $\exists$ inference, $A(s)$ may not occur in $\Gamma$.
(d) In a cut inference, neither $A$ nor $\overline{A}$ may occur in $\Gamma_1$ or $\Gamma_2$.

Note that the eigenvariable condition already prevents $A(b)$ from occurring in the side formulas of a $\forall$ inference.

**Lemma 1** *Let $P$ be a [tree-like] Tait proof. Then there is a [tree-like] Tait proof $P'$ satisfying the auxiliary condition proving the same conclusion as $P$. Furthermore, $|P'| \leq |P|$ and $h(P') \leq h(P)$.*

The proof of the lemma is straightforward using the fact that weakening inferences do not count towards proof size or height.

A *path* in a proof $P$ is a sequence of one or more cedents occurring in $P$, with the $(i+1)$st cedent a hypothesis of the inference inferring the $i$th cedent, for all $i$. A *branch* is a path that starts at the conclusion of $P$ and ends at an initial cedent.

Suppose $P$ is tree-like and satisfies the auxiliary condition. Also suppose a formula $A$ occurs in two cedents $C_1$ and $C_2$ in $P$, and let $A_1$ and $A_2$ denote the *occurrences* of $A$ in $C_1$ and $C_2$, respectively. We call $A_1$ a *direct ancestor* of $A_2$ (equivalently, $A_2$ is a *direct descendant* of $A_1$) provided there is a path in $P$ from

$C_2$ to $C_1$ such that the formula $A$ appears in every cedent in the path.[2] If $A_1$ is the principal formula of an inference, or occurs in an axiom, that we say $A_1$ is a place where $A_2$ is *introduced*. If $A_2$ is an auxiliary formula, then we say $A_2$ is the place where $A_1$ is *eliminated*. In view of the tree-like property of $P$, every formula occurring in $P$ either has a unique place where it is eliminated or has a direct descendant in the conclusion of $P$. However, due to the implicit use of contraction in the inference rules, formulas occurring in $P$ may be introduced in multiple places.

The notions of direct descendant and direct ancestor can be generalized to "descendant" and "ancestor" by tracking the flow of subformulas in a proof. If $\mathcal{I}$ is an $\wedge$, $\vee$, $\exists$, or $\forall$ inference, then the principal formula of $\mathcal{I}$ is the (only) *immediate descendant* of each auxiliary formula of $\mathcal{I}$. Then, the "descendant" relation is the reflective, transitive closure of the union of the immediate descendant and direct descendant relations. Namely, a formula $A'$ occurring in $P$ is a *descendant* of a formula $A$ occurring in $P$ iff there is a sequence of formula occurrences in $P$, starting with $A$ and ending with $A'$ such that each formula in the sequence is the immediate descendant or a direct descendant of the previous formula in the sequence. We also call $A$ an *ancestor* of $A'$.

The definitions of descendant and ancestor apply to formulas that appear in cedents. Similar notions also apply to subformulas. Suppose $A$ and $B$ are formulas appearing in cedents with $B$ a descendant of $A$. Let $C$ be a subformula of $A$. We wish to define a unique subformula $D$ of $B$, such that $C$ *corresponds to* $D$. This unique subformula is intended to be defined in the obvious way, with each subformula in an upper cedent of an inference corresponding to a subformula in the lower sequent. Assume $P$ is tree-like and satisfies the auxiliary condition. The "corresponds" relation is defined by taking the reflexive, transitive closure of the following conditions.

- The formula $A(s)$ in an $\exists$ inference corresponds to the subformula $A(x)$ in the lower sequent.
- In a $\forall$ inference, the formula $A(b)$ corresponds to the subformula $A(x)$.
- In an $\wedge$ or $\vee$ inference, the formulas $A$ and $B$ in the upper cedent(s) correspond to the subformulas $A$ and $B$ shown in the lower cedent. Except for an $\vee$ inference in which $A$ and $B$ are the same formula, the auxiliary formula corresponds to the subformula denoted $A$ in the lower cedent. That is, in this case, the $\vee$ inference is treated as if it were defined as

$$\frac{A, \Gamma}{A \vee B, \Gamma}$$

- If $C$ is a subformula of a side formula, namely of a formula $A$ in $\Gamma$, $\Gamma_1$, or $\Gamma_2$ in Fig. 1, then $C$ corresponds to the same subformula of the occurrence of $A$ in the lower cedent.
- If $A$ and $B$ appear in the upper and lower cedent of an inference and $A$ corresponds to $B$ and if $C$ is the $i$th subformula of $A$, then $C$ corresponds to

[2]This definition works because $P$ satisfies the auxiliary condition.

the $i$th subformula $D$ of $B$, where the subformulas of $C$ and $D$ are ordered (say) according to the left-to-right positions of their principal connectives.

It is often convenient to assume proofs use free variables in a controlled fashion. The following definition is slightly weaker than the usual definition, but suffices for our purposes.

**Definition** A proof $P$ is in *free variable normal form* provided that each free variable $b$ is used at most once as an eigenvariable, and provided that when $b$ is used as an eigenvariable for inference $\mathcal{I}$, then $b$ appears in $P$ only above $\mathcal{I}$ (that is, each occurrence of $b$ occurs in a cedent reachable from $\mathcal{I}$ by some path in $P$). The variables $c$ that appear in $P$ but are not used as eigenvariables are called the *parameter variables* of $P$.

Any tree-like proof $P$ may be put into free variable normal form without increasing its size or height; furthermore, this can be done while enforcing the auxiliary condition.

## 2.2 The Basic Constructions

This section describes the basic ideas and constructions used for the cut-elimination results obtained in Sects. 3 and 4.

The first important tool is a generalization of the well-known inversion lemmas for the outermost $\forall$ and $\vee$ connectives of a formula. Assume we have a tree-like proof $P$, in free variable normal form, that ends with the cedent $\Gamma, A \vee B$. Then there is a proof $P'$ of $\Gamma, A, B$, with $P'$ also tree-like, and with $|P'| \leq |P|$ and $h(P') \leq h(P)$. Similarly, if $P$ ends with $\Gamma, (\forall x)A(x)$ and $t$ is any term, then there is a proof $P''$ of $\Gamma, A(t)$, with $P''$ also tree-like and satisfying the same conditions on its size and height. The proofs are quite simple: $P'$ is obtained from $P$ by replacing all direct ancestors of $A \vee B$ with $A, B$ and removing all $\vee$ inferences that introduce a direct ancestor of $A \vee B$. Likewise, if $t$ does not contain any eigenvariables of $P$, then $P''$ is formed by replacing all direct ancestors of $(\forall x)A(x)$ with $A(t)$, and removing the $\forall$ inferences that introduce these direct ancestors and replacing their eigenvariables with $t$.

Iterating this construction allows us to formulate an inversion lemma that works for the entire set of outermost $\vee$ and $\forall$ connectives. If $B$ is a subformula of $A$, we call $B$ an $\vee\forall$*-subformula* of $A$ if every connective of $A$ containing $B$ in its scope is an $\vee$ or a $\forall$. Similarly, a connective $\vee$ or $\forall$ is said to be $\vee\forall$*-outermost* if it is not in the scope of any $\exists$ or $\wedge$ connective. Let $P$ be a tree-like proof of $\Gamma, A$, and let $B_1, \ldots, B_k$ enumerate the minimal $\vee\forall$-subformulas of $A$ in left-to-right order. The subformulas $B_i$ are called the $\vee\forall$*-components* of $A$. Note that each $B_i$ is atomic or has as outermost connective an $\wedge$ or an $\exists$.

**Lemma 2** *Let $P$, $A$, $B_1, \ldots, B_k$ be as above. Let $\sigma$ be any substitution mapping free variables to terms. Then there is a proof $P'$ of $\Gamma\sigma, B_1\sigma, \ldots, B_k\sigma$ such that $P'$ is tree-like and $|P'| \leq |P|$ and $h(P') \leq h(P)$.*

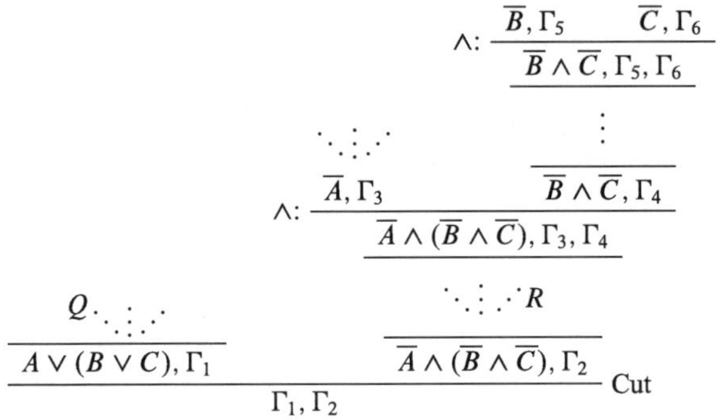

**Fig. 2** A simple example of $\vee$ cut to be eliminated. $Q$ and $R$ are the subproofs deriving the hypotheses of the cut

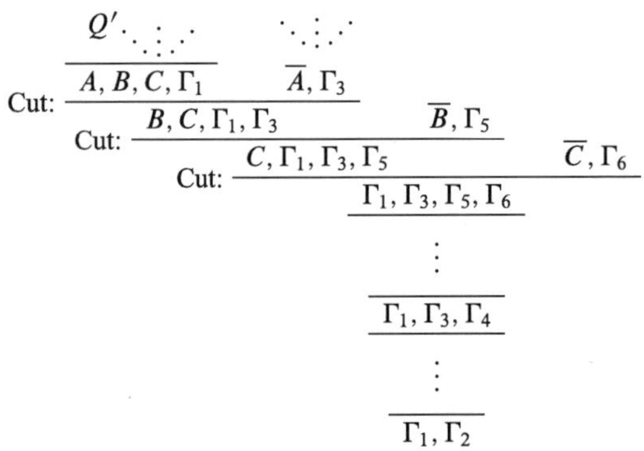

**Fig. 3** The proof $P'$ obtained after eliminating the cut of Fig. 2

The lemma is proved by iterating the inversion lemmas for $\vee$ and $\forall$.

Section 3 will give the details how to simplify cuts in a propositional Tait calculus proof by removing all outermost $\vee$ (or, all outermost $\wedge$) connectives from cut formulas. As a preview, we give the idea of the proof, which depends on the inversion lemma for $\vee$. Namely, suppose the proof $P$ ends with a cut on the formula $A \vee (B \vee C)$, as shown in Fig. 2. The right cut formula, in the final line of the subproof $R$, is in the dual form $\overline{A} \wedge (\overline{B} \wedge \overline{C})$ of course. Now suppose that in the subproof $R$ there are the two pictured $\wedge$ inferences that introduce the formulas $(\overline{B} \wedge \overline{C})$ and then $\overline{A} \wedge (\overline{B} \wedge \overline{C})$.

By the inversion lemma for $\vee$, the proof $Q$ can be transformed into a proof $Q'$ of $A, B, C, \Gamma_1$ with no increase in size or height. The cut in $P$ can thus be removed by replacing the $\wedge$ inferences in $R$ with cuts to obtain the proof $P'$ shown in Fig. 3. Note that this has replaced the $\wedge$ inference introducing $\overline{B} \wedge \overline{C}$ with two cuts, one on $B$ and one on $C$, and replaced the $\wedge$ inference introducing $\overline{A} \wedge (\overline{B} \wedge \overline{C})$ with a cut on $A$. Overall, two $\wedge$ inferences and one cut inference in $P$ have been replaced by three cut inferences in $P'$. More generally, due to contractions, there can be

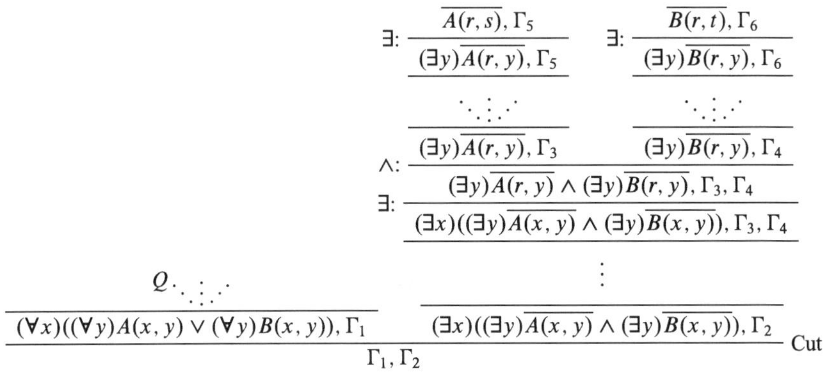

**Fig. 4** A simple example of cuts using $\vee$ and $\exists$ to be eliminated

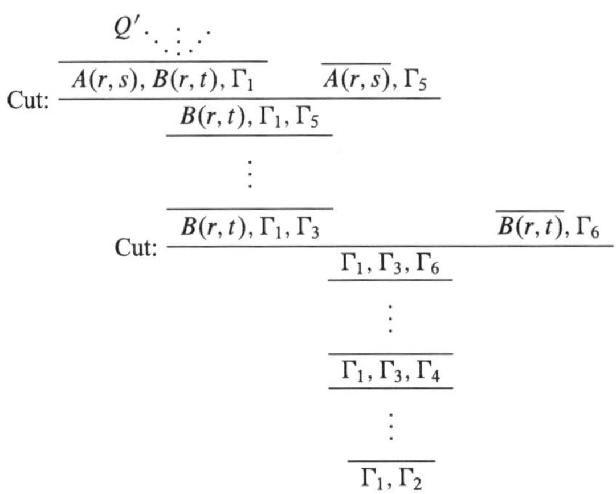

**Fig. 5** The results of eliminating the cuts in Fig. 4

$k_1 \geq 1$ inferences in $P$ that introduce $\overline{B} \wedge \overline{C}$, and $k_2 \geq 1$ inferences that introduce $\overline{A} \wedge (\overline{B} \wedge \overline{C})$: these $k_1 + k_2$ many $\wedge$ inferences and the cut inference in $P$ are replaced by $2k_1 + k_2$ many cut inferences in $P'$. Thus the size of $P'$ is no more than twice the size of $P$. The catch though is that $P'$ may now be dag-like rather than tree-like.

Finally, it should be noted that $P'$ is obtained from $P$ by moving the subproof $Q'$ and the subproof deriving $\overline{A}, \Gamma_3$ "rightward and upward" in the proof. This is crucial in allowing us to remove multiple cuts at once. Intuitively, the final cut of $P$ plus all the cuts that lie in the subproofs $Q$ or $R$ can be simplified in parallel without any unwanted "interference" between the different cuts.

Figure 4 shows a proof $P$ from which the outermost $\vee$ and $\forall$ (dually, $\wedge$ and $\exists$) connectives can be removed from cut formulas. The left subproof $Q$ can be inverted to give a proof $Q'$ of $A(r,s), B(r,t), \Gamma_1$, and this is used to form the proof $P'$ shown in Fig. 5. In this simple example, an $\wedge$ inference, three $\exists$ inferences, and the cut inference are replaced by just two cut inferences. As in the $\vee$ example, the proof $P'$ is formed by moving (instantiations of) subproofs of $P$ rightward. In particular, the subproof in $P$ ending with $\overline{(\exists y)A(r,y)}, \Gamma_3$ has become a proof of $B(r,t), \Gamma_1, \Gamma_3$ and has been moved rightward in the proof so as to be cut against $\overline{B(r,t)}, \Gamma_6$.

The general case of removing quantifiers is more complicated, however. For instance, there might be multiple places where the formula $\overline{(\exists y)A(x,y)}$ is introduced, using $k_1$ terms $s_1, \ldots, s_{k_1}$. Likewise, there could be $k_2$ terms $t_j$ used for introducing the formula $\overline{(\exists y)B(x,y)}$, and $k_3$ terms $r_\ell$ for introducing the $(\exists x)$. In this case we would need $k_1k_2k_3$ many inversions of $Q$, namely, proofs $Q_{i,j,\ell}$ of $A(r_\ell, s_i), B(r_\ell, t_j), \Gamma_1$ for all $i \leq j_1$, $j \leq k_2$, and $\ell \leq k_3$. The result is that $P'$ can have size exponential in the size of $P$; there is, however, still a succinct description of $P'$ which can be obtained directly from $P$. This will be described in Sect. 4.

## 3 Eliminating Like Propositional Connectives

This section describes how to eliminate an outermost block of propositional connectives from cut formulas. The construction applies to proofs in first-order logic.

**Definition** Suppose $B$ is a subformula occurring in $A$. Then $B$ is an $\vee$-*subformula* of $A$ iff $B$ occurs in the scope of only $\vee$ connectives. The notion of $\wedge$-*subformula* is defined similarly.

An $\vee$-*component* (resp., $\wedge$-*component*) of $A$ is a minimal $\vee$-subformula (resp., $\wedge$-subformula) of $A$.

**Definition** An $\vee/\wedge$-*component of a cut formula in* $P$ is an $\vee$-component of a left cut formula in $P$ or an $\wedge$-component of a right cut formula in $P$.

**Theorem 3** *Let $P$ be a tree-like Tait calculus proof of $\Gamma$. Then there is a dag-like proof $P'$, also of $\Gamma$, such that each cut formula of $P'$ is an $\vee/\wedge$-component of a cut formula of $P$, and such that $|P'| \leq 2 \cdot |P|$ and hence $h(P') \leq 2 \cdot |P|$. Furthermore, given $P$ as input, the proof $P'$ can be constructed by a polynomial time algorithm.*

Note that $P'$ is obtained by simplifying *all* the cut formulas in $P$ that have outermost connective $\wedge$ or $\vee$.

Without loss of generality, by Lemma 1, $P$ satisfies the auxiliary condition. The construction of $P'$ depends on classifying the formulas appearing in $P$ according to how they descend to cut formulas. For this, each formula $B$ in $P$ can be put into exactly one of the following categories ($\alpha$)–($\gamma$).

($\alpha$) $B$ has a left cut formula $A$ as a descendant and corresponds to an $\vee$-subformula of $A$, or

($\beta$) $B$ has a right cut formula $A$ as a descendant and corresponds to an $\land$-subformula of $A$, or

($\gamma$) Neither ($\alpha$) nor ($\beta$) holds.

**Definition** Let $B$ be an occurrence of a formula in $P$, and suppose $B$ is in category ($\beta$) with a cut formula $A$ as a descendant. The formula $A$ is a conjunction $\bigwedge_{i=1}^{k} C_i$ of its $k \geq 1$ many $\land$-components (parentheses are suppressed in the notation). The formula $B$ is a subconjunction of $A$ of the form $\bigwedge_{i=m}^{\ell} C_i$ where $1 \leq m \leq \ell \leq k$. The $\land$-components of $A$ to the right of $B$ are $C_{\ell+1}, \ldots, C_k$. The negations of these, namely $\overline{C}_{\ell+1}, \ldots, \overline{C}_k$, are called the *pending implicants for* $B$.

Each formula $B$ in $P$ will be replaced by a cedent denoted $*(B)$. For $B$ in category ($\alpha$), $*(B)$ is the cedent consisting of the $\lor$-components of $B$. For $B$ in category ($\beta$), $*(B)$ is the (possibly empty) cedent containing the pending implicants for $B$. For $B$ in category ($\gamma$), $*(B)$ is the cedent containing just the formula $B$.

**Definition** The *jump target* of a category ($\beta$) formula $B$ in $P$ is the first cut or $\land$ inference below the occurrence of $B$ which has some descendant of $B$ as an auxiliary formula in its right upper cedent. The jump target will be either:

$$\frac{\overline{D}, \Gamma_1 \qquad D, \Gamma_2}{\Gamma_1, \Gamma_2} \quad \text{or} \quad \frac{C, \Gamma_1 \qquad D, \Gamma_2}{C \land D, \Gamma_1, \Gamma_2} \tag{1}$$

where the formula $D$ is either equal to $B$ (a direct descendant of $B$) or is of the form $((\cdots(B \land B_1) \land \cdots \land B_{k-1}) \land B_k)$ with $k \geq 1$ (since only $\land$ inferences can operate on $B$ until reaching the jump target). The left upper cedent of the jump target (that is, $\overline{D}, \Gamma_1$ or $C, \Gamma_1$) is called the *jump target cedent*. The auxiliary formula of the left upper cedent, that is $\overline{D}$ or $C$, is called the *jump target formula*.

We shall consistently suppress parentheses when forming disjunctions and conjunctions. For instance, the formula $((\cdots(B \land B_1) \land \cdots \land B_{k-1}) \land B_k)$ above would typically be written as just $B \land B_1 \land \cdots \land B_k$. It should be clear from the context what the possible parenthesizations are.

**Lemma 4** *Suppose $B$ is category ($\beta$) formula in $P$. Let $C_1, \ldots, C_k$ be as above, so $B = \bigwedge_{i=m}^{\ell} C_i$ and the pending implicants of $B$ are $\overline{C}_{\ell+1}, \ldots, \overline{C}_k$. Consider $B$'s jump target, namely one of the inferences shown in (1), and let $E$ be the jump target formula, that is, either $\overline{D}$ or $C$. Then $*(E)$ is equal to the cedent $\overline{C}_m, \ldots, \overline{C}_k$.*

*Proof* If the jump target of $B$ is a cut inference, then $D$ is $\bigwedge_{i=1}^{k} C_i$. In this case, $E = \overline{D}$ is the formula $\overline{C}_1 \lor \cdots \lor \overline{C}_k$, and $m = 1$. It follows that $E$ is category ($\alpha$), and $*(E) = \overline{C}_1, \ldots, \overline{C}_k$, so the lemma holds. On the other hand, if the jump target is an $\land$ inference, then $D$ equals $C_m \land \cdots \land C_r$ for some $r \leq k$, and $E = C$ equals $\bigwedge_{i=j}^{m-1} C_i$ for some $1 \leq j < m$. In this case, $E$ is category ($\beta$), and $*(E)$ again equals $\overline{C}_m, \ldots, \overline{C}_k$. □

*Proof (of Theorem 3)* The cedents of $P'$ are formed by modifying each cedent $\Delta$ of $P$ to form a new cedent $\Delta^*$, called the $*$-*translation* of $\Delta$. A formula $B$ *occurs* *in or below* $\Delta$ if it is in $\Delta$ or is in some cedent below $\Delta$ in $P$. For each $\Delta$ in $P$, the cedent $\Delta^*$ is defined to include the formulas $*(B)$ for all formulas $B$ which occur in or below $\Delta$.

Theorem 3 is proved by showing that the cedents $\Delta^*$ can be put together to form a valid proof $P'$. This requires making the following modifications to $P$: (1) For any inference in $P$ that introduces an $\land$-component of a right cut formula, we must insert at that point in $P'$ a cut on that $\land$-component using (the $*$-translation of) its jump target cedent. (2) When forming $P'$, we remove from $P$ every $\land$ inference that introduces an $\land$-subformula of a right cut formula, every $\lor$ inference that introduces an $\lor$-subformula of a left cut formula, and every cut inference of $P$. (3) Weakening inferences are added as needed. These changes are described in detail below, where we describe how to combine the cedents $\Delta^*$ to form the proof $P'$. We consider separately each possible kind of inference in $P$.

For the first case, consider the case where $\Delta$ is an initial cedent $B, \overline{B}$. (Surprisingly, this is the hardest case of the proof.) Our goal is to show how the cedent $\Delta^*$ is derived in $P'$. As a first subcase, suppose neither $B$ nor $\overline{B}$ is in category ($\beta$), so neither descends to an $\land$-component of a right cut formula. Since $B$ is atomic, and $B$ and $\overline{B}$ are each in category ($\alpha$) or ($\gamma$), we have $*(B) = B$ and $*(\overline{B}) = \overline{B}$, respectively. The cedent $\Delta^*$ is equal to $B, \overline{B}, \Lambda$, where $\Lambda$ contains the formulas $*(E)$ for all formulas $E$ that occur below the cedent $B, \overline{B}$. The proof $P'$ merely derives $B, \overline{B}, \Lambda$ from $B, \overline{B}$ by a weakening inference. (Recall that weakening inferences do not count towards the size or height of proofs.)

For the second subcase, suppose exactly one of $B$ and $\overline{B}$ are in category ($\beta$). Without loss of generality, we may assume $B$ is of category ($\beta$), and $\overline{B}$ is not. The formula $B$ descends to a right cut formula $\bigwedge_{i=1}^{k} B_i$, and corresponds uniquely to one of its $\land$-components $B_\ell$. We have $1 \le \ell \le k$, and $\overline{B}_{\ell+1}, \ldots, \overline{B}_k$ are the pending implicants of $B = B_\ell$. Since $\overline{B}$ is atomic and not in category ($\beta$), $*(\overline{B}) = \overline{B} = \overline{B}_\ell$. Thus, $\Delta^*$ is equal to

$$\overline{B}_{\ell+1}, \ldots, \overline{B}_k, \overline{B}_\ell, \Lambda. \tag{2}$$

As before, the cedent $\Lambda$ is the set of *-translations of formulas that appear below $\Delta$ in $P$.

The jump target for $B$ has the form

$$\text{Cut: } \frac{\overline{D}, \Gamma_1 \qquad D, \Gamma_2}{\Gamma_1, \Gamma_2} \qquad \text{or} \qquad \land\text{: } \frac{C, \Gamma_1 \qquad D, \Gamma_2}{C \land D, \Gamma_1, \Gamma_2} \tag{3}$$

By Lemma 4, the $*$-translation of the upper left cedent has the form

$$\overline{B}_\ell, \ldots, \overline{B}_k, \Lambda' \tag{4}$$

where $\Lambda'$ contains the formulas $*(E)$ for all formulas $E$ occurring in or below the lower cedent of the inference (3). Of course, $\Lambda' \subseteq \Lambda$. Thus, in $P'$, the cedent $\Delta^*$ is derived from the cedent (4) by a weakening inference.

In the third subcase, both $B$ and $\overline{B}$ are in category ($\beta$). As in the previous subcase, $*(B)$ has the form $\overline{B}_{\ell+1}, \ldots, \overline{B}_k$, and the $*$-translation of its jump target cedent has the form

$$\overline{B}_{\ell}, \ldots, \overline{B}_k, \Lambda'$$

with $B_\ell = B$. Likewise, $*(\overline{B})$ has the form $B'_{\ell'+1}, \ldots, B'_{k'}$ and the $*$-translation of $\overline{B}$'s jump target cedent has the form

$$\overline{B}'_{\ell'}, \ldots, \overline{B}'_{k'}, \Lambda''$$

where $B'_{\ell'} = \overline{B}$. These two cedents combine with a cut on the formula $B$ to yield the inference

$$\frac{\overline{B}_{\ell}, \ldots, \overline{B}_k, \Lambda' \qquad \overline{B}'_{\ell'}, \ldots, B'_{k'}, \Lambda''}{\overline{B}_{\ell+1}, \ldots, \overline{B}_k, \overline{B}'_{\ell'+1}, \ldots, \overline{B}'_{k'}, \Lambda', \Lambda''}$$

Since $\Lambda', \Lambda'' \subseteq \Lambda$, the cedent $\Delta^*$ is derivable with one additional weakening inference. This completes the argument for the case of an initial cedent. Note that in the first two subcases, the initial cedent is eliminated, while bypassing a cut or $\wedge$ inference. In the third subcase, the initial cedent is replaced with a cut inference on an atomic formula.

For the second case of the proof of Theorem 3, consider a weakening inference

$$\frac{\Gamma}{\Gamma, \Delta}$$

in $P$. Here, the upper and lower sequents have exactly the same $*$-translations; that is, $\Gamma^*$ is the same as $(\Gamma, \Delta)^*$. Thus the weakening inference can be omitted in $P'$.

Now consider the case of an $\wedge$ inference in $P$:

$$\frac{A, \Gamma_1 \qquad B, \Gamma_2}{A \wedge B, \Gamma_1, \Gamma_2}$$

For the first subcase, suppose that $A \wedge B$ is in category ($\alpha$) or ($\gamma$), so $*(A \wedge B)$ is just $A \wedge B$. In this case, $A$ and $B$ are both in category ($\gamma$), so also $*(A) = A$ and $*(B) = B$. The $*$-translation of the $\wedge$ inference thus becomes

$$\frac{A, \Lambda, A \wedge B \qquad B, \Lambda, A \wedge B}{A \wedge B, \Lambda}$$

for suitable $\Lambda$, and this is still a valid inference. (The formula $A \wedge B$ appears in the upper cedents since the $*$-translations of the cedents $A, \Gamma_1$ and $B, \Gamma_1$ must contain $*(A \wedge B) = A \wedge B$.)

As the second subcase, suppose $A \wedge B$ is category ($\beta$), and thus $A$ and $B$ are also category ($\beta$). Expressing the formula $B$ as a conjunction of its $\wedge$-components yields $B = B_1 \wedge B_2 \wedge \cdots \wedge B_k$ for $k \geq 1$. Let the pending implicants of $A \wedge B$ be $\overline{C}_1, \ldots, \overline{C}_\ell$ with $\ell \geq 0$. The formula $B$ has the same pending implicants as $A \wedge B$.

Similarly, $*(A)$ is $\overline{B}_1, \ldots, \overline{B}_k, \overline{C}_1, \ldots, \overline{C}_\ell$. Thus the $*$-translations of the cedents in the $\wedge$ inference become

$$\frac{\overline{B}_1, \ldots, \overline{B}_k, \overline{C}_1, \ldots, \overline{C}_\ell, \Lambda \qquad \overline{C}_1, \ldots, \overline{C}_\ell, \Lambda}{\overline{C}_1, \ldots, \overline{C}_\ell, \Lambda}$$

for suitable $\Lambda$. The dashed line is used to indicate that this is no longer a valid inference. However, since the lower cedent is the same as the upper right cedent, this inference can be completely omitted in $P'$.

Next consider the case of a cut inference in $P$:

$$\frac{A, \Gamma_1 \qquad \overline{A}, \Gamma_2}{\Gamma_1, \Gamma_2}$$

Clearly, $A$ is of category ($\alpha$), and $\overline{A}$ is of category ($\beta$). Since $\overline{A}$ has no pending implicants, $*(\overline{A})$ is the empty cedent; thus, the $*$-translation of the three cedents has the form

$$\frac{*(A), \Lambda \qquad \Lambda}{\Lambda}$$

The cut inference therefore can be completely omitted in $P'$.

Now consider the case of an $\vee$ inference in $P$:

$$\frac{A, B, \Gamma}{A \vee B, \Gamma}$$

There are three subcases to consider. First, if $A \vee B$ is in category ($\gamma$), then so are $A$ and $B$. The $*$-translation of the two cedents has the form

$$\frac{A, B, \Lambda, A \vee B}{A \vee B, \Lambda} \tag{5}$$

This of course is a valid inference, and remains in this form in $P'$.

The second subcase is when $A \vee B$ is category ($\alpha$). Expressing $A$ and $B$ as disjunctions of their $\vee$-components yields $A = A_1 \vee \cdots \vee A_k$ and $B = B_1 \vee \cdots \vee B_\ell$ with $k, \ell \geq 1$. The $*$-translation of the $\vee$ inference is

$$\frac{A_1, \ldots, A_k, B_1, \ldots, B_k, \Lambda}{A_1, \ldots, A_k, B_1, \ldots, B_k, \Lambda}$$

and so this inference can be omitted in $P'$.

The third subcase is when $A \vee B$ is category ($\beta$). In this subcase, $A$ and $B$ are both category ($\gamma$). We have $*(A) = A$ and $*(B) = B$. And, $*(A \vee B)$ is $\overline{C}_1, \ldots, \overline{C}_k$, where the $\overline{C}_i$'s are the pending implicants of $A \vee B$, with $k \geq 0$. Thus, the $*$-translation of the cedents in the $\vee$ inference has the form

$$\frac{A, B, \Lambda, \overline{C}_1, \ldots, \overline{C}_k}{\overline{C}_1, \ldots, \overline{C}_k, \Lambda}$$

Of course, this is not a valid inference. Note that the formulas $\overline{C}_i$ must be included in the upper sequent since they are part of $*(A \vee B)$. From Lemma 4, the upper left sequent of the jump target of $A \vee B$ has $*$-translation of the form

$$\overline{A \vee B}, \overline{C}_1, \ldots, \overline{C}_k, \Lambda',$$

where $\Lambda' \subseteq \Lambda$. The following inferences are used in $P'$ to replace the $\vee$ inference:

$$\text{Cut: } \frac{\overline{A \vee B}, \overline{C}_1, \ldots, \overline{C}_k, \Lambda' \qquad \dfrac{A, B, \Lambda, \overline{C}_1, \ldots, \overline{C}_k}{A \vee B, \Lambda, \overline{C}_1, \ldots, \overline{C}_k}}{\overline{C}_1, \ldots, \overline{C}_k, \Lambda} \tag{6}$$

This cut is permitted in $P'$ since $A \vee B$ is an $\wedge$-component of a right cut formula in $P$. Note that the $\vee$ inference in $P$ has been replaced in $P'$ with two inferences, namely a cut and an $\vee$ inference.

Now consider the case of a $\forall$ inference in $P$

$$\frac{A(b), \Gamma}{(\forall x)A(x), \Gamma}$$

This case is handled similarly to the case of an $\vee$ inference. The formula $A(b)$ is category $(\gamma)$, so $*(A(b)) = A(b)$. If the formula $(\forall x)A(x)$ is category $(\alpha)$ or $(\gamma)$, then $*((\forall x)A(x)) = (\forall x)A(x)$. In this case, the $*$-translation of the $\forall$ inference gives

$$\frac{A(b), \Lambda, (\forall x)A(x)}{(\forall x)A(x), \Lambda}$$

for suitable $\Lambda$. This is still a valid inference, and is used as is in $P'$. Suppose, on the other hand, that $(\forall x)A(x)$ is category $(\beta)$. In this case, the $*$-translation of the $\forall$ inference has the form

$$\frac{A(b), \Lambda, \overline{C}_1, \ldots, \overline{C}_k}{\overline{C}_1, \ldots, \overline{C}_k, \Lambda}$$

where $\overline{C}_1, \ldots, \overline{C}_k$ are the pending implicants of $(\forall x)A(x)$. Note this is not a valid inference. By Lemma 4, the $*$-translation of the upper left cedent of the jump target of $(\forall x)A(x)$ is equal to

$$(\exists x)\overline{A(x)}, \overline{C}_1, \ldots, \overline{C}_k, \Lambda',$$

where $\Lambda' \subseteq \Lambda$. The following inferences are used in $P'$ to replace the $\forall$ inference:

$$\text{Cut: } \frac{(\exists x)\overline{A(x)}, \overline{C}_1, \ldots, \overline{C}_k, \Lambda' \qquad \dfrac{A(b), \Lambda, \overline{C}_1, \ldots, \overline{C}_k}{(\forall x)A(x), \Lambda, \overline{C}_1, \ldots, \overline{C}_k}}{\overline{C}_1, \ldots, \overline{C}_k, \Lambda} \tag{7}$$

Note that since $P$ is in free variable normal form, the variable $b$ does not appear in the lower cedent of the new $\forall$ inference. The $\forall$ inference in $P$ has been replaced in $P'$ with two inferences: a cut and a $\forall$ inference.

The case of an $\exists$ inference in $P$ is handled in exactly the same way as a $\forall$ inference. We omit the details.

The above completes the construction of $P'$ from $P$. By construction, the inferences in $P'$ are valid. To verify that $P'$ is globally a valid proof, we need to ensure that it is acyclic, so there is no chain of inferences that forms a cycle. This follows immediately from the fact that the inferences in $P'$ respect the postorder traversal of $P$. In particular, the upper left cedent of the jump target of a formula $B$ comes before the cedent containing $B$ in the postorder traversal of $P$. Therefore, $P'$ is well founded.

It is clear that $P'$ can be constructed in polynomial time from $P$. The size of $P'$ can be bounded as follows. First, each initial sequent in $P$ can add at most one cut inference to $P'$. Each $\wedge$ inference in $P$ can become at most one $\wedge$ inference in $P'$. Each $\vee$, $\forall$, and $\exists$ inference in $P$ can become up to two inferences in $P'$. Each cut in $P$ is replaced, at least locally, by zero inferences in $P'$. Let $n_{\mathrm{Ax}}$, $n_{\mathrm{Cut}}$, $n_\wedge$, $n_\vee$, $n_\forall$, and $n_\exists$ denote the numbers of initial sequents, cuts, $\wedge$, $\vee$, $\forall$, and $\exists$ inferences in $P$. Then $|P|$ equals $n_{\mathrm{Cut}} + n_\wedge + n_\vee + n_\forall + n_\exists$, and $|P'|$ is bounded by $n_{\mathrm{Ax}} + n_\wedge + 2(n_\vee + n_\forall + n_\forall + n_\exists)$. Since w.l.o.g. there is at least one cut in $P$ and since $n_{\mathrm{Ax}} = n_{\mathrm{Cut}} + n_\wedge + 1$, it follows that $|P'| \leq 2 \cdot |P|$. Q.E.D. Theorem 3. □

## 4 Eliminating Like Quantifiers

We next show how to eliminate the outermost block of quantifiers from cut formulas.

**Definition** An $\exists$-*subformula* (resp., $\forall$-*subformula*) of $A$ is a subformula that is contained in the scope of only $\exists$ (resp., $\forall$) quantifiers. An $\exists$-*component* (resp., $\forall$-*component*) of $A$ is a minimal $\exists$ or $\forall$-subformula (respectively). A $\forall/\exists$-*component of a cut formula in* $P$ is a $\forall$-component of a left cut formula in $P$ or an $\exists$-component of a right cut formula in $P$.

**Theorem 5** *Let $P$ be a tree-like Tait calculus proof of $\Gamma$. Then there is a dag-like proof $P'$, also of $\Gamma$, such that each cut formula of $P'$ is a $\forall/\exists$-component of a cut formula of $P$, and such that $|P'| \leq 4^{|P|/5} \leq (1.32)^{|P|}$ and $h(P') \leq |P|$. As a consequence of the height bound, $P'$ can also be expressed as a tree-like proof of size $\leq 2^{|P|}$. Similarly, $h(P') \leq 2^{h(P)}$.*

Without loss of generality, $P$ is in free variable normal form and satisfies the auxiliary condition. Each formula $B$ in $P$ can be put in one of the following categories ($\alpha$)–($\gamma$):

($\alpha$) $B$ has a left cut formula $A$ as a descendant and corresponds to a $\forall$-subformula of $A$, or

($\beta$) $B$ has a right cut formula $A$ as a descendant and corresponds to an $\exists$-subformula of $A$, or

($\gamma$) Neither ($\alpha$) nor ($\beta$) holds.

**Definition** An $\exists$ inference as shown in Fig. 1 is *critical* if the auxiliary formula $A(s)$ does not have an $\exists$ as its outermost connective. The formula $A(s)$ is also referred to as $\exists$-*critical*. If $A(s)$ is furthermore of category ($\beta$), then the $\exists$-*jump target* of $A(s)$ is the cut inference which has a descendant of $A(s)$ as a (right) cut formula. The $\exists$-*jump target cedent* of $A(s)$ is the upper left cedent of the jump target of $A(s)$. This is also referred to as the $\exists$-jump target cedent of the cedent $\Delta$ containing $A(s)$.

We now come to the crucial new definition for handling cut elimination of outermost like quantifiers. The intuition is that we want to trace, through the proof $P$, a possible branch in the proof $P'$. Along with this traced out path, we also need to keep a partial substitution assigning terms to variables: this substitution will track the needed term substitution for forming the corresponding cedent in $P'$. First we define an "$\exists$-path" and then we define the associated substitution.

**Definition** A cut inference is called *to-be-eliminated* if the outermost connective of the cut formula is a quantifier. An $\exists$-*path* $\pi$ through $P$ consists of a sequence of cedents $\Delta_1, \Delta_2, \ldots, \Delta_m$ from $P$ such that $\Delta_1$ is the endsequent of $P$ and such that for each $i < m$, one of the following holds:

- $\Delta_i$ is the lower cedent of a to-be-eliminated cut inference, and $\Delta_{i+1}$ is its right upper cedent, or
- $\Delta_i$ is the lower cedent of an inference other than a to-be-eliminated cut, and $\Delta_{i+1}$ is an upper cedent of the same inference, or
- $\Delta_i$ is the upper cedent of an $\exists$-critical inference, and $\Delta_{i+1}$ is the $\exists$-jump target cedent of $\Delta_i$.

The $\exists$-path is said to *lead to* $\Delta_m$.

It is easy to verify that, for $\Delta_i$ in $\pi$, the $\exists$-path $\pi$ contains every cedent in $P$ below $\Delta_i$.

The cedents in an $\exists$-path are in reverse postorder from $P$. The effect of an $\exists$-path is to repeatedly traverse up to an $\exists$-critical inference—always going rightward at to-be-eliminated cuts—and then jump back down to the associated $\exists$-jump target cedent. The most important information needed to specify the $\exists$-path is the subsequence of cedents $\Delta_{i_1}, \Delta_{i_2}, \ldots, \Delta_{i_k}$, $i_1 < i_2 < \cdots < i_k$ which are $\exists$-critical and for which $\Delta_{i_\ell+1}$ is the $\exists$-jump target cedent of $\Delta_{i_\ell}$. The entire $\exists$-path can be uniquely reconstructed from this subsequence plus knowledge of the last cedent $\Delta_m$ in $\pi$.

There is a substitution $\sigma_\pi$ associated with the $\exists$-path $\pi = \langle\Delta_1, \ldots, \Delta_m\rangle$. The domain of $\sigma_\pi$ is the set of free variables appearing in or below $\Delta_m$ *plus* the set of outermost universally quantified variables occurring in the category ($\alpha$) formulas in $\Delta_m$.

**Definition** The definition of $\sigma_\pi$ is by induction on the length of $\pi$. First, let $(\forall x_i)\cdots(\forall x_\ell)A$ be a formula in $\Delta_m$ in category $(\alpha)$ with $i \le \ell$ such that $A$ does not have outermost connective $\forall$. Since it is in category $(\alpha)$, this formula has the form $(\forall x_i)\cdots(\forall x_\ell)A(b_1,\ldots,b_{i-1},x_i,\ldots,x_\ell)$, and has a descendant of the form $(\forall x_1)\cdots(\forall x_\ell)A(x_1,\ldots,x_\ell)$ which is the left cut formula of a cut inference. Since the cut is to-be-eliminated, $\pi$ must reach the upper left cedent by way of a "jump" from an $\exists$-critical cedent $\Delta_i \in \pi$. As pictured, the associated $\exists$-critical formula must have the form $\overline{A(s_1,\ldots,s_\ell)}$:

$$
\dfrac{\begin{array}{c}(\forall x_i)\cdots(\forall x_\ell)A(b_1,\ldots,b_{i-1},x_i,\ldots,x_\ell),\Gamma\\ \vdots\\ (\forall x_1)\cdots(\forall x_\ell)A(x_1,\ldots,x_\ell),\Gamma_1\end{array}\qquad \begin{array}{c}\dfrac{\overline{A(s_1,\ldots,s_\ell)},\Gamma'}{(\exists x_\ell)\overline{A(s_1,\ldots,s_{\ell-1},x_\ell)},\Gamma'}\\ \vdots\\ (\exists x_1)\cdots(\exists x_\ell)\overline{A(x_1,\ldots,x_\ell)},\Gamma_2\end{array}}{\Gamma_1,\Gamma_2}
$$

Note that the terms $s_1,\ldots,s_\ell$ are uniquely determined by $\pi$, since they are found by following the path from the upper right cedent of the cut inference to the cedent $\Delta_i$, and setting the $s_i$'s to be the terms used for $\exists$ inferences acting on the descendants of $A(\vec{s})$.

Let $\pi'$ be $\pi$ truncated to end at $\overline{A(\vec{s})},\Gamma$. The substitution $\sigma_\pi$ is defined to map the bound variables $x_i,\ldots,x_\ell$ to the terms $s_i\sigma_{\pi'},\ldots,s_\ell\sigma_{\pi'}$. (Strictly speaking, the substitution $\sigma_\pi$ acts on the occurrences of variables, since the same variable may be used in multiple quantifiers and in different formulas; this is suppressed in the notation, however.)

For $b$ a free variable appearing in or below $\Delta_m$, the value $\sigma_\pi(b)$ is defined as follows. If there is a $\forall$ inference, below $\Delta_m$,

$$\frac{A(b),\Gamma}{(\forall x)A(x),\Gamma}$$

that uses $b$ as an eigenvariable, and if $(\forall x)A(x)$ is category $(\alpha)$, then define $\sigma_\pi(b)$ to equal the value of $\sigma_{\pi'}(a)$, where $\pi'$ is $\pi$ truncated to end at the lower cedent of the $\forall$ inference. For example, in the proof displayed above, $\sigma_\pi(b_i) = s_i$. Otherwise, if there is no such $\forall$ inference, define $\sigma_\pi(b) = b$.

**Definition** Let $A$ be a formula appearing in a cedent $\Delta$ of $P$. Let $\pi$ be an $\exists$-path leading to $\Delta$. Then $*_\pi(A)$ is defined as follows:

- If $A$ is in category $(\alpha)$ and has the form $A = (\forall x_1)\cdots(\forall x_\ell)B$ with $\ell > 0$ and $B$ not starting with a $\forall$ quantifier, then define $*_\pi(A)$ to be the formula $B\sigma_\pi$, namely the formula obtained by replacing each $x_i$ with $\sigma_\pi(x_i)$ and each free variable $b$ with $\sigma_\pi(b)$.
- If $A$ is in category $(\beta)$ and has outermost connective $\exists$, then $*_\pi(A)$ is the empty cedent.
- Otherwise $*_\pi(A)$ is the formula $A\sigma_\pi$, namely obtained by replacing each free variable $b$ with $\sigma_\pi(b)$.

For $A$ appearing below $\Delta$, we define $*_\pi(A)$ to equal $*_{\pi'}(A)$ where $\pi'$ is $\pi$ truncated to end at the cedent $\Delta'$ containing $A$. The $*_\pi$*-translation*, $*_\pi(\Delta)$, of $\Delta$ is the cedent containing exactly the formulas $*_\pi(A)$ for $A$ appearing in or below $\Delta$ in $P$.

We can now give the proof of Theorem 5. The proof $P'$ will be formed from the cedents $*_\pi(\Delta)$ where $\Delta$ ranges over the cedents of $P$, and $\pi$ ranges over the $\exists$-paths leading to $\Delta$. The inferences in $P'$ will respect the postordering of $P$, and $P'$ will be a dag.

As before, we must show how to connect up the cedents $*_\pi(\Delta)$ to make $P'$ into a valid proof. The argument again splits into cases based on the type of inference used to infer $\Delta$ in $P$. The cases of initial cedents, $\vee$ inferences, $\wedge$ inferences, and weakenings are all immediate. These inferences remain valid after their cedents are replaced by their $*_\pi$-translations, since initial cedents contain only atomic formulas, and since the $*_\pi$-translations respect propositional connectives.

Consider the case where $\Delta$ is inferred by a $\forall$ inference in $P$:

$$\frac{A(b), \Gamma}{(\forall x)A(x), \Gamma}$$

The $\exists$-path $\pi$ ends at the lower cedent $\Delta$. Define $\pi'$ to be the $\exists$-path that extends $\pi$ by one step to the upper cedent $\Delta'$. If $(\forall x)A(x)$ and $A(b)$ are not in category $(\alpha)$, then $\sigma_{\pi'}(b) = b$ and the inference is still valid since the $*_{\pi'}$-/$*_\pi$-translations of $A(b)$ and $(\forall x)A(x)$ are equal to $C(b)$ and $(\forall x)C(x)$ for $C$ defined by $C(b) = A(b)\sigma_\pi = A(b)\sigma_{\pi'}$. Thus, in this case, the result is still a valid $\forall$ inference. Otherwise, $A(b)$ and $(\forall x)A(x)$ are both in category $(\alpha)$. In this case, $*_\pi(A(b)) = *_\pi((\forall x)A(x))$; the $\forall$ inference has equal upper and lower cedents and is just omitted from $P'$.

Now consider the case where $\Delta$ is inferred in $P$ with an $\exists$ inference:

$$\frac{A(s), \Gamma}{(\exists x)A(x), \Gamma}$$

Define $\pi'$ as in the previous case. If $A(s)$ and $(\exists x)A(x)$ are not in category $(\beta)$, then the $*_\pi$-translation leaves the quantifier on $x$ untouched, and the $*_{\pi'}$-/$*_\pi$-translation of the inference is still a valid inference in $P'$. Otherwise, both formulas are in category $(\beta)$. If $A(s)$ has an $\exists$ as its outmost connective, then $*_{\pi'}(A(s))$ and $*_\pi((\exists x)A(x))$ are both empty, and the $*_{\pi'}$- and $*_\pi$-translations (respectively) of the upper and lower cedents are identical, and the $\exists$ inference can be omitted in $P'$. If $A$ does not have an $\exists$ as its outermost connective, then the $*_{\pi'}$-/$*_\pi$-translations of the cedents in the inference are

$$\frac{*_{\pi'}(A(s)), \Lambda}{\Lambda}$$

where $\Lambda$ contains the formulas $*_\pi(B)$ for all formulas $B$, other than $A(s)$, which occur in or below $\Delta$ in $P$. The upper left cedent of the $\exists$-jump target of $A(s)$ has the form

$$\Gamma_1, (\forall x_1)\cdots(\forall x_\ell)\overline{A(x_1, \ldots, x_\ell)},$$

where $x = x_\ell$ and $A(s) = A(s_1, \ldots, s_\ell)$ with $s$ corresponding to the term $s_\ell$. Let $\pi''$ be the $\exists$-path that extends $\pi'$ by the addition of this upper left cedent. The $*_{\pi''}$-translation of the upper left cedent has the form

$$\Lambda_1, \overline{A(s_1, \ldots, s_\ell)\sigma_{\pi''}}.$$

Here $\Lambda_1 \subseteq \Lambda$, and $\overline{A(s_1, \ldots, s_\ell)\sigma_{\pi''}}$ is the same as $\overline{*_{\pi'}(A(s))}$. Hence, a cut inference gives

$$\frac{\Lambda_1, \overline{A(s_1, \ldots, s_\ell)\sigma_{\pi''}} \qquad *_{\pi'}(A(s)), \Lambda}{\Lambda}$$

The $\exists$ inference in $P$ is thus replaced with a cut inference in $P'$, but on a formula of lower complexity than the cut in $P$.

Finally consider the case of a cut inference in $P$ as shown in Fig. 1 with left cut formula $A$ and right cut formula $\overline{A}$. First suppose it is not a to-be-eliminated cut. Let $\pi_1$ and $\pi_2$ be the $\exists$-paths which extend $\pi$ by one step to include the upper left or right cedent of the cut, respectively. Then $*_{\pi_1}(A)$ and $*_{\pi_2}(\overline{A})$ are complements of each other, and the cut remains valid in $P'$. Otherwise, the cut is to-be-eliminated, and $\pi_2$ is again a valid $\exists$-path. The right cut formula $\overline{A}$ is category $(\beta)$ and has outermost connective $\exists$. Thus $*_{\pi_2}(\overline{A})$ is the empty cedent, so the $*_{\pi_2}$-translation of the right upper cedent and the $*_\pi$-translation of the lower cedent are identical. In this case, the cut can be removed completely from $P'$.

The above completes the construction of $P'$. The next lemma will be used to bound its size.

**Lemma 6** *Let $\Delta$ be a cedent in $P$. The number of $\exists$-paths $\pi$ to $\Delta$ in $P$ is $\leq (1.32)^{|P|}$.*

*Proof* Recall that an $\exists$-path $\pi$ to $\Delta$ can be uniquely characterized by its final cedent $\Delta_m = \Delta$ and its subsequence $\Delta_{i_1}, \ldots, \Delta_{i_k}$ of cedents which are $\exists$-critical and have $\Delta_{i_\ell+1}$ the $\exists$-jump target cedent of $\Delta_{i_\ell}$. We will bound the number $N$ of ways to select the $\exists$-critical cedents in this subsequence. For this, we group the $\exists$-critical cedents of $P$ according to their $\exists$-jump target. Let there be $m$ many to-be-eliminated cut inferences in $P$, and suppose that the $i$th such cut has $n_i$ many $\exists$-critical cedents associated with it. The $i$th cut also has at least one $\forall$ inference associated with it that introduces a $\forall$ quantifier in its left cut formula. Therefore $|P| \geq \sum_{i=1}^m (n_i + 2)$. Each $\exists$-path $\pi$ can jump from at most one of the $n_i$ $\exists$-critical cedents associated with the $i$th cut. It follows that there are at most $\prod_{i=1}^m (n_i + 1)$ many $\exists$-paths; namely, there are at most $n_i + 1$ choices for which one, if any, of $i$th cut's associated $\exists$-critical cedents are included in $\pi$.

To upper bound the value $N = \prod_{i=1}^m (n_i + 1)$, take the logarithm, and upper bound $\sum_{i=1}^m \ln(n_i + 1)$ subject to $\sum_{i=1}^m (n_i + 2) \leq |P|$. For integer values of $x$, $(\ln x)/(x + 1)$ is maximized at $x = 4$. Thus, $\ln N \leq |P| \cdot (\ln 4)/5$; that is, $N \leq |P| \cdot 4^{|P|/5} \leq (1.32)^{|P|}$. □

The size bound of Theorem 5 follows immediately from the lemma. Namely, $P'$ contains at most one cedent for each path to each cedent $\Delta$ in $P$, and thus $|P'| \leq |P| \cdot (1.32)^{|P|}$. The height bound $h(P') \leq |P|$ follows from the construction of $P$, since paths $\pi$ traverse cedents of $P$ in reverse postorder, and each $\wedge$, $\vee$, $\exists$, $\forall$, and cut inference along $\pi$ contributes at most inference to $P'$. (Note that cuts contribute an inference only when used as a jump target.) Q.E.D. Theorem 5

The proof $P'$ was constructed in a highly uniform way from $P$. Indeed, $P'$ can be generated with a polynomial time algorithm $f$ that operates as follows: $f$ takes as input a string $w$ of length $\leq |P|$ many bits, and outputs whether the string $w$ is an index for a cedent $\Delta_w$ in $P'$, and if so, $f$ also outputs: (a) the cedent $\Delta_w$ with terms specified as dags, and (b) what kind of inference is used to derive $\Delta_w$, and (c) the index $w'$ or indices $w', w''$ of the cedent(s) from which $\Delta_w$ is inferred in $P'$. For (a), note that the cedent $\Delta_w$ can be written out in polynomial length only if terms are written as dags (that is, circuits) rather than as trees (that is, as formulas). This is because the iterated application of substitutions may cause the terms $\sigma_\pi(b)$ to be exponentially big when written out as formulas instead of as circuits. Also note that, although some inferences in $P'$ become trivial and are omitted in $P'$, we can avoid using REP inferences in $P'$ by the simple convention that indices $w$ that would lead to REP inferences are taken to not be valid indices. (An example of this would be a $w$ encoding an $\exists$-path leading to a to-be-eliminated cut.)

This means of course that there is a polynomial space algorithm that lists out the proof $P'$.

## 5 Eliminating and/Exists and or/Forall Blocks

This section gives an algorithm for eliminating outermost blocks of $\vee/\forall$ (equivalently, $\wedge/\exists$) connectives from cut formulas, where the $\vee$ and $\forall$ (resp., $\wedge$ and $\exists$) connectives can be arbitrarily interspersed.

**Definition** A subformula $B$ of $A$ is an $\vee\forall$-*subformula* of $A$ if $B$ is in the scope of only $\vee$ and $\forall$ connectives. The $\vee\forall$-*components* of $A$ are the minimal $\vee\forall$-subformulas of $A$. The $\wedge\exists$-*subformulas* and $\wedge\exists$-*components* of $A$ are defined similarly.

An $\vee\forall/\wedge\exists$-*component* of a cut formula in $P$ is either an $\vee\forall$-component of a left cut formula of $P$ or an $\wedge\exists$-component of a right cut formula of $P$.

**Theorem 7** *Let $P$ be a tree-like Tait calculus proof of $\Gamma$. Then there is a dag-like proof $P'$, also of $\Gamma$, such that each cut formula of $P'$ is an $\wedge\exists/\vee\forall$-component of a non-atomic cut formula of $P$, and such that $|P'| \leq 4^{|P|/5} \leq (1.32)^{|P|}$ and $h(P') \leq |P|$. Consequently, $P'$ can also be expressed as a tree-like proof of size $\leq 2^{|P|}$.*

Note that all cuts in $P$ are simplified in $P'$. The atomic cuts in $P$ are eliminated when forming $P'$. However, new cuts are added on $\wedge\exists/\vee\forall$-components of cuts in $P$, and some of these might be cuts on atomic formulas. If all cut formulas in $P$ are atomic, then $P'$ is cut free.

W.l.o.g., $P$ is in free variable normal form and satisfies the auxiliary condition. Each formula $B$ in $P$ can be put in one of the following categories ($\alpha$)–($\gamma$):

($\alpha$) $B$ has a left cut formula $A$ as a descendant and corresponds to an $\vee\forall$-subformula of $A$, or
($\beta$) $B$ has a right cut formula $A$ as a descendant and corresponds to an $\wedge\exists$-subformula of $A$, or
($\gamma$) Neither ($\alpha$) nor ($\beta$) holds.

**Definition** The *jump target* of a category ($\beta$) formula $B$ occurring in $P$ is the first cut or $\wedge$ inference below the cedent containing $B$ that has some descendant of $B$ as the auxiliary formula $D$ in its right upper cedent. The jump target will again be of the form (1). Its right auxiliary formula $D$ has a unique subformula $B'$ which corresponds to $B$. $B'$ occurs only in the scope of $\exists$ connectives and $\wedge$ connectives, and only in the first argument of $\wedge$ connectives. (The last part holds since otherwise the jump target would be an $\wedge$ inference higher in the proof.) The *jump target cedent* is defined as before.

Suppose a category ($\beta$) formula $B$ has descendant $D$ as the right auxiliary formula of its jump target. Let the $\wedge\exists$-components of $D$ be $D_m, \ldots, D_k$ in left-to-right order. The $\wedge\exists$-components of $B$ in left-to-right order can be listed as $B_m, \ldots, B_\ell$, with each $B_i$ corresponding to $D_i$, with $1 \leq m \leq \ell \leq k$. The formulas $\overline{D}_{\ell+1}, \ldots, \overline{D}_k$ are the *pending implicants* of $B$. The *pending quantifiers* of $B$ are the quantifiers $(\exists x)$ which appear to the right of the subformula $D_\ell$ in $D$ and are outermost connectives of $\wedge\exists$-subformulas of $D$. Let $B'$ be the subformula of $D$ that corresponds to $B$; the *current quantifiers* of $B$ are the quantifiers $(\exists x)$ in $D$ which contain $B'$ in their scope.

The pending implicants of $B$ will be used similarly as in the proof of Theorem 3, but first we need to define $\wedge\exists$-paths and substitutions $\sigma_\pi$ similarly to the proof of Theorem 5. Now, $\sigma_\pi$ must also map the pending quantifier variables to terms

**Definition** An upper cedent $\Delta$ of an $\wedge$ or $\exists$ inference is *critical* if the auxiliary formula in $\Delta$ is either atomic or has outermost connective $\vee$ or $\forall$.

**Definition** A cut inference in $P$ is *non-atomic* if its cut formulas are not atomic. An *$\wedge\exists$-path* $\pi$ through $P$ consists of a sequence $\Delta_1, \ldots, \Delta_m$ of cedents from $P$ such that $\Delta_1$ is the end cedent of $P$ and such that, for each $i < m$, one of the following holds:

- $\Delta_i$ is the lower cedent of non-atomic cut inference, and $\Delta_{i+1}$ is its right upper cedent, or
- $\Delta_i$ is the lower cedent of an inference other than a non-atomic cut, and $\Delta_{i+1}$ is an upper cedent of the same inference, or
- $\Delta_i$ is a critical upper cedent of an $\wedge$ or $\exists$ inference with auxiliary formula $A$, and $\Delta_{i+1}$ is the jump target cedent of $A$.

The next definition of $\sigma_\pi$ is more difficult than in the proof of Theorem 5 because the substitution has to act also on the pending implicants of category ($\beta$).

**Definition** Let $\pi$ be $\wedge\exists$-path as above. The domain of the substitution $\sigma_\pi$ is: the free variables appearing in or below $\Delta_m$, the variables of the $\vee\forall$-outermost quantifiers of each category ($\alpha$) formula in $\Delta_m$, and the variables of the pending quantifiers of each category ($\beta$) formula in $\Delta_m$.[3] The definition of $\sigma_\pi$ is defined by induction on the length of $\pi$. For $\pi$ containing just the end cedent, $\sigma_\pi$ is the identity mapping with domain the parameter variables of $P$. Otherwise, let $\pi'$ be the initial part of $\pi$ up through the next-to-last cedent $\Delta_{m-1}$ of $\pi$, and suppose $\sigma_{\pi'}$ is already defined. There are several cases to consider.

(a) Suppose $\Delta_{m-1}$ and $\Delta_m$ are the lower cedent and an upper cedent of some inference other than a $\forall$ inference. The $\sigma_\pi$ is same as $\sigma_{\pi'}$.
(b) Suppose $\Delta_{m-1}$ and $\Delta_m$ are the lower cedent and an upper cedent of a $\forall$ inference as shown in Fig. 1. If the principal formula $(\forall x)A(x)$ is category ($\alpha$), then $\sigma_\pi$ extends $\sigma_{\pi'}$ by letting $\sigma_\pi(b) = \sigma_{\pi'}(x)$ where $(\forall x)$ is the quantifier introduced by the $\forall$ inference. Otherwise, $\sigma_\pi(b) = b$. And, $\sigma_\pi$ is equal to $\sigma_{\pi'}$ for all other variables in its domain.
(c) Otherwise, $\Delta_m$ is the jump target cedent of $\Delta_{m-1}$. Suppose the jump target is an $\wedge$ inference

$$\frac{C, \Gamma_1 \qquad D, \Gamma_2}{C \wedge D, \Gamma_1, \Gamma_2}$$

For $b$ a free variable in $C, \Gamma_1$, the value $\sigma_\pi(b)$ is defined to equal $\sigma_{\pi'}(b)$. Similarly, for any pending quantifier $(\exists x)$ of any category ($\beta$) formula in $\Gamma_1$ and for any $\vee\forall$-outermost quantifier $(\forall x)$ of any category ($\alpha$) formula in $\Gamma_1$, set $\sigma_\pi(x) = \sigma_{\pi'}(x)$.

We also must define the action of $\sigma_\pi$ on the pending quantifiers of the category ($\beta$) formula $C$. Let $D_1$ be the first (leftmost) $\wedge\exists$-component of $D$. The cedent $\Delta_{m-1}$ has the form $B_1, \Gamma_3$ where $B_1$ is an ancestor of $D$ and corresponds to $D_1$. Write $D_1 = D_1(x_1, \ldots, x_j)$ where $(\exists x_1), \ldots, (\exists x_j)$ are the current quantifiers for $D_1$. Then $B_1 = B_1(s_1, \ldots, s_j)$ where the $s_i$'s are the terms used for $\exists$ inferences acting on descendants of $B_1$. The $(\exists x_i)$'s are pending quantifiers of $C$, and $\sigma_\pi(x_i)$ is defined to equal $s_i\sigma_{\pi'}$. The rest of the pending quantifiers of $C$ are the pending quantifiers of $B_1$ in the cedent $\Delta_{m-1}$: for these variables, $\sigma_\pi$ is defined to equal the value of $\sigma_{\pi'}$.
(d) Suppose that $\Delta_m$ is the left upper cedent of the jump target of $\Delta_{m-1}$, and the jump target is a cut inference

$$\frac{\overline{D}, \Gamma_1 \qquad D, \Gamma_2}{\Gamma_1, \Gamma_2}$$

[3] As before, strictly speaking, a variable might be quantified at multiple places, and $\sigma$ acts on variables according to how they are bound by a quantifier, but we suppress this in the notation.

Let $\pi'$ be as before, and set $\sigma_\pi(b) = \sigma_{\pi'}(b)$ for all free variables of the lower cedent. For any pending quantifier $(\exists x)$ of any category $(\beta)$ formula in $\Gamma_1$ and for any $\vee\forall$-outermost quantifier $(\forall x)$ of any category $(\alpha)$ formula in $\Gamma_1$, set $\sigma_\pi(x) = \sigma_{\pi'}(x)$. Now, let $D_1 = D_1(x_1, \ldots, x_j)$ and $B_1 = B_1(s_1, \ldots, s_j)$ as in the previous case. Consider any $\vee\forall$-outermost quantifier $(\forall y)$ of $\overline{D}$. If $y$ is one of the $x_i$'s, define $\sigma_\pi(y) = s_i\sigma_{\pi'}$. Otherwise, $(\exists y)$ is a pending quantifier of $D_1$, and a pending quantifier of $B_1$ in $\Delta_{m-1}$, and we define $\sigma_\pi(y) = \sigma_{\pi'}(y)$.

**Definition** Suppose $A$ is a formula occurring in cedent $\Delta$ in $P$, and $\pi$ is an $\wedge\exists$-path leading to $\Delta$. The formula $*_\pi(A)$ is defined as follows:

- If $A$ is category $(\beta)$, then $*_\pi(A)$ is the cedent containing the formulas $B\sigma_\pi$ for each pending implicant $B$ of $A$.
- If $A$ is category $(\alpha)$, then $*_\pi(A)$ is the cedent containing $B\sigma_\pi$ for each $\vee\forall$-component $B$ of $A$.
- Otherwise $*_\pi(A)$ is $A\sigma_\pi$.

The notation $*_\pi(A)$ is extended to apply also to $A$ appearing in a cedent $\Delta'$ below the cedent $\Delta$. Let $\pi'$ be the initial subsequence of $\pi$ leading to $\Delta'$. Then define $*_\pi(A) = *_{\pi'}(A)$. The $*_\pi$-translation of $\Delta$ consists of the formulas $*_\pi(A)$ such that $A$ appears in or below $\Delta$ in $P$.

The next lemma is analogous to Lemma 4.

**Lemma 8** *Suppose $B$ is a category $(\beta)$ formula in a cedent $\Delta$ in $P$, and let $\pi$ be an $\wedge\exists$-path to $\Delta$. Also suppose $B$ does not have outermost connective $\wedge$ or $\exists$. Let $\overline{C}_1, \ldots, \overline{C}_m$ be the pending implicants of $B$. Let $\Delta'$ be $B$'s jump target cedent, and $E$ be the auxiliary formula in $\Delta'$. Then there is an $\wedge\exists$-path $\pi'$ to $\Delta'$ such that $*_{\pi'}(E)$ equals the cedent $\overline{B}\sigma_\pi, \overline{C}_1\sigma_\pi, \ldots, \overline{C}_m\sigma_\pi$.*

*Proof* The jump target of $B$ is either a cut or an $\wedge$ inference as shown in (1), with $B$ corresponding to the first $\wedge\exists$-component $C_0$ of $D$. The remaining $\wedge\exists$-components of $D$ are $C_1, \ldots, C_r$ where $0 \le r \le m$. Of course, their negations are (some of the) pending implicants of $B$.

Suppose the jump target is a non-atomic cut inference. Then we have $r = m$. Since $B$ does not have outermost connective $\wedge$ or $\exists$ and since the cut formula $D$ is non-atomic, $B$ is not the same as $D$. Consider the lowest direct descendant of $B$; it appears in a cedent $\Delta''$, and is the auxiliary formula of an $\exists$ inference, or the left auxiliary formula of an $\wedge$ inference. In either case, $\Delta''$ is critical. Let $\pi''$ be the $\wedge\exists$-path consisting of the initial part of $\pi$ to $\Delta''$. Set $\pi'$ to be the $\wedge\exists$-path that follows $\pi''$ and then jumps from $\Delta''$ to the upper left cedent $\Delta'$ of the jump target. The left cut formula $E$ is equal to $\overline{D}$, and the $\vee\forall$-components of $E$ are $\overline{C}_0, \ldots, \overline{C}_m$. The cedent $*_{\pi'}(E)$ consists of the formulas $\overline{C}_i\sigma_{\pi'}$. For $i = 0$, $\sigma_{\pi'}$ was defined so that $C_0\sigma_{\pi'} = B\sigma_\pi$. Likewise, for $i > 0$, we have $C_i\sigma_{\pi'} = C_i\sigma_{\pi''}$. Also, by cases (a) and (b) of the definition of $\sigma_\pi$, we have $C_i\sigma_{\pi''} = C_i\sigma_\pi$. Thus the lemma holds.

Second, suppose the jump target is a cut on an atomic formula. The right cut formula is equal to $B$ of course; the left cut formula $E$ is equal to $\overline{B}$. Letting $\pi'$ be as above, $*_{\pi'}(E)$ is equal to $\overline{B}\sigma_{\pi'} = \overline{B}\sigma_\pi$ as desired.

Now suppose the jump target is an $\wedge$ inference, as in (1), where $E = C$. If $D$ is atomic, then $D$ is a direct descendant of $B$ (possibly even the same occurrence as $B$). In this case, let $\Delta''$ be the cedent containing $D$ (the upper right cedent of the $\wedge$ inference), let $\pi''$ be the initial part of $\pi'$ leading to $\Delta''$, and let $\pi'$ be $\pi''$ plus the upper left cedent $\Delta'$. (Note that $\Delta'$ is the jump target cedent of $D$.) Then, the pending implicants of $C$ in $\Delta'$ are $\overline{D} = \overline{B}$ and $\overline{C}_1, \ldots, \overline{C}_m$. We have $D\sigma_{\pi'} = D\sigma_{\pi''} = D\sigma_\pi$ and also $C_i\sigma_{\pi'} = C_i\sigma_{\pi''} = C_i\sigma_\pi$, so the lemma holds. Now suppose $D$ is not atomic. Then define $\pi'$, $\pi''$, and $\Delta''$ exactly as in the case above where jump target of $B$ was a cut inference. The pending implicants of $C$ are $\overline{C}_0, \ldots, \overline{C}_m$, and, as before, we have $C_0\sigma_{\pi'} = \overline{B}\sigma_{\pi''} = B\sigma_\pi$ and $C_i\sigma_{\pi'} = C_i\sigma_{\pi''} = C_i\sigma_\pi$, satisfying the conditions of the lemma. □

*Proof (of Theorem 7)* The proof combines the constructions from the proofs of the two previous theorems. For each cedent $\Delta$ in $P$ and each $\wedge\exists$-path leading to $\Delta$, form the cedent $\Delta^\pi$ as the $*_\pi$-translation of $\Delta$. Our goal is to show that these cedents can be combined to form a valid proof $P'$. The proof splits into cases to handle the different kinds of inferences in $P$ separately. In each case, we have a cedent $\Delta$ and an $\wedge\exists$-path $\pi$ leading to $\Delta$, and need to show how $\Delta^\pi$ is derived in $P'$.

For the first case, consider an initial cedent $\Delta$ of the form $B, \overline{B}$ in $P$. As the first subcase, suppose neither $B$ nor $\overline{B}$ is category $(\beta)$. Then $\Delta^\pi$ is the cedent $B\sigma_\pi, \overline{B}\sigma_\pi, \Lambda$ where $\Lambda$ is the cedent of formulas $*_\pi(E)$ for $E$ a formula appearing below $\Delta$ in $P$. This is obtained in $P'$ by applying a weakening to the initial cedent $B\sigma_\pi, \overline{B}\sigma_\pi$.

For the second subcase, suppose $B$ is category $(\beta)$ and $\overline{B}$ is not. The formula $B$ has a right cut formula as descendant, and corresponds to the $\ell$th $\wedge\exists$-component $D_\ell$ of $D$. Let the pending implicants of $B$ be $\overline{D}_{\ell+1}, \ldots, \overline{D}_k$. By Lemma 8, there is an $\wedge\exists$-path $\pi'$ to the upper left cedent $\Delta'$ of the jump target such that the auxiliary formula $E$ in $\Delta'$ has $*_{\pi'}(E)$ equal to $\overline{B}\sigma_\pi, \overline{D}_\ell\sigma_\pi, \ldots, \overline{D}_k\sigma_\pi$. Thus, $\Delta^\pi$ and $(\Delta')^{\pi'}$ are

$$\overline{B}\sigma_\pi, \overline{D}_{\ell+1}\sigma_\pi, \ldots, \overline{D}_k\sigma_\pi, \Lambda$$

and

$$\overline{B}\sigma_\pi, \overline{D}_{\ell+1}\sigma_\pi, \ldots, \overline{D}_k\sigma_\pi, \Lambda'$$

where $\Lambda' \subseteq \Lambda$. In $P'$, the first cedent is derived from the second by a weakening inference.

In the third subcase, both $B$ and $\overline{B}$ are category $(\beta)$. We have $*_\pi(B)$ still equal to $\overline{D}_{\ell+1}\sigma_\pi, \ldots, \overline{D}_k\sigma_\pi$, and now $*_\pi(\overline{B})$ is equal to $\overline{D}''_{\ell''+1}\sigma_\pi, \ldots, \overline{D}''_{k''}\sigma_\pi$ with the

$\overline{D}''_i$'s the $k''$ pending implicants of $\overline{B}$. Using Lemma 8 twice, we have $\wedge\exists$-paths $\pi'$ and $\pi''$ leading to cedents $\Delta'$ and $\Delta''$ such that $(\Delta')^{\pi'}$ and $(\Delta'')^{\pi''}$ (respectively) are

$$\overline{B}\sigma_\pi, \overline{D}_{\ell+1}, \ldots, \overline{D}_k\sigma_\pi, \Lambda'$$

and

$$B\sigma_\pi, \overline{D}''_{\ell''+1}, \ldots, \overline{D}''_{k''}\sigma_\pi, \Lambda''$$

where $\Lambda', \Lambda'' \subseteq \Lambda$. In $P'$, using a cut and then a weakening gives $\Delta^\pi$ as desired.

Second, consider the (very simple) case where the cedent $\Delta$ is inferred by a weakening inference

$$\frac{\Delta'}{\Delta}$$

where $\Delta \subset \Delta'$. The path $\pi$ to $\Delta$ can be extended by one more cedent to be a path $\pi'$ to the cedent $\Delta'$. The cedents $\Delta^\pi$ and $(\Delta')^{\pi'}$ are identical. Thus the weakening inference in $P$ is just omitted in $P'$.

Now consider the case where $\Delta$ is the lower cedent of an $\wedge$ inference in $P$:

$$\frac{A, \Gamma_1 \qquad B, \Gamma_2}{A \wedge B, \Gamma_1, \Gamma_2}$$

Let $\Delta_1$ and $\Delta_2$ be the left and right upper cedents, respectively, and let $\pi_1$ and $\pi_2$ be the $\wedge\exists$-paths obtained by adding $\Delta_1$ or $\Delta_2$, respectively, to the end of $\pi$. First, suppose $A \wedge B$ is category $(\alpha)$ or $(\gamma)$, so $*_\pi(A \wedge B)$ is $(A \wedge B)\sigma_\pi$. Then $A$ and $B$ are both category $(\gamma)$, and $*_{\pi_1}(A) = A\sigma_{\pi_1} = A\sigma_\pi$ and $*_{\pi_2}(B) = B\sigma_{\pi_2} = B\sigma_\pi$. Thus, in $P'$, the $\wedge$ inference becomes

$$\frac{A\sigma_\pi, \Lambda, (A \wedge B)\sigma_\pi \qquad B\sigma_\pi, \Lambda, (A \wedge B)\sigma_\pi}{(A \wedge B)\sigma_\pi, \Lambda}$$

and this is still a valid $\wedge$ inference.

For the second subcase, suppose $A \wedge B$, thus $A$ and $B$, are category $(\beta)$. The formula $B$ in $\Delta_2$ has the same pending implicants $\overline{C}_1, \ldots, \overline{C}_\ell$ as the formula $A \wedge B$ in $\Delta$. Also, $\overline{C}_i\sigma_{\pi_2} = \overline{C}_i\sigma_\pi$. Thus $\Delta^\pi$ is the same as $(\Delta_2)^{\pi_2}$. This means that the $\wedge$ inference can be omitted in $P'$.

Next consider the case where $\Delta$ is the lower cedent of a cut in $P$:

$$\frac{A, \Gamma_1 \qquad \overline{A}, \Gamma_2}{\Gamma_1, \Gamma_2}$$

Let $\Delta_1$ and $\Delta_2$ be the left and right upper cedents, respectively, and $\pi_1$ and $\pi_2$ be the extensions of $\pi$ to $\Delta_1$ and $\Delta_2$. The occurrence of $\overline{A}$ is category $(\beta)$ of course, and $*_{\pi_2}(\overline{A})$ is the empty cedent. Thus, the cedents $\Delta^\pi$ and $(\Delta_2)^{\pi_2}$ are identical, and the cut inference may be omitted from $P'$.

Next consider the case where $\Delta$ is the lower cedent of an $\vee$ inference:

$$\frac{A, B, \Gamma}{A \vee B, \Gamma}$$

In this, and the remaining cases, let $\Delta'$ be the upper cedent of the inference, and let $\pi'$ be $\pi$ extended to the cedent $\Delta'$. For the $\vee$ inference, $\sigma_{\pi'}$ is identical to $\sigma_\pi$. As a first subcase, suppose $A \vee B$ is category $(\gamma)$, and thus $A$ and $B$ are as well. In this subcase, $*_\pi(A \vee B) = (A \vee B)\sigma_\pi$, $*_{\pi'}(A) = A\sigma_\pi$, and $*_{\pi'}(B) = B\sigma_\pi$. The $*_\pi$-translation of the two cedents thus forms a valid $\vee$ inference in $P'$.

The second subcase is when $A \vee B$, $A$, and $B$ are category $(\alpha)$. Letting $A_1, \ldots, A_k$ be the $\vee\forall$-components of $A$, and $B_1, \ldots, B_{k'}$ be those of $B$, the $*_\pi$-translation of the $\vee$ inference has the form

$$\frac{A_1\sigma_\pi, \ldots, A_k\sigma_\pi, B_1\sigma_\pi, \ldots, B_{k'}\sigma_\pi, \Lambda}{A_1\sigma_\pi, \ldots, A_k\sigma_\pi, B_1\sigma_\pi, \ldots, B_{k'}\sigma_\pi, \Lambda}$$

and this can be omitted from $P'$.

The third subcase is when $A \vee B$ is category $(\beta)$. Then $A$ and $B$ are category $(\gamma)$, and $*(A) = A\sigma_\pi$ and $*(B) = B\sigma_\pi$. Also, $*(A \vee B)$ is $\overline{C}_1\sigma_\pi, \ldots, \overline{C}_k\sigma_\pi$, where the $\overline{C}_i$'s are the pending implicants of $A \vee B$. Thus, the $*_\pi$-translation of the cedents in the $\vee$ inference has the form

$$\frac{A\sigma_\pi, B\sigma_\pi, \Lambda, \overline{C}_1\sigma_\pi, \ldots, \overline{C}_k\sigma_\pi}{\overline{C}_1\sigma_\pi, \ldots, \overline{C}_k\sigma_\pi, \Lambda} \tag{8}$$

Of course, this is not a valid inference. Let $\Delta''$ be the upper left cedent of the jump target of $A \vee B$. From Lemma 8, there is an $\wedge\exists$-path $\pi''$ leading to $\Delta''$ so that the $*_{\pi''}$-translation of $\Delta''$ is

$$(\overline{A \vee B})\sigma_\pi, \overline{C}_1\sigma_\pi, \ldots, \overline{C}_k\sigma_\pi, \Lambda'$$

where $\Lambda' \subseteq \Lambda$. In $P'$, this cedent and the upper cedent of (8) are combined with an $\vee$ inference and a cut to yield the lower cedent of (8), similarly to what was done in (6).

Now consider the case where $\Delta$ is the lower cedent of a $\forall$ inference

$$\frac{A(b), \Gamma}{(\forall x)A(x), \Gamma}$$

First suppose $(\forall x)A(x)$ is category $(\gamma)$, so $*_\pi((\forall x)A(x)) = (\forall x)A(x)\sigma_\pi = (\forall x)A(x)\sigma_{\pi'}$. The formula $A(b)$ is category $(\gamma)$ and $\sigma_{\pi'}(b) = b$, thus $*_{\pi'}(A(b)) = A(b)\sigma_\pi$. The $\forall$ inference of $P$ becomes

$$\frac{A(b)\sigma_\pi, \Lambda, (\forall x)A(x)\sigma_\pi}{(\forall x)A(x)\sigma_\pi, \Lambda}$$

and this forms a valid $\forall$ inference in $P'$.

For the second subcase, suppose that $(\forall x)A(x)$ is category $(\beta)$. Hence, $A(b)$ is category $(\gamma)$. This case is similar to the third subcase for $\vee$ inferences above. We have $*_\pi((\forall x)A(x))$ equal to $\overline{C}_1\sigma_\pi, \ldots, \overline{C}_k\sigma_\pi$ where the $\overline{C}_i$'s are the pending implicants of $(\forall x)A(x)$. And, $*_\pi(A(b))$ equals $A(b)\sigma_\pi$; note $\sigma_\pi(b) = b$. Thus, the $*_{\pi'}$-/$*_\pi$-translation of the cedents in the $\forall$ inference has the form

$$\frac{A(b)\sigma_\pi, \Lambda, \overline{C}_1\sigma_\pi, \ldots, \overline{C}_k\sigma_\pi}{\Lambda, \overline{C}_1\sigma_\pi, \ldots, \overline{C}_k\sigma_\pi} \tag{9}$$

which is not a valid inference. Let $\Delta''$ be the upper left cedent of the jump target of $(\forall x)A(x)$. By Lemma 8, there is an $\wedge\exists$-path $\pi''$ leading to $\Delta''$ so that the $*_{\pi''}$-translation of $\Delta''$ is

$$\overline{(\forall x)A(x)}\sigma_\pi, \overline{C}_1\sigma_\pi, \ldots, \overline{C}_k\sigma_\pi, \Lambda'$$

where $\Lambda' \subseteq \Lambda$. In $P'$, this cedent and the upper cedent of (9) are combined with an $\forall$ inference and a cut to yield the lower cedent of (9), similarly to what was done in (7).

For the third subcase, suppose that $(\forall x)A(x)$ is category $(\alpha)$, so $A(b)$ is also category $(\alpha)$. By definition, $\sigma_{\pi'}(b) = \sigma_\pi(x)$. Thus, $*_{\pi'}(A(b)) = A(b)\sigma_{\pi'} = A(x)\sigma_\pi$. Also, $*_\pi((\forall x)A(x)) = A(x)\sigma_\pi$. Therefore, in $P'$, the $\forall$ inference becomes trivial with $\Delta^\pi$ and $(\Delta')^{\pi'}$ equal to each other; so, this inference is omitted from $P'$.

Finally, consider the case where $\Delta$ is the lower cedent of an $\exists$ inference

$$\frac{A(s), \Gamma}{(\exists x)A(x), \Gamma}$$

Note that $\sigma_{\pi'}$ is the same as $\sigma_\pi$. For the first subcase, suppose $(\exists x)A(x)$ is either category $(\alpha)$ or $(\gamma)$, so $A(s)$ is category $(\gamma)$. This gives $*_{\pi'}(A(s)) = A(s)\sigma_{\pi'} = A(s)\sigma_\pi$. And, since its outermost connective is $\exists$, $*_\pi((\exists x)A(x)) = (\exists x)A(x)\sigma_\pi$. The $\exists$ inference in $P$ becomes, in $P'$,

$$\frac{A(s)\sigma_\pi, \Lambda, (\exists x)A(x)\sigma_\pi}{(\exists x)A(x)\sigma_\pi, \Lambda}$$

which is a valid $\exists$ inference.

For the second subcase, suppose $(\exists x)A(x)$ and hence $A(s)$ are category $(\beta)$. The two formulas have the same pending implicants, $\overline{C}_1, \ldots, \overline{C}_k$, for $k \geq 0$. Thus, $*_{\pi'}(A(s))$ and $*_\pi((\exists x)A(x))$ are both equal to the cedent $\overline{C}_1\sigma_\pi, \ldots, \overline{C}_k\sigma_\pi$. That is to say, $\Delta^\pi$ and $(\Delta')^{\pi'}$ are identical, and thus the $\exists$ inference can be omitted from $P'$.

The above completes the construction of $P'$ from $P$. The discussion at the end of the proof of Theorem 5 applies equally well to the $P'$ just constructed, and $P'$ is again polynomial time uniform. □

## 6 Bounds on Eliminating All Cuts

This section gives bounds on eliminating all cuts from a proof. The bound obtained has the form $2^{|P|}_{d+O(1)}$, where $d$ is the maximum quantifier alternation of cut formulas in $P$. The first-order formula classes $\Sigma_i$ and $\Pi_i$ are defined as usual by counting alternations of quantifiers, allowing propositional connectives to appear arbitrarily. Namely, $\Sigma_0 = \Pi_0$ is the set of quantifier free formulas; and, using Bachus-Naur notation,

$$\Sigma_i ::= \Sigma_{i-1} | \Pi_{i-1} | \Sigma_i \wedge \Sigma_i | \Sigma_i \vee \Sigma_i | \neg \Pi_i | (\exists x) \Sigma_i$$

$$\Pi_i ::= \Pi_{i-1} | \Sigma_{i-1} | \Pi_i \wedge \Pi_i | \Pi_i \vee \Pi_i | \neg \Sigma_i | (\forall x) \Pi_i$$

The *alternating quantifier depth (aqd)* of a cut is the minimum $i > 0$ such that one cut formula is in $\Sigma_i$ and the other is in $\Pi_i$. The *alternation quantifier depth* of a proof $P$, denoted $\mathsf{aqd}(P)$, is the maximum aqd of any cut in $P$.

**Theorem 9** *Let $P$ be a tree-like proof, and let $d = \mathsf{aqd}(P)$. There is a cut free proof $P'$ with the same end cedent as $P$ with the size of $P'$ bounded by $|P'| \le 2^{|P|}_{d+O(1)}$.*

The proof of the theorem depends only on Theorem 5, not on Theorems 3 and 7. We also use upper bounds on eliminating cuts on quantifier free formulas as can be found in [5, 7, 17].

*Proof* It is helpful to briefly review the well-known fact that the size of formulas appearing in the tree-like proof $P$ can be bounded by the number of inferences in $P$ plus the size of the formulas in the end cedent of $P$. For this, recall that any formula $B$ appearing in $P$ has a unique descendent $A$ such that $A$ either is a cut formula or is in the end cedent of $P$. In addition, $B$ corresponds to a unique subformula $C$ of $A$. Let $C$ be a non-atomic subformula of a formula $D$ in $P$ which has a cut formula as descendant. If there is some ancestor $B$ of $D$ such that $B$ corresponds to $C$ and such that $B$ is a principal formula of a logical inference, then leave $C$ unchanged. If there is no such ancestor $D$, then mark $C$ for deletion. Now replace every maximal subformula $C$ in $P$ marked for deletion with an arbitrary atomic formula, say with $d{=}d$ for $d$ some new free variable. The proof remains a valid proof (since only atomic formulas are allowed in initial cedents), and its end cedent is unchanged. Clearly, in the resulting proof, every cut formula has number of logical connectives bounded by the total number of $\wedge$, $\vee$, $\exists$, and $\forall$ inferences in $P$. Without loss of generality, we assume this is true of the proof $P$ itself.

The main step in proving Theorem 9 is to convert $P$ into a proof in which all cuts are in prenex form. As a preliminary step, we show that we may assume w.l.o.g. that no cut formula in $P$ has multiple quantifiers on the same bound variable, or in other words, that the bound variables in a cut formula are distinct. Towards this end, for each cut inference in $P$, with formulas $A$ and $\overline{A}$ as its cut formulas, rename the bound variables in $A$ so that the quantifiers in $A$ use distinct bound variables. This also renames the bound variables of $\overline{A}$ of course. Furthermore, if $B$ is a formula with descendent $A$ or $\overline{A}$, this induces a renaming of the bound variables in $B$ according to the renaming of bound variables in the subformula of $A$ or $\overline{A}$ that corresponds to $B$. By applying these renamings to all such formulas $B$, and repeating for all cuts in $P$, we obtain a proof with the same end cedent as $P$ such that bound variables are never reused in cut formulas.[4] So, we may assume w.l.o.g. that $P$ satisfies this property.

Now, for each cut in $P$, with cut formulas $A$ and $\overline{A}$, choose an arbitrary prenex form $A'$ for $A$ so that the aqd of $A'$ is $\leq \mathsf{aqd}(P)$. The formula $A'$ is obtained by choosing an ordering of the quantifiers in $A$ which respects the scope of the quantifiers, and then using standard prenex operations to move the quantifiers out to the front of the formula in the chosen order. The prenex form $(\overline{A})'$ of $\overline{A}$ is chosen with the same ordering and thus equals $\overline{A'}$.

Let $B$ be any formula in $P$ with a cut formula $A$ as descendent. The quantifiers of $A$ are ordered as just discussed to form its prenex form $A'$. Since $B$ corresponds to a subformula of $A$, this induces an ordering on the quantifiers of $B$; the prenex form $B'$ of $B$ is defined using this induced ordering. On the other hand, if $B$ has a descendent in the end cedent of $P$, the formula $B'$ is defined to be equal to $B$. For any cedent $\Delta$ in $P$, define $\Delta'$ to contain exactly the formulas $B'$ for $B \in \Delta$.

The proof $P'$ will contain the cedents $\Delta'$ for all $\Delta \in P$. However, the $\wedge$ and $\vee$ inferences in $P$ may no longer be valid in $P'$. Cuts, weakenings, and quantifier inferences of $P$ do remain valid in $P'$. In addition, since only atomic formulas are allowed initial cedents, the initial cedents of $P$ are unchanged in $P'$.

In order to make $P'$ a valid proof, we must replace the $\wedge$ and $\vee$ inferences of $P$ with some new subproofs and cuts. The next lemma gives the key construction needed for this.

**Lemma 10** *Let $B \wedge C$ be the principal formula of an $\wedge$ inference in $P$ with a cut formula as descendent. The auxiliary formulas of the inference are $B$ and $C$. Let $B'$, $C'$, and $(B \wedge C)'$ be their prenex forms in $P'$. Then the cedent*

$$\overline{B'}, \overline{C'}, (B \wedge C)' \tag{10}$$

*has a cut free proof of length linear in the lengths of $B$ and $C$. Similarly, if $B \vee C$ is the principal formula of an $\vee$ inference of $P$, then the cedents*

$$\overline{B'}, (B \vee C)' \qquad \text{and} \qquad \overline{C'}, (B \vee C)' \tag{11}$$

*have cut free proofs of length linear in the lengths of $B$ and $C$.*

*Proof* Let $B'$ and $C'$ have the forms $\mathcal{Q}_1 B_0$ and $\mathcal{Q}_2 C_0$ where $\mathcal{Q}_1$ and $\mathcal{Q}_2$ denote blocks of zero or more quantifiers and where $B_0$ and $C_0$ are quantifier free. The

[4] The same construction could also rename bound variables in the end cedent of $P$, but this would then change the end cedent.

formula $(B \wedge C)'$ or $(B \vee C)'$ will have the form $\mathcal{Q}(B_0 \wedge C_0)$ or $\mathcal{Q}(B_0 \vee C_0)$. Here the quantifier block $\mathcal{Q}$ is obtained by arbitrarily interleaving (or, "shuffling") the two blocks $\mathcal{Q}_1$ and $\mathcal{Q}_2$.

We claim that, for any quantifier blocks $\mathcal{Q}_1$ and $\mathcal{Q}_2$, and any block $\mathcal{Q}$ obtained as a shuffle of $\mathcal{Q}_1$ and $\mathcal{Q}_2$, the cedents (10) and (11) have tree-like, cut free proofs with size equal to the number of logical connectives in the cedents being proved. This is proved by induction on the number of quantifiers in $\mathcal{Q}$.

The base case of the induction is when $\mathcal{Q}$ is empty, and $B$ and $C$ are quantifier free. As is well known (and easy to verify) there are proofs of the cedents $\overline{B}_0, B_0$ and $\overline{C}_0, C_0$ with sizes equal to twice the number of logical connectives in $B_0$ and $C_0$, respectively. These two cedents plus a single $\wedge$ or $\vee$ inference suffices to derive any of the cedents in (10) or (11).

For the induction step, suppose that $\mathcal{Q}$ contains at least one quantifier. The first quantifier can have the form $(\exists x)$ or $(\forall x)$ and is also the first quantifier of either $\mathcal{Q}_1$ or $\mathcal{Q}_2$. For instance, suppose $(\exists x)$ is the outermost quantifier of $\mathcal{Q}$ and $\mathcal{Q}_1$. Writing $B_0 = B_0(x)$ to show the occurrences of the bound variable $x$, and replacing occurrences of $x$ with a new free variable $a$, the induction hypothesis gives derivations of the cedents

$$\overline{\mathcal{Q}_1^- B_0(a)}, \mathcal{Q}^-(B_0(a) \vee C_0) \quad \text{and} \quad \overline{\mathcal{Q}_2 C_0}, \mathcal{Q}^-(B_0(a) \vee C_0)$$

or

$$\overline{\mathcal{Q}_1^- B_0(a)}, \overline{\mathcal{Q}_2 C_0}, \mathcal{Q}^-(B_0(a) \wedge C_0)$$

where $\mathcal{Q}_1^-$ and $\mathcal{Q}^-$ are the blocks $\mathcal{Q}_1$ and $\mathcal{Q}$ minus the first quantifier $\exists x$. For the $\vee$ case, the derivation

$$\frac{\overline{\mathcal{Q}_2 C_0}, \mathcal{Q}^-(B_0(a) \vee C_0)}{\overline{\mathcal{Q}_2 C_0}, (\exists x)\mathcal{Q}^-(B_0(x) \vee C_0)}$$

gives the desired derivation of $\overline{\mathcal{Q}_2 C_0}, \mathcal{Q}(B_0 \vee C_0)$; and the derivation

$$\frac{\dfrac{\overline{\mathcal{Q}_1^- B_0(a)}, \mathcal{Q}^-(B_0(a) \vee C_0)}{\overline{\mathcal{Q}_1^- B_0(a)}, (\exists x)\mathcal{Q}^-(B_0(x) \vee C_0)}}{(\forall x)\overline{\mathcal{Q}_1^- B_0(x)}, (\exists x)\mathcal{Q}^-(B_0(x) \vee C_0)}$$

gives the desired derivation of $\overline{\mathcal{Q}_1 B_0}, \mathcal{Q}(B_0 \vee C_0)$. Note that the second inference is a $\forall$ inference; by the assumption of distinctness of bound variables, the eigenvariable $a$ does not appear in $C_0$.

A similar argument works for the $\wedge$ case. The cases where outermost quantifier of $\mathcal{Q}$ is $(\forall x)$ are also similar. □

We can now complete the proof of Theorem 9. The proof $P'$ is formed from the cedents $\Delta'$ defined above. Using the cedents $\Delta'$ maintains the validity of all inferences except for some of the $\vee$ and $\wedge$ inferences. In $P'$ these inferences become

$$\frac{B', \Gamma_1' \qquad C', \Gamma_2'}{(B \wedge C)', \Gamma_1, \Gamma_2} \quad \text{and} \quad \frac{B', C', \Gamma}{(B \vee C)', \Gamma}$$

and these are no longer valid if their principal formula contains quantifiers and has a cut formula as descendent. However, the $\wedge$ inference can be simulated by using two cuts against the cedent $\overline{B'}, \overline{C'}, (B \wedge C)'$ given by Lemma 10. Likewise, the $\vee$ inference can be simulated by using two cuts with the cedents $\overline{B'}, (B \vee C)'$ and $\overline{C'}, (B \vee C)'$. This process replaces one inference in $P$ with two cuts in $P'$; in addition, $P'$ must contain the derivations of the cedents as given by Lemma 10. Since the formulas $(B \wedge C)'$ and $(B \vee C)'$ have cut formulas as descendents, their sizes are bounded by $|P|$ as discussed at the beginning of the proof. Therefore, the size of $|P'|$ can be bounded by $|P'| \leq 3|P|^2$, since the size of the proofs from Lemma 10 is strictly less than $3|P|$.

The proof $P'$ has all cut formulas in $\Sigma_d$ or $\Pi_d$, where $d = \mathsf{aqd}(P)$. It suffices to assume $d > 0$. Applying Theorem 5 $d$ times gives a tree-like proof $P''$ with the same end cedent, in which all cut formulas are quantifier free, with $h(P'') \leq 2_{d-1}^{3|P|^2}$. Now, applying Theorem 8 of [5] and the discussion from the end of Sect. 4 of [5], we get a proof $P'''$ of the same end cedent with height bounded by $h(P''') \leq 2^{|P|}2_{d-1}^{3|P|^2}$, such that all cut formulas in $P'''$ are atomic. Then, applying Lemma 7 of [5], we get another proof $P''''$ again with the same end cedent, which is cut free, and has height bounded by $2^{h(P''')+1}$. In particular, the size of $P''''$ is less than $2_2^{h(P''')+1}$.

Therefore, $|P''''| < 2_{d+2}^{|P|}$, at least for $|P| > 7$. For $d > 0$, this gives $|P''''| < 2_{d+2}^{|P|}$ for $|P| > 7$. This completes the proof of Theorem 9. □

The size bound on $P'''$ is not optimal; we expect that even $2_{d+1}^{|P|}$ might work.

**Acknowledgements** We thank G. Mints for encouraging comments, and the referee for helpful comments and corrections. Supported in part by NSF grants DMS-1101228 and CCF-1213151 and by a grant from the Simons Foundation (#208717 to Sam Buss).

## References

1. K. Aehlig, A. Beckmann, On the computational complexity of cut reduction. Ann. Pure Appl. Log. **161**, 711–736 (2010)
2. M. Baaz, A. Leitsch, Cut normal forms and proof complexity. Ann. Pure Appl. Log. **97**, 127–177 (1999)
3. A. Beckmann, S.R. Buss, Separation results for the size of constant-depth propositional proofs. Ann. Pure Appl. Log. **136**, 30–55 (2005)
4. A. Beckmann, S.R. Buss, Characterization of definable search problems in bounded arithmetic via proof notations, in *Ways of Proof Theory* (Ontos Verlag, Frankfurt, 2010), pp. 65–134

5. A. Beckmann, S.R. Buss, Corrected upper bounds for free-cut elimination. Theor. Comput. Sci. **412**, 5433–5445 (2011)
6. S.R. Buss, An introduction to proof theory, in *Handbook of Proof Theory*, ed. by S.R. Buss (North-Holland, Amsterdam, 1998), pp. 1–78
7. P. Gerhardy, Refined complexity analysis of cut elimination, in *Proceedings of 17th Workshop on Computer Science Logic (CSL)*. Lecture Notes in Computer Science 2803 (Springer, 2003), pp. 212–225
8. J.-Y. Girard, *Proof Theory and Logical Complexity* (Bibliopolis, Naples, 1987)
9. J. Krajíček, Lower bounds to the size of constant-depth propositional proofs. J. Symb. Log. **59**, 73–86 (1994)
10. J. Krajíček, Interpolation theorems, lower bounds for proof systems, and independence results for bounded arithmetic. J. Symb. Log. **62**, 457–486 (1997)
11. G. Kreisel, G.E. Mints, S.G. Simpson, The use of abstract language in elementary metamathematics: some pedagogical examples, in *Logic Colloquium*. Lecture Notes in Mathematics, vol. 453 (Springer, Berlin, 1975), pp. 38–131
12. G.E. Mints, Finite investigations of transfinite derivations. J. Sov. Math. **10**, 548–596 (1978). Russian original in Zapiski Nauchnykh Seminarov LOMI Steklova AN SSSR **49**, 67–121 (1975)
13. V.P. Orevkov, Upper bound on the lengthening of proofs by cut elimination. J. Sov. Math. **34**, 1810–1819 (1986). Original Russian version in Zap. Nauchn. Sem. L.O.M.I. Steklov **137**, 87–98 (1984)
14. V.P. Orevkov, Applications of cut elimination to obtain estimates of proof lengths. Sov. Math. Dokl. **36**, 292–295 (1988). Original Russian version in Dokl. Akad. Nauk. **296**(3), 539–542 (1987)
15. A.A. Razborov, On provably disjoint NP-pairs. Technical report RS-94-36, Basic Research in Computer Science Center, Aarhus, Denmark, 1994. http://www.brics.dk/index.html
16. A.S. Troelstra, H. Schwichtenberg, *Basic Proof Theory*, 2nd edn. Tracts in Theoretical Computer Science, vol. 43 (Cambridge University Press, Cambridge, 2000)
17. W. Zhang, Cut elimination and automatic proof procedures. Theor. Comput. Sci. **91**, 265–284 (1991)
18. W. Zhang, Depth of proofs, depth of cut-formulas, and complexity of cut formulas. Theor. Comput. Sci. **129**, 193–206 (1994)

# Spector's Proof of the Consistency of Analysis

**Fernando Ferreira**

## 1 Introduction

The editors of this volume asked me to present and discuss Clifford Spector's proof of the consistency of analysis. It is only fitting that, in a volume dedicated to Gerhard Gentzen, known for his epoch-making consistency proof of Peano arithmetic PA, Spector's proof of consistency of analysis is discussed. Gentzen's approach to consistency proofs has been systematically developed and generalized by the German school of proof theory (Schütte, Pohlers, Buchholz, Jäger, Rathjen, etc.) and others. For all its successes (and there were many), the approach is still very far from providing a proof of the consistency of full second-order arithmetic $\mathsf{PA}_2$ (analysis). There are quite serious difficulties in analyzing systems above $\Pi^1_2$-comprehension. In the words of Michael Rathjen in [32], the more advanced analyses "tend to be at the limit of human tolerance." How is it, then, that Spector was able to provide a proof of the consistency of analysis? What kind of proof is it? Spector's proof follows quite a different blueprint from Gentzen's. It does not reduce $\mathsf{PA}_2$ to finististic arithmetic together with the postulation of the well-ordering of a sufficiently long primitive recursive ordinal notation system. Instead, it reduces (in a finitary manner) the consistency of analysis to the consistency of a certain quantifier-free finite-type theory. The epistemological gain, if there is one, rests in the evidence for the consistency of Spector's quantifier-free theory.

The proof of Spector was published posthumously in 1962 (Spector died young of acute leukemia). It is a descendant of Gödel's interpretation of PA in 1958, in which it was shown that PA is interpretable in Gödel's quantifier-free finite-type theory T. In the last paragraph of his paper [11], Gödel writes that "it is clear that, starting from the same basic idea, one can also construct systems that are

F. Ferreira (✉)
Departamento de Matemática, Faculdade de Ciências, Universidade de Lisboa, Lisboa, Portugal
e-mail: fjferreira@fc.ul.pt

 
R. Kahle, M. Rathjen (eds.), *Gentzen's Centenary*,
DOI 10.1007/978-3-319-10103-3_11

much stronger than T, for example by admitting transfinite types or the sort of inference that Brouwer used in proving the 'fan theorem'." Spector took up the latter suggestion. The Brouwerian kind of inference that Gödel is presumably referring to is the bar theorem (a corollary of which is the 'fan theorem'). Brouwer's justification of the bar theorem is object of controversy (see [3] for Brouwer's own rendition and [37] for a modern defense and references) and not really formulated in a workable form. Following Stephen Kleene's enunciation in [21], the bar theorem is nowadays admitted in intuitionistic mathematics in the form of an axiom scheme known as bar induction. Spector follows this approach and advances two moves: he generalizes bar induction to finite types and, in a bold stroke, introduces a corresponding principle of definition known as (Spector's) bar recursion. In his own words, "bar recursion is a principle of definition and bar induction a corresponding principle of proof." One cannot but think of a parallel with ordinary recursion and ordinary induction. Spector's quantifier-free finite-type theory adjoins to Gödel's T new constants for the bar recursors and accepts the pertinent equations that characterize them. He is then able to show that analysis is interpretable in this extension of Gödel's T.

This paper is organized as follows. In the next section, we review Gödel's *dialectica* interpretation of 1958. We describe a direct interpretation of PA into T, instead of Gödel's own which relies on the interpretation of Heyting arithmetic accompanied by a double negation translation of classical logic into intuitionistic logic. The direct interpretation is very simple and was first described by Joseph Shoenfield in his well-known textbook [34]. Section 3 and 5 introduce bar recursion. In the first of these sections, we discuss bar recursion from the set theoretic point of view. As opposed to standard treatments of bar recursion, we take some time doing this. We have in mind the reader unfamiliar with bar recursion but comfortable with the basics of set theory. One of the aims of this paper is to explain Spector's proof to a logician not trained in proof theory or constructive mathematics. Two set-theoretic discussions are made. The first focuses on well-founded trees and their ordinal heights. The second has the advantage of immediately drawing attention to the principle of dependent choices, a principle which plays an important role in the discussions of bar recursion. Armed with the set-theoretic understanding, in Sect. 5 we finally discuss bar recursion from an intuitionistic point of view.

The interim Sect. 4 introduces Spector's quantifier-free theory with the bar recursive functionals of finite type. It also briefly mentions the two main models of this theory. Sections 6 and 7 are the heart of the paper. They present the interpretation of analysis into Spector's theory. The original proof is based on the interpretation of the so-called classical principle of numerical double negation shift (principle $F$ in Spector's paper), and this is sufficient to interpret full second-order comprehension. The technical matter boils down to solving a certain system of equations in finite-type theory, and the bar recursive functionals permit the construction of a solution. The solution of these equations is *ad hoc* (Paulo Oliva was, nevertheless, able to find a nice motivation for it in [31]). The interpretation of bar induction is more natural because bar induction and bar recursion go hand in hand, in a way similar to that of induction and recursion in Gödel's *dialectica* interpretation (see the discussion in [7]). Moreover, it provides additional information. We owe to William Howard in [14] the interpretation of bar induction into Spector's theory. Our paper develops Howard's strategy *directly* for the classical setting.

The paper includes a short appendix. It discusses a sort of perplexity caused by the existence of the *term model* of Spector's theory, a structure whose infinite numerical sequences are all recursive. How can bar recursion hold in such a classical structure when models of bar recursion are usually associated with producing non-recursive objects? The answer lies in the failure of quantifier-free choice in the term model and reveals a little of the subtlety of Spector's interpretation.

The main body of the paper finishes with an epilogue in which Spector's consistency proof is briefly assessed. We hope that this writing is able to convey to the uninitiated a little of the depth and beauty of Spector's proof of consistency, and also that the expert finds some interest in the paper.

## 2 Gödel's *Dialectica* Interpretation of 1958

David Hilbert did not precisely define what finitary mathematics is, but a very influential thesis of William Tait [39] identifies finitism with the quantifier-free system of primitive recursive arithmetic. This theory concerns only one sort of objects: the natural numbers. These are, in the Hilbertian terms as exposed by Gödel in his 1958 paper, "in the last analysis spatiotemporal arrangements of elements whose characteristics other than their identity or nonidentity are irrelevant." Gödel considers an *extension* of finitism (the work concerns, as its title says, "a hitherto unutilized extension of the finitary standpoint"), viz. a certain quantifier-free, *many-sorted*, theory. Its "axioms (...) are formally almost the same as those of primitive recursive number theory, the only exception being that the variables (other than those on which induction is carried out), as well as the defined constants, can be of any finite type over the natural numbers" (quoted from [11]). The variables are supposed to range over the so-called computable functionals of finite type (a primitive notion for Gödel). This is the crux of the extension: the requirement that the value of the variables be concrete ("spatiotemporal arrangements") is dropped, and certain *abstracta* are accepted.

The current literature has some very clear descriptions and explanations of Gödel's theory $\mathsf{T}$. Easily available sources are Avigad and Feferman's survey in [1] and Kohlenbach's monograph [25]. The latter source includes a detailed treatment of Spector's bar recursive interpretation (different from the one presented here). In the present section, we briefly highlight the main features of $\mathsf{T}$ but the reader is referred to the above sources for details and pointers to the literature. The quantifier-free language $\mathcal{T}$ of $\mathsf{T}$ has infinitely many sorts (variable ranges), one for each finite type over the natural numbers. These types are syntactic expressions defined inductively: $\mathsf{N}$ (the base type) is a finite type; if $\tau$ and $\sigma$ are finite types, then $\tau \to \sigma$ is a finite type. These are all the types. It is useful to have the following (set-theoretic) interpretation in mind: the base type $\mathsf{N}$ is the type constituted by the natural numbers

$\mathbb{N}$, whereas $\tau \to \sigma$ is the type of all (total) set-theoretic functions of objects of type $\tau$ to objects of type $\sigma$. To ease reading, we often omit brackets and associate the arrows to the right. For example, $\mathrm{N} \to \mathrm{N} \to \mathrm{N}$ means $\mathrm{N} \to (\mathrm{N} \to \mathrm{N})$. $\mathcal{T}$ has a denumerable set of variables $x^\sigma$, $y^\sigma$, $z^\sigma$, etc. for each type $\sigma$. When convenient, we omit the type scripts. There are two kinds of constants:

(a) *Logical constants* or *combinators*. For each pair of types $\sigma, \tau$ there is a logical constant $\Pi_{\sigma,\tau}$ of type $\sigma \to \tau \to \sigma$. For each triple of types $\delta, \sigma, \tau$ there is a logical constant $\Sigma_{\delta,\sigma,\tau}$ of type $(\delta \to \sigma \to \tau) \to (\delta \to \sigma) \to (\delta \to \tau)$.
(b) *Arithmetical constants*. The constant 0 of type N. The *successor* constant $S$ of type $\mathrm{N} \to \mathrm{N}$. For each type $\sigma$, there is a *recursor* constant $R_\sigma$ of type $\mathrm{N} \to \sigma \to (\sigma \to \mathrm{N} \to \sigma) \to \sigma$.

Constants and variables of type $\sigma$ are terms of type $\sigma$. If $t$ is a term of type $\sigma \to \tau$ and $q$ is a term of type $\sigma$, then one can form a new term, denoted by $App(t, q)$, of type $\tau$ ($t$ is said to be applied to $q$). These are all the terms. We write $tq$ or $t(q)$ for $App(t, q)$. We also write $t(q, r)$ instead of $(t(q))(r)$. In general, $t(q, r, \ldots, s)$ stands for $(\ldots((t(q))(r))\ldots)(s)$.

The intended meaning of these constants is given by certain identities. There are the identities for the combinators: $\Pi(x, y)$ is $x$ and $\Sigma(x, y, z)$ is $x(z, yz)$. The identities for the combinators make possible the definition of *lambda terms* within Gödel's T: given a term $t^\sigma$ and a variable $x^\tau$, there is a term $q^{\tau\to\sigma}$ (denoted by the lambda notation $\lambda x.t$) whose variables are all those of $t$ other than $x$, such that, for every term $s$ of type $\tau$, one has the identity between $qs$ and $t[s/x]$ (the notation '$[s/x]$' indicates the substitution of the variable $x$ by the term $s$ in the relevant expression). For the recursors, we have the following identities: $R(0, y, z)$ is $y$ and $R(Sx, y, z)$ is $z(R(x, y, z), x)$. These identities formulate definitions by recursion.

We have been speaking loosely about identities because there are subtle issues concerning the treatment of equality in functional interpretations: consult [40] and [1] for discussions. (These issues surface because extensional equality suffers from a serious shortcoming with respect to the *dialectica* interpretation, viz: the axiom of extensionality, i.e., the postulation that extensional equality enjoys substitution *salva veritate* fails to be interpretable. This was shown by Howard in [15].) We adopt the following minimal treatment: there is only the symbol for equality between terms of the base type N, and the formulas of $\mathcal{T}$ are defined as Boolean combinations of equalities of the form $t = q$, where $t$ and $q$ are terms of type N. How are the identities for the combinators and recursors to be formulated within this framework? They give rise to certain axiom schemes. For instance, the axioms for the recursors are given by the equivalences $A[R(0, y, z)/w] \leftrightarrow A[y/w]$ and $A[R(Sx, y, z)/w] \leftrightarrow A[z(R(x, y, z), x)/w]$, where $A$ is any formula of $\mathcal{T}$ with a distinguished variable $w$.

The axioms of T are the axioms of *classical* propositional calculus, the axioms of equality $x = x$ and $x = y \wedge A[x/w] \to A[y/w]$ ($A$ is any formula of $\mathcal{T}$, and $x$, $y$ and $w$ are of type N, of course), the schemata coming from the identities of combinators and recursors and, finally, the usual arithmetical axioms for the constants 0 and $S$, namely: $Sx \neq 0$ and $Sx = Sy \to x = y$. There are also two rules. The rule of substitution that allows to infer $A[s^\sigma/x]$ from $A$ and the rule of induction that, from $A(0)$ and $A(x^{\mathrm{N}}) \rightarrow A(Sx)$ allows the inference of $A(x)$ (in both rules, $A$ can be any formula of $\mathcal{T}$).

We have described Gödel's quantifier-free, many-sorted, system $\mathsf{T}$. Gödel showed that it is possible to interpret Heyting arithmetic (and, hence, Peano arithmetic) into $\mathsf{T}$ in a finitistic way. This result entails that the consistency of $\mathsf{PA}$ is finitistically reducible to the consistency of a natural extension of finitism. In the sequel, we describe Gödel's result. We formulate a direct interpretation of an extension (to finite types) of $\mathsf{PA}$ into $\mathsf{T}$. The extension, which we denote by $\mathsf{PA}^{\omega}$, is a quantifier version of $\mathsf{T}$. Its language $\mathcal{L}^{\omega}$ is obtained from $\mathcal{T}$ by adding quantifiers for each type. Formulas of $\mathcal{L}^{\omega}$ can now be constructed in the usual way by means of quantification. Note that the quantifier-free fragment of $\mathcal{L}^{\omega}$ is constituted exactly by the formulas of $\mathcal{T}$. $\mathsf{PA}^{\omega}$ is formulated in classical logic. Its axioms consist of the universal closures of the axioms of $\mathsf{T}$ and the induction scheme constituted by the universal closures of

$$A(0) \wedge \forall x^{\mathrm{N}}(A(x) \rightarrow A(Sx)) \rightarrow \forall x A(x)$$

where $A$ can be any formula of $\mathcal{L}^{\omega}$. There is (now) no (need for the) substitution rule nor (the) induction rule. $\mathsf{PA}^{\omega}$ can be considered an extension of first-order arithmetic $\mathsf{PA}$ because both sum and product can be defined using the recursors. As an aside, it is now possible to *define* equality $x =_{\sigma} y$ in higher types by $\forall F^{\sigma \rightarrow 0}(Fx = Fy)$. With this Leibnizian definition, we have the usual properties of equality (reflexivity, symmetry, transitivity and substitution *salva veritate*, but *not* that it coincides with extensional equality).

We are now ready to define an interpretation of $\mathsf{PA}^{\omega}$ into $\mathsf{T}$. As noted in the introduction, this interpretation is due to Shoenfield in [34]. Like all functional interpretations, it consists of a trade-off between quantifier complexity and higher types. Since the logic is classical, we may assume that the primitive logical connectives are disjunction, negation and universal quantifications.

**Definition** To each formula $A$ of the language $\mathcal{L}^{\omega}$ we assign formulas $A^{\mathrm{S}}$ and $A_{\mathrm{S}}$ so that $A^{\mathrm{S}}$ is of the form $\forall \underline{x} \exists \underline{y} A_{\mathrm{S}}(\underline{x}, \underline{y})$, with $A_{\mathrm{S}}(\underline{x}, \underline{y})$ a quantifier-free formula of $\mathcal{L}^{\omega}$, according to the following clauses:

1. $A^{\mathrm{S}}$ and $A_{\mathrm{S}}$ are simply $A$, for atomic formulas $A$.

If we have already interpretations of $A$ and $B$ given by $\forall \underline{x} \exists \underline{y} A_{\mathrm{S}}(\underline{x}, \underline{y})$ and $\forall \underline{z} \exists \underline{w} B_{\mathrm{S}}(\underline{z}, \underline{w})$ (respectively), then we define:

2. $(A \vee B)^{\mathrm{S}}$ is $\forall \underline{x}, \underline{z} \exists \underline{y}, \underline{w}(A_{\mathrm{S}}(\underline{x}, \underline{y}) \vee B_{\mathrm{S}}(\underline{z}, \underline{w}))$.
3. $(\neg A)^{\mathrm{S}}$ is $\forall \underline{f} \exists \underline{x} \neg A_{\mathrm{S}}(\underline{x}, \underline{f}\,\underline{x})$.
4. $(\forall u A(u))^{\mathrm{S}}$ is $\forall u \forall \underline{x} \exists \underline{y} A_{\mathrm{S}}(\underline{x}, \underline{y}, u)$.

In the above, the underlined variables denote tuples of variables (possibly empty). In the sequel, we omit the underlining. For example, $(\neg A)^{\mathrm{S}}$ is written as $\forall f \exists x \neg A_{\mathrm{S}}(x, fx)$. The formulas $A_{\mathrm{S}}$ are the matrices of $A^{\mathrm{S}}$. For instance, $(\neg A)_{\mathrm{S}}$ is $\neg A_{\mathrm{S}}(x, fx)$. There is a principle of choice that plays a fundamental role in

Shoenfield's interpretation. It is the quantifier-free axiom of choice in all finite types, denoted by $\mathsf{AC}^{\omega}_{\mathrm{qf}}$:

$$\forall x^{\sigma} \exists y^{\tau} A_{\mathrm{qf}}(x, y) \rightarrow \exists f^{\sigma \rightarrow \tau} \forall x A_{\mathrm{qf}}(x, fx)$$

where $\sigma$ and $\tau$ are any types and $A_{\mathrm{qf}}$ is a quantifier-free formula. This principle is called the *characteristic principle* of Shoenfield's interpretation because of the following result:

**Proposition (Characterization of Shoenfield's Interpretation)** *For any formula $A$ of $\mathcal{L}^{\omega}$, the theory $\mathsf{PA}^{\omega} + \mathsf{AC}^{\omega}_{\mathrm{qf}}$ proves the equivalence $A \leftrightarrow A^{\mathrm{S}}$.*

The proposition is easy to prove by induction on the complexity of $A$. All the clauses of Shoenfield's translation, with the exception of negation, give rise to classically equivalent formulas. The choice principle $\mathsf{AC}^{\omega}_{\mathrm{qf}}$ is exactly what is needed to deal with the negation clause. We are now ready to state Gödel's result of 1958 in the form that is most convenient for us:

**Theorem (After Gödel and Shoenfield)** *Let $A$ be a sentence of $\mathcal{L}^{\omega}$. If $\mathsf{PA}^{\omega} + \mathsf{AC}^{\omega}_{\mathrm{qf}} \vdash A$, then there are closed terms $t$ (of appropriate types) of $\mathcal{T}$ such that $\mathsf{T} \vdash A_{\mathrm{S}}(x, tx)$.*

The proof is not difficult, but it is delicate at some points. One works with a suitable axiomatization of classical logic (the one given by Shoenfield in [34] is specially convenient) and with the usual axioms of arithmetic (it is simpler to work with an induction rule instead). It can be shown that the axioms are interpretable and that the rules of inference preserve the interpretation. Roughly, the logical part of the calculus is dealt by the combinators whereas the recursors are used to interpret induction. The quantifier-free axiom of choice is interpretable (essentially) because of the way that the clause of negation is defined. The remaining axioms are universal and, therefore, trivially interpretable. This is an obviously finitistic proof.

## 3 What is Bar Recursion? Set-Theoretic Considerations

Let $C$ and $D$ be non-empty sets, and let $F : C^{<\mathbb{N}} \mapsto D$, $G : C^{<\mathbb{N}} \times D^{C} \mapsto D$ and $Y : C^{\mathbb{N}} \mapsto \mathbb{N}$ be given functions (here, $C^{<\mathbb{N}}$ denotes the set of all finite sequences of elements of $C$). We introduce some notation. First, we distinguish an element $0_C$ of $C$. Given $s \in C^{<\mathbb{N}}$, denote by $|s|$ the length of $s = \langle s_0, s_1, \ldots, s_{|s|-1} \rangle$; if $s, t \in C^{<\mathbb{N}}$, $s * t$ is the concatenation of $s$ with $t$. For $i \leq |s|$, let $s|_i$ be the sequence $\langle s_0, \ldots, s_{i-1} \rangle$. To each finite sequence $s \in C^{<\mathbb{N}}$, we denote by $\hat{s}$ the infinite sequence of $C^{\mathbb{N}}$ which prolongs $s$ by zeroes. More precisely: $\hat{s}(i) = s_i$, for $i < |s|$; $\hat{s}(i) = 0_C$, for $i \geq |s|$. Finally, for $x \in C^{\mathbb{N}}$ and $i$ a natural number, $\overline{x}(i)$ is the finite sequence $\langle x(0), \ldots, x(i-1) \rangle$.

A function $B$ from $C^{<\mathbb{N}}$ to $D$ is defined by *bar recursion* from $F$, $G$ and $Y$ if it satisfies the following equality:

$$B(s) = \begin{cases} F(s) & \text{if } \exists i \leq |s| \, (Y(\widehat{s|_i}) \leq i), \\ G(s, \lambda w. B(s * \langle w \rangle)) & \text{otherwise.} \end{cases}$$

(The knowledgeable reader will notice that the above definition is slightly different from Spector's definition. The present definition has the advantage of having the functional $Y$ directly related to a certain *tree*—as will be discussed below.) The above specification does not always define a total function. Take, for instance, the functional $Y : \mathbb{N}^{\mathbb{N}} \mapsto \mathbb{N}$ given by

$$Y(x) = \begin{cases} 0 & \text{if } \forall k \, (x(k) \neq 0), \\ i+1 & \text{if } x(i) = 0 \wedge \forall k < i \, (x(k) \neq 0). \end{cases}$$

Then, with appropriate $F$ and $G$, we could consider

$$B(s^{\mathbb{N}^{<\mathbb{N}}}) = \begin{cases} 0 & \text{if } \exists i \leq |s| \, (Y(\widehat{s|_i}) \leq i), \\ 1 + B(s * \langle 1 \rangle) & \text{otherwise} \end{cases}$$

but it is easy to argue that $B$ is not defined on the empty sequence $\langle \rangle$.

There is a simple condition on the function $Y$ whose validity ensures that $B$ is always defined. Consider the *tree*

$$T_Y := \{s \in C^{<\mathbb{N}} : \forall i \leq |s| (Y(\widehat{s|_i}) > i)\}.$$

(This set is a tree because whenever $s * t \in T_Y$ then $s \in T_Y$.) We say that the tree $T_Y$ is *well founded* if $\forall x \in C^{\mathbb{N}} \exists i \, (\overline{x}(i) \notin T_Y)$, i.e., $\forall x \in C^{\mathbb{N}} \exists i \, (Y(\widehat{\overline{x}(i)}) \leq i)$. We call the latter condition, *Spector's condition* for $Y$.

**Theorem** *Let $Y : C^{\mathbb{N}} \mapsto \mathbb{N}$ be given. The function $Y$ satisfies Spector's condition if, and only if, there is a map hgt from $T_Y$ into the ordinals such that, whenever $s$ is a strict subsequence of $t$, then $hgt(t) < hgt(s)$.*

*Proof* Suppose that there is an order inverting map *hgt* as above. Let $x \in C^{\mathbb{N}}$ be given and assume, in view of a contradiction, that $\forall i \, (\overline{x}(i) \in T_Y)$. Since, for each natural number $i$, $\overline{x}(i)$ is a strict subsequence of $\overline{x}(i+1)$, then $hgt(\overline{x}(i+1)) < hgt(\overline{x}(i))$. This gives an infinite descending sequence of ordinals, a contradiction.

Now, let us assume that $Y$ satisfies Spector's condition, i.e., that the tree $T_Y$ is well founded. Suppose, in order to reach a contradiction, that there is no order inverting map from $T_Y$ into the ordinals. This assumption implies that $T_Y \neq \emptyset$. Given $s \in T_Y$, let $T_Y/s$ be $\{t \in C^{<\mathbb{N}} : s * t \in T_Y\}$. Note that $T_Y/s$ is a non-empty tree. Consider the subset $T \subseteq T_Y$ constituted by the finite sequences $s \in T_Y$ such that $T_Y/s$ is a tree for which there is no order inverting map to the ordinals. Since $\langle \rangle \in T_Y$ and $T_Y = T_Y/\langle \rangle$, we have $\langle \rangle \in T$. Moreover, it is clear that $T$ is a *subtree* of $T_Y$. We claim that $T$ has no endnodes, i.e., we show that if $s \in T$ then there is $w \in C$ such that $s * \langle w \rangle \in T$. Suppose not. Then there is $s \in T$ such that, for each $w \in C$, we can find an ordinal $\alpha_w$ and an order inverting map $h_w$ from $T_Y/(s * \langle w \rangle)$

into $\alpha_w$. Let $\alpha = \sup_{w \in C}(\alpha_w + 1)$ (the axiom of replacement of Zermelo-Fraenkel set theory is being used in this argument), and define

$$h(t) := \begin{cases} \alpha & \text{if } t = \langle\rangle, \\ h_w(q) & \text{if } t = \langle w \rangle * q \end{cases}$$

for all $t \in T_Y/s$. By construction, $h$ is an order inverting function from $T_Y/s$ into the ordinals. This is a contradiction.

Now, since $T$ is a non-empty tree without endnodes, then $T$ has an infinite path, i.e., there is a function $x : N \mapsto T$ such that, for all natural numbers $i$, $\overline{x}(i) \in T$. This path is actually also a path through $T_Y$, contradicting Spector's condition for $Y$. □

If $Y$ satisfies Spector's condition, the above theorem permits to justify bar recursive definitions by *transfinite recursion*. In order to see this, note that if $s \in T_Y$, $w \in C$ and $s * \langle w \rangle \in T_Y$, then we have $hgt(s * \langle w \rangle) < hgt(s)$. So $B(s)$ is defined by $G(s, \lambda w.B(s * \langle w \rangle))$, an operation that only uses values of $B$ at points of $T_Y$ of smaller ordinal height than $s$ (the points outside $T_Y$ pose no problem).

There are two important conditions that easily ensure Spector's condition for the function $Y$. One is the *continuity condition*:

$$\forall x \in C^{\mathbb{N}} \exists k \in \mathbb{N} \forall y \in C^{\mathbb{N}} (\overline{y}(k) = \overline{x}(k) \to Y(x) = Y(y)).$$

The other is the (weaker) *bounding condition*:

$$\forall x \in C^{\mathbb{N}} \exists n \in \mathbb{N} \forall i \in \mathbb{N}\ Y(\widehat{\overline{x}(i)}) < n.$$

As we will briefly discuss in the next section, these two conditions are related to important structures for bar recursion. However, the bounding condition seems to be more fundamental (see [8]).

We saw that bar recursion is a form of definition by transfinite recursion on well-founded trees. We used set-theoretic arguments at will. The existence of the bar recursive functionals for well-founded $Y$ is, nevertheless, amenable to a more elementary set theoretic treatment. It is sufficient to be able to form certain subsets of $Z \subseteq C^{<\mathbb{N}} \times D$ and to use the following principle of dependent choices:

$$\forall s \in C^{<\mathbb{N}} \exists w \in C\ A(s, s * \langle w \rangle) \to \exists x \in C^{\mathbb{N}} \forall i \in \mathbb{N}\ A(\overline{x}(i), \overline{x}(i+1))$$

for suitable predicates $A$. (This principle ensures that there are "enough" infinite sequences around.) Let us briefly see why this is so.

Suppose that $Y$ satisfies Spector's condition. A set $Z \subseteq C^{<\mathbb{N}} \times D$ is a *partial bar function*, and we write $\mathbb{P}(Z)$, if $Z$ is a partial function (i.e., whenever $(s, d) \in Z$ and $(s, d') \in Z$ then $d = d'$) and, for all $s \in C^{<\mathbb{N}}$ with $s \in \mathrm{dom}(Z)$, either $s \notin T_Y \wedge Z(s) = F(s)$ or

$$s \in T_Y \wedge \forall w \in C\ (s * \langle w \rangle \in \mathrm{dom}(Z) \wedge Z(s) = G(s, \lambda w.Z(s * \langle w \rangle))).$$

We claim that if $\mathbb{P}(Z)$ and $\mathbb{P}(W)$ then $\mathbb{P}(Z \cup W)$. First, observe that it is easy to argue that if $s \in \mathrm{dom}(Z) \cap \mathrm{dom}(W)$ and $Z(s) \neq W(s)$ then there exists $w \in C$ such that $s * \langle w \rangle \in \mathrm{dom}(Z) \cap \mathrm{dom}(W)$ and $Z(s * \langle w \rangle) \neq W(s * \langle w \rangle)$. Now, if $Z$

and $W$ are not compatible at a certain given sequence $s \in C^{<\mathbb{N}}$, then (by the above observation) there must exist an infinite path $x \in C^N$ such that $\overline{x}(|s|) = s$ and, for all natural numbers $i \geq |s|$, $\overline{x}(i) \in \mathrm{dom}(Z) \cap \mathrm{dom}(W)$ and $Z(\overline{x}(i)) \neq W(\overline{x}(i))$. Of course, the existence of this path requires the principle of dependent choices. Clearly, we have $\forall i\, (\overline{x}(i) \in T_Y)$ and this contradicts the well-foundedness of $T_Y$.

Let $U := \bigcup\{Z : \mathbb{P}(Z)\}$. By the discussion above, it is clear that $\mathbb{P}(U)$. Note, also, that $C^{<\mathbb{N}} \setminus T_Y \subseteq \mathrm{dom}(U)$. If we show that $U$ is a total function, i.e., defined for every $s \in C^{<\mathbb{N}}$, then $U$ is the bar functional that we want. The following fact is easy to prove: if $s \in C^{<\mathbb{N}}$ and, for all $w \in C$, $s * \langle w \rangle \in \mathrm{dom}(U)$, then $s \in \mathrm{dom}(U)$. If $s \notin T_Y$, there is nothing to prove. If $s \in T_Y$, then $\mathbb{P}(U \cup \{(s, G(s, \lambda w.U(s * \langle w \rangle)))\})$. By the maximality of $U$, $s \in \mathrm{dom}(U)$. Now, to see that $U$ is a total function assume, in order to get a contradiction, that there is a sequence $s \in C^{<\mathbb{N}}$ such that $s \notin \mathrm{dom}(U)$. Using the above fact and dependent choices, it is easy to obtain $x \in C^{\mathbb{N}}$ such that, for all natural numbers $i$, if $i \geq |s|$, then $\overline{x}(i) \notin \mathrm{dom}(U)$. This entails $\forall i\, (\overline{x}(i) \in T_Y)$, contradicting the well-foundedness of $T_Y$.

## 4 Spector's Quantifier-Free Theory for Bar Recursion

In [36], Spector introduces a logic-free theory of computable functionals of finite type (called $\Sigma_4$ in Spector' paper). In this section, we describe a quantifier-free variant of $\Sigma_4$ building on Gödel's quantifier-free theory $\mathsf{T}$ described in Sect. 2. The terms of the language of Spector's theory include the terms of $\mathcal{T}$ together with those obtained by term application from new constants $B_{\sigma,\tau}$ of type

$$(\sigma^{<\mathrm{N}} \to \tau) \to (\sigma^{<\mathrm{N}} \to \tau^{\sigma} \to \tau) \to ((\mathrm{N} \to \sigma) \to \mathrm{N}) \to \sigma^{<\mathrm{N}} \to \tau,$$

one for each pair of types $\sigma, \tau$. We are casually using the type $\sigma^{<\mathrm{N}}$ of finite sequences of elements of type $\sigma$ even though this is not a primitive type of our language. It is nevertheless possible to deal with finite sequences via a pair consisting of an infinite sequence and a natural number (whose intended meaning is to signal the truncation of the infinite sequence at the length of the given natural number). We will not worry about these technical issues in here. Let us denote the extended quantifier-free language by $\mathcal{T}_{\mathsf{BR}}$. Its formulas are built as in Gödel's $\mathsf{T}$, only now with new terms coming from the bar constants. The theory $\mathsf{T} + \mathsf{BR}$ includes the rule of induction (and substitution) for the new formulas and the quantifier-free bar axioms (naturally) associated with the following equality:

$$B_{\sigma,\tau}(F, G, Y)(s) = \begin{cases} F(s) & \text{if } \exists i \leq |s| (Y(\widehat{s|_i}) \leq i), \\ G(s, \lambda w. B_{\sigma,\tau}(F, G, Y)(s * \langle w \rangle)) & \text{otherwise} \end{cases}$$

where, of course, $F$, $G$, $Y$ and $s$ are variables of types $\sigma^{<\mathrm{N}} \to \tau$, $\sigma^{<\mathrm{N}} \to \tau^{\sigma} \to \tau$, $(\mathrm{N} \to \sigma) \to \mathrm{N}$ and $\sigma^{<\mathrm{N}}$, respectively. As noted in the previous section, the above definition of bar recursion is not quite the same as Spector's. It is easy to see that our theory is included in Spector's theory. We do not know if they are the same theory.

We saw that the set-theoretical structure is not a model of $\mathsf{T} + \mathsf{BR}$. This is one of the main differences between Gödel's $\mathsf{T}$ and Spector's $\mathsf{T} + \mathsf{BR}$: the former, but not the latter, has the usual set-theoretic interpretation. This is due to the fact that definitions by ordinary number recursion are always available set-theoretically whereas definitions by bar-recursion depend on a certain well-foundedness condition (Spector's condition). Spector's theory enjoys the astonishing property that *every* functional whose type is of the form $(\mathsf{N} \to \sigma) \to \mathsf{N}$ automatically satisfies Spector's condition (cf. Kreisel's trick in Sect. 6). The property of well-foundedness is *unconditionally* associated with certain functionals, unlike in ordinary settings where one must always explicitly hypothesize conditions for it. This feature is related to certain intuitionistic ideas according to which uniform continuity automatically holds for real-valued functions defined on a closed bounded interval.

The first rigorous proofs that certain structures are models of Spector's $\mathsf{T} + \mathsf{BR}$ only appeared in the early seventies. If we put aside the term model (see [42] or [30]), Bruno Scarpellini's proof in [33] that the structure of all sequentially continuous functionals is a model of $\mathsf{T} + \mathsf{BR}$ is, to my knowledge, the first such rigorous proof. Spector's condition of the pertinent functionals is assured by the continuity condition mentioned in the previous section. Together with the fact that *all* functions with domain the (discrete topological) space of the natural numbers are sequentially continuous, it ensures—as we saw—that bar recursive functionals can be defined (one has also to check that they are sequentially continuous). Anne Troelstra shows in [42] that the structures $\mathrm{ICF}^{\omega}$ and $\mathrm{ECF}^{\omega}$ of the intensional (respectively, extensional) continuous functionals are models of $\mathsf{T} + \mathsf{BR}$ (it can be proven that the extensional structure is isomorphic to Scarpellini's model—cf. [19]). These structures, based on continuity assumptions, are natural to consider because they flow from the very intuitionistic ideas that were at the source of Spector's interpretation (see the next section). In 1985, Marc Bezem presents a quite different model. Bezem's structure [2] uses the so-called strongly majorizable functionals and admits discontinuous functionals. Spector's condition of the pertinent functionals is assured by the bounding condition mentioned in the previous section. Since *all* infinite sequences of (strongly) majorizable functionals are, themselves, strongly majorizable (i.e., are in Bezem's model), bar recursive functionals can be defined (of course, one must also check that the functionals so obtained are strongly majorizable). All rigorous proofs that some structures are models of Spector's theory appeared quite some years after 1962. This fact is a source of amazement for me, and it tells much about the spell of intuitionism among some logicians at the time.

As in the case of Gödel's $\mathsf{T}$, we can extend the quantifier-free language of the theory $\mathsf{T} + \mathsf{BR}$ to a quantificational language $\mathcal{L}^{\omega}_{\mathsf{BR}}$ and consider the corresponding quantificational theory $\mathsf{PA}^{\omega} + \mathsf{AC}^{\omega}_{\mathrm{qf}} + \mathsf{BR}$. This theory consists of $\mathsf{PA}^{\omega} + \mathsf{AC}^{\omega}_{\mathrm{qf}}$, allowing now for the new formulas in the schemata of induction and quantifier-free choice, together with the new bar axioms.

**Theorem (Soundness Theorem for Bar Recursion)** *Let $A$ be a sentence of the language of $\mathcal{L}^{\omega}_{\mathsf{BR}}$. If $\mathsf{PA}^{\omega} + \mathsf{AC}^{\omega}_{\mathsf{qf}} + \mathsf{BR} \vdash A$, then there are closed terms (of appropriate types) of $\mathcal{T}_{\mathsf{BR}}$ such that $\mathsf{T} + \mathsf{BR} \vdash A_{\mathsf{S}}(x, tx)$.*

*Proof* The proof of the soundness of Shoenfield's interpretation needs hardly any additional work because the new bar axioms are universal closures of quantifier-free formulas and, hence, are automatically interpreted (by themselves). □

## 5 What is Bar Recursion? Brouwerian Considerations

Spector's paper is entitled "Provably recursive functionals of analysis: a consistency proof of analysis by an extension of principles formulated in current intuitionistic mathematics." According to Georg Kreisel in p. 161 of [29], the long title incorporates contributions by Spector, Gödel and Kreisel himself. Be that as it may, the catchword 'extension' is common to the title of Gödel's paper of 1958. Spector's paper, like Gödel's, tries to reduce the consistency of a classical theory to the acceptance of an *extension* of a certain foundational framework: Hilbert's finitism in Gödel's paper, Brouwer's intuitionism in Spector's case. Furthermore, both extensions share a similar pattern: they follow Gödel's cherished idea of "gain(ing) knowledge abstractly by means of notions of higher type" (quoted from Gödel's [11]).

A form of bar induction commonly accepted in intuitionistic mathematics is *monotone bar induction*. In the following, we formulate this principle in the language of finite-type arithmetic $\mathcal{L}^{\omega}$. The type $\sigma$ and the formulas $P$ and $Q$ below are unrestricted (note that, in Brouwerian intuitionism, $\sigma$ must be the type N of the natural numbers): If

Hyp1. $\forall x^{\mathrm{N}\to\sigma} \exists k^{\mathrm{N}} P(\overline{x}(k))$
Hyp2. $\forall s^{\sigma^{<\mathrm{N}}} \forall i \leq |s| (P(s|_i) \to P(s))$
Hyp3. $\forall s^{\sigma^{<\mathrm{N}}} (P(s) \to Q(s))$
Hyp4. $\forall s^{\sigma^{<\mathrm{N}}} (\forall w^{\sigma} Q(s * \langle w \rangle) \to Q(s))$

then $Q(\langle\rangle)$.

It is easy to argue that this principle is set-theoretically true (contrast this fact with bar recursion). Suppose that $Q(\langle\rangle)$ is false. Then, by Hyp4, there is $w_0$ such that $\neg Q(\langle w_0 \rangle)$. By Hyp 4 again, there is $w_1$ with $\neg Q(\langle w_0, w_1 \rangle)$. We can continue this process and get a sequence $w$ of elements of type $\sigma$ such that $\forall k \in \mathbb{N} \neg Q(\overline{w}(k))$. Of course, a form of dependent choices is needed to arrive at this conclusion. By Hyp3, $\forall k \in \mathbb{N} \neg P(\overline{w}(k))$. This contradicts Hyp1. Notice that the monotonicity condition Hyp2 was not used in the argument. Even though Hyp2 is not needed to justify *classically* the principle of bar induction, without Hyp2 the principle is not intuitionistically acceptable (because it would entail the lesser limited principle of omniscience, a weaker form of excluded middle also rejected by the intuitionists: cf. exercise 4.8.11 in [41]).

Together with a continuity argument, the above principle of bar induction proves (intuitionistically) the existence of the bar recursive functionals. Take $F$, $G$ and $Y$ as in the previous section. Let $P(s^{\sigma^{<N}})$ be $\exists i \leq |s|(Y(\widehat{s|_i}) \leq i)$ and define $Q(s^{\sigma^{<N}})$ by:

$$\exists B \forall t^{\sigma^{<N}} [(P(s*t) \wedge B(s*t) = F(s*t)) \vee \\ (\neg P(s*t) \wedge B(s*t) = G(s*t, \lambda w^{\sigma}.B(s*t*\langle w \rangle)))]$$

where the variables have appropriate types. It is clear that Hyp2 and Hyp3 hold. The verification of Hyp4 uses an intuitionistically admissible form of choice. Hyp1 is true by appealing to the continuity of the functional $Y$ (this is the only place in the argument which is not set-theoretically sound). Therefore, we can conclude $Q(\langle\rangle)$, i.e., that there exists the bar functional $B(F, G, Y)$. It is also not difficult to prove (by bar induction) that this functional is unique.

We have shown that bar recursion of type $\sigma$ reduces intuitionistically to bar induction of the same type (with the aid of a principle of continuity). Can bar recursion of finite type be constructively justified? The matter was taken up in a seminar on the foundations of analysis led by Kreisel at Stanford in the summer of 1963, and a report [27] circulated. The answer was that "for the precise formulation in this report of constructive principles implicit in known intuitionistic mathematics, the answer is negative by a wide margin (...)." What are these principles? They "concern primarily functionals of finite and transfinite types, free choice sequences, and generalized inductive definitions." The story of the accomplishments of the seminar and of the ensuing work over the next years is long-winded. To cut through the fog, I believe that it is fair to say that the proof-theoretic strength of the principles of intuitionistic mathematics considered by Kreisel lies at the level of the theory $\mathsf{ID}_1$ of non-iterated monotone inductive definitions. They are enough to justify bar recursion of type N (i.e., when $\sigma$ is N) and perhaps (by slightly stronger theories) also of type N $\to$ N (see Sect. 7 of [18]), but not more. The results were certainly disillusioning. Kreisel confides in [28] that "when I originally considered the extension of [Spector] to analysis I believed that the *particular* notion of functional of finite type there described could be proved by intuitionistic methods to satisfy [the functional interpretation of analysis]. Put differently, I thought that the *existing* intuitionistic theory of free choice sequences, especially if one uses the formally powerful continuity axioms, was of essentially the same proof theoretic strength as full classical analysis!" (italics as in the original). As we now know, $\mathsf{ID}_1$ has the proof-theoretic strength of $\Pi^1_1$-comprehension without set parameters and it is a far cry from full second-order comprehension.

Note, however, that the answer is negative as measured against *existing* intuitionistic theory. The title of Spector's paper explicitly mentions an *extension* of principles formulated in current intuitionistic mathematics. We believe that the benefits of Spector's consistency proof have to be judged on its own terms: against the intuitionistic plausibility of the extension proposed. In the Stanford report, Kreisel writes that, according to Gödel, "if one finds Brouwer's argument for the bar theorem conclusive then one should accept the generalization in Spector's paper." (Interestingly, it is added that "nothing much was intended to follow from this because [Gödel] does not find Brouwer's argument conclusive." In the same vein, Spector says in his paper that the bar theorem is itself questionable and in need of a suitable foundation.) I see in Gödel's opinion the implication that a conclusive argument for the bar theorem would generalize to bar induction in finite types. It would be expedient if the specialists who are convinced by Brouwer's argument could give their assessment of its possible generalization to higher types. There is also the other leg of the argument, the one regarding the continuity condition. In an intuitionistic setting, continuity is a consequence of Brouwer's doctrine about choice sequences. Their analogue in Spector's framework are sequences of higher type functionals (vis-à-vis sequences of concrete natural numbers). Do choice sequences of higher-order *abstracta* make sense for the 'creating subject'?

## 6 Bar Recursion Entails Bar Induction (in the Presence of Quantifier-Free Choice)

We prove a result that is preparatory for interpreting analysis in Spector's $\mathsf{T}+\mathsf{BR}$, viz. that a certain simplified form of bar induction is a consequence of $\mathsf{PA}^{\omega}+\mathsf{AC}^{\omega}_{\mathrm{qf}}+\mathsf{BR}$. The next proposition is instrumental in showing this. It says that, in the presence of $\mathsf{BR}$, functionals of type $Y^{(\mathrm{N}\to\sigma)\to\mathrm{N}}$ automatically satisfy Spector's condition:

**Proposition (Kreisel's trick [26])** *For any type* $\sigma$, *the theory* $\mathsf{PA}^{\omega}+\mathsf{BR}$ *proves the sentence* $\forall Y^{(\mathrm{N}\to\sigma)\to\mathrm{N}}\forall x^{\mathrm{N}\to\sigma}\exists i^{\mathrm{N}}\,(Y(\widehat{\overline{x}(i)})\leq i)$.

*Proof* Fix $Y$ and $x$. Define $W:\sigma^{<\mathrm{N}}\to\mathrm{N}$ by bar recursion in the following way:

$$W(s):=\begin{cases}0 & \text{if } \exists i\leq |s|\,(Y(\widehat{s|_i})\leq i),\\ 1+W(s*\langle x(|s|)\rangle) & \text{otherwise}\end{cases}$$

and let $h(k):=W(\overline{x}(k))$. By definition, it is clear that

$$h(k)=\begin{cases}0 & \text{if } \exists i\leq k\,(Y(\widehat{\overline{x}(i)})\leq i),\\ 1+h(k+1) & \text{otherwise.}\end{cases}$$

Let $k$ be given. Clearly, if $h(k)\neq 0$ and $i\leq k$, then $h(0)=i+h(i)$. In particular, if $h(k)\neq 0$, $h(0)=k+h(k)$. Instantiating $k$ by $h(0)$, we can conclude that if $h(h(0))\neq 0$ then $h(0)=h(0)+h(h(0))$. Therefore, $h(h(0))=0$. By definition of $h$, we conclude that $\exists i\leq h(0)\,(Y(\widehat{\overline{x}(i)})\leq i)$. □

Given a type $\sigma$ and an existential formula $P(s)$ with a distinguished variable $s$ of type $\sigma^{<\mathrm{N}}$, we consider the following simplified version of monotone bar induction, denoted by $\mathsf{BI}^{-}_{\exists}$: From the three hypotheses

H1. $\forall x^{\mathrm{N}\to\sigma}\exists k^{N}\,P(\overline{x}(k))$

H2. $\forall s^{\sigma^{<\mathrm{N}}}\forall i\leq |s|\,(P(s|_i)\to P(s))$

H3. $\forall s^{\sigma^{<\mathrm{N}}}(\forall w^{\sigma}\,P(s*\langle w\rangle)\to P(s))$

one can conclude $P(\langle\rangle)$.

**Theorem (After Howard)** *The theory* $\mathsf{PA}^\omega + \mathsf{AC}^\omega_{\mathrm{qf}} + \mathsf{BR}$ *proves* $\mathsf{BI}^-_\exists$.

*Proof* Let $P(s)$ be the existential statement $\exists a^\tau P_{\mathrm{qf}}(s,a)$, where $s$ has type $\sigma^{<\mathrm{N}}$ and $P_{\mathrm{qf}}$ is a quantifier-free formula. Assume the hypotheses of bar-induction. By the first hypothesis, $\forall x \exists k, a P_{\mathrm{qf}}(\overline{x}(k), a)$. By $\mathsf{AC}^\omega_{\mathrm{qf}}$, there are functionals $Y : (\mathrm{N} \to \sigma) \to \mathrm{N}$ and $H : (\mathrm{N} \to \sigma) \to \tau$ such that

$\tilde{\mathrm{H}}1.\quad \forall x\, P_{\mathrm{qf}}(\overline{x}(Yx), Hx).$

By the second hypothesis, $\forall s \forall i \leq |s| \forall a \exists b (P_{\mathrm{qf}}(s|_i, a) \to P_{\mathrm{qf}}(s,b))$. Hence, by $\mathsf{AC}^\omega_{\mathrm{qf}}$, there is a functional $F : \sigma^{<\mathrm{N}} \to \mathrm{N} \to \tau \to \tau$ such that

$\tilde{\mathrm{H}}2.\quad \forall s \forall i \leq |s| \forall a\, (P_{\mathrm{qf}}(s|_i, a) \to P_{\mathrm{qf}}(s, F(s,i,a))).$

By $\mathsf{AC}^\omega_{\mathrm{qf}}$, it is easy to see that $\forall s, f^{\sigma\to\tau} \exists w, b\, (P_{\mathrm{qf}}(s * \langle w\rangle, fw) \to P_{\mathrm{qf}}(s,b))$ is equivalent to the last hypothesis of bar induction. Using $\mathsf{AC}^\omega_{\mathrm{qf}}$ to witness $w$ and $b$ and then disregarding the witness of $w$, there is $G : \sigma^{<\mathrm{N}} \to (\sigma \to \tau) \to \tau$ such that

$\tilde{\mathrm{H}}3.\quad \forall s, f\, (\forall w P_{\mathrm{qf}}(s * \langle w\rangle, fw) \to P_{\mathrm{qf}}(s, G(s,f))).$

Let us define, by bar-recursion, the following functional:

$$B(s) := \begin{cases} F(s, Y(\widehat{s|_{i_0}}), H(\widehat{s|_{i_0}})) & \text{if } \exists i \leq |s|\, (Y(\widehat{s|_i}) \leq i), \\ G(s, \lambda w. B(s * \langle w\rangle)) & \text{otherwise} \end{cases}$$

where $i_0$ is the least number $i$ such that $Y(\widehat{s|_i}) \leq i$.

We claim that, for all $s$ of type $\sigma^{<\mathrm{N}}$, if $\exists i \leq |s|\, Y(\widehat{s|_i}) \leq i$, then $P_{\mathrm{qf}}(s, Bs)$. In fact, by $\tilde{\mathrm{H}}1$, we have $P_{\mathrm{qf}}(\overline{\widehat{s|_{i_0}}}(Y(\widehat{s|_{i_0}})), H(\widehat{s|_{i_0}}))$. Since $Y(\widehat{s|_{i_0}}) \leq i_0 \leq |s|$, the finite sequence $\overline{\widehat{s|_{i_0}}}(Y(\widehat{s|_{i_0}}))$ is actually the sequence $s|_{Y(\widehat{s|_{i_0}})}$. Hence, $P_{\mathrm{qf}}(s|_{Y(\widehat{s|_{i_0}})}, H(\widehat{s|_{i_0}}))$. Using $\tilde{\mathrm{H}}2$, we get $P_{\mathrm{qf}}(s, F(s, Y(\widehat{s|_{i_0}}), H(\widehat{s|_{i_0}})))$, that is, $P_{\mathrm{qf}}(s, Bs)$.

Secondly, we claim that, for all $s$ of type $\sigma^{<\mathrm{N}}$, if $\forall i \leq |s|\, Y(\widehat{s|_i}) > i$, then

$\forall w P_{\mathrm{qf}}(s * \langle w\rangle, B(s * \langle w\rangle)) \to P_{\mathrm{qf}}(s, Bs).$

To see this, suppose that $\forall w P_{\mathrm{qf}}(s * \langle w\rangle, B(s * \langle w\rangle))$. Let $f := \lambda w. B(s * \langle w\rangle)$. With this notation, we have $\forall w P_{\mathrm{qf}}(s * \langle w\rangle, fw)$. By $\tilde{\mathrm{H}}3$, we conclude that $P_{\mathrm{qf}}(s, G(s,f))$, that is, $P_{\mathrm{qf}}(s, Bs)$.

Of course, the above two claims entail that, for every $s$ of type $\sigma^{<\mathrm{N}}$,

$\forall w P_{\mathrm{qf}}(s * \langle w\rangle, B(s * \langle w\rangle)) \to P_{\mathrm{qf}}(s, Bs).$

Therefore,

$\forall s\, [\neg P_{\mathrm{qf}}(s, Bs) \to \exists w \neg P_{\mathrm{qf}}(s * \langle w\rangle, B(s * \langle w\rangle))].$

By $\mathsf{AC}^\omega_{\mathrm{qf}}$, there is a functional $T : \sigma^{<\mathrm{N}} \to \sigma$ such that

$\forall s\, [\neg P_{\mathrm{qf}}(s, Bs) \to \neg P_{\mathrm{qf}}(s * \langle Ts\rangle, B(s * \langle Ts\rangle))].$

We now define, by recursion, a functional $z$ of type $N \to \sigma$ according to the following clause: $z(k) = T(\langle z(0), z(1), \ldots, z(k-1)\rangle)$. Note that $z(0) = T(\langle\rangle)$. By construction, we have

$$\neg P_{\mathrm{qf}}(\bar{z}(k), B(\bar{z}(k))) \to \neg P_{\mathrm{qf}}(\bar{z}(k+1), B(\bar{z}(k+1))),$$

for every natural number $k$.

Suppose, in order to reach a contradiction, that the conclusion of bar-induction fails, i.e., that $\neg P(\langle\rangle)$. Therefore, $\forall a \neg P_{\mathrm{qf}}(\langle\rangle, a)$. In particular, $\neg P_{\mathrm{qf}}(\bar{z}(0), B(\bar{z}(0)))$. By induction, we get $\neg P_{\mathrm{qf}}(\bar{z}(k), B(\bar{z}(k)))$, for all $k^{\mathrm{N}}$. This contradicts the fact that, by Kreisel's trick, there is $i^{\mathrm{N}}$ such that $Y(\widehat{\bar{z}(i)}) \leq i$ and, hence, by the first claim above, that $P_{\mathrm{qf}}(\bar{z}(i), B(\bar{z}(i)))$. □

We finish this section with a discussion concerning equality. At a certain point of the previous argument, we apparently used the axiom of extensionality. The finite sequences $\overline{\widehat{s|_{i_0}}}(Y(\widehat{s|_{i_0}}))$ and $s|_{Y(\widehat{s|_{i_0}})}$ are extensionally equal. As a result, we used the implication

$$P_{\mathrm{qf}}(\overline{\widehat{s|_{i_0}}}(Y(\widehat{s|_{i_0}})), H(\widehat{s|_{i_0}})) \to P_{\mathrm{qf}}(s|_{Y(\widehat{s|_{i_0}})}, H(\widehat{s|_{i_0}})),$$

which amounts to a substitution *salva veritate*. Actually, by carefully defining the notion of finite sequence, this implication can be justified without any extensionality assumptions, and the theorem above is correct as it stands (i.e., based on the *minimal* theory $\mathsf{PA}^{\omega}$). There are alternatives, though: one way out is not to worry about the precise definition of finite sequences and simply admit in our theory the universal statements $j \leq i \leq |s| \wedge \Phi(\overline{\widehat{s|_i}}(j)) \to \Phi(s|_j)$. Granted, this is an *ad hoc* maneuver (but quite an admissible one). A more systematic way of getting the desired universal statements is to include in our base theory a so-called weak extensionality rule. This is the choice of Spector in his original paper. In [25], Kohlenbach follows this route and the reader is directed to this reference for a thorough discussion of this rule.

## 7 The Interpretation of Analysis

Analysis, a.k.a. full second-order arithmetic $\mathsf{PA}_2$, is the extension of first-order arithmetic $\mathsf{PA}$ to a language $\mathcal{L}^2$ with a new sort of (second-order) variables for sets of natural numbers, a new kind of atomic formulas taking the form '$t \in X$', where $t$ is a first-order term and $X$ is a second-order variable, and whose axioms include the full comprehension scheme:

$$\exists X \forall x (x \in X \leftrightarrow A(x))$$

where $A$ is any formula of $\mathcal{L}^2$ (first and second-order parameters are allowed). Induction in $\mathsf{PA}_2$ can be stated by the single axiom

$$\forall X(0 \in X \wedge \forall x(x \in X \to Sx \in X) \to \forall x(x \in X)).$$

Given that we have full comprehension, induction actually applies to every formula of the second-order language. The language $\mathcal{L}^2$ can be embedded into $\mathcal{L}^\omega$ by letting the number variables run over arguments of type N, letting the set variables run over variables of type N $\to$ N subjected to a process of normalization (so that they take values in $\{0, 1\}$), and by interpreting $t \in X$ by $X(t) = 0$.

We need three definitions within $\mathcal{L}^\omega$:

**Definition** The principle of full numerical comprehension $\mathsf{CA}^{\mathsf{N}}$ is the following scheme:

$$\exists f^{\mathsf{N}\to\mathsf{N}}\forall x^{\mathsf{N}}(fx = 0 \leftrightarrow A(x))$$

where $A$ is any formula.

**Definition** The principle of dependent choices $\mathsf{DC}^\omega$ is the following scheme:

$$\forall x^\sigma \exists y^\sigma A(x, y) \to \forall u^\sigma \exists f^{\mathsf{N}\to\sigma}(f0 = u \wedge \forall k A(fk, f(k+1))),$$

where $\sigma$ is any type and $A$ is any formula. The restriction of the above principle to universal formulas $A$ is denoted by $\mathsf{DC}^\omega_\forall$.

**Definition** The principle of numerical choice $\mathsf{AC}^{\mathsf{N},\omega}$ is the following scheme:

$$\forall k \exists x^\sigma A(k, x) \to \exists f^{\mathsf{N}\to\sigma}\forall k\, A(k, fk),$$

where $\sigma$ is any type and $A$ is any formula. The restriction of this principle when $\sigma$ is the type of natural numbers N is denoted by $\mathsf{AC}^{\mathsf{N},\mathsf{N}}$.

**Proposition (Easy Facts)**

1. $\mathsf{PA}^\omega + \mathsf{BI}^-_\exists \vdash \mathsf{DC}^\omega_\forall$.
2. $\mathsf{PA}^\omega + \mathsf{AC}^\omega_{\mathsf{qf}} + \mathsf{DC}^\omega_\forall \vdash \mathsf{DC}^\omega$.
3. $\mathsf{PA}^\omega + \mathsf{DC}^\omega \vdash \mathsf{AC}^{\mathsf{N},\omega}$.
4. $\mathsf{PA}^\omega + \mathsf{AC}^{\mathsf{N},\mathsf{N}} \vdash \mathsf{CA}^{\mathsf{N}}$.

*Proof* Let $A(x^\sigma, y^\sigma)$ be a universal formula such that $\forall x \exists y A(x, y)$. Fix $u^\sigma$. We must show that there is $f : \mathsf{N} \to \sigma$ such that $f0 = u$ and $\forall k A(fk, f(k+1))$. Take $v^\sigma$ such that $A(u, v)$ and define the following existential formula $P(s^{\sigma^{<\mathsf{N}}})$:

$$P(s) := \exists i \leq |s| \neg A((\langle u, v\rangle * s)_i, (\langle u, v\rangle * s)_{i+1}).$$

By the choice of $v$, it is clear that $\neg P(\langle\rangle)$. We claim that the hypotheses H2 and H3 of $\mathsf{BI}^-_\exists$ hold for $P$. This is straightforward for H2 and not so difficult to verify for H3. Suppose that, for a given $s^{\sigma^{<\mathsf{N}}}$ one has $\forall w P(s * \langle w\rangle)$. This means that, for all $w^\sigma$, either

$$\exists i \leq |s| \neg A((\langle u, v\rangle * s * \langle w\rangle)_i, (\langle u, v\rangle * s * \langle w\rangle)_{i+1})$$

or $\neg A(((\langle u, v\rangle * s * \langle w\rangle)_{|s|+1}, (\langle u, v\rangle * s * \langle w\rangle)_{|s|+2})$. The first disjunct is equivalent to $P(s)$ whereas the second is equivalent to $\neg A(z, w)$, where $z = v$ if $|s| = 0$ and $z = s_{|s|-1}$ otherwise. By the arbitrariness of $w$, one has $P(s) \vee \forall w \neg A(z, w)$. This entails $P(s)$ and, therefore, the verification of H3 is finished.

By $\mathsf{BI}_{\exists}^{-}$, we must conclude that H1 fails for $P$. Therefore,

$$\exists x^{\mathrm{N}\to\sigma} \forall k \forall i \leq k\ A(((\langle u, v\rangle * \overline{x}(k))_i, (\langle u, v\rangle * \overline{x}(k))_{i+1}).$$

It is now clear that the function

$$f(k) := \begin{cases} u & \text{if } k = 0, \\ v & \text{if } k = 1, \\ x(k-1) & \text{if } k \geq 2 \end{cases}$$

satisfies $f0 = u$ and $\forall k A(fk, f(k+1))$.

We have just proved the first easy fact. (This argument is basically in the appendix of [14].) In order to prove the second fact, take $A(x^\sigma, y^\sigma)$ any formula such that $\forall x \exists y A(x, y)$ and fix $u^\sigma$. By Proposition 2, the formula $A(x, y)$ is equivalent to $\forall w \exists z A_{\mathrm{S}}(w, z, x, y)$ for $w$ and $z$ of appropriate types. Using $\mathsf{AC}^{\omega}_{\mathrm{qf}}$, $A(x, y)$ is equivalent to $\exists h \forall w A_{\mathrm{S}}(w, hw, x, y)$. By hypothesis, and inserting a dummy variable $g$ (of the same type as $h$), we have

$$\forall x, g \exists y, h \forall w A_{\mathrm{S}}(w, hw, x, y).$$

Since $A_{\mathrm{S}}$ is quantifier-free, we are in the conditions of application of $\mathsf{DC}^{\omega}_{\forall}$. Therefore, there are $f^{\mathrm{N}\to\sigma}$ and appropriate $l$ such that $f(0) = u$ and

$$\forall k \forall w A_{\mathrm{S}}(w, l(k+1)w, fk, f(k+1)).$$

We conclude that $\forall k A(fk, f(k+1))$.

Let us now consider the third fact. Suppose that $\forall k \exists x^\sigma A(k, x)$. Fix $u$ such that $A(0, u)$. Clearly, $\forall k, x \exists n, y\ (n = k + 1 \wedge A(n, y))$. By $\mathsf{DC}^\omega$, there are $f^{\mathrm{N}\to\sigma}$ and $g^{\mathrm{N}\to\mathrm{N}}$ such that $f(0) = u$, $g(0) = 0$ and $\forall k(g(k+1) = g(k) + 1 \wedge A(g(k+1), f(k+1)))$. It is clear, by induction, that $\forall k(gk = k)$. It easily follows that $\forall k A(k, fk)$.

The proof of the fourth fact is well known. Take an arbitrary formula $A(k^{\mathrm{N}})$. Clearly, $\forall k \exists n((n = 0 \wedge A(k)) \vee (n = 1 \wedge \neg A(k)))$. By $\mathsf{AC}^{\mathrm{N,N}}$, there is $f^{\mathrm{N}\to\mathrm{N}}$ such that, for all $k^{\mathrm{N}}$, $A(k)$ if, and only if, $fk = 0$. □

Howard's theorem of the previous section, together the above facts, entails that $\mathsf{DC}^\omega$ is a consequence of the theory $\mathsf{PA}^\omega + \mathsf{AC}^{\omega}_{\mathrm{qf}} + \mathsf{BR}$. We highlight the following result:

**Corollary** *The theory* $\mathsf{PA}^\omega + \mathsf{AC}^{\omega}_{\mathrm{qf}} + \mathsf{BR}$ *proves* $\mathsf{CA}^{\mathrm{N}}$.

By the above corollary, $\mathsf{PA}_2$ can be considered a subtheory of $\mathsf{PA}^\omega + \mathsf{AC}^{\omega}_{\mathrm{qf}} + \mathsf{BR}$. For ease of reading, in the next theorem we identify a formula of $\mathcal{L}^2$ with its translation into $\mathcal{L}^\omega$.

**Theorem (After Spector)** *Let $A$ be a sentence of the language of second-order arithmetic. If* $\mathsf{PA}_2 \vdash A$*, then there are closed terms $t$ (of appropriate types) of* $\mathcal{T}_{\mathsf{BR}}$ *such that* $\mathsf{T} + \mathsf{BR} \vdash A_{\mathrm{S}}(x, tx)$.

*Proof* Suppose that $\mathsf{PA}_2 \vdash A$. By the discussion above, $\mathsf{PA}^{\omega} + \mathsf{AC}^{\omega}_{\mathrm{qf}} + \mathsf{BR} \vdash A$. By the soundness theorem of Sect. 4, the result follows. □

Note that the above proof is finitistic. Therefore, by considering the formula $A$ to be $0 = 1$, this theorem shows that the consistency of analysis is finitistically reducible to the consistency of Spector's quantifier-free theory $\mathsf{T} + \mathsf{BR}$.

Let us finish this section with some remarks concerning subsystems of $\mathsf{T} + \mathsf{BR}$. The restriction of $\mathsf{T}+\mathsf{BR}$ to bar recursion of type N is very-well understood (see the next section), and it has played a fruitful role in the foundations of mathematics. For instance, Kohlenbach gave in [23] a particularly perspicuous analysis of arithmetical comprehension based on the bar-recursor $B_{\mathrm{N},\mathrm{N}\to\mathrm{N}}$.

## 8 Epilogue

Spector's consistency proof is a beautiful and sophisticated piece of work. It provides a surprising way of replacing the comprehension principles of analysis by forms of transfinite recursion. But, as a *consistency proof*, does it command any epistemological conviction? In Sect. 5, we defended that the most promising argument for a possible epistemological gain provided by Spector's proof is still the original intended one: to rely on an *extension* of the principles of Brouwerian intuitionism by considering the generalization of bar induction to finite types. It is nevertheless an almost universal conviction that Brouwer's argument for the bar theorem is inconclusive, let alone its possible extension to higher types. With no conclusive arguments for the extension, nothing much is attained.

Other readings of Spector's proof are possible. Spector's proof reduces the comprehension principles of analysis to the termination of some effective processes (viz, to the normalization of the closed terms of $\mathcal{T}_{\mathsf{BR}}$). This is no mean achievement. The postulation of the normalization of the closed terms of $\mathcal{T}_{\mathsf{BR}}$ is sufficient to prove (modulo some weak arithmetic) the consistency of analysis. Proofs of normalization for the terms of $\mathcal{T}_{\mathsf{BR}}$ do exist in the literature (they, in fact, guarantee the existence of the term model). The first such proof is, to my knowledge, due to Tait in [38]. Of necessity (by Gödel's second incompleteness theorem), these proofs use proof-theoretic power stronger than the power of analysis. Tait's proof is not ordinal informative. However, the situation is different for some subclasses of $\mathcal{T}_{\mathsf{BR}}$. There are proofs of the normalization of the terms of Gödel's $\mathsf{T}$ by the method of assigning to them ordinals less than $\epsilon_0$ (see [40] for references), therefore providing another route to Gentzen's proof of the consistency of $\mathsf{PA}$. Moreover, Helmut Vogel and

Howard gave in [44], [16] and [17] a detailed ordinal analysis of bar recursion of type N. As far as we are aware, ordinal analyses of stronger forms of bar recursion have not been pursued.

The most important benefits of Spector's proof probably lie elsewhere. Not in consistency proofs but in applications to the extraction of computational information from ordinary proofs of mathematics. The methods of Kohlenbach's proof mining (conveniently reported in [25]) can be applied to full second-order arithmetic because of the work of Spector. Kohlenbach, as a matter of course, works with systems with full second-order comprehension. In more recent studies, bar recursion has also been extended to new types, used to interpret—for instance—abstract normed spaces (see [24] and [10]). Even though the uses of bar recursion have not yet shown up in an essential way in the analyses of ordinary mathematical proofs, the situation can—in principle—change. Kohlenbach's methods are also deeply interwoven with questions of uniformity (i.e., the obtaining of bounds independent from some parameters), including a set-theoretic false uniform boundedness principle. These methods are possible within analysis because of the majorizability of the bar recursive constants. Majorizability considerations have played an important role in the removal of ideal elements (conservation results). The paramount example is the elimination of weak König's lemma (fan theorem) for theories without arithmetical comprehension: see [35] and [22]. The bounded functional interpretation of Ferreira and Oliva [9] can be seen as a thorough exploitation of majorizability properties. It was first defined for arithmetic but it extends to analysis via bar recursion (cf. [5] and [4]). The relations between functional interpretations, majorizability, uniformity results, elimination of ideal elements, extraction of computational information and the role of some classically false principles constitute a fascinating topic in the foundations of mathematics. My paper [6] includes a general discussion on these issues.

# Appendix

Let $T_K$ be a Kleene binary tree, i.e., a primitive recursive infinite tree of finite binary sequences with no infinite recursive path (Kleene introduced such an example in [20]). The form of bar induction described in Sect. 6 can be used to prove that $T_K$ has an infinite path. This can be seen by considering the existential statement

$$P(s^{\{0,1\}^{<\mathrm{N}}}) := \exists l^{\mathrm{N}}(l \geq |s| \wedge \forall t^{\{0,1\}^{<\mathrm{N}}}(|s * t| = l \rightarrow s * t \notin T_K)).$$

(The universal quantification on the finite binary sequence $t$ can be considered bounded because the length of $t$ does not exceed $l$.) Both H2 and H3 hold but, since Kleene's tree is infinite, the conclusion $P(\langle\rangle)$ fails. Therefore, H1 must fail and this readily entails that there is an infinite path through $T_K$. By the choice of $T_K$, this infinite Boolean sequence is not recursive. A close inspection of the proof of Howard's theorem in Sect. 6 shows that the amount of quantifier-free choice needed to prove the required bar induction is just $\mathsf{AC}_{\mathsf{qf}}^{\mathrm{N}\to\{0,1\},\mathrm{N}}$ (the meaning of this notation should be clear). This amount of choice justifies the existence of a functional $Y$ of type $(\mathrm{N} \to \{0,1\}) \to \mathrm{N}$ with the property that $\forall x^{\mathrm{N}\to\{0,1\}}(\overline{x}(Yx) \notin T_K)$, and this fact is sufficient to pull the proof through.

It is the *combination* of bar recursion and quantifier-free choice that is responsible for the introduction of non-recursive Boolean sequences in models of $\mathsf{PA}^{\omega}$ (via forms of bar induction or, what is classically the same thing, via dependent choices). The lack of just one of these ingredients may result in the failure of introducing non-recursive sequences. For instance, the structure $\mathrm{HRO}^{\omega}$ of the hereditarily recursive operations is a case where $\mathsf{AC}_{\mathsf{qf}}^{\mathrm{N}\to\{0,1\},\mathrm{N}}$ is available but $\mathsf{BR}$ fails. On the other hand, the term model is a case where $\mathsf{BR}$ holds but $\mathsf{AC}_{\mathsf{qf}}^{\mathrm{N}\to\{0,1\},\mathrm{N}}$ fails. Both structures only have recursive Boolean sequences.

Interestingly, the soundness theorem for bar recursion applies to the theory with the combination of quantifier-free choice and bar recursion (see the end of Sect. 4). Hence, bar induction is available and the theory proves the existence of non-recursive Boolean sequences (of course, only a very restricted form of bar recursion is needed for obtaining infinite paths through infinite binary recursive trees but, as we saw, unrestricted bar recursion even proves full second-order comprehension). From the soundness theorem, one easily shows that $\mathsf{PA}^{\omega}+\mathsf{AC}_{\mathsf{qf}}^{\omega}+\mathsf{BR}$ is conservative over $\mathsf{PA}^{\omega} + \mathsf{BR}$ with respect to sentences which, in prenex normal form, have quantifier prefix $\forall\exists$. The existence of an infinite path through Kleene's tree $T_K$ is a statement of quantifier prefix $\exists\forall$, and the conservation result does not apply. Spector's interpretation is subtle indeed.

**Acknowledgements** I would like to thank Reinhard Kahle and Michael Rathjen, editors of this volume, for inviting me to contribute to this book dedicated to Gerhard Gentzen. The comments of an anonymous referee and an interesting series of discussions with Ulrich Kohlenbach concerning the contents of this paper have certainly improved it. I have also benefited from correspondence with Solomon Feferman and William Howard. The latter urged me to mention some computability issues, especially since the termination of effective processes was a fundamental issue for Gentzen. I gratefully acknowledge these comments and discussions but, of course, the responsibility for the choice of material and for the writing of the paper is fully mine. The financial support of FCT-Fundação para a Ciência e a Tecnologia [UID/MAT/04561/2013, PTDC/MAT/104716/2008 and PTDC/FIL-FCI/109991/2009] was, of course, very welcomed. Finally, I want to thank Faculdade de Ciências da Universidade de Lisboa for conceding me a sabbatical leave in the academic year of 2012/2013.

## References

1. J. Avigad, S. Feferman, Gödel's functional ("Dialectica") interpretation, in *Handbook of Proof Theory*, ed. by S.R. Buss. Studies in Logic and the Foundations of Mathematics, vol. 137 (North Holland, Amsterdam, 1998), pp. 337–405
2. M. Bezem, Strongly majorizable functionals of finite type: a model for bar recursion containing discontinuous functionals. J. Symb. Log. **50**, 652–660 (1985)

3. L.E.J. Brouwer, Über Definitionsbereiche von Funktionen. Math. Ann. **93**, 60–75 (1927). English translation in [43], 457–463
4. P. Engrácia, The bounded functional interpretation of bar induction. Ann. Pure Appl. Log. **163**(9), 1183–1195 (2012)
5. P. Engrácia, F. Ferreira, The bounded functional interpretation of the double negation shift. J. Symb. Log. **75**, 759–773 (2010)
6. F. Ferreira, A most artistic package of a jumble of ideas. Dialectica **62**, 205–222 (2008). Special Issue: Gödel's *dialectica* Interpretation. Guest editor: Thomas Strahm
7. F. Ferreira, A short note on Spector's proof of consistency of analysis, in *How the World Computes*, ed. by B. Cooper, A. Dawar, B. Löwe Lecture Notes in Computer Science, vol. 7318 (Springer, New York, 2012) pp. 222–227
8. F. Ferreira, U. Kohlenbach, (2015, in preparation)
9. F. Ferreira, P. Oliva, Bounded functional interpretation. Ann. Pure Appl. Log. **135**, 73–112 (2005)
10. P. Gerhardy, U. Kohlenbach, General logical metatheorems for functional analysis. Trans. Am. Math. Soc. **360**, 2615–2660 (2008)
11. K. Gödel, Über eine bisher noch nicht benützte Erweiterung des finiten Standpunktes. Dialectica **12**, 280–287 (1958). Reprinted with an English translation in [13], pp. 240–251
12. K. Gödel, On an extension of finitary mathematics which has not yet been used (1972). Published in [13], pages 271–280. Revised version of [11].
13. K. Gödel, *Collected Works, Vol. II*. ed. by S. Feferman et al. (Oxford University Press, Oxford, 1990)
14. W.A. Howard, Functional interpretation of bar induction by bar recursion. Compos. Math. **20**, 107–124 (1968)
15. W.A. Howard, Hereditarily majorizable functionals of finite type, 1973. Published in [42], pages 454–461
16. W.A. Howard, Ordinal analysis of simple cases of bar recursion. J. Symb. Log. **46**, 17–30 (1980)
17. W.A. Howard, Ordinal analysis of bar recursion of type zero. Compos. Math. **42**, 105–119 (1981)
18. W.A. Howard, G.Kreisel. Transfinite induction and bar induction of types zero and one, and the role of continuity in intuitionistic analysis. J. Symb. Log. **31**, 325–358 (1966)
19. M. Hyland, Filter spaces and continuous functionals. Ann. Math. Log. **16**, 101–143 (1979)
20. S. Kleene, Recursive functions and intuitionistic mathematics, in *Proceedings of the International Congress of Mathematicians, Cambridge 1950*, vol. 1 (American Mathematics Society, 1952), pp. 679–685
21. S. Kleene, R. Vesley, *The Foundations of Intuitionistic Mathematics* (Amsterdam, North Holland, 1965)
22. U. Kohlenbach, Effective bounds from ineffective proofs in analysis: an application of functional interpretation and majorization. J. Symb. Log. **57**, 1239–1273 (1992)
23. U. Kohlenbach, On the no-counterexample interpretation. J. Symb. Log. **64**, 1491–1511 (1999)
24. U. Kohlenbach, Some logical metatheorems with applications in functional analysis. Trans. Am. Math. Soc. **357**, 89–128 (2005)
25. U. Kohlenbach, *Applied Proof Theory: Proof Interpretations and their Use in Mathematics*. Springer Monographs in Mathematics (Springer, Berlin, 2008)
26. G. Kreisel, Interpretation of analysis by means of constructive functionals of finite types, in *Constructivity in Mathematics* ed. by A. Heyting (Amsterdam, North Holland, 1959), pp. 101–128
27. G. Kreisel, Reports of the seminar on the foundations of analysis. Technical report, Stanford University, 1963. Mimeographed report in two parts
28. G. Kreisel, A survey of proof theory. J. Symb. Log. **3**, 321–388 (1968)
29. G. Kreisel, Gödel's excursions into intuitionistic logic, in *Gödel Remembered* (Bibliopolis, Napoli, 1987), pp. 65–186

30. H. Luckhardt, *Extensional Gödel Functional Interpretation: A Consistency Proof of Classical Analysis*, Lecture Notes in Mathematics, vol. 306 (Springer, Berlin, 1973)
31. P. Oliva, Understanding and using Spector's bar recursive interpretation of classical analysis, in *Logical Approaches to Computational Barriers – Proceedings of Computability in Europe 2006*, ed. by B. Löwe A. Beckmann, U. Berger, J. Tucker. Lecture Notes in Computer Science, vol. 3988 (Springer, New York, 2006) pp. 423–434
32. M. Rathjen, The realm of ordinal analysis, in *Sets and Proofs*, ed by S.B. Cooper, J.K. Truss London Mathematical Society Lecture Notes Series, vol. 258 (Cambridge University Press, Cambridge, 1999), pp. 219–280
33. B. Scarpellini, A model for bar recursion of higher types. Compos. Math. **23**, 123–153 (1971)
34. J.R. Shoenfield, *Mathematical Logic*. (Addison-Wesley Publishing Company, Wokingham, 1967). Republished in 2001 by AK Peters.
35. W. Sieg, Fragments of arithmetic. Ann. Pure Appl. Log. **28**, 33–71 (1985)
36. C. Spector, Provably recursive functionals of analysis: a consistency proof of analysis by an extension of principles in current intuitionistic mathematics, in *Recursive Function Theory: Proceedings of Symposia in Pure Mathematics* vol 5, ed. by F.D.E. Dekker (American Mathematical Society, Providence, Rhode Island 1962), pp. 1–27
37. G. Sundholm, M. van Atten, The proper explanation of intuitionistic logic: on Brouwer's demonstration of the Bar Theorem, in *One Hundred Years of Intuitionism (1907–2007)*, ed. by M. van Atten, P. Boldini, M. Bourdeau, G. Heinzmann. Publications des Archives Henri Poincaré / Publications of the Henri Poincaré Archives (Birkhäuser Basel, 2008), pp. 60–77
38. W. Tait, Normal form theorem for bar recursive functions of finite type, in *Proceedings of the 2nd Scandinavian Logic Symposium*. Studies in Logic and the Foundations of Mathematics (Elsevier, Amsterdam, 1971), pp. 353–367
39. W. Tait, Finitism. J. Philos. **78**, 524–546 (1981)
40. A.S. Troelstra, Introductory note to [11] and [12] (1990). Published in [13], pages 214–241
41. A.S. Troelstra, D. van Dalen, *Constructivism in Mathematics. An Introduction*, volume 121 of *Studies in Logic and the Foundations of Mathematics* (Amsterdam, North Holland, 1988)
42. A.S. Troelstra (ed.), *Metamathematical Investigation of Intuitionistic Arithmetic and Analysis*, Lecture Notes in Mathematics, vol. 344 (Springer, Berlin, 1973)
43. J. van Heijenoort (ed.), *From Frege to Gödel* (Harvard University Press, Cambridge, 1967)
44. H. Vogel, Über die mit dem Bar-Rekursor vom Typ 0 definierbaren Ordinalzahlen. Archiv für mathematische Logik und Grundlagenforschung **19**, 165–173 (1978/1979)

# Climbing Mount $\varepsilon_0$

**Herman Ruge Jervell**

**Abstract** Gentzen showed in his Habilitation that transfinite induction up to any ordinal $< \varepsilon_0$ is provable in first order arithmetic—and made a constructive justification of how to reach any ordinal $< \varepsilon_0$. In his analysis Gentzen used ordinals in Cantor normal form. We shall look at ordinals as given by finite trees and then see how the climbing up to $\varepsilon_0$ can be justified there with methods from first order arithmetic, and methods to use where we climb above it.

## 1 Gödel's Incompleteness

Gödel showed with his incompleteness theorem [2] that in any reasonable elementary theory of arithmetic $\mathcal{A}$ there are undecidable sentences $G$ with

$$\nvdash_{\mathcal{A}} G \textit{ and } \nvdash_{\mathcal{A}} \neg G.$$

The Gödel sentence $G$ can be written as

$$\neg\mathbf{Provable}(\mathcal{A}, 0 = 1).$$

There are usually three requirements of the theory $\mathcal{A}$

**Language:** The language should be rich enough to code formulas, proofs and provability in the system. Gödel had to include $+$ and $\times$ to be able to do the coding. Provability is coded in such a way that "something is provable" can be coded as an existential sentence—a sentence with existential quantifiers outermost and only connectives and bounded quantifiers inside.

**Strength:** The formal system should be strong enough to derive all true existential sentences. Usually we include in the system induction over quantifier free formulas—a fairly weak requirement.

**Consistency:** The system is consistent and $\omega$-consistent.

---

H.R. Jervell (✉)
Department of Informatics, University of Oslo, Ole Johan Dahls hus, Gaustadalleen, 0373 Oslo, Norway
e-mail: herman@ifi.uio.no

 
R. Kahle, M. Rathjen (eds.), *Gentzen's Centenary*,
DOI 10.1007/978-3-319-10103-3_12

These requirements are quite robust—we can make Gödel theorems for any reasonable elementary theory of a data structure.

One problem is that the undecidable sentences do not give any information about the reasonable theory—the requirements are only sufficient to simulate provability within the system. For another theory $\mathcal{T}$ of some other datastructure we use a similar Gödel sentence $\neg$**Provable**$(\mathcal{T}, 0 = 1)$. The only difference is that we simulate provability in $\mathcal{T}$.

## 2 Gentzen's Result

In Gödel's argument there is no analysis of the theory $\mathcal{A}$ involved. This was done by Gentzen for first order arithmetic $\mathcal{PA}$—also called Peano arithmetic. We can write Gentzen's result in a simplified way as

**Theorem 1 (Gentzen)** *Let TI denote transfinite induction. Then in first order arithmetic, $\mathcal{A}$, for any ordinal*

$$(\vdash_{\mathcal{PA}} TI\,\alpha) \Leftrightarrow \alpha < \varepsilon_0.$$

Gentzen proved $\Rightarrow$ with his 1936 proof of the consistency of first order arithmetic [3]. Then with his 1943 Habilitation [4] he showed $\Leftarrow$. Gentzen's result gives an analysis of first order arithmetic. The scheme derived from the Gödel sentence is $TI\,\varepsilon_0$ and Gentzen's result combines a logical statement about provability on the left-hand side and a combinatorial statement on the right-hand side. It was the first and still one of the best mathematical incompleteness in first order arithmetic. Compare it with [10].

## 3 Schemes and Formulas

In the formulation of Gentzen's result we must distinguish between schemes and formulas. The Gödel sentence is a formula and not a scheme. From Gentzen's work we get a particular formula $G$ and where we cannot prove $\vdash_{\mathcal{PA}} TI\varepsilon_0 G$—and hence not the scheme $\vdash_{\mathcal{PA}} TI\varepsilon_0$. On the other hand we can prove the scheme up to $\varepsilon_0$ by a climbing argument. There we require two transformations—one of ordinals $\alpha \mapsto \alpha^\dagger$ and the other one of formulas $F \mapsto F^*$ such that the following argument can be carried through

- Assume we have $TI_{PA}\alpha$
- Let $F$ be any formula
- We can prove that $TI_{PA}\alpha^\dagger F$ follows from $TI_{PA}\alpha F^*$
- By assumption $TI_{PA}\alpha F^*$ and hence $TI_{PA}\alpha^\dagger F$
- Hence we get the scheme $TI_{PA}\alpha^\dagger$

Gentzen's theorem talks about "any ordinal $\alpha$". This is made precise by using ordinal notations. For ordinals $< \varepsilon_0$ we can do this as follows:

$\omega$: The order type of natural numbers.

$\omega^\omega$: The order type of $\omega$-sequences of natural numbers with finite support ordered by inverse lexicographical ordering. Finite support means that the sequences are 0 except for finitely many places.

$\omega^{\omega^\omega}$: The order type of $\omega^\omega$-sequences of natural numbers with finite support ordered by inverse lexicographical ordering.

...

...

It is straightforward to represent this within arithmetic and also to take the limit of these ordertypes. The limit is the ordertype $\varepsilon_0$.

Gentzen's result gives the Gödel sentence for first order arithmetic—we can remember it as the scheme $TI(\varepsilon_0)$.

The proof of Gentzen's result is quite involved. For the $\Rightarrow$-part we use some kind of cut elimination. There are essentially two proofs

- Schütte showed [11] that we could embed a derivation in first order arithmetic into a derivation in $\omega$-arithmetic of height $< \omega^2$. In $\omega$-arithmetic we can perform the ordinary cut-elimination and we end up with a cut-free proof of length $< \varepsilon_0$. Novikov had similar results earlier [9].
- Gentzen considered the Endstück of a derivation of a quantifier free sentence in first order arithmetic. In such a derivation one could perform cut-elimination with a process controlled by ordinals $< \varepsilon_0$. Annika Kanckos has given an improved version of this proof in the present volume [7].

We shall here consider the $\Leftarrow$-part.

# 4 Formal Theories of Arithmetic

In 1923 Thoralf Skolem published his investigations in a formal system for arithmetic [12]. There he introduced a system based on the following

- there is an underlying datastructure—like the unary numbers with 0 $s$ and $<$,
- we can build and name new terms using primitive recursion,
- the logic is first order predicate logic with equality,
- we have the basic axioms for the datastructure,
- we have as axioms the recursion equations for the primitive recursive terms,
- we have induction for quantifier free formulas.

This is clearly a general construction—we can build such systems for all usual datastructures. We call them primitive recursive arithmetic $\mathcal{PRA}$. A better name could have been Skolem arithmetic.

$\mathcal{PRA}$ is the usual starting system for proof theoretic investigations. It is sufficiently strong to do all the codings required for Gödels incompleteness theorem.

Gentzen analyzed what is called first order arithmetic or Peano arithmetic $\mathcal{PA}$. It consists of $\mathcal{PRA}$ and in addition induction over arbitrary formulas—not only the quantifier free ones.

## 5 Finite Trees as Ordinals

The ordinal notations give a way to break down an ordinal into smaller ordinals. This is important in the proof of Gentzen's theorem. Gentzen used in his proof a variant of iterated lexicographical orderings mentioned above. We shall here give a proof using finite trees. The trees are built up from the empty tree, $\bullet$, using finite and ordered branching.

From Kruskals theorem [8] we know that the finite trees are well quasi ordered with respect to topological embedding. This means that any linear extension of the finite trees ordered by topological embedding is a well ordering. And in such a linear extension we can talk about the ordinal of a tree.

In previous papers [5, 6] we have shown how to order finite trees with ordered branchings. We have the finite trees defined by

$$\mathcal{T} :: \bullet|(\mathcal{T})|(\mathcal{T},\mathcal{T})|(\mathcal{T},\mathcal{T},\mathcal{T})|\cdots$$

and write them with the root down. Here is the tree $((\bullet, (\bullet, \bullet)))$

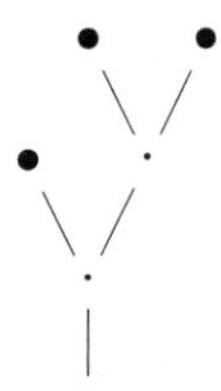

For a finite tree $\alpha \in \mathcal{T}$ we write $\langle\alpha\rangle$ for the sequence of immediate subtrees. The branchings are ordered from left to right. And then have the following abbreviations

$\alpha \leq \langle\beta\rangle$: $\alpha$ is either equal to or $<$ some element in $\langle\beta\rangle$.

$\langle\alpha\rangle < \beta$: All elements from $\langle\alpha\rangle$ are $< \beta$.

$\langle\alpha\rangle < \langle\beta\rangle$: The sequence $\langle\alpha\rangle$ is $<$ the sequence $\langle\beta\rangle$ in the inverse lexicographical ordering. In this ordering we have as the first priority the length of the two sequences and for sequences of equal length we look at the rightmost place where they differ.

We define the ordering of trees

$$\boxed{\alpha < \beta \Leftrightarrow \alpha \leq \langle\beta\rangle \vee (\langle\alpha\rangle < \beta \wedge \langle\alpha\rangle < \langle\beta\rangle)}$$

We have developed this theory in previous papers [5, 6]. Let us note the following—proved in primitive recursive arithmetic $\mathcal{PRA}$

- The ordering is decidable.
- The ordering is a total linear ordering.
- The least element in the ordering is the empty tree—$\bullet$.
- The equality in the ordering is the ordinary equality of (ordered) trees.
- Successor is given by $\lambda x.(x)$—we tack on a unary branch below the root.
- Monotonicity: For the usual tree functions $\alpha < \beta \Rightarrow F(\alpha) < F(\beta)$.
- Embedding: If tree $\alpha$ can be topologically embedded in $\beta$, then $\alpha < \beta$.
- Approximations: To each tree $\alpha$ we can build the set of all trees $< \alpha$ in a simple way. Below we give examples of these approximations.

It may be easier to grasp the ordering by looking at a decision tree for the ordering. Say we want to decide the ordering between $\alpha$ and $\beta$

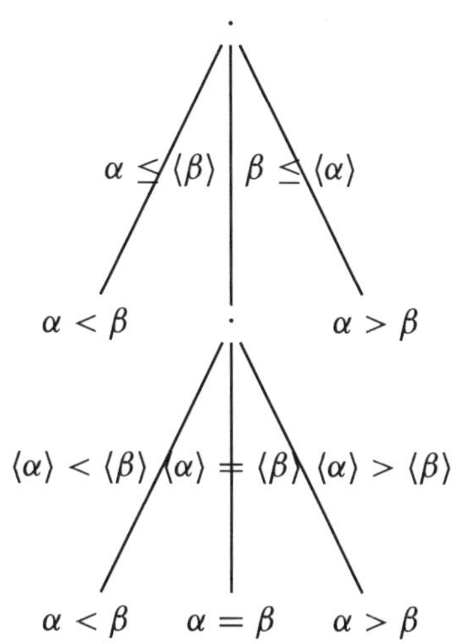


We start at the top—and first look at whether $\alpha \leq \langle\beta\rangle$ or $\beta \leq \langle\alpha\rangle$. In those cases we get either $\alpha < \beta$ or $\alpha > \beta$. Else we go through the middle branch and look at the inverse lexicographical ordering of $\langle\alpha\rangle$ and $\langle\beta\rangle$. The ordering itself is an extension of an embedding relation using also inverse lexicographical ordering.

By Kruskals theorem it follows immediately that the ordering is a well ordering. And the tree ordering has an ordinal. This ordinal is what is called the small Veblen ordinal. It also follows that to any ordinal less than the small Veblen ordinal there is a unique finite tree corresponding to it. But to get more insight we must look at the approximations of the finite trees.

## 6 Approximating Trees

The approximation theorem [5]—proved in $\mathcal{PRA}$—gives a natural way of breaking down a tree.

For the tree $(\alpha, \beta)$ we construct the following approximating set $\mathcal{D}$

- $\bullet \in \mathcal{D}$.
- For each $\alpha^- < \alpha$ we have $(\alpha^-, \beta) \in \mathcal{D}$.
- $\mathcal{D}$ is closed under
  - $\lambda x.(x)$,
  - $\lambda x.(x, \beta^-)$ for any $\beta^- < \beta$.

There are similar constructions for arbitrary finite branching [5].

We then have

$$\gamma < (\alpha, \beta) \Leftrightarrow \gamma \in \mathcal{D}.$$

For the case that $\beta = \bullet = 0$ there are no $\beta^- < \beta$ and the last condition does not apply. We only require that $\mathcal{D}$ is closed under the successor $\lambda x.(x)$.

Using the approximation theorem we can calculate some ordinals

$$\begin{aligned}
\bullet &= 0\\
(\bullet) &= 1\\
((\bullet)) &= 2\\
(\bullet, \bullet) &= \omega\\
((\bullet), \bullet) &= \omega \cdot 2\\
((\bullet, \bullet), \bullet) &= \omega^2\\
(\bullet, (\bullet)) &= \omega^\omega\\
(\bullet, (\bullet, \bullet)) &= \omega^{(\omega^\omega)}
\end{aligned}$$

In [5] we proved that for trees with unary and binary branching we have

$$\begin{aligned}
(\alpha) &= \alpha + 1\\
(\alpha, \bullet) &= \omega \cdot (1 + \alpha)\\
(\bullet, \alpha) &= \omega^{\omega^\alpha}
\end{aligned}$$

For trees with more branching we get a similar result but have to take fix points of the functions on the right into consideration. Using this we get [5]

$$
\begin{aligned}
(\bullet, \bullet, \bullet) &= \varepsilon_0 \\
((\bullet), \bullet, \bullet) &= \varepsilon_1 \\
(\bullet, (\bullet), \bullet) &= \kappa_0 \ \textit{the first critical } \varepsilon\textit{-number} \\
(\bullet, \bullet, (\bullet)) &= \Gamma_0
\end{aligned}
$$

In this paper we are first interested in how to climb up to $\varepsilon_0$. This is analyzed here by giving appropriate proofs of transfinite inductions in first order arithmetic.

## 7 Transfinite Induction

We define the following for predicates $F$ in the language of first order logic

$$
\begin{aligned}
\mathit{PROG}(F) &: \forall x.(\forall y < x.Fy \to Fx) && \textit{the predicate } F \textit{ is progressive} \\
\mathit{TI}(F, \alpha) &: \mathit{PROG}(F) \to \forall x < \alpha.Fx && \textit{transfinite induction up to } \alpha \\
\mathit{TI}\,\alpha &: \mathit{TI}(F, \alpha) && \textit{for any predicate } F
\end{aligned}
$$

Observe the following simple facts about transfinite inductions

$$
\begin{aligned}
\alpha < \beta \wedge \mathit{TI}\,\beta &\Rightarrow \mathit{TI}\,\alpha \\
\mathit{TI}\,\alpha &\Rightarrow \mathit{TI}\,(\alpha)
\end{aligned}
$$

## 8 Climbing Up to $\omega^\omega$

We are going to prove

$$\mathit{TI}\,\alpha \Rightarrow \mathit{TI}\,(\alpha, \bullet).$$

We define addition of a natural number to a tree by

$$
\begin{aligned}
(\beta)^0 &= \beta, \\
(\beta)^{n+1} &= ((\beta)^n).
\end{aligned}
$$

Or as a picture

For the climbing up to $(\alpha, \bullet)$ the crucial observation is that for $\alpha > 0$

$$\gamma < (\alpha, \bullet) \Leftrightarrow \exists \beta < \alpha . \exists n < \omega . \gamma = ((\beta, \bullet))^n .$$

We can write this out in a more perspicuous way

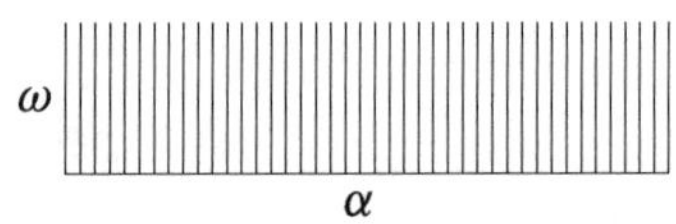


We have a comb of length $\alpha$ and each tooth is of length $\omega$. To compare two points there we first check which tooth they are on and if they are on the same tooth we check how high up they are on the tooth. We get $(\alpha, \bullet) \approx \omega \cdot (1 + \alpha)$.

We have already *TI* $(\bullet, \bullet)$. This is just a version of ordinary induction. We assume

$$PROG\ F \text{ and } TI\ \alpha \text{ where } \alpha > 0$$

and want to prove

$$\forall \beta < \alpha . \forall n < \omega . F\ (\beta)^n .$$

The quantifiers are proved by *TI* $\alpha$ and *TI* $\omega$, respectively.
Assume $\beta < \alpha$ and $\forall \gamma < \beta . \forall n < \omega . F\ (\gamma)^n$.
Assume $n < \omega$ and $\forall m < n . F\ (\beta)^m$.
Then $\forall x < (\beta)^n . F\ x$ by the two assumptions about $F$.
By $PROG\ F : F\ (\beta)^n$ and we get $PROG\ \lambda n < \omega . F\ (\beta)^n$.
By $TI\ \omega : \forall n < \omega . F\ (\beta)^n$ and we get $PROG\ \lambda \beta < \alpha . \forall n < \omega . F\ (\beta)^n$.
By $TI\ \alpha : \forall \beta < \alpha . \forall n < \omega . F\ (\beta)^n$ and we are done.
We have just used $\Pi_1$ inductions in the proof—and it can be done within $\mathcal{PRA}$.
This gives the $\Leftarrow$-part of the well known theorem

**Theorem 2** *Let TI denote transfinite induction. Then in* $\mathcal{PRA}$ *for any ordinal* $\alpha$

$$\vdash TI\ \alpha \Leftrightarrow \alpha < \omega^{\omega} .$$

## 9 Reaching $\omega^\omega$

We have climbed up to any ordinal $< \omega^\omega$ but have not reached $\omega^\omega$ itself. Let us see how this is done. We explain this with a similarity between defining the primitive recursive functions and defining the Ackermann function [1]. At the core of the Ackermann function is the higher order iteration function

$$it : \mathcal{N} \to ((\mathcal{N} \to \mathcal{N}) \to (\mathcal{N} \to \mathcal{N}))$$

defined by

$$it\, n\, f = f^n.$$

Gerhard Gentzen used in his Habilitation [4] a similar construction to define the limits up to $\omega^\omega$ and—as we shall see in the next section—to any ordinal $< \varepsilon_0$. We want to iterate a construction involving the function $\lambda\alpha\,.\,(\alpha, \bullet)$. We introduce the more general notation

$$(\gamma)^0_\beta = \gamma,$$
$$(\gamma)^{n+1}_\beta = ((\gamma)^n_\beta, \beta).$$

Here we are interested in $(\alpha)^n_0$. Given a predicate $F$ with $PROG\, F$. We then introduce the first order variant of operation to be iterated

$$F^\star : \forall y.(\forall z < y\, Fz \to \forall z < (y)^1_0\, Fz).$$

Assume $F^\star$. We then get using iteration of the $\to$ for any $n < \omega$

$$\forall y.(\forall z < y\, Fz \to \forall z < (y)^n_0\, Fz).$$

Or using $TI\,\omega$ with a $\Pi_2$ induction formula

$$\forall y.\forall n < \omega\,.\,(\forall z < y\, Fz \to \forall z < (y)^n_0\, Fz).$$

Note that

$$\alpha < \omega^\omega \leftrightarrow \exists n < \omega\,.\alpha < (0)^n_0.$$

And we have proved

$$TI\,\omega \Rightarrow TI\,\omega^\omega.$$

## 10 Climbing Up to $\varepsilon_0$

Note that $(\bullet, \alpha) \approx \omega^{\omega^\alpha}$ [5]. Now we want to prove

$$TI\,\alpha \Rightarrow TI\,(\bullet, \alpha).$$

We use the function $(\alpha)^n_\beta$ introduced in the previous section. Here is a picture of $(\alpha)^3_\beta$

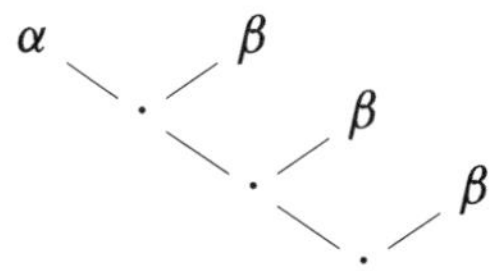


We use the approximation for $\alpha > 0$

$$\beta < (\bullet, \alpha) \leftrightarrow \exists \alpha^- < \alpha \,.\, \exists n < \omega \,.\, \beta < (\bullet)^n_{\alpha^-}.$$

Assume we have a predicate $F$ with *PROG* $F$. As before we introduce a new predicate but now with an extra argument

$$F^\star x : \forall y.(\forall z < y \,.\, Fz \rightarrow \forall z < (y)^1_x \,.\, Fz).$$

We can iterate the conditional more times and we get for any $n < \omega$

$$F^\star x \rightarrow \forall y.(\forall z < y \,.\, Fz \rightarrow \forall z < (y)^n_x \,.\, Fz).$$

Using $TI\,\omega$ we also get

$$F^\star x \rightarrow \forall y.\forall n < \omega.(\forall z < y \,.\, Fz \rightarrow \forall z < (y)^n_x \,.\, Fz).$$

Using this we get *PROG* $F^\star$—and by using $TI\,\alpha$.

$$\forall z < (\bullet, \alpha) \,.\, Fz.$$

So

$$TI\,\alpha \Rightarrow TI\,(\bullet, \alpha).$$

Let us look closer at the steps.

- The new predicate $F^\star$ has a higher quantifier complexity than $F$.
- If $F$ is $\Pi_n$, then $F^\star$ is $\Pi_{n+1}$.

- In the crucial iteration we needed the extra quantifiers—we first picked a $y$ to get $(y)_x^1$, and then substituted this for $y$ to get $(y)_x^2$ and so on. We used both elimination and introduction in the $\forall y$ repeatedly.

So now we are able to climb up to any ordinal $< \varepsilon_0$ using $\omega$-induction with more and more complicated induction formulae.

## 11 Problems with $\varepsilon_0$

There are some stumbling blocks in reaching $\varepsilon_0$

- we have used the schema $TI\,\alpha$ instead of the formula $TI\,\alpha\,F$. This is no problem. We can easily replace the schema with appropriate formulae. Below we shall not bother with this replacement where it is obvious how it is done,
- to get higher $TI\,\beta$ we iterate the process which takes a proof of $\vdash TI\,\alpha$ into a proof of $\vdash TI\,(\bullet,\alpha)$. There is no uniformity in the process,
- we really want to iterate the proof of the conditional $\vdash TI\,\alpha \rightarrow TI\,(\bullet,\alpha)$—in other words we want the axiom and not the rule.

We see that whenever we iterate the rule $\vdash TI\,\alpha \Rightarrow \vdash TI\,(\bullet,\alpha)$ we need more and more quantifiers in the inductionformulae.

## 12 Coding Logic

The trick now is to consider codes of proofs instead of proofs. We have

- a standard coding of terms, formulae, proofs, etc.—this is done within Skolem arithmetic.
- To each formula $F$ we have a $\Sigma_1$-formula $\Box F$ meaning "there is a coded proof of the code of $F$".

And we have

**Theorem 3 (Formal $\Sigma_1$ completeness)** *For proofs in systems which contain $\mathcal{PRA}$, and for $\Sigma_1$-formulae $F$ the following is provable in $\mathcal{PRA}$*

$$\vdash F \rightarrow \Box F.$$

We always have

$$\vdash F \Rightarrow \vdash \Box F$$

for any formula $F$. Formal $\Sigma_1$ completeness provides a formal analogue for $\Sigma_1$-formulae. It doesn't hold for arbitrary $\Pi_1$-formulae. The converse, $\Box F \rightarrow F$, called

$\Sigma_1$-reflection, is also not provable (take $F$ to be $0{=}1$ and use Gödel's theorem). This gives the idea for a hierarchy of systems. We define systems for arithmetic $\mathcal{H}_i$ by

- the language of the systems is the language of $\mathcal{PRA}$,
- $\mathcal{H}_0$ is ordinary first order arithmetic $\mathcal{PA}$,
- $\Box_i$ is provability within $\mathcal{H}_i$,
- $\mathcal{H}_{i+1}$ is $\mathcal{H}_i$ extended with axioms $\Box_i F \to F$ for all formulae $F$.

## 13 Reaching $\boldsymbol{\varepsilon_0}$

We get arbitrary close to $\varepsilon_0$ by iterating $\alpha \mapsto (\bullet, \alpha)$. Now we have shown in $\mathcal{H}_0$ the rule

$$\vdash TI\,\alpha \Rightarrow \vdash TI\,(\bullet, \alpha).$$

We can transform this into a proof in $\mathcal{H}_0$ of the $\Sigma_1$-formula

$$\vdash \Box(TI\,\alpha \to TI\,(\bullet, \alpha)).$$

But then we have in $\mathcal{H}_1$

$$\vdash TI\,\alpha \to TI\,(\bullet, \alpha).$$

Then using the same trick as in proving $TI\,\omega^\omega$ above, we get in $\mathcal{H}_1$

$$\vdash TI\,\varepsilon_0.$$

## 14 Up Through the $\boldsymbol{\varepsilon}$-Numbers

As before we use the approximation theorem to write down the set $\mathcal{E}$ of all trees below $(\alpha, \bullet, \bullet)$

- $\bullet$ is in $\mathcal{E}$.
- $(\alpha^-, \bullet, \bullet)$ is in $\mathcal{E}$ for any $\alpha^- < \alpha$.
- $\mathcal{E}$ is closed under the functions
  - $\lambda x.(x)$,
  - $\lambda x.\lambda y.(x, y)$.

And we see that we in fact get the $\varepsilon$-numbers. For example, we get

$$\varepsilon_1 = ((\bullet), \bullet, \bullet).$$

The climbing in the $\varepsilon$-numbers is easy. We work within $\mathcal{H}_1$. There we can express the transformation

$$\vdash TI\, \alpha \Rightarrow \vdash TI\, (\alpha, \bullet, \bullet)$$

as a conditional and then use the same arguments as in climbing up to $\omega^\omega$. This can be done within $\mathcal{H}_1$. Reaching $\omega^\omega$ will then correspond to reaching $(\bullet, (\bullet), \bullet)$—the first critical $\varepsilon$-number.

## 15 Up Through the Critical $\varepsilon$-Numbers

We enumerate the fix points of the $\varepsilon$-numbers with $(\bullet, \alpha, \bullet)$. The set $\mathcal{F}$ of all trees below it is given by

- $\bullet$ is in $\mathcal{F}$.
- $\mathcal{F}$ is closed under the functions
  - $\lambda x.(x)$,
  - $\lambda x.\lambda y.(x, y)$,
  - $\lambda x.(x, \alpha^-, \bullet)$ for any $\alpha^- < \alpha$.

Within $\mathcal{H}_1$ we do the iteration as we did for $\varepsilon_0$ to climb up to

$$(\bullet, \varepsilon_0, \bullet) = (\bullet, (\bullet, \bullet, \bullet), \bullet).$$

To get beyond that we had to use the higher $\mathcal{H}_i$.

**Theorem 4** *Assume* $\vdash TI\, \alpha$ *is done in* $\mathcal{H}_i$. *Then* $\vdash TI\, (\bullet, \alpha, \bullet)$ *in* $\mathcal{H}_{i+1}$.

## 16 Reaching $\Gamma_0$

$\Gamma_0$ is the first fix point of the critical $\varepsilon$-numbers. So

$$\Gamma_0 = (\bullet, \bullet, (\bullet)) =$$

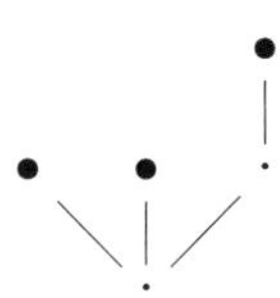

or as the limit of

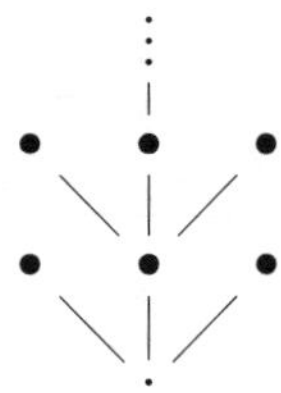

We get it by iterating the operation

$$\alpha \mapsto (\bullet, \alpha, \bullet).$$

Now we have shown

$$\vdash_{\mathcal{H}_i} TI\,\alpha \Rightarrow \vdash_{\mathcal{H}_{i+1}} TI\,(\bullet, \alpha, \bullet).$$

Now we find a system where we can prove

$$\vdash \Box(TI\,\alpha \rightarrow TI\,(\bullet, \alpha, \bullet)).$$

And then we can reach $\Gamma_0$ and prove

$$TI\,\Gamma_0.$$

## 17 Going Beyond $\Gamma_0$

There is no reason to stop at $\Gamma_0$. It is the limit to how far we get with the $\mathcal{H}_i$'s. We must look out for other systems which can capture the iterations which we have used so far. It is easy to write down the approximations for the $\Gamma$-numbers $(\bullet, \bullet, \alpha)$. It is the set $\mathcal{G}$ of all trees given by

- $\bullet$ is in $\mathcal{G}$.
- $\mathcal{G}$ is closed under the functions
  - $\lambda x.\,(x)$,
  - $\lambda x.\lambda y.\,(x, y)$,
  - $\lambda x.\lambda y.\,(x, y, \alpha^-)$ for any $\alpha^- < \alpha$.

But to analyze it and go further we need more systems.

## References

1. W. Ackermann, Zum Hilbertschen Aufbau der reellen Zahlen. Math. Ann. **99**, 118–133 (1928)
2. K. Gödel, Über formal unentscheidbare Sätze der Principia Mathematica und verwandter Systeme I. Monatshefte für Mathematik und Physik **38**, 173–198 (1931)
3. G. Gentzen, Die Widerspruchsfreiheit der reinen Zahlentheorie. Math. Ann. **112**, 493–565 (1936)
4. G. Gentzen, Beweisbarkeit und Unbeweisbarkeit von Anfängsfallen der transfiniten Induktion in der reinen Zahlentheorie. Math. Ann. **119**, 140–161 (1943)
5. H.R. Jervell, Finite trees as ordinals, in *New Computational Paradigms, CIE 2005* (Springer, Heidelberg, 2005), pp. 211–220
6. H.R. Jervell, Constructing ordinals, in *Constructivism: Mathematics, Logic, Philosophy and Linguistics* (2006), pp. 5–20. ISBN 2-84174-399-3
7. A. Kanckos, A direct Gentzen-style consistency proof for Heyting arithmetic, in *Gentzen's Centenary: The Quest for Consistency*, ed. by R. Kahle, M. Rathjen (Springer, Heidelberg, 2015)
8. J.B. Kruskal, Well-quasi-ordering, the tree theorem, and Vazsonyi's conjecture. Trans. Am. Math. Soc. **95**, 210–225 (1960)
9. P.S. Novikov, Sur quelques théorèmes d'existence. Comptes Rendus (Doklady) de l'Academie des Sciences de l'URSS **23**, 438–440 (1939)
10. J. Paris, L. Harrington, A mathematical incompleteness in peano arithmetic, in *Handbook of Mathematical Logic*, ed. by J. Barwise (North-Holland, Amsterdam, 1977), pp. 1133–1142
11. K. Schütte, Beweistheoretische Erfassung der unendlichen Induktion in der Zahlentheorie. Math. Ann. **122**, 369–389 (1950)
12. T. Skolem, Begründung der elementären Arithmetik durch die rekurrierende Denkweise ohne Anwendung scheinbarer Veränderlichen mit unendlichem Ausdehnungsbereich. Skrifter, Videnskabsakademiet i Kristiania (1923), pp. 1–38

# Semi-Formal Calculi and Their Applications

**Wolfram Pohlers**

## 1 Introduction

By a semi-formal system we understand a proof system which includes infinitary inference rules. The use of inference rules with infinitely many premises was already suggested by David Hilbert in his paper "Die Grundlegung der elementaren Zahlentheorie" [6] and was later systematically used by Kurt Schütte in his work on proof theory. The heigths of proof trees in a semi-formal system are canonically measured by ordinals. Therefore, in contrast to Gentzen's original approach, ordinals enter proof theoretic research via semi-formal systems in a completely canonical way.

In this paper we will, however, not talk about applications of semi-formal systems to proof theory but indicate that there are also many applications outside of ordinal analysis. Our intention is to show that semi-formal systems provide a versatile tool that can be applied to obtain a series of results in a very uniform way. These applications will comprise results about the structure theory of infinite countable structures. The key property for these applications is the Boundedness Theorem, a variant of a theorem that has already been proved by Gerhard Gentzen in his seminal paper [3]. Gentzen himself suspected there that this theorem has further reaching applications (cf. Note 5.13 below). In Sect. 5 we try to indicate some of such possible applications.

I am indebted to the anonymous referee for pointing out several gaps and typos in the first version and giving a series of helpful hints to improve the paper.

W. Pohlers (✉)
Westfälische Wilhelms—Universität, Institut für Mathematische Logik und Grundlagenforschung Einsteinstrasse 62, D-48149 Münster, Germany
e-mail: pohlers@uni-muenster.de

 
R. Kahle, M. Rathjen (eds.), *Gentzen's Centenary*,
DOI 10.1007/978-3-319-10103-3_13

## 2 Abstract Logical Languages

### 2.1 Syntax

To obtain as much generality as possible, we start with a very general notion of logical language.

**Definition 2.1** The basic symbols of a logical language comprise

- Constants for individuals and predicates of finite arity,
- Variables for individuals and predicates of finite arity,
- Symbols for functions mapping tuples of individuals to individuals,
- Logical operators. Every operator possesses an arity which is an index set $I$. Some of the operators may bind variables.
- Terms are built up from constants and variables for individuals by function symbols in the familiar way.
- Atomic formulas are built up from terms by constants or variables for predicates.
- If $\langle F_\iota \mid \iota \in I\rangle$ is a sequence of **L**-formulas and $\mathcal{O}$ is a logical operator of arity $I$, then $(\mathcal{O}\langle F_\iota \mid \iota \in I\rangle)$ is a well-formed **L**-formula. If $\mathcal{O}$ binds a variable $x$ then $x$ is no longer free (i.e. bound) in $(\mathcal{O}\langle F_\iota \mid \iota \in I\rangle)$

We will use the language **L** in Tait style. That is, there is no logical operation for negation. Instead we require that for every predicate variable $X$, for every predicate constant $P$ and every logical operation $\mathcal{O}$ there are the dual symbols $X^c$, $P^c$ and the dual operation $\mathcal{O}^c$. For every **L**-formula $F$ its *dual formula* $F^c$ is inductively defined by

- $(Xt_1,\dots,t_n)^c = (X^c t_1,\dots,t_n)$ and $(Pt_1,\dots,t_n)^c = (P^c t_1,\dots,t_n)$,
- $((X^c t_1,\dots,t_n))^c = (Xt_1,\dots,t_n)$ and $((P^c t_1,\dots,t_n))^c = (Pt_1,\dots,t_n)$,
- $(\mathcal{O}\langle F_\iota \mid \iota \in I\rangle)^c = \mathcal{O}^c\langle (F_\iota)^c \mid \iota \in I\rangle$,
- $(\mathcal{O}^c\langle F_\iota \mid \iota \in I\rangle)^c = \mathcal{O}\langle (F_\iota)^c \mid \iota \in I\rangle$.

We divide the formulas of **L** into three *types*

- Formulas without type,
- Formulas in $\bigwedge$-type,
- Formulas in $\bigvee$-type,

satisfying the requirement that all formulas without type are atomic.[1]

**Definition 2.2** We call an abstract logical language **L** *decorated* if every formula $F$ of **L** is equipped with a *characteristic sequence* $\mathrm{CS}(F)$ which is a sequence of subformulas of $F$ satisfying the conditions:

[1]Observe, however, that atomic formulas may well belong to a type. Cf., e.g., Definition 2.10 below.

(C0) If $\mathrm{CS}(F) = \langle F_\iota | \iota \in I \rangle$, then $\mathrm{CS}(F^c) = \langle F_\iota^c | \iota \in I \rangle$,
(C1) there is a complexity measure $\mathrm{rnk}(F)$ such that $\mathrm{rnk}(G)$ is less than $\mathrm{rnk}(F)$ for all formulas in $G \in \mathrm{CS}(F)$,
(C2) the formulas in $\mathrm{CS}(F)$ must not contain free variables that are not free in $F$.

**Definition 2.3** An abstract language is *countable* iff the set of its terms and formulas is countable (and thus every characteristic sequence in a decorated language contains at most countably many formulas).

Let $\mathbb{A}$ be an admissible structure in the sense of Barwise [1] and **L** a decorated language that is $\Delta_1$-definable in $\mathbb{A}$. The admissible fragment $\mathbf{L}_{\mathbb{A}}$ of **L** consists of

- the set $\{t | t$ is an **L**-term and $t \in \mathbb{A}\}$,
- the set $\{F | F$ is an **L**-formula, $F \in \mathbb{A}$ and $\mathrm{CS}(F) \in \mathbb{A}\}$

and satisfies the condition that

- $F \in \mathbf{L}_{\mathbb{A}}$ entails $\mathrm{CS}(F) \in \mathbf{L}_{\mathbb{A}}$.

## 2.2 Semantics for L

Let **L** be an abstract language. The signature of **L** is the set of function—and predicate constants of **L** together with their arities. An **L**-structure $\mathfrak{M}$ is a non void set $|\mathfrak{M}|$ together with a function $\cdot^{\mathfrak{M}}$ that matches the signature of **L**. That is

- $c^{\mathfrak{M}} \in |\mathfrak{M}|$ for every constant $c$ of **L**,
- for every $k$-ary function symbol $f$ of **L** the interpretation $f^{\mathfrak{M}}$ is a function $f^{\mathfrak{M}}: |\mathfrak{M}|^k \longrightarrow |\mathfrak{M}|$,
- for every $k$-ary predicate constant $P$ of **L** it is $P^{\mathfrak{M}} \subseteq |\mathfrak{M}|^k$, where $(P^c)^{\mathfrak{M}}$ is supposed to be the complement of $P^{\mathfrak{M}}$.

An $\mathfrak{M}$-assignment is a map $\Phi$ that assigns

- an element $\Phi(x) \in |\mathfrak{M}|$ to every individual variable $x$,
- a set $\Phi(X) \subseteq |\mathfrak{M}|^k$ for every predicate variable of arity $k$ with the proviso that $\Phi(X^c)$ is the complement of $\Phi(X)$.

**Definition 2.4** Let **L** be a decorated abstract language, $\mathfrak{M}$ an **L**-structure and $\Phi$ an $\mathfrak{M}$-assignment. For every **L**-term $t$ we then obtain in the canonical way an evaluation $t^{\mathfrak{M}}[\Phi] \in |\mathfrak{M}|$. We define the relation $\mathfrak{M} \models F[\Phi]$ saying that the structure $\mathfrak{M}$ satisfies the formula $F$ under the assignment $\Phi$ inductively by the following clauses:

- If $F$ has no type, then $F$ is an atomic formula $(Rt_1, \ldots, t_n)$ or $(Xt_1, \ldots, t_n)$ and we define $\mathfrak{M} \models F[\Phi]$ if the tuple $(t_1^{\mathfrak{M}}[\Phi], \ldots t_n^{\mathfrak{M}}[\Phi])$ is in $R^{\mathfrak{M}}$ or $\Phi(X)$, respectively.

- If $F$ is in $\bigwedge$-type, we define
  - $\mathfrak{M} \models F[\Phi]$ iff $\mathfrak{M} \models G[\Phi]$ for *all* $G \in \mathrm{CS}(F)$.
- If $F$ is in $\bigvee$-type, we define
  - $\mathfrak{M} \models F[\Phi]$ iff $\mathfrak{M} \models G[\Phi]$ for *some* $G \in \mathrm{CS}(F)$.

## *2.3 The Verification Calculus*

**Definition 2.5** Let $\mathbf{L}$ be a decorated abstract logical language. Then there is a canonical verification calculus $\models_{\mathbf{L}}^{\alpha} \Delta$ for finite sequences $\Delta$ of $\mathbf{L}$-formulas and ordinals $\alpha$ which is defined by the following clauses. We write $\Delta, F$ instead of $\Delta \cup \{F\}$.

(Ax) If $\{F, F^c\} \subseteq \Delta$ for a formula $F$ without type, then $\models_{\mathbf{L}}^{\alpha} \Delta$ holds true for all ordinals $\alpha$.

($\bigwedge$) If $F \in \Delta \cap \bigwedge$-type, $\models_{\mathbf{L}}^{\alpha_G} \Delta, G$ and $\alpha_G < \alpha$ holds true for all $G \in \mathrm{CS}(F)$, then $\models_{\mathbf{L}}^{\alpha} \Delta$.

($\bigvee$) If $F \in \Delta \cap \bigvee$-type and $\models_{\mathbf{L}}^{\alpha_0} \Delta, \Gamma$ holds true for some finite subset $\Gamma \subseteq \mathrm{CS}(F)$, then $\models_{\mathbf{L}}^{\alpha} \Delta$ holds true for all $\alpha > \alpha_0$.

We call the formula(s) $F$ [and $F^c$ in a clause (Ax)] the *main-formula(s)* of the respective clause.

There are some simple observations that follow directly from the definition of the verification calculus.

**Lemma 2.6 (Monotonicity)** $\models_{L}^{\alpha} \Delta$, *$\alpha \leq \beta$ and $\Delta \subseteq \Gamma$ imply* $\models_{L}^{\beta} \Gamma$.

**Lemma 2.7 ($\bigwedge$-inversion)** *$F \in \bigwedge$-type and $\models_{L}^{\alpha} \Delta, F$ imply $\models_{L}^{\alpha} \Delta, G$ for all $G \in$ CS$(F)$*

**Lemma 2.8 ($\bigvee$-exportation)** *If $F \in \bigvee$-type and CS$(F)$ is finite, then $\models_{L}^{\alpha} \Delta, F$ implies $\models_{L}^{\alpha} \Delta, CS(F)$.*

The proofs of all these lemmas are straightforward by induction on $\alpha$.

**Lemma 2.9 (Tautology)** *Let rnk$(F)$ denote the complexity of $F$. Then $\models_{L}^{2 \cdot rnk(F)} \Delta, F, F^c$ holds true for all finite sets $\Delta$ of $\mathbf{L}$-formulas.*

*Proof* Straightforward by induction on the complexity $\mathrm{rnk}(F)$ of the formula $F$. □

## 2.4 The Language $L_{\mathfrak{M}}$

Not all logical languages are decorated. Examples are first order languages in the familiar sense.[2] To decorate a first order language we have to equip every formula with a characteristic sequence. It is obvious how to do that for formulas whose outmost logical operation is $\wedge$ and $\vee$. It is less obvious for the first order quantifiers. Defining $\mathrm{CS}((\mathsf{Q}x)F(x)) = \langle F(x)\rangle$ would violate (C2) in Definition 2.2.

We call an abstract language *semi-decorated* if all but its first order formulas possess a characteristic sequence.

In order to decorate a semi-decorated language $\mathbf{L}$ we need a structure that matches the signature of $\mathbf{L}$.

**Definition 2.10 (The Language $\mathbf{L}_{\mathfrak{M}}$)** Let $\mathbf{L}$ be a semi-decorated language and $\mathfrak{M}$ an $\mathbf{L}$-structure. We extend $\mathbf{L}$ to the language $\mathbf{L}_{\mathfrak{M}}$ by adding a constant $\underline{m}$ for every element $m \in |\mathfrak{M}|$. This allows us to dispense with free individual variables. Therefore there are only closed terms in $\mathbf{L}_{\mathfrak{M}}$. For simplicity we identify $\mathbf{L}_{\mathfrak{M}}$-terms that obtain the same values in $\mathfrak{M}$. Therefore we may w.l.o.g. assume that all terms have the form $\underline{m}$ for some $m \in |\mathfrak{M}|$.

Any $\mathbf{L}$-structure $\mathfrak{M}$ canonically extends to an $\mathbf{L}_{\mathfrak{M}}$-structure $\mathfrak{M}_{\mathfrak{M}}$ by defining $\underline{m}^{\mathfrak{M}_{\mathfrak{M}}} := m$ for all $m \in |\mathfrak{M}|$. We mostly identify $\mathfrak{M}$ and $\mathfrak{M}_{\mathfrak{M}}$. Only if we want to emphasize that we are talking about $\mathbf{L}_{\mathfrak{M}}$-formulas we will write $\mathfrak{M}_{\mathfrak{M}}$.

To fully decorate $\mathbf{L}_{\mathfrak{M}}$ we make the following definitions.

- If $G(\vec{x})$ is an $\mathbf{L}$-formula in $\bigwedge$-type ($\bigvee$-type), where $\vec{x} = x_1, \ldots, x_n$ are all the individual variables occurring in $G$, then for every tuple $\vec{m} \in |\mathfrak{M}|^n$ the formula $G(\underline{\vec{m}})$ belongs to $\bigwedge$-type ($\bigvee$-type) of $\mathbf{L}_{\mathfrak{M}}$.
- For every decorated $\mathbf{L}$-formula $F$ we translate $\mathrm{CS}(F)$ into the language $\mathbf{L}_{\mathfrak{M}}$ by replacing formulas $G(\vec{x}) \in \mathrm{CS}(F)$ by the sequence $\langle G(\underline{\vec{m}}) \mid \vec{m} \in |\mathfrak{M}|^n\rangle$, i.e. $\mathrm{CS}(F)$ in the extended language becomes $\langle G(\underline{\vec{m}}) \mid G(\vec{x}) \in \mathrm{CS}(F) \wedge \vec{m} \in |\mathfrak{M}|^n\rangle$. Defining $\mathrm{rnk}(G(\underline{\vec{m}})) := \mathrm{rnk}(G(\vec{x}))$ we extend the complexity measure to the language $\mathbf{L}_{\mathfrak{M}}$.
- We extend the $\bigwedge$-type beyond the $\bigwedge$-type of $\mathbf{L}$ by including all sentences that are in the diagram of $\mathfrak{M}$, i.e., the true atomic sentences of $\mathbf{L}_{\mathfrak{M}}$ and all formulas the outmost logical symbol of which is $\wedge$ or $\forall$.
- We extend the $\bigvee$-type beyond the $\bigvee$-type of $\mathbf{L}$ by including all false atomic sentences and all formulas whose outmost logical symbol is $\vee$ or $\exists$.
- All atomic formulas have empty characteristic sequences.
- It is $\mathrm{CS}(F_1 \circ \cdots \circ F_n) = \langle F_1, \ldots, F_n\rangle$ if $\circ$ is one of the operators $\wedge$ of $\vee$.
- It is $\mathrm{CS}((\mathsf{Q}x)F(x)) = \langle F(\underline{s}) \mid s \in |\mathfrak{M}|\rangle$ for $\mathsf{Q} \in \{\forall, \exists\}$.[3]

[2] That is, languages that comprise a complete set of Boolean operations, first order variables and the first order quantifiers $\forall$ and $\exists$ (either explicitly or by definition) together with a set of non-logical symbols and—possibly—a set of free predicate variables.

[3] This ensures that the semantics defined in Definition 2.4 coincides with the usual semantics for first order languages.

**Theorem 2.11** *Let* **L** *be a semi-decorated language and* $\mathfrak{M}$ *an* **L**-*structure. Then* $\models^{\alpha}_{\mathbf{L}_{\mathfrak{M}}} \Delta$ *implies* $\mathfrak{M} \models \bigvee \Delta[\Phi]$ *for all* $\mathfrak{M}$-*assignments* $\Phi$.

*Proof* We prove the lemma by induction on $\alpha$.

If the last clause in the definition of $\models^{\alpha}_{\mathbf{L}_{\mathfrak{M}}} \Delta$ is according to (Ax), then there is an atomic formula such that $\{F, F^c\} \in \Delta$. This immediately implies $\mathfrak{M} \models \bigvee \Delta[\Phi]$ for any $\mathfrak{M}$-assignment $\Phi$.

If the last clause is according to $(\bigwedge)$, then there is a formula $F \in \bigwedge$-type $\cap\, \Delta$. If $\mathrm{CS}(F) = \emptyset$, then $F$ is an atomic sentence in the diagram of $\mathfrak{M}$. Hence $\mathfrak{M} \models \bigvee \Delta[\Phi]$ for any assignment $\Phi$. If $F \simeq \bigwedge \langle G_\iota | \iota \in I\rangle$,[4] we have premises $\models^{\alpha_\iota}_{\mathbf{L}_{\mathfrak{M}}} (\bigvee \Delta \vee G_\iota)[\Phi]$ for $\alpha_\iota < \alpha$. By induction hypothesis we thus either have $\mathfrak{M} \models \bigvee \Delta[\Phi]$ or $\mathfrak{M} \models G_\iota[\Phi]$ for all $\iota \in I$. In the latter case we get $\mathfrak{M} \models F[\Phi]$ and, since $F \in \Delta$ also $\mathfrak{M} \models \bigvee \Delta[\Phi]$.

The remaining case follows similarly from the induction hypothesis. □

Let $F$ be an **L**-formula which does not contain logical operators that bind predicate variables but contains the predicate variables $X_1, \dots, X_n$ free. We then call the formula $(\forall X_1) \dots (\forall X_n) F(X_1, \dots, X_n)$ a $\Pi^1_1$-sentence of **L**. We clearly have $\mathfrak{M} \models (\forall X_1) \dots (\forall X_n) F(X_1, \dots, X_n)$ in the sense of full second order logic iff $\mathfrak{M} \models F[\Phi]$ for every $\mathfrak{M}$-assignment $\Phi$. Therefore we obtain the next theorem as an immediate consequence of Theorem 2.11.

**Theorem 2.12** *Let* **L** *be a semi-decorated language and* $\mathfrak{M}$ *an* **L**-*structure. Then* $\models^{\alpha}_{\mathbf{L}_{\mathfrak{M}}} F(X_1, \dots, X_n)$ *implies* $\mathfrak{M} \models (\forall X_1) \dots (\forall X_n) F(X_1, \dots, X_n)$.

There is also a trivial inversion of Theorem 2.11.

**Theorem 2.13** *Let* **L** *be a semi-decorated language,* $\mathfrak{M}$ *an* **L**-*structure and* $F$ *an* $\mathbf{L}_{\mathfrak{M}}$-*sentence that is valid in* $\mathfrak{M}$, *i.e.,* $\mathfrak{M} \models F$. *Then there is an* $\alpha \leq rnk(F)$ *such that* $\models^{\alpha}_{\mathbf{L}_{\mathfrak{M}}} F$.

*Proof* Straightforward by induction on $\mathrm{rnk}(F)$. □

*Remark 2.14* Theorem 2.13 is in fact a triviality which is essentially a remake of the truth definition for $F$. Nonetheless there are situations where verifications of sentences yield useful information. This is, e.g., the case in the ordinal analyses of subsystems of set theory. (cf. [10, 11]).

Here we will, however, not talk about ordinal analyses. The first challenge is to check if the opposite direction in Theorem 2.12 also holds true. We are going to show that in a more general setting.

---

[4] We write $F \simeq \bigwedge \langle G_\iota | \iota \in I\rangle$ to indicate that $F$ is in $\bigwedge$-type and $\mathrm{CS}(F) = \langle G_\iota | \iota \in I\rangle$. Similarly we use $F \simeq \bigvee \langle G_\iota | \iota \in I\rangle$.

## 2.5 L-*Consequences*

**Definition 2.15 (L-*consequence*)** Let **L** be an abstract language, $F$ an **L**-formula and $T$ a set of **L**-formulas. We say that $F$ is **L**-derivable from $T$ or, synonymously, that $F$ is an **L**-consequence of $T$, if for every **L**-structure $\mathfrak{M}$ and every $\mathfrak{M}$-assignment $\Phi$, that satisfies all the formulas in $T$, we also have $\mathfrak{M} \models F[\Phi]$. We denote **L**-consequence by $T \models_{\mathbf{L}} F$.

Now let **L** be a decorated language. To extend the definition of a verification calculus to a verification calculus $T \overset{\alpha}{\underset{\mathbf{L}}{\models}} \Delta$ for **L**-consequences we keep the old rules. i.e.,

- $\overset{\alpha}{\underset{\mathbf{L}}{\models}} \Delta$ extends to $T \overset{\alpha}{\underset{\mathbf{L}}{\models}} \Delta$

and augment it by a theory rule

($T$-rule) if $T \overset{\alpha_0}{\underset{\mathbf{L}}{\models}} \Delta, F^c$ then $T, F \overset{\alpha}{\underset{\mathbf{L}}{\models}} \Delta$ for all $\alpha > \alpha_0$.

The ($T$-rule) looks a bit weird at first sight. One would expect a rule saying that $T \overset{\alpha}{\underset{\mathbf{L}}{\models}} \Delta$ iff $T \cap \Delta$ is not void. Such a rule is in fact permissible. With the help of Lemma 2.9 and the theory rule we get easily

$$\textit{if} \quad F \in \Delta \cap T \quad \textit{then} \quad T \overset{2\cdot \mathrm{rnk}(F)+1}{\underset{\mathbf{L}}{\models}} \Delta. \tag{1}$$

As a triviality we remark

$$\overset{\alpha}{\underset{\mathbf{L}}{\models}} \Delta \Leftrightarrow \emptyset \overset{\alpha}{\underset{\mathbf{L}}{\models}} \Delta. \tag{2}$$

**Lemma 2.16** *Let* **L** *be a (semi-)decorated language and* $\mathfrak{M}$ *an* **L**-*structure. Then* $\mathbf{L}_{\mathfrak{M}}$ *is a decorated language and* $T \overset{\alpha}{\underset{\mathbf{L}_{\mathfrak{M}}}{\models}} \Delta$ *implies* $\mathfrak{M} \models \bigvee \Delta[\Phi]$ *for all* $\mathfrak{M}$-*assignments* $\Phi$ *that satisfy all formulas in* $T$.

*Proof* The proof is essentially that of Theorem 2.11. We assume $\mathfrak{M} \models G[\Phi]$ for all $G \in T$ and show $\mathfrak{M} \models \bigvee \Delta[\Phi]$ by induction on $\alpha$. The only new case is that of a theory rule $T_0 \overset{\alpha_0}{\underset{\mathbf{L}_{\mathfrak{M}}}{\models}} \Delta, F^c \Rightarrow T_0, F \overset{\alpha}{\underset{\mathbf{L}_{\mathfrak{M}}}{\models}} \Delta$. Since $F \in T$ we have $\mathfrak{M} \models F[\Phi]$ by hypothesis which implies $\mathfrak{M} \not\models F^c[\Phi]$. By induction hypothesis we also have $\mathfrak{M} \models (\bigvee \Delta \vee F^c)[\Phi]$. Hence $\mathfrak{M} \models \bigvee \Delta[\Phi]$. □

A possibility to read Lemma 2.16 is

$$T \overset{\alpha}{\underset{\mathbf{L}_{\mathfrak{M}}}{\models}} F \Rightarrow T \models_{\mathbf{L}_{\mathfrak{M}}} F \tag{3}$$

although the latter notion is still quite narrow[5] and needs to be more elaborated. (That will be done in Sect. 4.3).

[5]It only says that every $\mathfrak{M}$-assignment $\Phi$ that satisfies all formulas in $T$ also satisfies $F$ in $\mathfrak{M}$.

Nevertheless there is the question for the opposite direction in (3) which we are going to address in the next section.

# 3 Search Trees

## 3.1 The Syntactical Main Lemma

In the following section we concentrate on countable decorated languages. In such languages the characteristic sequences $\mathrm{CS}(F) = \langle G_\iota | \iota \in I\rangle$ are countable. We may therefore assume that $I$ is always a countable index set and thus carries a well-ordering of order-type $\omega$.

**Definition 3.1** Let $\mathbf{L}$ be a countable decorated language, $T$ a countable and $\Delta$ a finite sequence of $\mathbf{L}$-formulas. We fix an enumeration of the formulas in $T$ and define the search tree $\mathsf{S}^{\mathbf{L}}_{T,\Delta}$ together with a label function $\Delta\colon \mathsf{S}^{\mathbf{L}}_{T,\Delta} \longrightarrow \mathbf{L}^{<\omega}$ by the following clauses [we write $\Delta_s$ instead of $\Delta(s)$].

(root) $\langle\,\rangle \in \mathsf{S}^{\mathbf{L}}_{T,\Delta}$ and $\Delta_{\langle\,\rangle} := \Delta$.

For the following clauses we assume $s \in \mathsf{S}^{\mathbf{L}}_{T,\Delta}$ such that $\Delta_s$ is not a logical axiom according to (Ax).

The *redex* of a finite sequence $\Delta_s$ of $\mathbf{L}$-formulas is the leftmost formula that possesses a type. We obtain the *reduced sequence* $\Delta^r_s$ by discharging the redex in $\Delta_s$.

(id) If $\Delta_s$ has no redex, then $s^\frown\langle 0\rangle \in \mathsf{S}^{\mathbf{L}}_{T,\Delta}$ and $\Delta_{s^\frown\langle 0\rangle} := \Delta_s, F_\iota{}^c$ where $F_\iota$ is the first formula in $T$ that has not occurred in $\bigcup_{t\sqsubseteq s} \Delta_t$.

($\bigwedge$) If $F$ is the redex of $\Delta_s$ and $F \simeq \bigwedge\langle G\iota | \iota \in I\rangle$, then $s^\frown\langle\iota\rangle \in \mathsf{S}^{\mathbf{L}}_{T,\Delta}$ for all $\iota \in I$ and $\Delta_{s^\frown\langle\iota\rangle} := \Delta^r_s, G_\iota$.

($\bigvee$) If $F$ is the redex of $\Delta_s$ and $F \simeq \bigvee\langle G\iota | \iota \in I\rangle$, then $s^\frown\langle 0\rangle \in \mathsf{S}^{\mathbf{L}}_{T,\Delta}$ and $\Delta_{s^\frown\langle 0\rangle} := \Delta^r_s, G, F, F_\iota{}^c$ where $G$ is the first formula in $\mathrm{CS}(F)$ and $F_\iota$ the first formula in $T$ that have not yet occurred in $\bigcup_{t\sqsubseteq s} \Delta_t$.

**Lemma 3.2 (Syntactical Main Lemma)** *Let $\mathbf{L}$ be a decorated language, $T$ a countable and $\Delta$ a finite set of $\mathbf{L}$-formulas. If the search tree $\mathsf{S}^{\mathbf{L}}_{T,\Delta}$ is well-founded of ordertype $\alpha$, then there is a subset $T_0$ of $T$ such that $T_0 \,\substack{\alpha\\ \models\\ \mathbf{L}}\, \Delta$.*

*If $\mathbf{L}$ is an $\mathbb{A}$-admissible fragment $\mathbf{L}_{\mathbb{A}}$ and the set $T$ is $\Sigma$-definable in $\mathbb{A}$, then $T_0$ and $\alpha$ can be chosen $\mathbb{A}$-finite.*

*Proof* Let $\mathsf{S}^{\mathbf{L}}_{T,\Delta}$ be well-founded and $s$ a node in $\mathsf{S}^{\mathbf{L}}_{T,\Delta}$. Let $|s|$ denote the ordertype of $s$ in $\mathsf{S}^{\mathbf{L}}_{T,\Delta}$. By induction on $|s|$ we then easily obtain the existence of a set $T_s \subseteq T$ such that $T_s \,\substack{|s|\\ \models\\ \mathbf{L}}\, \Delta_s$.

To prove the addendum we show that for all $s \in \mathsf{S}^{\mathbf{L}}_{T,\Delta}$ there is a set $T_s \in \mathbb{A}$ and an ordinal $\alpha_s \in \mathbb{A}$ such that $T_s \mathrel{\substack{|\alpha_s \\ \overline{\overline{\mathbf{L}}}}} \Delta_s$. The only delicate case is clause $(\bigwedge)$ in the definition of the search tree. There we obtain by induction hypothesis for all $\iota \in I$ a set $T_\iota \subseteq T$ and an ordinal $\alpha_\iota \in \mathbb{A}$ such that $T_\iota \in \mathbb{A}$ and $T_\iota \mathrel{\substack{|\alpha_\iota \\ \overline{\overline{\mathbf{L}}}}} \Delta_s, G_\iota$. Since $I \in \mathbb{A}$ we obtain by $\Sigma$-collection and union a set $T_s \in \mathbb{A}$ such that $T_s \mathrel{\substack{|\alpha_\iota \\ \overline{\overline{\mathbf{L}}}}} \Delta_s, G_\iota$ for all $\iota \in I$ and by $\Sigma$-collection and $\Delta_0$-separation an ordinal $\alpha_s$ such that $\alpha_\iota < \alpha_s$ for all $\iota \in I$. By a clause $(\bigwedge)$ it then follows $T_s \mathrel{\substack{|\alpha_s \\ \overline{\overline{\mathbf{L}}}}} \Delta_s$. □

## 3.2 The Semantical Main Lemma

The semantic counterpart of the Syntactical Main Lemma is the Semantical Main Lemma which needs an **L**-structure in its formulation.

**Lemma 3.3 (Semantical Main Lemma)** *Let* **L** *be a (semi-)decorated language, $\mathfrak{M}$ a countable* **L***-structure, $T$ a countable and $\Delta$ a finite set of $\mathbf{L}_{\mathfrak{M}}$-formulas. If the search tree $\mathsf{S}^{\mathbf{L}_{\mathfrak{M}}}_{T,\Delta}$ is not well-founded, then there is an $\mathfrak{M}$-assignment $\Phi$ that verifies all the formulas in $T$ but falsifies the formulas in $\Delta$.*

*Proof* Let $f$ be an infinite path in $\mathsf{S}^{\mathbf{L}_{\mathfrak{M}}}_{T,\Delta}$ and $\Delta_f := \bigcup_{s \in f} \Delta_s$. Since $f$ is infinite we have $F^c$ in $\Delta_f$ for every $F \in T$. We define an assignment

$$\Phi(X) := \{(\underline{m}_1, \ldots, \underline{m}_k) \mid (X^c \underline{m}_1, \ldots, \underline{m}_k) \quad \text{does not occur in} \quad \Delta_f\}$$

and prove $\mathfrak{M} \not\models F[\Phi]$ for all $F \in \Delta_f$ by induction on the complexity of $F$. For atomic formulas $(X\underline{m}_1, \ldots, \underline{m}_k)$ we cannot have $(X^c\underline{m}_1, \ldots, \underline{m}_k) \in \Delta_f$ since otherwise $f$ would be finite. Hence $\mathfrak{M} \not\models F[\Phi]$. If $F$ is an atomic sentence, it cannot be in the diagram of $\mathfrak{M}$ since in this case $f$ would be finite. For formulas $F \simeq \bigwedge \langle G_\iota \mid \iota \in I\rangle$ there is at least one $G_\iota$ in $\Delta_f$. Hence $\mathfrak{M} \not\models G_\iota[\Phi]$ which implies $\mathfrak{M} \not\models F[\Phi]$. For formulas $F \simeq \bigvee \langle G_\iota \mid \iota \in I\rangle$ we secured by the definition of the search tree that all $G_\iota$ occur in $\Delta_f$. Hence $\mathfrak{M} \not\models G_\iota[\Phi]$ for all $\iota \in I$ by induction hypothesis and this implies $\mathfrak{M} \not\models F[\Phi]$. □

From the Syntactical- and Semantical Main Lemma we obtain also the opposite direction in (3).

**Lemma 3.4** *Let* **L** *be a countable (semi-)decorated language and $\mathfrak{M}$ a countable* **L***-structure and $F$ an* **L***-formula. For a countable set $T$ of $\mathbf{L}_{\mathfrak{M}}$-formulas, we then have $T \models_{\mathbf{L}_{\mathfrak{M}}} F$ iff there is a subset $T_0$ of $T$ and a countable ordinal $\alpha$ such that $T_0 \mathrel{\substack{|\alpha \\ \overline{\overline{\mathbf{L}_{\mathfrak{M}}}}}} F$.*

*If $\mathbf{L}_{\mathfrak{M}}$ is an $\mathbb{A}$-admissible fragment and $T$ is $\Sigma$-definable in $\mathbb{A}$, then $\alpha$ and $T_0$ can be chosen $\mathbb{A}$-finite.*

*Proof* One direction is (3). For the other direction assume $T_0 \not\models^{\alpha}_{\mathbf{L}_{\mathfrak{M}}} F$ for all subsets $T_0 \subseteq T$ and all countable ordinals $\alpha$ (all $T_0 \in \mathbb{A}$ and all $\alpha \in \mathbb{A}$ in case that $\mathbf{L}_{\mathfrak{M}}$ is an $\mathbb{A}$-admissible fragment). By the Syntactical Main Lemma $\mathsf{S}^{\mathbf{L}_{\mathfrak{M}}}_{T,F}$ cannot be well-founded. Therefore there is by the Semantical Main Lemma an $\mathfrak{M}$-assignment which satisfies all the formulas in $T$ but falsifies $F$. Hence $T \not\models_{\mathbf{L}_{\mathfrak{M}}} F$. □

As a consequence of Lemma 3.4 we obtain a completeness theorem for $\Pi^1_1$-sentences.

**Theorem 3.5** *Let* **L** *be a countable (semi-)decorated language and $F$ an* **L***-formula containing at most the predicate variables $X_1, \ldots, X_n$. For any countable* **L***-structure $\mathfrak{M}$ we get $\mathfrak{M} \models (\forall X_1) \ldots (\forall X_n) F$ iff there is countable ordinal $\alpha$ such that $\models^{\alpha}_{\mathbf{L}_{\mathfrak{M}}} F$.*

*Proof* One direction is Theorem 2.12. The opposite direction follows immediately from the Semantical Main Lemma. □

We use the verification calculus to define the $\mathbf{L}_{\mathfrak{M}}$ truth-complexity of a $\Pi^1_1$-sentence.

**Definition 3.6** For a $\Pi^1_1$-sentence $(\forall X_1) \ldots (\forall X_n) F$ we define

$$\mathrm{tc}_{\mathbf{L}_{\mathfrak{M}}}((\forall X_1) \ldots (\forall X_n) F) := \begin{cases} \min\{\alpha \mid \models^{\alpha}_{\mathbf{L}_{\mathfrak{M}}} F\} & \text{if this exists,} \\ \overline{\overline{|\mathfrak{M}|}}^{+} & \text{otherwise} \end{cases}$$

where $\overline{\overline{A}}$ denotes the cardinality of a set $A$ and $\kappa^+$ the first cardinal bigger than $\kappa$.

Using this notion Theorem 3.5 can be reread as

**Theorem 3.7** *Let* **L** *be a countable semi-decorated language and $\mathfrak{M}$ a countable* **L***-structure. Then $\mathfrak{M} \models F$ holds true for a $\Pi^1_1$-sentence $F$ iff $\mathrm{tc}_{\mathbf{L}_{\mathfrak{M}}}(F) < \omega_1$, where $\omega_1$ denotes the first uncountable ordinal.*

## 3.3 *The Term-Model*

In order to fully decorate a semi-decorated language **L** we need an **L**-structure. To deal with pure logic, i.e., logic not talking about special structures, we need a most general structure. The following section will show that the *term-model* for a semi-decorated language provides such a most general structure.

**Definition 3.8** Let **L** be a semi-decorated language with identity symbol $\equiv$. Its term-model $\mathfrak{T}_{\mathbf{L}}$ is defined by

- The domain $|\mathfrak{T}_{\mathbf{L}}|$ is the set $\{t \mid t \text{ is an } \mathbf{L}\text{-term}\}$.
- For any individual constant $c$ we define $c^{\mathfrak{T}_{\mathbf{L}}} := c$.
- For a function symbol we put $f^{\mathfrak{T}_{\mathbf{L}}}(t_1, \ldots, t_n) = (f t_1, \ldots, t_n)$.

- $\equiv^{\mathfrak{T}_{\mathbf{L}}} := \{(s, s) \mid s \in |\mathfrak{T}_{\mathbf{L}}|\}$.
- The other relation constants remain uninterpreted i.e., they are treated as relation variables.[6]

*Remark 3.9* There is a, in fact important, difference between the language $\mathbf{L}$ and $\mathbf{L}_{\mathfrak{T}_{\mathbf{L}}}$ (which we are going to denote more briefly by $\mathbf{L}_{\mathfrak{T}}$). According to our agreement to dispense with free individual variables in languages $\mathbf{L}_{\mathfrak{M}}$ there are no free individual variables in $\mathbf{L}_{\mathfrak{T}}$. The term-model $\mathfrak{T}_{\mathbf{L}}$ is clearly also an $\mathbf{L}$-structure. Therefore also $\mathbf{L}$-formulas can be interpreted in $\mathfrak{T}_{\mathbf{L}}$. The meanings of the $\mathbf{L}$-formula $F(x)$ and that of the $\mathbf{L}_{\mathfrak{T}}$-formula $F(\underline{x})$ are, however, different. Provided that $F(\underline{x})$ does not contain additional predicate variables it has a fixed truth value in $\mathfrak{T}_{\mathbf{L}}$ whereas $F(x)$ needs an assignment for $x$ to obtain a truth value in $\mathfrak{T}_{\mathbf{L}}$ (cf. Lemma 3.13 below).

Observe moreover that the characteristic sequences of the formulas in the language $\mathbf{L}_{\mathfrak{T}}$ only depend on $\mathbf{L}$. Therefore we commonly write $T \models^{\alpha}_{\mathbf{L}} \Delta$ instead of $T \models^{\alpha}_{\mathbf{L}_{\mathfrak{T}}} \Delta$ even for semi-decorated languages $\mathbf{L}$.

Similarly the search tree for sequences of $\mathbf{L}_{\mathfrak{T}}$ formulas only depends on $\mathbf{L}$. Therefore we also write $\mathsf{S}^{\mathbf{L}}_{T,\Delta}$ instead of $\mathsf{S}^{\mathbf{L}_{\mathfrak{T}}}_{T,\Delta}$.

**Definition 3.10** Let $\mathbf{L}$ be a semi-decorated language and $\mathfrak{M}$ an $\mathbf{L}$-structure. By an $\mathfrak{M}$*-translation* we understand a map $\phi$ that transfers $\mathbf{L}_{\mathfrak{T}}$-formulas into $\mathbf{L}_{\mathfrak{M}}$-formulas. To every constant $\underline{t}$ we assign an element $\phi(\underline{t})$ of $|\mathfrak{M}|$. By $F^{\phi}$ we denote the $\mathbf{L}_{\mathfrak{M}}$-formula that is obtained from $F$ by replacing every constant $\underline{t}$ by $\underline{\phi(\underline{t})}$. For a set $T$ of $\mathbf{L}_{\mathfrak{T}}$-formulas we denote by $T^{\phi}$ the set $\{F^{\phi} \mid F \in T\}$.

We call a translation $\phi$ *adequate* if

$$\mathrm{CS}(F^{\phi}) = \langle G^{\phi} \mid G \in \mathrm{CS}(F)\rangle$$

holds true for every formula $F$ the outmost logical operation of which is not a first order quantifier,

**Lemma 3.11 (Substitution Lemma)** *Let* $\mathbf{L}$ *be a semi-decorated language and* $\mathfrak{M}$ *be an* $\mathbf{L}$*-structure. Then* $T \models^{\alpha}_{\mathbf{L}} \Delta$ *implies* $T^{\phi} \models^{\alpha}_{\mathbf{L}_{\mathfrak{M}}} \Delta^{\phi}$ *for all adequate* $\mathfrak{M}$*-translations* $\phi$.

*Proof* The proof is a straightforward induction on $\alpha$ using the fact that $\phi$ is adequate. The only case that needs some care is that of $(\bigwedge)$-rule whose critical formula is a first order formula $(\forall x)F(x)$. There we have the premises $T \models^{\alpha_t}_{\mathbf{L}} \Delta, F(\underline{t})$ for all $\mathbf{L}$-terms $t$. W.l.o.g. we may assume that $T \cup \Delta$ does not contain all $\mathbf{L}$-terms. Then we obtain $T \models^{\alpha_t}_{\mathbf{L}} \Delta, F(\underline{t})$ for a term $t$ not occurring in $T \cup \Delta$. By induction hypothesis we get $T^{\psi} \models^{\alpha_t}_{\mathbf{L}_{\mathfrak{M}}} \Delta^{\psi}, F(\underline{t})^{\psi}$ for all adequate translations $\psi$ which coincide with $\phi$

[6] Which means that there are many term-models according to the interpretation of the relation constants.

but for the value $\psi(\underline{t})$. That means, however, that we have $T^{\phi} \mathrel{\substack{\alpha_t\\ \models\\ \mathbf{L}_{\mathfrak{M}}}} \Delta^{\phi}, F^{\phi}(\underline{m})$ for all $m \in |\mathfrak{M}|$. Since $\alpha_t < \alpha$ and $\mathrm{CS}((\forall x)F(x)^{\phi}) = \langle F^{\phi}(\underline{m}) | \, m \in |\mathfrak{M}|\rangle$ we get $T^{\phi} \mathrel{\substack{\alpha\\ \models\\ \mathbf{L}_{\mathfrak{M}}}} \Delta^{\phi}, (\forall x)F(x)^{\phi}$ by an inference $(\bigwedge)$. □

As a corollary of Lemma 3.11 we obtain

**Corollary 3.12** *Let* **L** *be a semi-decorated language,* $\Delta$ *and* $\Gamma$ *finite and* $T$ *an arbitrary set of* $\mathbf{L}_{\mathfrak{T}}$*-formulas. Then* $T \mathrel{\substack{\alpha\\ \models\\ \mathbf{L}}} \Delta, \Gamma(\underline{t})$ *for an* $\mathbf{L}_{\mathfrak{T}}$*-term* $\underline{t}$ *that does neither occur in* $T$ *nor in* $\Delta$ *implies* $T \mathrel{\substack{\alpha\\ \models\\ \mathbf{L}}} \Delta, \Gamma(\underline{s})$ *for any* **L***-term* $s$.

*Proof* Just apply the Substitution Lemma to the structure $\mathfrak{T}_{\mathbf{L}}$. □

**Lemma 3.13** *Let* **L** *be a semi-decorated language. For every* **L***-formula* $F$ *there is a canonical* $\mathbf{L}_{\mathfrak{T}}$*-formula* $\underline{F}$ *which is obtained from* $F$ *by "underlining" all* **L***-terms occurring in* $F$*. For an* **L***-structure* $\mathfrak{M}$ *and an* $\mathfrak{M}$*-assignment* $\Phi$ *we obtain an adequate translation from* $\mathbf{L}_{\mathfrak{T}}$ *into* $\mathbf{L}_{\mathfrak{M}}$ *by defining* $\chi(\underline{t}) := t^{\mathfrak{M}}[\Phi]$ *such that*

$$\mathfrak{M} \models F[\Phi] \quad \textit{iff} \quad \mathfrak{M}_{\mathfrak{M}} \models \underline{F}^{\chi}[\Phi]. \tag{4}$$

*Conversely we obtain for every* $\mathbf{L}_{\mathfrak{T}}$*-formula* $F$ *an* **L***-formula* $\overline{F}$ *by stripping all the underlines. By putting* $\overline{\Phi}(x) := x$ *we may extend any* $\mathfrak{T}_{\mathfrak{T}_{\mathbf{L}}}$*-assignment* $\Phi$ *for the language* $\mathbf{L}_{\mathfrak{T}}$ *to an* $\mathfrak{T}_{\mathbf{L}}$*-assignment* $\overline{\Phi}$ *for the language* **L** *such that* $t^{\mathfrak{T}_{\mathbf{L}}}[\overline{\Phi}] = t^{\mathfrak{T}_{\mathfrak{T}_{\mathbf{L}}}}$ *for any* **L***-term* $t$ *and*

$$\mathfrak{T}_{\mathbf{L}} \models \overline{F}[\overline{\Phi}] \quad \textit{iff} \quad \mathfrak{T}_{\mathfrak{T}_{\mathbf{L}}} \models F[\Phi]. \tag{5}$$

*Proof* The proof is straightforward. Observe, however, that $\mathbf{L}_{\mathfrak{M}}$ is a decorated language which means that the satisfaction relation in an $\mathbf{L}_{\mathfrak{M}}$-structure is defined according to Definition 2.4. In the case that $F$ is a formula $(\forall x)G(x)$ we therefore have to argue as follows: We have $\mathfrak{M} \models (\forall x)G(x)[\Phi]$ iff $\mathfrak{M} \models F(x)[\Psi]$ for all $\mathfrak{M}$-assignments that coincide with $\Phi$ but for $\Psi(x)$. By induction hypothesis this is equivalent to $\mathfrak{M}_{\mathfrak{M}} \models \underline{G(x)}^{\chi}[\Psi]$. By choice of $\Psi$ this holds true iff we have $\mathfrak{M}_{\mathfrak{M}} \models \underline{G}(\underline{m})^{\chi}[\Phi]$ for all $\overline{m \in |\mathfrak{M}|}$ and, since $\mathrm{CS}((\forall x)\underline{G}(x)) = \langle \underline{G}(\underline{m}) | \, m \in |\mathfrak{M}|\rangle$, this is equivalent to $\mathfrak{M}_{\mathfrak{M}} \models \underline{(\forall x)G(x)}^{\chi}[\Phi]$. A similar argument is needed in the proof of the second claim. □

As a first application of term-models we obtain a generalization of Lemma 3.4.

**Theorem 3.14 (Completeness Theorem)** *Let* **L** *be a countable semi-decorated language and* $T$ *a countable set of* **L***-formulas. Then we have* $T \models_{\mathbf{L}} F$ *iff there is a subset* $T_0 \subseteq T$ *and a countable ordinal* $\alpha$ *such that* $\underline{T}_0 \mathrel{\substack{\alpha\\ \models\\ \mathbf{L}}} \underline{F}$.

*If $\mathbf{L}_{\mathfrak{T}}$ is an $\mathbb{A}$-admissible fragment and $\underline{T}$ a $\Sigma$-definable set of $\mathbf{L}_{\mathfrak{T}}$-formulas, then $T_0$ and $\alpha$ can be chosen $\mathbb{A}$-finite.*

*Proof* Assume first $\underline{T} \models^{\alpha}_{\mathbf{L}} \underline{F}$, let $\mathfrak{M}$ be an $\mathbf{L}$-structure and $\Phi$ an $\mathfrak{M}$-assignment that satisfies all the formulas in $T$. By the Substitution Lemma (Lemma 3.11) we get $\underline{T}^{\Phi} \models^{\alpha}_{\mathbf{L}_{\mathfrak{M}}} \underline{F}^{\Phi}$ and by Lemma 2.16 $\mathfrak{M}_{\mathfrak{M}} \models \underline{F}^{\Phi}[\Phi]$, i.e., $\mathfrak{M} \models F[\Phi]$ by (4). Hence $T \models_{\mathbf{L}} F$.

For the opposite direction assume $\underline{T_0} \not\models^{\alpha}_{\mathbf{L}} \underline{F}$ for all countable ordinals $\alpha$ and all $T_0 \subseteq T$ ($\alpha, T_0 \in \mathbb{A}$ for the addendum). By the Syntactical Main Lemma $\mathsf{S}^{\mathbf{L}}_{\underline{T},\underline{F}}$ cannot be well-founded and by the Semantical Main Lemma there is a $\mathfrak{T}_{\mathfrak{T}_{\mathbf{L}}}$-assignment $\Phi$ that verifies all formulas in $\underline{T}$ but falsifies $\underline{F}$. By (5) $\overline{\Phi}$ then verifies all formulas in $T$ and falsifies $F$. Hence $T \not\models_{\mathbf{L}} F$. □

Let $\mathbf{L}$ be a semi-decorated language and $\mathfrak{M}$ an adequate $\mathbf{L}$-structure. We call an $\mathbf{L}$-formula $F$ valid in $\mathfrak{M}$, if $\mathfrak{M} \models F[\Phi]$ for all $\mathfrak{M}$-assignments $\Phi$. An $\mathbf{L}$-formula $F$ is valid iff it is valid in all $\mathbf{L}$-structures.

As a corollary to the Completeness Theorem we obtain

**Corollary 3.15** *An $\mathbf{L}$ formula $F$ is valid iff $\models F$ and this holds true iff there is a countable ordinal $\alpha$ such that $\models^{\alpha}_{\mathbf{L}} \underline{F}$.*

As another immediate consequence of the Completeness Theorem we obtain the permissibility of the cut-rule in the verification calculus.

**Theorem 3.16 (Permissibility of Cuts)** *Let $\mathbf{L}$ be a decorated language. If $\models^{\alpha}_{\mathbf{L}} \Delta, F$ and $\models^{\beta}_{\mathbf{L}} \Delta, F^c$ there is a countable ordinal $\gamma$ such that $\models^{\gamma}_{\mathbf{L}} \Delta$.*

*Proof* From $\models^{\alpha}_{\mathbf{L}} \Delta, F$ and $\models^{\beta}_{\mathbf{L}} \Delta, F^c$ we obtain by the correctness direction of Theorem 3.14 $\models \bigvee \Delta \vee F$ and $\models \bigvee \Delta \vee F^c$. Hence $\models \bigvee \Delta$ and by the completeness direction of Theorem 3.14 it follows $\models^{\gamma}_{\mathbf{L}} \bigvee \Delta$ for some countable ordinal $\gamma$. The claim now follows by $\bigvee$-exportation (Lemma 2.8). □

The next theorem shows that the term-model can in fact be regarded as the most general model.

**Theorem 3.17** *Let $\mathbf{L}$ be a semi-decorated language. An $\mathbf{L}$-formula is valid iff it is valid in the term-model of $\mathbf{L}$.*

*Proof* If $F$ is valid, then $F$ is also valid in the term-model. If $F$ is not valid, then by Corollary 3.15 we have $\not\models^{\alpha}_{\mathbf{L}} \underline{F}$ for all countable ordinals $\alpha$. By the Syntactical Main Lemma the search tree $\mathsf{S}^{\mathbf{L}}_{\underline{F}}$ cannot be well-founded. By the Semantical Main Lemma there is an $\mathfrak{T}_{\mathfrak{T}_{\mathbf{L}}}$-assignment $\Phi$ that falsifies $\underline{F}$. By (5) $\overline{\Phi}$ is a $\mathfrak{T}_{\mathbf{L}}$-assignment that falsifies $F$. Hence $F$ is not valid in the term-model. □

### 3.4 Admissible Fragments

Let $\mathbf{L} = \mathbf{L}_{\mathbb{A}}$ be a countable semi-decorated admissible fragment. If the $\mathbf{L}$-terms form a set in $\mathbb{A}$, which is at least the case if $\mathbb{A}$ satisfies infinity,[7] the language $\mathbf{L}_{\mathfrak{T}}$ of the term-model is a decorated admissible language. Therefore we can apply Theorem 3.14 to reobtain a famous theorem. Our tacit hypothesis for the rest of this section is that the $\mathbf{L}$-terms form a set in $\mathbb{A}$.

**Theorem 3.18 (Barwise Compactness Theorem)** *Let $\mathbf{L}_{\mathbb{A}}$ be an $\mathbb{A}$-admissible fragment of a countable semi-decorated language $\mathbf{L}$ and $T$ a set of $\mathbf{L}_{\mathbb{A}}$-formulas that is $\Sigma$-definable in $\mathbb{A}$. Then $T \models_{\mathbf{L}_{\mathbb{A}}} F$ iff there is an $\mathbb{A}$ finite subset $T_0 \subseteq T$ such that $T_0 \models_{\mathbf{L}_{\mathbb{A}}} F$.*

*Proof* From $T \models_{\mathbf{L}_{\mathbb{A}}} F$ and the fact that $\mathbf{L}_{\mathfrak{T}}$ is an $\mathbb{A}$-fragment we get by Theorem 3.14 an $\mathbb{A}$-finite set $\underline{T}_0 \subseteq \underline{T}$ such that $\underline{T}_0 \overset{\alpha}{\underset{\mathbf{L}_{\mathbb{A}}}{\models}} \underline{F}$ for some ordinal $\alpha \in \mathbb{A}$. Again by Theorem 3.14 this entails $T_0 \models_{\mathbf{L}_{\mathbb{A}}} F$. □

Summing up our hitherto obtained results we get two theorems.

**Theorem 3.19 (Barwise Completeness Theorem)** *Let $\mathbf{L}$ be a countable $\mathbb{A}$-admissible fragment of a semi-decorated language and $T$ a set of $\mathbf{L}$-formulas that is $\Sigma$-definable in $\mathbb{A}$. Then the following claims are equivalent.*

1. $T \models_{\mathbf{L}} F$.
2. $(\exists\alpha)[T \overset{\alpha}{\underset{\mathbf{L}}{\models}} F]$.
3. $\mathbb{A} \models (\exists\alpha)[T \overset{\alpha}{\underset{\mathbf{L}}{\models}} F]$.

**Theorem 3.20** *Let $\mathbf{L}$ be a countable $\mathbb{A}$-admissible fragment of a semi-decorated language and $T$ a set of $\mathbf{L}$-formulas that is $\Sigma$-definable on $\mathbb{A}$. Then $T$ is consistent iff every $\mathbb{A}$-finite subset of $T$ is consistent.*

*Proof* For the non-trivial direction assume that $T$ is inconsistent. This is the case iff there is a formula $F$ in $T$ such that $T \models_{\mathbf{L}} F^c$. By Theorem 3.18 there is an $\mathbb{A}$-finite subset $T_0 \subseteq T$ such that $T_0 \models_{\mathbf{L}} F^c$. Then $T_0 \cup \{F\}$ is still $\mathbb{A}$-finite and inconsistent. □

## 4 Applications to Logic

In the following section we study the applicability of the previous section to logics. We do that in an exemplary way and start with the most natural test case, first order logic.

[7] Cf. [1, III, Proposition 1.4].

## 4.1 First Order Logic

First order languages are semi-decorated languages. They share the special feature that their verification calculi can be replaced by finite verification calculi. This fact is due to Corollary 3.12 of the Substitution Lemma.

**Definition 4.1** Let **L** be a first order language (with free predicate variables) and $\mathfrak{T}$ its term-model. If we define the complexity $\text{rnk}(F)$ in the familiar way, we obtain $\mathbf{L}_{\mathfrak{T}}$ as a fully decorated language. We modify the verification calculus to a calculus $T \left|\frac{\alpha}{\mathbf{L}}\right. \Delta$ by keeping the rules (Ax), $(\bigvee)$ and the (T-rule) and modifying $(\bigwedge)$ to

($\bigwedge$) If $F \in \Delta \cap \bigwedge$-type, $T \left|\frac{\alpha_G}{\mathbf{L}}\right. \Delta, G, \alpha_G < \alpha$ for all $G \in \text{CS}(F)$ and the outmost logical symbol of $F$ is not a universal quantifier, then $T \left|\frac{\alpha}{\mathbf{L}}\right. \Delta$ holds true.

($\forall$) If $(\forall x)F(x) \in \Delta$ and $T \left|\frac{\alpha_0}{\mathbf{L}}\right. \Delta, F(\underline{t})$ for some **L**-term $t$ such that $\underline{t}$ does not occur in the formulas of $T \cup \Delta$, then $T \left|\frac{\alpha}{\mathbf{L}}\right. \Delta$ holds true for all $\alpha > \alpha_0$.

It is immediate from Definition 4.1 that $\alpha$ in $T \left|\frac{\alpha}{\mathbf{L}}\right. \Delta$ can always be chosen finite.

**Lemma 4.2** *Let* $\mathsf{L}$ *be a first order language and* $T$ *an arbitrary,* $\Delta$ *a finite set of* $\mathsf{L}_{\mathfrak{T}}$*-formulas. Then* $T \models^{\alpha}_{\mathsf{L}} \Delta$ *and* $T \left|\frac{\alpha}{\mathsf{L}}\right. \Delta$ *are equivalent.*

*Proof* Clearly $T \models^{\alpha}_{\mathsf{L}} \Delta$ implies $T \left|\frac{\alpha}{\mathsf{L}}\right. \Delta$. For the opposite direction we show $T \left|\frac{\alpha}{\mathsf{L}}\right. \Delta \Rightarrow T \models^{\alpha}_{\mathsf{L}} \Delta$ by induction on $\alpha$. The only critical case is that of an inference according to ($\forall$). Here we have the premise $T \left|\frac{\alpha_0}{\mathsf{L}}\right. \Delta, F(\underline{t})$ for some term $t$ such that $\underline{t}$ neither occurs in $T$ nor in $\Delta$. Hence $T \models^{\alpha_0}_{\mathsf{L}} \Delta, F(\underline{t})$ by induction hypothesis. From Corollary 3.12 we then get $T \models^{\alpha_0}_{\mathsf{L}} \Delta, F(\underline{s})$ for all $\mathsf{L}$-terms $s$. Using an inference $(\bigwedge)$ this yields $T \models^{\alpha}_{\mathsf{L}} \Delta$. □

*Remark 4.3* The modified verification calculus $\left|\frac{\alpha}{\mathsf{L}}\right. \Delta$ with empty set $T$ is the familiar cut-free one-sided sequent calculus à la Tait.

Combining Theorem 3.14 and Lemma 4.2 we obtain the completeness theorem for first order logic.

**Theorem 4.4 (Correctness and Completeness for First Order Logic)** *Let* $\mathsf{L}$ *be a first order language and* $T$ *be a countable set of* $\mathsf{L}$*-formulas. Then* $T \models_L F$ *holds true iff there is a finite subset* $T_0 \subseteq T$ *and a finite ordinal* $n$ *such that* $T \left|\frac{n}{\mathsf{L}}\right. F$.

The fraternal twin of first order completeness is the Compactness Theorem for first order logic.

**Theorem 4.5** *A countable set of first order formulas is consistent iff all its finite subsets are consistent.*

*Proof* The proof is that of Theorem 3.20. Here we can, however, replace $\mathbb{A}$-finite by finite. □

*Remark 4.6* It is well known that the Compactness—and the Completeness Theorem for first order logic hold also for arbitrary, not only for countable sets $T$. The restriction "countable" is the price we have to pay for the generality of our approach. For admissible fragments, for example, the hypothesis "countable" is indispensable.

As a consequence of Theorem 3.17 we get the term-model as the most general model for first order logic.

**Theorem 4.7** *A first order formula is valid iff it is valid in the term-model.*

Combining Theorems 3.5 and 4.4 and Lemma 4.2 we obtain

**Theorem 4.8** *Every logically valid $\Pi^1_1$-sentence has a finite truth complexity.*

## 4.2 Second Order Logics

The language $\mathbf{L}^2$ of second order logic above $\mathbf{L}$ is obtained from $\mathbf{L}$ by allowing quantifications over predicate variables. To simplify notations we restrict ourselves to unary predicate variables.[8] We then talk about second order quantifiers. To obtain the semantics for $\mathbf{L}^2$-languages we start with an $\mathbf{L}$-structure $\mathfrak{M} = (|\mathfrak{M}|, \dots)$ and extend it to an $\mathbf{L}^2$-structure $\mathfrak{M}^2 = (|\mathfrak{M}|, \mathsf{S}, \dots)$ by adding a domain $\mathsf{S} \subseteq \mathrm{Pow}(|\mathfrak{M}|)$ for the second order quantifiers. An $\mathfrak{M}^2$-assignment $\Phi$ assigns an element $\Phi(x) \in |\mathfrak{M}|$ to every first order (i.e. individual) variable $x$ and a set $\Phi(X) \in \mathsf{S}$ to every second order (i.e. predicate) variable $X$. We say that $\mathfrak{M} \models (\mathsf{Q}X)F(X)[\Phi]$ iff $\mathfrak{M}^2 \models F(X)[\Psi]$ for all ($\mathsf{Q} = \forall$) or some ($\mathsf{Q} = \exists$) $\mathfrak{M}^2$-assignments $\Psi$ which coincide with $\Phi$ but for the value $\Psi(X)$.

Due to different semantics there are three kinds of second order logic.

- *Weak second order logic* which is characterized by the fact that the domain of the second order quantifiers may be any subset of $\mathrm{Pow}(|\mathfrak{M}|)$.
- *Simple second order logic* for which the domain of the second order quantifiers is the class of all $\mathbf{L}^2$-definable subsets of $|\mathfrak{M}|$.
- *Full second order logic* for which the domain of the second order quantifiers is the full power set of $|\mathfrak{M}|$.

### 4.2.1 Weak Second Order Logic

To obtain a term model for weak second order logic $\mathbf{L}^2$ we have to extend the term-model $\mathfrak{T}_{\mathbf{L}}$ by the second order domain $\mathsf{S} := \{X \mid X \text{ is a second order variable}\}$ to the

[8]For full generality we need a domain $\mathsf{S}^k \subseteq \mathrm{Pow}\big(|\mathfrak{M}|\big)^k$ for every quantifier on $k$-ary predicate variables. The restriction to unary predicate variables is in fact only a matter of simplifying notations.

term-model $\mathfrak{T}^2_{\mathbf{L}}$. If $\mathbf{L}$ is countable and semi-decorated, the language $\mathbf{L}^2_{\mathfrak{T}^2_{\mathbf{L}}}$ becomes a countable decorated language extending $\mathbf{L}_{\mathfrak{T}_{\mathbf{L}}}$ by putting formulas $(\forall X)F(X)$ into $\bigwedge$-type and dually formulas $(\exists X)F(X)$ into $\bigvee$-type and defining

- $\mathrm{CS}((\mathsf{Q}X)F(X)) := \langle F(\underline{X}) |\, X \quad \text{is a second order variable}\rangle$.

Therefore all the results of Sect. 3.1 and Definition 4.1 carry over to weak second order logic.[9] We leave it to the reader to transfer the results.

### 4.2.2 Simple Second Order Logic

To obtain a term model for simple second order logic $\mathsf{s}\mathbf{L}^2$ over a semi-decorated logic $\mathbf{L}$ we have to extend the term model $\mathfrak{T}_{\mathbf{L}}$ by the set $\mathsf{S} := \{\{x \mid F(x)\} |\, F \text{ is an } \mathbf{L}^2\text{-formula}\}$ and consequently

- $\mathrm{CS}((\mathsf{Q}X)F(X)) = \langle F(\underline{\{x \mid G(x)\}}) |\, G \text{ is an } \mathbf{L}^2\text{-formula}\rangle$.

But here we also have to decorate the formulas $s \in \underline{\{x \mid F(x)\}}$ and their duals $s \notin \underline{\{x \mid F(x)\}}$ that are new in the language of the term-model. We put

- $\mathrm{CS}(s \in \underline{\{x \mid F\}(x)}) := \langle F(s)\rangle$.

Observe that simple second order logic is already pretty strong. It allows for full comprehension. If we denote the term-model for $\mathsf{s}\mathbf{L}^2$ by $\mathfrak{T}^2$, we get

$$\mathfrak{T}^2 \models (\exists X)(\forall x)[(Xx) \leftrightarrow F(x)] \tag{6}$$

for every $\mathbf{L}^2$-formula $F(x)$.

Although $\mathfrak{T}^2$ is a countable structure the decoration via $\mathfrak{T}^2$ violates—at least prima facie—condition (C1). It is not immediately clear how to define a complexity that satisfies (C1). It follows from the work of Takeuti [14] that the results of Sect. 3 cannot be easily extended to $\mathsf{s}\mathbf{L}^2$. According to [14] cut eliminability as in Theorem 3.16 for $\mathsf{s}\mathbf{L}^2$ would imply the consistency of second order arithmetic with full comprehension. A theory the ordinal analysis of which is still one of the biggest challenges.

This, however, does not exclude the possibility that results for $\mathsf{s}\mathbf{L}^2$ similar to those in Sect. 3 can be obtained using stronger means. The only result for which condition (C1) is essential is the Semantical Main Lemma. Schütte in [12] used search trees to show that cut elimination for simple type theory is equivalent to the fact that every partial valuation for simple type theory is extensible to a total valuation and Tait used this result in [13] to prove cut-elimination for simple second order logic by showing that partial valuations can be extended to valuations in $\omega$-

[9] This is of course not surprising because weak first order logic is in fact nothing but a two sorted first order logic.

models. It seems possible that similar techniques may be used to prove a variation of the Semantical Main Lemma. We have not yet checked that.

Also an attempt to use Girards "canditats de reducibilité" [4] might yield some results similar to those in Sect. 3.

#### 4.2.3 Full Second Order Logic

The only possibility to decorate full second order logic via "term-models" we can imagine is to define $\mathrm{CS}((\mathsf{Q}X)F(X)) := \langle F(\underline{S}) | \; S \subseteq |\mathfrak{T}|\rangle$. This, however, yields not a countable decorated language. Therefore there is, as far as we can see it now, no way to treat full second order languages. There are, however, logics in between first order (or weak second order) logic and full second order logic which are feasible and are treated in the next sections.

### *4.3 $\mathfrak{M}$-Logic*

A possible way to read Lemma 2.16 is Eq. (3) in which we find the notion $T \models_{\mathbf{L}_{\mathfrak{M}}} F$. This notion is not yet well elaborated. To elaborate this notion further we have to specify the notion of an $\mathbf{L}_{\mathfrak{M}}$-structure. Clearly $\mathfrak{M}_{\mathfrak{M}}$ is an $\mathbf{L}_{\mathfrak{M}}$-structure. To obtain a wider class of $\mathbf{L}_{\mathfrak{M}}$-structures we extend the language $\mathbf{L}_{\mathfrak{M}}$ by a new unary predicate symbol $\mathsf{M}$, the intended meaning of which is the domain $|\mathfrak{M}|$ of the structure $\mathfrak{M}$, and allowing free individual variables. Call this extended language $\mathbf{L}^{+}_{\mathfrak{M}}$.

**Definition 4.9** Let $\mathbf{L}$ be a semi-decorated language and $\mathfrak{M}$ an $\mathbf{L}$-structure. An $\mathbf{L}^{+}_{\mathfrak{M}}$-structure $\mathfrak{N}$ is called an $\mathbf{L}$-$\mathfrak{M}$-structure iff

- $M^{\mathfrak{N}} - |\mathfrak{M}|$,
- $\underline{m}^{\mathfrak{N}} = m$ for all $m \in |\mathfrak{M}|$,
- $\mathfrak{M}_{\mathfrak{M}}$ is a substructure of $\mathfrak{N} \restriction \mathbf{L}_{\mathfrak{M}}$.

We say that a formula $F$ is a consequence of a set $T$ of $\mathbf{L}^{+}_{\mathfrak{M}}$-formulas in $\mathfrak{M}$-logic (written as $T \models^{\mathbf{L}}_{\mathfrak{M}} F$) iff for every $\mathbf{L}$–$\mathfrak{M}$-structure $\mathfrak{N}$ and every $\mathfrak{N}$-assignment $\Phi$ that verifies all the formulas in $T$ we also have $\mathfrak{N} \models F[\Phi]$.

Commonly the basic language $\mathbf{L}$ is the first order language $\mathsf{L}(\mathfrak{M})$ of a given structure. Instead of $\mathsf{L}(\mathfrak{M})$-$\mathfrak{M}$-structures we then talk about $\mathfrak{M}$-structures and write $T \models_{\mathfrak{M}} F$ instead of $T \models^{\mathsf{L}(\mathfrak{M})}_{\mathfrak{M}} F$.

Although $\mathbf{L}_{\mathfrak{M}}$ is a fully decorated language we must not regard $\mathbf{L}^{+}_{\mathfrak{M}}$ as fully decorated. It is not even semi-decorated since there are new formulas $(\mathsf{M}t)$ and $(\mathsf{M}^{c}t)$ which are not yet decorated with characteristic sequences. These formulas—

let us write $(t \in \mathsf{M})$ and $(t \notin \mathsf{M})$ instead of $(\mathsf{M}t)$ and $\mathsf{M}^c t)$—are not regarded as atomic but we put

- all formulas $(t \in \mathsf{M})$ in $\bigvee$-type and dually
- formulas $(t \notin \mathsf{M})$ in $\bigwedge$-type,

define

- $\mathrm{CS}(t \in \mathsf{M}) := \langle t \equiv \underline{m} \mid m \in |\mathfrak{M}|\rangle$.

and assign a complexity larger than $\mathrm{rnk}(t \equiv \underline{m})$ to the formulas $t \in \mathsf{M}$ and $t \notin \mathsf{M}$.[10]

Having made these definitions we may regard $\mathbf{L}_{\mathfrak{M}}^{+}$ as a semi-decorated language. To fully decorate it we need an $\mathbf{L}_{\mathfrak{M}}^{+}$-structure $\mathfrak{N}$. According to Definition 2.10 the language $(\mathbf{L}_{\mathfrak{M}}^{+})_{\mathfrak{N}}$ is then a fully decorated language.

To deal with "pure $\mathfrak{M}$-logic" we have to introduce a term-model for $\mathbf{L}_{\mathfrak{M}}^{+}$. Observe that there are constants $\underline{m}$ for $m \in |\mathfrak{M}|$ in $\mathbf{L}_{\mathfrak{M}}^{+}$. Sometimes it is useful to strip the underlines. We therefore denote by $\overline{s}$ the $\mathbf{L}_{\mathfrak{M}}^{+}$-term that is obtained from $s$ by stripping all the underlines. Clearly $\overline{s}$ is not an $\mathbf{L}_{\mathfrak{M}}^{+}$-term.

**Definition 4.10** Let $\mathbf{L}$ be an abstract language and $\mathfrak{M}$ an $\mathbf{L}$-structure. The term-model $\mathfrak{T}_{\mathbf{L}_{\mathfrak{M}}}$ for $\mathbf{L}_{\mathfrak{M}}^{+}$ is defined by the following items.

- The domain is the set of all $\mathbf{L}_{\mathfrak{M}}^{+}$-terms.
- $f^{\mathfrak{T}_{\mathbf{L}_{\mathfrak{M}}}}(t_1, \ldots, t_k) := (f t_1, \ldots, t_k)$.
- $\equiv^{\mathfrak{T}_{\mathbf{L}_{\mathfrak{M}}}} := \{(s, s) \mid s \text{ is an } \mathbf{L}_{\mathfrak{M}}^{+}\text{-term}\}$.
- For every $m \in |\mathfrak{M}|$ we define $\underline{m}^{\mathfrak{T}_{\mathbf{L}_{\mathfrak{M}}}} := m$.
- It is $\mathsf{M}^{\mathfrak{T}_{\mathbf{L}_{\mathfrak{M}}}} = \{s \mid \overline{s} \in |\mathfrak{M}|\}$ and $\mathsf{M}^{c\,\mathfrak{T}_{\mathbf{L}_{\mathfrak{M}}}} = \{s \mid \overline{s} \notin |\mathfrak{M}|\}$.
- The diagram of $\mathfrak{T}_{\mathbf{L}_{\mathfrak{M}}}$ comprises all sentences of the form $(R\underline{m}_1, \ldots, \underline{m}_k)$ for which $(Rm_1, \ldots, m_k)$ is in the diagram of $\mathfrak{M}$.
- Relation constants different from $\equiv$ and $\mathsf{M}$ remain uninterpreted and are treated as relation variables. Observe, however, the extension of the diagram above.[11]

Using the standard complexity definition for $\mathrm{rnk}(F)$ the language $\mathbf{L}_{\mathfrak{T}_M}$ (which we will briefly call $\mathbf{L}_{\mathfrak{T}}^{\mathfrak{M}}$) is now fully decorated[12] and thus defines a verification calculus $T \models^{\alpha}_{\mathbf{L}_{\mathfrak{T}}^{\mathfrak{M}}} \Delta$ (in case that $\mathbf{L}$ is the first order language of $\mathfrak{M}$ we briefly write $T \models^{\alpha}_{\mathfrak{M}} \Delta$). We can transfer the notions and results of Sect. 3.3 to the term-model of $\mathfrak{M}$-logic. For translations in Definition 3.10 we clearly have to require that $\underline{m}^{\phi} = m$ for all $m \in |\mathfrak{M}|$. Transferring the results we get the following theorems.

---

[10] $\mathrm{rnk}(t \in \mathsf{M}) = \mathrm{rnk}(t \notin \mathsf{M}) = 1$ will work in most cases, especially in the case that the basis language is the first order language of $\mathfrak{M}$.

[11] According to this extension $\mathsf{M}$ and $\mathsf{M}^c$ are interpreted by the definition of $\mathrm{CS}(t \in \mathsf{M})$ and $\mathrm{CS}(t \notin \mathsf{M})$, respectively.

[12] Observe the peculiarity that for $m \in |\mathfrak{M}|$ we have a constant $\underline{m}$ in $\mathbf{L}_{\mathfrak{M}}^{+}$ and thus a constant $\underline{\underline{m}}$ in $\mathbf{L}_{\mathfrak{T}}^{\mathfrak{M}}$. The interpretation of $\underline{\underline{m}}$ in the extended model $(\mathfrak{T}_{\mathbf{L}_{\mathfrak{M}}})_{\mathfrak{T}_{\mathbf{L}_{\mathfrak{M}}}}$ is $\underline{m}$ and the interpretation of $\underline{m}$ in $\mathfrak{T}_{\mathbf{L}_{\mathfrak{M}}}$ is $m$. Thus $(R\underline{\underline{m}}_1, \ldots, \underline{\underline{m}}_k)$ belongs to the diagram of $(\mathfrak{T}_{\mathbf{L}_{\mathfrak{M}}})_{\mathfrak{T}_{\mathbf{L}_{\mathfrak{M}}}}$ iff $(R\underline{m}_1, \ldots, \underline{m}_k)$ belongs to the diagram of $\mathfrak{T}_{\mathfrak{M}}$ iff $(Rm_1, \ldots, m_k)$ belongs to the diagram of $\mathfrak{M}$.

**Theorem 4.11** *Let* **L** *be a semi-decorated countable language,* $\mathfrak{M}$ *a countable* **L**-*structure and* $T$ *a countable set of* $\mathbf{L}_{\mathfrak{M}}$-*formulas. Then* $T \models^{\mathbf{L}}_{\mathfrak{M}} F$ *holds true iff there is a subset* $T_0 \subseteq T$ *and a countable ordinal* $\alpha$ *such that* $T_0 \underset{\mathbf{L}^{\mathfrak{M}}_{\mathfrak{T}}}{\overset{\alpha}{\models}} F$.

For a structure $\mathfrak{M}$ let $\mathbb{HYP}_{\mathfrak{M}}$ denote the least admissible structure above $\mathfrak{M}$ and define $O(\mathfrak{M}) := o(\mathbb{HYP}_{\mathfrak{M}})$ as the least ordinal which does not belong to $\mathbb{HYP}_{\mathfrak{M}}$.[13] Since $|\mathfrak{M}| \in \mathbb{HYP}_{\mathfrak{M}}$ the language $\mathsf{L}^{\mathfrak{M}}_{\mathfrak{T}} := \mathsf{L}(\mathfrak{M})^{\mathfrak{M}}_{\mathfrak{T}}$ is a $\mathbb{HYP}_{\mathfrak{M}}$-admissible fragment. Therefore we can refine Theorem 4.11 to

**Theorem 4.12** *Let* $\mathfrak{M}$ *be a countable structure and* $T$ *a countable set of* $\mathbf{L}_{\mathfrak{M}}$-*formulas that is* $\Sigma$-*definable in* $\mathbb{HYP}_{\mathfrak{M}}$. *Then* $T \models_{\mathfrak{M}} F$ *holds true iff there is an ordinal* $\alpha < O(\mathfrak{M})$ *and an* $\mathbb{HYP}_{\mathfrak{M}}$-*finite subset* $T_0$ *of* $T$ *such that* $T_0 \underset{\mathbf{L}_{\mathfrak{M}}}{\overset{\alpha}{\models}} F$.

The standard application for $\mathfrak{M}$-logic is $\omega$-logic where the basic structure is the standard structure $\mathbb{N}$ of natural numbers. In this case Theorem 4.12 reads that $T \models_{\mathbb{N}} F$ for a $\Pi^1_1$-definable set of formulas holds true iff there is an ordinal $\alpha$ below $\omega_1^{CK}$—the first ordinal that cannot be represented by a recursive well-ordering on the natural numbers—and a subset $T_0 \subseteq T$ that is $\Delta^1_1$-definable such that $T_0 \underset{\mathbf{L}_{\mathbb{N}}}{\overset{\alpha}{\models}} F$. The calculus $\underset{\mathbf{L}_{\mathbb{N}}}{\overset{\alpha}{\models}} \Delta$ corresponds to a cut-free one-sided sequent calculus with $\omega$-rule. Theorem 4.12, or rather Theorem 3.5 applied to $\mathsf{L}(\mathbb{N})_{\mathbb{N}}$, the extended first order logic of the standard structure of natural numbers, is then a version of the $\omega$-completeness theorem which is originally due to Henkin [5] and Orey [9].

## 4.4 *Infinitary Logic* $\mathsf{L}_{\kappa,\omega}$

A logic that is essentially stronger than first order logic but is still in some sense treatable is infinitary logic $\mathsf{L}_{\kappa,\omega}$. The language $\mathsf{L}_{\kappa,\omega}$ is an extension of first order logic that allows for the formation of infinitely long disjunctions and conjunctions provided that they contain only a finite number of free variables. To be more precise the formation rules besides those of first order logic are

- If $\langle F_\iota \mid \iota < \lambda < \kappa\rangle$ is sequence of $\mathsf{L}_{\kappa,\omega}$-formulas which contains only a finite number of free individual variables, then $\bigwedge \langle F_\iota \mid \iota < \lambda < \kappa\rangle$ and $\bigvee \langle F_\iota \mid \iota < \lambda < \kappa\rangle$ are $\mathsf{L}_{\kappa,\omega}$-formulas.

The semantics for $\mathsf{L}_{\kappa,\omega}$ is the obvious extension of the semantics for first order logic.

Putting infinite conjunctions in $\bigwedge$-type and infinite disjunctions in $\bigvee$-type and defining

- $\mathrm{CS}(\mathcal{O}\langle F_\iota \mid \iota < \lambda\rangle) := \langle F_\iota \mid \iota < \lambda\rangle$ where $\mathcal{O}$ is one of the operators $\bigwedge$ or $\bigvee$.

$\mathsf{L}_{\kappa,\omega}$ becomes a semi-decorated language. For any $\mathsf{L}$-structure $\mathfrak{M}$ we can rewrite Theorems 2.12 and 2.13 as

[13] Cf. [1, II Sect. 5].

**Theorem 4.13** *Let $F$ be an $(\mathsf{L}_{\kappa,\omega})_{\mathfrak{M}}$-formula whose second order variables are all in the list $X_1, \ldots, X_k$ then $\models^{\alpha}_{(\mathsf{L}_{\kappa,\omega})_{\mathfrak{M}}} F$ implies $\mathfrak{M} \models (\forall X_1) \ldots (\forall X_k) F(X_1, \ldots, X_k)$.*

and

**Theorem 4.14** *A $(\mathsf{L}_{\kappa,\omega})_{\mathfrak{M}}$-sentence $F$ is valid in an $\mathsf{L}$-structure $\mathfrak{M}$ iff there is an ordinal $\alpha \le rnk(F)$ such that $\models^{\alpha}_{(\mathsf{L}_{\kappa,\omega})_{\mathfrak{M}}} F$.*

Theorem 3.5 now reads as

**Theorem 4.15** *Let $\mathfrak{M}$ be a countable $\mathsf{L}$-structure and $F$ be an $(\mathsf{L}_{\omega_1,\omega})_{\mathfrak{M}}$-formula whose second order variables all occur in the list $X_1, \ldots, X_k$. Then we have $\mathfrak{M} \models (\forall X_1) \ldots (\forall X_k) F(X_1, \ldots, X_k)$ iff there is a countable ordinal $\alpha$ such that $\models^{\alpha}_{(\mathsf{L}_{\omega_1,\omega})_{\mathfrak{M}}} F$.*

The hypothesis of countability in Theorem 4.15 is indispensable. There are counterexamples for languages $\mathsf{L}_{\kappa,\omega}$ with $\kappa > \omega_1$ (cf., e.g., [10, Exercise 8.2.3]).

Clearly all the other results about countable admissible fragments of $\mathsf{L}_{\omega_1,\omega}$ and $\mathsf{L}_{\omega_1,\omega}$-consequences carry over. We leave the exact formulation to the reader.

## 4.5 *Logic for Inductive Definitions*

There are more examples for logical languages that are semi-decorable. We will not be able to treat all of them (probably we do not even know all of them). Examples which are certainly important are languages for inductive definitions and their generalizations, languages with the game quantifier $\Game$. Here we will only treat a language for one non-iterated inductive definition. We will not treat languages with the game quantifier. This is an interesting topic with wide reaching consequences as indicated in [7] which we leave for further studies.

### 4.5.1 A Brief Recapitulation of Inductive Definitions

To describe logics for inductive definitions we need some facts about inductive definitions which we will sketch roughly. For a profound study of inductive definitions visit [8] or (better and) [1].

Let **L** be a (semi-) decorated language and $F(X, x_1, \ldots, x_n)$ an $X$-positive **L**-formula[14] where $X$ is an $n$-ary predicate variable. This defines a monotone operator

$$\Gamma_F : \mathrm{Pow}(|\mathfrak{M}|^n) \longrightarrow \mathrm{Pow}(|\mathfrak{M}|^n)$$

[14] An **L**-formula is $X$ positive if the dual variable $X^c$ does not occur in the Tait version of **L**.

on every **L**-structure $\mathfrak{M}$ by setting

$$\Gamma_F(S) := \{(m_1, \ldots, m_n) \in |\mathfrak{M}|^n \,|\, \mathfrak{M} \models F[S, m_1, \ldots, m_n]\}.$$

We call such an operator a positively definable inductive definition (or positive inductive definition for short) of $\mathfrak{M}$. The operator $\Gamma_F$ possesses a least fixed-point (*the* fixed-point) $I_F$ satisfying

$$\Gamma_F(I_F) = I_F \quad \text{and} \quad (\forall X)[\Gamma_F(X) \subseteq X \to I_F \subseteq X].$$

The fixed-point is thus $\Pi^1_1$-definable by the formula

$$I_F = \{\vec{x} \,|\, (\forall X)[(\forall \vec{y})[F(X, \vec{y}) \to (X\vec{y})] \to (X\vec{x})]\}. \tag{7}$$

If **L** is the first order language $\mathsf{L}(\mathfrak{M})$ of a structure $\mathfrak{M}$ we call the slices of fixed-points of $\mathsf{L}(\mathfrak{M})$ positively definable inductive definitions the *positive-inductively definable relations* of $\mathfrak{M}$.

The *stages* of an inductive definition $\Gamma_F$ are recursively defined by

$$I_F^\alpha := \Gamma_F(I_F^{<\alpha}) \quad \text{for} \quad I_F^{<\alpha} := \bigcup_{\xi<\alpha} I_F^\xi.$$

Because of the monotonicity of the operator $\Gamma_F$ the stages of a monotone inductive definition are increasing, i.e.,

$$\alpha < \beta \Rightarrow I_F^\alpha \subseteq I_F^\beta.$$

By cardinality reasons there is therefore an ordinal $\kappa < \overline{\overline{|\mathfrak{M}|}}^+$ such that $I_F^{<\kappa} = I_F^\kappa$. The least such ordinal is the *closure ordinal* of the inductive definition $\Gamma_F$, denoted by $\|F\|$. For an element $\vec{n} \in I_F$ we define its *inductive norm* by

$$|\vec{n}|_F := \min\{\xi \,|\, \vec{n} \in I_F^\xi\}.$$

It is then easy to see that

$$I_F = I_F^{\|F\|} \quad \text{and} \|F\| := \sup\{|\vec{n}|_F + 1 \,|\, \vec{n} \in I_F\}. \tag{8}$$

### 4.5.2 The Language $\mathbf{L}_{\mathrm{ID}}$

Let **L** be a (semi-)decorated language. For every $X$-positive formula $F(X, x_1, \ldots, x_n)$ we introduce an $n$-ary predicate constant $I_F$. Again we write $t \in I_F$ instead of $(I_F t)$ and $(t \notin I_F)$ instead of $(I_F{}^c t)$. An $\mathbf{L}_{\mathrm{ID}}$-structure is an

**L**-structure in which $I_F$ is interpreted as the fixed-point of the monotone operator $\Gamma_F$ defined by $F(X, x_1, \ldots, x_n)$.

To (semi-)decorate $\mathbf{L}_{\text{ID}}$ we put formulas of the form $(\vec{t} \in I_F)$ into $\bigvee$-type and dually all formulas $(\vec{t} \notin I_F)$ in $\bigwedge$-type and define

- $\text{CS}(\vec{t} \in I_F) = \langle F(I_F, \vec{t}\,)\rangle$.

Using the standard complexity definition for formulas this obviously violates condition (C1).

However, we easily see that for any $\mathbf{L}_{\text{ID}}$-structure $\mathfrak{M}$ we still have

$$\mathfrak{M} \models \vec{t} \in I_F[\Psi] \quad \text{iff} \quad \mathfrak{M} \models G[\Psi] \quad \text{for some} \quad G \in \text{CS}(\vec{t} \in I_F)$$

and

$$\mathfrak{M} \models \vec{t} \notin I_F[\Psi] \quad \text{iff} \quad \mathfrak{M} \models G[\Psi] \quad \text{for all} \quad G \in \text{CS}(\vec{t} \notin I_F).$$

This entails that the verification calculus $\models^{\alpha}_{\mathbf{L}_{\text{ID}}} \Delta$ is sound for all $\mathbf{L}_{\text{ID}}$-structures.

To obtain also completeness for countable $\mathbf{L}_{\text{ID}}$-structures $\mathfrak{M}$ we transfer Sect. 3.1 to the language $\mathbf{L}_{\text{ID}}$. In doing so we observe that condition (C1) is only needed in the proof of the Semantical Main Lemma. We thus have to check the Semantical Main Lemma for countable $\mathbf{L}_{\text{ID}}(\mathfrak{M})$. So we modify the proof of the Semantical Main Lemma (Lemma 3.3) as follows.

Let $\mathfrak{M}$ be a countable **L**-structure and $f$ an infinite path in the search tree $\mathsf{S}^{\mathbf{L}_{\text{ID}}(\mathfrak{M})}_{T,\Delta}$. We define an $\mathfrak{M}$-assignment $\Phi$ as in the proof of Lemma 3.3 and interpret the new constants $I_F$ by $I_F^{<\alpha}$ for all countable ordinals $\alpha$. By $G^{\alpha}$ we denote that all constants $I_F$ in $G$ are interpreted by $I_F^{<\alpha}$.

By main-induction on $\alpha$ and side induction on the complexity of the formula $G$ we then prove that for all formulas $G \in \Delta_f$ we get $\mathfrak{M} \not\models G^{\alpha}[\Phi]$ for all $\alpha$. The only new cases are that the redex is a formula $\vec{t} \in I_F$ or $\vec{t} \notin I_F$. In the first case the formula $F(I_F, \vec{t}\,)$ belongs to $\Delta_f$ and we get by induction hypothesis $\mathfrak{M} \not\models F(I_F^{<\xi}, \vec{t}\,)[\Phi]$ for all $\xi < \alpha$. This entails $\mathfrak{M} \not\models \vec{t} \in I_F^{\xi}[\Phi]$ for all $\xi < \alpha$. Hence $\mathfrak{M} \not\models \vec{t} \in I_F^{<\alpha}[\Phi]$.

In the second case the formula $F(I_F, \vec{t}\,)^c$ belongs to $\Delta_f$. By induction hypothesis we therefore have $\mathfrak{M} \models F(I_F^{<\xi}, \vec{t}\,)[\Phi]$ for some $\xi < \alpha$. Hence $\mathfrak{M} \models \vec{t} \in I_F^{\xi} \subseteq I_F^{<\alpha}$ and we have $\mathfrak{M} \not\models \vec{t} \notin I_F^{<\alpha}$.

Since $I_F = I_F^{<\|F\|}$ we may choose $\alpha = \|F\|$ to make $\mathfrak{M}$ an $\mathbf{L}_{\text{ID}}$-structure for which we have $\mathfrak{M} \not\models G$ for all $G \in \Delta$. □

Summing up we have shown that the Semantical Main Lemma holds true for the language $\mathbf{L}_{\text{ID}}$ which implies that all the results of Sect. 3 also hold for the language $\mathbf{L}_{\text{ID}}$.

# 5 Applications to Structure Theory

**Definition 5.1** Let $\mathfrak{M}$ be a structure. By $\mathsf{L}(\mathfrak{M})$ we denote the first order language of $\mathfrak{M}$. We already assumed some familiarity with the usual complexity hierarchy of formulas. As a reminder: The $\Delta^0_0$-formulas are obtained from atomic formulas by the Boolean operations and bounded quantification. A $\Sigma^0_n$-formula has a $\Delta^0_0$-matrix in front of which there are $n$ alternating blocks of quantifiers starting with a block of existential quantifier. $\Pi^0_n$-formulas are the dual of $\Sigma^0_n$-formulas. A formula $F$ is said to be $\Delta^0_n$ in $\mathfrak{M}$ if there are a $\Sigma^0_n$-formula $F_\Sigma$ and a $\Pi^0_n$-formula $F_\Pi$ such that

$$\mathfrak{M} \models (\forall x_1)\dots(\forall x_n)[F_\Pi \leftrightarrow F_\Sigma \leftrightarrow F]$$

where all the free individual variables occurring in $F$ are listed in $x_1, \dots, x_n$.

By $\Delta^1_0$ we denote the set of all first order formulas. $\Pi^1_1$-formulas are $\Delta^1_0$-formulas preceded by a block of second order universal quantifier. $\Sigma^1_1$-formulas are the dual of $\Pi^1_1$-formulas.

Theorem 3.5 shows the close connection between semi-formal systems and $\Pi^1_1$-sentences. In this section we will show that there are applications to the structure theory of countable structures concerning the $\Pi^1_1$-sentences that are valid over this structure.

Observe, however, that by validity of a $\Pi^1_1$-sentence over a structure $\mathfrak{M}$ we understand validity in the sense of full second order logic. The second order quantifies are supposed to vary over the full subset of the domain of $\mathfrak{M}$. In this sense our structures are all $\beta$-structures, i.e., $\Pi^1_1$-sentences are absolute for $\mathfrak{M}$.

There is an obvious observation connecting $\Pi^1_1$-relations over an infinite countable structure and relations that are $\Sigma_1$-definable in $\mathbb{HYP}_\mathfrak{M}$. First we observe that the relation $\models^{\alpha}_{\mathsf{L}_\mathfrak{M}} \Delta$ is definable by $\Sigma$-recursion in $\mathbb{HYP}_\mathfrak{M}$. Therefore we obtain

$$\mathfrak{M} \models (\forall X)F(X) \quad \text{iff} \quad \mathbb{HYP}_\mathfrak{M} \vdash (\exists\alpha)\big[\models^{\alpha}_{\mathsf{L}_\mathfrak{M}} F(X)\big] \tag{9}$$

from which we immediately obtain the following theorem.

**Theorem 5.2** *Any $\Pi^1_1$-relation that is valid over an infinite countable structure $\mathfrak{M}$ is $\Sigma^0_1$-definable in $\mathbb{HYP}_\mathfrak{M}$.*

As a corollary to Theorem 5.2 we obtain that the relations that are $\Delta^1_1$-definable over $\mathfrak{M}$ are already members of $\mathbb{HYP}_\mathfrak{M}$.

*Remark 5.3* The opposite direction of Theorem 5.2 holds true even without the hypothesis of countability. For countable structures $\mathfrak{M}$ there is also a proof of the opposite direction using semi-formal systems. Since this is the weaker result we just give a rough indication how such a proof could work. Spelling out the details would leads us too far away.

The key point is to define a semi-decorated language for the relativized constructible hierarchy $\mathsf{L}_\alpha(\mathfrak{M})$.[15] Since $\mathbb{HYP}_\mathfrak{M} = \mathsf{L}_\alpha(\mathfrak{M})$ for $\alpha = O(\mathfrak{M})$ and $\mathfrak{M}$ is countable we have an enumeration $\mathcal{T}$ of all the elements of $\mathsf{L}_\alpha(\mathfrak{M})$ and may use it to define $\mathrm{CS}((\mathsf{Q}x)F(x)) = \langle F(s) | s \in \mathcal{T}\rangle$ for $\mathsf{L}_{\mathbb{HYP}_\mathfrak{M}}$-formulas the quantifier of which is supposed to range over $\mathbb{HYP}_\mathfrak{M}$. Using search trees we prove a completeness theorem saying that $\mathbb{HYP}_\mathfrak{M} \models F(X)$ iff $\mathbb{HYP}_\mathfrak{M} \models \big[(\exists\beta) \models^{\beta}_{\mathsf{L}_{\mathbb{HYP}_\mathfrak{M}}} F(X)\big]$ for $\Sigma^0_1$-sentences $F$ (this is needed to secure that $\beta$ is an element of $\mathbb{HYP}_\mathfrak{M}$. For $\Pi$-formulas this would fail.). Now we need a translation of the languages to construct a formula $F'(X)$ which allows us to convert $\models^{\beta}_{\mathsf{L}_{\mathbb{HYP}_\mathfrak{M}}} F(X)$ into $\models^{\beta'}_{\mathsf{L}_\mathfrak{M}} F'(X)$.[16] Using the Correctness Theorem for $\models^{\alpha}_{\mathsf{L}_\mathfrak{M}}$ this finally implies $\mathfrak{M} \models (\forall X)F'(X)$.

We will not pursue these aspects of structure theory further but concentrate on characteristic ordinals of structures in which semi-formal systems have some advantages.

## 5.1 *Characteristic Ordinals of a Structure*

**Definition 5.4** Let $\mathfrak{M}$ be a structure. There is a series of ordinals that are characteristic for $\mathfrak{M}$. By an ordering we mean a binary transitive relation that is irreflexive.

- $\sigma^i_n(\mathfrak{M})$ ($\pi^i_n(\mathfrak{M})$, $\delta^i_n(\mathfrak{M})$) is the supremum of the ordertypes of well-founded orderings that are $\Sigma^i_n$– ($\Pi^i_n$–, $\Delta^i_1$–) definable in $\mathfrak{M}$, where $i \in \{0, 1\}$ and $n \in \{0, 1\}$ in case that $i = 1$.
- $\delta^1_0(\mathfrak{M})$ ($= \sigma^1_0(\mathfrak{M}) = \pi^1_0(\mathfrak{M})$) is the supremum of all ordertypes of well-founded orderings that are $\mathsf{L}(\mathfrak{M})$-definable in $\mathfrak{M}$.
- $o(\mathfrak{M})$ is the least ordinal that does not belong to $\mathfrak{M}$. This definition only makes sense for structures in which ordinals are definable. An ordinal $\alpha$ is *admissible* iff $\alpha = o(\mathbb{A})$ for an admissible structure $\mathbb{A}$.
- $O(\mathfrak{M}) = o(\mathbb{HYP}_\mathfrak{M})$ where $\mathbb{HYP}_\mathfrak{M}$ is the least admissible structure $\mathbb{A}$ for which we have $|\mathfrak{M}| \in \mathbb{A}$ (cf. [1, II.5.8] for details). The ordinal $O(\mathfrak{M})$ is the next admissible above $o(\mathfrak{M})$.
- $\pi^\mathfrak{M}$ is the supremum of the truth-complexities of $\Pi^1_1$-sentences that are valid in $\mathfrak{M}$.

*Note 5.5* The ordinal which is genuinely connected to semi-formal systems is $\pi^\mathfrak{M}$. According to our definition of the truth-complexity[17] we always get $\pi^\mathfrak{M} = \overline{\overline{|\mathfrak{M}|}}^{+}$

---

[15] Cf. [10, Sect. 11.9] for the definition of a semi-decorated language for the constructible hierarchy $\mathsf{L}_\alpha$.

[16] Cf. [1] for the use of extended first order- and extended $\Pi^1_1$-formulas.

[17] Since we are working with the first order logic $\mathsf{L}(\mathfrak{M})$ of $\mathfrak{M}$ as basis language we denote from now on truth complexity briefly by $\mathrm{tc}_\mathfrak{M}(F)$ instead of $\mathrm{tc}_{\mathsf{L}(\mathfrak{M})_\mathfrak{M}}(F)$.

for structures for which the verification calculus is not complete. So $\pi^{\mathfrak{M}}$ is only an interesting ordinal for countable structures. It is still unclear to us if there is a modification of the definition of $\pi^{\mathfrak{M}}$ which is also meaningful for uncountable structures. For example

$$\pi^{\mathfrak{M}} := \sup(\{\mathrm{tc}_{\mathfrak{M}}(F) \mid \mathfrak{M} \models F\} \cap \overline{\overline{|\mathfrak{M}|}}^{+}).$$

Section 5.4 hides a hint that this might be a sensible definition.

In the following we study the connection of $\pi^{\mathfrak{M}}$ to the other characteristic ordinals on countable structures. First notice that $\pi^{\mathfrak{M}}$ is apparently a strict supremum. Whenever we have $\models^{\alpha}_{\mathsf{L}(\mathfrak{M})} F$ we obviously obtain $\models^{\alpha+1}_{\mathsf{L}(\mathfrak{M})} F \vee G$ for any $\mathsf{L}(\mathfrak{M})$-formula $G$.

*Remark 5.6* As an aside we want to mention that the characteristic ordinals for *structures* are closely connected to proof theoretic ordinals for *axiom systems*.

Because of how we interpret second order quantification $\mathfrak{M}$ is always a $\beta$-structure, i.e., a structure that is absolute for $\Pi^1_1$-sentences. Therefore we can replace the phrase "$\prec$-is well-founded" by "$\mathfrak{M} \models Wf(\prec)$", e.g.,

$$\delta^i_n = \sup\{\mathrm{otyp}(\prec) \mid \prec \text{ is a } \Delta^i_n\text{-definable ordering and } \mathfrak{M} \models Wf(\prec)\}.$$

For a theory $T$ we define analogously

$$\|T\| := \sup\{\mathrm{otyp}(\prec) \mid \prec \text{ is a } \Delta^0_0\text{-definable ordering and } T \vdash Wf(\prec)\}.$$

The ordinal $\|T\|$ is known as the proof-theoretic ordinal of $T$.

Similarly we may define

$$\|T\|_{\Pi^1_1} := \sup\{\mathrm{tc}\big((\forall X)F\big) \mid T \vdash F\}.$$

The computation of these ordinals is called an ordinal analysis of $T$. Proof theoretic ordinals are always ordinals below $\omega_1^{CK}$ while their model-theoretic counterparts are commonly much bigger. For "standard" axiom systems $T$ comprising Peano arithmetic directly or via interpretation the ordinals $\|T\|$ and $\|T\|_{\Pi^1_1}$ coincide. Therefore it makes no sense to define proof theoretical ordinals, e.g. $\|T\|_{\Delta^0_n}$ analogously to $\delta^0_n(\mathfrak{M})$, etc. There is, however, a notion of a $\Pi^0_2$-ordinal of a theory $T$ which is much more sophisticated and needs a notation systems for ordinals. We will not mention any details, since ordinal analyses are not among the aims of this paper.

## 5.2 The Boundedness Theorem

The central theorem that connects characteristic ordinals and semi-formal systems is the Boundedness Theorem which goes back to Gentzen [3] although Gentzen did not work with semi-formal systems. An (essential) improvement of the theorem is due to Beckmann [2].

Because of the centrality of the Boundedness Theorem and to improve the readability of the paper we are going to reprove it although there is a detailed proof in [10] which, however, uses slightly different notations.

To prepare the Boundedness Theorem we recall the fact that every element in the field of a well-founded ordering $\prec$ possesses an ordertype $\mathrm{otyp}_{\prec}(x)$ which is the strict supremum of the ordertypes of all its predecessors.

**Definition 5.7** Let $\prec$ be a binary well-founded relation. For a set $X \subseteq \mathit{field}(\prec)$ let $O(X) := \{\mathrm{otyp}_{\prec}(x) \mid x \in X\}$. Let $en_X^c$ enumerate the complement of $O(X)$. Define

$$R_{\prec}^{\alpha}(X) := X \cup \{y \in \mathit{field}(\prec) \mid \mathrm{otyp}_{\prec}(y) \leq en_X^c(\alpha)\} \tag{10}$$

and let

$$R_{\prec}^{<0}(X) := \{X\} \quad \text{and} \quad R_{\prec}^{<\alpha}(X) := \bigcup_{\xi<\alpha} R_{\prec}^{\xi}(X) \quad \text{for} \quad \alpha > 0.$$

We easily get

$$en_{X\cup\{x\}}^c(\alpha) \leq en_X^c(\alpha + 1), \tag{11}$$

hence

$$R_{\prec}^{\alpha}(X \cup \{x\}) \subseteq R_{\prec}^{\alpha+1}(X) \cup \{x\}. \tag{12}$$

The inequality (11) is trivial for $x \in X$. For $x \notin X$ let $\mathrm{otyp}_{\prec}(x) = en_X^c(\eta)$. Then $en_{X\cup\{x\}}^c(\eta) = en_X^c(\eta + 1)$.

If $\prec$ is transitive, then there is also a recursive characterization of the set $R_{\prec}^{\alpha}(X)$.

**Lemma 5.8** *Let $\prec$ be a well-founded ordering. Then*

$$R_{\prec}^{\alpha}(X) = R_{\prec}^{<\alpha}(X) \cup \{y \mid (\forall x \prec y)[x \in R_{\prec}^{<\alpha}(X)]\}.$$

*Proof* For the direction from left to right let $y \in R_{\prec}^{\alpha}(X)$. The claim is trivial for $y \in X$. So assume $y \notin X$, $\mathrm{otyp}_{\prec}(y) \leq en_X^c(\alpha)$ and $x \prec y$. If $x \in X$, then we trivially have $x \in R_{\prec}^{<\alpha}(X)$. If $x \notin X$, then $\mathrm{otyp}_{\prec}(x) = en_X^c(\beta)$ for some $\beta < \alpha$. Hence $x \in R_{\prec}^{<\alpha}(X)$.

For the opposite direction we show

$$(\forall x \prec y)[x \in R_{\prec}^{<\alpha}(X)] \Rightarrow \mathrm{otyp}_{\prec}(y) \leq en_X^c(\alpha) \tag{i}$$

by $\prec$-induction. For $z \prec y$ we get $(\forall x \prec z)[x \in R_{\prec}^{<\alpha}(X)]$ by transitivity of $\prec$. Hence $\mathrm{otyp}_{\prec}(z) \leq en_X^c(\alpha)$ by induction hypothesis. If $z \notin X$, we get $\mathrm{otyp}_{\prec}(z) < en_X^c(\alpha)$ by $z \in R_{\prec}^{<\alpha}(X)$. If $z \in X$, we cannot have $\mathrm{otyp}_{\prec}(z) = en_X^c(\alpha)$, hence $\mathrm{otyp}_{\prec}(z) < en_X^c(\alpha)$. So we have $\mathrm{otyp}_{\prec}(z) < en_X^c(\alpha)$ for all $z \prec y$ and thus $\mathrm{otyp}_{\prec}(y) \preceq en_X^c(\alpha)$. □

We will now study well-founded binary relations $\prec$. Well-foundedness is expressed by the formula

$$(\forall x)[(\forall y \prec x)(y \in X) \to x \in X] \to (\forall x \in \mathit{field} \prec)(x \in X).$$

The premise in this implication expresses that the relation $\prec$ is progressive what we are going to abbreviate by $Prog(X, \prec)$. In Tait style well-foundedness of the binary relation $\prec$ is thus expressed by

$$Wf(\prec) :\Leftrightarrow Prog(X, \prec)^c \vee (\forall x \in \mathit{field} \prec)(x \in X).$$

**Lemma 5.9 (Boundedness Lemma)** *Let $\mathfrak{M}$ be a countable structure, $\prec$ an $\mathsf{L}(\mathfrak{M})$ definable well-founded ordering and $\Delta(X)$ a finite set of $X$-positive $\mathsf{L}(\mathfrak{M})$-formulas. If*

$$\models_{\mathfrak{M}}^{\alpha} Prog(X, \prec)^c, s_1 \notin X, \ldots, s_n \notin X, \Delta(X),$$

*then*

$$\mathfrak{M} \models \bigvee \Delta(X)[\Phi]$$

*for any $\mathfrak{M}$ assignment $\Phi$ that assigns the set $R_{\prec}^{<\alpha}(\{s_1^{\mathfrak{M}}[\Phi], \ldots, s_n^{\mathfrak{M}}[\Phi]\})$ to $X$.*

*Proof* The proof is by induction on $\alpha$. The cases that the main-formula of the last clause belong to $\Delta(X)$ follow directly from the induction hypothesis and the $X$-positivity of the formulas in $\Delta(X)$.

If the main formula of the last clause is $Prog(X, \prec)^c$, we have the premise

$$\models_{\mathfrak{M}}^{\alpha_0} Prog(X, \prec)^c, (\forall y \prec s)(y \in X) \wedge s \notin X, s_1 \notin X, \ldots, s_n \notin X, \Delta(X)$$

for some $\mathsf{L}(\mathfrak{M})$-term $s$. By $\bigwedge$-inversion (Lemma 2.7) this entails

$$\models_{\mathfrak{M}}^{\alpha_0} Prog(X, \prec)^c, (\forall y \prec s)(y \in X), s_1 \notin X, \ldots, s_n \notin X, \Delta(X) \tag{i}$$

and

$$\mathop{\models}\limits_{\mathfrak{M}}^{\alpha_0} Prog(X,\prec)^c, s \notin X, s_1 \notin X, \ldots, s_n \notin X, \Delta(X). \tag{ii}$$

Now assume $\mathfrak{M} \not\models \bigvee \Delta[\Phi]$. Applying the induction hypothesis to (i) we get

$$(\forall y \prec s^{\mathfrak{M}}[\Phi])[y \in R_{\prec}^{<\alpha_0}(\{s_1^{\mathfrak{M}}[\Phi], \ldots, s_n^{\mathfrak{M}}[\Phi]\})]$$

which by Lemma 5.8 implies

$$s^{\mathfrak{M}}[\Phi] \in R_{\prec}^{\alpha_0}(\{s_1^{\mathfrak{M}}[\Phi], \ldots, s_n^{\mathfrak{M}}[\Phi]\}). \tag{iii}$$

Applying the induction hypothesis to (ii) implies

$$\mathfrak{M} \models \bigvee \Delta(X)[R_{\prec}^{<\alpha_0}(\{s_1^{\mathfrak{M}}[\Phi], \ldots, s_n^{\mathfrak{M}}[\Phi], s^{\mathfrak{M}}[\Phi]\})]$$

which by (12) implies

$$\mathfrak{M} \models \bigvee \Delta(X)[R_{\prec}^{\leq\alpha}(\{s_1^{\mathfrak{M}}[\Phi], \ldots, s_n^{\mathfrak{M}}[\Phi]\}) \cup \{s^{\mathfrak{M}}[\Phi]\}]. \tag{iv}$$

However, (iii) and (iv) show that we in fact have $\mathfrak{M} \models \bigvee \Delta(X)[\Phi]$. □

An immediate consequence of the Boundedness Lemma is the Boundedness Theorem.

**Theorem 5.10 (Boundedness Theorem)** *Let $\mathfrak{M}$ be a countable structure and $\prec$ an $\mathsf{L}(\mathfrak{M})$-definable well-founded ordering. Then* $\mathrm{otyp}(\prec) \leq tc_{\mathfrak{M}}(Wf(\prec))$.

*Proof* Let $\alpha = \mathrm{tc}_{\mathfrak{M}}(Wf(\prec))$. If $\alpha = \omega_1$, we are done. Otherwise we obtain

$$\mathop{\models}\limits_{\mathfrak{M}}^{\alpha} Prog(X,\prec)^c, (\forall x \in field(\prec))(x \in X)$$

by $\bigvee$-exportation (Lemma 2.8). This entails $(\forall x \in field(\prec))(x \in R_{\prec}^{\leq\alpha}(\emptyset))$ by the Boundedness Lemma. Hence $\mathrm{otyp}_{\prec}(x) < en_{\emptyset}^{c}(\alpha)$ for all $x$ in the field of $\prec$. Since $en_{\emptyset}^{c}(\alpha) = \alpha$ this entails $\mathrm{otyp}_{\prec}(x) < \alpha$ for all $x \in field(\prec)$. □

It is easy to see that Theorem 5.10 can be extended to $\Sigma_1^1$-definable well-founded orderings. If $m \prec n$ is defined by the $\Sigma_1^1$-sentence $\mathfrak{M} \models (\exists Y)G(Y, \underline{m}, \underline{n})$, we have $n \in field(\prec)$ iff there is a set $S \subseteq |\mathfrak{M}|$ such that

$$(\exists x)[G(S, \underline{n}, x) \vee G(S, x, \underline{n})].$$

From $\models^{\alpha}_{\mathfrak{M}} Prog(X, \prec)^c, (\forall x \in field(\prec))(x \in X)$ we then get by Lemma 5.9

$$\neg((\exists x)[G(S, \underline{n}, x) \vee G(S, x, \underline{n})]) \vee \underline{n} \in R_{\prec}^{<\alpha}(\emptyset)$$

for every $n$. Hence $\mathrm{otyp}(\prec) \leq \alpha$.

There is even an extension of the Boundedness Theorem to definable classes of well-founded orderings.

**Theorem 5.11 (Extended Boundedness Theorem)** *Let $\mathfrak{M}$ be a countable structure and $\mathcal{P}$ a $\Sigma^1_1$-definable class of well-founded orderings on $|\mathfrak{M}|$. Then* $\sup\{\mathrm{otyp}(\prec) \mid \prec \in \mathcal{P}\} \leq \pi^{\mathfrak{M}}$.

*Proof* Let $Y \in \mathcal{P}$ iff $(\exists Z)G(Y, Z)$. Then $(\forall Y)[Y \in \mathcal{P} \to Wf(Y)]$ is a $\Pi^1_1$-sentence. If $\mathfrak{M} \models (\forall Y)[Y \in \mathcal{P} \to Wf(Y)]$, there is an $\alpha < \pi^{\mathfrak{M}}$ such that

$$\models^{\alpha}_{\mathfrak{M}} G(Y, Z)^c, Prog(X, Y)^c, (\forall x \in field(Y))[x \in X]. \qquad \text{(i)}$$

If $\prec$ is a well-ordering in $\mathcal{P}$ there is a set $S \subseteq |\mathfrak{M}|$ such that $\mathfrak{M} \models G(\prec, S)$. By the Boundedness Lemma we get from (i)

$$\mathfrak{M} \models G(\prec, S)^c \vee (\forall x \in field(\prec))[x \in R_{\prec}^{<\alpha}(\emptyset)].$$

Hence $\mathrm{otyp}(\prec) \leq \alpha < \pi^{\mathfrak{M}}$ for all $\prec \in \mathcal{P}$. □

**Corollary 5.12** *If $\prec$ is a well-ordering that is $\Sigma^1_1$-definable on a countable structure $\mathfrak{M}$, then* $\mathrm{otyp}(\prec)$ *is less than* $\pi^{\mathfrak{M}}$.

*Proof* Let $\mathcal{P} := \{Y \mid Y$ is a linear order and $(\forall x)(\forall y)[(x, y) \in Y \to x \prec y]\}$. Then $\mathcal{P}$ is a $\Sigma^1_1$-definable class and $\prec \in \mathcal{P}$. Thus $\mathrm{otyp}(\prec) < \pi^{\mathfrak{M}}$ by the proof of the theorem. □

*Note 5.13* The Boundedness Lemma is based on the results in Sect. 3 of [3]. Gentzen's result in our terminology is $\mathrm{otyp}(\prec) \leq \omega^{\mathrm{tc}_{\mathbb{N}}(Wf(\prec))}$ for a well-ordering $\prec$. This can easily be improved to $\mathrm{otyp}(\prec) \leq 2^{\mathrm{tc}_{\mathbb{N}}(Wf(\prec))}$ (which makes not too much difference for infinite ordinals). The sharper bound $\mathrm{tc}_{\mathbb{N}}(Wf(\prec))$ is due to Beckmann.

The hypothesis of transitivity in the Boundedness Lemma is indispensable. Without transitivity Lemma 5.8 fails (cf. [10, Exercise 6.7.6]). In Lemma 5.26 below we have to deal with relations that are not transitive and get the weaker bound $2^{\alpha}$.

Apparently Gentzen himself already observed that the Boundedness Theorem might have further reaching consequences. In his paper [3] he writes on page 159 *"Diese Ergebnisse lassen sich noch wesentlich verschärfen; ich hoffe später einmal näheres hierüber veröffentlichen zu können."* which roughly means *"these results can still be essentially sharpened; I hope to be able to say more about it in a later publication"*.

In many cases the bound given by the Boundedness Theorem is the sharp bound. To clarify that we prove a lemma.

**Lemma 5.14** *Let $\prec$ be a well-ordering that is $\Sigma^1_1$-definable in a countable structure $\mathfrak{M}$ such that the sentences $(\underline{m} \prec \underline{n})^c$ in the diagram of $\mathfrak{M}$ uniformly have a truth complexity $\leq \beta$. Then we get*

$$\models^{\alpha_n}_{\mathfrak{M}} Prog(X, \prec)^c, \underline{n} \in X$$

*for* $\alpha_n := \beta + 4 \cdot (\text{otyp}_{\prec}(n) + 1)$.

*Proof* We use $\prec$-induction. By induction hypothesis and Theorem 2.13 we have

$$\models^{\alpha_m}_{\mathfrak{M}} Prog(X, \prec)^c, (\underline{m} \prec \underline{n})^c, \underline{m} \in X$$

for all $m \in |\mathfrak{M}|$. Using a clause $(\bigvee)$ and $(\bigwedge)$ we obtain

$$\models^{\beta_n}_{\mathfrak{M}} Prog(X, \prec)^c, (\forall x)[(x \prec \underline{n}) \rightarrow x \in X]$$

for $\beta_n = \beta + 4 \cdot \text{otyp}_{\prec}(n) + 2$. By clause (Ax) we have

$$\models^{0}_{\mathfrak{M}} Prog(X, \prec)^c, (\forall x)[(x \prec \underline{n}) \rightarrow x \in X], \underline{n} \notin X, \underline{n} \in X$$

and obtain by a clause $(\bigwedge)$

$$\models^{\beta_n+1}_{\mathfrak{M}} Prog(X, \prec)^c, (\forall x)[(x \prec \underline{n}) \rightarrow x \in X] \wedge \underline{n} \notin X, \underline{n} \in X$$

and finally by a clause $(\bigvee)$

$$\models^{\beta_n+2}_{\mathfrak{M}} Prog(X, \prec)^c, \underline{n} \in X. \qquad \square$$

As a consequence of Lemma 5.14 we obtain

**Theorem 5.15** *Let $\prec$ be a well-ordering that is $\Sigma^1_1$-definable on a countable structure $\mathfrak{M}$ and assume $tc_{\mathfrak{M}}((\underline{m} \prec \underline{n})^c)$ is uniformly less or equal to $\beta$ for all atomic sentences $(m \prec n)^c$ that are valid in $\mathfrak{M}$. If the ordertype of $\prec$ is a limit ordinal, then* $\text{otyp}(\prec) \leq tc_{\mathfrak{M}}(Wf(\prec)) \leq \beta + \text{otyp}(\prec)$.

*Proof* Let $\alpha := \text{tc}_{\mathfrak{M}}(Wf(\prec))$. Then we have

$$\models^{\alpha}_{\mathfrak{M}} Prog(X, \prec)^c, (\forall x \in field(\prec))[x \in X],$$

hence $\text{otyp}(\prec) \leq \alpha$ by the Boundedness Theorem 5.10. Since $\text{otyp}(\prec)$ is a limit ordinal we obtain from Lemma 5.14 $\alpha \leq \beta + \text{otyp}(\prec)$. $\square$

If the relation $\prec$ is $\mathsf{L}(\mathfrak{M})$-definable, then $\mathrm{tc}_{\mathfrak{M}}((\underline{m} \prec \underline{n})^c) \leq \mathrm{rnk}(\underline{m} \prec \underline{n})$ holds true for all $(m \prec n)^c$ in the diagram of $\mathfrak{M}$ by Theorem 2.13. Since all first order sentences in $\mathsf{L}(\mathfrak{M})_{\mathfrak{M}}$ have finite rank we get the following corollary.

**Corollary 5.16** *Let $\mathfrak{M}$ be a countable structure. For any $\mathsf{L}(\mathfrak{M})$-definable well-ordering of limit ordertype we have* $\mathrm{otyp}(\prec) = tc_{\mathfrak{M}}(Wf(\prec))$.

## 5.3 Definable Ordinals, the Next Admissible and $\pi^{\mathfrak{M}}$

There is a series of consequences of the Boundedness Theorem for the characteristic ordinals of countable structures.

**Theorem 5.17** *Let $\mathfrak{M}$ be an infinite countable structure then*

$$\delta_0^0(\mathfrak{M}) \leq \delta_n^0(\mathfrak{M}) \leq \delta_0^1(\mathfrak{M}) \leq \sigma_1^1(\mathfrak{M}) \leq \pi^{\mathfrak{M}} \leq O(\mathfrak{M}).$$

*Proof* The first three inequalities are obvious. The fourth one follows from the Boundedness Theorem (Theorem 5.10). The last one follows from Theorem 4.12. □

A natural question to ask is under which condition the $\leq$ in Theorem 5.17 can be sharpened to $=$.

A reinspection of Definition 3.1 shows that for a finite sequence $\Delta$ of $\mathsf{L}(\mathbb{N})$-formulas the search-tree $\mathsf{S}_{\Delta}^{\mathsf{L}(\mathbb{N})}$ can be defined by course-of-values recursion and thus is primitive-recursively definable. This entails that the tree ordering on a well-founded search tree $\mathsf{S}_{\Delta}^{\mathsf{L}(\mathbb{N})}$ is primitive-recursively definable. Therefore we obtain for the standard structure of arithmetic

**Theorem 5.18** $\delta_0^0(\mathbb{N}) = \pi^{\mathbb{N}}$. *Hence* $\delta_0^0(\mathbb{N}) = \delta_0^1(\mathbb{N}) = \sigma_1^1(\mathbb{N}) = \pi^{\mathbb{N}}$.

*Proof* By Theorem 5.17 it suffices to show $\pi^{\mathbb{N}} \leq \delta_0^0(\mathbb{N})$. If $\alpha < \pi^{\mathbb{N}}$, then there is a $\Pi_1^1$-sentence $(\forall X)F(X)$ that is valid in $\mathbb{N}$ such that $\alpha \leq \mathrm{tc}_{\mathbb{N}}((\forall X)F(X))$. Then $\mathsf{S}_{F(X)}^{\mathsf{L}(\mathbb{N})}$ is well-founded and $\alpha$ is less than or equal to the ordertype of $\mathsf{S}_{F(X)}^{\mathsf{L}(\mathbb{N})}$. Since $\mathsf{S}_{F(X)}^{\mathsf{L}(\mathbb{N})}$ is primitive-recursively definable its ordertype is less than $\delta_0^0(\mathbb{N})$. Hence $\pi^{\mathbb{N}} \leq \delta_0^0(\mathbb{N})$. □

We did not check which are the minimal conditions under which the search tree becomes definable in a structure. We certainly need some kind of coding machinery.

A structure is acceptable in the sense of [8] if it allows for an elementarily definable coding machinery. In an acceptable structure the search tree becomes elementarily definable. By the same argument as in the proof of Theorem 5.18 we therefore obtain the next theorem.

**Theorem 5.19** *Let $\mathfrak{M}$ be an acceptable structure. Then $\delta_0^1(\mathfrak{M}) = \pi^{\mathfrak{M}}$. Hence $\delta_0^1(\mathfrak{M}) = \sigma_1^1(\mathfrak{M}) = \pi^{\mathfrak{M}}$.*

The relation between $\pi^{\mathfrak{M}}$ and $O(\mathfrak{M})$ is established by the next theorem. According to [1] we say that a relation is $\mathbb{A}$-recursively definable on an admissible structure $\mathbb{A}$ iff it is $\Delta_1^0$-definable on $\mathbb{A}$.

**Theorem 5.20** *Let $\mathfrak{M}$ be an infinite countable structure that allows for a $\mathbb{HYP}_{\mathfrak{M}}$-recursively definable pairing function. Then $\sigma_1^1(\mathfrak{M}) = \pi^{\mathfrak{M}} = O(\mathfrak{M})$.*

*Proof* The $\leq$-direction is Theorem 5.17. For the proof of the opposite inequality we have to borrow results from [1]. By [1, VI.4.12] we know that the structure $\mathbb{HYP}_{\mathfrak{M}}$ is projectible into $\mathfrak{M}$ . Thus by [1, V.5.9.] we have

$$O(\mathfrak{M}) = \sup\{\mathrm{otyp}(\prec)|\ \prec\subseteq|\mathfrak{M}|^2\wedge\prec\in\mathbb{HYP}_{\mathfrak{M}}\wedge Wf(\prec)\}.$$

By [1, IV.3.4] we obtain that every relation $\prec$ in $\mathbb{HYP}_{\mathfrak{M}}$ is $\Sigma_1^1$-definable (even $\Delta_1^1$-definable) in $\mathfrak{M}$. Hence $O(\mathfrak{M}) \leq \sigma_1^1(\mathfrak{M}) \leq \pi^{\mathfrak{M}}$. □

**Corollary 5.21** *Let $\mathfrak{M}$ be an acceptable structure. Then $\delta_0^1(\mathfrak{M}) = \delta_1^1(\mathfrak{M}) = \sigma_1^1(\mathfrak{M}) = \pi^{\mathfrak{M}} = O(\mathfrak{M})$.*

Recall that an ordinal $\alpha$ is admissible if $\alpha = o(\mathbb{A})$ for an admissible structure $\mathbb{A}$.

**Corollary 5.22** *The Kleene ordinal $\omega_1^{CK}$ is the first admissible ordinal above $\omega$.*

*Proof* For the proof just recall that $\omega_1^{CK} = \delta_1^0(\mathbb{N})$ and apply Corollary 5.21. □

*Remark 5.23* It follows from Theorem 5.20 that for structures $\mathfrak{M}$ that allow for a $\mathbb{HYP}_{\mathfrak{M}}$-recursive pairing the ordinal $\pi^{\mathfrak{M}}$ is admissible and thus an ordinal with strong closure properties, including the closure under powers to the basis 2. As we will see later (cf. Lemma 5.28 below) closure under 2-powers is of some importance.

If $\mathfrak{M}$ is the trivial structure with equality as the only relation symbol, we apparently have $\pi^{\mathfrak{M}} = \omega$. This ordinal is closed under powers to the basis 2. We do not know if there are structures whose $\Pi_1^1$-ordinal is not closed under 2-powers. Our conjecture is that this holds true for almost all structures.

## 5.4 $\pi^{\mathfrak{M}}$ and Inductive Definitions

We already touched inductive definitions in Sect. 4.5.1. Here we start with an arbitrary structure $\mathfrak{M}$ and regard the semi-decorated first order language $\mathsf{L}(\mathfrak{M})$. Putting $|\vec{n}|_F := \overline{\overline{|\mathfrak{M}|}}^{+}$ for $\vec{n} \notin I_F$ we obtain the *stage comparison relations*

$$\vec{m} \preceq^* \vec{n} :\Leftrightarrow |\vec{m}|_F \leq |\vec{n}|_G,$$
$$\vec{m} \prec^* \vec{n} :\Leftrightarrow |\vec{m}|_F < |\vec{n}|_G.$$

The most important theorem in the theory of inductive definitions is the Stage Comparison Theorem which says that the stage comparison relations are positive-inductively definable in $\mathfrak{M}$.

Due to this theorem there is a characteristic ordinal which is very important for the structure theory of the sets that are positive-inductively definable in $\mathfrak{M}$.

**Definition 5.24** Let $\mathfrak{M}$ be a structure. We define the ordinal $\kappa^{\mathfrak{M}}$ of $\mathfrak{M}$ as the supremum of the closure ordinals of positive inductions that are $\mathsf{L}(\mathfrak{M})$-definable in $\mathfrak{M}$ (cf. [8]).

The ordinal $\kappa^{\mathfrak{M}}$ is characteristic for the structure theory of $\mathfrak{M}$ in so far that a positive-inductively definable relation is hyperelementary (i.e. the relation and its complement are inductively definable) iff its closure ordinal is less than $\kappa^{\mathfrak{M}}$. Details are in [8] and [1]

*Remark 5.25* Again there is a proof theoretic counterpart of the ordinal $\kappa^{\mathfrak{M}}$. For a theory in the language of inductive logic (cf. Sect. 4.5) we define

$$\kappa^T := \sup\{|n|_F \mid T \vdash n \in I_F\}.$$

To establish the connection between $\kappa^{\mathfrak{M}}$ and $\pi^{\mathfrak{M}}$ we have to modify the Boundedness Theorem. We introduce the abbreviations

$$Cl_F(X) :\Leftrightarrow (\forall \vec{x})[F(X,\vec{x}) \rightarrow \vec{x} \in X],$$

$$F_{\vec{t}_1,\ldots,\vec{t}_n}(X,\vec{x}) :\Leftrightarrow F(X,\vec{x}) \vee \vec{x} = \vec{t}_1 \vee \cdots \vee \vec{x} = \vec{t}_n$$

and prove

$$\vec{t} \in I_F^{\alpha} \Rightarrow I_{F_{\vec{t}}}^{\beta} \subseteq I_F^{\alpha+\beta} \tag{13}$$

by induction on $\beta$. From $\vec{x} \in I_{F_{\vec{t}}}^{\beta}$ we get $F(I_{F_{\vec{t}}}^{<\beta},\vec{x}) \vee \vec{x} = \vec{t}$. If $\vec{x} = \vec{t} \in I_F^{\alpha} \subseteq I_F^{\alpha+\beta}$, we are done. Otherwise we have $F(I_{F_{\vec{t}}}^{<\beta},\vec{x})$ and obtain by induction hypothesis and $X$-positivity $F(I_F^{<\alpha+\beta},\vec{x})$ which implies $\vec{x} \in I_F^{\alpha+\beta}$. □

Now we prove a variation of the boundedness lemma.

**Lemma 5.26 (Stage Lemma)** *Let $F(X,\vec{x})$ be an $X$-positive formula and $\Delta$ a finite set of $X$-positive $\mathsf{L}(\mathfrak{M})$-formulas. Then $\models^{\alpha}_{\mathsf{L}(\mathfrak{M})} Cl_F(X)^c, \vec{t}_1 \notin X, \ldots, \vec{t}_n \notin X, \Delta(X)$ implies $\bigvee \Delta[I_{F_{\vec{t}_1,\ldots,\vec{t}_n}}^{<2^{\alpha}}]$.*

*Proof* The proof is by induction on $\alpha$. The cases that the main-formula of the last clauses belongs to $\Delta(X)$ follow easily from the induction hypothesis and the $X$-positivity of the formulas in $\Delta(X)$. Let $\Gamma := \{\vec{t}_1 \notin X, \ldots, \vec{t}_n \notin X\}$. If the main-formula is $Cl_F(X)^c$, we have the premise

$$\overset{\alpha_0}{\underset{\mathsf{L}(\mathfrak{M})}{\models}} Cl_F(X)^c, F(X,\vec{t}\,) \wedge \vec{t} \notin X, \Gamma, \Delta(X) \tag{i}$$

for some tuple $\vec{t}$ of $\mathsf{L}(\mathfrak{M})$-terms. From (i) we get by $\bigwedge$-inversion

$$\overset{\alpha_0}{\underset{\mathsf{L}(\mathfrak{M})}{\models}} Cl_F(X)^c, F(X,\vec{t}\,), \Gamma, \Delta(X) \tag{ii}$$

and

$$\overset{\alpha_0}{\underset{\mathsf{L}(\mathfrak{M})}{\models}} Cl_F(X)^c, \Gamma, \vec{t} \notin X, \Delta(X). \tag{iii}$$

Assuming $\neg \bigvee \Delta[I^{<2^\alpha}_{F_{\vec{t}_1,\dots,\vec{t}_n}}]$ we get $\vec{t} \in I^{2^{\alpha_0}}_{F_{\vec{t}_1,\dots,\vec{t}_n}}$ from (ii) and $\bigvee \Delta[I^{<2^{\alpha_0}}_{F_{\vec{t}_1,\dots,\vec{t}_n,\vec{t}}}]$ from (iii) by induction hypothesis. Using (13) and the $X$-positivity of the formulas in $\Delta(X)$ we thus have $\bigvee \Delta[I^{<2^{\alpha_0}+2^{\alpha_0}}_{F_{\vec{t}_1,\dots,\vec{t}_n}}]$. Since $2^{\alpha_0} + 2^{\alpha_0} \leq 2^\alpha$ this contradicts our assumption. □

By (7) we have $\vec{n} \in I_F$ iff $(\forall X)[Cl_F(X) \to \vec{n} \in X]$ and define

$$\mathrm{tc}_{\mathfrak{M}}(\vec{n} \in I_F) := \mathrm{tc}_{\mathfrak{M}}((\forall X)[Cl_F(X) \to \underline{\vec{n}} \in X]) \tag{14}$$

and

$$\mathrm{tc}_{\mathfrak{M}}(I_F) := \sup\left(\{\mathrm{tc}_{\mathfrak{M}}(\vec{n} \in I_F) + 1 \,|\, n \in |\mathfrak{M}|\} \cap \overline{\overline{|\mathfrak{M}|}}^{+}\right). \tag{15}$$

The counterpart of the Boundedness Theorem is the following Stage Theorem.

**Theorem 5.27 (Stage Theorem)** *Let $F(X,\vec{x})$ be an $X$-positive $\mathsf{L}(\mathfrak{M})$-formula. Then $|\vec{t}|_F < 2^{tc_{\mathfrak{M}}(t \in I_F)}$ holds true for all $\vec{t} \in I_F$. Hence $\|F\| \leq 2^{tc_{\mathfrak{M}}(I_F)}$.*

*Proof* Let $\vec{t} \in I_F$ and $\alpha := \mathrm{tc}_{\mathfrak{M}}(\vec{t} \in I_F)$. Then we have $\overset{\alpha}{\underset{\mathsf{L}(\mathfrak{M})}{\models}} Cl_F(x)^c, \vec{t} \in X$. By the Stage Lemma it follows $\mathfrak{M} \models \vec{t} \in I^{<2^\alpha}_F$. Hence $|\vec{t}|_F < 2^\alpha$. □

As a corollary to Theorem we get

**Corollary 5.28** *For any countable structure $\mathfrak{M}$ we have $\kappa^{\mathfrak{M}} \leq 2^{\pi^{\mathfrak{M}}}$. If $\pi^{\mathfrak{M}}$ is closed under powers to the basis 2 this means $\kappa^{\mathfrak{M}} \leq \pi^{\mathfrak{M}}$.*[18]

There is also an "inversion" of Lemma 5.26.

**Lemma 5.29** *Let $F(X,n)$ be an $X$-positive $\mathsf{L}(\mathfrak{M})$-formula without further relation variables. If $\mathfrak{M} \models F(I^{<\alpha}_G, \underline{n})$, then $\overset{\xi}{\underset{\mathsf{L}(\mathfrak{M})}{\models}} Cl_G(X)^c, F(X,n)$ holds true for $\xi = \omega \cdot \alpha + rnk(F)$.*

[18]Cf. Remark 5.23.

*Proof* We prove the lemma by main induction on $\alpha$ with side induction on $\mathrm{rnk}(F)$. Observe first, that $\mathrm{rnk}(F)$ is always finite.

If $F$ is a formula in which $X$ does not occur, then we get the claim by Theorem 2.13.

For composed formulas $F(X,\underline{n})$ we obtain the claim directly from the side induction hypothesis.

If finally $F(I_G^{<\alpha},\underline{n})$ is the formula $\underline{n} \in I_G^{<\alpha}$, then there is a $\beta < \alpha$ such that $\mathfrak{M} \models \underline{n} \in I_G^{\beta}$, i.e., $\mathfrak{M} \models G(I_G^{<\beta},\underline{n})$. By main induction hypothesis we thus have $\models^{\gamma}_{\mathsf{L}(\mathfrak{M})} Cl_G(X)^c, G(X,\underline{n})$ for $\gamma = \omega \cdot \beta + \mathrm{rnk}(G(X,\underline{n}))$. Together with axiom (Ax) $\models^{0}_{\mathsf{L}(\mathfrak{M})} Cl_G(X)^c, \underline{n} \notin X, \underline{n} \in X$ we get $\models^{\omega\cdot(\beta+1)}_{\mathsf{L}(\mathfrak{M})} Cl_G(X)^c, \underline{n} \in X$. □

Since $\mathrm{rnk}(\underline{n} \in X) = 0$ we obtain as a corollary to Lemma 5.29 and Theorem 5.27

$$\mathrm{tc}\big(\underline{\vec{n}} \in I_F\big) \leq \omega \cdot |\vec{n}|_F \leq \omega \cdot 2^{\mathrm{tc}\left(\vec{n} \in I_F\right)} \tag{16}$$

and

$$\mathrm{tc}\big(I_F\big) \leq \omega \cdot \|F\| \leq \omega \cdot 2^{\mathrm{tc}\left(I_F\right)}. \tag{17}$$

Up to now we know that $\kappa^{\mathfrak{M}} \leq \pi^{\mathfrak{M}}$. Inequality (17) is not sufficient to establish also the opposite inequality since we do not know if every $\Pi^1_1$-sentence that is valid in $\mathfrak{M}$ is already expressible as a fixed point of an $X$-positive operator. Therefore we have to study the connection between $\Pi^1_1$-relations and positive-inductively definable relations on $\mathfrak{M}$ further. We will, however, not go too much into details. The reader who wants to know more is advised to consult [1, Sect. VI].

It follows from Definition 2.5 that the relation $\models^{\alpha}_{\mathfrak{M}} \Delta$ is inductively defined and $\alpha$ an upper bound for the stage of this inductive definition. However, it is not clear that this is an inductive definition over an arbitrary structure $\mathfrak{M}$. What is lacking is enough coding machinery to express formulas and their characteristic sequences in $\mathfrak{M}$. In acceptable structures $\mathfrak{M}$ this coding machinery is available. However, the hypothesis of acceptability is even too strong.

It is clear that $\models^{\alpha}_{\mathfrak{M}} \Delta$ is an inductive definition over the structure $\mathbb{HF}_{\mathfrak{M}}$, the *hereditarily finite sets over* $\mathfrak{M}$ (cf. [1, Sect. II. 2]). But even this is too strong. Barwise in [1, II. Definition 2.7] introduces the notion of an extended first order language over $\mathfrak{M}$ as a sublanguage $\mathsf{L}(\mathbb{HF}_{\mathfrak{M}})$. This language allows quantifiers over urelements (i.e., the elements of $|\mathfrak{M}|$), bounded quantifiers and unbounded existential quantifiers and is closed under the positive Boolean operations $\vee$ and $\wedge$. It is then easy to see that $\models^{\alpha}_{\mathfrak{M}} \Delta$ is inductively definable by an $X$-positive formula that is extended first order over $\mathfrak{M}$. Barwise (cf. [1, IV. Definition 3.7]) calls that *extended inductively definable* on $\mathfrak{M}$ or *inductive** on $\mathfrak{M}$.

By Theorem 3.5 we therefore obtain the following theorem.

**Theorem 5.30** *Let $\mathfrak{M}$ be a countable infinite structure. Then every $\Pi^1_1$-relation on $\mathfrak{M}$ is extended inductive on $\mathfrak{M}$.*

Let $\kappa^{*\mathfrak{M}}$ denote the supremum of the closure ordinals of all inductive* definitions on $\mathfrak{M}$. Then we have the following theorem.

**Theorem 5.31** *Let $\mathfrak{M}$ be a countable infinite structure. Then $\kappa^{\mathfrak{M}} \leq \pi^{\mathfrak{M}} \leq \kappa^{*\mathfrak{M}}$.*

*Proof* From $\mathfrak{M} \models (\forall X)F(X)$ we get by Theorem 3.5 $\models^{\alpha}_{\mathfrak{M}} F(X)$ for $\alpha = \mathrm{tc}_{\mathfrak{M}}(F(X))$. Since $\alpha$ is an upper bound for the stage in the inductive definition of $\models^{\alpha}_{\mathfrak{M}} F(X)$ we have $\alpha \leq \kappa^{*\mathfrak{M}}$. □

According to [1, VI. Theorem 3.11] we have

$$O(\mathfrak{M}) = \kappa^{*\mathfrak{M}},$$

hence

$$\kappa^{\mathfrak{M}} \leq \pi^{\mathfrak{M}} \leq O(\mathfrak{M}) = \kappa^{*\mathfrak{M}}$$

for countable infinite structures $\mathfrak{M}$. By [1, VI. Theorem 4.1] the relations which are inductive* on a structure $\mathfrak{M}$ coincide with the inductive relations on $\mathfrak{M}$ if $\mathfrak{M}$ has an inductive pairing. Hence $\kappa^{\mathfrak{M}}$ equals $\kappa^{*\mathfrak{M}}$ on such structures and we get the following theorem.

**Theorem 5.32** *For an infinite countable structure $\mathfrak{M}$ with inductive pairing we have $\kappa^{\mathfrak{M}} = \sigma^1_1(\mathfrak{M}) = \pi^{\mathfrak{M}} = O(\mathfrak{M})$.*

*Proof* Since $\kappa^{\mathfrak{M}}$ equals the supremum of the ordertype of all well-founded orderings which are coinductively definable on $\mathfrak{M}$ and every coinductive relation is $\Sigma^1_1$-definable we have

$$\kappa^{\mathfrak{M}} \leq \sigma^1_1(\mathfrak{M}) \leq \pi^{\mathfrak{M}} \leq O(\mathfrak{M}) = \kappa^{*\mathfrak{M}} = \kappa^{\mathfrak{M}}$$

and thus equality. □

## References


1. J. Barwise, *Admissible Sets and Structures*. Perspectives in Mathematical Logic (Springer, Berlin, 1975)
2. A. Beckmann, W. Pohlers, Application of cut–free infinitary derivations to generalized recursion theory. Ann. Pure Appl. Log. **94**, 1–19 (1998)
3. G. Gentzen, Beweisbarkeit und Unbeweisbarkeit von Anfangsfällen der transfiniten Induktion in der reinen Zahlentheorie Math. Ann. **119**, 140–161 (1943)
4. J.-Y. Girard, Une extension de l'interpretation de Gödel a l'analyse et son application a l'elimination des coupures dans l'analyse et la theorie des types, in *Proceedings of the 2nd*

*Scandinavian Logic Symposium*, ed. by J.E. Fenstad. Studies in Logic and the Foundations of Mathematics, vol. 63 (North-Holland Publishing Company, Amsterdam, 1971), pp. 63–92
5. L. Henkin, A generalization of the notion of $\omega$-consistency. J. Symb. Log. **19**, 183–196 (1954)
6. D. Hilbert, Die Grundlegung der elementaren Zahlenlehre. Math. Ann. **104**, 485–494 (1931)
7. M. Möllerfeld, Systems of inductive definitions. Ph.D. thesis, Münster, 2003
8. Y.N. Moschovakis, *Elementary Induction on Abstract Structures*. Studies in Logic and the Foundations of Mathematics, vol. 77 (North-Holland Publishing Company, Amsterdam, 1974)
9. S. Orey, On $\omega$-consistency and related properties. J. Symb. Log. **21**, 246–252 (1956)
10. W. Pohlers, *Proof Theory. The First Step into Impredicativity*. Universitext (Springer, Berlin, 2009)
11. M. Rathjen, An ordinal analyis of stability. Arch. Math. Log. **48/2**, 1–62 (2005)
12. K. Schütte, Syntactical and semantical properties of simple type theory. J. Symb. Log. **25**, 305–326 (1960)
13. W.W. Tait, A non constructive proof of Gentzen's Hauptsatz for second order predicate logic. Bull. Am. Math. Soc. **72**, 980–983 (1966)
14. G. Takeuti, On a generalized logic calculus. Jpn. J. Math. **24**, 149–156 (1953)

# Part IV
# Developments

# Proof Theory for Theories of Ordinals III: $\Pi_N$-Reflection

**Toshiyasu Arai**

**Abstract** This paper deals with a proof theory for a theory $\mathrm{T}_N$ of $\Pi_N$-reflecting ordinals using a system $Od(\Pi_N)$ of ordinal diagrams. This is a sequel to the previous one (Arai, Ann Pure Appl Log 129:39–92, 2004) in which a theory for $\Pi_3$-reflecting ordinals is analysed proof-theoretically.

## 1 Prelude

This is a sequel to the previous ones [3, 4]. Namely our aim here is to give finitary analyses of finite proof figures in a theory for $\Pi_N$-reflecting ordinals, [11] via cut-eliminations as in Gentzen–Takeuti's consistency proofs, [7, 12]. Throughout this paper $N$ denotes a positive integer such that $N \geq 4$.

Let $T$ be a theory of ordinals. Let $\Omega$ denote the (individual constant corresponding to the) ordinal $\omega_1^{CK}$. We say that $T$ is a $\Pi_2^\Omega$-*sound theory* if

$$\forall \Pi_2\, A(T \vdash A^\Omega \Rightarrow A^\Omega).$$

**Definition 1.1** (**$\Pi_2^\Omega$-Ordinal of a Theory**) Let $T$ be a $\Pi_2^\Omega$-sound and recursive theory of ordinals. For a sentence $A$ let $A^\alpha$ denote the result of replacing unbounded quantifiers $Qx$ ($Q \in \{\forall, \exists\}$) in $A$ by $Qx < \alpha$. Define the $\Pi_2^\Omega$-*ordinal* $|T|_{\Pi_2^\Omega}$ *of* $T$ by

$$|T|_{\Pi_2^\Omega} := \inf\{\alpha \leq \omega_1^{CK} : \forall \Pi_2 \text{ sentence } A(T \vdash A^\Omega \Rightarrow A^\alpha)\} < \omega_1^{CK}.$$

Roughly speaking, the aim of proof theory for theories $T$ of ordinals is to describe the ordinal $|T|_{\Pi_2^\Omega}$. This gives $\Pi_2^\Omega$-ordinal of an equivalent theory of sets, cf. [3].

T. Arai (✉)
Graduate School of Science, Chiba University, 1-33, Yayoi-cho, Inage-ku, Chiba 263-8522, Japan
e-mail: tosarai@faculty.chiba-u.jp



R. Kahle, M. Rathjen (eds.), *Gentzen's Centenary*,
DOI 10.1007/978-3-319-10103-3_14

Let $\mathrm{KP}\Pi_N$ denote the set theory for $\Pi_N$-reflecting universes. $\mathrm{KP}\Pi_N$ is obtained from the Kripke–Platek set theory with the Axiom of Infinity by adding the axiom: for any $\Pi_N$ formula $A(u)$

$$A(u) \to \exists z(u \in z \,\&\, A^z(u)).$$

In [6] we introduced a recursive notation system $Od(\Pi_N)$ of ordinals, which we studied first in [1]. An element of the notation system is called an *ordinal diagram* (henceforth abbreviated by *o.d.*). The system is designed for proof theoretic study of theories of $\Pi_N$-reflection. We [6] showed that for each $\alpha < \Omega$ in $Od(\Pi_N)$ $\mathrm{KP}\Pi_N$ proves that the initial segment of $Od(\Pi_N)$ determined by $\alpha$ is a well ordering.

Let $\mathrm{T}_N$ denote a theory of $\Pi_N$-reflecting ordinals. The aim of this paper is to show an upper bound theorem for the ordinal $|\mathrm{T}_N|_{\Pi_2^\Omega}$:

**Theorem 1.1** $\forall \Pi_2\, A(\mathrm{T}_N \vdash A^\Omega \Rightarrow \exists\alpha \in Od(\Pi_N) \mid \Omega\ A^\alpha)$.

Combining Theorem 1.1 with the result in [6] mentioned above yields the:

**Theorem 1.2** $|\mathrm{KP}\Pi_N|^\Omega_{\Pi_2} = |\mathrm{T}_N|^\Omega_{\Pi_2} =$ the order type of $Od(\Pi_N) \mid \Omega$.

Proof theoretic study for $\Pi_N$-reflecting ordinals via ordinal diagrams was first obtained in a handwritten note [2].

For an alternative approach to ordinal analyses of set theories, see Rathjen's papers [8–10].

Let us mention the contents of this paper.

In Sect. 2 a preview of our proof-theoretic analysis for $\Pi_N$-reflection is given. As in [4] inference rules $(c)^\sigma_{\alpha_1}$ are added to analyse an inference rule ($\Pi_N$-rfl) saying the universe of the theory $\mathrm{T}_N$ is $\Pi_N$-reflecting. A *chain* is defined to be a consecutive sequence of rules $(c)$.

In Sect. 2.1 we expound that chains have to merge each other for a proof theoretic analysis of $\mathrm{T}_N$ for $N \geq 4$. An ordinal diagram in the system $Od(\Pi_N)$ defined in [6] may have its $Q$ part, which has to obey complicated requirements. In Sect. 2.2 we explain what parts correspond to the $Q$ part in proof figures.

In Sect. 3 the theory $\mathrm{T}_N$ for $\Pi_N$-reflecting ordinals is defined. In Sect. 4 let us recall briefly the system $Od(\Pi_N)$ of ordinal diagrams (abbreviated by o.d.'s) in [6].

In Sect. 5 we extend $\mathrm{T}_N$ to a formal system $\mathrm{T}_{Nc}$. The language is expanded so that individual constants $c_\alpha$ for o.d.'s $\alpha \in Od(\Pi_N) \mid \pi$ are included. Inference rules $(c)^\sigma_{\alpha_1}$ are added. *Proofs* in $\mathrm{T}_{Nc}$ defined in Definition 5.8 are proof figures enjoying some provisos and obtained from given proofs in $\mathrm{T}_N$ by operating rewriting steps. Some lemmata for proofs are established. These are needed to verify that rewritten proof figures enjoy these provisos. To each proof $P$ in $\mathrm{T}_{Nc}$ an o.d. $o(P) \in Od(\Pi_N) \mid \Omega$ is attached. Then the Main Lemma 5.1 is stated as follows: If $P$ is a proof in $\mathrm{T}_{Nc}$, then the endsequent of $P$ is true.

In Sect. 6 the Main Lemma 5.1 is shown by a transfinite induction on $o(P) \in Od(\Pi_N) \mid \Omega$.

This paper relies heavily on the previous ones [3, 4].

**General Conventions** Let $(X, <)$ be a quasiordering. Let $F$ be a function $F : X \ni \alpha \mapsto F(\alpha) \subseteq X$. For subsets $Y, Z \subset X$ of $X$ and elements $\alpha, \beta \in X$, put

1. $\alpha \leq \beta \Leftrightarrow \alpha < \beta$ or $\alpha = \beta$,
2. $Y \mid \alpha = \{\beta \in Y : \beta < \alpha\}$,
3. $Y < Z :\Leftrightarrow \exists \beta \in Z \forall \alpha \in Y(\alpha < \beta)$,
4. $Y < \beta :\Leftrightarrow Y < \{\beta\} \Leftrightarrow \forall \alpha \in Y(\alpha < \beta)$; $\alpha < Z :\Leftrightarrow \{\alpha\} < Z$,
5. $Z \leq Y :\Leftrightarrow \forall \beta \in Z \exists \alpha \in Y(\beta \leq \alpha)$,
6. $\beta \leq Y :\Leftrightarrow \{\beta\} \leq Y \Leftrightarrow \exists \alpha \in Y(\beta \leq \alpha)$; $Z \leq \alpha :\Leftrightarrow Z \leq \{\alpha\}$,
7. $F(Y) = \bigcup\{F(\alpha) : \alpha \in Y\}$.

## 2 A Preview of Proof-Theoretic Analysis

In this section a preview of our proof-theoretic analysis for $\Pi_N$-reflection is given.

Let us recall briefly the system $Od(\Pi_N)$ of o.d.'s in [6]. The main constructor in $Od(\Pi_N)$ is to form an o.d. $d_\sigma^q\alpha$ from a symbol $d$ and o.d.'s in $\{\sigma, \alpha\} \cup q$, where $\sigma$ denotes a recursively Mahlo ordinal and $q = Q(d_\sigma^q\alpha)$ a finite sequence of quadruples of o.d.'s called *Q part* of $d_\sigma^q\alpha$. By definition we set $d_\sigma^q\alpha < \sigma$. Let $\gamma \prec_2 \delta$ denote the transitive closure of the relation $\gamma = d_\delta^q\alpha$ for some $q$ and $\alpha$, and $\preceq_2$ its reflexive closure. Then the set $\{\tau : \sigma \prec_2 \tau\}$ is finite and linearly ordered by $\prec_2$ for each $\sigma$.

An o.d. of the form $\rho = d_\sigma^q\alpha$ is introduced in proof figures only when an inference rule ($\Pi_N$-rfl) for $\Pi_N$-reflection is resolved by using an inference rule $(c)_\rho$.

$q$ in $\rho = d_\sigma^q\alpha$ includes some data $st_i(\rho), rg_i(\rho)$ for $2 \leq i < N$. $st_{N-1}(\rho)$ is an o.d. less than $\varepsilon_{\pi+1}$ and $rg_{N-1}(\rho) = \pi$, while $st_i(\rho), rg_i(\rho)$ for $i < N-1$ may be undefined. If these are defined, then we write $rg_i(\rho) \downarrow$, etc. and $\kappa = rg_i(\rho)$ is an o.d. such that $\rho \prec_i \kappa$, where $\gamma \prec_i \delta$ is a transitive closure of the relation $pd_i(\gamma) = \delta$ on o.d.'s such that $\prec_{i+1} \subseteq \prec_i$ and $\prec_2$ is the same as one mentioned above. $q$ also includes data $pd_i(\rho)$. $st_{N-1}(\rho)$ is defined so that

$$\gamma \prec_{N-1} \rho \Rightarrow st_{N-1}(\gamma) < st_{N-1}(\rho). \tag{1}$$

In Sect. 2.2 we explain what parts correspond to the $Q$ part in proof figures.

A theory $\mathrm{T}_N$ for $\Pi_N$-reflection is formulated in Tait's logic calculus, i.e., one-sided sequent calculus and $\Gamma, \Delta \ldots$ denote a *sequent*, i.e., a finite set of formulae. $\mathrm{T}_N$ has the inference rule ($\Pi_N$-rfl):

$$\frac{\Gamma, A \quad \neg\exists z A^z, \Gamma}{\Gamma} \; (\Pi_N\text{-rfl})$$

where $A \equiv \forall x_N \exists x_{N-1} \cdots Q x_1 B$ with a bounded formula $B$.

So ($\Pi_N$-rfl) says $A \to \exists z A^z$.[1]

To deal with the inference rule ($\Pi_N$-rfl) we introduce new inference rules $(c)^\sigma_\rho$ and $(\Sigma_i)^\sigma$ $(1 \le i \le N)$ as in [4]:

$$\frac{\Gamma, \Lambda^\sigma}{\Gamma, \Lambda^\rho}\ (c)^\sigma_\rho$$

where $\Lambda$ is a set of $\Pi_N$-sentences as above, $\Lambda^\sigma = \{A^\sigma : A \in \Lambda\}$, the side formulae $\Gamma$ consists solely of $\Sigma^\sigma_1$-sentences and $\rho$ is of the form $d^q_\sigma \alpha$.

$$\frac{\Gamma, \neg A^\sigma \quad A^\sigma, \Lambda}{\Gamma, \Lambda}\ (\Sigma_i)^\sigma$$

where $A$ is a $\Sigma_i$ sentence. Although this rule $(\Sigma_i)^\sigma$ is essentially a (cut) inference, we need to distinguish between this and (cut) to remember that a ($\Pi_N$-rfl) was resolved.

When we apply the rule $(c)^\sigma_\rho$ it must be the case:

$$\text{any instance term } \beta < \sigma \text{ for the existential quantifiers } \exists x_{N-i} < \sigma (i\text{:odd})$$
$$\text{in } A^\sigma \equiv \forall x_N < \sigma \exists x_{N-1} < \sigma \cdots Q x_1 < \sigma B \text{ is less than } \rho. \tag{2}$$

As in [4] an inference rule ($\Pi_N$-rfl) is resolved by forming a succession of rules $(c)$'s, called a *chain*, which grows downwards in proof figures. We have to pinpoint, for each $(c)$, the unique chain, which describes how to introduce the $(c)$. To retain the uniqueness of the chain, i.e., not to branch or split a chain, we have to be careful in resolving rules with two uppersequents. Our guiding principles are:

**(ch1)** For any $\dfrac{A^\sigma}{A^\tau}\ (c)^\sigma_\tau$ with $\tau = d^q_\sigma \alpha$, if an o.d. $\beta$ is substituted for an existential quantifier $\exists y < \sigma$ in $A^\sigma$, i.e., $\beta$ is a realization for $\exists y < \sigma$, then $\beta < \tau$, cf. (2), and

**(ch2)** Resolving rules having several uppersequents must not branch a chain.

## 2.1 Merging Chains

As contrasted with [4] for $\Pi_3$-reflection we have to merge chains here. Let us explain this phenomenon.

We omit side formulae in this subsection.

[1]For simplicity we suppress the parameter. Correctly $\forall u(A(u) \to \exists z(u < z \,\&\, A^z(u)))$.

1. First resolve a $(\Pi_N\text{-rfl})$ in the left figure, and resolve the $(\Sigma_N)^\sigma\ J_0$ to the right figure with a $\Sigma_{N-1}\ A_1$:

$$\dfrac{\dfrac{A}{A^\sigma}\,(c)^\pi_\sigma \qquad \neg A^\sigma}{\phantom{A}}\,(\Sigma_N)^\sigma\ J_0 \qquad\qquad \dfrac{\dfrac{\dfrac{A}{A^\sigma} \quad \neg A^\sigma, \neg A_1^\sigma}{\neg A_1^\sigma} \qquad \dfrac{A_1}{A_1^\sigma}\,(c)^\pi_\sigma\ I_0}{\phantom{A}}\ J_1\ (\Sigma_{N-1})^\sigma$$

with $A \equiv \forall x_N \exists x_{N-1} \forall x_{N-2} A_3$, $\sigma = d^\alpha_\pi \alpha$, where $A_3 \equiv \exists x_{N-3} A_4$ is a $\Sigma_{N-3}$-formula and $\alpha$ denotes the o.d. attached to the uppersequent $A$ of $(c)^\pi_\sigma$.

2. Second resolve a $(\Pi_N\text{-rfl})$ above the $(c)^\pi_\sigma\ I_0$ and a $(\Sigma_N)$ as in 1):

$$\dfrac{\dfrac{\dfrac{\dfrac{\neg A_1^\sigma \quad \dfrac{\begin{array}{c}P_1\\ A_1, B\end{array}}{A_1^\sigma, B^\sigma}\,(c)^\pi_\sigma\ \tilde{I}_0}{B^\sigma}\,(\Sigma_{N-1})^\sigma}{B^\tau} \qquad \dfrac{\neg A_1^\sigma \quad \dfrac{A_1, \neg B^\tau, \neg B_1^\tau}{A_1^\sigma, \neg B^\tau, \neg B_1^\tau}}{\neg B^\tau, \neg B_1^\tau}}{\neg B_1^\tau} \qquad \dfrac{\dfrac{\neg A_1^\sigma \quad \dfrac{A_1, B_1}{A_1^\sigma, B_1^\sigma}\,(c)^\pi_\sigma}{B_1^\sigma}\,J_1}{B_1^\tau}\,(c)^\sigma_\tau\ I_1}{\phantom{A}}$$

*Fig.* 1

with a $\tau = d^\nu_\sigma \beta$ and a $\Sigma_{N-1}\ B_1 \equiv \exists y_{N-1} \forall y_{N-2} B_3$, where $\nu$ denotes the o.d. attached to the subproof $P_1$ ending with the uppersequent $A_1, B$ of $(c)^\pi_\sigma\ \tilde{I}_0$.
After that resolve the $(\Sigma_{N-1})^\sigma\ J_1$:

$$\dfrac{\dfrac{\dfrac{\dfrac{\neg A_1^\sigma \quad \dfrac{\begin{array}{c}P_2\\ A_1, B\end{array}}{A_1^\sigma, B^\sigma}\,\tilde{I}_0}{B^\sigma}}{B^\tau} \qquad \neg B^\tau, \neg B_1^\tau}{\neg B_1^\tau} \qquad \dfrac{\dfrac{\dfrac{\neg A_1^\sigma \quad \dfrac{A_1, B_1, A_2}{A_1^\sigma, B_1^\sigma, A_2^\sigma}}{B_1^\sigma, A_2^\sigma} \qquad \dfrac{\dfrac{A}{A^\sigma} \quad \neg A^\sigma, \neg A_2^\sigma}{\neg A_2^\sigma}\,J_0'}{B_1^\sigma}}{B_1^\tau}}{\phantom{A}}$$

Then resolve the $(\Sigma_N)^\sigma\ J_0'$ :

$$\dfrac{\dfrac{\dfrac{\dfrac{\neg A_1^\sigma \quad \dfrac{\begin{array}{c}P_3\\ A_1, B\end{array}}{A_1^\sigma, B^\sigma}\,\tilde{I}_0}{B^\sigma}}{B^\tau} \qquad \neg B^\tau, \neg B_1^\tau}{\neg B_1^\tau} \qquad \dfrac{\dfrac{B_1^\sigma, A_2^\sigma \qquad \dfrac{\neg A_2^\sigma, \neg \tilde{A}_1^\sigma \quad \dfrac{\begin{array}{c}(\Pi_N\text{-rfl})\ H\\ \vdots\\ \tilde{A}_1\end{array}}{\tilde{A}_1^\sigma}\,(c)^\pi_\sigma\ I_0'}{\neg A_2^\sigma}\,(\Sigma_{N-2})^\sigma\ J_2}{B_1^\sigma}}{B_1^\tau}\,(c)^\sigma_\tau\ I_1}{\phantom{A}}\ K$$

3. Thirdly resolve a $(\Pi_N\text{-rfl})\ H$ above the $(c)^\pi_\sigma\ I_0'$. One cannot resolve the $(\Pi_N\text{-rfl})\ H$ by introducing a $(c)^\sigma_\rho$ with $\rho < \tau$. Let me explain the reason.

Suppose that we introduce a new $(c)^{\sigma}_{\rho}\, I'_1$ with $\rho = d^{\eta}_{\sigma}\gamma$ immediately above the $(\Sigma_{N-2})^{\sigma}\, J_2$ as in [4]. Then the new $(c)^{\sigma}_{\rho}\, I'_1$ is introduced after the $(c)^{\sigma}_{\tau}\, I_1$ and so $\rho = d^{\eta}_{\sigma}\gamma < \tau$. Hence a new $(\Sigma_N)^{\rho}\, K'$ is introduced below the $(\Sigma_{N-1})^{\tau}\, K$:

$$
\dfrac{\dfrac{\neg B^{\tau}_1 \qquad \dfrac{\dfrac{B^{\sigma}_1, A^{\sigma}_2 \qquad \dfrac{\dfrac{\neg A^{\sigma}_2, \neg\tilde{A}^{\sigma}_1 \qquad \dfrac{\tilde{A}_1, D}{\tilde{A}^{\sigma}_1, D^{\sigma}}\,(c)^{\pi}_{\sigma}}{\neg A^{\sigma}_2, D^{\sigma}}}{\neg A^{\sigma}_2, D^{\rho}}\,(c)^{\sigma}_{\rho}\, I'_1}{B^{\sigma}_1, D^{\rho}}\,J_2}{B^{\tau}_1, D^{\rho}}\,(c)^{\sigma}_{\tau}\, I_1}{D^{\rho}}\,K \qquad \neg D^{\rho}}{}\,K' \qquad \textit{Fig.2}
$$

with $D \equiv \forall z_N \exists z_{N-1} \forall z_{N-2} D_3$. Nevertheless this does not work, because $\neg A_2 \equiv \exists x_{N-3} \forall x_{N-4} \neg A_4$ is a $\Sigma_{N-2}$ sentence with $N - 2 \geq 2$. Namely the principle **(ch1)** may break down for the $(c)^{\sigma}_{\rho}\, I'_1$ since any o.d. $\delta < \sigma$, i.e., possibly $\delta \geq \rho$ may be an instance term for the existential quantifier $\exists x_{N-3}$ in $A_2 \equiv \forall x_{N-2} \exists x_{N-3} A_4$ and may be substituted for the variable $x_{N-3}$ in $\neg A^{\sigma}_2$. Only we know that such a $\delta$ is less than $\sigma$ and comes from the left upper part of $J_2$.

4. Therefore the chain for $H$ has to connect or merge with the chain $I_0 - I_1$ for $B$:

$$
\dfrac{\dfrac{\dfrac{\dfrac{\neg A^{\sigma}_1 \qquad \dfrac{\begin{matrix} P_4 \\ A_1, B \end{matrix}}{A^{\sigma}_1, B^{\sigma}}\,\tilde{I}_0}{\dfrac{B^{\sigma}}{B^{\tau}}\,I'_1} \qquad \neg B^{\tau}, \neg B^{\tau}_1}{\neg B^{\tau}_1} \qquad \dfrac{\dfrac{B^{\sigma}_1, A^{\sigma}_2 \qquad \dfrac{\dfrac{\dfrac{A}{A^{\sigma}}\,I''_0 \qquad \neg A^{\sigma}, \neg A^{\sigma}_2, \neg\tilde{A}^{\sigma}_1}{\neg A^{\sigma}_2, \neg\tilde{A}^{\sigma}_1}\,J'_0 \qquad \dfrac{\tilde{A}_1, D}{\tilde{A}^{\sigma}_1, D^{\sigma}}\,I'_0}{\neg A^{\sigma}_2, D^{\sigma}}}{B^{\sigma}_1, D^{\sigma}}\,J_2}{B^{\tau}_1, D^{\tau}}\,I_1}{D^{\tau}}}{D^{\rho}}\,(c)^{\tau}_{\rho}\, I_2
$$

*Fig.* 3

with $\rho = d^{\eta}_{\tau}\gamma$ and a $(\Sigma_N)^{\rho}$ with the cut formula $D^{\rho}$ follows this figure as in *Fig.* 2, where $\eta$ denotes the o.d. attached to the uppersequent $\tilde{A}_1, D$ of $(c)^{\pi}_{\sigma}\, I'_0$. $(\Sigma_{N-2})^{\sigma}\, J_2$ is a merging point for chains $I_0 - I_1$ and $I'_0 - I_1 - I_2$.

The principle **(ch1)** for the new $(c)^{\tau}_{\rho}\, I_2$ will be retained for the simplest case $N = 4$ as in [4]. The problem is that the proviso (1) may break down: it may be the case $\nu = st_{N-1}(\tau) \leq st_{N-1}(\rho) = \eta$ since we cannot expect the upper part of $(c)^{\pi}_{\sigma}\, I'_0$ is simpler than the one of $(c)^{\pi}_{\sigma}\, \tilde{I}_0$.

In other words a new succession $I'_0 - I_1 - I_2$ of collapsing starts. This is required to resolve $\Sigma^{\sigma}_{N-2}$ sentence $\neg A^{\sigma}_2$ $(N - 2 \geq 2)$ and hence $\sigma$ has to be $\Pi_{N-1}$-reflecting.

If this chain $I'_0 - I_1 - I_2$ would grow downwards as in $\Pi_3$-reflection, i.e., in a chain $I'_0 - I_1 - I_2 - \cdots - I_n$, $I_n$ would come only from the upper part of $I'_0$, then the proviso (1) would suffice to kill this process. But the whole process may be iterated: in *Fig.* 3 another succession $I''_0 - I_1 - I_2 - I_3$ may arise by resolving the $(\Sigma_N)^\sigma\ J'_0$.

Nevertheless still we can find a reducing part, that is, the upper part of the $(c)^\tau_\rho\ I_2$: the upper part of the $(c)^\tau_\rho\ I_2$ becomes simpler in the step $I_2 - I_3$. Furthermore in the general case $N > 4$ merging processes could be iterated, vz. the merging point $(\Sigma_{N-2})^\sigma\ J_2$ may be resolved into a $(\Sigma_{N-3})^{\rho_1}$, which becomes a new merging point to analyse a $\Sigma_{N-3}$ sentence $A_3^{\rho_1}$ where $\rho_1 \preceq \rho$ is a $\Pi_{N-2}$-reflecting and so on. Therefore in $Od(\Pi_N)$ the $Q$ part of an o.d. may consist of several factors:

$$(\tau, \alpha, q = \{\nu_i, \kappa_i, \tau_i : i \in In(\rho)\}) \mapsto d^q_\tau \alpha = \rho$$

with $\kappa_{N-1} = rg_{N-1}(\rho) = \pi$. $In(\rho)$ denotes a set such that

$$N - 1 \in In(\rho) \subseteq \{i : 2 \leq i \leq N - 1\}.$$

We set for $i \in In(\rho)$:

$$st_i(\rho) = \nu_i, rg_i(\rho) = \kappa_i, pd_i(\rho) = \tau_i.$$

If $i \notin In(\rho)$, set

$$pd_i(\rho) = pd_{i+1}(\rho), st_i(\rho) \simeq st_i(pd_i(\rho)), rg_i(\rho) \simeq rg_i(pd_i(\rho)).$$

Also these are defined so that $pd_2(\rho) = \tau$ for $\rho = d^q_\tau \alpha$.

For the o.d. $\rho = d^q_\tau \gamma$ in *Fig.* 3, $In(\rho) = \{N-2, N-1\}, st_{N-1}(\rho) = \eta$, $pd_{N-1}(\rho) = \sigma, rg_{N-2}(\rho) = \tau = pd_{N-2}(\rho), st_{N-2}(\rho) = \gamma = st_2(\gamma)$.

Thus $\nu_i = st_i(\rho)$ corresponds to the upper part of a $(c)^{rg_i(\rho)}$ while $\tau_{N-1} = pd_{N-1}(\rho)$ indicates that the first, i.e., uppermost merging point for a chain ending with a $(c)_\rho$ is a rule $(\Sigma_{N-2})^{\tau_{N-1}}$, e.g., the rule $J_2$ in *Fig.* 3. Note that $st_{N-1}(\rho) = \eta < st_{N-1}(pd_{N-1}(\rho))$, cf. (1). $\kappa_i = rg_i(\rho)$ is an o.d. such that there exists a $(c)^{\kappa_i}$ in the chain for $(c)_\rho$. We will explain how to determine the rule $(c)^{rg_i(\rho)}$, i.e., the point to which we direct our attention in Sect. 2.2.

The case $In(\rho) = \{N-1\}$ corresponds to the case when a $(c)^{pd_{N-1}(\rho)}_\rho$ is introduced without merging points, i.e., as a resolvent of a ($\Pi_N$-rfl) above the top of the chain whose bottom is a $(c)_{pd_{N-1}(\rho)}$. The case $In(\rho) = \{N-2, N-1\}$ corresponds to the case when a $(c)^{pd_2(\rho)}_\rho$ $(pd_2(\rho) = pd_{N-2}(\rho))$ is introduced with a merging point $(c)^{pd_{N-1}(\rho)}$.

In *Fig.* 3 a new succession with a merging point $(c)^\tau_\rho\ I_2$ arises by resolving a $(\Sigma_N)^\tau$ below the $(c)^\sigma_\tau\ I'_1$, i.e., $\tilde{I}_0 - I'_1 - I_2 - I_3\ (c)^\rho_\kappa$ for a $\kappa$ with a $\lambda = st_{N-1}(\kappa)$. But in this case we have

$$\lambda = st_{N-1}(\kappa) < st_{N-1}(\tau) = \nu.$$

$st_{N-1}(\kappa)$ corresponds to the upper part $P_1$ of a $(c)^{\pi}_{\sigma}\ \tilde{I}_0$ in *Fig.* 1, when the $(c)^{\sigma}_{\tau}$ was originally introduced. This part $P_1$ is unchanged up to *Fig.* 3:

$P_1 = P_2 = P_3 = P_4$. Roughly speaking, $\tilde{I}_0 - I'_1 - I_3$ can be regarded as a $\Pi_{N-1}$ resolving series $I_0 - I_1 - I_3$. This prevents the new merging points from going downwards unlimitedly.

## 2.2 The *Q* Part of an Ordinal Diagram

In this subsection we explain how to determine the $Q$ part $q$ of $\rho = d^q_{\sigma}\alpha$ from a proof figure when an inference rule $(c)^{\sigma}_{\rho}$ is introduced.

In general such a $(c)^{\sigma}_{\rho}$ is formed when we resolve an inference rule $(\Pi_N\text{-rfl})\ H$:

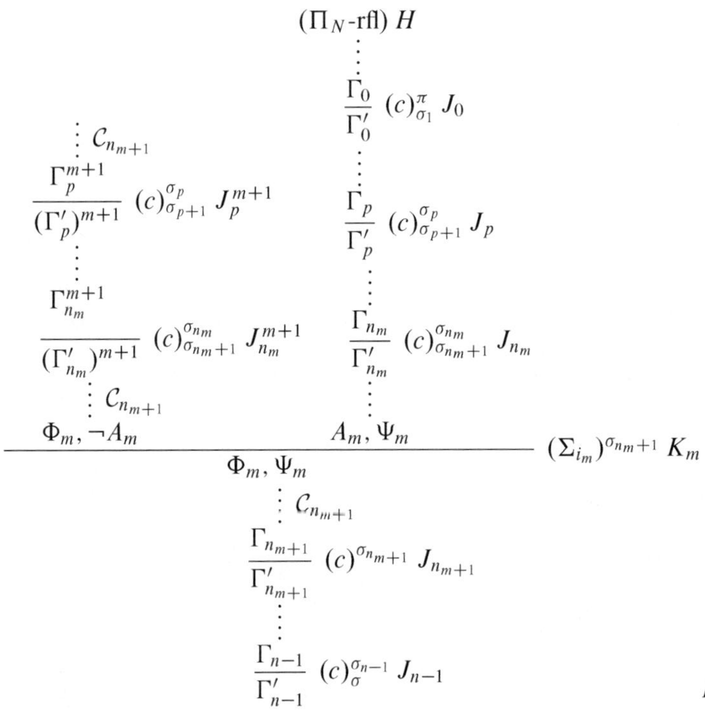

*Fig.* 4

where $\mathcal{R} = J_0, \ldots, J_{n-1}$ denotes a series of rules $(c)^{\sigma_p}_{\sigma_{p+1}}\ J_p$ with $\pi = \sigma_0, \sigma = \sigma_n$. $(\Pi_N\text{-rfl})\ H$ is resolved into a $(c)^{\sigma}_{\rho}\ J_n$ and a $(\Sigma_N)^{\rho}$ below $J_{n-1}$.

This series $\mathcal{R}$ is divided into intervals $\{\mathcal{R}_m = J_{n_{m-1}+1}, \ldots, J_{n_m} : m \leq l\}$ with an increasing sequence $n_{-1} + 1 = 0 \leq n_0 < n_1 < \cdots < n_l = n - 1\ (l \geq 0)$ of numbers so that

1. $\mathcal{R}_0 = J_0, \ldots, J_{n_0}$ is a chain $\mathcal{C}_{n_0}$ leading to $J_{n_0}$.
2. For $m < l$ $\mathcal{R}_{m+1} = J_{n_m+1}, \ldots, J_{n_{m+1}}$ is a tail of a chain $\mathcal{C}_{n_{m+1}} = J^{m+1}_0, \ldots, J^{m+1}_{n_m}, J_{n_m+1}, \ldots, J_{n_{m+1}}$ leading to $J_{n_{m+1}}$ such that the chain $\mathcal{C}_{n_{m+1}}$

passes through the left side of an inference rule $(\Sigma_{i_m})^{\sigma_{n_m+1}} K_m$ with $2 \le i_m < N-1$, $J_{n_m}$ is above the right uppersequent $A_m, \Psi_m$ and $J_{n_m}^{m+1}$ is above the left uppersequent $\Phi_m, \neg A_m$ of $K_m$, resp. $A_m$ is a $\Sigma_{i_m}$ sentence. Each rule $J_p^{m+1}$ for $p \le n_m$ is again an inference rule $(c)_{\sigma_{p+1}}^{\sigma_p}$. $K_m$ will be a merging point of chains $\mathcal{C}_{n_{m+1}}$ and a new chain $\mathcal{C}_\rho = J_0, \ldots, J_{n-1}, J_n$ leading to $(c)_\rho\, J_n$.
3. There is no such a merging point below $J_{n-1}$, viz. there is no $(\Sigma_k)^\sigma$ with $1 < k < N-1$ such that $J_{n-1}$ is in the right upper part of the inference rule and there exists a chain passing through its left side.

Set $N-1 \in In(\rho), rg_{N-1}(\rho) = \pi$ and $st_{N-1}(\rho)$ is the o.d. attached to the upper part of $(c)^\pi\, J_0$, where by the upper part we mean the part after resolving $(\Pi_N\text{-rfl})\, H$.

First consider the case $l = 0$, i.e., there is no merging point for the new chain $\mathcal{C}_\rho$ leading to the new $J_n$. Then set $In(\rho) = \{N-1\}$ and $pd_{N-1}(\rho) = \sigma$.

Suppose $l > 0$ in what follows. Then set $pd_{N-1}(\rho) = \sigma_{n_0+1}$, i.e., $pd_{N-1}(\rho)$ is the superscript of the first uppermost merging point $(\Sigma_{i_0})^{\sigma_{n_0+1}} K_0$.

In any cases we have $st_{N-1}(\rho) < st_{N-1}(pd_{N-1}(\rho))$, cf. (1). $st_i(\rho)$ always corresponds to the upper part of a $(c)^{rg_i(\rho)}$ in the chain $\mathcal{C}_\rho$ for $i \in In(\rho)$.

### The Simplest Case $N = 4$

Here suppose $N = 4$ and we determine the $Q$ part of $\rho$. First set $2 \in In(\rho)$, vz. $In(\rho) = \{2, 3\}$ and $pd_2(\rho) = \sigma$. It remains to determine the o.d. $rg_2(\rho)$. In other words to specify a rule $(c)^{\sigma_q}\, J_q$ with $rg_i(\rho) = \sigma_q$.

Note that $i_m = 2$ for any $m$ with $0 < m \le l$ since $2 \le i_m < N-1 = 3$ in this case. There are two cases to consider. First suppose there is a $p < n$ such that

1. $p > n_0$, i.e., $\sigma_{p+1} \prec_2 \sigma_{n_0+1} = pd_3(\rho)$ and
2. $2 \in In(\sigma_{p+1})$, i.e., there was a merging point of the chain leading to $(c)_{\sigma_{p+1}}\, J_p$.

Then pick the minimal $q$ satisfying these two conditions, viz. the uppermost rule $(c)_{\sigma_{q+1}}^{\sigma_q}\, J_q$ below the first uppermost merging point $(\Sigma_{i_0})^{\sigma_{n_0+1}} K_0$ with $2 \in In(\sigma_{q+1})$. Then set

**Case 1** $rg_2(\rho) = rg_2(\sigma_{q+1})$.

Otherwise set

**Case 2** $rg_2(\rho) = \sigma = pd_2(\rho)$.

Consider the first case **Case 1** $rg_2(\rho) = rg_2(\sigma_{q+1}) \neq pd_2(\rho)$. From the definition we see $rg_2(\rho) = rg_2(\sigma_{q+1}) = pd_2(\sigma_{q+1}) = \sigma_q$. We have $\sigma_q = rg_2(\rho) \preceq_3 pd_3(\rho) = \sigma_{n_0+1}$. This follows from the minimality of $q$, i.e., $\forall t[n_0 < t < q \to 2 \notin In(\sigma_{t+1})]$ and hence $\forall t[n_0 < t < q \to \sigma_t = pd_2(\sigma_{t+1}) = pd_3(\sigma_{t+1})]$.

Furthermore $q$ is minimal, i.e, $\sigma_q$ is maximal in the following sense:

$$\forall t[n_0 < t < n(\leftrightarrow pd_2(\rho) = \sigma \preceq_2 \sigma_{t+1} \prec_2 pd_3(\rho)) \,\&\, rg_2(\sigma_{t+1}) \downarrow$$
$$\rightarrow rg_2(\sigma_{t+1}) \preceq_2 \sigma_q] \tag{3}$$

In general we have the following fact.

**Proposition 2.1** *Let $\mathcal{C} = J_0, \ldots, J_{n-1}$ be a chain leading to a $(c)^{\sigma_{n-1}}_{\sigma_n} J_{n-1}$. Each $J_p$ is a rule $(c)^{\sigma_p}_{\sigma_{p+1}}$ with $\sigma_0 = \pi$. Suppose that $2 \in In(\sigma_n)$ and the chain passes through the left side of a $(\Sigma_2)^{\sigma_p} K$ for a $p$ with $0 < p < n$ so that $J_{p-1}$ is in the left upper part of $K$ and $J_p$ is below $K$. Then $\sigma_q = rg_2(\sigma_n) \preceq_2 \sigma_p$, i.e., $q \geq p$.*

$$\begin{array}{ccc}
\vdots\ \mathcal{C} & & \\
\dfrac{\Gamma_{p-1}}{\Gamma'_{p-1}}\ (c)^{\sigma_{p-1}}_{\sigma_p}\ J_{p-1} & & \\
\vdots & & \vdots \\
\Phi, \neg A^{\sigma_p} & & A^{\sigma_p}, \Psi \\
\hline
& \Phi, \Psi & \\
\end{array}\ (\Sigma_2)^{\sigma_p}\ K$$
$$\begin{array}{c}
\vdots\ \mathcal{C} \\
\dfrac{\Gamma_p}{\Gamma'_p}\ (c)^{\sigma_p}\ J_p \\
\vdots \\
\dfrac{\Gamma_{n-1}}{\Gamma'_{n-1}}\ (c)^{\sigma_{n-1}}_{\sigma}\ J_{n-1}
\end{array}$$

This means that when in *Fig.* 4 a $(\Sigma_3)^{\sigma_t} K^3$ $(0 < t \leq n)$ in the new chain $\mathcal{C}_\rho = J_0, \ldots, J_{n-1}, J_n$ leading to $(c)^{\sigma}_{\rho} J_n$ is to be resolved into a $(\Sigma_2)^{\sigma_t} K^2$, then $t \leq q$, i.e., $rg_2(\rho) = \sigma_q \preceq_2 \sigma_t$. In other words any $(\Sigma_3)^{\sigma_t}$ with $q < t \leq n$, equivalently $(\Sigma_3)^{\sigma_t}$ which is below $(c)^{\sigma_q} J_q$ has to wait to be resolved, until the chain $\mathcal{C}_\rho$ will disappear by inversion.

For example consider, in *Fig.* 4, an inference rule $(\Sigma_3)^{\sigma_t} K^3$ for $t = n_{m+1} + 1$. Its right cut formula is a $\Sigma_3^{\sigma_t}$ sentence $C^{\sigma_t}$ and a descendent of a $\Sigma_3$ sentence $C$: a series of sentences from $C$ to $C^{\sigma_t}$ are in the chain $\mathcal{C}_{n_{m+1}} = J_0^{m+1}, \ldots, J_{n_m}^{m+1}, J_{n_m+1}, \ldots, J_{n_{m+1}}$ leading to $J_{n_{m+1}}$. Then the chain $\mathcal{C}_{n_{m+1}}$ passes through the left side of the inference rule $(\Sigma_{i_m})^{\sigma_{n_m+1}} K_m$ and hence $K^3$ will not be resolved until $K_m$ will be resolved and its right upper part will disappear since we always perform rewritings of proof figure on the rightmost branch. But then the chain $\mathcal{C}_\rho$ will disappear by inversion since it passes through the right side of $K_m$. In this way we see Proposition 2.1, cf. Lemma 5.7 in Sect. 5 for a full statement and a detailed proof.

Equation (3) is seen from Proposition 2.1 and the minimality of $q$. Thus we have shown, cf. the conditions $(\mathcal{D}^{\mathcal{Q}}.1)$ for $Od(\Pi_4)$ in [6] or Sect. 4,

$$rg_2(\sigma) = rg_2(pd_2(\rho)) \preceq_2 rg_2(\rho) \preceq_3 pd_3(\rho)$$

and

$$\forall t[rg_2(pd_2(\rho)) \preceq_2 \sigma_t \prec_2 \sigma_q \Rightarrow rg_2(\sigma_t) \preceq_2 \sigma_q].$$

Furthermore we have

$$st_2(\rho) < st_2(\sigma_{p+1}) < \sigma_q^+ \tag{4}$$

for the maximal $p$, vz. for the latest $(c)_{\sigma_{p+1}} J_p$ with $rg_2(\sigma_{p+1}) = \sigma_q \,\&\, 2 \in In(\sigma_{p+1})$.

Let $m < l$ denote the number such that $n_m < q \leq n_{m+1}$, i.e., $J_q$ is a member of the tail $\mathcal{R}_{m+1} = J_{n_m+1}, \ldots, J_{n_{m+1}}$ of the chain $\mathcal{C}_{n_{m+1}}$. Then from Proposition 2.1 we see that $J_p$ is also a member of $\mathcal{R}_{m+1}$ and further that $J_q$ is a member of a chain $\mathcal{C}_p$ leading to $J_p$. Thus the upper part of $(c)^{\sigma_q} J_q$ corresponding to $st_2(\rho)$ is a result of performing several non-void rewritings to the upper part of a $(c)^{\sigma_q}$ which determined $st_2(\sigma_{p+1})$ when $(c)_{\sigma_{p+1}} J_p$ was introduced originally. This yields (4).

Thus we have established the conditions $(\mathcal{D}^Q.1)$ in [6] or Sect. 4 for the newly introduced $\rho$.

Why we choose such a $\sigma_q$ as $rg_2(\rho)$? First introducing $\sigma_q = rg_2(\rho)$ is meant to express the fact that $\sigma_q$ is (iterated) $\Pi_3$-reflecting and it is responsible to $\Sigma_2^{\sigma_q}$ sentences occurring above a $(c)^{\sigma_q}$. Therefore even if there exists a $\sigma_{p+1}$ above $pd_3(\rho)$, i.e., $p \leq n_0$ such that $2 \in In(\sigma_{p+1})$, we ignore these in determining $rg_2(\rho)$. Second in the **Case 1** the reason why we chose $\sigma_q$ as the uppermost one is explained by Proposition 2.1: any $(\Sigma_3)^{\sigma_t}$ in the new chain $\mathcal{C}_\rho$ will not be resolved for $q < t \leq n$ until the chain $\mathcal{C}_\rho$ will disappear by inversion. Hence any $\sigma_{q_1}$ with $rg_2(\sigma_{p_1+1}) = \sigma_{q_1} \prec_2 \sigma_q$ for some $p_1 \leq n$ will not be $rg_2(\kappa)$ for $\kappa \prec_2 \rho$ in the future. This means that a collapsing series $\{(c)_\kappa : rg_2(\kappa) = \sigma_{q_1}\}$ expressing the fact that $\sigma_{q_1}$ is $\Pi_3$-reflecting is killed by introducing $\rho$ such that $\rho \prec_2 \sigma_{q_1} \prec_2 \sigma_q = rg_2(\rho)$. Therefore once we introduce such a $\rho$, then we can ignore $rg_2(\sigma_{p_1+1}) = \sigma_{q_1}$ between $rg_2(\rho)$ and $\rho$.

## The General Case $N > 4$

Here suppose $N > 4$ and we determine the $Q$ part of $\rho$, i.e., determine the set $In(\rho)$ and o.d.'s $pd_i(\rho), rg_i(\rho)$ for $i \in In(\rho)$ by referring *Fig.* 4.

First set $i_0 \in In(\rho)$ where $i_0$ denotes the number such that the first merging point is a $(\Sigma_{i_0})^{\sigma_{n_0+1}} K_0$. Now let us assume inductively that for $k_0 \geq 0$ we have specified merging points $\{K_{m_k} : k \leq k_0\}$ so that $0 = m_0 < \cdots < m_{k_0}$, $N - 1 > i_{m_0} > \cdots > i_{m_{k_0}} \geq 2$ and $\forall m \forall k < k_0[m_k < m < m_{k+1} \rightarrow i_m \geq i_{m_k}]$, and have set $\{i_{m_k} : k \leq k_0\} \subseteq In(\rho)$. Namely $K_{m_0}, \ldots, K_{m_{k_0}}$ is a series of merging points going downwards with decreasing indices $i_{m_k}$ and $K_{m_k}$ is the uppermost merging point with $i_{m_k} < i_{m_{k-1}}$ $(i_{m_{-1}} := N - 1)$.

If there exists an $m < l$ such that $m_{k_0} < m \,\&\, i_{m_{k_0}} > i_m \geq 2$, then let $m$ denote the minimal one, vz. the uppermost merging point $K_m$ below the latest one $K_{m_{k_0}}$ with $i_{m_{k_0}} > i_m$, and set $i_m \in In(\rho)$. Otherwise set

$$In(\rho) = \{i_{m_k} : k \leq k_0\} \cup \{N-1\}.$$

This completes a description of the set

$$\begin{aligned} In(\rho) &= \{N-1 = i_{m_{-1}}\} \cup \{i_{m_k} : 0 \leq k \leq k_1\} \\ &= \{N-1 = i_{m_{-1}} > i_{m_0} > \cdots > i_{m_{k_1}}\}. \end{aligned}$$

Observe that for $i < N-1$

$$i \in In(\rho) \Leftrightarrow \exists m < l[i_m = i \,\&\, \forall p < m(i_p \geq i)].$$

Now set $pd_{i_{m_k}}(\rho) = \sigma_{n_{m_{k+1}}+1}$ for $-1 \leq k \leq k_1$ with $m_{k_1+1} := l$, vz. the merging point $K_{m_k}$ chosen for $i_{m_k} \in In(\rho)$ is a $(\Sigma_{i_{m_k}})^{pd_{i_{m_{k-1}}}(\rho)}$ for $0 \leq k \leq k_1$ and $pd_2(\rho) = pd_{i_{m_{k_1}}}(\rho) = \sigma_{n_l+1} = \sigma_n = \sigma$. Observe that for any $i$ with $2 \leq i \leq N-1$ there exists an $m(i) \leq l$ such that $pd_i(\rho) = \sigma_{n_{m(i)}+1}$ and this $m(i)$ is the minimal $m$ for which $i_m < i$.

It remains to determine the o.d.'s $rg_i(\rho)$ for $N-1 \neq i = i_{m_k} \in In(\rho)$. As in the case $N = 4$ there are two cases to consider. First suppose there is a $p < n$ such that

1. $\rho \prec_i \sigma_{p+1} \prec_i \sigma_{n_{m_k}+1} = pd_{i_{m_{k-1}}}(\rho) = pd_{i+1}(\rho)$ and
2. $i \in In(\sigma_{p+1})$.

Then pick the minimal $p$ satisfying these two conditions, vz. the uppermost rule $(c)_{\sigma_{p+1}} J_p$ below the merging point $(\Sigma_i)^{pd_{i+1}(\rho)} K_{m_k}$ with $\sigma_{p+1} \prec_i pd_{i+1}(\rho) \,\&\, i \in In(\sigma_{p+1})$. Then set

**Case 1** $rg_i(\rho) := \sigma_q := rg_i(\sigma_{p+1})$.

Otherwise set

**Case 2** $rg_i(\rho) = pd_i(\rho)$.

In general we have the following fact.

**Proposition 2.2** *Let $\mathcal{C} = J_0, \ldots, J_{n-1}$ be a chain leading to a $(c)^{\sigma_{n-1}}_{\sigma_n} J_{n-1}$. Each $J_p$ is a rule $(c)^{\sigma_p}_{\sigma_{p+1}}$ with $\sigma_0 = \pi$. Suppose that the chain passes through the left side of a $(\Sigma_j)^{\sigma_p} K$ for a $p$ with $0 < p < n$ and a $j \geq i$ so that $J_{p-1}$ is in the left upper part of $K$ and $J_p$ is below $K$. Then $\sigma_n \prec_i \sigma_p$ and if further $N-1 \neq i \in In(\sigma_n)$, then $\sigma_q = rg_i(\sigma_n) \preceq_i \sigma_p$, cf. the figure in Proposition 2.1.*

Let us explain this Proposition 2.2 using the new chain $\mathcal{C}_\rho = J_0, \ldots, J_{n-1}, J_n$ leading to $(c)^{\sigma}_{\rho} J_n$, cf. *Fig.* 4. When a $(\Sigma_{j+1})^{\sigma_t} K^{j+1}$ $(0 < t \leq n)$ in the new chain

$\mathcal{C}_\rho$ is to be resolved, a $(\Sigma_j)^{\sigma_s} K^j$ is introduced at a point below $K^{j+1}$. The point and $s \geq t$ is determined as the lowest position as far as we can lower a rule $(\Sigma_j)^{\sigma_t}$, cf. Definition 5.5 in Sect. 5. For example when $K^{j+1}$ is the rule $(\Sigma_{i_m})^{\sigma_{n_m+1}} K_m$ in *Fig.* 4, let $m_1$ denote the minimal $m_1$ such that $i_{m_1} < i_m$ and we introduce a new $(\Sigma_{i_m-1})^{\sigma_{n_{m_1}+1}}$ $(s = n_{m_1+1})$ between the rules $(c)_{\sigma_{n_{m_1}+1}} J_{n_{m_1}}$ and $(\Sigma_{i_{m_1}})^{\sigma_{n_{m_1}+1}} K_{m_1}$. Observe that the new $(\Sigma_{i_m-1})$ together with $(\Sigma_{i_{m_2}}) K_{m_2}$ $(m < m_2 < m_1)$ by inversion will be merging points for the next chain leading to a $(c)^\rho$.

Let us consider the case when the $(\Sigma_{i_m-1})^{\sigma_{n_{m_1}+1}}$ is the rule $(\Sigma_j)^{\sigma_p} K$ in Proposition 2.2: $j = i_m - 1$ & $p = n_{m_1+1}$. Also put $pd_i(\rho) = \sigma_{n_{m(i)}+1}$, where $m(i)$ denotes the minimal $m(i)$ such that $i_{m(i)} < i$. Then $i \leq j = i_m - 1$. By Proposition 2.3 below we see that $i_m < i_{m_3}$ for any $m_3 < m$, i.e., any merging point $(\Sigma_{i_{m_3}}) K_{m_3}$ above $(\Sigma_{i_m}) K^{j+1} = K_m$ has larger index since we are assuming that $K_m$ is to be resolved. Therefore $m(i) \geq m_1$, i.e., the merging point $(\Sigma_{i_{m(i)}})^{\sigma_{n_{m(i)}+1}} K_{m(i)}$ determining $pd_i(\rho)$ is equal to or below the merging point $(\Sigma_{i_{m_1}})^{\sigma_{n_{m_1}+1}} K_{m_1}$. In the former case we have $pd_i(\rho) = \sigma_{n_{m(i)}+1} = \sigma_{n_{m_1}+1} = \sigma_p$ and hence $\rho \prec_i \sigma_p$. In the latter case we have $i_{m_3} \geq i$ for $m_1 \leq m_3 < m(i)$. Thus we see $\rho \prec_i \sigma_p$ inductively. This shows the first half of Proposition 2.2.

Now assume $N - 1 \neq i \in In(\sigma_n)$ and show $rg_i(\rho) \preceq_i \sigma_p$. Consider the **Case 1**, vz. $\sigma_q = rg_i(\rho) \neq pd_i(\rho)$. Let $p_0$ denote the minimal $p_0$ such that $\rho \prec_i \sigma_{p_0+1} \prec_i pd_{i+1}(\rho)$ and $i \in In(\sigma_{p_0+1})$. By the definition we have $\sigma_q = rg_i(\rho) = rg_i(\sigma_{p_0+1})$.

Let $m(i+1) < m(i)$ denote the number such that $pd_{i+1}(\rho) = \sigma_{n_{m(i+1)}+1}$. Then by $i \in In(\rho)$ we have $i_{m(i+1)} = i \leq i_{m_1}$, i.e., $m_1 \leq m(i+1) < m(i)$ and hence $pd_{i+1}(\rho) \leq \sigma_{n_{m_1}+1} = \sigma_p$. On the other hand we have $\rho \prec_i \sigma_q = rg_i(\rho)$ by the definition and $\rho \prec_i \sigma_p$ by the first half of the Proposition 2.2. Hence it suffices to show $\sigma_q \leq \sigma_p$ since the set $\{\tau : \rho \prec_i \tau\}$ is linearly ordered by $\prec_i$. Now we see $\sigma_q = rg_i(\rho) = rg_i(\sigma_{p_0+1}) \preceq_i pd_{i+1}(\rho)$ inductively, i.e., by using Proposition 2.2 for smaller parts. Thus we get $\sigma_q \preceq_i pd_{i+1}(\rho) \leq \sigma_p$. This shows the second half of Proposition 2.2.

Further we have the following fact.

**Proposition 2.3** *Let $\mathcal{C} = J_0, \ldots, J_{n-1}$ be a chain leading to a $(c)^{\sigma_{n-1}}_{\sigma_n} J_{n-1}$. Each $J_k$ is a rule $(c)^{\sigma_k}_{\sigma_{k+1}}$ with $\sigma_0 = \pi$. Suppose that the chain $\mathcal{C}$ passes through the left side of a $(\Sigma_j)^{\sigma_p} K^{lw}$ for a $p$ with $0 < p < n$ so that $J_{p-1}$ is in the left upper part of $K^{lw}$ and $J_p$ is below $K^{lw}$. Let $\mathcal{D} = I_0, \ldots, I_{m-1}$ $(m \geq p)$ be a chain leading to a $(c)^{\sigma_{m-1}}_{\sigma_m} I_{m-1}$. Each $I_k$ is a rule $(c)^{\tau_k}_{\tau_{k+1}}$ such that $\tau_k = \sigma_k$ for $0 \leq k < \min\{n, m\}$. Suppose that the chain $\mathcal{D}$ passes through the left side of a $(\Sigma_i)^{\sigma_k} K^{up}$ for a $k$ with $0 < k < p$ so that $I_{k-1}$ is in the left upper part of $K^{up}$ and $I_k$ is below $K^{up}$. Further assume the rule $(c)_{\sigma_p} I_{p-1}$ is in the right upper part of $(\Sigma_j)^{\sigma_p} K^{lw}$ and $i \leq j$.*

*Then the upper $K^{up}$ foreruns the lower $K^{lw}$, i.e., analyses of $K^{up}$ have to precede ones of $K^{lw}$.*

Let us explain Proposition 2.3 by referring *Fig.* 4: $\mathcal{C}$ is the new chain $\mathcal{C}_\rho$, $K^{lw}$ is the new $(\Sigma_{i_m-1})^{\sigma_{n_{m_1}+1}}$ which is resulted from $(\Sigma_{i_m})^{\sigma_{n_m+1}} K_m$ with $m = l - 1$, i.e., the resolved rule $K_{l-1}$ is the lowest merging point. Then $K^{lw}$ is a $(\Sigma_{i_m-1})^\sigma$ with $m_1 = l$. Further $\mathcal{D}$ is the chain $\mathcal{C}_{n_{m+1}} = J_0^{m+1}, \ldots, J_{n_m}^{m+1}, J_{n_m+1}, \ldots, J_{n_{m+1}}$

leading to the last member $(c)_\sigma\ J_{n-1}\ (n-1 = n_{m+1} = n_l)$ of the series $\mathcal{R}$. Then the last member $(c)_\sigma\ J_{n-1}$ is in the right upper part of $(\Sigma_{i_m-1})^\sigma\ K^{lw}$. Let $I$ be a rule $(\Sigma_{i+1})^\tau$ such that the chain $\mathcal{D}$ passes through its right side. Suppose the rule $I$ in the chain $\mathcal{D}$ is resolved and produces a $(\Sigma_i)^{\sigma_k}\ K^{up}$ for a $k$ with $0 < k < n$ so that the chain $\mathcal{D}$ passes through the left side of $K^{up}$.

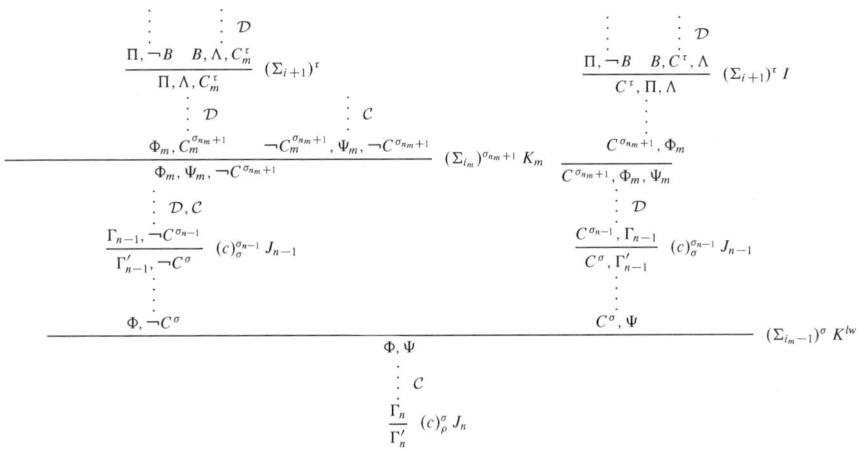

*Fig. 5*

where $\neg A_m \equiv C_m^{\sigma_{n_m+1}} \equiv \forall x < \sigma_{n_m+1} C_0(x)$ and $C^{\sigma_{n_m+1}} \equiv C_0(\alpha)$ for an $\alpha < \sigma_{n_m+1}$.

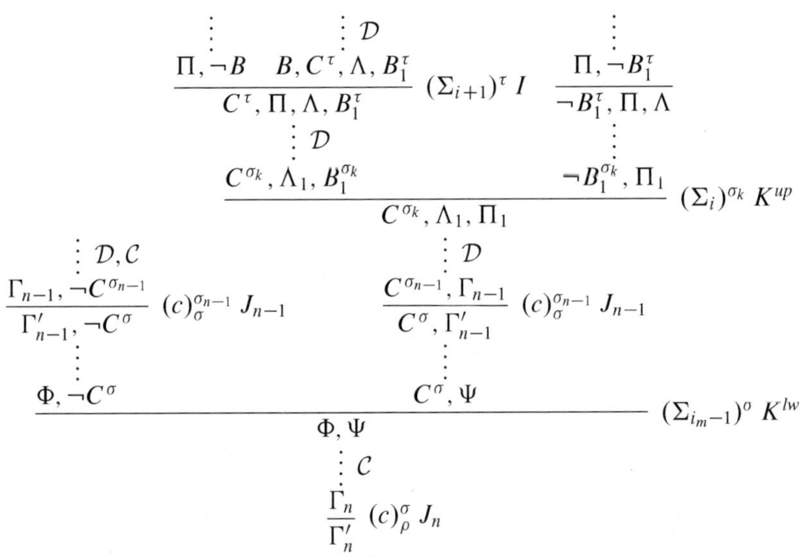

*Fig. 6*

We show, in *Fig.* 6, no ancestor of the right cut formula $C^\sigma$ of $K^{lw}$ is in the right upper part of $K^{up}$ in order to see that $K^{up}$ foreruns $K^{lw}$. It suffices to see that, in *Fig.* 5, no ancestor of the right cut formula $C^\sigma$ of $K^{lw}$ is in the left upper part of the resolved rule $(\Sigma_{i+1})^\tau\ I$. Any ancestor of the right cut formula $C^\sigma$ of $K^{lw}$ comes from the left cut formula $\neg A_m \equiv C_m^{\sigma_{n_m+1}}$ of $(\Sigma_{i_m})^{\sigma_{n_m+1}}\ K_m$ and any ancestor of the latter is in the chain $\mathcal{D}$, which in turn passes through the right side of $(\Sigma_{i+1})^\tau\ I$. Thus any ancestor of the right cut formula $C^\sigma$ of $K^{lw}$ is in the right upper part of $I$ in *Fig.* 5, a fortiori, in the left upper part of $K^{up}$ in *Fig.* 6. This shows Proposition 2.3.

For full statements and proofs of Propositions 2.2, 2.3, see Lemmata 5.7, the proviso **(uplw)** in Definition 5.8 in Sect. 5 and the case **M7.2** in Sect. 6.

From Propositions 2.2, 2.3 we see that the conditions $(\mathcal{D}^Q.1)$ for $Od(\Pi_N)$ in [6] or Sect. 4 are enjoyed with respect to the $Q$ part of $\rho$ as for the case $N = 4$. A set-theoretic meaning and a wellfoundedness proof of $Od(\Pi_N)$ are derived from these conditions on o.d.'s as we saw in [5, 6].

Consider a rule $(\Sigma_j)$ in the chain $\mathcal{C}_\rho$ for $j \geq i \in In(\rho)$ which is below $(\Sigma_{i_{m(i)}})^{pd_i(\rho)}\ K_{m(i)}\ (i_{m(i)} < i)$. Then from Proposition 2.3 we see that analyses of such a $(\Sigma_j)$ have to follow ones of the rule $(\Sigma_{i_{m(i)}})^{pd_i(\rho)}\ K_{m(i)}$. Thus when such a reversal happens, the lower rule with greater indices $(j > i_{m(i)})$ is dead and we can ignore it. The o.d. $pd_i(\rho)$ and the rule $(c)_{pd_i(\rho)}\ J_{n_{m(i)}}$ is the predecessor of the o.d. $\rho$ and the rule $(c)_\rho$ with respect to $i$: any member $(c)_\kappa$ of the chain $\mathcal{C}_\rho$ with $\rho < \kappa < pd_i(\rho)$ is irrelevant to the fact that $pd_i(\rho)$ and $rg_i(\rho)$ are iterated $\Pi_i$-reflecting. But the member may be relevant to $\Pi_j$-reflection for $j < i$. This motivates the definitions of $In(\rho)$ and $pd_i(\rho)$. A series $\kappa_n \prec_i \kappa_{n-1} \prec_i \cdots \prec_i \kappa_0$ expresses a possible stepping down for the fact that $\kappa_0$ is an iterated $\Pi_i$-reflecting ordinal. Degrees of iterations are measured by an ordinal $\nu < \kappa^+$ with $\kappa = rg_i(\kappa_0), \nu = st_i(\kappa_0)$ (and by predecessors of $rg_i(\kappa_0)$) as we saw in [5, 6]. Therefore we search only for o.d.'s $\sigma_{p+1}$ with $\rho \prec_i \sigma_{p+1}$ in determining the o.d. $rg_i(\rho) = rg_i(\sigma_{p+1})$.

In the **Case 1** the reason why we chose $\sigma_q$ as the uppermost one is explained by Propositions 2.2, 2.3 as in the case $N = 4$.

Now details follow.

## 3 The Theory $\mathrm{T}_N$ for $\Pi_N$-Reflecting Ordinals

In this section a theory $\mathrm{T}_N$ of $\Pi_N$-reflecting ordinals is defined.

Let $T_0$ denote the base theory defined in [3]. $\mathcal{L}_1$ denotes the language of $T_0$. Recall that $\mathcal{L}_1 = \mathcal{L}_0 \cup \{R^{\mathcal{A}}, R^{\mathcal{A}}_< : \mathcal{A} \text{ is a } \Delta_0 \text{ formula in } \mathcal{L}_0 \cup \{X\}\}$ with $\mathcal{L}_0 = \{0, 1, +, -, \cdot, q, r, \max, j, ()_0, ()_1, =, <\}$. $R^{\mathcal{A}}, R^{\mathcal{A}}_<$ are predicate constants for inductively defined predicates. The axioms and inference rules in $T_0$ are designed for this language $\mathcal{L}_1$.

The *language* $\mathcal{L}(\mathrm{T}_N)$ of the theory $\mathrm{T}_N$ is defined to be $\mathcal{L}_1 \cup \{\Omega\}$ with an individual constant $\Omega$.

The *axioms* of $\mathrm{T}_N$ are the same as for the theory $T_3$ in [4], i.e., are obtained from those of $T_{22}$ in [3] by deleting the axiom $\Gamma, Ad(\Omega)$. Thus the axioms $\Gamma, \Lambda_f$ for the closure of $\Omega$ under the function $f$ in $\mathcal{L}_0$ are included as mathematical axiom in $\mathrm{T}_N$.

The *inference rules* in $\mathrm{T}_N$ are obtained from $T_0$ by adding the following rules ($\Pi_N$-rfl) and ($\Pi_2^\Omega$-rfl).

$$\frac{\Gamma, A \quad \neg\exists z(t_0 < z \land A^z), \Gamma}{\Gamma}\ (\Pi_N\text{-rfl})$$

where $A \equiv \forall x_N \exists x_{N-1} \cdots Q x_1 B(x_N, x_{N-1}, \dots, x_1, t_0)$ is a $\Pi_N$ formula.

$$\frac{\Gamma, A^\Omega \quad \neg\exists z(t < z < \Omega \land A^z), \Gamma \quad \Gamma, t < \Omega}{\Gamma}\ (\Pi_2^\Omega\text{-rfl})$$

where $A \equiv \forall x \exists y B(x, y, t)$ is a $\Pi_2$ formula.

Concepts related to proof figures, principal or auxiliary formulae, pure variable condition, branch, etc. are defined exactly as in Section 2 of [3].

## 4 The System $Od(\Pi_N)$ of Ordinal Diagrams

In this section first let us recall briefly the system $Od(\Pi_N)$ of ordinal diagrams (abbreviated by o.d.'s) in [6].

Let $0, \varphi, \Omega, +, \pi$ and $d$ be distinct symbols. Each o.d. in $Od(\Pi_N)$ is a finite sequence of these symbols. $\varphi$ is the Veblen function. $\Omega$ denotes the first recursively regular ordinal $\omega_1^{CK}$ and $\pi$ the first $\Pi_N$-reflecting ordinal.

The set $Od(\Pi_N)$ is classified into subsets $R, SC, P$ according to the intended meanings of o.d.'s. $P$ denotes the set of additive principal numbers, $SC$ the set of strongly critical numbers and $R$ the set of recursively regular ordinals (less than or equal to $\pi$). If $\pi > \sigma \in R$, then $\sigma^+$ denotes the next recursively regular diagram to $\sigma$.

Recall that $K\alpha$ denotes the finite set of o.d.'s defined as follows.

1. $K0 = \emptyset$.
2. $K(\alpha_1 + \cdots + \alpha_n) = \bigcup\{K\alpha_i : 1 \le i \le n\}$.
3. $K\varphi\alpha\beta = K\alpha \cup K\beta$.
4. $K\alpha = \{\alpha\}$ otherwise, i.e., $\alpha \in SC$.

**Definition 4.1** 1. $\mathcal{D}_\sigma(\alpha) \subseteq \mathcal{D}_\sigma$.

(a) $\mathcal{D}_\sigma(\alpha) = \emptyset$ if $\alpha \in \{0, \Omega, \pi\}$.
(b) $\mathcal{D}_\sigma(\alpha) = \mathcal{D}_\sigma(K\alpha)$ if $\alpha \notin SC$.
(c) If $\alpha \in \mathcal{D}_\tau$,

$$\mathcal{D}_\sigma(\alpha) = \begin{cases} \mathcal{D}_\sigma(\{\tau\} \cup c(\alpha)) & \text{if } \tau > \sigma, \\ \{\alpha\} \cup \mathcal{D}_\sigma(c(\alpha)) & \text{if } \tau = \sigma, \\ \mathcal{D}_\sigma(\tau) & \text{if } \tau < \sigma. \end{cases}$$

2. $\mathcal{B}_\sigma(\alpha) = \max\{b(\beta) : \beta \in \mathcal{D}_\sigma(\alpha)\}$.
3. $\mathcal{B}_{>\sigma}(\alpha) = \max\{\mathcal{B}_\tau(\alpha) : \tau > \sigma\}$.

For an o.d. $\alpha$ set

$$\alpha^+ = \min\{\sigma \in R \cup \{\infty\} : \alpha < \sigma\}.$$

For $\sigma \in R$, $\mathcal{D}_\sigma \subseteq SC$ denotes the set of o.d.'s of the form $\rho = d_\sigma^q \alpha$ with a (possibly empty) list $q$, where the following condition has to be met:

$$\mathcal{B}_{>\sigma}(\{\sigma, \alpha\} \cup q) < \alpha \tag{5}$$

$\alpha$ is the *body* of $d_\sigma^q \alpha$.

If $q$ is not empty, then $d_\sigma^q \alpha \in \mathcal{D}^Q$ by definition. Its *Q part* $Q(d_\sigma^q \alpha) = q = \overline{\nu \kappa \tau j}$ denotes a sequence of quadruples $\nu_m \kappa_m \tau_m j_m$ of length $l + 1$ $(0 \le l)$ such that

1. $2 \le j_0 < j_1 < \cdots < j_l = N - 1$,
2. $\kappa_l = \pi, \kappa_m \in R \mid \pi \ (m < l) \,\&\, \sigma \preceq \kappa_m \ (m \le l)$,
3. $\nu_l \in Od(\Pi_N)$,

$$\sigma = \pi \Rightarrow \nu_l \le \alpha \tag{6}$$

and

$$m < l \Rightarrow \nu_m < \kappa_m^+, \tag{7}$$

4. $\tau_0 = \sigma, \tau_m \in \{\pi\} \cup \mathcal{D}^Q, \sigma \preceq \tau_m \ (m \le l)$ and

$$\tau_l = \pi \Rightarrow \sigma = \pi. \tag{8}$$

From $q = Q(\rho)$ define

1. $in_j(\rho) = st_j(\rho) rg_j(\rho)$ (a pair) and $pd_j(\rho)$: Given $j$ with $2 \le j < N$, put $m = \min\{m \le l : j \le j_m\}$.
2. $pd_j(\rho) = \tau_m$.
3. $\exists m \le l (j = j_m)$: Then $st_j(\rho) = \nu_m$, $rg_j(\rho) = \kappa_m$.
4. Otherwise: $in_j(\rho) = in_j(pd_j(\rho)) = in_j(\tau_m)$. If $in_j(\tau_m) = \emptyset$, then set $st_j(\rho) \uparrow$, $rg_j(\rho) \uparrow$.
5. $In(\rho) = \{j_m : m \le l\}$.

Observe that

$$\pi < \beta \in q = Q(\rho) \Rightarrow \beta = \nu_l = st_{N-1}(\rho). \tag{9}$$

The relation $\alpha \prec_i \beta$ is the transitive closure of the relation $pd_i(\alpha) = \beta$.

In [6] we impose several conditions on a diagram of the form $\rho = d_\sigma^q \alpha$ to be in $Od(\Pi_N)$. For $\alpha \in Od(\Pi_N), q \subseteq Od(\Pi_N) \,\&\, \sigma \in R \setminus \{\Omega\}, \rho = d_\sigma^q \alpha \in Od(\Pi_N)$ if the following conditions are fulfilled besides (5);

($\mathcal{D}^Q$.1) Assume $i \in In(\rho)$. Put $\kappa = rg_i(\rho)$. Then

($\mathcal{D}^Q$.11) $in_i(rg_i(\rho)) = in_i(pd_{i+1}(\rho))$, $rg_i(\rho) \preceq_i pd_{i+1}(\rho)$ and $pd_i(\rho) \neq pd_{i+1}(\rho)$ if $i < N-1$.
Also $pd_i(\rho) \preceq_i rg_i(\rho)$ for any $i$.

($\mathcal{D}^Q$.12) One of the following holds:

($\mathcal{D}^Q$.12.1) $rg_i(\rho) = pd_i(\rho) \,\&\, \mathcal{B}_{>\kappa}(st_i(\rho)) < b(\alpha_1)$ with $\rho \preceq \alpha_1 \in \mathcal{D}_\kappa$.

($\mathcal{D}^Q$.12.2) $rg_i(\rho) = rg_i(pd_i(\rho)) \,\&\, st_i(\rho) < st_i(pd_i(\rho))$.

($\mathcal{D}^Q$.12.3) $rg_i(pd_i(\rho)) \prec_i \kappa \,\&$
$\forall \tau (rg_i(pd_i(\rho)) \preceq_i \tau \prec_i \kappa \to rg_i(\tau) \preceq_i \kappa) \,\&\, st_i(\rho) < st_i(\sigma_1)$ with

$$\sigma_1 = \min\{\sigma_1 : rg_i(\sigma_1) = \kappa \,\&\, pd_i(\rho) \prec_i \sigma_1 \prec_i \kappa\}$$

and such a $\sigma_1$ exists.

($\mathcal{D}^Q$.2)

$$\forall \kappa \leq rg_i(\rho)(K_\kappa st_i(\rho) < \rho) \tag{10}$$

for $i \in In(\rho)$.

We set $Q(d_\sigma \alpha) = \emptyset$, i.e., $d_\sigma^\emptyset \alpha = d_\sigma \alpha$.

The order relation $\alpha < \beta$ on $\mathcal{D}_\sigma$ is defined through finite sets $K_\tau \alpha$ for $\tau \in R, \alpha \in Od(\Pi_N)$, and the latter is defined through the relation $\alpha \prec \beta$, which is the transitive closure of the relation $\alpha \in \mathcal{D}_\beta$. Thus $\alpha \prec_2 \beta \Leftrightarrow \alpha \prec \beta$.

For $\rho = d_\tau^q \alpha$ $c(\rho) = \{\tau, \alpha\} \cup q$ and

$$K_\sigma \rho = \begin{cases} K_\sigma(\{\tau\} \cup c(\rho)) = K_\sigma\{\tau, \alpha\} \cup q, & \sigma < \tau, \\ K_\sigma \tau, & \tau < \sigma \,\&\, \tau \not\prec \sigma. \end{cases}$$

The following Proposition 4.1 is shown in [6].

**Proposition 4.1** *1. The finite set $\{\tau : \sigma \prec_i \tau\}$ is linearly ordered by $\prec_i$.*
*In the following assume $\kappa = rg_i(\rho) \downarrow$.*

*2. $\rho \prec_i rg_i(\rho)$.*
*3. $\rho \prec_i \sigma \prec_i \tau \,\&\, in_i(\rho) = in_i(\tau) \Rightarrow in_i(\rho) = in_i(\sigma)$.*
*4. $\rho \prec_i \tau \prec_i rg_i(\rho) \Rightarrow rg_i(\tau) \preceq_i rg_i(\rho)$.*

**Definition 4.2** For o.d.'s $\alpha, \sigma$ with $\sigma \in R$,

$$\mathcal{K}_\sigma(\alpha) := \max K_\sigma \alpha.$$

The following lemmata are seen as in [3].

**Lemma 4.1** *Suppose $\mathcal{B}_{>\kappa}(\alpha_i) < \alpha_i$ for $i = 0, 1$, and $\alpha_0 < \alpha_1$. Then*

$$\tau > \kappa \Rightarrow d_\tau \alpha_i \in Od(\Pi_N) \,\&\, d_\tau \alpha_0 < d_\tau \alpha_1.$$

**Lemma 4.2** *For $\alpha, \beta, \sigma \in Od(\Pi_N)$ with $\sigma \in R|\pi$ assume $\forall \tau < \pi[\mathcal{B}_\tau(\beta) \leq \mathcal{B}_\tau(\alpha)]$, and put $\gamma = \max\{\mathcal{B}_\pi(\beta), \mathcal{B}_{>\sigma}(\{\sigma, \alpha\})\} + \omega^\beta$. Then $\mathcal{B}_{>\sigma}(\{\sigma, \gamma, \gamma + \mathcal{K}_\sigma(\alpha)\}) < \gamma$, and hence (5) is fulfilled for $d_\sigma \gamma, d_\sigma(\gamma + \mathcal{K}_\sigma(\alpha)) \in Od(\Pi_N)$.*

## 5 The System $T_{Nc}$

In this section we extend $T_N$ to a formal system $T_{Nc}$. The *universe* $\pi(T_N)$ of the theory $T_N$ is defined to be the o.d. $\pi \in Od(\Pi_N)$. The language is expanded so that individual constants $c_\alpha$ for o.d.'s $\alpha \in Od(\Pi_N) \mid \pi$ are included. Inference rules $(c)^\sigma$ are added. To each proof $P$ in $T_{Nc}$ an o.d. $o(P) \in Od(\Pi_N) \mid \Omega$ is attached. *Chains* are defined to be a consecutive sequence of rules $(c)$. *Proofs* in $T_{Nc}$ defined in Definition 5.8 are proof figures enjoying some provisos and obtained from given proofs in $T_N$ by operating rewriting steps. Some lemmata for proofs are established. These are needed to verify that rewrited proof figures enjoy these provisos.

The *language* $\mathcal{L}_{Nc}$ of $T_{Nc}$ is obtained from the language $\mathcal{L}(T_N)$ by adding individual constants $c_\alpha$ for each o.d. $\alpha \in Od(\Pi_N)$ such that $1 < \alpha < \pi \,\&\, \alpha \neq \Omega$. We identify the constant $c_\alpha$ with the o.d. $\alpha$.

In what follows $A, B, \ldots$ denote formulae in $\mathcal{L}_{Nc}$ and $\Gamma, \Delta, \ldots$ sequents in $\mathcal{L}_{Nc}$.

The *axioms* of $T_{Nc}$ are obtained from those of $T_N$ as in [3].

Complexity measures $\deg(A), \mathrm{rk}(A)$ of formulae $A$ are defined as in [3] by replacing the universe $\pi(T_{22}) = \mu$ by $\pi(T_N) = \pi$.

Also the sets $\Delta_0^\sigma, \Sigma_i^\sigma$ of formulae are defined as in [3]. Recall that for a bounded formula $A$ and a multiplicative principal number $\alpha \leq \pi$, we have $A \in \Delta^\alpha \Leftrightarrow \deg(A) < \alpha$.

**Definition 5.1**

$$\deg_N(A) := \begin{cases} \deg(A) + N - 1 & \text{if } A \text{ is a bounded formula,} \\ \deg(A) & \text{otherwise.} \end{cases}$$

Note that

$$\deg_N(A) \notin \{\alpha + i : i < N - 1, \alpha < \pi \text{ is a limit o.d.}\}.$$

The *inference rules* of $\mathrm{T}_{Nc}$ are obtained from those of $\mathrm{T}_N$ by adding the following rules $(h)^\alpha$ ($\alpha \in \{\alpha : \pi \leq \alpha < \pi + \omega\} \cup \{0, \Omega\}$), $(c\Pi_2)^\Omega_{\alpha_1}$, $(c\Sigma_1)^\Omega_{\alpha_1}$, $(c\Pi_N)^\sigma_\tau$, $(c\Sigma_{N-1})^\sigma_\tau$ for each $\sigma \in R \subseteq Od(\Pi_N) \,\&\, \sigma \neq \Omega$ and $(\Sigma_i)^\sigma$ for each $\sigma \in R \subseteq Od(\Pi_N) \,\&\, \sigma \notin \{\Omega, \pi\}$ and $i = 1, 2, \ldots, N$. The rule $(h)^\alpha$, $(c\Pi_2)^\Omega_{\alpha_1}$ and $(c\Sigma_1)^\Omega_{\alpha_1}$ are the same as in [4]. We write $(w)$ for $(h)^0$.

1.

$$\frac{\Gamma, A^\sigma}{\Gamma, A^\tau}\ (c\Pi_N)^\sigma_\tau$$

where

(a) $A \equiv \forall x_N \exists x_{N-1} \cdots Q x_1 B$ is a $\Pi_N$-sentence with a $\Delta^\tau$-matrix $B$,
(b) $\tau \in \mathcal{D}_\sigma$ with the *body* $\alpha = b(\tau)$ of the rule and
(c) the formula $A^\tau$ in the lowersequent is the *principal formula* of the rule and the formula $A^\sigma$ in the uppersequent is the *auxiliary formula* of the rule, resp. Each formula in $\Gamma$ is a *side formula* of the rule.

2.

$$\frac{\Gamma, \Lambda^\sigma}{\Gamma, \Lambda^\tau}\ (c\Sigma_{N-1})^\sigma_\tau$$

where

(a) $\Lambda$ is a nonempty set of unbounded $\Pi_N$-sentences with $\Delta^\tau$-matrices.
(b) $\tau \in \mathcal{D}_\sigma$ with the *body* $\alpha = b(\tau)$ of the rule and
(c) each formula in $\Gamma$ is a *side formula* of the rule.

3.

$$\frac{\Gamma, \neg A^\sigma \quad A^\sigma, \Lambda}{\Gamma, \Lambda}\ (\Sigma_i)^\sigma$$

where $1 \leq i \leq N$ and $A^\sigma$ is a genuine $\Sigma^\sigma_i$-sentence, i.e., $A^\sigma \in \Sigma^\sigma_i$ and $A^\sigma \notin \Pi^\sigma_{i-1} \cup \Sigma^\sigma_{i-1}$.
$A^\sigma$ $[\neg A^\sigma]$ is said to be the *right* [*left*] *cut formula* of the rule $(\Sigma_i)^\sigma$, resp.

The rules $(c\Pi_2)^\Omega$ and $(c\Pi_N)$ are *basic rules* but not the rules $(h)^\alpha$, $(c\Sigma_1)^\Omega$, $(c\Sigma_{N-1})^\sigma$ and $(\Sigma_i)^\sigma$.

A *preproof* in $\mathrm{T}_{Nc}$ is a proof in $\mathrm{T}_{Nc}$ in the sense of [3], i.e., a proof tree built from axioms and inference rules in $\mathrm{T}_{Nc}$. The underlying tree Tree$(P)$ of a preproof $P$ is a tree of finite sequences of natural numbers such that each occurrence of a sequent or an inference rule receives a finite sequence. The root (empty sequence) () is attached to the endsequent, and in an inference rule

$$\frac{a * (0,0) : \Lambda_0 \quad \cdots \quad a * (0,n) : \Lambda_n}{a : \Gamma}\ (r)\, a * (0)$$

where $(r)$ is the name of the inference rule. Finite sequences are denoted by Roman letters $a, b, c, \ldots, I, J, K, \ldots$. Roman capitals $I, J, K, \ldots$ denote exclusively inference nodes. We will identify the attached sequence $a$ with the occurrence of a sequent or an inference rule.

Let $P$ be a preproof and $\gamma < \pi + \omega$ an o.d. in $Od(\Pi_N)$. For each sequent $a : \Gamma$ ($a \in \text{Tree}(P)$), we assign the *height* $\text{h}_\gamma(a; P) < \pi + \omega$ of the node *a with the base height* $\gamma$ in $P$ as in [3] except we replace $\pi(\text{T}_{22}) = \mu$ by $\pi(\text{T}_N) = \pi$ and replace $\deg(A)$ by $\deg_N(A)$.

Then the *height* $\text{h}(a; P)$ of $a$ in $P$ is defined to be the height with the base height $\gamma = 0$:

$$\text{h}(a; P) := \text{h}_0(a; P).$$

A pair $(P, \gamma)$ of a preproof $P$ and an o.d. $\gamma$ is said to be *height regulated* if it enjoys the conditions in [3], or equivalently in [4, Definition 5.4]. For the rules $(\Sigma_i)^\sigma$, this requires the condition: If $a : \Gamma$ is the lowersequent of a rule $(\Sigma_i)^\sigma$ $a * (0)$ $(1 \leq i \leq N)$ in $P$, then $\text{h}_\gamma(a; P) \leq \sigma + i - 2$ if $i = N - 1, N$. Otherwise $\text{h}_\gamma(a; P) \leq \sigma + i - 1$.

Therefore for the uppersequent $a * (0, k) : \Lambda$ of a $(\Sigma_i)^\sigma$ we have $\text{h}_\gamma(a * (0, k); P) = \sigma + i - 1$. Note that this implies that there are no nested rules $(\Sigma_i)^\sigma$, i.e., there is no $(\Sigma_i)^\sigma$ below any $(\Sigma_i)^\sigma$ for $i \geq N - 1$.

A preproof is height regulated iff $(P, 0)$ is height regulated.

Let $P$ be a preproof and $\gamma < \pi + \omega$. Assume that $(P, \gamma)$ is height regulated. Then the o.d $o_\gamma(a; P) \in O(\Pi_N)$ assigned to each node $a$ in the underlying tree $\text{Tree}(P)$ of $P$ is defined exactly as in [4].

Furthermore for $\tau \in R \cap Od(\Pi_N)$, o.d.'s $B_{\tau,\gamma}(a; P), Bk_{\tau,\gamma}(a; P) \in O(\Pi_N)$ are assigned to each sequent node $a$ such that $\text{h}_\gamma(a; P) \leq \tau \in R$ as in [4]. Namely

$$B_{\tau,\gamma}(a; P) := \begin{cases} \pi \cdot o_\gamma(a; P) \\ \quad \text{if } \text{h}_\gamma(a; P) = \tau = \pi, \\ \max\{\mathcal{B}_\pi(o_\gamma(a; P)), \mathcal{B}_{>\tau}(\{\tau\} \cup (a; P))\} + \omega^{o_\gamma(a;P)} \\ \quad \text{if } \text{h}_\gamma(a; P) < \pi. \end{cases}$$

$$Bk_{\tau,\gamma}(a; P) := B_{\tau,\gamma}(a; P) + \mathcal{K}_\tau(a; P),$$

$B_\tau(a; P)$ $[Bk_\tau(a; P)]$ denotes $B_{\tau,0}(a; P)$ $[Bk_{\tau,0}(a; P)]$, resp.

Then propositions and lemmata (Rank Lemma 7.3, Inversion Lemma 7.9, etc.) in Section 9 of [3] and Replacement Lemma 5.15 in [4] hold also for $\text{T}_{Nc}$.

Lemma 4.2 yields $o_\gamma(a; P) \in Od(\Pi_N)$ for each node $a \in \text{Tree}(P)$ if $(P, \gamma)$ is height regulated and $\gamma < \pi + \omega$.

**Definition 5.2** Let $\mathcal{T}$ be a branch in a preproof $P$ and $J$ a rule $(\Sigma_i)^\sigma$.

1. *Left branch:* $\mathcal{T}$ is a *left branch* of $J$ if
   (a) $\mathcal{T}$ starts with a lowermost sequent $\Gamma$ such that $h(\Gamma) \geq \pi$,
   (b) each sequent in $\mathcal{T}$ contains an ancestor of the left cut formula of $J$ and
   (c) $\mathcal{T}$ ends with the left uppersequent of $J$.
2. *Right branch:* $\mathcal{T}$ is a *right branch* of $J$ if
   (a) $\mathcal{T}$ starts with a lowermost sequent $\Gamma$ such that $\Gamma$ is a lowersequent of a basic rule whose principal formula is an ancestor of the right cut formula of $J$ and
   (b) $\mathcal{T}$ ends with the right uppersequent of $J$.

*Chains* in a preproof are defined as in Definition 6.1 of [4] when we replace $((c\Pi_3), (\Sigma_3))$, $((c\Sigma_2), (\Sigma_2))$ by $((c\Pi_N), (\Sigma_N))$, $((c\Sigma_{N-1}), (\Sigma_{N-1}))$. For definitions related to chains such as *starting with, top, branch* of a chain, *passing through*, see Definition 6.1 of [4]. Also *rope sequence* of a rule, the *end* of a rope sequence and the *bar* of a rule are defined as in Definition 6.2 of [4]. Moreover a *chain analysis* for a preproof together with the *bottom* of a rule is defined as in Definition 6.3 of [4].

**Definition 5.3** *Q part of a chain and the i-origin.*

1. Let $\mathcal{C} = J_0, J_0', \ldots, J_n, J_n'$ be a chain starting with a $(c)_\sigma$ $J_n$. Put
   (a) $In(\mathcal{C}) := In(J_n) := In(\sigma)$.
   (b) $in_i(\mathcal{C}) := in_i(J_n) := in_i(\sigma)$ for $2 \leq i < N$.
   (c) $st_i(\mathcal{C}) := st_i(J_n) := st_i(\sigma)$, $rg_i(\mathcal{C}) := rg_i(J_n) := rg_i(\sigma)$
   where $st_i(\mathcal{C}) \uparrow \; \& \; rg_i(\mathcal{C}) \uparrow$ if $st_i(\sigma) \uparrow \; \& \; rg_i(\sigma) \uparrow$.
   (d) $J_k$ is the *i-origin* of the chain $\mathcal{C}$ or the rule $J_n$ if $J_k$ is a rule $(c)^\kappa$ with $\kappa = rg_i(\sigma) \downarrow$.
   (e) $J_k$ is the *i-predecessor* of $J_n$, denoted by $J_k = pd_i(J_n)$ or
   *i predecessor* of the chain $\mathcal{C}$, denoted by $J_k = pd_i(\mathcal{C})$ if $J_k$ is a rule $(c)_\rho$ with $\rho = pd_i(\sigma)$.

**Definition 5.4** *Knot and rope.*
Assume that a chain analysis for a preproof $P$ is given and by a chain we mean a chain in the chain analysis.

1. *i-knot:* Let $K$ be a rule $(\Sigma_i)^\sigma$ $(1 \leq i \leq N-2)$. We say that $K$ is an *i-knot* if there are an uppermost rule $(c)^\sigma$ $J_{lw}$ below $K$ and a chain $\mathcal{C}$ such that $J_{lw}$ is a member of $\mathcal{C}$ and $\mathcal{C}$ passes through the left side of $K$.
   The rule $J_{lw}$ is said to be the *lower rule* of the $i$-knot $K$. The member $(c)_\sigma$ $J_{ul}$ of the chain $\mathcal{C}$ is the *upper left rule* of $K$ and a rule $(c)_\sigma$ $J_{ur}$ which is above the right uppersequent of $K$ is an *upper right rule* of $K$ if such a rule $(c)_\sigma$ $J_{ur}$ exists.

$$
\dfrac{\begin{array}{cc} \vdots\; \mathcal{C} & \vdots \\ \Gamma, \neg A^\sigma & A^\sigma, \Lambda \end{array}}{\Gamma, \Lambda}\;(\Sigma_i)^\sigma\; K
$$
$$
\begin{array}{c} \vdots \\ \dfrac{\Delta}{\Delta'} \;\text{uppermost } (c)^\sigma\; J_{lw} \in \mathcal{C} \\ \vdots \end{array}
$$

2. A rule is a *knot* if it is an $i$-knot for some $i > 1$.

   *Remark* Note that a 1-knot $(\Sigma_1)$ is not a knot by definition.

3. Let $K$ be a knot, $J_{lw}$ the lower rule of $K$ and $J_{ur}$ an upper right rule of $K$. Then we say that $K$ is a *knot of* $J_{ur}$ *and* $J_{lw}$.
4. Let $\mathcal{C}_n = J_0, \dots, J_n$ be a chain starting with $J_n$ and $K$ a knot. $K$ is a *knot for the chain* $\mathcal{C}_n$ or *the rule* $J_n$ if

   (a) the lower rule $J_{lw}$ of $K$ is a member $J_k$ $(k < n)$ of $\mathcal{C}_n$,
   (b) $\mathcal{C}_n$ passes through the right side of $K$, and
   (c) for any $k < n$ the chain $\mathcal{C}_k$ starting with $J_k$ does not pass through the right side of $K$.

   The knot $K$ is a merging rule of the chain $\mathcal{C}_n$ and the chain $\mathcal{C}_k$ starting with the lower rule $J_{lw} = J_k$.

$$
\dfrac{\begin{array}{cc} \vdots\; \mathcal{C}_k & \vdots\; \mathcal{C}_n \\ \Gamma, \neg A^\sigma & A^\sigma, \Lambda \end{array}}{\Gamma, \Lambda}\;(\Sigma_i)^\sigma\; K
$$
$$
\begin{array}{c} \vdots \\ \dfrac{\Delta_k}{\Delta'_k} \;\text{uppermost } (c)^\sigma\; J_{lw} = J_k \in \mathcal{C}_n \\ \vdots \\ \dfrac{\Delta_n}{\Delta'_n}\; J_n \end{array}
$$

5. A series $\mathcal{R}_{J_0} = J_0, \dots, J_{n-1}$ $(n \geq 1)$ of rules $(c)$ is said to be the *rope starting with* $J_0$ if there is an increasing sequence of numbers (uniquely determined)

$$
0 \leq n_0 < n_1 < \cdots < n_l = n - 1 \; (l \geq 0) \tag{11}
$$

   for which the following hold:

   (a) each $J_{n_m}$ is the bottom of $J_{n_{m-1}+1}$ for $m \leq l$ $(n_{-1} = -1)$,
   (b) there is an uppermost knot $K_m$ such that $J_{n_m}$ is an upper right rule and $J_{n_m+1}$ is the lower rule of $K_m$ for $m < l$, and
   (c) there is no knot whose upper right rule is $J_{n_l} = J_{n-1}$.

We say that the rule $J_{n-1}$ is the *edge of the rope* $\mathcal{R}_{J_0}$ or *the edge of the rule* $J_0$. For a rope the increasing sequence of numbers (11) is called the *knotting numbers* of the rope.

*Remark* These knots $K_m$ are uniquely determined for a *proof* defined below.

6. Let $K_{-1}$ be an $i_{-1}$-knot ($i_{-1} \geq 1$) and $J_0$ the lower rule of $K_{-1}$. The *left rope* ${}_{K_{-1}}\mathcal{R}$ *of* $K_{-1}$ is inductively defined as follows:

   (a) Pick the lowermost rule $(c)$ $J_{n_0}$ such that the chain $\mathcal{C}$ starting with $J_{n_0}$ passes through the left side of the $i_{-1}$-knot $K_{-1}$ and $J_0$ is a member of $\mathcal{C}$. Let ${}_0\mathcal{R} = I_0, \ldots, I_q$ be the part of the chain $\mathcal{C}$ with $J_0 = I_0$ & $J_{n_0} = I_q$.
   (b) If there exists an uppermost knot $K_0$ such that $J_{n_0}$ is an upper right rule of $K_0$, then ${}_{K_{-1}}\mathcal{R}$ is defined to be a concatenation:

$$ {}_{K_{-1}}\mathcal{R} = {}_0\mathcal{R}^\frown {}_{K_0}\mathcal{R} $$

   where ${}_{K_0}\mathcal{R}$ denotes the left rope of $K_0$.
   (c) Otherwise. Set:

$$ {}_{K_{-1}}\mathcal{R} = {}_0\mathcal{R}. $$

Therefore for the left rope ${}_{K_{-1}}\mathcal{R} = J_0, \ldots, J_{n-1}$ of $K_{-1}$ there exists a uniquely determined increasing sequence of numbers (11) such that:

(a) each $J_{n_m}$ is the lowermost rule $(c)$ such that the chain $\mathcal{C}$ starting with $J_{n_m}$ passes through the left side of the $i_{m-1}$-knot $K_{m-1}$ and $J_{n_{m-1}+1}$ is a member of $\mathcal{C}$ ($n_{-1} = -1$) for $m \leq l$,
(b) there is an $i_m$-knot $K_m$ ($i_m > 1$) such that $J_{n_m}$ is an upper right rule and $J_{n_m+1}$ is the lower rule of $K_m$ for $m < l$, and
(c) there is no knot whose upper right rule is $J_{n_l} = J_{n-1}$. ( $K_{-1}$ is the $i_{-1}$-knot whose lower rule is $J_0$.)

These numbers (11) are called the *knotting numbers* of the left rope and each knot $K_m$ ($m < l$) a *knot for the left rope.*
By the *left rope* ${}_{J_0}\mathcal{R}$ of the lower rule $J_0$ of $K_{-1}$ we mean the left rope ${}_{K_{-1}}\mathcal{R}$ of $K_{-1}$.

When a rule $(\Sigma_{i+1})^\sigma$ $K$ ($0 < i < N$) is resolved, we introduce a new rule $(\Sigma_i)^{\sigma_{n_{m(i+1)}+1}}$ at a sequent $\Phi$, which is defined to be the *resolvent* of $K$ and a $\sigma_{n_{m(i+1)}+1} \preceq \sigma$ defined as follows.

**Definition 5.5** *Resolvent*
Let $K$ be a rule $(\Sigma_{i+1})^\sigma$ ($0 < i < N$). The *resolvent* of the rule $K$ is a sequent $a : \Phi$ defined as follows: let $K'$ denote the lowermost rule $(\Sigma_{i+1})^\sigma$ below or equal to $K$ and $b : \Psi$ the lowersequent of $K'$.

**Case 1** The case when there exists an $(i+1)$-knot $(\Sigma_{i+1})^\sigma$ which is between an uppersequent of $K$ and $b : \Psi$: Pick the uppermost such knot $(\Sigma_{i+1})^\sigma$ $K_{-1}$ and let ${}_{K_{-1}}\mathcal{R} = J_0, \ldots, J_{n-1}$ denote the left rope of $K_{-1}$. Each $J_p$ is a rule $(c)^{\sigma_p}_{\sigma_{p+1}}$. Let

$$0 \leq n_0 < n_1 < \cdots < n_l = n - 1 \; (l \geq 0) \tag{11}$$

be the knotting numbers of the left rope ${}_{K_{-1}}\mathcal{R}$ and $K_m$ an $i_m$-knot $(\Sigma_{i_m})^{\sigma_{n_m+1}}$ of $J_{n_m}$ and $J_{n_m+1}$ for $m < l$. Put

$$m(i+1) = \max\{m : 0 \leq m \leq l \,\&\, \forall p \in [0, m)(i + 1 \leq i_p)\}. \tag{12}$$

Then the resolvent $a : \Phi$ is defined to be the uppermost sequent $a : \Phi$ below $J_{n_{m(i+1)}}$ such that $h(a; P) < \sigma_{n_{m(i+1)}+1} + i$.

**Case 2** Otherwise: Then the resolvent $a\Phi$ is defined to be the sequent $b : \Psi$.

**Definition 5.6** Let $J$ and $J'$ be rules in a preproof such that both $J$ and $J'$ are one of rules $(\Sigma_i)$ $(1 \leq i \leq N-1)$ and $J$ is above the right uppersequent of $J'$. We say that $J$ *foreruns* $J'$ if any right branch $\mathcal{T}$ of $J'$ *is left to* $J$, i.e., there exists a merging rule $K$ such that $\mathcal{T}$ passes through the left side of $K$ and the right uppersequent of $K$ is equal to or below the right uppersequent of $J$.

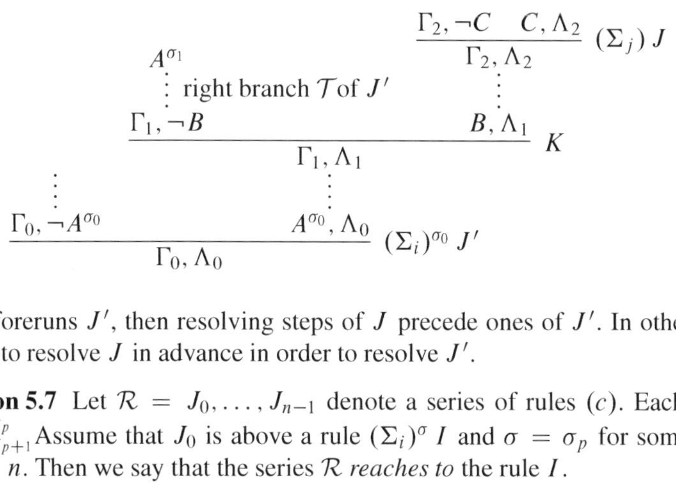


If $J$ foreruns $J'$, then resolving steps of $J$ precede ones of $J'$. In other words, we have to resolve $J$ in advance in order to resolve $J'$.

**Definition 5.7** Let $\mathcal{R} = J_0, \ldots, J_{n-1}$ denote a series of rules $(c)$. Each $J_p$ is a rule $(c)^{\sigma_p}_{\sigma_{p+1}}$Assume that $J_0$ is above a rule $(\Sigma_i)^\sigma\, I$ and $\sigma = \sigma_p$ for some $p$ with $0 < p \leq n$. Then we say that the series $\mathcal{R}$ *reaches to* the rule $I$.

In a *proof* defined in the next definition, if a series $\mathcal{R} = J_0, \ldots, J_{n-1}$ reaches to the rule $(\Sigma_i)^\sigma\, I$, then either $\mathcal{R}$ passes through $I$ in case $p < n$, or the subscript $\sigma_n$ of the last rule $(c)^{\sigma_{n-1}}_{\sigma_n}\, J_{n-1}$ is equal to $\sigma$, i.e., $J_{n-1}$ is a lowermost rule $(c)$ above $I$.

**Definition 5.8** *Proof*

Let $P$ be a preproof. Assume a chain analysis for $P$ is given. The preproof *$P$ together with the chain analysis* is said to be a *proof* in $\mathrm{T}_{Nc}$ if it satisfies the following conditions besides the conditions **(pure)**, **(h-reg)**, **(c:side)**, **(c:bound)**, **(next)**, **(h:bound)**, **(ch:pass)** (a chain passes through only rules $(c)$, $(h)$, $(\Sigma_i)$ $(i < N)$), **(ch:left)**, which are the same as in [4]:

**(st:bound)** Let $\mathcal{C}$ be a chain, $i \in In(\mathcal{C})$ and $a : \Gamma$ be the uppersequent of the $i$ origin of the chain $\mathcal{C}$. Then

**(st:bound1)** Let $i = N - 1$. Then

$$o(a; P) \leq st_{N-1}(\mathcal{C}).$$

**(st:bound2)** Let $i < N - 1$ and $\kappa = rg_i(\mathcal{C})$. Then for an $\alpha$

$$st_i(\mathcal{C}) = d_{\kappa}+\alpha$$

and

$$B_{\kappa}(a; P) \leq \alpha.$$

**(ch:link)** *Linking chains:* Let $\mathcal{C} = J_0, J_0', \ldots, J_n, J_n'$ and $\mathcal{D} = I_0, I_0', \ldots, I_m, I_m'$ be chains such that $J_i$ is a rule $(c)^{\tau_i}_{\tau_i+1}$ and $I_i$ a rule $(c)^{\sigma_i}_{\sigma_i+1}$. Assume that branches of these chains intersect. Then one of the following three types must occur (cf. [4] for **Type1 (segment)** and **Type2 (jump)**):

**Type1 (segment)**: One is a part of the other, i.e.,

$$n \leq m \,\&\, J_i = I_i$$

or vice versa.

Assume that there exists a merging rule $K$ such that $\mathcal{C}$ passes through the left side of $K$ and $\mathcal{D}$ the right side of $K$. Then by **(ch:left)** the merging rule $K$ is a $(\Sigma_l)^{\tau_j}$ for some $j \leq n$ and some $l$ with $1 \leq l \leq N - 2$.

**Type2 (jump)**: The case when there is an $i \leq m$ so that

1. $J'_{j-1}$ is above $K$ and $J_j$ is below $K$,
2. $I_i$ is above $K$,
3. $I'_i$ is below $J'_n$ and
4. $\sigma_{i+1} < \tau_{n+1}$.

**Type3 (merge)**: The case when $\tau_j = \sigma_j$. Then it must be the case:

1. $l > 1$,
2. $I'_{j-1}$ and $J'_{j-1}$ are rules $(\Sigma_{N-1})^{\tau_j}$ above $K$, and
3. $n < m \,\&\, J_{j+k} = I_{j+k} \,\&\, J'_{j+k} = I'_{j+k}$ for any $k$ with $j \leq j + k \leq n$.

That is to say, $\mathcal{C}$ and $\mathcal{D}$ share the part from $J_j = I_j$ to $J_n = I_n$ , the right chain $\mathcal{D}$ has to be longer $n < m$ than the left chain $\mathcal{C}$ and the merging rule $K$ is not a rule $(\Sigma_1)$.

If **Type2 (jump)** or **Type3 (merge)** occurs for chains $\mathcal{C}$ and $\mathcal{D}$, then we say that $\mathcal{D}$ *foreruns* $\mathcal{C}$, since the resolving of the chain $\mathcal{D}$ precedes the resolving of the chain $\mathcal{C}$.

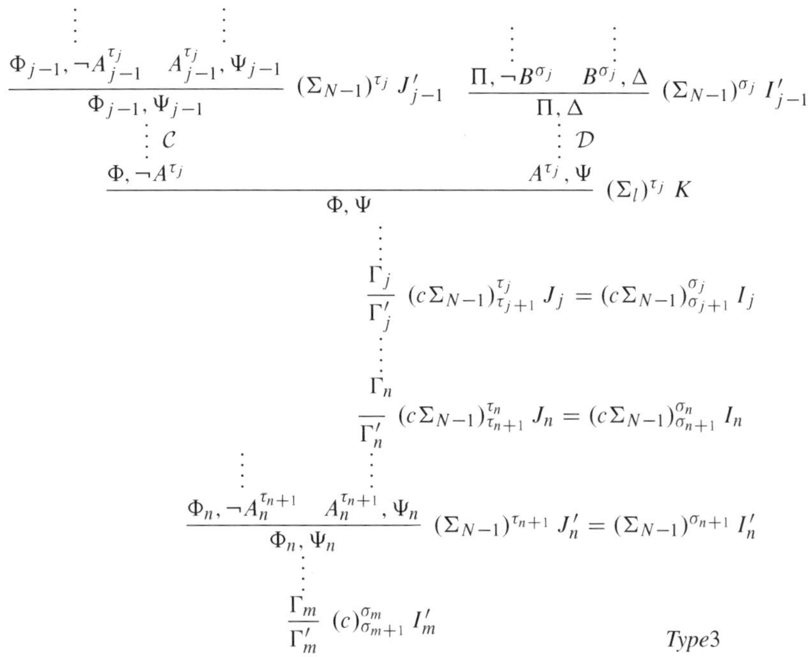

*Type3*

**(ch:Qpt)** Let $\mathcal{C} = J_0, \ldots, J_n$ be a chain with a $(c)^{\sigma_p}_{\sigma_{p+1}}\ J_p\ (p \leq n)$ and put $\rho = \sigma_{n+1}$. Then by **(ch:link)** there exists a uniquely determined increasing sequence of numbers

$$0 \leq n_0 < n_1 < \cdots < n_l = n - 1\ (l \geq 0) \tag{11}$$

such that for each $m < l$ there exists an $i_m$-knot $(\Sigma_{i_m})^{\sigma_{n_m+1}}\ K_m\ (2 \leq i_m \leq N-2)$ for the chain $\mathcal{C}$. (The $i_m$-knot $K_m$ is the merging rule of the chain $\mathcal{C}$ and the chain starting with the rule $J_{n_m+1}$, cf. **Type3 (merge)**.) These numbers are called the *knotting numbers* of the chain $\mathcal{C}$.

Then $pd_i(\rho), In(\rho), rg_i(\rho)$ have to be determined as follows:

1. For $2 \leq i < N$,

$$pd_i(\rho) = \sigma_{n_{m(i)}+1}$$

with

$$m(i) = \max\{m : 0 \leq m \leq l \ \&\ \forall p \in [0, m)(i \leq i_p)\} \tag{13}$$

that is to say,

$$J_{n_{m(i)}} = pd_i(J_n).$$

2. For $2 \leq i < N - 1$

$$\begin{aligned} i \in In(\mathcal{C}) = In(\rho) &\Leftrightarrow \exists p \in [0, m(i))(i_p = i) \\ &\Leftrightarrow \exists p \in [0, l)(i_p = i \,\&\, \forall q < p(i_q > i)) \\ &\Leftrightarrow m(i) > m(i+1) = \min\{m < l : i_m = i\}. \end{aligned}$$

And by the definition $N - 1 \in In(\mathcal{C}) = In(\rho)$.

3. For $i \in In(\mathcal{C}) \,\&\, i \neq N - 1$,

   (a) The case when there exists a $q$ such that

   $$\exists p[n_{m(i)} \geq p \geq q > n_{m(i+1)} \,\&\, \rho \prec_i \sigma_{p+1} \,\&\, \sigma_q = rg_i(\sigma_{p+1})]. \qquad (14)$$

   Then

   $$rg_i(\rho) = \sigma_q$$

   where $q$ denotes the minimal $q$ satisfying (14).

   (b) Otherwise.

   $$rg_i(\rho) = pd_i(\rho) = \sigma_{n_{m(i)}+1}.$$

**(lbranch)** Any left branch of a $(\Sigma_i)$ is the rightmost one in the left upper part of the $(\Sigma_i)$.

**(forerun)** Let $J^{lw}$ be a rule $(\Sigma_j)^\sigma$. Let $\mathcal{R}_{J_0} = J_0, \ldots, J_{n-1}$ denote the rope starting with a $(c)$ $J_0$. Assume that $J_0$ is above the right uppersequent of $J^{lw}$ and the series $\mathcal{R}_{J_0}$ reaches to the rule $J^{lw}$. Then there is no merging rule $K$, *cf.* the figure below, such that

1. the chain $\mathcal{C}_0$ starting with $J_0$ passes through the right side of $K$, and
2. a right branch $\mathcal{T}$ of $J^{lw}$ passes through the left side of $K$.

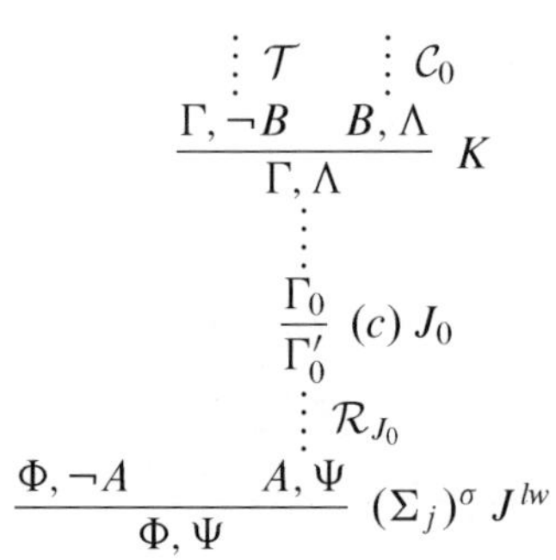

**(uplw)** Let $J^{lw}$ be a rule $(\Sigma_j)^{\sigma}$ and $J^{up}$ an $i$-knot $(\Sigma_i)^{\sigma_0}$ $(1 \leq i, j \leq N)$. Let $J_0$ denote the lower rule of $J^{up}$. Assume that the left rope ${}_{J^{up}}\mathcal{R} = J_0, \ldots, J_{n-1}$ of $J^{up}$ reaches to the rule $J^{lw}$. Then

**(uplwl)** if $J^{up}$ is above the left uppersequent of $J^{lw}$, then $j < i < N$.

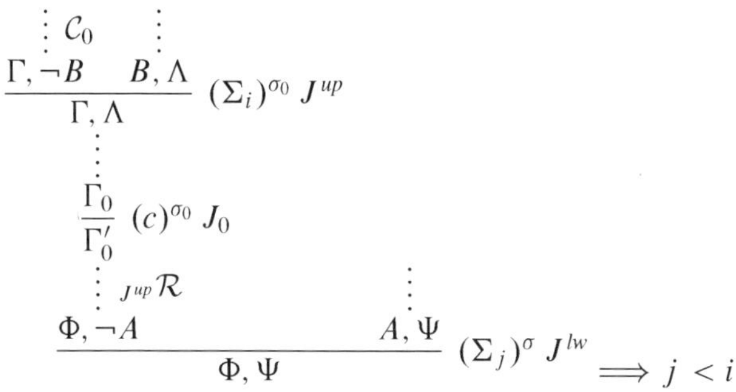

where $\mathcal{C}_0$ denotes the chain starting with $J_0$, and

**(uplwr)** if $J^{up}$ is above the right uppersequent of $J^{lw}$ and $i \leq j \leq N$, then the rule $(\Sigma_i)^{\sigma_0}$ $J^{up}$ foreruns the rule $(\Sigma_j)^{\sigma}$ $J^{lw}$, *cf.* Proposition 2.3 in Sect. 2.2. In other words if there exists a right branch $\mathcal{T}$ of $J^{lw}$ as shown in the following figure, then $j < i$.

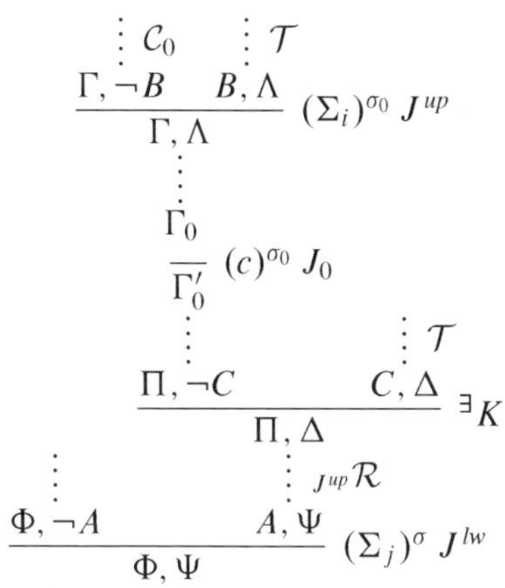

## 5.1 Decipherment

These provisos for a preproof to be a proof are obtained by inspection to rewrite proof figures. We decipher only additional provisos from [4].

**(ch:link)** Now a new type of linking chains, **Type3 (merge)** enters, cf. Sect. 2.1.

For a chain $\mathcal{D} = I_0, I'_0, \ldots, I_m, I'_m$ and a member $I_n$ $(n < m)$ of $\mathcal{D}$ let $\mathcal{C} = J_0, J'_0, \ldots, J_n, J'_n$ denote the chain starting with $J_n = I_n$. Then there are two possibilities:

**Type1 (segment)** $\mathcal{C}$ is a part $I_0, \ldots, I_n$ of $\mathcal{D}$ and hence the tops $I_0$ and $J_0$ are identical.

**Type3 (merge)** The branch of $\mathcal{C}$ is left to the branch of $\mathcal{D}$.

**(st:bound), (ch:Qpt)** By these provisos we see that an o.d. $\rho$ is in $Od(\Pi_N)$ for a newly introduced rule $(c)_\rho$, cf. Propositions 2.2, 2.3 in Sect. 2.2, Lemma 5.8 below and the case **M5.2** in Sect. 6.

**(uplwl)** By the proviso we see that a preproof $P'$ which is resulted from a proof $P$ is again a proof with respect to the proviso **(ch:Qpt)**, cf. Lemma 5.7.2.

**(uplwr), (forerun), (lbranch)** By these provisos we see that a preproof $P'$ which is resulted from a proof $P$ by resolving a rule $(\Sigma_{i+1})$ is again a proof with respect to the provisos **(forerun)** and **(uplw)**, cf. Proposition 2.3 in Sect. 2.2, the case **M7.2** in Sect. 6, Lemmas 5.5 and 5.4.

In the following any sequent and any rule are in a fixed proof.

As in the previous paper [4] we have the following lemmata. Lemma 5.1 follows from the provisos **(h-reg)** and **(ch:link)** in Definition 5.8, Lemma 5.2 from **(h-reg)** and **(c:bound1)** and Lemma 5.3 from **(h-reg)**.

**Lemma 5.1** *Let $J$ be a rule $(c)_\sigma$ and $J'$ the trace $(\Sigma_{N-1})^\sigma$ of $J$. Let $J_1$ be a rule $(c)^\sigma$ below $J'$. If there exists a chain $\mathcal{C}$ to which both $J$ and $J_1$ belong, then $J_1$ is the uppermost rule $(c)^\sigma$ below $J$ and there is no rule $(c)$ between $J'$ and $J_1$.*

**Lemma 5.2** *Let $J_{top}$ be a rule $(c)^\pi$. Let $\Phi$ denote the bar of $J_{top}$. Assume that the branch $\mathcal{T}$ from $J_{top}$ to $\Phi$ is the rightmost one in the upper part of $\Phi$. Then no chain passes through $\Phi$.*

**Lemma 5.3** *Let $J$ be a rule $(c)$ and $b : \Phi$ the bar of the rule $J$. Then there is no (cut) $I$ with $b \subset I \subset J$ nor a right uppersequent of a $(\Sigma_N)$ $I$ with $b \subset I * (1) \subseteq J$ between $J$ and $b : \Phi$.*

The following lemma is used to show that a preproof $P'$ which results from a proof $P$ by resolving a rule $(\Sigma_j)$ $J^{lw}$ is again a proof with respect to the proviso **(uplwl)**, cf. the Claim 6.6 in the case **M7** in the next subsection.

**Lemma 5.4** *Let $J^{lw}$ be a rule $(\Sigma_j)$. Assume that there exists a right branch $\mathcal{T}$ of $J^{lw}$ such that $\mathcal{T}$ is the rightmost one in the upper part of $J^{lw}$. Then there is no i-knot $(\Sigma_i)$ $J^{up}$ above the right uppersequent of $J^{lw}$ such that $i \leq j$ and the left rope ${}_{J^{up}}\mathcal{R}$ of $J^{up}$ and $J^{up}$ reaches to $J^{lw}$.*

*Proof* Suppose such a rule $J^{up}$ exists. By **(uplwr)** the rule $J^{up}$ foreruns $J^{lw}$. Thus the branch $\mathcal{T}$ would not be the rightmost one.

$$\dfrac{\Gamma, \neg A \qquad \dfrac{\dfrac{\Psi, \neg B \quad B, \Phi}{\Psi, \Phi}\,(\Sigma_i)\, J^{up}}{A, \Lambda}}{\Gamma, \Lambda}\,(\Sigma_j)\, J^{lw}$$

□

The following lemma is used to show that a preproof $P'$ which results from a proof $P$ by resolving a $(\Sigma_{i+1})$ is a proof with respect to the proviso **(uplwr)**, and to show a newly introduced rule $(\Sigma_i)$ in such a $P'$ does not split any chain, cf. the Claim 6.6 in the case **M7**.

**Lemma 5.5** *Let $J$ be a rule $(\Sigma_{i+1})^{\sigma_0}$ $(0 < i < N)$ and $b : \Phi$ the resolvent of $J$. Assume that the branch $\mathcal{T}$ from $J$ to $b$ is the rightmost one in the upper part of $b$. Then every chain passing through $b$ passes through the right side of $J$.*

*Proof* Let $a * (0)$ denote the lowermost rule $(\Sigma_{i+1})^{\sigma_0}$ below or equal to $J$, and $a : \Psi$ the lowersequent of $a * (0)$. The sequent $a : \Psi$ is the uppermost sequent below $J$ such that $h(a; P) < \sigma_0 + i$ by **(h-reg)**.

**Case 2.** $b = a$: If a chain passes through $a$ and a left side of a $(\Sigma_{i+1})^{\sigma_0}$ $K_{-1}$ with $a \subset K_{-1} \subseteq J$, then the chain would produce an $(i + 1)$-knot $K_{-1}$.

**Case 1.** Otherwise: Then there exists an $(i + 1)$-knot $(\Sigma_{i+1})^{\sigma_0}$ with $a \subset K_{-1} \subseteq J$. Let $(\Sigma_{i+1})^{\sigma}$ $K_{-1}$ denote the uppermost such knot and ${}_{K_{-1}}\mathcal{R} = J_0, \ldots, J_{n-1}$ the left rope of $K_{-1}$. Each $J_p$ is a rule $(c)^{\sigma_p}_{\sigma_{p+1}}$. Let

$$0 \leq n_0 < n_1 < \cdots < n_l = n - 1 \; (l \geq 0) \tag{11}$$

be the knotting numbers of the left rope ${}_{K_{-1}}\mathcal{R}$ and $K_m$ an $i_m$-knot $(\Sigma_{i_m})^{\sigma_{n_m+1}}$ of $J_{n_m}$ and $J_{n_m+1}$ for $m < l$. Put

$$m(i + 1) = \max\{m : 0 \leq m \leq l \;\&\; \forall p \in [0, m)(i + 1 \leq i_p)\}. \tag{12}$$

Then the resolvent $b : \Phi$ is the uppermost sequent $b : \Phi$ below $J_{n_{m(i+1)}}$ such that

$$h(b; P) < \sigma_{n_{m(i+1)}+1} + i.$$

Put

$$m = m(+1), \sigma = \sigma_{n_m+1}.$$

Assume that there is a chain $\mathcal{C}$ passing through $b$. As in **Case 2** it suffices to show that the chain $\mathcal{C}$ passes through the right side of $K_{-1}$. Assume that this is not the case. Let $(c)^{\rho}_{\rho'}$ $K$ denote the lowermost member of $\mathcal{C}$ which is above $b$.

**Claim 5.1** *$K$ is on the branch $\mathcal{T}$.*

*Proof of Claim 5.1* Assume that this is not the case. Then we see that there exists a merging rule $(\Sigma_j)^{\rho'}\ I$ and a member $(c)^{\rho'}\ K'$ of $\mathcal{C}$ such that the chain $\mathcal{C}$ passes through the left side of $I$. $K' \subset b \subset I$ and hence $h(K'; P) = \rho' \leq \sigma$. We see $\rho' = \sigma$ from (h-reg).

Suppose $m = l$. Then by the definition of the left rope ${}_{K_{-1}}\mathcal{R}$, the rule $(\Sigma_j)^{\rho'}\ I$ is not a knot, i.e., $j = 1$. But then $h(I * (1); P) = h(K'; P) = \sigma$, and hence $I \subset b$. A contradiction. Therefore $m < l$ and $i_m \leq i$. This means $K_m * (1) \subseteq b \subset I$. On the other hand, we have $1 \leq i < j$ by $b \subset I$, and by Lemma 5.1 $K'$ is the uppermost rule $(c)^\sigma$ below $(\Sigma_j)^\sigma\ I$. Therefore $(\Sigma_j)^{\rho'}\ I$ would be a knot below $J_{n_m}$. On the other side, $K_m$ is the uppermost knot below $J_{n_m}$. This is a contradiction.

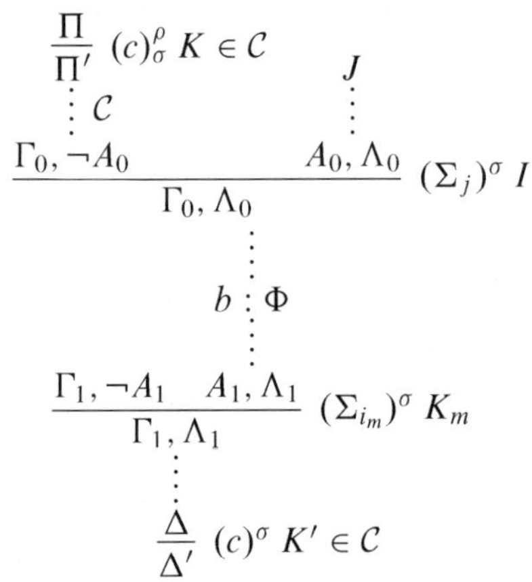

□

Then as in the proof of Lemma 7.13 of [4] we see that $K = J_{n_m}$, i.e., $(c)^{\rho}_{\rho'}\ K$ and $(c)^{\sigma_{nm}}_{\sigma}\ J_{n_m}$ coincide. Consider the chain $\mathcal{C}_m$ starting with $(c)^{\sigma_{nm}}_{\sigma}\ J_{n_m}$. Then by **(ch:link)** either $\mathcal{C}_m$ is a segment of $\mathcal{C}$ by **Type1(segment)**, or $\mathcal{C}$ foreruns $\mathcal{C}_m$ by **Type3(merge)**. Since $(c)^{\sigma_{nm}}_{\sigma}\ J_{n_m}$ is the lowest one such that $\mathcal{C}_m$ passes through the left side of $K_{m-1}$ and $J_{n_{m-1}+1}$ is a member of $\mathcal{C}_m$, **Type1(segment)** does not occur. In **Type3(merge)** $K_{m-1}$ has to be the merging rule of $\mathcal{C}_m$ and $\mathcal{C}$ since, again, $(c)^{\sigma_{nm}}_{\sigma}\ J_{n_m}$ is the lowest one, and the branch $\mathcal{T}$ is the rightmost one. Therefore $\mathcal{C}$ passes through the right side of $K_{m-1}$. If $m = 0$, then we are done. Otherwise we see the chain $\mathcal{C}$ and the chain $\mathcal{C}_{m-1}$ starting with $(c)^{\sigma_{n_{m-1}}}_{\sigma}\ J_{n_{m-1}}$ has to share the rule $(c)^{\sigma_{n_{m-1}}}_{\sigma}\ J_{n_{m-1}}$. As above we see that $\mathcal{C}$ passes through the right side of $K_{m-2}$, and so forth. □

**Lemma 5.6** *Let $\mathcal{C} = J_0, \ldots, J_n$ be a chain with rules $(c)^{\sigma_p}_{\sigma_{p+1}}\ J_p$ for $p \leq n$, and $(\Sigma_j)^{\sigma_p}\ K\ (p < n)$ a rule such that $\mathcal{C}$ passes through the right side of $K$ and the chain $\mathcal{C}_p$ stating with $J_p$ passes through the left side of $K$. Further let $\mathcal{R} = {}_K\mathcal{R} = J_p, \ldots, J_{q-1}\ (q \leq n)$ denote the left rope of the $j$-knot $K$. Then the chain $\mathcal{C}_q$ starting with $J_q$ is a part of the chain $\mathcal{C} = \mathcal{C}_n$, $\mathcal{C}_q \subseteq \mathcal{C}$. Therefore any knot for the chain $\mathcal{C}$ is* *below $J_q$ and $q < n$, and in particular, if $K$ is a knot for the chain $\mathcal{C}$, then $b = n$, cf. Definition 5.4.3.*

*Proof* Suppose $q < n$. By Definition 5.4.6 there is no knot of $J_{q-1}$ and $J_q$. Let $I_q$ denote a knot such that the chain $\mathcal{C}_{q-1}$ starting with $J_{q-1}$ passes through the left side of $I_q$. $c \subseteq b$. From the definition of a left rope we see that the chain $\mathcal{C}_q$ starting with $J_q$ does not pass through the left side of the knot $I_q$. Therefore by **(ch:link) Type1 (segment)** the chain $\mathcal{C}_q$ must be a part of the chain $\mathcal{C}$, $\mathcal{C}_q \subset \mathcal{C}$, i.e., the top of the chain $\mathcal{C}_q$ is the top $J_0$ of $\mathcal{C}$.

$$
\begin{array}{c}
\dfrac{\begin{array}{cc} \vdots\ \mathcal{C}_p & \vdots\ \mathcal{C}_q \subset \mathcal{C} \\ \Phi_p, \neg A_p & A_p, \Psi_p \end{array}}{\Phi_p, \Psi_p}\ (\Sigma_j)^{\sigma_p}\ K \\
\vdots \\
\dfrac{\Gamma_p}{\Gamma'_p}\ (c)^{\sigma_p}\ J_p \\
\begin{array}{cc} \vdots\ \mathcal{C}_{q-1} & \vdots\ \mathcal{R} \end{array} \\
\dfrac{\Phi_{q-1}, \neg A_{q-1} \quad A_{q-1}, \Psi_{q-1}}{\Phi_{q-1}, \Psi_{q-1}}\ I_q \\
\vdots\ \mathcal{R} \\
\dfrac{\Gamma_{q-1}}{\Gamma'_{q-1}}\ (c)_{\sigma_q}\ J_{q-1} \\
\vdots\ \mathcal{C}_q \\
\dfrac{\Gamma_q}{\Gamma'_q}\ (c)^{\sigma_q}\ J_q \\
\vdots\ \mathcal{C} \\
\dfrac{\Gamma_n}{\Gamma'_n}\ J_n
\end{array}
$$

□

The following Lemma 5.7 is a preparation for Lemma 5.8. From Lemma 5.8 we see that an o.d. $\rho$ is in $Od(\Pi_N)$ for a newly introduced rule $(c)_\rho$, cf. the case **M5.2** in the next subsection.

In the following Lemma 5.7, $J$ denotes a rule $(c)_\rho$ and $\mathcal{C} = J_0, \ldots, J_n$ the chain starting with $J_n = J$. Each $J_p$ is a rule $(c)^{\sigma_p}_{\sigma_{p+1}}$ for $p \le n$ with $\sigma_{n+1} = \rho$.

$K$ denotes a rule $(\Sigma_j)^{\sigma_a}$ ($j \le N-2, 0 < a \le n$) such that the chain $\mathcal{C}$ passes through $K$. If $\mathcal{C}$ passes through the left side of $K$, then $j \le N-2$ holds by **(ch:left)**.

$J_{a-1}$ denotes the lowermost member $(c)_{\sigma_a}$ of $\mathcal{C}$ above $K$, $K \subset J_{a-1}$.

Let

$$0 \le n_0 < n_1 < \cdots < n_l = n - 1 \; (l \ge 0) \tag{11}$$

be the knotting numbers of the chain $\mathcal{C}$, cf. **(ch:Qpt)**, and $K_m$ an $i_m$-knot $(\Sigma_{i_m})^{\sigma_{n_m+1}}$ of $J_{n_m}$ and $J_{n_m+1}$ for $m < l$. Let $m(i)$ denote the number

$$m(i) = \max\{m : 0 \le m \le l \,\&\, \forall p \in [0, m)(i \le i_p)\}. \tag{13}$$

**Lemma 5.7 (cf. Proposition 2.2 in Sect. 2.2.)**

1. *Let $m \le m(i)$. Then*

$$i \le i_{m-1} \,\&\, \sigma_{n_m+1} \prec_i \sigma_{n_{m-1}+1}.$$

2. *Assume that $\mathcal{C}$ passes through the left side of the rule $K$, i.e., $K * (0) \subset J_{a-1}$. Then $J_{a-1}$ is the upper left rule of $K$. Let $i \le j$.*
   *(a)* $\rho \prec_i \sigma_a$,
   *and hence*
   *(b) the $i$-predecessor of $J$ is equal to or below $J_{a-1}$, and*
   *(c) if $K_p$ is an $i_p$-knot $(\Sigma_{i_p})$ for the chain $\mathcal{C}$ above $K$, then $j < i_p$.*

$$\cfrac{\begin{matrix}\vdots & \vdots\, \mathcal{C}\\ \Phi_p, \neg A_p & A_p, \Psi_p\end{matrix}}{\Phi_p, \Psi_p}\,(\Sigma_{i_p})^{\sigma_{n_p+1}}\, K_p$$

$$\vdots$$

$$\cfrac{\Gamma_{a-1}}{\Gamma'_{a-1}}\,(c)_{\sigma_a}\, J_{a-1}$$

$$\cfrac{\begin{matrix}\vdots\, \mathcal{C} & & \vdots\\ \Phi, \neg A & & A, \Psi\end{matrix}}{\Phi, \Psi}\,(\Sigma_j)^{\sigma_a}\, K$$

$$\vdots$$

$$\cfrac{\Gamma_n}{\Gamma'_n}\,(c)^{\sigma_n}_{\rho}\, J_n = J \qquad \Longrightarrow \rho \prec_j \sigma_a \,\&\, j < i_p$$

3. *Let $J_{b-1}$ be a member of $\mathcal{C}$ such that $\rho \prec_i \sigma_b$ for an $i$ with $2 \le i \le N - 2$. Let $\mathcal{C}_{b-1}$ denote the chain starting with $J_{b-1}$. Assume that the chain $\mathcal{C}_{b-1}$ intersects $\mathcal{C}$ of* **Type3 (merge)** *in* **(ch:link)** *and $(\Sigma_j)$ $K$ is the merging rule of $\mathcal{C}_{b-1}$ and $\mathcal{C}$. Then $i \le j$.*

$$
\begin{array}{c}
\dfrac{\begin{array}{cc} \vdots\, \mathcal{C}_{b-1} & \vdots\, \mathcal{C} \\ \Phi, \neg A & A, \Psi \end{array}}{\Phi, \Psi}\ (\Sigma_j)\ K \\
\vdots \\
\dfrac{\Gamma_{b-1}}{\Gamma'_{b-1}}\ (c)_{\sigma_b}\ J_{b-1} \\
\vdots\, \mathcal{C} \\
\dfrac{\Gamma_n}{\Gamma'_n}\ (c)_{\sigma_{n+1}}\ J_n \qquad \& \, \sigma_{n+1} \prec_i \sigma_b \Longrightarrow i \leq j
\end{array}
$$

4. *Assume that $\mathcal{C}$ passes through the left side of the rule $K$, i.e., $K * (0) \subset J_{a-1}$. Let $i \leq j$.*
*Assume that the $i$-origin $J_q$ of $\mathcal{C}$ is not below $K$, i.e., $\sigma_q = rg_i(\rho) \downarrow \Rightarrow q < a$. Then*

$$\forall b \in (a, n+1]\{\rho \preceq_i \sigma_b \prec_i \sigma_a \rightarrow i \notin In(\sigma_b)\}$$

*and hence*

$$\forall b \in (a, n+1]\{\rho \preceq_i \sigma_b \prec_i \sigma_a \rightarrow in_i(J) = in_i(J_{b-1}) = in_i(J_{a-1}) \text{, i.e.,} \\ in_i(\rho) = in_i(\sigma_b) = in_i(\sigma_a)\}.$$

*In particular by Lemma 5.7.2 we have*

$$\rho \prec_i \sigma_a \,\&\, in_i(J) = in_i(J_{a-1}) \text{, i.e., } in_i(\rho) = in_i(\sigma_a).$$

5. *Assume that $\mathcal{C}$ passes through the left side of the rule $K$. Let $J_{b-1}$ be a member of $\mathcal{C}$ such that $J_{b-1}$ is below $K$, i.e., $a < b$, and assume that $\sigma_{n+1} \preceq_i \sigma := \sigma_b$ for an $i \leq j$. If $\sigma_q = rg_i(\sigma) \downarrow \Rightarrow q < a$, then*

$$\forall d \in (a, b]\{\sigma \preceq_i \sigma_d \prec_i \sigma_a \rightarrow i \notin In(\sigma_d)\}$$

*and*

$$\sigma \prec_i \sigma_a.$$

*Hence*

$$\forall d \in (a, b]\{\sigma \preceq_i \sigma_d \prec_i \sigma_a \rightarrow in_i(\sigma_d) = in_i(\sigma_a)\} \,\&\, in_i(\sigma) = in_i(\sigma_a).$$

*The following figure depicts the case* $\sigma_q = rg_i(\sigma)\downarrow$*:*

$$
\begin{array}{ccc}
\vdots\ \mathcal{C} & & \\
\dfrac{\Gamma_q}{\Gamma'_q}\ (c)^{rg_i(\sigma)}\ J_q & & \\
\vdots\ \mathcal{C} & & \vdots \\
\Phi, \neg A & & A, \Psi \\
\hline
& \Phi, \Psi & (\Sigma_j)^{\sigma_a}\ K \\
& \vdots & \\
& \dfrac{\Gamma_{b-1}}{\Gamma'_{b-1}}\ (c)_\sigma\ J_{b-1} & \\
& \vdots\ \mathcal{C} & \\
& \dfrac{\Gamma_n}{\Gamma'_n}\ (c)_{\sigma_{n+1}}\ J_n &
\end{array}
$$

6. *Assume that the chain* $\mathcal{C}$ *passes through the left side of the rule* $K$. *For an* $i \leq j$ *assume that there exists a* $q$ *such that*

$$\exists p[n \geq p \geq q \geq a \,\&\, \rho \preceq_i \sigma_{p+1} \,\&\, \sigma_q = rg_i(\sigma_{p+1})].$$

*Pick the minimal such* $q_0$ *and put* $\kappa = \sigma_{q_0}$. *Then*

(a) $\forall d \in (a, q_0]\{\sigma_{q_0} \preceq_i \sigma_d \prec_i \sigma_a \rightarrow i \notin In(\sigma_d)\}$ *and* $in_i(J_{a-1}) = in_i(J_{q_0-1})$, *i.e.,* $in_i(\sigma_a) = in_i(\kappa)$ *and* $\kappa \preceq_i \sigma_a$.

(b) $\forall t[\rho \preceq_i \sigma_t \prec_i \kappa \Rightarrow rg_i(\sigma_t) \preceq_i \kappa]$.

7. *Assume that* $\mathcal{C}$ *passes through the left side of the rule* $K$. *Let* $J_{b-1}$ *be a member of* $\mathcal{C}$ *such that* $J_{b-1}$ *is below* $K$, *i.e.,* $a < b$ *and* $\rho \preceq_i \sigma := \sigma_b$ *for an* $i \leq j$. *Suppose* $rg_i(\sigma)\downarrow$ *and put* $\sigma_q = rg_i(\sigma)$. *If the member* $(c)^{\sigma_q}\ J_q$ *is below* $K$, *i.e.,* $a \leq q$, *then for* $st_i(\sigma) = d_{\sigma_q^+}\alpha$, *cf.* **(st:bound)**,

$$B_{\sigma_q}(c; P) \leq \alpha$$

*for the uppersequent* $c : \Gamma_q$ *of the rule* $J_q$.

*Proof* First we show Lemmata 5.7.1 and 5.7.2 simultaneously by induction on the number of sequents between $K$ and $J$.

*Proof of Lemma 5.7.1* By the definition of the number $m(i)$ we have $i \leq i_{m-1}$. Since the chain $\mathcal{C}_{n_m}$ starting with $J_{n_m}$ passes through the left side of the $i_{m-1}$-knot $K_{m-1}$, we have the assertion $\sigma_{n_m+1} \prec_i \sigma_{n_{m-1}+1}$ by IH on Lemma 5.7.2.

$$\begin{array}{c} \vdots\ \mathcal{C}_{n_m} \\ \dfrac{\Phi_{m-1}, \neg A_{m-1} \quad A_{m-1}, \Psi_{m-1}}{\Phi_{m-1}, \Psi_{m-1}}\ (\Sigma_{i_{m-1}})^{\sigma_{n_{m-1}+1}}\ K_{m-1} \\ \vdots \\ \dfrac{\Gamma_{n_m}}{\Gamma'_{n_m}}\ (c)_{\sigma_{n_m+1}}\ J_{n_m} \end{array}$$

This shows Lemma 5.7.1. □

*Proof of Lemma 5.7.2* By **(ch:Qpt)** we have

$$pd_i(\rho) = \sigma_{n_{m(i)}+1} \text{ and } J_{n_{m(i)}} = pd_i(J).$$

**Claim 5.2** *$a \leq n_{m(i)} + 1$, i.e., $J_{n_{m(i)}+1} \subset K$.*

*Proof of Claim 5.2* If $m(i) = l$, then $a \leq n = n_l + 1$. Assume

$$m(i) < l \neq 0 \,\&\, a > n_{m(i)} + 1.$$

Then the $i_{m(i)}$-knot $(\Sigma_{i_{m(i)}})^{\sigma_{n_{m(i)}+1}} K_{m(i)}$ is above the left uppersequent of $K$, $K * (0) \subset K_{m(i)}$, and $j \geq i > i_{m(i)}$. Consider the left rope ${}_{K_{m(i)}}\mathcal{R} = J_{n_{m(i)}+1}, \ldots, J_{b-1}$ of the knot $K_{m(i)}$ for the chain $\mathcal{C}$. Then by Lemma 5.6 we have $b = n$. Therefore ${}_{K_{m(i)}}\mathcal{R}$ reaches to the rule $K$. Thus by **(uplwl)** we have $i \leq j < i_{m(i)}$. This is a contradiction. □

By Claim 5.2 we have Lemma 5.7.2b.

**Case 1** $a = n_{m(i)} + 1$: This means that the $i$-predecessor $J_{n_{m(i)}}$ of $J$ is the rule $J_{a-1}$, and $pd_i(\rho) = \sigma_a$.

**Case 2** $a < n_{m(i)} + 1$: This means that $J_{n_{m(i)}} \subset K$. Put

$$m_1 = \min\{m \leq m(i) : a < n_m + 1\}. \tag{15}$$

Then $J_{n_{m_1}}$ is the uppermost rule $J_{n_m}$ below $K$. The chain $\mathcal{C}_{n_{m_1}}$ starting with $J_{n_{m_1}}$ passes through the left side of the knot $(\Sigma_{i_{m_1-1}})^{\sigma_{n_{m_1-1}+1}} K_{m_1-1}$. If $K \subset K_{m_1-1}$, then $\mathcal{C}_{n_{m_1}}$ passes through the left side of $K$.

$$\begin{array}{c} \vdots\ \mathcal{C}_{n_{m_1}} \qquad \vdots \\ \dfrac{\Phi, \neg A \qquad A, \Psi}{\Phi, \Psi}\ (\Sigma_j)^{\sigma_a}\ K \\ \vdots \\ \dfrac{\Gamma_{n_{m_1}}}{\Gamma'_{n_{m_1}}}\ (c)_{\sigma_{n_{m_1}+1}}\ J_{n_{m_1}} \end{array}$$

And by the minimality of $m_1$, if $K_{m_1-1} \subset K$, then $J_{a-1} = J_{n_{m_1-1}}$, i.e., $a = n_{m_1-1} + 1$.

$$
\dfrac{\dfrac{\dfrac{\dfrac{\vdots}{\Gamma_{a-1}}\;J_{a-1}=J_{n_{m_1-1}}}{\Gamma'_{a-1}}}{\vdots\;\mathcal{C}}\!\!\!\!\quad \dfrac{\vdots}{\Phi, \neg A} \qquad\qquad \dfrac{\vdots}{A, \Psi}}{\Phi, \Psi}\;(\Sigma_j)^{\sigma_a}\;K
$$

$$
\dfrac{\begin{matrix}\vdots\;\mathcal{C}_{n_{m_1}} \\ \Phi_{m_1-1}, \neg A_{m_1-1}\end{matrix} \quad \begin{matrix}\vdots\;\mathcal{C} \\ A_{m_1-1}, \Psi_{m_1-1}\end{matrix}}{\Phi_{m_1-1}, \Psi_{m_1-1}}\;(\Sigma_{i_{m_1-1}})^{\sigma_a}\;K_{m_1-1}
$$

$$
\begin{matrix}\vdots\;\mathcal{C}_{n_{m_1}} \\ \dfrac{\Gamma_{n_{m_1}}}{\Gamma'_{n_{m_1}}}\;(c)_{\sigma_{n_{m_1}+1}}\;J_{n_{m_1}}\end{matrix}
$$

By Lemma 5.7.1 we have $pd_i(\rho) = \sigma_{n_{m(i)}+1} \preceq_i \sigma_{n_{m_1}+1}$. Once again by IH we have $\sigma_{n_{m_1}+1} \preceq_i \sigma_a$. Thus we have shown Lemma 5.7.2a, $\rho \prec_i \sigma_a$. □

*Proof of Lemma 5.7.2c* $j < i_p$: This is seen from **(uplwl)** as in the proof of the Claim 5.2 since in this case we have $l \neq 0$.

A proof of Lemma 5.7.2 is completed. □

*Proof of Lemma 5.7.3* The chain $\mathcal{C}_{b-1}$ passes through the left side of $K$ and $\mathcal{C}$ the right side of $K$. By **(ch:Qpt)** we have

$$pd_i(\sigma_{n+1}) = \sigma_{n_{m(i)}+1} \text{ and } J_{n_{m(i)}} = pd_i(J_n).$$

If $K$ is a $K_m$ for an $m < l$, then the assertion $i \leq j = i_m$ follows from (13) since $b - 1 \leq n_{m(i)}$ by $\sigma_{n+1} \prec_i \sigma_b$, and hence $m < m(i)$.

Otherwise let $m \leq m(i)$ denote the number such that $n_m > b - 1 > n_{m-1}$, i.e., $J_{b-1}$ is between $K_{m-1}$ and $J_{n_m}$. Then $K$ is below $K_{m-1}$ and the rule $K$ is the merging rule of $\mathcal{C}_{b-1}$ and the chain $\mathcal{C}_{n_m}$ starting with $J_{n_m}$, i.e., $\mathcal{C}_{n_m}$ passes through the right side of $K$.

$$
\dfrac{\dfrac{\begin{matrix}\vdots\ \mathcal{C}_{b-1}\\ \Phi, \neg A\end{matrix} \qquad \begin{matrix}\dfrac{\begin{matrix}\vdots\ \mathcal{C}_{n_m}\\ \Phi_{m-1}, \neg A_{m-1}\end{matrix} \quad \begin{matrix}\vdots\ \mathcal{C}\\ A_{m-1}, \Psi_{m-1}\end{matrix}}{\Phi_{m-1}, \Psi_{m-1}}\ K_{m-1}\\ \vdots\ \mathcal{C}_{n_m}, \mathcal{C}\\ A, \Psi\end{matrix}}{\Phi, \Psi}\ (\Sigma_j)\, K}{}
$$

$$
\begin{matrix}\vdots\\ \dfrac{\Gamma_{b-1}}{\Gamma'_{b-1}}\ (c)_{\sigma_b}\ J_{b-1}\\ \vdots\\ \dfrac{\Gamma_{n_m}}{\Gamma'_{n_m}}\ (c)_{\sigma_{n_m+1}}\ J_{n_m}\end{matrix}
$$

By IH it suffices to show that $\sigma_{n_m+1} \prec_i \sigma_b$ and this follows from

$$\sigma_{n+1} \prec_i \sigma_{n_m+1} \tag{16}$$

since the set $\{\tau : \sigma_{n+1} \prec_i \tau\}$ is linearly ordered by $\prec_i$, Proposition 4.1.1. Now (16) follows from (13) and Lemma 5.7.2a, i.e,

$$\sigma_{n+1} \prec_i pd_i(\sigma_{n+1}) = \sigma_{n_{m(i)}+1} \prec_i \cdots \prec_i \sigma_{n_{m-1}+1} \prec_i \sigma_{n_m+1}.$$

This shows Lemma 5.7.3. □

*Proof of Lemma 5.7.4* by induction on the number of sequents between $K$ and $J$.
By Claim 5.2 we have $a \leq n_{m(i)} + 1$.

**Case 1** $a = n_{m(i)} + 1$: This means that the $i$-predecessor $J_{n_{m(i)}}$ of $J$ is the rule $J_{a-1}$ and $pd_i(\rho) = \sigma_a$. By Lemma 5.7.2c we have $i_p > j \geq i$ for any $p < m(i)$. On the other side by **(ch:Qpt)**

$$i \in In(\mathcal{C}) = In(\rho) \Leftrightarrow \exists p \in [0, m(i))(i_p = i). \tag{17}$$

Hence $i \notin In(\rho)$. Thus $in_i(\sigma_a) = in_i(pd_i(\rho)) = in_i(\rho)$.

**Case 2** $a < n_{m(i)} + 1$: This means that $J_{n_{m(i)}}$ is below $K$. Let $m_1$ denote the number (15) defined in the proof of Lemma 5.7.2.

**Claim 5.3** *For each $m \in (m_1, m(i)]$ the $i$-origin of $J_{n_m}$ is not below $K_{m-1}$, $i < i_{m-1}$, $i \notin In(\rho)$, $\sigma_{n_m+1} \prec_i \sigma_{n_{m-1}+1}$ and $\forall b \in (n_{m-1}+1, n_m+1]\{\sigma_{n_m+1} \preceq_i \sigma_b \prec_i \sigma_{n_{m-1}+1} \rightarrow i \notin In(\sigma_b)\}$ and*

$$\forall b \in (n_{m-1}+1, n_m+1]\{\sigma_{n_m+1} \preceq_i \sigma_b \prec_i \sigma_{n_{m-1}+1} \rightarrow i \notin In(\sigma_b)\}$$

$$\forall b \in (n_{m-1}+1, n_m+1]\{\sigma_{n_m+1} \preceq_i \sigma_b \prec_i \sigma_{n_{m-1}+1} \rightarrow$$

$$in_i(J_{n_m}) = in_i(J_{b-1}) = in_i(J_{n_{m-1}}),\ \text{i.e.,}$$

$$in_i(\sigma_{n_m+1}) = in_i(\sigma_b) = in_i(\sigma_{n_{m-1}+1})\} \tag{18}$$

*Proof of Claim 5.3* First we show $i < i_{m-1}$. By Lemma 5.7.1 we have $i \leq i_{m-1}$. Assume $i = i_{m-1}$ for some $m \in (m_1, m(i)]$. Pick the minimal such $m_2$. Then by **(ch:Qpt)**, (17) we have $i \in In(\rho)$ and hence $rg_i(\rho) \downarrow$. By Lemma 5.7.2c we have

$$p < m_1 \Rightarrow i_p > j \geq i. \tag{19}$$

Here $p < m_1$ means that $K_p$ is above $K$. Thus by **(ch:Qpt)**

$$m(i+1) = \max\{m : 0 \leq m \leq l \,\&\, \forall p \in [0, m)(i+1 \leq i_p)\} = m_2 - 1 \geq m_1,$$

i.e.,

$$J_{n_{m_2-1}} = pd_{i+1}(J) \,\&\, \sigma_{n_{m_2-1}+1} = pd_{i+1}(\rho) \,\&\, n_{m_2-1} \geq n_{m_1} \geq a.$$

On the other hand by **(ch:Qpt)** we have for the $i$-origin $J_q$ of $\mathcal{C}$, i.e., $\sigma_q = rg_i(\rho)$,

$$n_{m_2-1} = n_{m(i+1)} < q \leq n_{m(i)} + 1.$$

Thus $J_q$ is below $J_{n_{m_2-1}}$ and hence by $a \leq n_{m_2-1}$ the $i$-origin $J_q$ is below $K$. This contradicts our hypothesis.

$$\begin{array}{c}
\dfrac{\Phi, \neg A \quad A, \Psi}{\Phi, \Psi}\; K \\
\vdots \\
\dfrac{\Gamma_{n_{m_2-1}}}{\Gamma'_{n_{m_2-1}}}\; (c)_{pd_{i+1}(\rho)}\; J_{n_{m_2-1}} \\
\vdots \\
\dfrac{\Phi_{m_2-1}, \neg A_{m_2-1} \quad A_{m_2-1}, \Psi_{m_2-1}}{\Phi_{m_2-1}, \Psi_{m_2-1}}\; (\Sigma_i)\; K_{m_2-1} \\
\vdots \\
\dfrac{\Gamma_q}{\Gamma'_q}\; (c)^{rg_i(\rho)}\; J_q \\
\vdots \\
\dfrac{\Gamma_{n_{m(i)}}}{\Gamma'_{n_{m(i)}}}\; (c)_{pd_i(\rho)}\; J_{n_{m(i)}}
\end{array}$$

Thus we have shown $i < i_{m-1}$ for any $m \in (m_1, m(i)]$. From this, (19) and (17) we see $i \notin In(\rho)$ and hence

$$in_i(\rho) = in_i(pd_i(\rho)) = in_i(\sigma_{n_{m(i)}+1}).$$

In particular, if $rg_i(\rho)\downarrow$, then $\sigma_q = rg_i(\rho) = rg_i(\sigma_{n_{m(i)}+1})$, i.e., the $i$-origin of $J_{n_{m(i)}}$ equals to the $i$-origin $J_q$ of $J$. Therefore the $i$-origin of $J_{n_{m(i)}}$ is not below $K$ and a fortiori not below the $i_{m(i)-1}$-knot $K_{m(i)-1}$. Also the chain starting with $J_{n_{m(i)}}$ passes through the left side of $K_{m(i)-1}$ and $i \leq i_{m(i)-1}$. Thus by IH we have (18) for $m = m(i)$. We see similarly that for each $m \in (m_1, m(i)]$ the $i$-origin of $J_{n_m}$ is not below $K_{m-1}$ and (18).

Thus we have shown Claim 5.3. □

From Claim 5.3 we see

$$\forall b \in (n_{m_1}+1, n+1]\{\rho \preceq_i \sigma_b \prec_i \sigma_{n_{m_1}+1} \to i \notin In(\sigma_b)\}$$

and hence

$$\forall b \in (n_{m_1}+1, n+1]\{\rho \preceq_i \sigma_b \prec_i \sigma_{n_{m_1}+1} \to$$
$$in_i(J) = in_i(J_{b-1}) = in_i(J_{n_{m_1}}), \text{ i.e., } in_i(\rho) = in_i(\sigma_b) = in_i(\sigma_{n_{m_1}+1})\}.$$

Further

$$\forall m \in (m_1, m(i)]\{\rho \prec_i \sigma_{n_m+1} \prec_i \sigma_{n_{m-1}+1} \preceq_i \sigma_{n_{m_1}+1}\}.$$

Once more by IH we have, cf. figures in the proof of Lemma 5.7.2, **Case 2.**, $\sigma_{n_{m_1}+1} \preceq_i \sigma_a$ and

$$\forall b \in (a, n_{m_1}+1]\{\sigma_{n_{m_1}+1} \preceq_i \sigma_b \prec_i \sigma_a \to i \notin In(\sigma_b)\}$$

and hence

$$\forall b \in (a, n_{m_1}+1]\{\sigma_{n_{m_1}+1} \preceq_i \sigma_b \prec_i \sigma_a \to$$
$$in_i(J_{n_{m_1}}) = in_i(J_{b-1}) = in_i(J_{a-1}), \text{ i.e., } in_i(\sigma_{n_{m_1}+1}) = in_i(\sigma_b) = in_i(\sigma_a)\}.$$

Thus we have shown Lemma 5.7.4. □

*Proof of Lemma 5.7.5* by induction on the number of sequents between $K$ and $J_{b-1}$.

Let $\mathcal{C}_b = I_0, \ldots, I_{b-1}$ denote the chain starting with $J_{b-1} = I_{b-1}$. Each rule $I_p$ is again a rule $(c)^{\sigma_p}_{\sigma_{p+1}}$. Chains $\mathcal{C}_b$ and $\mathcal{C}$ intersect in a way described as **Type1 (segment)** or **Type3 (merge)** in **(ch:link)**. If the chain $\mathcal{C}_b$ passes through the left side of $K$, then the $i$-origin $I_q$ of $\mathcal{C}_b$ is above $K$ if it exists, and hence the assertion follows from Lemma 5.7.4.

Otherwise there exists a merging rule $(\Sigma_l)^{\sigma_c}$ $I$ below $K$ such that the chain $\mathcal{C}_b$ passes through the left side of $I$ and $\mathcal{C}$ the right side of $I$.

$$\dfrac{\begin{array}{cc}\vdots\;\mathcal{C} & \vdots\\ \Phi,\neg A & A,\Psi\end{array}}{\Phi,\Psi}\;K$$

$$\vdots$$

$$\dfrac{\Gamma_{c-1}}{\Gamma'_{c-1}}\;(c)_{\sigma_c}\;J_{c-1}$$

$$\dfrac{\begin{array}{cc}\vdots\;\mathcal{C}_b & \vdots\;\mathcal{C}\\ \Pi,\neg B & B,\Delta\end{array}}{\Pi,\Delta}\;(\Sigma_l)^{\sigma_c}\;I$$

$$\vdots$$

$$\dfrac{\Gamma_{b-1}}{\Gamma'_{b-1}}\;(c)_{\sigma}\;J_{b-1}$$

Then by Lemma 5.7.3 we have $i \leq l$. The $i$-origin $I_q$ of $\mathcal{C}_b$ is not below $I$. Therefore by Lemma 5.7.4 we have

$$\forall d \in (c,b]\{\sigma \preceq_i \sigma_d \prec_i \sigma_c \to i \notin In(\sigma_d)\}$$

and

$$\sigma \prec_i \sigma_c.$$

Hence

$$\forall d \in (c,b]\{\sigma \preceq_i \sigma_d \prec_i \sigma_c \to in_i(\sigma_d) = in_i(\sigma_c)\}.$$

In particular

$$in_i(\sigma_c) = in_i(\sigma)\;\&\;\sigma \prec_i \sigma_c. \tag{20}$$

Now consider the member $J_{c-1}$ of $\mathcal{C}$. $J_{c-1}$ is again below $K$, $\sigma_{n+1} \preceq_i \sigma_c$ and $rg_i(\sigma_c) \simeq rg_i(\sigma)$ by (20). Thus by IH we have

$$\forall d \in (a,c]\{\sigma_c \preceq_i \sigma_d \prec_i \sigma_a \to i \notin In(\sigma_d)\}$$

and

$$\sigma_c \prec_i \sigma_a.$$

Therefore

$$\forall d \in (a,c]\{\sigma_c \preceq_i \sigma_d \prec_i \sigma_a \to in_i(\sigma_d) = in_i(\sigma_a)\}.$$

This shows Lemma 5.7.5. □

*Proof of Lemma 5.7.6* Pick a $p_0$ so that $n \geq p_0 \geq q_0$, $\rho \preceq_i \sigma_{p_0+1}$ and $rg_i(\sigma_{p_0+1}) = \sigma_{q_0} = \kappa$.

$$
\begin{array}{c}
\vdots \\
\dfrac{\Gamma_{a-1}}{\Gamma'_{a-1}}\ (c)_{\sigma_a}\ J_{a-1} \\
\vdots\ \mathcal{C} \\
\dfrac{\Phi, \neg A \qquad\qquad \overset{\vdots}{A, \Psi}}{\Phi, \Psi}\ (\Sigma_j)^{\sigma_a}\ K \\
\vdots \\
\dfrac{\Gamma_{q_0}}{\Gamma'_{q_0}}\ (c)^{\kappa}\ J_{q_0}\ (rg_i(\sigma_{p_0+1}) = \kappa) \\
\vdots \\
\dfrac{\Gamma_{p_0}}{\Gamma'_{p_0}}\ (c)_{\sigma_{p_0+1}}\ J_{p_0} \\
\vdots \\
\dfrac{\Gamma_{n}}{\Gamma'_{n}}\ (c)_{\rho}\ J_{n}
\end{array}
$$

Lemma 5.7.6a. First note that $\rho \preceq_i \sigma_{p_0+1} \prec_i rg_i(\sigma_{p_0+1}) = \kappa$ by Proposition 4.1.2 (or by the proviso **(ch:Qpt)**) and hence $\rho \prec_i \kappa$. Thus the assertion follows from Lemma 5.7.5 and the minimality of $q_0$.

Lemma 5.7.6b. Suppose $rg_i(\sigma_t) \not\preceq_i \kappa$ for a $t$ with $\rho \preceq_i \sigma_t \prec_i \kappa$. Put $\sigma_b = rg_i(\sigma_t)$. Then by Propositions 4.1.1 and 4.1.2 we have $\kappa = \sigma_{q_0} \prec_i \sigma_b$ and $b < q_0 < t \;\&\; q_0 \geq a$. Hence by the minimality of $q_0$ we have $b < a$.

$$
\begin{array}{c}
\vdots \\
\dfrac{\Gamma_{b}}{\Gamma'_{b}}\ (c)^{\sigma_b}\ J_{b}\ (\sigma_b = rg_i(\sigma_t)) \\
\vdots\ \mathcal{C} \\
\dfrac{\Phi, \neg A \qquad\qquad \overset{\vdots}{A, \Psi}}{\Phi, \Psi}\ (\Sigma_j)^{\sigma_a}\ K \\
\vdots \\
\dfrac{\Gamma_{q_0}}{\Gamma'_{q_0}}\ (c)^{\kappa}\ J_{q_0} \\
\vdots \\
\dfrac{\Gamma_{t-1}}{\Gamma'_{t-1}}\ (c)_{\sigma_t}\ J_{t-1}
\end{array}
$$

Thus by Lemma 5.7.5 we have

$$in_i(\sigma_a) = in_i(\sigma_t).$$

From this and Lemma 5.7.6a we have

$$in_i(\sigma_a) = in_i(\sigma_t) = in_i(\kappa) \,\&\, \sigma_t \prec_i \kappa \preceq_i \sigma_a. \tag{21}$$

**Case 1** $t \leq p_0$: Then $\sigma_{p_0+1} \prec_i \sigma_t \prec_i \kappa = rg_i(\sigma_{p_0+1})$ by Proposition 4.1.1. By Proposition 4.1.4 we would have $\sigma_b = rg_i(\sigma_t) \preceq_i \kappa$. Thus this is not the case. Alternatively we can handle this case without appealing Proposition 4.1.4 as follows. Let $p_0$ denote the minimal $p_0$ such that

$$n \geq p_0 \geq q_0 \,\&\, \rho \preceq_i \sigma_{p_0+1} \,\&\, rg_i(\sigma_{p_0+1}) = \sigma_{q_0} = \kappa.$$

Then by **(ch:Qpt)** we have $\kappa = rg_i(\sigma_{p_0+1}) = pd_i(\sigma_{p_0+1})$ and hence this is not the case, i.e., $p_0 < t$.

**Case 2** $p_0 < t$: Then $\sigma_t \preceq_i \sigma_{p_0+1} \prec_i \kappa$. By (21) and Proposition 4.1.3, or by Lemma 5.7.5 we would have $in_i(\sigma_{p_0+1}) = in_i(\kappa)$. In particular $\kappa = rg_i(\sigma_{p_0+1}) = rg_i(\kappa)$ but $rg_i(\kappa)$ is a proper subdiagram of $\kappa$. This is a contradiction.

This shows Lemma 5.7.6b. □

*Proof of Lemma 5.7.7* by induction on the number of sequents between $K$ and $J_{b-1}$.

**Case 1** $J_q$ is the $i$-origin of $J_{b-1}$, i.e., $J_q$ is a member of the chain starting with $(c)_\sigma\, J_{b-1}$: By the proviso **(st:bound)** we can assume $i \not\in In(\sigma)$. Then $in_i(\sigma) = in_i(pd_i(\sigma)) = in_i(J_p)$ with $J_p = pd_i(J_{b-1}) \,\&\, a \leq b-1 < p$ by Lemma 5.7.2. In particular $rg_i(\sigma) = rg_i(pd_i(\sigma))$. IH and $st_i(pd_i(\sigma)) = st_i(\sigma)$ yields the lemma.

**Case 2** Otherwise: First note that $\sigma_{n+1} \neq \sigma$ and $\sigma_{n+1} \prec_i \sigma$. By **(ch:Qpt)** we have

$$pd_i(\sigma_{n+1}) = \sigma_{n_{m(i)}+1} \text{ and } J_{n_{m(i)}} = pd_i(J_n).$$

Also $\sigma_{n_{m(i)}+1} \preceq_i \sigma$ and hence $\sigma_{n_{m(i)}+1} \leq \sigma$. Let $m_1$ denote the number such that

$$m_1 = \min\{m : \sigma_{n_m+1} \leq \sigma\} \leq m(i).$$

Then the rule $(c)_\sigma\, J_{b-1}$ is a member of the chain $\mathcal{C}_{n_{m_1}}$ starting with $J_{n_{m_1}}$ and $J_{b-1}$ is below $(\Sigma_{i_{m_1-1}})\, K_{m_1-1}$. Also the chain $\mathcal{C}_{n_{m_1}}$ passes through the left side of the knot $K_{m_1-1}$. By $m_1 \leq m(i)$ and Lemma 5.7.1 we have $\sigma_{n_{m(i)}+1} \preceq_i \sigma_{n_{m_1}}$ and hence

$$i \leq i_{m_1-1} \,\&\, \sigma_{n_{m_1}} \preceq_i \sigma. \tag{22}$$

**Case 2.1** $J_q$ is below $K_{m_1-1}$, i.e., $\sigma_q \leq \sigma_{n_{m_1-1}+1}$, i.e., $n_{m_1-1} < q$: By IH and (22) we get the assertion.

**Case 2.2** Otherwise: By Lemma 5.7.5 and (22) we have

$$in_i(\sigma) = in_i(\sigma_{n_{m_1}+1}) \,\&\, \sigma \prec_i \sigma_{n_{m_1}+1}.$$

Hence $st_i(\sigma) = st_i(\sigma_{n_{m_1}+1}) \,\&\, rg_i(\sigma) = rg_i(\sigma_{n_{m_1}+1})$. IH and $st_i(\sigma_{n_{m_1}+1}) = st_i(\sigma)$ yield the lemma. □

**Lemma 5.8** *Let $\mathcal{R} = J_0, \ldots, J_{n-1}$ denote the rope starting with a top $(c)^{\pi}$ $J_0$. Each $J_p$ is a rule $(c)^{\sigma_p}_{\sigma_{p+1}}$. Let*

$$0 \leq n_0 < n_1 < \cdots < n_l = n - 1 \; (l \geq 0) \tag{11}$$

*be the knotting numbers of the rope $\mathcal{R}$, and $K_m$ an $i_m$-knot $(\Sigma_{i_m})^{\sigma_{n_m+1}}$ of $J_{n_m}$ : and $J_{n_m+1}$ for $m < l$. For $2 \leq i < N$ let $m(i)$ denote the number*

$$m(i) = \max\{m : 0 \leq m \leq l \,\&\, \forall p \in [0, m)(i \leq i_p)\}. \tag{13}$$

*Note that $i_m \leq N - 2$ by* **(p10)**. *Also put* (cf. **(ch:Qpt)**)

*1.*

$$pd_i = \sigma_{n_{m(i)}+1}.$$

2.

$$i \in In \Leftrightarrow \exists p \in [0, m(i))(i_p = i).$$

*3. For $i \in In$ $(i \neq N - 1)$,*

**Case 1** *The case when there exists a $q$ such that*

$$\exists p[n_{m(i)} \geq p \geq q > n_{m(i+1)} \,\&\, pd_i \preceq_i \sigma_{p+1} \,\&\, \sigma_q = rg_i(\sigma_{p+1})]. \tag{23}$$

*Then*

$$rg_i = \sigma_q$$

*where $q$ denotes the minimal $q$ satisfying (23).*

**Case 2** *Otherwise.*

$$rg_i = pd_i = \sigma_{n_{m(i)}+1}.$$

*1. For each $i \in In$ we have*

*(a) $in_i(rg_i) = in_i(pd_{i+1})\,\&\,pd_i \preceq_i rg_i \preceq_i pd_{i+1}\,\&\,pd_i \neq pd_{i+1}$.*
*(b) $\forall t[rg_i(pd_i) \preceq_i \sigma_t \prec_i rg_i \Rightarrow rg_i(\sigma_t) \preceq_i rg_i]$.*
*(c) Either $rg_i = pd_i$ or $rg_i(pd_i) \preceq_i rg_i$.*

*2. Assume $i \in In\,\&\,\sigma_q := rg_i \neq pd_i$, i.e.,* **Case 1** *occurs. Then*

$$B_{\sigma_q}(c; P) \leq \alpha$$

*for the uppersequent $c : \Gamma_q$ of the rule $J_q$, $st_i(\sigma_{p+1}) = d_{\sigma_q^+}\alpha$ and $p$ denotes a number such that*

$$n_{m(i)} \geq p \geq q > n_{m(i+1)}\,\&\,pd_i = \sigma_{n_{m(i)}+1} \preceq_i \sigma_{p+1}\,\&\,\sigma_q = rg_i(\sigma_{p+1}).$$

*Proof* Lemma 5.8.1.

Let $i \in In$, and put $\sigma_{q0} = rg_i\,\&\,\sigma_{p0} = pd_i\,\&\,\sigma_r = pd_{i+1}$. By the definition we have $p_0 = n_{m(i)} + 1\,\&\,r = n_{m(i+1)} + 1$, $m(i) > m(i+1)\,\&\,i_{m(i+1)} = i$, $p_0 \leq q_0 \leq r$ and $\sigma_{p0} \preceq_i \sigma_{q0}$. Also

$$\forall p \in [m(i+1), m(i))(i \leq i_p).$$

From this and Lemma 5.7.1 we see

$$\forall p \in [m(i+1), m(i))(\sigma_{n_{p+1}+1} \prec_i \sigma_{n_p+1}). \tag{24}$$

On the other hand, we have by the definition of $rg_i = \sigma_{q0}$

$$\neg\exists q < q_0 \exists p[p_0 - 1 \leq p \leq q > r - 1\,\&\,pd_i \preceq_i \sigma_{p+1}\,\&\,\sigma_q = rg_i(\sigma_p)]. \tag{25}$$

**Case 2**. Then $pd_i = rg_i$, i.e., $p_0 = q_0$, and Lemma 5.8.1b vacuously holds. Lemma 5.8.1a, $in_i(\sigma_{q0}) = in_i(\sigma_r)\,\&\,\sigma_{q0} \preceq_i \sigma_r$, follows from (24) and Lemma 5.7.4 with (25).

**Case 1**. Let $m$ denote the number such that

$$m(i) \geq m > m(i+1)\,\&\,n_m \geq q_0 > n_{m-1} \tag{26}$$

i.e., the rule $(c)^{\sigma_{q0}}\,J_{q_0}$ is a member of the chain $C_{n_m}$ starting with $J_{n_m}$.

**Claim 5.4** *Let $p_1$ denote the minimal $p_1$ such that $\sigma_{p0} \preceq_i \sigma_{p_1+1}$ and $\sigma_{q0} = rg_i(\sigma_{p_1+1})$. Then $p_1 \leq n_m\,\&\,\sigma_{n_m+1} \preceq_i \sigma_{p_1+1}$.*

$$
\cfrac{\begin{array}{cc} \vdots\; \mathcal{C}_{n_m} & \vdots \\ \Phi_{m-1}, \neg A_{m-1} & A_{m-1}, \Psi_{m-1} \end{array}}{\Phi_{m-1}, \Psi_{m-1}}\; K_{m-1}
$$

$$
\begin{array}{c} \vdots \\ \dfrac{\Gamma_{q_0}}{\Gamma'_{q_0}}\; (c)^{rg_i(\sigma_{p_1+1})}\; J_{q_0} \\ \vdots \\ \dfrac{\Gamma_{p_1}}{\Gamma'_{p_1}}\; (c)_{\sigma_{p_1+1}}\; J_{p_1} \\ \vdots \\ \dfrac{\Gamma_{n_m}}{\Gamma'_{n_m}}\; (c)_{\sigma_{n_m+1}}\; J_{n_m} \end{array}
$$

*Proof of Claim 5.4* Let $m_1$ denote the number such that

$$
m(i) \geq m_1 > m(i+1)\,\&\, n_{m_1} \geq p_1 > n_{m_1-1}.
$$

Then by (24), $pd_i \preceq_i \sigma_{n_{m_1}+1}$ and $pd_i \preceq_i \sigma_{p_1+1}$ we have $\sigma_{n_{m_1}+1} \preceq_i \sigma_{p_1+1}$. It remains to show $m = m_1$. Assume $m < m_1$. Then by Lemma 5.7.5 and $q_0 < n_{m_1-1}+1$ we would have $in_i(\sigma_{n_{m_1-1}+1}) = in_i(\sigma_{p_1+1})$ and hence $rg_i(\sigma_{n_{m_1-1}+1}) = rg_i(\sigma_{p_1+1}) = \sigma_{q_0}$. This contradicts the minimality of $p_1$ by (24).

$$
\cfrac{\begin{array}{cc} \vdots\; \mathcal{C}_{n_m} & \vdots \\ \Phi_{m-1}, \neg A_{m-1} & A_{m-1}, \Psi_{m-1} \end{array}}{\Phi_{m-1}, \Psi_{m-1}}\; K_{m-1}
$$

$$
\begin{array}{c} \vdots \\ \dfrac{\Gamma_{q_0}}{\Gamma'_{q_0}}\; (c)^{rg_i(\sigma_{p_1+1})}\; J_{q_0} \\ \vdots \\ \dfrac{\Gamma_{n_m}}{\Gamma'_{n_m}}\; J_{n_m} \end{array}
$$

$$
\cfrac{\begin{array}{cc} \vdots\; \mathcal{C}_{n_{m_1}} & \vdots \\ \Phi_{m_1-1}, \neg A_{m_1-1} & A_{m_1-1}, \Psi_{m_1-1} \end{array}}{\Phi_{m_1-1}, \Psi_{m_1-1}}\; (\Sigma_{i_{m_1-1}})^{\sigma_{n_{m_1-1}}}\; K_{m_1-1}
$$

$$
\begin{array}{c} \vdots \\ \dfrac{\Gamma_{p_1}}{\Gamma'_{p_1}}\; (c)_{\sigma_{p_1+1}}\; J_{p_1} \\ \vdots \\ \dfrac{\Gamma_{n_{m_1}}}{\Gamma'_{n_{m_1}}}\; J_{n_{m_1}} \end{array}
$$

This shows Claim 5.4. □

By the minimality of $q_0$ and Claim 5.4 $q_0$ is the minimal $q$ such that

$$\exists p[n_m \geq p \geq q \geq n_{m-1} + 1 \,\&\, \sigma_{n_m+1} \preceq_i \sigma_{p+1} \,\&\, \sigma_q = rg_i(\sigma_{p+1})].$$

Hence by Lemma 5.7.6 we have

$$in_i(\sigma_{n_{m-1}+1}) = in_i(\sigma_{q_0}) \,\&\, \sigma_{q_0} \preceq_i \sigma_{n_{m-1}+1} \tag{27}$$

and

$$\forall t[\sigma_{n_m+1} \preceq_i \sigma_t \prec_i \sigma_{q_0} \Rightarrow rg_i(\sigma_t) \preceq_i \sigma_{q_0}]. \tag{28}$$

Lemma 5.8.1. By (27) it suffices to show that

$$in_i(\sigma_{n_{m-1}+1}) = in_i(pd_{i+1}) \,\&\, \sigma_{n_{m-1}+1} \preceq_i pd_{i+1}.$$

This follows from (24) and Lemma 5.7.4 with (25).
Lemma 5.8.1b and 5.8.1c. In view of (28) it suffices to show the

**Claim 5.5** $\sigma_{p_0} \preceq_i \sigma_t \prec_i \sigma_{n_m+1} \Rightarrow rg_i(\sigma_t) \preceq_i \sigma_{q_0}$.

*Proof of Claim 5.5.* By induction on $t$ with $p_0 > t > n_m + 1$.

Let $m_2 \geq m$ denote the number such that $n_{m_2+1} \geq t > n_{m_2}$. Then the chain $\mathcal{C}_{n_{m_2+1}}$ starting with $J_{n_{m_2+1}}$ passes through the left side of the rule $(\Sigma_{i_{m_2}})^{\sigma_{n_{m_2}+1}}\ K_{m_2}$.

$$\dfrac{\begin{array}{cc} \vdots\ \mathcal{C}_{n_{m_2+1}} & \vdots \\ \Phi_{m_2}, \neg A_{m_2} & A_{m_2}, \Psi_{m_2} \end{array}}{\Phi_{m_2}, \Psi_{m_2}}\ (\Sigma_{i_{m_2}})^{\sigma_{n_{m_2}+1}}\ K_{m_2}$$

$$\vdots$$

$$\dfrac{\Gamma_{t-1}}{\Gamma'_{t-1}}\ (c)_{\sigma_t}\ J_{t-1}$$

$$\vdots$$

$$\dfrac{\Gamma_{n_{m_2+1}}}{\Gamma'_{n_{m_2+1}}}\ (c)_{\sigma_{n_{m_2+1}+1}}\ J_{n_{m_2+1}}$$

We have

$$\sigma_{n_{m_2+1}+1} \preceq_i \sigma_t \prec_i \sigma_{n_{m_2}+1} \tag{29}$$

by (24). Put $\sigma_b = rg_i(\sigma_t)$. It suffices to show $b \geq q_0$.

First consider the case when $b \leq n_{m_2}$. Then by (29) and Lemma 5.7.5 we have $in_i(\sigma_t) = in_i(\sigma_{n_{m_2}+1})$ and $rg_i(\sigma_t) = rg_i(\sigma_{n_{m_2}+1})$. Thus IH when $m_2 > m$ and (28) when $m_2 = m$ yield $b \geq q_0$.

Next suppose $b > n_{m_2}$. Let $q_1 \leq b$ denote the minimal $q \leq b$ such that

$$\exists p[n_{m_2+1} \geq p \geq q \geq n_{m_2} + 1 \,\&\, \sigma_{n_{m_2+1}} \preceq_i \sigma_{p+1} \,\&\, \sigma_q = rg_i(\sigma_{p+1})].$$

The pair $(p, q) = (t - 1, b)$ enjoys this condition.

Then by Lemma 5.7.6 we have $\sigma_{q_1} \preceq_i \sigma_{n_{m_2}+1}$. Thus $\sigma_b \preceq_i \sigma_{q_1} \preceq_i \sigma_{n_{m_2}+1} \preceq_i \sigma_{n_m+1} \preceq_i \sigma_{q_0}$. This shows Claim 5.5. □

Lemma 5.8.2.

First observe that as in (22) in the proof of Lemma 5.7.7,

$$\forall m \leq m(i)[\sigma_{n_{m(i)}+1} \preceq_i \sigma_{n_m+1}] \tag{30}$$

Put

$$m_1 = \min\{m : p \leq n_m\}$$
$$m_2 = \min\{m : q \leq n_m\}.$$

Then the rule $J_p$ $[J_q]$ is a member of the chain $\mathcal{C}_{n_{m_1}}$ $[\mathcal{C}_{n_{m_2}}]$ starting with $J_{n_{m_1}}$ [starting with $J_{n_{m_2}}$], resp. and $m(i + 1) < m_2 \leq m_1 \leq m(i)$.

**Claim 5.6 (cf. Claim 5.4.)** *There exists a* $p_0$ *such that*

$$in_i(\sigma_{p+1}) = in_i(\sigma_{p_0+1}) \,\&\, \sigma_{n_{m_2}+1} \preceq_i \sigma_{p_0+1} \,\&\, n_{m_2} \geq p_0 > n_{m_2-1},$$

*i.e., the rule* $J_{p_0}$ *is a member of the chain* $\mathcal{C}_{n_{m_2}}$ *starting with* $J_{n_{m_2}}$.

*Proof of Claim 5.6* 1. The case $m_1 = m_2$: By (30) and $\sigma_{n_{m(i)}+1} \preceq_i \sigma_{p+1}$ we have

$$\sigma_{n_{m_1}+1} \preceq_i \sigma_{p+1}. \tag{31}$$

Set $p_0 = p$.

2. The case $m_2 < m_1$: By (31) and Lemma 5.7.5 we have

$$in_i(\sigma_{p+1}) = in_i(\sigma_{n_{m_1-1}+1}) = \cdots = in_i(\sigma_{n_{m_2}+1})$$

and

$$\sigma_{p+1} \prec_i \sigma_{n_{m_1-1}+1} \prec_i \cdots \prec_i \sigma_{n_{m_2}+1}.$$

Set $p_0 = n_{m_2}$.

This shows the Claim 5.6. □

By Claim 5.6 and Lemma 5.7.7 we conclude $rg_i(\sigma_{p+1}) = rg_i(\sigma_{p_0+1})$ and

$$B_{\sigma_q}(c; P) \leq b(st_i(\sigma_{p_0+1})) = b(st_i(\sigma_{p+1})) = \alpha. \qquad \square$$

**Main Lemma 5.1** *If $P$ is a proof, then the endsequent of $P$ is true.*

In the next section we prove the Main Lemma 5.1 by a transfinite induction on $o(P) \in Od(\Pi_N) \mid \Omega$.

Assuming the Main Lemma 5.1 we see Theorem 1.1 as in [4], i.e., attach $(h)^{\pi}$, $(c\Pi_2)^{\Omega}$ and $(h)^{\Omega}$ as last rules to a proof $P_0$ of $A^{\Omega}$ in $\mathrm{T}_N$.

$$\begin{array}{l}
\vdots \; P_0 \\
\dfrac{\dfrac{\dfrac{A^{\Omega}}{A^{\Omega}}\,(h)^{\pi}}{A^{\alpha}}\,(c\Pi_2)^{\Omega}_{\alpha}}{A^{\alpha}}\,(h)^{\Omega}
\end{array} \qquad P$$

## 6 Proof of Main Lemma

Throughout this section $P$ denotes a proof with a chain analysis in $\mathrm{T}_{Nc}$ and $r : \Gamma_{rdx}$ the redex of $P$.

**M1**. The case when $r : \Gamma_{rdx}$ is a lowersequent of an explicit basic rule $J$.
**M2**. The case when $r : \Gamma_{rdx}$ is a lowersequent of an $(ind)$ $J$.
**M3**. The case when the redex $r : \Gamma_{rdx}$ is an axiom.

These are treated as in [3,4].
By virtue of **M1-3** we can assume that the redex $r : \Gamma_{rdx}$ of $P$ is a lowersequent of a rule $J = r * (0)$ such that $J$ is one of the rules ($\Pi_2^{\Omega}$-rfl), ($\Pi_N$-rfl) or an implicit basic rule.

**M4**. $J$ is a ($\Pi_2^{\Omega}$-rfl). As in [3] introduce a $(c)^{\Omega}_{d_{\Omega}\alpha}$ and a (cut).
**M5**. $J$ is a ($\Pi_N$-rfl).
**M5.1**. There is no rule $(c)^{\pi}$ below $J$.

$$\begin{array}{c}
\vdots \qquad\qquad \vdots \\
\dfrac{\Gamma, A \quad \neg\exists z(t < z \wedge A^z), \Gamma}{r : \Gamma}\,J \\
\vdots \\
\dfrac{a : \Phi}{a_0 : \Lambda}\,(h)^{\pi}
\end{array} \qquad P$$

where $a : \Phi$ denotes the uppermost sequent below $J$ such that $h(a; P) = \pi$. The sequent $a_0 : \Lambda$ is the lowersequent of the lowermost $(h)^{\pi}$.

Let $P'$ be the following:

$$
\dfrac{\dfrac{\dfrac{\dfrac{\vdots}{\Gamma}}{\Gamma, A}\,(w)}{\dfrac{a_1 : \Phi, A}{\Phi, A^\sigma}\,(c\Pi_N)^\pi_\sigma\, J_0}\;\; (h)^\pi \;\; \Lambda, A^\sigma \qquad \dfrac{\vdots\; z := \sigma}{\dfrac{\neg A^\sigma, \Gamma}{\neg A^\sigma, \Gamma}\,(w)} \cdots \dfrac{\dfrac{\neg A^\sigma, \Phi}{\neg A^\sigma, \Phi}\,(w)}{\neg A^\sigma, \Lambda}\,(h)^\pi}{a_0 : \Lambda}\,(\Sigma_{N-1})^\sigma\, J_0' \qquad P'
$$

where the o.d. $\sigma$ in the new $(c\Pi_N)^\pi_\sigma\ J_0$ is defined to be

$$\sigma = d^q_\pi \alpha \text{ with } q = \nu\pi\pi N - 1, \nu = o(a_1; P') \text{ and } \alpha = \pi \cdot o(a_1; P') + \mathcal{K}_\pi(a; P).$$

Namely $In(\sigma) = \{N - 1\}$, $st_{N-1}(\sigma) = \nu$ and $pd_{N-1}(\sigma) = rg_{N-1}(\sigma) = \pi$.

Then as in [4] we see that $\Phi \subseteq \Delta^\sigma$, $\alpha < Bk_\pi(a; P) \,\&\, \sigma < o(a_0; P)$, $\sigma \in Od(\Pi_N)$ and $o(a_0; P') < o(a_0; P)$. Hence $o(P') < o(P)$. Moreover in $P$, no chain passes through $a_0 : \Lambda$, and the new $(\Sigma_N)^\sigma\ J_0'$ does not split any chain.

**M5.2**. There exists a rule $(c)^\pi\ J_0$ below $J$.

Let $\mathcal{R} = J_0, \ldots, J_{n-1}$ denote the rope starting with $J_0$. The rope $\mathcal{R}$ need not to be a chain as contrasted with [4]. Each rule $J_p$ is a $(c)^{\sigma_p}_{\sigma_{p+1}}$. Put $\sigma = \sigma_n$.

$$
\begin{array}{l}
\dfrac{\Gamma, A \qquad \neg\exists z(t < z \wedge A^z), \Gamma}{r : \Gamma}\; J \\
\qquad\vdots \\
\dfrac{\Gamma_0}{a_0 \Gamma_0'}\; (c)^\pi_{\sigma_1}\ J_0 \\
\qquad\vdots \\
\dfrac{a_i : \Gamma_i}{\Gamma_i'}\; (c)^{\sigma_i}_{\sigma_{i+1}}\ J_i \\
\qquad\vdots \\
\dfrac{a_{n-1} : \Gamma_{n-1}}{\Gamma_{n-1}'}\; (c)^{\sigma_{n-1}}_{\sigma}\ J_{n-1} \\
\qquad\vdots \\
\dfrac{\qquad}{a_n : \Gamma_n}\; (\Sigma_{N-1})^\sigma\ J_{n-1}' \\
\qquad\vdots \\
a : \Phi \qquad\qquad\qquad P
\end{array}
$$

where $a_n : \Gamma_n$ denotes the lowersequent of the trace $(\Sigma_{N-1})^{\sigma}\ J'_{n-1}$ of $J_{n-1}$, and $a : \Phi$ the bar of the rule $(c)_{\sigma}\ J_{n-1}$. Let $(\Sigma_{N-1})^{\sigma_i+1}\ J'_i$ denote the trace of $J_i$ for $0 \leq i < n$. Put

$$h := \mathrm{h}(a; P).$$

By Lemma 5.2 there is no chain passing through the bar $a : \Phi$.
Let $P'$ be the following:

$$\dfrac{\begin{array}{c} \vdots \\ \Gamma, A \\ \vdots \\ \dfrac{a^l_0 : \Gamma_0, A}{\Gamma'_0, A^{\sigma_1}}\ J^l_0 \\ \vdots \\ \dfrac{a^l_i : \Gamma_i, A^{\sigma_i}}{\Gamma'_i, A^{\sigma_i+1}}\ J^l_i \\ \vdots \\ \dfrac{a^l_{n-1} : \Gamma_{n-1}, A^{\sigma_{n-1}}}{\Gamma'_{n-1}, A^{\sigma}}\ J^l_{n-1} \\ \vdots \\ \hline \dfrac{a^l_n : \Gamma_n, A^{\sigma}}{\Gamma_n, A^{\rho}}\ (c\Pi_N)^{\sigma}_{\rho}\ J_n \\ \vdots \\ \Phi, A^{\rho} \end{array} \qquad \begin{array}{c} \vdots \\ \neg A^{\rho}, \Gamma \\ \vdots \\ \dfrac{a^r_0 : \neg A^{\rho}, \Gamma_0}{\neg A^{\rho}, \Gamma'_0}\ J^r_0 \\ \vdots \\ \dfrac{a^r_i : \neg A^{\rho}, \Gamma_i}{\neg A^{\rho}, \Gamma'_i}\ J^r_i \\ \vdots \\ \dfrac{a^r_{n-1} : \neg A^{\rho}, \Gamma_{n-1}}{\neg A^{\rho}, \Gamma'_{n-1}}\ J^r_{n-1} \\ \vdots \\ \hline \dfrac{a^r_n : \neg A^{\rho}, \Gamma_n}{\neg A^{\rho}, \Gamma_n}\ (w)\ J^r_n \\ \vdots \\ \neg A^{\rho}, \Phi \end{array}}{a : \Phi}\ (\Sigma_N)^{\rho}\ J'_n \qquad P'$$

For the proviso **(lbranch)** in $P'$, any ancestor of the left cut formula of the new $(\Sigma_N)^{\rho}\ J'_n$ is a genuine $\Pi^{\tau}_N$-formula $A^{\tau}$ for a $\tau$ with $\rho \preceq \tau$. The formula $A^{\tau}$ is not in the branch $\mathcal{T}$ from $r : \Gamma$ to $a : \Phi$ in $P$ since no genuine $\Pi^{\tau}_N$-formula with $\tau > \Omega$ is on the rightmost branch $\mathcal{T}$. Therefore any left branch of the new $(\Sigma_N)^{\rho}\ J'_n$ is the rightmost one in the left upper part of the $J'_n$ in $P'$.

In $P'$, a new chain $J^l_0, \ldots, J^l_{n-1}, J_n$ starting with the new $J_n$ is in the chain analysis for $P'$ and $\rho = d^q_{\sigma}\alpha \in \mathcal{D}_{\sigma}$ is determined as follows:

$$b(\rho) = \alpha =$$

$$\max\{\mathcal{B}_{\pi}(o(a^l_n; P')), \mathcal{B}_{>\sigma}(\{\sigma\} \cup (a_n; P))\} + \omega^{o(a^l_n; P')} + \max\{\mathcal{K}_{\sigma}(a_n; P), \mathcal{K}_{\sigma}(h)\},$$

$$rg_{N-1}(\rho) = \pi \text{ and } st_{N-1}(\rho) = o(a^l_0; P').$$

Let

$$0 \le n_0 < n_1 < \cdots < n_l = n - 1 \, (l \ge 0) \tag{11}$$

be the knotting numbers of the rope $\mathcal{R}$ and $K_m$ an $i_m$-knot $(\Sigma_{i_m})^{\sigma_{n_m+1}}$ of $J_{n_m}$ and $J_{n_m+1}$ for $m < l$. Let $m(i)$ denote the number

$$m(i) = \max\{m : 0 \le m \le l \,\&\, \forall p \in [0, m)(i \le i_p)\}. \tag{13}$$

Then $pd_i(\rho), In(\rho), rg_i(\rho), st_i(\rho)$ are determined so as to enjoy the provisos **(ch:Qpt)** and **(st:bound)**.

1. $pd_i(\rho) = \sigma_{n_{m(i)}+1}$ for $2 \le i < N$. Note that $pd_i(\rho) \ne \pi = \sigma_0$ since $n_0 \ge 0$, cf. the condition (8) in Sect. 4 which says that $pd_{N-1}(\rho) = \pi \Leftrightarrow \sigma = \pi$.
2. $N - 1 \in In(\rho)$ and $i \in In(\rho) \Leftrightarrow \exists p \in [0, m(i))(i_p = i)$ for $2 \le i < N - 1$.
3. Let $i \in In(\rho) \,\&\, i \ne N - 1$. $q$ denotes a number determined as follows.

   **Case 1** The case when there exists a $q$ such that

   $$\exists p[n_{m(i)} \ge p \ge q > n_{m(i+1)} \,\&\, \rho \prec_i \sigma_{p+1} \,\&\, \sigma_q = rg_i(\sigma_{p+1})]. \tag{14}$$

   Then $q$ denotes the minimal $q$ satisfying (14). Note that $\rho \prec_i \sigma_{p+1}$ is equivalent to $pd_i(\rho) = \sigma_{n_{m(i)}+1} \preceq_i \sigma_{p+1}$.

   **Case 2** Otherwise. Then set $q = n_{m(i)} + 1$.

   In each case set $rg_i(\rho) = \sigma_q := \kappa$ for the number $q$, and $st_i(\rho) = d_{\kappa^+}\alpha_i$ for

   $$\alpha_i = B_\kappa(a^l_q; P')$$

   where $a^l_q$ denotes the uppersequent $\Gamma_q, A^{\sigma_q}$ of $J^l_q$ in the left upper part of $(\Sigma_N)^\rho \, J^l_n$ in $P'$.
   By Lemma 4.2 we have $\mathcal{B}_{>\kappa^+}(\alpha_i) \subset \mathcal{B}_{>\kappa}(\alpha_i) < \alpha_i$, and hence $st_i(\rho) \in Od(\Pi_N)$.

Obviously the provisos **(ch:Qpt)** and **(st:bound)** are enjoyed for the new chain $J^l_0, \ldots, J^l_{n-1}, J_n$.
Observe that, cf. (9) in Sect. 4,

$$\pi < \beta \in q = Q(\rho) \Rightarrow \beta = st_{N-1}(\rho).$$

**Claim 6.1** $\rho = d^q_\sigma\alpha \in Od(\Pi_N)$.

*Proof of Claim 6.1* (5) $\mathcal{B}_{>\sigma}(\{\sigma, \alpha\} \cup q) < \alpha$: By Lemma 4.2 we have $\mathcal{B}_{>\sigma}(\{\sigma, \alpha\}) < \alpha$. It suffices to see $\mathcal{B}_{>\sigma}(q) < \alpha$. By the definition we have $\{pd_i(\rho), rg_i(\rho) : i \in In(\rho)\} \subset \{\sigma_p, \sigma^+_p : p \le n\}$. On the other hand, we have $\mathcal{B}_{>\sigma}(\{\sigma_p, \sigma^+_p : p \le n\}) \subset \mathcal{B}_{>\sigma}(\sigma)$.

We have $\mathcal{B}_{>\sigma}(st_{N-1}(\rho)) \subset \mathcal{B}_{>\sigma}(\alpha)$. Finally for $st_i(\rho) = d_\kappa + \alpha_i$ with $i < N - 1$, we have $\mathcal{B}_{>\sigma}(st_i(\rho)) \subset \mathcal{B}_{>\sigma}(\{\sigma, \alpha_i\}) \cup \{\alpha_i\}$, and $\mathcal{B}_{>\sigma}(\{\sigma, \alpha_i\}) \subset \mathcal{B}_{>\sigma}(\{\sigma, \alpha\})$ and $\alpha_i < \alpha$.
($\mathcal{D}^{\mathcal{Q}}$.12):

**Case 2** This corresponds to ($\mathcal{D}^{\mathcal{Q}}$.12.1), $\kappa = rg_i(\rho) = pd_i(\rho)$. Let $\alpha_1$ denote the diagram such that $\rho \preceq \alpha_1 \in \mathcal{D}_\kappa$. Then

$$\alpha_1 = \sigma_{n_{m(i)}+2}\,(pd_i(\rho) = \sigma_{n_{m(i)}+1} \,\&\, \sigma_{n+1} = \rho).$$

We have by Lemma 4.2 and **(c:bound2)**,

$$\mathcal{B}_{>\sigma}(B_\kappa(a_{n_{m(i)}+1}; P)) < B_\kappa(a_{n_{m(i)}+1}; P) \leq b(\alpha_1).$$

On the other, hand we have for $st_i(\rho) = d_\kappa + \alpha_i$

$$\mathcal{B}_{>\kappa}(st_i(\rho)) \subset \mathcal{B}_{>\kappa}(\{\kappa, \alpha_i\}) \leq \mathcal{B}_{>\sigma}(B_\kappa(a_{n_{m(i)}+1}; P)).$$

Thus $\mathcal{B}_{>\kappa}(st_i(\rho)) < b(\alpha_1)$.
**Case 1** This corresponds to ($\mathcal{D}^{\mathcal{Q}}$.12.2), $rg_i(\rho) = rg_i(pd_i(\rho))$ or to ($\mathcal{D}^{\mathcal{Q}}$.12.3), $rg_i(pd_i(\rho)) \prec_i \kappa$ by Lemma 5.8.1. Let $p$ denote the maximal $p$ such that

$$rg_i(\sigma_{p+1}) = \sigma_q = rg_i(\rho) \,\&\, pd_i(\rho) \preceq_i \sigma_{p+1}.$$

$st_i(\rho) < st_i(pd_i(\rho))$ for the case ($\mathcal{D}^{\mathcal{Q}}$.12.2) and $st_i(\rho) < st_i(\sigma_{p+1})$ for the case ($\mathcal{D}^{\mathcal{Q}}$.12.3) follow from Lemma 5.8.2 since for $rg_i(\sigma_{p+1}) = \sigma_q = rg_i(\rho)$

$$b(st_i(\rho)) = B_{\sigma_q}(a_q^l; P') < B_{\sigma_q}(a_q; P) \leq b(st_i(\sigma_{p+1}))$$

and hence by Lemmata 4.1 and 4.2

$$st_i(\rho) < st_i(\sigma_{p+1}).$$

($\mathcal{D}^{\mathcal{Q}}$.11) and ($\mathcal{D}^{\mathcal{Q}}$.12.3): These follow from Lemma 5.8.1.

($\mathcal{D}^{\mathcal{Q}}$.2): $\forall \tau \leq rg_i(\rho)(K_\tau st_i(\rho) < \rho)$. For $\tau \leq \kappa = rg_i(\rho)$ and $st_i(\rho) = d_\kappa + \alpha_i$, we have $K_\tau(st_i(\rho)) = K_\tau\{\kappa, \alpha_i\} \leq \mathcal{K}_\tau(a_q^l; P') < \rho$ as in the case **M6.2** in [4]. □

As in the case **M6.2** in [4] we see that $o(P') < o(P)$.

We have to verify that $P'$ is a proof. The provisos other than **(uplwl)** are seen to be satisfied as in the case **M5.2** of [4]. For the proviso **(forerun)** see Claim 6.3 in the subcase **M7.2** below. It suffices to see that $P'$ enjoys the proviso **(uplwl)** when the lower rule $J^{lw}$ is the new $(\Sigma_N)^\rho$ $J'_n$. For example, the left rope ${}_{K_m}\mathcal{R}$ of the $i_m$-knot $(\Sigma_{i_m})^{\sigma_{n_m+1}}$ $K_m$ of $J_{n_m}$ and $J_{n_m+1}$ ends with the rule $(c)_\sigma$ $J_{n-1}$. We show the following claim.

**Claim 6.2** *Any left rope* ${}_{J^{up}}\mathcal{R}$ *of a knot* $J^{up}$ *in the left upper part of the new* $(\Sigma_N)^\rho\ J'_n$ *does not reach to* $J'_n$.

*Proof of Claim 6.2* Consider the original proof $P$. By Lemma 5.2 there is no chain passing through the bar $a : \Phi$ and hence it suffices to see that there is no rule $(c)^\sigma_\rho$ above $a : \Phi$. First observe that we have $\rho < \tau$ for any rules $(c)_\tau$ and $(\Sigma_i)^\tau$ which are between $(\Pi_N\text{-rfl})\ J$ and $a : \Phi$. Thus there is no rule $(c)^\sigma_\rho$ on the branch $\mathcal{T}_0$ from $(c)^\pi\ J_0$ to $a : \Phi$. Consider another branch $\mathcal{T}$ above $a : \Phi$ and suppose that there is a rule $(c)^\sigma_\rho\ I$ on $\mathcal{T}$. We can assume that the merging rule $K$ of $\mathcal{T}$ and $\mathcal{T}_0$ is below $J_0$ and hence the rule $K$ is a $(\Sigma_i)^\tau$. By the proviso **(h-reg)** (cf. Definition 5.4.4 in [4]) we have $\tau \leq \sigma$, i.e., $K$ is between $(c)^{\sigma_{n-1}}_\sigma\ J_{n-1}$ and $a : \Phi$. Then we have seen $\rho < \tau$ and hence the trace $(\Sigma_{N-1})^\rho\ I_0$ of $(c)^\sigma_\rho\ I$ is below $K$ by the proviso **(h-reg)**. Therefore the chain stating with the trace $I_0$ passes through the left side of $K$. This is impossible by the proviso **(ch:left)**.

$$
\dfrac{\dfrac{\dfrac{\Psi_2}{\Psi'_2}\,(c)^\sigma_\rho\ I}{\Psi_1, \neg C^\tau}\,\vdots\,\mathcal{T} \qquad \dfrac{\dfrac{\Gamma_{n-1}}{\Gamma'_{n-1}}\,(c)^{\sigma_{n-1}}_\sigma\ J_{n-1}}{C^\tau, \Phi_1}\,\vdots\,\mathcal{T}_0}{\Psi_1, \Phi_1}\,(\Sigma_i)^\tau\ K
$$

$$
\dfrac{\dfrac{\Psi_0, \neg B^\rho \qquad B^\rho, \Phi_0}{\Psi_0, \Phi_0}\,(\Sigma_{N-1})^\rho\ I_0}{a : \Phi}\,\vdots\,\mathcal{T}, \mathcal{T}_0 \qquad\qquad P
$$

In what follows we assume that $r * (0) = J$ is a basic rule. Let $v * (0) = I$ denote the vanishing cut of $r * (0) = J$. $v * (0) = I$ is either a $(\Sigma_i)$ or a $(cut)$.

**M6.** $I$ is a $(\Sigma_N)^\sigma$.

$$
\dfrac{\Gamma, A^\sigma \qquad \dfrac{\dfrac{\alpha < \sigma, \Lambda_0 \qquad \neg A^\sigma_{N-1}(\alpha), \Lambda_0}{\exists x < \sigma \neg A^\sigma_{N-1}(x), \Lambda_0}\,(b\exists)\ J}{\neg A^\sigma, \Lambda}\,\vdots}{v : \Gamma, \Lambda}\,(\Sigma_N)^\sigma\ I \qquad\qquad P
$$

where $A \equiv \forall x A_{N-1}(x)$ is a $\Pi_N$ formula.

Assuming $\alpha < \sigma$ let $P'$ be the following:

$$
\dfrac{\dfrac{\Gamma, A^{\sigma} \qquad \begin{array}{c} \vdots \\ \dfrac{\neg A^{\sigma}_{N-1}(\alpha), \Lambda_0}{\neg A^{\sigma}, \Lambda_0, \neg A^{\sigma}_{N-1}(\alpha)}\,(w) \\ \vdots \\ \neg A^{\sigma}, \Lambda, \neg A^{\sigma}_{N-1}(\alpha) \end{array}}{\Gamma, \Lambda, \neg A^{\sigma}_{N-1}(\alpha)} \qquad \begin{array}{c} \vdots \\ \Gamma, A^{\sigma}_{N-1}(\alpha) \end{array}}{v : \Gamma, \Lambda}\,(\Sigma_{N-1})^{\sigma}\, I_{N-1} \qquad P'
$$

where the preproof ending with $\Gamma, A^{\sigma}_{N-1}(\alpha)$ is obtained from the left upper part of $I$ in $P$ by inversion.

As in the case **M6** of [4] we see that $o(v; P') < o(v; P)$.

For the proviso **(lbranch)** in $P'$, cf. the case **M5.2**. We verify that $P'$ is a proof with respect to the proviso **(uplw)**.

The proviso **(uplwl)** when the lower rule $J^{lw}$ is the new $(\Sigma_{N-1})^{\sigma}\, I_{N-1}$: Consider the original proof $P$. By Lemma 5.3 no left rope in the right upper part of $(\Sigma_N)^{\sigma}\, I$ reaches to $I$. Also by **(uplwl)** with the lower rule $J^{lw} = I$ there is no left rope of an $i$-knot $J^{up}$ reaching to $I$.

The proviso **(uplwr)** when the lower rule $J^{lw}$ is the new $(\Sigma_{N-1})^{\sigma}\, I_{N-1}$: As above there is no left rope of an $i$-knot $J^{up}$ reaching to $I$.

The proviso **(uplwr)** when the upper rule $J^{up}$ is the $(\Sigma_{N-1})^{\sigma}\, I_{N-1}$: $(\Sigma_{N-1})^{\sigma}\, I_{N-1}$ is not an $(N-1)$-knot since there is no chain passing through $(\Sigma_N)^{\sigma}\, I$ by **(ch:pass)**.

For the proviso **(forerun)** see Claim 6.3 in the subcase **M7.2** below.

**M7.** $I$ is a $(\Sigma_{i+1})^{\sigma}$ with $1 \leq i < N-1$.
Then $J$ is either an $(\exists)$ or a $(b\exists)$. Let $u_0 : \Psi$ denote the uppermost sequent below $I$ such that $h(u_0; P) < \sigma + i$. Also let $u : \Phi$ denote the resolvent of $I$, cf. Definition 5.5.

**M7.1** $u_0 = u$.

$$
\begin{array}{c}
\dfrac{\begin{array}{c} \vdots \\ \Gamma, \neg A^{\sigma}_{i+1} \end{array} \qquad \begin{array}{c} \dfrac{\begin{array}{c}\vdots \\ \alpha < \tau, \Lambda_0\end{array} \quad \begin{array}{c}\vdots \\ A^{\tau}_{i}(\alpha), \Lambda_0\end{array}}{A^{\tau}_{i+1}, \Lambda_0}\, x \\ \vdots \\ A^{\sigma}_{i+1}, \Lambda \end{array}}{\Gamma, \Lambda}\,(\Sigma_{i+1})^{\sigma}\, I \\
\vdots \\
u : \Psi
\end{array} \qquad P
$$

where $A_{i+1} \equiv \exists y A_i(y)$ is a $\Sigma_{i+1}$ formula. Also if $x$ is an $(\exists)$, then $\tau = \pi$ and the left upper part of the true sequent $\alpha < \tau, \Lambda_0$ is absent. Anyway $\sigma \preceq \tau$.

Assuming $\alpha < \tau$ and then $\alpha < \sigma$ by **(c:bound)**, let $P'$ be the following:

$$
\dfrac{
\dfrac{
\begin{array}{c}\vdots\\ \Gamma, \neg A^{\sigma}_{i+1}\end{array}
\quad
\begin{array}{c}\vdots\\ \dfrac{A^{\tau}_{i}(\alpha), \Lambda_0}{A^{\tau}_{i+1}, A^{\tau}_{i}(\alpha), \Lambda_0}\,(w)\\ \vdots\\ A^{\sigma}_{i+1}, A^{\sigma}_{i}(\alpha), \Lambda\end{array}
}{
\begin{array}{c}A^{\sigma}_{i}(\alpha), \Gamma, \Lambda\\ \vdots\\ \Psi, A^{\sigma}_{i}(\alpha)\end{array}
}
\qquad
\begin{array}{c}\vdots\\ \dfrac{\Gamma, \neg A^{\sigma}_{i}(\alpha)}{\Gamma, \Lambda, \neg A^{\sigma}_{i}(\alpha)}\,(w)\\ \vdots\\ \neg A^{\sigma}_{i}(\alpha), \Psi\end{array}
}{u : \Psi}\,(\Sigma_i)^{\sigma} \qquad P'
$$

It is easy to see that $o(u; P') < o(u; P)$. For the proviso **(lbranch)** in $P'$, cf. the case **M5.2**. To see that $P'$ is a proof with respect to the provisos **(forerun)**, **(uplw)**, cf. the subcase **M7.2** below.

**M7.2** Otherwise.

Let $K$ denote the lowermost rule $(\Sigma_{i+1})^{\sigma}$ below or equal to $I$. Then $u_0 : \Psi$ is the lowersequent of $K$ by **(h-reg)**. There exists an $(i+1)$-knot $(\Sigma_{i+1})^{\sigma}$ which is between an uppersequent of $I$ and $u_0 : \Psi$. Pick the uppermost such knot $(\Sigma_{j+1})^{\sigma}$ $K_{-1}$ and let ${}_{K_{-1}}\mathcal{R} = J_0, \dots, J_{n-1}$ denote the left rope of $K_{-1}$. Each $J_p$ is a rule $(c)^{\sigma_p}_{\sigma_{p+1}}$ with $\sigma = \sigma_0$. Let

$$0 \le n_0 < n_1 < \cdots < n_l = n - 1 \; (l \ge 0) \tag{11}$$

be the knotting numbers of the left rope ${}_{K_{-1}}\mathcal{R}$ and $K_m$ an $i_m$-knot $(\Sigma_{i_m})^{\sigma_{n_m+1}}$ of $J_{n_m}$ and $J_{n_m+1}$ for $m < l$. Put

$$m(i+1) = \max\{m : 0 \le m \le l \,\&\, \forall p \in [0, m)(i + 1 \le i_p)\}. \tag{12}$$

Then the resolvent $u : \Phi$ is the uppermost sequent $u : \Phi$ below $J_{n_{m(i+1)}}$ such that

$$h(u; P) < \sigma_{n_{m(i+1)}+1} + i.$$

In the following figure of $P$ the chain $\mathcal{C}_{n_{m+1}}$ starting with $J_{n_{m+1}}$ passes through the left side of $K_m$.

$$
\dfrac{\alpha < \tau, \Lambda_0 \quad A_i^{\tau}(\alpha), \Lambda_0}{A_{i+1}^{\tau}, \Lambda_0}\ J
$$

$$
\dfrac{\begin{array}{c}\vdots\ \mathcal{T}_l\\ \Gamma, \neg A_{i+1}^{\sigma}\end{array} \qquad \begin{array}{c}\vdots\\ A_{i+1}^{\sigma}, \Lambda\end{array}}{v : \Gamma, \Lambda}\ (\Sigma_{i+1}^{\sigma})\ I
$$

$$
\vdots\ \mathcal{T}_1
$$

$$
\dfrac{\Gamma_0}{\Gamma_0'}\ (c)_{\sigma_1}^{\sigma}\ J_0
$$

$$
\vdots
$$

$$
\dfrac{\Gamma_{n_m}}{\Gamma_{n_m}'}\ (c)_{\sigma_{n_m+1}}^{\sigma_{n_m}}\ J_{n_m}
$$

$$
\dfrac{\begin{array}{c}\vdots\ \mathcal{C}_{n_{m+1}}\\ \Pi_m, \neg B_m\end{array} \qquad \begin{array}{c}\vdots\\ B_m, \Delta_m\end{array}}{\Pi_m, \Delta_m}\ (\Sigma_{i_m})^{\sigma_{n_m+1}}\ K_m
$$

$$
\vdots
$$

$$
\dfrac{\Gamma_{n_m+1}}{\Gamma_{n_m+1}'}\ (c)_{\sigma_{n_m+2}}^{\sigma_{n_m+1}}\ J_{n_m+1}
$$

$$
\vdots
$$

$$
\dfrac{\Gamma_{n_{m+1}}}{\Gamma_{n_{m+1}}'}\ (c)_{\sigma_{n_{m+1}}+1}^{\sigma_{n_{m+1}}}\ J_{n_{m+1}}
$$

$$
\vdots
$$

$$
\dfrac{\Gamma_{n_{m(i+1)}}}{\Gamma_{n_{m(i+1)}}'}\ (c)_{\sigma_{n_{m(i+1)}}+1}^{\sigma_{n_{m(i+1)}}}\ J_{n_{m(i+1)}}
$$

$$
\vdots\ \mathcal{T}_1
$$

$$
u : \Phi \qquad\qquad P
$$

Assuming $\alpha < \tau$ and then $\alpha < \sigma_n \leq \sigma_{n_{m(i)}+1}$, let $P'$ be the following:

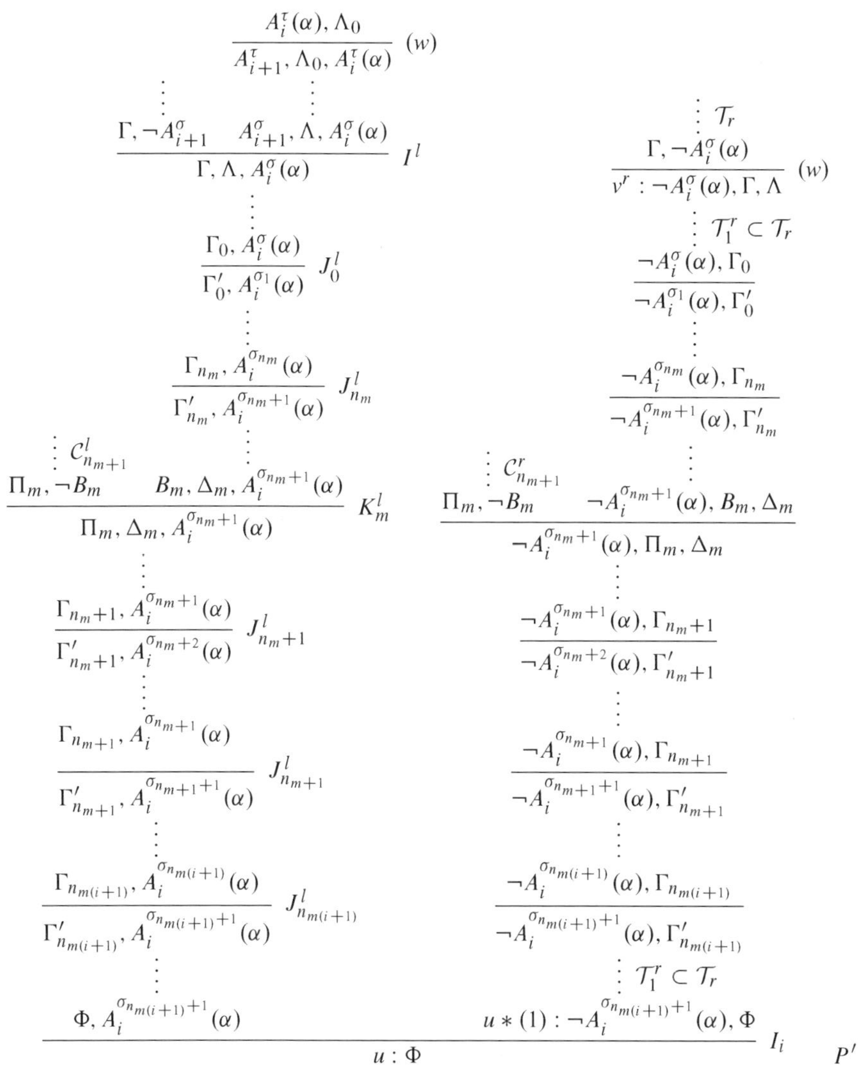

Here $I_i$ denotes a $(\Sigma_i)^{\sigma_{n_{m(i+1)}+1}}$.

It is straightforward to see $o(u; P') < o(u; P)$. We show $P'$ is a proof.

First by Lemma 5.5, in $P$ every chain passing through the resolvent $u : \Phi$ passes through the right side of $I$ and, by inversion, the right upper part of $I$ disappears in $P'$. Hence the new $(\Sigma_i)^{\sigma_{n_{m(i+1)}+1}}$ $I_i$ does not split any chain. For the proviso **(lbranch)** in $P'$, cf. the case **M5.2**.

**Claim 6.3** *The proviso* **(forerun)** *holds for the lower rule $J^{lw} = I_i$ in $P'$.*

*Proof of Claim 6.3* Consider a right branch $\mathcal{T}_r$ of $I_i$. We show that there is no rule $K$ such that $\mathcal{T}_r$ passes through the left side of $K$ and $h(a; P') < \pi$ with the lowersequent $a$ of $K$. The assertion follows from this and **(h-reg)**. The ancestors of the right cut formula $\neg A_i^{\sigma_{n_{m(i+1)}+1}}(\alpha)$ of $I_i$ comes from the left cut formula $\neg A_{i+1}^{\sigma}$ of $I$ in $P$. Let $\mathcal{T}_1^r$ denote the branch in $P'$ from the lowersequent $v^r : \neg A_i^{\sigma}(\alpha), \Gamma, \Lambda$ of the new $(w)$ to the right uppersequent $u * (1) : \neg A_i^{\sigma_{n_{m(i+1)}+1}}(\alpha), \Phi$ of $I_i$. Also let $\mathcal{T}_l$ denote a (the) left branch of $I$ in $P$. There exists a (possibly empty) branch $\mathcal{T}_0$ such that $\mathcal{T}_r = \mathcal{T}_0{}^\frown\mathcal{T}_l{}^\frown\mathcal{T}_1^r$. By **(lbranch)** any left branch $\mathcal{T}_l$ of $I$ is the rightmost one in the left upper part of $I$. Therefore there is no rule $K$ such that $\mathcal{T}_r$ passes through the left side of $K$ and $h(a; P') < \pi$ with the lowersequent $a$ of $K$. □

**Claim 6.4** *The proviso* **(uplwr)** *holds for the upper rule $J^{up} = I_i$ in $P'$.*

*Proof of Claim 6.4* Suppose that $I_i$ is a knot. Then there exists a chain $\mathcal{C}_1$ starting with an $I_1$ such that $\mathcal{C}_1$ passes through the left side of $I_i$. This chain comes from a chain in $P$ which passes through $u : \Phi$. Call the latter chain in $P$ $\mathcal{C}_1$ again. Further assume that, in $P'$, the left rope ${}_{I_i}\mathcal{R}$ of $I_i$ reaches to a rule $(\Sigma_j)^\kappa\ J^{lw}$ with $i \leq j$. Let $I_2$ denote the lower rule of $I_i$. We have to show $I_i$ foreruns $J^{lw}$. It suffices to show that, in $P$, any right branch $\mathcal{T}$ of $J^{lw}$ passes through the right side of $I$ if the branch $\mathcal{T}$ passes through $u : \Phi$. Since, by inversion, the right upper part of $I$ disappears in $P'$, for such a branch $\mathcal{T}$ there exists a unique branch $\mathcal{T}'$ corresponding to it in $P'$ so that $\mathcal{T}'$ passes through the left side of $I_i$ and hence $\mathcal{T}'$ is left to $I_i$.

$$
\dfrac{\vdots \qquad \dfrac{\dfrac{\vdots \qquad \vdots\ \mathcal{T}}{\Gamma, \neg A_{i+1}^{\sigma} \quad A_{i+1}^{\sigma}, \Lambda}\ (\Sigma_{i+1})^{\sigma}\ I}{\Gamma, \Lambda} \atop {\vdots \atop {u : \Phi \atop \vdots\ \mathcal{T}}}}{\Gamma_{lw}, \neg C_{lw} \qquad C_{lw}, \Lambda_{lw}} \ (\Sigma_j)^\kappa\ J^{lw}
$$

$$
\dfrac{\Gamma_{lw}, \neg C_{lw} \qquad C_{lw}, \Lambda_{lw}}{\Gamma_{lw}, \Lambda_{lw}} \qquad P
$$

$$
\dfrac{\vdots \qquad \vdots\ \mathcal{T}'}{\Gamma, \neg A_{i+1}^{\sigma} \quad A_{i+1}^{\sigma}, \Lambda, A_i^{\sigma}(\alpha)}\ I^l
$$

$$
\dfrac{\Gamma, \Lambda, A_i^{\sigma}(\alpha) \atop \vdots}{\Phi, A_i^{\sigma_{n_{m(i)}+1}}(\alpha) \qquad \neg A_i^{\sigma_{n_{m(i)}+1}}(\alpha), \Phi}\ I_i
$$

$$
\dfrac{\vdots \qquad u : \Phi \atop {\vdots\ \mathcal{T}'}}{\Gamma_{lw}, \neg C_{lw} \qquad C_{lw}, \Lambda_{lw}}\ J^{lw}
$$

$$
\Gamma_{lw}, \Lambda_{lw} \qquad P'
$$

**Case 1**. The case when, in $P$, there exists a member $I_3$ of the chain $\mathcal{C}_1$ such that $I_3$ is between $u : \Phi$ and $J^{lw}$, and the chain $\mathcal{C}_3$ starting with $I_3$ passes through the resolvent $u : \Phi$ in $P$: Then by Lemma 5.5 the chain $\mathcal{C}_3$ passes through the right side of $I$. The rope $\mathcal{R}_{I_3}$ starting with $I_3$ in $P$ corresponds to a part (a tail) of the left rope ${}_{I_i}\mathcal{R}$ in $P'$. Thus by the assumption the rope $\mathcal{R}_{I_3}$ also reaches to $J^{lw}$ in $P$. Hence by **(forerun)** there is no merging rule $K$ such that

1. the chain $\mathcal{C}_3$ starting with $I_3$ passes through the right side of $K$, and
2. the right branch $\mathcal{T}$ of $J^{lw}$ passes through the left side of $K$.

Therefore the right branch $\mathcal{T}$ of $J^{lw}$ passes through the right side of $I$ in $P$.

$$
\begin{array}{c}
\dfrac{\begin{array}{cc}\vdots & \vdots\ \mathcal{C}_3,\ \mathcal{T}' \\ \Gamma, \neg A^{\sigma}_{i+1} & A^{\sigma}_{i+1}, \Lambda, A^{\sigma}_i(\alpha)\end{array}}{\Gamma, \Lambda, A^{\sigma}_i(\alpha)}\ I^l \\
\vdots \qquad\qquad \vdots \\
\dfrac{\Phi, A_i^{\sigma_{n_{m(i)}+1}}(\alpha) \qquad \neg A_i^{\sigma_{n_{m(i)}+1}}(\alpha), \Phi}{u : \Phi}\ I_i \\
\vdots \\
\dfrac{\Delta_2}{\Delta_2'}\ I_2 \\
\vdots \\
\dfrac{\Delta_3}{\Delta_3'}\ I_3 \\
\begin{array}{cc}\vdots & \vdots\ \mathcal{R}_{I_3} \subset {}_{I_i}\mathcal{R}\end{array} \\
\dfrac{\Gamma_{lw}, \neg C_{lw} \qquad C_{lw}, \Lambda_{lw}}{\Gamma_{lw}, \Lambda_{lw}}\ J^{lw} \qquad\qquad P'
\end{array}
$$

**Case 2**. Otherwise: First we show the following claim:

**Claim 6.5** *In $P$, we have $m(i+1) < l$ for the number of knots $l$ in (11), and $I_2$ is the lower rule of the $i_{m(i+1)}$-knot $K_{m(i+1)}$. Let ${}_{K_{m(i+1)}}\mathcal{R}$ denote the left rope of $K_{m(i+1)}$ in $P$. Then ${}_{K_{m(i+1)}}\mathcal{R}$ reaches to $J^{lw}$.*

*Proof of Claim 6.5* In $P'$, the lower rule $I_2$ of the knot $I_i$ is a member of the chain $\mathcal{C}_1$ starting with $I_1$ and passing through the left side of $I_i$. Further $I_2$ is above $J^{lw}$ since the left rope ${}_{I_i}\mathcal{R}$ of $I_i$ is assumed to reach to $J^{lw}$ in $P'$, cf. Definition 5.7. Since we are considering when **Case 1** is not the case, in $P$, $I_1$ is below $J^{lw}$ and the chain $\mathcal{C}_2$ starting with $I_2$ does not pass through $u : \Phi$, and hence chains $\mathcal{C}_1$ and $\mathcal{C}_2$ intersect as **Type3 (merge)** in **(ch:link)**. In other words, there is a knot below $u : \Phi$ whose upper right rule is $(c)_{\sigma_{n_{m(i+1)}+1}} J_{n_{m(i+1)}}$. This means that the knot is the $i_{m(i+1)}$-knot $K_{m(i+1)}$. Thus we have shown that $m(i+1) < l$ and $I_2$ is the lower rule of the $i_{m(i+1)}$-knot $K_{m(i+1)}$.

Next we show that the left rope ${}_{K_{m(i+1)}}\mathcal{R}$ of $K_{m(i+1)}$ reaches to $J^{lw}$ in $P$. Suppose this is not the case. Let $(c)_{\kappa_4}\, I_4$ denote the lowest (last) member of the left rope ${}_{K_{m(i+1)}}\mathcal{R}$. Then $\kappa < \kappa_4$ for the rule $(\Sigma_j)^\kappa\, J^{lw}$. By $\kappa < \kappa_4$, the next member $(c)^{\kappa_4}\, I_5$ of the chain $\mathcal{C}_1$ is above $J^{lw}$. Since we are considering when **Case 1** is not the case, the chain $\mathcal{C}_5$ starting with $I_5$ does not pass through $u : \Phi$. By Definition 5.4.6 of left ropes and **(ch:link)** there would be a knot $K'$ whose lower rule is $I_5$ and whose upper right rule is $I_4$. This is a contradiction since $I_4$ is assumed to be the last member of the left rope ${}_{K_{m(i+1)}}\mathcal{R}$. This shows Claim 6.5.

In the following figure note that $u : \Phi$ is above $K_{m(i+1)}$ by **(h-reg)** and the definition of the resolvent $u : \Phi$.

$$
\dfrac{\begin{array}{c}\vdots\ \mathcal{C}_2 = \mathcal{C}_{n_{m(i+1)+1}}\\ \Pi_{m(i+1)}, \neg B_{m(i+1)}\end{array} \qquad \begin{array}{c}\vdots\ \mathcal{C}_1\\ u : \Phi\\ \vdots\\ B_{m(i+1)}, \Delta_{m(i+1)}\end{array}}{\Pi_{m(i+1)}, \Delta_{m(i+1)}}\ (\Sigma_{i_{m(i+1)}})^{\sigma_{n_{m(i+1)+1}}}\ K_{m(i+1)}
$$

$$
\vdots
$$

$$
\dfrac{\Delta_2}{\Delta'_2}\ (c)^{\sigma_{n_{m(i+1)+1}}}\ I_2
$$

$$
\vdots
$$

$$
\dfrac{\Delta_4}{\Delta'_4}\ (c)_{\kappa_4}\ I_4
$$

$$
\dfrac{\begin{array}{c}\vdots\ \mathcal{C}_5\\ \Pi, \neg B\end{array} \quad \begin{array}{c}\vdots\\ B, \Delta\end{array}}{\Pi, \Delta}\ K'
$$

$$
\vdots
$$

$$
\dfrac{\Delta_5}{\Delta'_5}\ (c)^{\kappa_4}\ I_5
$$

$$
\dfrac{\begin{array}{c}\vdots\\ \Gamma_{lw}, \neg C_{lw}\end{array} \quad \begin{array}{c}\vdots\\ C_{lw}, \Lambda_{lw}\end{array}}{\Gamma_{lw}, \Lambda_{lw}}\ (\Sigma_j)^\kappa\ J^{lw}
$$

$$
\vdots\ \mathcal{C}_1
$$

$$
\dfrac{\Delta_1}{\Delta'_1}\ I_1
$$

$P$

$\square$

By Claim 6.5, **(uplwr)** and $i_{m(i+1)} \leq i \leq j$, $K_{m(i+1)}$ foreruns $J^{lw}$ in $P$. Therefore the right branch $\mathcal{T}$ of $J^{lw}$ is left to $K_{m(i+1)}$. Also by **(h-reg)** $K_{m(i+1)}$ is below $u : \Phi$. Hence $\mathcal{T}$ does not pass through $u : \Phi$ in this case. This shows Claim 6.4. In the following figure $\mathcal{C}_2$ denotes the chain starting with $I_2$.

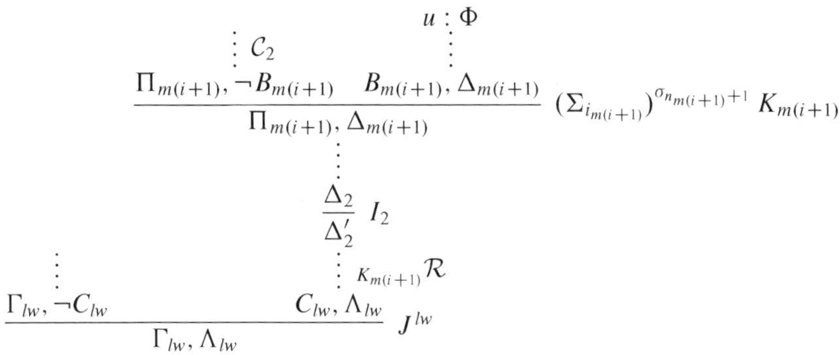

$P$

□

**Claim 6.6** *The proviso* **(uplw)** *holds for the lower rule* $J^{lw} = I_i$ *in* $P'$.

*Proof of Claim 6.6* Let $J^{up}$ be a $j$-knot $(\Sigma_j)$ above $I_i$. Let $H_0$ denote the lower rule of $J^{up}$. Assume that the left rope ${}_{J^{up}}\mathcal{R} = H_0, \ldots, H_{k-1}$ of $J^{up}$ reaches to the rule $I_i$. We show

$$i < j$$

even if $J^{up}$ is in the right upper part of $I_i$. Consider the corresponding rule $J^{up}$ in $P$.

**Case 1** Either $J^{up}$ is $I$ or between $I$ and $u : \Phi$: If either $J^{up}$ is $I$ or an $i_m$-knot $K_m$ with $m < m(i+1)$, then $i < i+1 = j$ or $i < i_m = j$ by (12), resp. Otherwise $J^{up}$ is between $K_{m-1}$ and $J_{n_m}$ for some $m$ with $0 \leq m \leq m(i+1)$. Then the rule $J^{up}$ is the merging rule of the chain $\mathcal{C}_{n_m}$ starting with $J_{n_m}$ and the chain $\mathcal{C}_{H_0}$ starting with $H_0$ so that $\mathcal{C}_{n_m}$ passes through the right side of $J^{up}$ and $\mathcal{C}_{H_0}$ the left side of $J^{up}$. Hence by **(ch:link) Type3 (merge)** the rule $H_{k-1}$ is above $J_{n_m}$ and the left rope ${}_{H_0}\mathcal{R}$ does not reach to $I_i$. Thus this is not the case.

$$
\begin{array}{c}
\vdots\ \mathcal{C}_{n_m} \\
\dfrac{\Pi_{m-1}, \neg B_{m-1} \quad B_{m-1}, \Delta_{m-1}}{\Pi_{m-1}, \Delta_{m-1}}\ K_{m-1} \\
\vdots\ \mathcal{C}_{H_q} \qquad \vdots\ \mathcal{C}_{n_m} \\
\dfrac{\Delta, \neg C \qquad C, \Psi}{\Delta, \Psi}\ J^{up} \\
\vdots \\
\dfrac{\Lambda_q}{\Lambda'_q}\ H_q \\
\vdots \\
\dfrac{\Lambda_{k-1}}{\Lambda'_{k-1}}\ H_{k-1} \\
\vdots \\
\dfrac{\Gamma_{n_m}}{\Gamma'_{n_m}}\ J_{n_m}
\end{array}
$$

where $H_q$ denotes the lowermost member of ${}_{H_0}\mathcal{R}$ such that the chain $\mathcal{C}_{H_q}$ starting with $H_q$ passes through the left side of $J^{up}$. By **(ch:link) Type3 (merge)** the rule $H_q$ is above $J_{n_m}$ and so on.

**Case 2** $J^{up}$ is in the right upper part of $I$: Then the left rope ${}_{H_0}\mathcal{R}$ reaches to $I$. Hence by Lemma 5.4, i.e., by **(uplwr)** we have $i < i + 1 < j$.

**Case 3** $J^{up}$ is in the left upper part of $I$: Then the left rope ${}_{H_0}\mathcal{R}$ reaches to $I$. Hence by **(uplwl)** we have $i < i + 1 < j$.

**Case 4** Otherwise: Then there exists a rule $K$ such that $J^{up}$ is in the left upper part of $K$ and $K$ is between $I$ and $\Phi$. By **(h-reg), (ch:pass)** $K$ is a rule $(\Sigma_p)^\kappa$. The left rope ${}_{H_0}\mathcal{R} = H_0, \ldots, H_{k-1}$ reaches to $K$. Hence by **(uplwl)** we have

$$p < j \tag{32}$$

$$
\begin{array}{c}
\begin{array}{cc}
\dfrac{\begin{array}{cc}\vdots & \vdots \\ \Delta, \neg C & C, \Psi\end{array}}{\Delta, \Psi}\ J^{up} &
\dfrac{\begin{array}{cc}\vdots & \vdots \\ \Gamma, \neg A^\sigma_{i+1} & A^\sigma_{i+1}, \Lambda\end{array}}{\Gamma, \Lambda}\ (\Sigma^\sigma_{i+1})\ I \\
\vdots & \vdots
\end{array} \\
\dfrac{\Gamma_K, \neg D \qquad\qquad D, \Lambda_K}{\Gamma_K, \Lambda_K}\ (\Sigma_p)^\kappa\ K \\
\vdots \\
\Phi
\end{array}
$$

**Case 4.1** $H_{k-1}$ is below $K$: Let $K'$ denote the uppermost knot such that $K'$ is equal to or below $K$, and there exists a member of ${}_{H_0}\mathcal{R}$ such that the chain starting with the member passes through the left side of $K'$. Let $H_q$ be the lowermost member of ${}_{H_0}\mathcal{R}$ such that the chain $\mathcal{C}_{H_q}$ starting with $H_q$ passes through the left side of $K'$. If there exists a member of ${}_{H_0}\mathcal{R}$ such that the chain starting with the member passes through the left side of $K$, then $K'$ is equal to $K$.

$$\begin{array}{c} J^{up} \\ \vdots\ \mathcal{C}_{H_q} \qquad\qquad \vdots \\ \dfrac{\Gamma_K, \neg D \qquad D, \Lambda_K}{\Gamma_K, \Lambda_K}\ (\Sigma_p)^{\kappa}\ K = K' \\ \vdots \\ \dfrac{\Delta_q}{\Delta'_q}\ H_q \end{array}$$

Otherwise $K'$ is below $K$ and it is a knot for the left rope ${}_{H_0}\mathcal{R}$. Let $H_{q-1}$ denote the lowermost member of ${}_{H_0}\mathcal{R}$ above $K$. Then $H_{q-1}$ is an upper right rule of the knot $K'$ and $K'$ is a rule $(\Sigma_{p'})^{\kappa}$ with

$$p' \leq p \tag{33}$$

by **(h-reg)**.

$$\begin{array}{c} J^{up} \\ \vdots \\ \dfrac{\Delta_{q-1}}{\Delta'_{q-1}}\ H_{q-1} \qquad\qquad I \\ \vdots \qquad\qquad\qquad \vdots \\ \dfrac{\Gamma_K, \neg D \qquad\qquad D, \Lambda_K}{\Gamma_K, \Lambda_K}\ (\Sigma_p)^{\kappa}\ K \\ \vdots\ \mathcal{C}_{H_q} \qquad\qquad \vdots \\ \dfrac{\Gamma_{K'}, \neg D' \qquad D', \Lambda_{K'}}{\Gamma_{K'}, \Lambda_{K'}}\ (\Sigma_{p'})^{\kappa}\ K' \\ \vdots \\ \dfrac{\Delta_q}{\Delta'_q}\ H_q \end{array}$$

By Lemma 5.1 the uppermost member of $\mathcal{C}_{H_q}$ below $K'$ is the lower rule of the knot $K'$. By (32), (33) and **Case 1** it suffices to show that the left rope ${}_{K'}\mathcal{R} = G_0, \ldots, G_{k_0}$ of $K'$ reaches to $I_i$, i.e., to show the last member $G_{k_0}$ is equal to or below the rule $H_{k-1}$. Then we will have $i < p' \leq p < j$.

Let $G_0 = H_{q_0}$ ($q_0 \geq q$) denote the lower rule of $K'$ and $G_{k_1}$ the lowermost member of ${}_{K'}\mathcal{R}$ such that the chain $\mathcal{C}_{G_{k_1}}$ starting with $G_{k_1}$ passes through the left side of $K'$. Then by **(ch:link)** $G_{k_1}$ is equal to or below $H_q$.

**Case 4.1.1** $G_{k_1} = H_q$: Then $G_{k_0} = H_{k-1}$, i.e., $G_{q_1 - q_0} = H_{q_1}$ for any $q_1$ with $q_0 \leq q_1 < k$.

$$
\begin{array}{c}
\vdots\ \mathcal{C}_{H_q} \qquad \vdots \\
\dfrac{\Gamma_{K'}, \neg D' \quad D', \Lambda_{K'}}{\Gamma_{K'}, \Lambda_{K'}}\ K' \\
\vdots \\
\dfrac{\Lambda_{q_0}}{\Lambda'_{q_0}}\ H_{q_0} = G_0 \\
\vdots \\
\dfrac{\Lambda_{q}}{\Lambda'_{q}}\ H_{q} = G_{k_1} \\
\vdots\ \mathcal{C}_{H_{q_1}} \qquad \vdots \\
\dfrac{\Gamma_{K_1}, \neg D_1 \quad D_1, \Lambda_{K_1}}{\Gamma_{K_1}, \Lambda_{K_1}}\ K_1 \\
\vdots \\
\dfrac{\Lambda_{q+1}}{\Lambda'_{q+1}}\ H_{q+1} = G_{q+1-q_0} \\
\vdots \\
\dfrac{\Lambda_{q_1}}{\Lambda'_{q_1}}\ H_{q_1} = G_{q_1-q_0}
\end{array}
$$

where $K_1$ is a knot of $H_{q+1} = G_{q+1-q_0}$ and $H_q = G_{k_1}$ with $q+1-q_0 = k_1+1$.

**Case 4.1.2** Otherwise: Then by **(ch:link)** $G_{k_1}$ is already below $H_{k-1}$.

$$\dfrac{\begin{array}{cc}\vdots\ \mathcal{C}_{H_q} & \vdots\ \mathcal{C}_{G_{k_1}}\\ \Gamma_{mg},\neg D_{mg} & D_{mg},\Lambda_{mg}\end{array}}{\Gamma_{mg},\Lambda_{mg}}\,K_{mg}$$

$$\vdots$$

$$\dfrac{\Gamma_{K'},\neg D' \quad D',\Lambda_{K'}}{\Gamma_{K'},\Lambda_{K'}}\,K'$$

$$\vdots$$

$$\dfrac{\Lambda_{q_0}}{\Lambda'_{q_0}}\,H_{q_0}=G_0$$

$$\vdots$$

$$\dfrac{\Lambda_q}{\Lambda'_q}\,H_q=G_{q-q_0}$$

$$\dfrac{\begin{array}{cc}\vdots\ \mathcal{C}_{H_{q_1}} & \vdots\\ \Gamma_{K_1},\neg D_1 & D_1,\Lambda_{K_1}\end{array}}{\Gamma_{K_1},\Lambda_{K_1}}\,K_1$$

$$\vdots$$

$$\dfrac{\Lambda_{q+1}}{\Lambda'_{q+1}}\,H_{q+1}=G_{q+1-q_0}$$

$$\vdots$$

$$\dfrac{\Lambda_{q_1}}{\Lambda'_{q_1}}\,H_{q_1}=G_{q_1-q_0}$$

$$\vdots$$

$$\dfrac{\Lambda_{k-1}}{\Lambda'_{k-1}}\,H_{k-1}=G_{k-1-q_0}$$

$$\vdots$$

$$\dfrac{\Lambda_{k_1+q_0}}{\Lambda'_{k_1+q_0}}\,G_{k_1}$$

where $K_{mg}$ is a merging rule of the chain $\mathcal{C}_{H_q}$ starting with $H_q$ and the chain $\mathcal{C}_{G_{k_1}}$ starting with $G_{k_1}$. Since the chain $\mathcal{C}_{H_{q+1}}$ starting with the lower rule $H_{q+1} = G_{q+1-q_0}$ of $K_1$ passes through the left side of $K_1$, $G_{k_1}$ is not equal to $H_{q+1}$ and hence is below $H_{q+1}$ and so on.

**Case 4.2** $H_{k-1}$ is above $K$: Then $H_{k-1}$ is a rule $(c)_{\sigma_{n_{m(i+1)}+1}}$ and $K$ is a rule $(\Sigma_p)^{\sigma_{n_{m(i+1)}+1}}$. Let $d : \Gamma_K, \neg D$ denote an uppersequent of $K$. By **(h-reg)** and the definition of the sequent $u : \Phi$ we have $\sigma_{n_{m(i+1)}+1} + i \leq h(d; P) \leq \sigma_{n_{m(i+1)}+1} + p - 1$. Thus by (32) we get $i \leq p - 1 < j$.

$$
\begin{array}{ccc}
J^{up} & & \vdots \\
\vdots & & \\
\dfrac{\Lambda_{k-1}}{\Lambda'_{k-1}}\ (c)_{\sigma_{n_{m(i+1)}}+1}\ H_{k-1} & & \dfrac{\Gamma_{n_{m(i+1)}}}{\Gamma'_{n_{m(i+1)}}}\ (c)^{\sigma_{n_{m(i+1)}}}_{\sigma_{n_{m(i+1)}}+1}\ J_{n_{m(i+1)}} \\
\vdots & & \vdots
\end{array}
$$

$$
\dfrac{d : \Gamma_K, \neg D \qquad D, \Lambda_K}{\Gamma_K, \Lambda_K}\ (\Sigma_p)^{\sigma_{n_{m(i+1)}}+1}\ K
$$

$$
\vdots\ \Phi\ \vdots
$$

$$
\dfrac{\Pi_{m(i+1)}, \neg B_{m(i+1)} \qquad B_{m(i+1)}, \Delta_{m(i+1)}}{\Pi_{m(i+1)}, \Delta_{m(i+1)}}\ (\Sigma_{i_{m(i+1)}})^{\sigma_{n_{m(i+1)}}+1}\ K_{m(i+1)}
$$

where the $i_{m(i+1)}$-knot $K_{m(i+1)}$ disappears when $m(i+1) = l$ in (12).

This shows Claim 6.6. □

**M8**. $I$ is a $(\Sigma_1)^\sigma$.

This is treated as in the case **M8** of [4].

Other cases are easy.

This completes a proof of the Main Lemma 5.1.

## References


1. T. Arai, A system $(O(\pi_n), <)$ of ordinal diagrams. Handwritten note (1994)
2. T. Arai, Proof theory for reflecting ordinals IV: $\Pi_N$-reflecting ordinals. Handwritten note (1994)
3. T. Arai, Proof theory for theories of ordinals I: recursively Mahlo ordinals. Ann. Pure Appl. Log. **122**, 1–85 (2003)
4. T. Arai, Proof theory for theories of ordinals II: $\Pi_3$-reflection. Ann. Pure Appl. Log. **129**, 39–92 (2004)
5. T. Arai, Iterating the recursively Mahlo operations, in *Proceedings of the Thirteenth International Congress of Logic Methodology, Philosophy of Science*, ed. by C. Glymour, W. Wei, D. Westerstahl (College Publications, London, 2009), pp. 21–35
6. T. Arai, Wellfoundedness proofs by means of non-monotonic inductive definitions II: first order operators. Ann. Pure Appl. Log. **162**, 107–143 (2010)
7. G. Gentzen, Neue Fassung des Widerspruchsfreiheitsbeweises für die reine Zahlentheorie, Forschungen zur Logik und zur Grundlegung der exakter Wissenschaften. Neue Folge **4**, 19–44 (1938)
8. M. Rathjen, Proof theory of reflection. Ann. Pure Appl. Log. **68**, 181–224 (1994)
9. M. Rathjen, An ordinal analysis of stability. Arch. Math. Log. **44**, 1–62 (2005)
10. M. Rathjen, An ordinal analysis of parameter free $\pi^1_2$-comprehension. Arch. Math. Log. **44**, 263–362 (2005)
11. W.H. Richter, P. Aczel, Inductive definitions and reflecting properties of admissible ordinals, in *Generalized Recursion Theory*, ed. by J.E. Fenstad, P.G. Hinman (North-Holland, Amsterdam, 1974), pp. 301–381
12. G. Takeuti, *Proof Theory*, 2nd edn. (North-Holland, Amsterdam, 1987). Reprinted from Dover Publications, 2013

# A Proof-Theoretic Analysis of Theories for Stratified Inductive Definitions

**Gerhard Jäger and Dieter Probst**

**Abstract** In this article we study subsystems $\mathsf{SID}_\nu$ of the theory $\mathsf{ID}_1$ in which fixed point induction is restricted to properly stratified formulas.

## 1 Introduction

Several years ago Leivant presented a conceptually interesting form of stratified complete induction that can be roughly described as follows: Let $T$ be some weak base theory about the natural numbers with the constant 0 and the unary successor function $S$. In the extension $T^*$ of $T$ we have, for every natural number $i$, a fresh unary relation symbols $N_i$ and the axiom

$$\forall x(N_i(x) \leftrightarrow (x = 0 \vee \exists y(x = S(y) \wedge N_i(y))),$$

stating that $N_i$ has the usual closure properties of the natural numbers. The new aspect is that in $T^*$ complete induction is restricted to a form of stratified induction. For all natural numbers $i$ and all formulas $A[u]$,

$$A[0] \wedge \forall x(A[x] \rightarrow A[S(x)]) \rightarrow \forall x(N_i(x) \rightarrow A[x]),$$

*provided that* $A[u]$ *does not contain relation symbols* $N_j$ *with* $j \geq i$. As a simple consequence, $T^*$ proves $\forall x(N_{i+1}(x) \rightarrow N_i(x))$, and so the sequence $(N_i : i \in \mathbb{N})$ provides finer and finer approximations of the "real" natural numbers.

Leivant is interested in this system in connection with Nelson's predicative arithmetic and his own research on systems tailored for representing feasible complexity classes. See Leivant [14, 15] for all details. Wainer and Williams [22] begin from a similar standpoint and analyze inductive definitions over a predicative arithmetic.

---

G. Jäger (✉) • D. Probst
Institut für Informatik und angewandte Mathematik, Universität Bern, Neubrückstrasse 10, 3012 Bern, Switzerland
e-mail: jaeger@iam.unibe.ch; probst@iam.unibe.ch



R. Kahle, M. Rathjen (eds.), *Gentzen's Centenary*,
DOI 10.1007/978-3-319-10103-3_15

In this article, we apply Leivant's idea in the context of inductive definitions: we study an analogous stratification of the fixed point induction axioms of the theory $\mathsf{ID}_1$ of one inductive definition, incontrovertibly one of the best studied theories at the borderline between predicative and impredicative proof theory. There exist numerous articles on $\mathsf{ID}_1$, good introductory texts are, for example, Buchholz et al. [7] and Pohlers [16].

The language $\mathcal{L}(\mathsf{ID}_1)$ of $\mathsf{ID}_1$ is the extension of the language $\mathcal{L}$ of first order arithmetic by a new (unary) relation symbol $\mathcal{P}^{\mathfrak{A}}$ for every $X$ positive arithmetic formula $\mathfrak{A}[X, x]$. The axioms of $\mathsf{ID}_1$ are the axioms of Peano arithmetic $\mathsf{PA}$ with the schema of complete induction for all $\mathcal{L}(\mathsf{ID}_1)$ formulas plus the fixed point axioms

$$\forall x(\mathcal{P}^{\mathfrak{A}}(x) \leftrightarrow \mathfrak{A}[\mathcal{P}^{\mathfrak{A}}, x]) \qquad (Fix)$$

and the axioms for fixed point induction

$$\forall x(\mathfrak{A}[B, x] \rightarrow B[x]) \rightarrow \forall x(\mathcal{P}^{\mathfrak{A}}(x) \rightarrow B[x]), \qquad (FI)$$

where $B[u]$ ranges over all $\mathcal{L}(\mathsf{ID}_1)$ formulas. Together, they formalize that $\mathcal{P}^{\mathfrak{A}}$ represents the least definable fixed point of the monotone operator $\Gamma_{\mathfrak{A}}$ defined by

$$\Gamma_{\mathfrak{A}}(M) := \{n \in \mathbb{N} : \mathbb{N} \models \mathfrak{A}[M, n]\}$$

for any $M \subseteq \mathbb{N}$. The proof-theoretic ordinal of $\mathsf{ID}_1$ is the Bachmann–Howard ordinal $\theta\varepsilon_{\Omega+1}0$.

Currently studied subsystems of $\mathsf{ID}_1$ with classical logic are obtained by weakening fixed point induction $(FI)$. For instance, in $\mathsf{ID}^*$ $(FI)$ is restricted to formulas positive in the fixed point constants, and $\widehat{\mathsf{ID}}_1$ is obtained by dropping $(FI)$ completely. Both systems are significantly weaker than $\mathsf{ID}_1$ and have the proof-theoretic ordinal $\varphi\varepsilon_0 0$; see, e.g., Aczel [2], Afshari and Rathjen [3], Feferman [9], Friedman [10], Probst [17].

In contrast to these approaches, the subsystems of $\mathsf{ID}_1$ studied in this article are obtained by a stratification of the fixed point axioms $(FI)$. The idea is simple: Let $\prec$ be a primitive recursive wellordering of order type $\nu$, given from outside. For any $X$ positive arithmetic $\mathfrak{A}[X, x]$ and any $\alpha \prec \nu$ we add a fresh unary relation symbol $\mathcal{P}^{\mathfrak{A}}_{\alpha}$ to the language $\mathcal{L}$ and consider the fixed point axiom

$$(\forall a \prec \nu)\forall x(\mathcal{P}^{\mathfrak{A}}_a(x) \leftrightarrow \mathfrak{A}[\mathcal{P}^{\mathfrak{A}}_a, x]), \qquad (Fix[\nu])$$

stating that any $\mathcal{P}^{\mathfrak{A}}_a$ with $a \prec \nu$ is a fixed point of $\mathfrak{A}[X, x]$. However, instead of $(FI)$ we only permit *stratified fixed point induction*

$$(\forall a \prec \nu)(\forall x(\mathfrak{A}[B, x] \rightarrow B[x]) \rightarrow \forall x(\mathcal{P}^{\mathfrak{A}}_a(x) \rightarrow B[x])), \qquad (SI[\nu])$$

where now the formula $B[u]$ must not speak about any $\mathcal{P}_b^{\mathfrak{A}}$ with $a \preceq b$. Call the resulting theory $\mathsf{SID}_\nu$. Analogous to the Leivant case, we now have $\forall x(\mathcal{P}_b^{\mathfrak{A}}(x) \to \mathcal{P}_a^{\mathfrak{A}}(x))$ for all $a \prec b \prec \nu$, provable in $\mathsf{SID}_\nu$, and the sequence of relations $(\mathcal{P}_a^{\mathfrak{A}} : a \prec \nu)$ thus provides a decreasing approximation of the least fixed point of the operator $\Gamma_{\mathfrak{A}}$.

Obviously, all $\mathsf{SID}_{1+\nu}$ contain $\widehat{\mathsf{ID}}_1$. However, for finite $\nu$, the theories $\mathsf{SID}_\nu$ are surprisingly weak. As shown in Ranzi and Strahm [18], $\mathsf{SID}_1, \mathsf{SID}_2, \dots$ have the same strength and are all proof-theoretically equivalent to the fixed point theory $\widehat{\mathsf{ID}}_1$.

The first interesting increase of strength happens when we move to $\mathsf{SID}_\omega$ and at the limit ordinals later. In the following we present a complete proof-theoretic analysis of the theories $\mathsf{SID}_\nu$ and obtain the following result: Given an ordinal $\nu > 0$, we write it as

$$\nu = \omega^{\alpha_m} + \dots + \omega^{\alpha_1} \quad \text{with} \quad \alpha_1 \leq \dots \leq \alpha_m,$$

let $\varepsilon(\nu)$ be the least $\varepsilon$-number greater than $\nu$ and define

$$\Lambda_\nu := \varphi\alpha_m(\dots(\varphi\alpha_1\varepsilon(\nu))\dots).$$

Then $\varphi\Lambda_\nu 0$ is the proof-theoretic ordinal of $\mathsf{SID}_\nu$. This implies, for example, that we need all stratifications less than $\varepsilon_0$ in order to reach the strength of $\widehat{\mathsf{ID}}_2$, i.e. $\mathsf{SID}_{<\varepsilon_0} \equiv \widehat{\mathsf{ID}}_2$.

Very much in the spirit of Gentzen-style proof theory, the lower bounds are established by carrying out wellordering proofs within the systems $\mathsf{SID}_\nu$. For determining the upper bounds, we use a combination of various forms of cut elimination and asymmetric interpretations. For obtaining the full picture and since it is needed for the general reduction, we also sketch the finite case, using an approach slightly different from that in [18].

## 2 Ordinal-Theoretic Preliminaries

Every theory $\mathsf{SID}_\nu$ is based on stratifications of inductive definitions along a primitive recursive wellordering of order type $\nu$. In order to concentrate on the essential proof-theoretic aspects of the theories $\mathsf{SID}_\nu$ and to make our approach as perspicuous as possible we fix a specific primitive recursive wellordering $\prec$ of order type $\Gamma_0$ right in advance and iterate stratifications along its initial segments.

It will become evident that $\prec$ and $\Gamma_0$ could be replaced by ordinal notations system generated, for example, from the ternary Veblen functions (cf. Jäger and Strahm [12]), the Veblen functions of all finite arities, Schütte's Klammersymbole (cf. Schütte [20]), Feferman's $\theta$ functions (cf. Aczel [1] and Buchholz [4]), or Buchholz's $\psi$ functions (cf. Buchholz [5]). The choice of $\prec$ and $\Gamma_0$ is only motivated by notational simplicity.

The standard notation system up to the Feferman–Schütte ordinal $\Gamma_0$ is provided by the usual Veblen hierarchy $(\varphi\alpha : \alpha \in On)$ of ordinal functions from $On$ to $On$, inductively defined as follows:

(1) $\varphi 0\beta \;:=\; \omega^\beta$,
(2) if $\alpha > 0$, then $\varphi\alpha$ enumerates $\{\xi \in On : (\forall\gamma < \alpha)(\varphi\gamma\xi = \xi)\}$.

$\Gamma_0$ is the least ordinal $\alpha$ with $\varphi\alpha 0 = \alpha$. We write

$$LI \;:=\; \{\omega\xi : 0 < \xi < \Gamma_0\} \qquad \text{and} \qquad AP \;:=\; \{\omega^\xi : \xi < \Gamma_0\}$$

for the sets of *limit numbers* and *additive principal numbers* less than $\Gamma_0$; the ordinals $\xi$ with $\omega^\xi = \xi$ are called *ε-numbers*. Also, we set for all ordinals $\alpha$,

$$\varepsilon_\alpha \;:=\; \varphi 1\alpha \quad \text{and} \quad \varepsilon(\alpha) \;=\; \text{least element of } \{\xi > \alpha : \omega^\xi = \xi\}.$$

Hence $\varepsilon(\alpha)$ is the least $\varepsilon$-number greater than $\alpha$. From now on we expect that the reader is familiar with ordinal computations and the basic properties of the Veblen hierarchy; all relevant details can be found, for example, in Pohlers [16] and Schütte [21]. In particular, there are two important decomposition properties:

(D.1) For any ordinal $\alpha > 0$ there are uniquely determined $\alpha_1, \dots, \alpha_m$ such that $\alpha = \omega^{\alpha_m} + \cdots + \omega^{\alpha_1}$ and $\alpha_1 \le \cdots \le \alpha_m$.
(D.2) For any $\varepsilon$-number $\alpha < \Gamma_0$ there are uniquely determined $\beta$ and $\gamma$ such that $\alpha = \varphi\beta\gamma$, $0 < \beta < \alpha$, and $\gamma < \alpha$.

Making use of these decompositions, we inductively assign a *fundamental sequence* $(\alpha[n] : n < \omega)$ to any limit number less than $\Gamma_0$:

(FS.1) If $\alpha = \omega^{\alpha_m} + \cdots + \omega^{\alpha_1}$ with $0 < \alpha_1 \le \cdots \le \alpha_m$ and $\omega^{\alpha_1} < \alpha$, then

$$\alpha[n] \;:=\; \omega^{\alpha_m} + \cdots + \omega^{\alpha_1}[n].$$

(FS.2) If $\alpha = \omega^{\beta+1}$, then $\alpha[n] := \omega^\beta n$.
(FS.3) If $\alpha = \omega^\beta$ with $\beta < \alpha$ and $\beta \in LI$, then $\alpha[n] := \omega^{\beta[n]}$.
(FS.4) If $\alpha = \varphi(\beta+1)0$, then $\alpha[0] := 0$ and $\alpha[n+1] := \varphi\beta\alpha[n]$.
(FS.5) If $\alpha = \varphi(\beta+1)(\gamma+1)$, then

$$\alpha[0] := \varphi(\beta+1)\gamma + 1 \quad \text{and} \quad \alpha[n+1] := \varphi\beta\alpha[n].$$

(FS.6) If $\alpha = \varphi\beta 0$ with $\beta \in LI$, then $\alpha[n] := \varphi\beta[n]0$.
(FS.7) If $\alpha = \varphi\beta(\gamma+1)$ with $\beta \in LI$, then

$$\alpha[0] := \varphi\beta\gamma + 1 \quad \text{and} \quad \alpha[n+1] := \varphi\beta[n]\alpha[n].$$

(FS.8) If $\alpha = \varphi\beta\gamma$ with $\gamma \in LI$, then $\alpha[n] := \varphi\beta\gamma[n]$.

As it is easy to check, for all $\alpha \in LI$ and $n < \omega$ we have $\alpha[n] < \alpha[n+1] < \alpha$ and $\alpha = \sup(\{\alpha[n] : n < \omega\})$.

Decomposition (D1) is used once more. Given $\nu > 0$, we write it as $\nu = \omega^{\alpha_m} + \cdots + \omega^{\alpha_1}$ with $\alpha_1 \leq \cdots \leq \alpha_m$. Depending on this presentation of $\nu$, we inductively define a sequence of ordinals $\nu_0, \ldots, \nu_m$ by

$$\nu_0 := \varepsilon(\nu) \qquad \text{and} \qquad \nu_{i+1} := \varphi\alpha_{i+1}\nu_i$$

and then set

$$\Lambda_\nu := \nu_m.$$

To finish the definition of $\Lambda$ we set $\Lambda_0 := 1$. As we will see, the ordinals $\Lambda_\nu$ play a crucial role in the proof-theoretic analysis of the theories $\mathsf{SID}_\nu$.

Simple computations show that, for example, $\Lambda_n = \varepsilon_0$ for all $0 < n < \omega$, $\Lambda_\omega = \varepsilon_{\varepsilon_0}$, $\Lambda_{\omega+\omega} = \varepsilon_{\varepsilon_{\varepsilon_0}}$, $\Lambda_{\varepsilon_\alpha} = \varphi\varepsilon_\alpha\varepsilon_{\alpha+1}$ for all $\alpha$, and $\Lambda_{\varphi\alpha\beta} = \varphi(\varphi\alpha\beta)\varepsilon_{\varphi\alpha\beta+1}$ for all $\alpha > 1$ and all $\beta$. The following lemma is immediate from the definition of $\Lambda_\nu$.

**Lemma 1** *If $\nu = \mu + k$ for some $k < \omega$, then $\varepsilon(\nu) = \varepsilon(\mu)$; if, in addition, $\mu > 0$, then $\Lambda_\nu = \Lambda_\mu$.*

In the textbooks by Pohlers and Schütte it is also explained in detail that there exists a primitive recursive wellordering $\prec$ on the natural numbers corresponding to the ordinals less than $\Gamma_0$; $m \preceq n$ is written iff $m \prec n$ or $m = n$.

Each natural number codes exactly one ordinal less than $\Gamma_0$, and given an $n \in \mathbb{N}$, we write $ot(n)$ for this ordinal; $ot(n)$ is called the *order type* of $n$ with respect to $\prec$. The inverse of $ot$, let us call it $nr$, assigns a natural number $nr(\alpha)$ to any $\alpha < \Gamma_0$. The sets $\{nr(\xi) : \xi \in LI\}$ and $\{nr(\xi) : \xi \in AP\}$ are primitive recursive subsets of the natural numbers.

Furthermore, for all ordinal functions $f$ on $(\Gamma_0, <)$ such as addition, multiplication, exponentiation, $\varepsilon$, $\varphi$, fundamental sequences, $\Lambda$, ... there exist primitive recursive functions $f_{code}$ acting on $(\mathbb{N}, \prec)$ that correspond to these ordinal operations. Without loss of generality we can assume that $ot(0) = 0$.

## 3 The Theories $\mathsf{SID}_\nu$

In the following we let $\mathcal{L}$ denote our language of first order arithmetic. It includes *number variables* $a, b, c, d, u, v, w, x, y, z$ (possibly with subscripts), symbols for all primitive recursive functions and relations as well as the unary relation symbol $W$. This relation symbol $W$ plays the role of an anonymous relation variable with no specific meaning. Its role will become clear in Definition 2 below. Furthermore, there is a symbol $\sim$ for forming negative literals. When dealing with primitive recursive functions and relations we often write the same expression for the primitive recursive function (relation) and for the associated function (relation) symbol.

The *number terms* $p, q, r, s, t$ (possibly with subscripts) of $\mathcal{L}$ are defined as usual; in particular, the numeral associated with the natural number $n$ is denoted by $\bar{n}$. The *positive literals* of $\mathcal{L}$ are all expressions of the form $R(s_1, \ldots, s_n)$ where $R$ is a symbol for an $n$-ary primitive recursive relation and all expressions $W(s)$. The *negative literals* of $\mathcal{L}$ are all expressions $\sim E$ such that $E$ is a positive literal of $\mathcal{L}$. Infix notation is used whenever convenient, and $(s = t)$ stands for $R_=(s, t)$ if $R_=$ is the symbol for primitive recursive equality.

The *formulas* $A, B, C, D$ (possibly with subscripts) of $\mathcal{L}$ are generated from the positive and negative literals of $\mathcal{L}$ by closing against disjunctions, conjunctions as well as existential and universal number quantifications. The negation $\neg A$ of an $\mathcal{L}$ formula $A$ is defined by making use of De Morgan's laws and the law of double negation; the remaining logical connectives are abbreviated as usual. We will often omit parentheses and brackets whenever there is no danger of confusion. Also, we frequently make use of the vector notation $\vec{\mathfrak{e}}$ as shorthand for a finite string $\mathfrak{e}_1, \ldots, \mathfrak{e}_n$ of expressions whose length is not important or evident from the context.

Suppose now that $\vec{a} = a_1, \ldots, a_n$ and $\vec{s} = s_1, \ldots, s_n$. Then $A[\vec{s}/\vec{a}]$ is the $\mathcal{L}$ formula that is obtained from the $\mathcal{L}$ formula $A$ by simultaneously replacing all free occurrences of the variables $\vec{a}$ by the $\mathcal{L}$ terms $\vec{s}$ (in order to avoid collision of variables, a renaming of bound variables may be necessary). If the $\mathcal{L}$ formula $A$ is written as $B[\vec{a}]$, then we often simply write $B[\vec{s}]$ instead of $A[\vec{s}/\vec{a}]$; variants of this notation will be self-explaining.

If $X$ is a fresh unary relation symbol, we let $\mathcal{L}(X)$ denote the extension of $\mathcal{L}$ by $X$; i.e. expressions of the forms $X(s)$ and $\sim X(s)$ are additional literals. Given a formula $A[X]$ of $\mathcal{L}(X)$ and a formula $B[u]$ of $\mathcal{L}$, we write $A[\{x{:}B[x]\}]$ to indicate the result of substituting $B[s]$ for each occurrence of $X(s)$ and $\neg B[s]$ for each occurrence of $\sim X(s)$ in $A[X]$ (again, bound variables are renamed if necessary). If $\mathcal{L}'$ is a language extending $\mathcal{L}$, then $\mathcal{L}'(X)$ is defined accordingly.

We will be interested in determining the proof-theoretic ordinals of the theories $\mathsf{SID}_\nu$. For this purpose we fix the auxiliary notions of progressiveness and transfinite induction. Given a primitive recursive relation $\lhd$, an $\mathcal{L}$ term $s$, and a formula $A[a]$ of $\mathcal{L}$ (or of some extension of $\mathcal{L}$ to be introduced later), we set:

$$Prog[\lhd, \{x{:}A[x]\}] \ := \ \forall x((\forall y \lhd x)A[y] \to A[x]),$$

$$TI[\lhd, \{x{:}A[x]\}] \ := \ Prog[\lhd, \{x{:}A[x]\}] \to \forall x A[x],$$

$$TI[\lhd, s, \{x{:}A[x]\}] \ := \ Prog[\lhd, \{x{:}A[x]\}] \to (\forall x \lhd s)A[x].$$

In the following we often work with the primitive recursive wellordering $\prec$ introduced in the previous section and thus, for instance, simply write $Prog[\{x{:}A[x]\}]$ for $Prog[\prec, \{x{:}A[x]\}]$ and $TI[s, \{x{:}A[x]\}]$ for $TI[\prec, s, \{x{:}A[x]\}]$.

**Definition 2** Let $T$ be a theory formulated in $\mathcal{L}$ or an extension of $\mathcal{L}$.

1. An ordinal $\alpha$ is called *provable in* $T$ iff there exists a primitive recursive wellordering $\lhd$ of order type $\alpha$ such that $T \vdash TI[\lhd, \{x{:}W(x)\}]$.
2. The *proof-theoretic ordinal* $|T|$ *of* $T$ is the least ordinal that is not provable in $T$.

We call an $\mathcal{L}(X)$ formula $X$ *positive* if it has no subformulas of the form $\sim X(s)$. An $X$ positive $\mathcal{L}(X)$ formula that contains at most the variable $x$ free is called an *inductive operator form*, and we let $\mathfrak{A}[X, x]$ range over such forms.

From now on $\nu$ always stands for an ordinal less than $\Gamma_0$, and $\bar{\nu}$ denotes the numeral $\overline{nr(\nu)}$ corresponding to the element $nr(\nu) \in OT$. For the formulation of the theories $\mathsf{SID}_\nu$ we add to the first order language $\mathcal{L}$ a new unary relation symbol $\mathcal{P}^{\mathfrak{A}}$ for every inductive operator form $\mathfrak{A}[X, x]$ and call this new language $\mathcal{L}_S$. We write $\mathcal{P}^{\mathfrak{A}}_s(t)$ for $\mathcal{P}^{\mathfrak{A}}(\langle s, t\rangle)$ and $\mathcal{P}^{\mathfrak{A}}_{\prec s}(t)$ for $(t = \langle (t)_0, (t)_1\rangle \wedge (t)_0 \prec s \wedge \mathcal{P}^{\mathfrak{A}}(t))$, where $\langle .,. \rangle$ denotes a primitive recursive pairing function with the associated primitive recursive projection functions $(.)_0$ and $(.)_1$. Also, $\mathfrak{A}[\mathcal{P}^{\mathfrak{A}}_a, b]$ and $\mathfrak{A}[\mathcal{P}^{\mathfrak{A}}_{\prec a}, b]$ are short for $\mathfrak{A}[\{x{:}\mathcal{P}^{\mathfrak{A}}_a(x)\}, b]$ and $\mathfrak{A}[\{x{:}\mathcal{P}^{\mathfrak{A}}_{\prec a}(x)\}, b]$, respectively.

We express the closure of an $\mathcal{L}_S$ formula $B[a]$ under the inductive operator form $\mathfrak{A}[X, x]$ by the formula

$$Cl_{\mathfrak{A}}[\{a{:}B[a]\}] \;:=\; \forall x(\mathfrak{A}[\{a{:}B[a]\}, x] \rightarrow B[x]).$$

For formulating stratified fixed point induction, a further shorthand notation is useful. Given a number variable $u$, we call an $\mathcal{L}_S$ formula $A$ *bounded by* $u$ iff all relation symbols $\mathcal{P}^{\mathfrak{A}}$ occur in $A$ only in the form $\mathcal{P}^{\mathfrak{A}}_{\prec u}(t)$ or $\sim\mathcal{P}^{\mathfrak{A}}_{\prec u}(t)$. More formally, $\mathcal{BL}_S(u)$ is the collection of $\mathcal{L}_S$ formulas inductively generated as follows:

($\mathcal{B}$.1) All atomic formulas of $\mathcal{L}$ as well as all formulas $\mathcal{P}^{\mathfrak{A}}_{\prec u}(t)$ and $\sim\mathcal{P}^{\mathfrak{A}}_{\prec u}(t)$ belong to $\mathcal{BL}_S(u)$ (for all inductive operator forms $\mathfrak{A}[X, x]$).
($\mathcal{B}$.2) If $A$ and $B$ belong to $\mathcal{BL}_S(u)$, then $(A \vee B)$ and $(A \wedge B)$ belong to $\mathcal{BL}_S(u)$.
($\mathcal{B}$.3) If $A$ belongs to $\mathcal{BL}_S(u)$ and $x$ is a number variable different from $u$, then $\exists x A$ and $\forall x A$ belong to $\mathcal{BL}_S(u)$.

Therefore, if $a$ is an element of $OT$, then the formulas in $\mathcal{BL}_S(a)$ are $\mathcal{L}_S$ formulas in which only stratifications less than $a$ play a role.

Every theory $\mathsf{SID}_\nu$ is formulated in the language $\mathcal{L}_S$ for stratified inductive definitions. Its axioms and rules of inference are the usual axioms and rules of inference of first order logic, the usual equality axioms formulated for all $\mathcal{L}_S$ formulas plus the following four classes of non-logical axioms.

**I. Peano axioms.** All axioms of Peano arithmetic $\mathsf{PA}$ with the schema of complete induction for all formulas of $\mathcal{L}_S$.

**II. Transfinite induction up to** $\nu$. For all formulas $A[u]$ of $\mathcal{L}_S$:

$$TI[\bar{\nu}, \{x{:}A[x]\}]. \qquad (TI[\nu])$$

**III. Fixed point axioms.** For all inductive operator forms $\mathfrak{A}[X,x]$:

$$(\forall a \prec \bar{\nu})\forall x(\mathcal{P}_a^{\mathfrak{A}}(x) \leftrightarrow \mathfrak{A}[\mathcal{P}_a^{\mathfrak{A}}, x]). \qquad (Fix[\nu])$$

**IV. Stratified fixed point induction.** For all inductive operator forms $\mathfrak{A}[X,x]$ and all formulas $B[u,v]$ from $\mathcal{BL}_S(u)$:

$$(\forall a \prec \bar{\nu})(Cl_{\mathfrak{A}}[\{x{:}B[a,x]\}] \rightarrow \forall x(\mathcal{P}_a^{\mathfrak{A}}(x) \rightarrow B[a,x])). \qquad (SI[\nu])$$

The theories $\mathsf{SID}_\nu$ have the important property that the stratifications of the fixed points form a weakly decreasing sequence of relations.

**Lemma 3** *For all inductive operator forms* $\mathfrak{A}[X,x]$ *we can prove in* $\mathsf{SID}_\nu$ *that*

$$(\forall a, b \prec \bar{\nu})\forall x(a \prec b \wedge \mathcal{P}_b^{\mathfrak{A}}(x) \rightarrow \mathcal{P}_a^{\mathfrak{A}}(x)).$$

*Proof* We define $B[u,v,w] := (v \prec u \wedge P_{\prec u}^{\mathfrak{A}}(\langle v, w\rangle))$. Obviously, this formula belongs to $\mathcal{BL}_S(u)$ and $\mathsf{SID}_\nu$ proves

$$a \prec b \prec \bar{\nu} \rightarrow \forall x(\mathcal{P}_a^{\mathfrak{A}}(x) \leftrightarrow B[b,a,x]).$$

In view of $(Fix[\nu])$ and $(SI[\nu])$ our assertion follows immediately. □

Besides the theories $\mathsf{SID}_\nu$, also their unions are of some interest. If $\nu > 0$, we write $\mathsf{SID}_{<\nu}$ for the union of the theories $\mathsf{SID}_\mu$ with $\mu < \nu$,

$$\mathsf{SID}_{<\nu} := \bigcup_{\mu<\nu} \mathsf{SID}_\mu.$$

In the following sections we will show that $|\mathsf{SID}_\nu| = \varphi\Lambda_\nu 0$. The theory $\mathsf{SID}_0$ contains neither fixed point axioms nor axioms for stratified fixed point induction and simply is a variant of Peano arithmetic $\mathsf{PA}$. The theories $\mathsf{SID}_\nu$ for $\nu > 0$ are more interesting. Here are some specific examples of theories and their proof-theoretic ordinals:

- $|\mathsf{SID}_1| = |\mathsf{SID}_{<\omega}| = \varphi\varepsilon_0 0$,
- $|\mathsf{SID}_\omega| = |\mathsf{SID}_{<\omega+\omega}| = \varphi\varepsilon_{\varepsilon_0} 0$ and $|\mathsf{SID}_{\omega+\omega}| = \varphi\varepsilon_{\varepsilon_{\varepsilon_0}} 0$,
- $|\mathsf{SID}_{<\omega^\omega}| = \varphi(\varphi\omega 0)0$ and $|\mathsf{SID}_{\omega^\omega}| = \varphi(\varphi\omega\varepsilon_0)0$,
- $|\mathsf{SID}_{<\varepsilon_0}| = \varphi(\varphi\varepsilon_0 0)0$ and $|\mathsf{SID}_{\varepsilon_0}| = \varphi(\varphi\varepsilon_0\varepsilon_1)0$,
- $|\mathsf{SID}_{<\Gamma_0}| = \Gamma_0$.

## 4 Lower Proof-Theoretic Bound for $\mathsf{SID}_\nu$

The lower bounds of the theories $\mathsf{SID}_\nu$ will be established by carrying out wellordering proofs within the theories $\mathsf{SID}_\nu$. To increase readability we shall use in our formal language $\mathcal{L}_S$ the ordinal-theoretic functions $f$ on $(\Gamma_0, <)$ introduced in Sect. 2 instead of their primitive recursive analogues $f_{code}$ on $(OT, \prec)$. We also write $\alpha$ instead of $\overline{nr(\alpha)}$ in terms and formulas of $\mathcal{L}_S$. Thus, for instance $s+t$, $\omega^s$, $\varphi\omega 0$ are to be considered as terms of $\mathcal{L}_S$ and $(\forall x \prec \omega^\omega)(\exists y \prec \omega)(x \prec \omega^y)$ is to be considered as a formula of $\mathcal{L}_S$. *LI* and *AP* are used as relation symbols for the sets of (the codes of) the limit numbers and additive principal numbers below $\Gamma_0$.

For the following considerations, the provably accessible parts of the relation $\prec$ play the decisive role. We only need the inductive operator form

$$\mathfrak{Ap}[X, x] \;:=\; (\forall y \prec x)X(y).$$

Then, given any $\mathcal{L}_S$ formula $B[u]$, the closure assertion $Cl_{\mathfrak{Ap}}[\{x\!:\!B[x]\}]$ simply means $Prog[\{x\!:\!B[x]\}]$. For all number terms $s$ and $t$ we introduce as abbreviations:

$$AC_s[t] \;:=\; \mathcal{P}^{\mathfrak{Ap}}_s(t),$$

$$AC_{\prec s}[t] \;:=\; \mathcal{P}^{\mathfrak{Ap}}_{\prec s}(t),$$

$$AC^s[t] \;:=\; (\forall x \prec s)AC_x[t].$$

Thus $AC^s$ describes the intersection of all stratifications of the inductive operator form $\mathfrak{Ap}[X, x]$ less than $s$. As we can easily conclude from Lemma 3 these intersections have the following property.

**Lemma 4** *We can prove in* $\mathsf{SID}_\nu$ *that*

$$\forall a(a \prec \bar{\nu} \;\rightarrow\; \forall x(AC^{a+1}[x] \;\leftrightarrow\; AC_a[x])).$$

Recall some notation: If the number terms $s$ and $t$ code the ordinal $\alpha$ and the number $n < \omega$, respectively, then $s[t]$ codes the $n$th component of the fundamental sequence of $\alpha$. The following useful observation is directly implied by the fixed point axioms.

**Lemma 5** *We can prove in* $\mathsf{SID}_\nu$*:*

1. $a \prec \bar{\nu} \wedge (\forall x \prec \omega)AC_a[s[x]]) \;\rightarrow\; AC_a[s]$.
2. $a \preceq \bar{\nu} \wedge (\forall x \prec \omega)AC^a[s[x]]) \;\rightarrow\; AC^a[s]$.

After these preparatory remarks we now turn to the wellordering proofs. We begin with considering a property of the stratifications $AC_a$ that will turn out to be central for what follows.

**Lemma 6** *We can prove in* $\mathsf{SID}_\nu$ *that*

$$a + 1 \prec \bar{\nu} \wedge AC_{a+1}[b] \to AC_a[\omega^b].$$

*Proof* Working in $\mathsf{SID}_\nu$, we fix an $a$ such that $a + 1 \prec \bar{\nu}$. Then define

$$A[x] := \forall z(AC_a[z] \to AC_a[z + \omega^x]).$$

We show

$$(\forall y \prec x)A[y] \to A[x]. \tag{1}$$

for an arbitrary $x$ by distinguishing the following cases.

(i) For $x = 0$ or $LI(x)$, assertion (1) is an immediate consequence of the closure properties of $AC_a$ or of the previous lemma.
(ii) Now assume that $x = y + 1$ for some $y$. Then complete induction yields

$$A[y] \wedge AC_a[z] \to (\forall e \prec \omega)AC_a[z + \omega^y e],$$

and we conclude

$$A[y] \wedge AC_a(z) \to AC_a[z + \omega^{y+1}].$$

This establishes (1) also in this case and finishes the proof of this auxiliary consideration. We further observe that the formula

$$B[u, v, w] := v \prec u \wedge \forall z(AC_{\prec u}[\langle v, z\rangle] \to AC_{\prec u}[\langle v, z + \omega^w\rangle])$$

belongs to $\mathcal{BL}_S(u)$ and that

$$\forall x(A[x] \leftrightarrow B[a + 1, a, x]). \tag{2}$$

From (1) we have $Prog[\{x : A[x]\}]$, i.e. $Cl_{\mathfrak{A}_\mathfrak{p}}[\{x : B[a + 1, a, x]\}]$, hence stratified fixed point induction implies

$$AC_{a+1}[b] \to B[a + 1, a, b]$$

and thus

$$AC_{a+1}[b] \to \forall z(AC_a[z] \to AC_a[z + \omega^b])$$

according to (2) and the definition of $A$. For $z = 0$ this is the assertion of our lemma. □

We continue with introducing a formula, depending on $\nu$, that describes a specific property of stratifications of the accessible parts with respect to the Veblen functions,

$$A_\nu[u] \ := \ \forall x \forall y (x + \omega^u \prec \bar{\nu} \wedge AC_{x+\omega^u}[y] \ \rightarrow \ AC^{x+\omega^u}[\varphi u y]).$$

The following two lemmas isolate some technical properties that will be needed in the proofs of Theorem 9 and Corollary 10 below.

**Lemma 7** *We can prove in* $\mathsf{SID}_\nu$ *that*

$$A_\nu[s] \wedge s \prec r \wedge t + \omega^r \preceq \bar{\nu} \ \rightarrow \ \forall x (AC^{t+\omega^r}[x] \rightarrow AC^{t+\omega^r}[\varphi s x]).$$

*Proof* Assume $A_\nu[s]$, $s \prec r$, and $t + \omega^r \preceq \bar{\nu}$ and pick an $x$ such that $AC^{t+\omega^r}[x]$. Now we distinguish the following cases:

(i) $r = p + 1$. Choose an arbitrary $u \prec \omega$. Then $t + \omega^p u + \omega^s \prec t + \omega^r$, and so $AC^{t+\omega^r}[x]$ implies $AC_{t+\omega^p u+\omega^s}[x]$. Using the assumption $A_\nu[s]$, we conclude that $AC^{t+\omega^p u+\omega^s}[\varphi s x]$, hence also $AC^{t+\omega^p u}[\varphi s x]$. So we have shown that

$$(\forall u \prec \omega) AC^{t+\omega^p u}[\varphi s x].$$

In view of Lemma 5 this implies $AC^{t+\omega^r}[\varphi s x]$.

(ii) $r \in LI$. Again we choose an arbitrary $u \prec \omega$ and observe that now $t + \omega^{r[u]} + \omega^s \prec t + \omega^r$. Because of $AC^{t+\omega^r}[x]$ we thus have $AC_{t+\omega^{r[u]}+\omega^s}[x]$, and the assumption $A_\nu[s]$ implies $AC^{t+\omega^{r[u]}+\omega^s}[\varphi s x]$, hence also $AC^{t+\omega^{r[u]}}[\varphi s x]$. This means that we have

$$(\forall u \prec \omega) AC^{t+\omega^{r[u]}}[\varphi s x],$$

and again a simple application of Lemma 5 yields $AC^{t+\omega^r}[\varphi s x]$.

Since for $r = 0$ nothing is to show, the proof of our assertion is complete. □

**Lemma 8** *We can prove in* $\mathsf{SID}_\nu$ *that*

$$(\forall z \prec r) A_\nu[z] \wedge 0 \prec r \wedge t + \omega^r \preceq \bar{\nu} \ \rightarrow \ Prog[\{x{:}AC^{t+\omega^r}[\varphi r x]\}].$$

*Proof* For any $a$ we show that under the assumptions $(\forall z \prec r) A_\nu[z]$, $0 \prec r$, and $t + \omega^r \preceq \bar{\nu}$,

$$(\forall x \prec a) AC^{t+\omega^r}[\varphi r x] \ \rightarrow \ AC^{t+\omega^r}[\varphi r a]. \tag{3}$$

For this purpose, fix an arbitrary $a$ and consider the fundamental sequence $(\varphi r a)[u]$, for $u \prec \omega$, of $\varphi r a$. By complete induction on $u$, making essential use of the previous lemma, we then show

$$(\forall x \prec a)AC^{t+\omega^r}[\varphi r x] \;\rightarrow\; (\forall u \prec \omega)AC^{t+\omega^r}[(\varphi r a)[u]]. \tag{4}$$

In this proof a case distinction with respect to $\varphi ra$ is carried through. We discuss one case and leave the others to the reader. So assume $r = p + 1$ and $a = b + 1$. Then $(\varphi ra)[0] = \varphi rb + 1$ and $(\varphi ra)[u + 1] = \varphi p(\varphi ra)[u]$. Clearly,

$$(\forall x \prec a)AC^{t+\omega^r}[\varphi r x] \;\rightarrow\; AC^{t+\omega^r}[(\varphi r a)[0]]. \tag{5}$$

Applying Lemma 7, we also obtain

$$AC^{t+\omega^r}[(\varphi r a)[u]] \;\rightarrow\; AC^{t+\omega^r}[(\varphi r a)[u+1]]. \tag{6}$$

Assertion (4) is immediate from (5) and (6) by complete induction. All other cases are similar.

So we have (4). But this completes the proof of our lemma since (3) is an immediate consequence of (4) and Lemma 5. □

**Theorem 9** *The theory* $\mathsf{SID}_\nu$ *proves* $Prog[\{z{:}A_\nu[z]\}]$.

*Proof* Given any $a$, we have to show in $\mathsf{SID}_\nu$ that

$$(\forall z \prec a)A_\nu[z] \;\rightarrow\; A_\nu[a].$$

If $a = 0$, then $A_\nu[a]$ follows immediately from Lemmas 4 and 6. So let $0 \prec a$ and assume $(\forall z \prec a)A_\nu[z]$. In view of the previous lemma we have

$$\forall x(x + \omega^a \prec \bar{\nu} \;\rightarrow\; Prog[\{z{:}AC^{x+\omega^a}[\varphi a z]\}]).$$

Since $AC^{x+\omega^a}[\varphi az]$ is (equivalent to) a formula bounded by $x + \omega^a$, stratified fixed point induction yields

$$\forall x(x + \omega^a \prec \bar{\nu} \;\rightarrow\; \forall y(AC_{x+\omega^a}[y] \rightarrow AC^{x+\omega^a}[\varphi a y])),$$

and this formula is equivalent to $A_\nu[a]$. This completes the proof of this lemma. □

**Corollary 10** *The theory* $\mathsf{SID}_\nu$ *proves*

$$t + \omega^r \preceq \bar{\nu} \wedge 0 \prec r \;\rightarrow\; Prog[\{x{:}AC^{t+\omega^r}[\varphi r x]\}].$$

*Proof* Since transfinite induction up to $\nu$ is available in $\mathsf{SID}_\nu$, the previous theorem gives us $(\forall z \prec \bar{\nu})A_\nu[z]$. By Lemma 8 this implies what we want. □

**Theorem 11** *Let* $\nu$ *be a limit number. If* $\alpha < \Lambda_\nu$, *then* $\mathsf{SID}_\nu$ *proves* $AC^{\omega}[\bar{\alpha}]$ *for the code* $\bar{\alpha}$ *of* $\alpha$.

*Proof* Since $\nu$ is a limit ordinal, there are uniquely determined ordinals $\sigma_1, \ldots, \sigma_m$ such that $\nu = \omega^{\sigma_m} + \cdots + \omega^{\sigma_1}$ and $1 \leq \sigma_1 \leq \cdots \leq \sigma_m$. Pick an arbitrary $\alpha < \Lambda_\nu$.

Then simple ordinal computation shows that there exists an ordinal $\beta < \varepsilon(\nu)$ for which

$$\alpha < \varphi\sigma_m(\varphi\sigma_{m-1}(\dots(\varphi\sigma_1\beta)\dots)). \tag{7}$$

Since $\mathsf{SID}_\nu$ comprises transfinite induction up to $\nu$, standard proof-theoretic techniques yield

$$TI[\bar{\beta}, \{x{:}A[x]\}] \tag{8}$$

for all $\mathcal{L}_S$ formulas $A[u]$ where $\bar{\beta}$ is the code of $\beta$.

For $i = 1, \dots, m$ we let $s_i$ be the code of $\sigma_i$, and for $j = 0, \dots, m-1$ we let $t_j$ be the code of $\omega^{\sigma_m} + \dots + \omega^{\sigma_{j+1}}$. Then we define, for $i = 1, \dots, m$ and $j = 1, \dots, m-1$:

$$B_i[u] := AC^{t_{i-1}}[\varphi s_i u] \qquad \text{and} \qquad C_j[u] := AC_{t_j}[u].$$

With these notations we have immediately that, for $i = 1, \dots, m-1$,

$$\forall x(B_i[x] \to C_i[\varphi s_i x]). \tag{9}$$

By Corollary 10 we also know that

$$Prog[\{x{:}B_i[x]\}] \tag{10}$$

for all $i = 1, \dots, m$. As in previous proofs we observe that every formula $B_{j+1}[u]$, for $j = 1, \dots, m-1$, is (equivalent to) a formula bounded by $t_j$ such that stratified fixed point induction implies

$$\forall x(C_i[x] \to B_{i+1}[x]) \tag{11}$$

for $i = 1, \dots, m-1$. Now we proceed as follows. From (8) and (10) we obtain $B_1[\bar{\beta}]$, and then iterative applications of (9) and (11) lead to

$$B_m[\varphi s_{m-1}(\dots(\varphi s_1\bar{\beta})\dots)], \text{ i.e. to } AC^{t_{m-1}}[\varphi s_m(\dots(\varphi s_1\bar{\beta})\dots)].$$

Since $\omega \preceq \omega^{s_m} = t_{m-1}$ and in view of (7) we finally conclude that $AC^{\omega}[\bar{\alpha}]$, as desired. □

Now the stage is set for determining the lower proof-theoretic bounds of $\mathsf{SID}_\nu$. For finite $\nu$, the situation is trivial. In the transfinite cases, the previous theorem and some methods of predicative proof theory do the job.

**Theorem 12** $\varphi\Lambda_\nu 0 \leq |\mathsf{SID}_\nu|$.

*Proof* $\mathsf{SID}_0$ is equivalent to Peano arithmetic PA, and $\Lambda_0 = \varepsilon_0$. If $\nu$ is finite and greater than 0, then $\mathsf{SID}_\nu$ contains the theory $\widehat{\mathsf{ID}}_1$, whose proof-theoretic ordinal is $\varphi\varepsilon_0 0$ (cf.,e.g., Aczel [2] or Feferman [9]), and $\Lambda_\nu = \varphi\varepsilon_0 0$. So the theorem is clear for $\nu < \omega$.

Let us turn to the more interesting situation and assume that $\nu = \mu + n$ for some limit number $\mu$ and some $n < \omega$. Then $\Lambda_\nu = \Lambda_\mu$, and given any ordinal $\alpha < \Lambda_\nu$, the previous theorem yields

$$\mathsf{SID}_\nu \vdash AC^\omega[\bar{\alpha}] \tag{12}$$

for the code $\bar{\alpha}$ of $\alpha$. Consider an arbitrary formula $A[u, v]$ belonging to $\mathcal{BL}_S(u)$. Because of the axioms about stratified fixed point induction we have

$$\mathsf{SID}_\nu \vdash (\forall a \prec \bar{\nu})(Prog[\{z{:}A[a, z]\}] \rightarrow \forall x(AC_a[x] \rightarrow A[a, x])). \tag{13}$$

In particular, if we write 1 for the code of the ordinal $1 \prec \omega$, then (13) implies

$$\mathsf{SID}_\nu \vdash \forall x(AC_1[x] \rightarrow TI[x, \{z{:}A[1, z]\}]).$$

Together with (12) we thus have

$$\mathsf{SID}_\nu \vdash TI[\bar{\alpha}, \{x{:}A[1, x]\}] \tag{14}$$

for all formulas $A[u, v]$ from $\mathcal{BL}_S(u)$.

The following is a standard result of predicative proof-theory, rephrased in the terminology of this article:

> Let $r$ be a closed number term. If $\mathsf{SID}_\nu \vdash TI[r, \{z{:}A[1, z]\}]$ for all formulas $A[u, v]$ from $\mathcal{BL}_S(u)$, then $\mathsf{SID}_\nu \vdash TI[\varphi r 0, \{z{:}B[z]\}]$ for all formulas $B[v]$ of $\mathcal{L}$.

This assertion is proved in detail in Buchholz [6] for the theory $\widehat{\mathsf{ID}}_1$, but since $\mathsf{SID}_\nu$ contains $\widehat{\mathsf{ID}}_1$ it transfers without problems. Applying this result to (14) yields

$$\mathsf{SID}_\nu \vdash TI[\varphi\bar{\alpha}0, \{z{:}B[z]\}]$$

for all $\mathcal{L}$ formulas $B[v]$, hence, in particular, $\mathsf{SID}_\nu \vdash TI[\varphi\bar{\alpha}0, \{x{:}W(x)\}]$. Therefore, $\varphi\alpha 0$ is provable in $\mathsf{SID}_\nu$.

We have shown that for any $\alpha < \Lambda_\nu$ the ordinal $\varphi\alpha 0$ is provable in $\mathsf{SID}_\nu$. This implies the assertion $\varphi\Lambda_\nu 0 \leq |\mathsf{SID}_\nu|$. □

# 5 Upper Proof-Theoretic Bound for $\mathsf{SID}_\nu$

In this section we establish the (sharp) upper proof-theoretic bounds of the theories $\mathsf{SID}_\nu$. Our strategy is similar to that used in Jäger et al. [13] for the proof-theoretic analysis of transfinitely iterated fixed point theories.

We first introduce auxiliary semiformal systems $\mathsf{H}_\nu$, in which the theories $\mathsf{SID}_\nu$ can be embedded. The ordinal analysis of these $\mathsf{H}_\nu$ via the methods of partial cut elimination and asymmetric interpretation finally yields the desired result. To simplify the notation, we assume from now on that in $\mathsf{SID}_\nu$ we work with one inductive operator form $\mathfrak{A}[X, x]$ only; the generalization of everything to a finite number of such forms is obvious.

The language $\mathcal{L}_\infty$ extends the basic first order language $\mathcal{L}$ by unary relation symbols $P^{\mathfrak{A}}_{<\alpha}$, $P^{\mathfrak{A}}_{\alpha}$, and $Q^{<\alpha}_{\mathfrak{A}}$ for all ordinals $\alpha < \Gamma_0$. In the systems $\mathsf{H}_\nu$ the relation symbols $P^{\mathfrak{A}}_{<\alpha}$ and $P^{\mathfrak{A}}_{\alpha}$ will be used to deal with the stratifications $\mathcal{P}^{\mathfrak{A}}_{\prec a}$ and $\mathcal{P}^{\mathfrak{A}}_{a}$ of the theories $\mathsf{SID}_\nu$, whereas the sets defined by the $Q^{<\alpha}_{\mathfrak{A}}$ represent the initial stages

$$I^{<\alpha}_{\mathfrak{A}} = \bigcup_{\xi<\alpha} \{n \in \mathbb{N} : \mathbb{N} \models \mathfrak{A}[I^{<\xi}_{\mathfrak{A}}, n]\}$$

of the inductive definition that is associated with the inductive operator form $\mathfrak{A}[X, x]$ up to the ordinals $\alpha < \Gamma_0$.

**Definition 13** The *formulas* $(A, B, C, A_0, B_0, C_0, \dots)$ of $\mathcal{L}_\infty$ together with their lengths are inductively generated as follows:

1. Every closed literal of $\mathcal{L}$ is an $\mathcal{L}_\infty$ formula of length 0.
2. If $t$ is a closed number term and $\alpha < \Gamma_0$, then $P^{\mathfrak{A}}_{<\alpha}(t)$ and $\sim P^{\mathfrak{A}}_{<\alpha}(t)$ are $\mathcal{L}_\infty$ formulas of length 1.
3. If $t$ is a closed number term and $\alpha < \Gamma_0$, then $P^{\mathfrak{A}}_{\alpha}(t)$ and $\sim P^{\mathfrak{A}}_{\alpha}(t)$ are $\mathcal{L}_\infty$ formulas of length 0.
4. If $t$ is a closed number term and $\alpha < \Gamma_0$, then $Q^{<\alpha}_{\mathfrak{A}}(t)$ and $\sim Q^{<\alpha}_{\mathfrak{A}}(t)$ are $\mathcal{L}_\infty$ formulas of length 0.
5. If $A$ is an $\mathcal{L}_\infty$ formula of length $m$ and if $B$ is an $\mathcal{L}_\infty$ formula of length $n$, then $(A \vee B)$ and $(A \wedge B)$ are $\mathcal{L}_\infty$ formulas of length $\max(m, n) + 1$.
6. If $A[0]$ is an $\mathcal{L}_\infty$ formula of length $m$, then $\exists x A[x]$ and $\forall x A[x]$ are $\mathcal{L}_\infty$ formulas of length $m + 1$.

$\mathcal{L}_\infty$ formulas of length 0 are called $\mathcal{L}_\infty$ *literals*. Clearly, as in the language $\mathcal{L}$, a literal of the form $\sim E$ acts as negation of the literal $E$, the negations $\neg A$ of arbitrary $\mathcal{L}_\infty$ formulas $A$ are defined by making use of De Morgan's laws plus the law of double negation, and the remaining logical connectives are abbreviated as usual.

For the later proof-theoretic considerations we assign two ordinals to any $\mathcal{L}_\infty$ formula $A$: its level $lev(A)$ provides a bound of the stratifications of the fixed point occurring in $A$, and its stage $stg(A)$ informs us about the maximal $\alpha$ for which $Q^{<\alpha}_{\mathfrak{A}}(t)$ or $\neg Q^{<\alpha}_{\mathfrak{A}}(t)$ is a subformula of $A$.

**Definition 14** The *level* $lev(A)$ and the *stage* $stg(A)$ of an $\mathcal{L}_\infty$ formula $A$ are inductively defined as follows:

1. If $A$ is a closed literal of $\mathcal{L}$, then $lev(A) := stg(A) := 0$.
2. If $A$ is of the form $P^{\mathfrak{A}}_{<\alpha}(t)$ or $\neg P^{\mathfrak{A}}_{<\alpha}(t)$, then
$$lev(A) := \alpha \text{ and } stg(A) := 0.$$
3. If $A$ is of the form $P^{\mathfrak{A}}_{\alpha}(t)$ or $\neg P^{\mathfrak{A}}_{\alpha}(t)$, then
$$lev(A) := \alpha + 1 \text{ and } stg(A) := 0.$$
4. If $A$ is of the form $Q^{<\alpha}_{\mathfrak{A}}(t)$ or $\neg Q^{<\alpha}_{\mathfrak{A}}(t)$, then
$$lev(A) := 0 \text{ and } stg(A) := \alpha.$$
5. If $A$ is of the form $(B \vee C)$ or $(B \wedge C)$, then
$$lev(A) := \max(lev(B), lev(C)) \text{ and } stg(A) := \max(stg(B), stg(C)).$$
6. If $A$ is of the form $\exists x B[x]$ or $\forall x B[x]$, then
$$lev(A) := lev(B[0]) \text{ and } stg(A) := stg(B[0]).$$

If $\nu < \Gamma_0$, we write $\mathcal{L}_\nu$ for the collection of all $\mathcal{L}_\infty$ formulas of levels less than or equal to $\nu$.

Observe that $\mathcal{L}_\infty$ formulas do not contain free number variables. As a consequence, any number term $t$ occurring in an $\mathcal{L}_\infty$ formula has a specific numerical value $t^{\mathbb{N}}$, and we denote two $\mathcal{L}_\infty$ literals as *numerically equivalent* iff they are syntactically identical modulo number terms of the same value.

Furthermore, we write $pair[t]$ iff the closed number term $t$ codes a pair, i.e. iff $t^{\mathbb{N}}$ is equal to $\langle (t^{\mathbb{N}})_0, (t^{\mathbb{N}})_1 \rangle$. Finally, we extend the function $ot$ mentioned in Sect. 2 to all closed number terms by setting $ot(t) := ot(t^{\mathbb{N}})$. Hence $ot(t)$ is the unique ordinal less than $\Gamma_0$ that is associated with the closed number term $t$ with respect to the wellordering $\prec$.

Every semiformal system $\mathsf{H}_\nu$ is formulated as a Tait-style calculus for finite subsets $(\Delta, \Pi, \Sigma, \Delta_0, \Pi_0, \Sigma_0, \ldots)$ of $\mathcal{L}_\nu$. If $\Delta \subseteq \mathcal{L}_\nu$ and $A \in \mathcal{L}_\nu$, then $\Delta, A$ is shorthand for $\Delta \cup \{A\}$; similarly for expressions such as $\Delta, A, B$ and $\Delta, \Pi, A$. Every system $\mathsf{H}_\nu$ comprises the following axioms and rules of inference.

**I. Axioms, group 1.** For all finite $\Delta \subseteq \mathcal{L}_\nu$, all numerically equivalent literals $A, B \in \mathcal{L}_\nu$, and all true literals $C$ of $\mathcal{L}$:
$$\Delta, \neg A, B \quad \text{and} \quad \Delta, C.$$

**II. Axioms, group 2.** For all finite $\Delta \subseteq \mathcal{L}_\nu$, all $\alpha \leq \nu$, all closed number terms $s$ such that $pair[s]$ is false, and all closed number terms $t$ such that $pair[t]$ is true and $\alpha \leq ot((t)_0)$:

$$\Delta,\ \neg P^{\mathfrak{A}}_{<\alpha}(s) \qquad \text{and} \qquad \Delta,\ \neg P^{\mathfrak{A}}_{<\alpha}(t).$$

**III. Induction axioms.** For all finite $\Delta \subseteq \mathcal{L}_\nu$, all $\alpha < \nu$, all closed number terms $t$, and all formulas $B[0] \in \mathcal{L}_\alpha$:

$$\Delta,\ \neg Cl_{\mathfrak{A}}[B],\ \neg P^{\mathfrak{A}}_{\alpha}(t),\ B[t].$$

**IV. Stage axioms.** For all finite $\Delta \subseteq \mathcal{L}_\nu$, all $\alpha \leq \beta < \Gamma_0$, and all closed number terms $s, t$ that have the same value:

$$\Delta,\ \neg Q^{<\alpha}_{\mathfrak{A}}(s),\ Q^{<\beta}_{\mathfrak{A}}(t).$$

**V. Fixed point rules, group 1.** For all finite $\Delta \subseteq \mathcal{L}_\nu$, all $\alpha < \beta \leq \nu$, and all closed number terms $t$ such that $pair[t]$ is true and $ot((t)_0) = \alpha$:

$$\frac{\Delta,\ P^{\mathfrak{A}}_{\alpha}((t)_1)}{\Delta,\ P^{\mathfrak{A}}_{<\beta}(t)} \qquad \text{and} \qquad \frac{\Delta,\ \neg P^{\mathfrak{A}}_{\alpha}((t)_1)}{\Delta,\ \neg P^{\mathfrak{A}}_{<\beta}(t)}.$$

**VI. Fixed point rules, group 2.** For all finite $\Delta \subseteq \mathcal{L}_\nu$, all $\alpha < \nu$, and all closed number terms $t$:

$$\frac{\Delta,\ \mathfrak{A}[P^{\mathfrak{A}}_{\alpha}, t]}{\Delta,\ P^{\mathfrak{A}}_{\alpha}(t)} \qquad \text{and} \qquad \frac{\Delta,\ \neg\mathfrak{A}[P^{\mathfrak{A}}_{\alpha}, t]}{\Delta,\ \neg P^{\mathfrak{A}}_{\alpha}(t)}.$$

**VII. Stage rules.** For all finite $\Delta \subseteq \mathcal{L}_\nu$, all $\alpha < \beta < \Gamma_0$, and all closed number terms $t$:

$$\frac{\Delta,\ \mathfrak{A}[Q^{<\alpha}_{\mathfrak{A}}, t]}{\Delta,\ Q^{<\beta}_{\mathfrak{A}}(t)} \qquad \text{and} \qquad \frac{\Delta,\ \neg\mathfrak{A}[Q^{<\xi}_{\mathfrak{A}}, t] \quad \text{for all } \xi < \beta}{\Delta,\ \neg Q^{<\beta}_{\mathfrak{A}}(t)}.$$

**VIII. Propositional rules.** For all finite $\Delta \subseteq \mathcal{L}_\nu$ and all $A, B \in \mathcal{L}_\nu$:

$$\frac{\Delta,\ A,\ B}{\Delta,\ A \vee B} \qquad \text{and} \qquad \frac{\Delta,\ A \qquad \Delta,\ B}{\Delta,\ A \wedge B}.$$

**IX. Quantifier rules.** For all finite $\Delta \subseteq \mathcal{L}_\nu$ and all $A[s] \in \mathcal{L}_\nu$:

$$\frac{\Delta,\ A[s]}{\Delta,\ \exists x A[x]} \qquad \text{and} \qquad \frac{\Delta,\ A[t] \quad \text{for all closed number terms } t}{\Delta,\ \forall x A[x]}.$$

**X. Cut rules.** For all finite $\Delta \subseteq \mathcal{L}_\nu$ and all $A \in \mathcal{L}_\nu$:

$$\frac{\Delta, A \qquad \Delta, \neg A}{\Delta},$$

where the formulas $A$ and $\neg A$ are called the *cut formulas* of this cut.

Before turning to the proof-theoretic analysis of the systems $\mathsf{H}_\nu$, we need some further auxiliary notions. We first fix the collection of those formulas that still may be used as cuts after partial cut elimination has been carried through.

**Definition 15** $Simp_\nu$ is defined to be the subset of $\mathcal{L}_\nu$ that comprises $\mathcal{L}_0$, all $\mathcal{L}_\infty$ formulas of levels less than $\nu$, and all elements of $\mathcal{L}_\nu$ of the form $P^{\mathfrak{A}}_\alpha(t)$ and $\neg P^{\mathfrak{A}}_\alpha(t)$ for $\alpha < \nu$.

According to this definition, $Simp_0 = \mathcal{L}_0$. Hence $Simp_\nu$ is an interesting set of formulas only for $\nu > 0$. Now, depending on $Simp_\nu$, we introduce a complexity measure for all formulas in $\mathcal{L}_\nu$ that measures their complexities "above" $Simp_\nu$. This is the measure to be used for partial cut elimination.

**Definition 16** The $\nu$-*rank* $rk_\nu(A)$ of an $A \in \mathcal{L}_\nu$ is inductively defined as follows:

1. If $A \in Simp_\nu$, then $rk_\nu(A) := 0$.
2. If $A$ is a formula $P^{\mathfrak{A}}_{<\nu}(t)$ or $\neg P^{\mathfrak{A}}_{<\nu}(t)$, then $rk_\nu(A) := 1$.
3. If $A$ does not belong to $Simp_\nu$ and is of the form $(B \vee C)$ or $(B \wedge C)$, then $rk_\nu(A) := \max(rk_\nu(B), rk_\nu(C)) + 1$.
4. If $A$ does not belong to $Simp_\nu$ and is of the form $\exists x B[x]$ or $\forall x B[x]$, then $rk_\nu(A) := rk_\nu(B[0]) + 1$.

Obviously, the $\nu$-rank of any $A \in \mathcal{L}_\nu$ is finite and less than or equal to the length of $A$, and $rk_\nu(A) = 0$ iff $A \in Simp_\nu$; also $rk_\nu(A) = rk_\nu(\neg A)$ for any $A \in \mathcal{L}_\nu$. Since $Simp_0 = \mathcal{L}_0$ we have $rk_0(A) = 0$ for all $A \in \mathcal{L}_0$.

**Definition 17** We define $\mathsf{H}_\nu \overset{\alpha}{\underset{(\sigma,m,n)}{\vdash}} \Delta$ for all finite $\Delta \subseteq \mathcal{L}_\nu$, all $m, n < \omega$, and all ordinals $\alpha, \sigma < \Gamma_0$ by induction on $\alpha$.

1. If $\Delta$ is an axiom of $\mathsf{H}_\nu$, then we have $\mathsf{H}_\nu \overset{\alpha}{\underset{(\sigma,m,n)}{\vdash}} \Delta$ for all $m, n < \omega$ and all $\alpha, \sigma < \Gamma_0$.
2. If $\mathsf{H}_\nu \overset{\alpha_i}{\underset{(\sigma,m,n)}{\vdash}} \Delta_i$ and $\alpha_i < \alpha$ for every premise $\Delta_i$ of a fixed point rule, a stage rule, a propositional rule, or a quantifier rule of $\mathsf{H}_\nu$, then we have $\mathsf{H}_\nu \overset{\alpha}{\underset{(\sigma,m,n)}{\vdash}} \Delta$ for the conclusion $\Delta$ of this rule.
3. Under the assumptions
   - $\mathsf{H}_\nu \overset{\alpha_0}{\underset{(\sigma,m,n)}{\vdash}} \Delta, A$ and $\alpha_0 < \alpha$,
   - $\mathsf{H}_\nu \overset{\alpha_1}{\underset{(\sigma,m,n)}{\vdash}} \Delta, \neg A$ and $\alpha_1 < \alpha$,
   - $stg(A) < \sigma$, $rk_\nu(A) < m$, and the length of $A$ is less than $n$

   we have $\mathsf{H}_\nu \overset{\alpha}{\underset{(\sigma,m,n)}{\vdash}} \Delta$.

In addition, $\mathsf{H}_\nu \overset{<\alpha}{\underset{(\sigma,m,n)}{\vdash}} \Delta$ means $\mathsf{H}_\nu \overset{\beta}{\underset{(\sigma,m,n)}{\vdash}} \Delta$ for some $\beta < \alpha$.

Thus $\mathsf{H}_\nu \mathrel{\vert\!\frac{\alpha}{(\sigma,m,n)}} \Delta$ states that there exists a proof of $\Delta$ in the system $\mathsf{H}_\nu$ whose depth is bounded by $\alpha$ such that any cut formula in this proof is of stage less than $\sigma$, of $\nu$-rank smaller than $m$, and of length smaller than $n$. Consequently, $\mathsf{H}_\nu \mathrel{\vert\!\frac{\alpha}{(\sigma,1,n)}} \Delta$ implies that there is a proof of $\Delta$ in $\mathsf{H}_\nu$ with all its cut formulas belonging to the set $Simp_\nu$.

It is easy to verify that the axioms and rules of $\mathsf{H}_\nu$ and the notion of $\nu$-rank are tailored in such a way that all cuts but the ones from $Simp_\nu$ can be eliminated. The following lemma, whose proof is left to the reader, is shown by standard proof-theoretic methods. You may also consult Ranzi and Strahm [18] for a similar result.

**Lemma 18** *For all ordinals $\alpha, \sigma < \Gamma_0$, all $m, n < \omega$, and all finite $\Delta \subseteq \mathcal{L}_\nu$ we have*

$$\mathsf{H}_\nu \mathrel{\vert\!\frac{\alpha}{(\sigma,m+2,n)}} \Delta \quad\Longrightarrow\quad \mathsf{H}_\nu \mathrel{\vert\!\frac{\omega^\alpha}{(\sigma,m+1,n)}} \Delta.$$

Since $\omega_n(\alpha) < \varepsilon(\alpha)$ for all $n < \omega$, the previous lemma immediately yields the following partial cut elimination result for the systems $\mathsf{H}_\nu$.

**Theorem 19 (Partial Cut Elimination)** *For all $\alpha, \sigma < \Gamma_0$, all $m, n < \omega$, and all finite $\Delta \subseteq \mathcal{L}_\nu$ we have*

$$\mathsf{H}_\nu \mathrel{\vert\!\frac{\alpha}{(\sigma,m+1,n)}} \Delta \quad\Longrightarrow\quad \mathsf{H}_\nu \mathrel{\vert\!\frac{<\varepsilon(\alpha)}{(\sigma,1,n)}} \Delta.$$

Our next aim is a reduction result for $\mathsf{H}_{\nu+1}$: if a finite $\Delta \subseteq \mathcal{L}_\nu$ is provable in $\mathsf{H}_{\nu+1}$, then it can already be proved in $\mathsf{H}_\nu$. To show this reduction theorem and several other properties of the systems $\mathsf{H}_\nu$, several auxiliary lemmas are needed.

**Lemma 20** *If $A$ and $B$ are numerically equivalent elements of $\mathcal{L}_\nu$ of length $k$, then we have $\mathsf{H}_\nu \mathrel{\vert\!\frac{2k}{(0,0,0)}} \neg A,\ B$.*

**Lemma 21** *Let $X$ be a fresh unary relation symbol and $C[X]$ an $X$ positive $\mathcal{L}(X)$ formula of length $k$ (in the usual sense); in addition, assume that $\Delta, A[u], B[u] \subseteq \mathcal{L}_\nu$, $\alpha, \sigma < \Gamma_0$, and $m, n < \omega$. If*

$$\mathsf{H}_\nu \mathrel{\vert\!\frac{\alpha}{(\sigma,m,n)}} \Delta,\ \neg A[t],\ B[t]$$

*for all closed number terms $t$, then we have*

$$\mathsf{H}_\nu \mathrel{\vert\!\frac{\alpha+2k}{(\sigma,m,n)}} \Delta,\ \neg C[\{x{:}A[x]\}],\ C[\{x{:}B[x]\}].$$

**Lemma 22** *Let $B[u]$ be any element of $\mathcal{L}_\nu$. Then we have for all $\alpha < \Gamma_0$ and all closed number terms $t$ that*

$$\mathsf{H}_\nu \mathrel{\vert\!\frac{\omega\alpha}{(0,0,0)}} \neg Cl_{\mathfrak{A}}[\{x{:}B[x]\}],\ \neg Q_{\mathfrak{A}}^{<\alpha}(t),\ B[t].$$

*Proof* We show this assertion by induction on $\alpha$. In the case $\alpha = 0$ we simply use the appropriate right stage rule. Given any $0 < \alpha < \Gamma_0$, the induction hypothesis yields

$$\mathsf{H}_\nu \vdash^{\omega\xi}_{(0,0,0)} \neg Cl_{\mathfrak{A}}[\{x{:}B[x]\}],\ \neg Q^{<\xi}_{\mathfrak{A}}(t),\ B[t].$$

for all $\xi < \alpha$ and all closed number terms $t$. Now we apply the previous lemma and conclude

$$\mathsf{H}_\nu \vdash^{<\omega\xi+\omega}_{(0,0,0)} \neg Cl_{\mathfrak{A}}[\{x{:}B[x]\}],\ \neg\mathfrak{A}[Q^{<\xi}_{\mathfrak{A}}, t],\ \mathfrak{A}[B, t].$$

Making use of Lemma 20 and some propositional rules we thus have

$$\mathsf{H}_\nu \vdash^{<\omega\xi+\omega}_{(0,0,0)} \neg Cl_{\mathfrak{A}}[\{x{:}B[x]\}],\ \neg\mathfrak{A}[Q^{<\xi}_{\mathfrak{A}}, t],\ \mathfrak{A}[B, t] \wedge \neg B[t],\ B[t].$$

By an existential quantification over $t$ we therefore obtain

$$\mathsf{H}_\nu \vdash^{<\omega\xi+\omega}_{(0,0,0)} \neg Cl_{\mathfrak{A}}[\{x{:}B[x]\}],\ \neg\mathfrak{A}[Q^{<\xi}_{\mathfrak{A}}, t],\ B[t].$$

Since $\omega\xi + \omega \leq \omega\alpha$ for all $\xi < \alpha$, a final application of a right stage rule implies what we want. □

We write $\mathcal{L}^-_{\nu+1}$ for the collection of all $A \in \mathcal{L}_{\nu+1}$ that do not contain subformulas of the form $P^{\mathfrak{A}}_{<\nu+1}(t)$ or $\neg P^{\mathfrak{A}}_{<\nu+1}(t)$. For the following considerations it is convenient (and sufficient) to restrict our attention to such formulas.

**Definition 23** Assume $\alpha, \beta < \Gamma_0$.

1. For any $A \in \mathcal{L}^-_{\nu+1}$ we define the set $Var_\nu(A, \alpha, \beta)$ of the *$(\alpha, \beta)$-variants of $A$ with respect to $\nu$* by induction on the build-up of $A$ as follows:
   (a) If $A \in \mathcal{L}_\nu$, then $A$ belongs to $Var_\nu(A, \alpha, \beta)$.
   (b) If $A$ is the formula $\neg P^{\mathfrak{A}}_\nu(t)$ and $\eta \leq \alpha$, then $\neg Q^{<\eta}_{\mathfrak{A}}(t)$ belongs to $Var_\nu(A, \alpha, \beta)$.
   (c) If $A$ is the formula $P^{\mathfrak{A}}_\nu(t)$ and $\beta \leq \xi$, then $Q^{<\xi}_{\mathfrak{A}}(t)$ belongs to $Var_\nu(A, \alpha, \beta)$.
   (d) If $A$ is the formula $(B_0 \vee B_1)$ and $C_i \in Var_\nu(B_i, \alpha, \beta)$ for $i = 0, 1$, then also $(C_0 \vee C_1)$ belongs to $Var_\nu(A, \alpha, \beta)$.
   (e) If $A$ is the formula $(B_0 \wedge B_1)$ and $C_i \in Var_\nu(B_i, \alpha, \beta)$ for $i = 0, 1$, then also $(C_0 \wedge C_1)$ belongs to $Var_\nu(A, \alpha, \beta)$.
   (f) If $A$ is the formula $\exists x B[x]$ and $C[0] \in Var_\nu(B[0], \alpha, \beta)$, then also $\exists x C[x]$ belongs to $Var_\nu(A, \alpha, \beta)$.
   (g) If $A$ is the formula $\forall x B[x]$ and $C[0] \in Var_\nu(B[0], \alpha, \beta)$, then also $\forall x C[x]$ belongs to $Var_\nu(A, \alpha, \beta)$.
2. Now let $\Delta$ be a finite subset of $\mathcal{L}^-_{\nu+1}$. We call a finite $\Pi \subseteq \mathcal{L}_\nu$ an $(\alpha, \beta)$-variant of $\Delta$ with respect to $\nu$ iff for every $A \in \Delta$ there exists a $B \in \Pi$ such that

$B \in Var_\nu(A, \alpha, \beta)$. The set of all $(\alpha, \beta)$-variants of $\Delta$ with respect to $\nu$ is denoted by $Var_\nu(\Delta, \alpha, \beta)$.

For $A \in \mathcal{L}^-_{\nu+1}$ any $(\alpha, \beta)$-variant of $A$ with respect to $\nu$ is an asymmetric interpretation of $A$: (i) every occurrence of a negative literal $\sim P^{\mathfrak{A}}_\nu(t)$ in $A$ is replaced by $\sim Q^{<\eta}_{\mathfrak{A}}(t)$ for some $\eta \leq \alpha$ and (ii) every occurrence of a positive literal $P^{\mathfrak{A}}_\nu(t)$ in $A$ is replaced by $Q^{<\xi}_{\mathfrak{A}}(t)$ for some $\xi \geq \beta$. This form of asymmetric interpretation is instrumental for the reduction of $\mathsf{H}_{\nu+1}$ to $\mathsf{H}_\nu$, as carried through now.

The method of asymmetric interpretation belongs to the standard repertoire of predicative proof theory. One of its first applications is in Schütte [21], but it has been used in numerous other contexts since then; cf., for example, Cantini [8], Jäger [11], or Rathjen [19]. Ranzi and Strahm [18] also proves a similar result.

**Lemma 24** *For all $\alpha, \beta, \sigma < \Gamma_0$, all $n < \omega$, all $\tau \geq \max(\sigma, \beta + \omega^\alpha)$, all finite $\Delta \subseteq \mathcal{L}^-_{\nu+1}$, and all $\Pi \in Var_\nu(\Delta, \beta, \beta + \omega^\alpha)$ we have that*

$$\mathsf{H}_{\nu+1} \left|\frac{\alpha}{(\sigma,1,n)}\right. \Delta \quad \Longrightarrow \quad \mathsf{H}_\nu \left|\frac{\omega\beta+\omega^{2\alpha}}{(\tau,n,n)}\right. \Pi.$$

*Proof* We proceed by induction on $\alpha$ and distinguish the following cases.

(i) If $\Delta$ is an axiom of group 1 or group 2, or a stage axiom, then our assertion is obvious. If $\Delta$ is an induction axiom, then $\Pi$ is either an induction axiom, or we simply apply Lemma 22.

(ii) $\Delta$ is the conclusion of a fixed point rule whose main formula belongs to $\mathcal{L}_\nu$, the conclusion of a propositional rule, the conclusion of a quantifier rule, or the conclusion of a cut whose cut formulas belong to $\mathcal{L}_\nu$. Then our assertion immediately follows from the induction hypothesis.

(iii) $\Delta$ is the conclusion of a left group 2 fixed point rule of the form

$$\frac{\Sigma,\ \mathfrak{A}[P^{\mathfrak{A}}_\nu, t]}{\Sigma,\ P^{\mathfrak{A}}_\nu(t)}.$$

In this case $\Pi$ is of the form $\Pi',\, Q^{<\gamma}_{\mathfrak{A}}(t)$ with $\Pi' \in Var_\nu(\Delta, \beta, \beta + \omega^\alpha)$ and $\beta + \omega^\alpha \leq \gamma$, and there exists a $\delta < \alpha$ such that

$$\mathsf{H}_{\nu+1} \left|\frac{\delta}{(\sigma,1,n)}\right. \Sigma,\ \mathfrak{A}[P^{\mathfrak{A}}_\nu, t].$$

Hence the induction hypothesis implies

$$\mathsf{H}_\nu \left|\frac{\omega\beta+\omega^{2\delta}}{(\tau,n,n)}\right. \Pi',\ \mathfrak{A}[Q^{<\beta+\omega^\delta}_{\mathfrak{A}}, t].$$

Since $\omega\beta + \omega^{2\delta} + 2 \leq \omega\beta + \omega^{2\alpha}$ and $\beta + \omega^\delta < \beta + \omega^\alpha \leq \gamma$, it follows

$$\mathsf{H}_\nu \vdash^{\omega\beta+\omega^{2\alpha}}_{(\tau,n,n)} \Pi',\ Q^{<\gamma}_{\mathfrak{A}}(t)$$

by a left stage rule.

(iv) $\Delta$ is the conclusion of a right group 2 fixed point rule of the form

$$\frac{\Sigma,\ \neg\mathfrak{A}[P^{\mathfrak{A}}_\nu, t]}{\Sigma,\ \neg P^{\mathfrak{A}}_\nu(t)}.$$

Now $\Pi$ is of the form $\Pi', \neg Q^{<\gamma}_{\mathfrak{A}}(t)$ with $\Pi' \in \mathit{Var}_\nu(\Delta, \beta, \beta+\omega^\alpha)$ and $\gamma \leq \beta$, and there exists a $\delta < \alpha$ such that

$$\mathsf{H}_{\nu+1} \vdash^{\delta}_{(\sigma,1,n)} \Sigma,\ \neg\mathfrak{A}[P^{\mathfrak{A}}_\nu, t].$$

If $\gamma = 0$, then our assertion is immediate by an application of a right stage rule. Otherwise, we obtain in view of the induction hypothesis that

$$\mathsf{H}_\nu \vdash^{\omega\beta+\omega^{2\delta}}_{(\tau,n,n)} \Pi',\ \neg\mathfrak{A}[Q^{<\xi}_{\mathfrak{A}}, t]$$

for all $\xi < \gamma$. Consequently, since $\omega\beta + \omega^{2\delta} + 2 \leq \omega\beta + \omega^{2\alpha}$, we have

$$\mathsf{H}_\nu \vdash^{\omega\beta+\omega^{2\alpha}}_{(\tau,n,n)} \Pi',\ \neg Q^{<\gamma}_{\mathfrak{A}}(t)$$

by a left stage rule.

(v) $\Delta$ is the conclusion of a cut of the form

$$\frac{\Delta,\ P^{\mathfrak{A}}_\nu(t) \qquad \Delta,\ \neg P^{\mathfrak{A}}_\nu(t)}{\Delta}.$$

Then there exist $\gamma, \delta < \alpha$ such that

$$\mathsf{H}_{\nu+1} \vdash^{\gamma}_{(\sigma,1,n)} \Delta,\ P^{\mathfrak{A}}_\nu(t), \tag{15}$$

$$\mathsf{H}_{\nu+1} \vdash^{\delta}_{(\sigma,1,n)} \Delta,\ \neg P^{\mathfrak{A}}_\nu(t). \tag{16}$$

From (15) we conclude with the induction hypothesis that

$$\mathsf{H}_\nu \vdash^{\omega\beta+\omega^{2\gamma}}_{(\tau,n,n)} \Pi,\ Q^{<\beta+\omega^\gamma}_{\mathfrak{A}}(t). \tag{17}$$

Since $\beta+\omega^\gamma+\omega^\delta < \beta+\omega^\alpha$, we have $\Pi \in \mathit{Var}_\nu(\Delta, \beta+\omega^\gamma, \beta+\omega^\gamma+\omega^\delta)$. Hence we obtain from (16) by means of the induction hypothesis that

$$\mathsf{H}_\nu \vdash^{\omega(\beta+\omega^\gamma)+\omega^{2\delta}}_{(\tau,n,n)} \Pi,\ \neg Q^{<\beta+\omega^\gamma}_{\mathfrak{A}}(t). \tag{18}$$

Since $\omega\beta + \omega^{2\gamma} < \omega(\beta + \omega^\gamma) + \omega^{2\delta} < \omega\beta + \omega^{2\alpha}$, $stg(Q_{\mathfrak{A}}^{<\beta+\omega^\gamma}(t)) < \tau$, and both, the $\nu$-rank and the length of $Q_{\mathfrak{A}}^{<\beta+\omega^\gamma}(t)$, are smaller than $n$, a cut applied to (17) and (18) yields

$$\mathsf{H}_\nu \overset{\omega\beta+\omega^{2\alpha}}{\underset{(\tau,n,n)}{\vdash}} \Pi.$$

Since now all possible cases have been covered, the proof of our lemma is completed. □

The following reduction theorem for the systems $\mathsf{H}_{\nu+1}$ is an immediate consequence of Theorem 19 and the previous lemma.

**Theorem 25 (Reduction)** *For all finite $\Delta \subseteq \mathcal{L}_\nu$, all $\alpha, \sigma < \Gamma_0$, and all $n < \omega$ we have that*

$$\mathsf{H}_{\nu+1} \overset{\alpha}{\underset{(\sigma,n,n)}{\vdash}} \Delta \quad \Longrightarrow \quad \mathsf{H}_\nu \overset{<\varepsilon(\alpha)}{\underset{(\sigma+\omega^\alpha,1,n)}{\vdash}} \Delta.$$

In a next step we turn to the reduction of systems $\mathsf{H}_\nu$ for limit ordinals $\nu$. We are interested in finding out what it means that a finite $\Delta \subseteq \mathcal{L}_\mu$ for $\mu < \nu$ is provable in $\mathsf{H}_\nu$. The following lemma gives the correct answer.

**Lemma 26** *For all $\alpha, \beta, \sigma < \Gamma_0$, all $\gamma$ with $0 < \gamma < \Gamma_0$, all $\delta < \omega^\gamma$, all $n < \omega$, and all finite $\Delta \subseteq \mathcal{L}_{\beta+\delta}$ we have that*

$$\mathsf{H}_{\beta+\omega^\gamma} \overset{\alpha}{\underset{(\sigma,1,n)}{\vdash}} \Delta \quad \Longrightarrow \quad \mathsf{H}_{\beta+\delta} \overset{\varphi\gamma\alpha}{\underset{(\varphi\gamma(\sigma+\alpha),1,n)}{\vdash}} \Delta.$$

*Proof* We prove this assertion by main induction on $\gamma$ and side induction on $\alpha$. If $\Delta$ is an axiom of $\mathsf{H}_{\beta+\omega^\gamma}$, then it is also an axiom of $\mathsf{H}_{\beta+\delta}$ and our claim is trivially satisfied. If $\Delta$ is the conclusion of a rule different from a cut rule, our claim is immediate from the induction hypothesis. Finally, if $\Delta$ is the conclusion of a cut rule, then there exist $\alpha_0, \alpha_1 < \alpha$ and a formula $A \in Simp_{\beta+\omega^\gamma}$ such that

$$\mathsf{H}_{\beta+\omega^\gamma} \overset{\alpha_0}{\underset{(\sigma,1,n)}{\vdash}} \Delta, A \qquad \text{and} \qquad \mathsf{H}_{\beta+\omega^\gamma} \overset{\alpha_1}{\underset{(\sigma,1,n)}{\vdash}} \Delta, \neg A. \tag{*}$$

Now we distinguish whether $\gamma = 1$, $\gamma$ is a successor ordinal greater than 1, or $\gamma$ is a limit ordinal.

(i) $\gamma = 1$. In this case $\delta$ is smaller than $\omega$. Since $A \in Simp_{\beta+\omega}$, we know that there exists a natural number $m \geq \delta$ for which $\Delta \subseteq \mathcal{L}_{\beta+m}$ and $A, \neg A \in \mathcal{L}_{\beta+m}$. By the side induction hypothesis we obtain from (*) that

$$\mathsf{H}_{\beta+m} \overset{\varphi\gamma\alpha_0}{\underset{(\varphi\gamma(\sigma+\alpha_0),1,n)}{\vdash}} \Delta, A \qquad \text{and} \qquad \mathsf{H}_{\beta+m} \overset{\varphi\gamma\alpha_1}{\underset{(\varphi\gamma(\sigma+\alpha_1),1,n)}{\vdash}} \Delta, \neg A,$$

and a cut yields

$$\mathsf{H}_{\beta+m} \overset{\eta}{\underset{(\xi,n,n)}{\vdash}} \Delta$$

for $\eta := \max(\varphi\gamma\alpha_0, \varphi\gamma\alpha_1) + 1$ and $\xi := \max(\varphi\gamma(\sigma + \alpha_0), \varphi\gamma(\sigma + \alpha_1))$. Finitely many applications of Theorem 25 therefore yield

$$\mathsf{H}_{\beta+\delta} \overset{\tau}{\underset{(\varphi(\gamma(\sigma+\alpha),1,n)}{\vdash}} \Delta$$

for some $\tau < \varphi\gamma\alpha$. This proves our claim for the case $\gamma = 1$.

(ii) $\gamma = \rho + 1$ for some $\rho \geq 1$. Now $\delta = \omega^\rho m + \zeta$ for some $m < \omega$ and $\zeta < \omega^\rho$. From $A \in Simp_{\beta+\omega^{\rho+1}}$ we now conclude that there exist a natural number $k > m$ such that $\Delta \subseteq \mathcal{L}_{\beta+\omega^\rho k}$ and $A, \neg A \in \mathcal{L}_{\beta+\omega^\rho k}$. Therefore the side induction hypothesis applied to (*) yields

$$\mathsf{H}_{\beta+\omega^\rho k} \overset{\varphi\gamma\alpha_0}{\underset{(\varphi\gamma(\sigma+\alpha_0)1,n)}{\vdash}} \Delta, A \qquad \text{and} \qquad \mathsf{H}_{\beta+\omega^\rho k} \overset{\varphi\gamma\alpha_1}{\underset{(\varphi\gamma(\sigma+\alpha_1)1,n)}{\vdash}} \Delta, \neg A,$$

and a cut gives us

$$\mathsf{H}_{\beta+\omega^\rho k} \overset{\eta}{\underset{(\xi,n,n)}{\vdash}} \Delta$$

for $\eta := \max(\varphi\gamma\alpha_0, \varphi\gamma\alpha_1) + 1$ and $\xi := \max(\varphi\gamma(\sigma + \alpha_0), \varphi\gamma(\sigma + \alpha_1)$. In view of Theorem 19 we have

$$\mathsf{H}_{\beta+\omega^\rho k} \overset{\tau}{\underset{(\xi,1,n)}{\vdash}} \Delta$$

for some $\tau < \varepsilon(\eta)$. For $i < \omega$ we now set

$$\tau_0 := \tau, \quad \tau_{i+1} := \varphi\rho\tau_i \qquad \text{and} \qquad \xi_0 := \xi, \quad \xi_{i+1} := \varphi\rho(\xi_i + \tau_i).$$

Then $k - m$ applications of the main induction hypothesis yield

$$\mathsf{H}_{\beta+\omega^\rho(k-1)} \overset{\tau_1}{\underset{(\xi_1,1,n)}{\vdash}} \Delta,$$

$$\mathsf{H}_{\beta+\omega^\rho(k-2)} \overset{\tau_2}{\underset{(\xi_2,1,n)}{\vdash}} \Delta,$$

$$\vdots$$

$$\mathsf{H}_{\beta+\omega^\rho(m+1)} \overset{\tau_{k-m-1}}{\underset{(\xi_{k-m-1},1,n)}{\vdash}} \Delta,$$

$$\mathsf{H}_{\beta+\omega^\rho m+\zeta} \overset{\tau_{k-m}}{\underset{(\xi_{k-m},1,n)}{\vdash}} \Delta.$$

In these reductions we have successively replaced the $\beta$ of the main induction hypothesis by

$$\beta + \omega^\rho(k-1),\ \beta + \omega^\rho(k-2),\ \ldots,\ \beta + \omega^\rho(m+1),\ \beta + \omega^\rho m.$$

In addition, observe that $\tau_i < \varphi\gamma\alpha$ and $\xi_i < \varphi\gamma(\sigma + \alpha)$ for all $i < \omega$. Hence we have shown that

$$\mathsf{H}_{\beta+\delta} \vdash^{\varphi\gamma\alpha}_{(\varphi\gamma(\sigma+\alpha),1,n)} \Delta,$$

as desired, finishing the treatment of this case.

(iii) $\gamma$ is a limit number. Because of $A \in Simp_{\beta+\omega^\gamma}$ we know that there exists a $\rho < \gamma$ satisfying $\Delta \subseteq \mathcal{L}_{\beta+\omega^\rho}$ and $A, \neg A \in \mathcal{L}_{\beta+\omega^\rho}$. Thus the side induction hypothesis applied to (*) asserts that

$$\mathsf{H}_{\beta+\omega^\rho} \vdash^{\varphi\gamma\alpha_0}_{(\varphi\gamma(\sigma+\alpha_0),1,n)} \Delta, A \qquad \text{and} \qquad \mathsf{H}_{\beta+\omega^\rho} \vdash^{\varphi\gamma\alpha_1}_{(\varphi\gamma(\sigma+\alpha_1),1,n)} \Delta, \neg A.$$

For $\eta := \max(\varphi\gamma\alpha_0, \varphi\gamma\alpha_1) + 1$ and $\xi := \max(\varphi\gamma(\sigma + \alpha_0), \varphi\gamma(\sigma + \alpha_1))$ we deduce by a cut that

$$\mathsf{H}_{\beta+\omega^\rho} \vdash^{\eta}_{(\xi,n,n)} \Delta.$$

As before, by making use of Theorem 19, we find a $\tau < \varepsilon(\eta)$ such that

$$\mathsf{H}_{\beta+\omega^\rho} \vdash^{\tau}_{(\xi,1,n)} \Delta$$

Now we are in the position to apply the main induction hypothesis and conclude

$$\mathsf{H}_{\beta+\delta} \vdash^{\varphi\rho\tau}_{(\varphi\rho(\xi+\tau),1,n)} \Delta.$$

Since $\varphi\rho\tau < \varphi\gamma\alpha$ and $\varphi\rho(\xi + \tau) < \varphi(\sigma + \alpha)$, this establishes our claim, finishing the proof of case (iii) and also the verification of our lemma. □

**Theorem 27** *Assume that $\Delta$ is a finite subset of $\mathcal{L}_0$ and $\mathsf{H}_\nu \vdash^{<\varepsilon(\nu)}_{(0,n,n)} \Delta$ for some $n < \omega$. Then there exist $\alpha, \beta < \Lambda_\nu$ such that $\mathsf{H}_0 \vdash^{\alpha}_{(\beta,1,n)} \Delta$.*

*Proof* We first observe that $\nu$ can be uniquely written as

$$\nu = \omega^{\nu_m} + \cdots + \omega^{\nu_1} + k$$

with $k < \omega$ and ordinals $\nu_1 \leq \cdots \leq \nu_m \leq \nu$. Set $\mu := \omega^{\nu_m} + \cdots + \omega^{\nu_1}$. By our assumptions we know that there exists an ordinal $\gamma < \varepsilon(\nu)$ for which

$$\mathsf{H}_{\mu+k} \vdash^{\gamma}_{(0,n,n)} \Delta.$$

In view of ($k$ applications of) Theorem 25 this implies

$$\mathsf{H}_{\mu} \vdash^{\delta}_{(\rho,1,n)} \Delta \tag{19}$$

for suitable $\delta, \rho < \varepsilon(\nu)$. By induction on $i$ we now define for all natural numbers $i$ such that $0 \leq i \leq m-1$:

$$\begin{aligned} \mu_0 &:= \varepsilon(\mu) \text{ and } \mu_{i+1} := \varphi\nu_{i+1}\mu_i, \\ \sigma_0 &:= \delta \quad \text{ and } \sigma_{i+1} := \varphi\nu_{i+1}\sigma_i, \\ \tau_0 &:= \rho \quad \text{ and } \tau_{i+1} := \varphi\nu_{i+1}(\tau_i + \sigma_i). \end{aligned}$$

A simple induction on $i$ then shows for all $i \leq m$ that

$$\sigma_i < \mu_i \qquad \text{and} \qquad \tau_i < \mu_i. \tag{20}$$

In view of Lemma 1 and the choice of $\mu$ we also know that

$$\Lambda_\nu = \Lambda_\mu = \mu_m. \tag{21}$$

It remains to apply Lemma 26 several times. More precisely, starting off from (19) and making use of Lemma 26 repeatedly we obtain

$$\begin{aligned} \mathsf{H}_{0+\omega^{\nu_m}+\ldots+\omega^{\nu_1}} &\Big|\frac{\sigma_0}{(\tau_0,1,n)}\ \Delta, \\ \mathsf{H}_{0+\omega^{\nu_m}+\ldots+\omega^{\nu_2}} &\Big|\frac{\sigma_1}{(\tau_1,1,n)}\ \Delta, \\ &\vdots \\ \mathsf{H}_{0+\omega^{\nu_m}} &\Big|\frac{\sigma_{m-1}}{(\tau_{m-1},1,n)}\ \Delta, \\ \mathsf{H}_0 &\Big|\frac{\sigma_m}{(\tau_m,1,n)}\ \Delta. \end{aligned}$$

Because of (20) and (21) the last line immediately gives us our assertion for $\alpha := \sigma_m$ and $\beta := \tau_m$. □

This theorem provides a reduction of the systems $\mathsf{H}_\nu$ to $\mathsf{H}_0$ with respect to all finite $\Delta \subseteq \mathcal{L}_0$. To complete the proof-theoretic analysis of the $\mathsf{H}_\nu$ we now turn to complete cut elimination for $\mathsf{H}_0$. To this end we first assign a rank $rk(A)$ to any $A \in \mathcal{L}_0$.

**Definition 28** The *rank* $rk(A)$ of an $A \in \mathcal{L}_0$ is inductively defined as follows:

1. If $A$ is a closed literal of $\mathcal{L}$, then $rk(A) := 0$.
2. If $A$ is of the form $P^{\mathfrak{A}}_{<0}(t)$ or $\neg P^{\mathfrak{A}}_{<0}(t)$, then $rk(A) := 1$.
3. If $A$ is of the form $Q^{<\alpha}_{\mathfrak{A}}(t)$ or $\neg Q^{<\alpha}_{\mathfrak{A}}(t)$, then $rk(A) := \omega\alpha$.
4. If $A$ is of the form $(B \vee C)$ or $(B \wedge C)$, then
$$rk(A) := \max(rk(B), rk(C)) + 1.$$
5. If $A$ is of the form $\exists x B[x]$ or $\forall x B[x]$, then $rk(A) := rk(B[0]) + 1$.

By straightforward induction on the build-up of the formulas in $\mathcal{L}_0$ we can easily verify a close relationship between their stages and ranks.

**Lemma 29** *If $A$ is a formula from $\mathcal{L}_0$ of length $n$ and $stg(A) = \alpha$, then $rk(A) < \omega\alpha + n$.*

If we restrict ourselves to the system $\mathsf{H}_0$, then (obviously) the ranks of the cut formulas are the appropriate parameters for measuring the complexities of cuts; levels and lengths of cut formula are no longer interesting. To make this precise, we introduce a slightly modified notion of derivability within $\mathsf{H}_0$.

**Definition 30** We define $\mathsf{H}_0 \overset{\alpha}{\underset{\sigma}{\vdash}} \Delta$ for all finite $\Delta \subseteq \mathcal{L}_0$ and all ordinals $\alpha, \sigma < \Gamma_0$ by induction on $\alpha$.

1. If $\Delta$ is an axiom of $\mathsf{H}_0$, then we have $\mathsf{H}_\nu \overset{\alpha}{\underset{\sigma}{\vdash}} \Delta$ for all $\alpha, \sigma < \Gamma_0$.
2. If $\mathsf{H}_0 \overset{\alpha_i}{\underset{\sigma}{\vdash}} \Delta_i$ and $\alpha_i < \alpha$ for every premise $\Delta_i$ of a stage rule, a propositional rule, or a quantifier rule or a cut of $\mathsf{H}_0$ whose cut formulas have rank less than $\sigma$, then we have $\mathsf{H}_0 \overset{\alpha}{\underset{\sigma}{\vdash}} \Delta$ for the conclusion $\Delta$ of this rule.

Thus $\mathsf{H}_0 \overset{\alpha}{\underset{0}{\vdash}} \Delta$ means that $\Delta$ is cut-free provable in the system $\mathsf{H}_0$. Our two methods of measuring derivations in $\mathsf{H}_0$ are, of course, closely linked. In particular, we can easily transform the former into the latter. The proof of the following lemma is trivial by induction on $\alpha$.

**Lemma 31** *For all $\alpha, \beta < \Gamma_0$, all $n < \omega$, and all finite $\Delta \subseteq \mathcal{L}_0$ we have that*

$$\mathsf{H}_0 \overset{\alpha}{\underset{(\beta,n,n)}{\vdash}} \Delta \quad \Longrightarrow \quad \mathsf{H}_0 \overset{\alpha}{\underset{\omega\beta+n}{\vdash}} \Delta.$$

The last step in our proof-theoretic analysis of the systems $\mathsf{H}_\nu$ is complete cut elimination for $\mathsf{H}_0$ with respect to our new derivability relation. However, it is easy to check that the assignment of ranks and the rules of inference are tailored such that the methods of predicative proof theory yield full cut elimination for $\mathsf{H}_0$. Therefore we omit the proof of the following theorem and refer to the standard literature, for example, Pohlers [16] or Schütte [21].

**Theorem 32** *For all $\alpha, \beta, \gamma < \Gamma_0$ and all finite $\Delta \subseteq \mathcal{L}_0$ we have that*

$$\mathsf{H}_0 \overset{\alpha}{\underset{\beta+\omega^\gamma}{\vdash}} \Delta \quad \Longrightarrow \quad \mathsf{H}_0 \overset{\varphi\gamma\alpha}{\underset{\beta}{\vdash}} \Delta.$$

From Theorem 27, Lemma 31, Theorem 32 and the fact that $\varphi\eta\xi < \varphi\Lambda_\nu 0$ for all $\eta < \Lambda_\nu$ and $\xi < \varphi\Lambda_\nu 0$ we deduce the following key result.

**Corollary 33** *Assume that $\Delta$ is a finite subset of $\mathcal{L}_0$ and $\mathsf{H}_\nu \overset{<\varepsilon(\nu)}{\underset{(0,n,n)}{\vdash}} \Delta$ for some $n < \omega$. Then there exists an $\alpha < \varphi\Lambda_\nu 0$ such that $\mathsf{H}_0 \overset{\alpha}{\underset{0}{\vdash}} \Delta$.*

This corollary brings us very close to the desired computation of the upper proof-theoretic bounds of the theories $\mathsf{SID}_\nu$. It now remains only to embed the theories

$\mathsf{SID}_\nu$ into the systems $\mathsf{H}_\nu$. The following lemma deals with transfinite induction within $\mathsf{H}_\nu$.

**Lemma 34** *For all $\alpha < \Gamma_0$, all closed number terms $t$ such that $ot(t) = \alpha$, and all $A[0] \in \mathcal{L}_\nu$ we have:*

1. $\mathsf{H}_\nu \,\overset{<\omega\alpha+\omega}{\underset{(0,0,0)}{\vdash}}\, \neg Prog[\prec, \{x{:}A[x]\}],\ A[t]$.
2. $\mathsf{H}_\nu \,\overset{\omega\alpha+\omega}{\underset{(0,0,0)}{\vdash}}\, Prog[\prec, \{x{:}A[x]\}] \rightarrow (\forall x \prec t)A[x]$.

*Proof* As to be expected, we prove the first assertion by induction on $\alpha$. Let $t$ be a closed term such that $ot(t) = \alpha$. We first consider all closed number terms $r$ for which $\alpha \leq ot(r)$. In this case $r \not\prec t$ is a true closed literal, hence

$$\mathsf{H}_\nu \,\overset{0}{\underset{(0,0,0)}{\vdash}}\, \neg Prog[\prec, \{x{:}A[x]\}],\ r \not\prec t,\ A[r]. \tag{22}$$

Secondly, if $s$ is a closed number term for which $ot(s) = \xi < \alpha$, then the induction hypothesis implies

$$\mathsf{H}_\nu \,\overset{\omega\xi+\omega}{\underset{(0,0,0)}{\vdash}}\, \neg Prog[\prec, \{x{:}A[x]\}],\ s \not\prec t,\ A[s]. \tag{23}$$

By some simple applications of propositional and quantifier rules, (22) and (23) yield

$$\mathsf{H}_\nu \,\overset{\omega\alpha+3}{\underset{(0,0,0)}{\vdash}}\, \neg Prog[\prec, \{x{:}A[x]\}],\ (\forall x \prec t)A[x].$$

Making use of Lemma 20, we can continue with

$$\mathsf{H}_\nu \,\overset{<\omega\alpha+\omega}{\underset{(0,0,0)}{\vdash}}\, \neg Prog[\prec, \{x{:}A[x]\}],\ (\forall x \prec t)A[x] \wedge \neg A[t],\ A[t]$$

and obtain

$$\mathsf{H}_\nu \,\overset{\omega\alpha+\omega}{\underset{(0,0,0)}{\vdash}}\, \neg Prog[\prec, \{x{:}A[x]\}],\ A[t]$$

by a further application of a quantifier rule, as desired. This completes the proof of the first assertion. The second assertion is a simple consequence of the first. □

The embedding of $\mathsf{SID}_\nu$ into $\mathsf{H}_\nu$ can now be defined straightforwardly: To each closed formula $A$ in the language $\mathcal{L}_S$ of $\mathsf{SID}_\nu$ we associate its interpretation $A^{(\nu)}$, given by replacing all subexpressions $\mathcal{P}^{\mathfrak{A}}(t)$ and $\sim\mathcal{P}^{\mathfrak{A}}(t)$ of $A$ by $P^{\mathfrak{A}}_{<\nu}(t)$ and $\neg P^{\mathfrak{A}}_{<\nu}(t)$, respectively. Based on this translation of the closed $\mathcal{L}_S$ formulas into formulas belonging to $\mathcal{L}_\nu$, we can formulate our embedding theorem.

**Theorem 35 (Embedding)** *Let $A$ be a closed formula of $\mathcal{L}_S$ and further assume $\mathsf{SID}_\nu \vdash A$. Then we have $\mathsf{H}_\nu \,\overset{<\varepsilon(\nu)}{\underset{(0,n,n)}{\vdash}}\, A^{(\nu)}$ for some $n < \omega$.*

*Proof* Since $\omega\nu + \omega < \varepsilon(\nu)$, the previous lemma takes care of the axioms about transfinite induction up to $\nu$. The translations of (the universal closures of) all other axioms of $\mathsf{SID}_\nu$ are obviously provable in $\mathsf{H}_\nu$ and closure under the translations of the inference rules of $\mathsf{SID}_\nu$ is guaranteed; always respecting the required bounds. □

The previous embedding theorem and Corollary 33 finally provide the reduction of $\mathsf{SID}_\nu$ with respect to all closed $\mathcal{L}$ formulas to the cut-free part of the system $\mathsf{H}_0$.

**Theorem 36 (Final Reduction)** *If $A$ is a closed $\mathcal{L}$ formula and $\mathsf{SID}_\nu \vdash A$, then there exists an $\alpha < \varphi\Lambda_\nu 0$ such that $\mathsf{H}_0 \,|\!\frac{\alpha}{0}\, A$.*

**Corollary 37** $|\mathsf{SID}_\nu| \;=\; \varphi\Lambda_\nu 0.$

*Proof* $\varphi\Lambda_\nu 0 \;\le\; |\mathsf{SID}_\nu|$ is Theorem 12. To show $|\mathsf{SID}_\nu| \;\le\; \varphi\Lambda_\nu 0$ assume that $\mathsf{SID}_\nu$ proves $TI[\lhd, \{x{:}W(x)\}]$ for some primitive recursive wellordering $\lhd$. By Theorem 33 we have an $\alpha \;<\; \varphi\Lambda_\nu 0$ such that $\mathsf{H}_0 \,|\!\frac{\alpha}{0}\, TI[\lhd, \{x{:}W(x)\}]$. Standard boundedness techniques as presented, for example, in Pohlers [16] and Schütte [21] then imply that the order-type of $\lhd$ is less than or equal to $\omega\alpha \;<\; \varphi\Lambda_\nu 0$. Hence every ordinal provable in $\mathsf{SID}_\nu$ is smaller than $\varphi\Lambda_\nu 0$. □

This finishes our proof-theoretic analysis of the theories $\mathsf{SID}_\nu$. Of course, this result also provides the proof-theoretic ordinals of the systems $\mathsf{SID}_{<\nu}$, namely

$$|\mathsf{SID}_{<\nu}| \;=\; \sup(\{\varphi\Lambda_\mu 0 : \mu < \nu\}).$$

The proof-theoretic ordinals of some important such theories have been mentioned in Sect. 3.

## References


1. P. Aczel, A new approach to the Bachmann method for describing large countable ordinals. Unpublished notes (1970)
2. P. Aczel, The strength of Martin-Löf's intuitionistic type theory with one universe. Unpublished notes (1976)
3. B. Afshari, M. Rathjen, A note on the theory of positive induction, $\mathsf{ID}_1^*$. Arch. Math. Log. **49**(2), 275–281 (2010)
4. W. Buchholz, Normalfunktionen und konstruktive Systeme von Ordinalzahlen, in *Proof Theory Symposion Kiel 1974* , ed. by J. Diller, G.H. Müller. Lecture Notes in Mathematics, vol. 500 (Springer, Berlin, 1975), pp. 4–25
5. W. Buchholz, A simplified version of local predicativity, in *Proof Theory*, ed. by P. Aczel, H. Simmons, S.S. Wainer (Cambridge University Press, New York, 1992), pp. 115–147
6. W. Buchholz, *Prädikative Beweistheorie*. Vorlesungsskript (Universität München, München, 2004/2005)
7. W. Buchholz, S. Feferman, W. Pohlers, W. Sieg, *Iterated Inductive Definitions and Subsystems of Analysis: Recent Proof-Theoretical Studies*. Lecture Notes in Mathematics, vol. 897 (Springer, Berlin, 1981)

8. A. Cantini, *Logical Frameworks for Truth and Abstraction. An Axiomatic Study*. Studies in Logic and the Foundations of Mathematics, vol. 135 (Elsevier, North-Holland, 1996)
9. S. Feferman, Iterated inductive fixed-point theories: application to Hancock's conjecture, in *The Patras Symposion*, ed. by G. Metakides (Elsevier, North-Holland, 1982), pp. 171–196
10. H. Friedman, Theories of inductive definition. Unpublished notes (1969)
11. G. Jäger, *Theories for Admissible Sets: A Unifying Approach to Proof Theory*. Studies in Proof Theory, Lecture Notes, vol. 2 (Bibliopolis, 1986)
12. G. Jäger, T. Strahm, Upper bounds for metapredicative Mahlo in explicit mathematics and admissible set theory. J Symb. Log. **66**(2), 935–958 (2001)
13. G. Jäger, R. Kahle, A. Setzer, T. Strahm, The proof-theoretic analysis of transfinitely iterated fixed point theories. J. Symb. Log. **64**(1), 53–67 (1999)
14. D. Leivant, *Peano Theories and the Formalization of Feasible Mathematics*. Extended summary ( Indiana University, Bloomington, 1993)
15. D. Leivant, Intrinsic theories and computational complexity, in *Logic and Computational Complexity*, ed. by D. Leivant. Lecture Notes in Computer Science, vol. 960 (Springer, Berlin, 1995), pp. 177–194
16. W. Pohlers, *Proof Theory. An Introduction*. Lecture Notes in Mathematics, vol. 1407 (Springer, Berlin, 1989)
17. D. Probst, The proof-theoretic analysis of transfinitely iterated quasi least fixed points. J. Symb. Log. **71**(3), 721–746 (2006)
18. F. Ranzi, T. Strahm, A note on the theory $\mathsf{SID}_{<\omega}$ of stratified induction. Math. Log. Q. **60**(6), 487–497 (2014)
19. M. Rathjen, Admissible proof theory and beyond, in *Logic, Methodology and Philosophy of Science - Proceedings of the Ninth International Congress*, ed. by D. Prawitz, B. Skyrms, D. Westerstahl. Studies in Logic and the Foundations of Mathematics, vol. 134 (Elsevier, Amsterdam, 1994), pp. 123–147
20. K. Schütte, Kennzeichnung von ordnungszahlen durch rekursiv erklärte funktionen. Math. Ann. **127**, 15–32 (1954)
21. K. Schütte, *Proof Theory*. Grundlehren der mathematischen Wissenschaften, vol. 225 (Springer, Berlin, 1977)
22. S.S. Wainer, R.S. Williams, Inductive definitions over a predicative arithmetic. Ann. Pure Appl. Log. **136**(1–2), 175–188 (2005)

# Classifying Phase Transition Thresholds for Goodstein Sequences and Hydra Games

**Frederik Meskens and Andreas Weiermann**

**Abstract** A classification of the phase transition thresholds behind the Kirby Paris style independence results about Goodstein sequences and hydras is given. Moreover earlier phase transition results by Kent and Hodgson are improved.

The article is intended to be suitable for teaching purposes and just requires basic familiarity with the standard classification of the provably recursive functions of PA and its fragments in terms of the Hardy functions.

## 1 Introduction

This article is part of a general program on classifying phase transitions for independence results for (sub-)systems of Peano Arithmetic. The basic idea is to investigate parameterized assertions $A_r$ (where $r$ is a rational number parameter) which are provable for small parameter values in a given formal system $S$ under consideration and which become unprovable in $S$ (but still remain true) for large parameter values. We moreover assume that $A_r$ is monotone in the sense that when $r$ ranges from small to large values there is only once a transition from $S$-provability to $S$-unprovability. Classifying the resulting threshold for the transition from provability to unprovability may shed light on the general question: *What makes a true assertion A unprovable from Peano Arithmetic?* For Friedman's miniaturization $FKT_r$ of Kruskal's theorem [10] the critical threshold value $\rho$ for the phase transition (in its logarithmic formulation a la [9]) has approximate value $0.639578\ldots$ (which is currently not known to be rational, irrational, algebraic or transcendental) [11].

F. Meskens • A. Weiermann (✉)
Gent Universitet, Vakgroep Wiskunde, Krijgslaan 281 Gebouw S22, 9000 Ghent, Belgium
e-mail: Frederik.Meskens@UGent.be; Andreas.Weiermann@UGent.be

 
R. Kahle, M. Rathjen (eds.), *Gentzen's Centenary*,
DOI 10.1007/978-3-319-10103-3_16

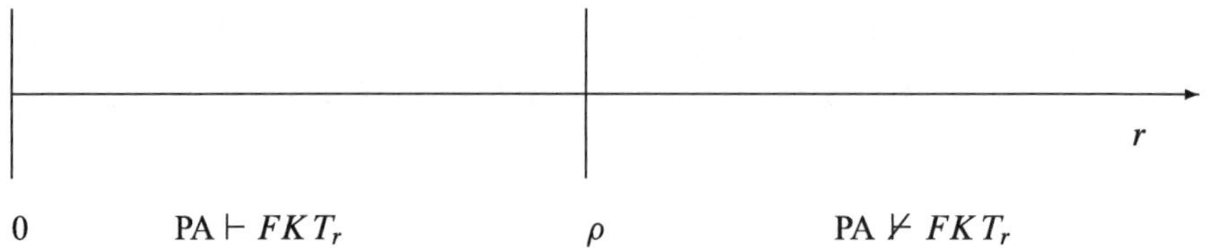


In a more general context, as for example in this article, we shall assume that $A$ depends on a number-theoretic function parameter $f$ (which is assumed to be elementary recursive).

We shall work under the assumption that $A_f$ is always true for any $f$ and provable if $f$ is very slow growing. Moreover we may assume similarly as above that if $A_f$ is provable in Peano Arithmetic (which will be abbreviated by PA) and $g$ is eventually dominated by $f$ then $A_g$ is provable in PA, too. Moreover we assume that $A_f$ becomes unprovable in PA if $f$ grows reasonably fast.

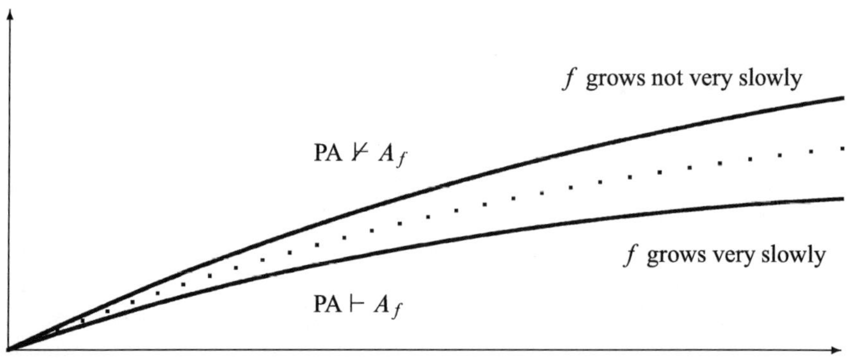


In this article we consider such phase transitions for the Kirby Paris principles about Goodstein sequences and Hydra games [7]. In case of Goodstein sequences preliminary results have already been obtained by Kent and Hodgson [6]. In this paper we improve their results and give simultaneously a treatment of the Hydra games, too. Moreover we investigate the situation for fragments of arithmetic.

Our results will finally show that in our context there will always be a sharp threshold function which can be expressed in terms of functions like $+, \cdot, \div, \exp_2, \log_2,\ \exp_2^x(y), \log_2^x(y)$ (where the upper index denotes the number of function iterations and the lower index denotes the base to which the function under consideration refers) and functions from the hierarchy $(H_\alpha^{-1})_{\alpha \leqslant \varepsilon_0}$. The hierarchy $(H_\alpha^{-1})_{\alpha \leqslant \varepsilon_0}$ consists in this context of a scale of slow growing functions which are given by inverse functions of functions from the Hardy hierarchy $(H_\alpha)_{\alpha \leqslant \varepsilon_0}$ of fast growing functions. In practice most functions of the form $H_\alpha^{-1}$ (for $\alpha$ reasonably large) grow so slow that they cannot be distinguished by computer calculations from constant functions. Nevertheless *PA* is able to prove the unboundedness of $H_\alpha^{-1}$ for $\alpha < \varepsilon_0$. When $\alpha$ increases the speed of $H_\alpha^{-1}$ becomes slower and slower and finally

$H_{\varepsilon_0}^{-1}$ grow so slow that *PA* is unable to prove its unboundedness (although $H_{\varepsilon_0}^{-1}$, in theory, still is unbounded).

If we are going to measure the phase transition in more general terms, then sharp threshold functions will no longer exist. Instead the fine structure of the threshold region will form a dense lattice which has similar properties as the lattice of honest elementary degrees considered by Kristiansen et al. [8].

This article requires only basic familiarity with the classification of the provably recursive functions of Peano arithmetic and is therefore a natural follow-up of the article *Classifying the provably total functions of* PA [12]. The classification of the provably recursive functions of PA is of course standard and can be found in many sources as, e.g., in [4] or [5]. What basically is needed for the phase transition results concerning PA studied in this paper is just the result that

$$\mathrm{PA} \nvdash \forall x \exists y H_{\varepsilon_0}(x) = y.$$

This article is also intended and, in fact, already proved useful for teaching purposes.[1]

## 2 Preliminaries

### 2.1 Definitions

To keep the paper short we assume the usual standard association between the Hydras and ordinal numbers which is, for example, explained in detail in the original source by Kirby and Paris [7] or in Buchholz's seminal paper [1]. A step in the hydra game for hydra $\alpha$ at time $x$ then corresponds in stepping down from $\alpha$ to $\alpha[x]$ with respect to the standard system of fundamental sequences for the ordinals below $\varepsilon_0$.

(If not mentioned otherwise positive integers are denoted by small Latin letters and ordinals not exceeding $\varepsilon_0$ will be denoted by small Greek letters.)

For simplicity we assume that hydras are represented by ordinals in Cantor normal form. Let us introduce some standard notations.

**Definition 1** For $0 < \alpha, \beta, \gamma < \varepsilon_0$ and $\omega \geqslant j \geqslant i \geqslant 2$, define

1. $\alpha =_{\mathrm{NF}} \omega^{\alpha_1} + \cdots + \omega^{\alpha_n}$ if $\alpha = \omega^{\alpha_1} + \cdots + \omega^{\alpha_n}$ and $\alpha_1 \geqslant \cdots \geqslant \alpha_n$.
2. $NF(\beta, \gamma)$ if $\beta =_{\mathrm{NF}} \omega^{\beta_1} + \cdots + \omega^{\beta_m}, \gamma =_{\mathrm{NF}} \omega^{\gamma_1} + \cdots + \omega^{\gamma_n}$ and $\beta_m \geqslant \gamma_1$.
3. $\alpha =_{\mathrm{CNF}} \omega^{\alpha_1} m_1 + \cdots + \omega^{\alpha_n} m_n$ if $\alpha = \omega^{\alpha_1} m_1 + \cdots + \omega^{\alpha_n} m_n, \alpha_1 > \cdots > \alpha_n$ and $m_1, \ldots, m_n \in \mathbb{N} \setminus \{0\}$.
4. If $\alpha =_{\mathrm{CNF}} \omega^{\alpha_1} m_1 + \cdots + \omega^{\alpha_n} m_n$, then we define the maximal coefficient as $mc(\alpha) := \max\{mc(\alpha_i), m_i | i = 1, \ldots, n\}$ with $mc(0) := 0$.

[1] The second author has used the main results of this paper repeatedly during lectures on selected chapters from proof theory.

5. If $a = i^{a_1}m_1 + \cdots + i^{a_n}m_n$ with $a_1 > \cdots > a_n$ and $i > m_k > 0$, then $a(i \mapsto j) := j^{a_1(i \mapsto j)}m_1 + \cdots + j^{a_n(i \mapsto j)}m_n$, else $a(i \mapsto j) := a$.

Since we will work several times with exponential towers and iterated logarithms it is convenient to agree also on the following notations.

**Definition 2** For $\alpha, \beta < \varepsilon_0$ and nonnegative integers $h$ we define:

6. $\alpha_0(\beta) := \beta, \alpha_{h+1}(\beta) := \alpha^{\alpha_h(\beta)}$,
   $\omega_h := \omega_h(1), 2_h := 2_h(1)$,
7. $|0| := 1, |i| := \lceil \log_2(i+1) \rceil$ if $i > 0$,
   $|i|_h := i$ if $h = 0, |i|_{h+1} := ||i|_h|$.

Then, of course, $|i|$ is the binary length of $i$ and $|i|_h$ stands for the $h$-times iterated binary length function.

Let us moreover recall the standard assignment of fundamental sequences for ordinals below $\varepsilon_0$.

**Definition 3** If $\alpha \in \{0, 1\}$ then $\alpha[k] := 0$, if $1 < \alpha < \varepsilon_0$ then write $\alpha =_{\mathrm{NF}} \omega^{\alpha_1} + \cdots + \omega^{\alpha_n}$ and

$$\alpha[k] := \begin{cases} \omega^{\alpha_1} + \cdots + \omega^{\alpha_{n-1}} & \text{if } \alpha_n = 0, \\ \omega^{\alpha_1} + \cdots + \omega^{\alpha_n - 1} \cdot (k+1) & \text{if } \alpha_n \notin \mathrm{Lim}, \\ \omega^{\alpha_1} + \cdots + \omega^{\alpha_n[k]} & \text{if } \alpha_n \in \mathrm{Lim}. \end{cases}$$

Finally put $\varepsilon_0[k] := \omega_k$. (As common we write Lim for the class of limit ordinals.)

For technical reasons (to guarantee the so-called Bachmann condition [2]) we put $\omega^{\alpha+1}[k] = \omega^\alpha \cdot (k+1)$. Our results will also hold (modulo some obvious modifications) for the choice $\omega^{\alpha+1}[k] = \omega^\alpha \cdot k$ and not too exotic variations thereof.

To fix the context we moreover recall some well-known definitions and lemmata from subrecursive hierarchy theory. In order to find the phase transition for the Hydra game ($Q$-steps) we will consider a more complex and a simpler version. The simpler version will be a Friedman style slowly well orderedness assertion with respect to a norm provided by the maximal coefficient ($mc(\cdot)$) [10] and the more complex corresponds to the termination of the Goodstein sequences ($P$-steps). In the sequel let $f$ denote a given weakly increasing elementary recursive function $f : \mathbb{N} \to \mathbb{N}$.

**Definition 4 (Predecessor Operations)** For ordinals $\alpha$ and $\lambda \in \mathrm{Lim}$ define:

$$P_x^f(0) = 0, \qquad P_x^f(\alpha+1) = \alpha, \qquad P_x^f(\lambda) = P_x^f(\lambda[f(x)]),$$
$$Q_x^f(0) = 0, \qquad Q_x^f(\alpha+1) = \alpha, \qquad Q_x^f(\lambda) = \lambda[f(x)].$$

If $f(i) = id(i) = i$ we write $R_x(\alpha)$ instead of $R_x^f(\alpha)$ for $R \in \{P, Q\}$; and if there is no confusion possible, we omit the brackets as well.

The stepping down relation on the ordinal numbers defined by $P$ and $Q$ is written as follows:

**Definition 5** For $R \in \{P, Q\}$, define:

1. $\alpha \succ_f^{R,0} \alpha$.
2. $\alpha \succ_f^{R,n} \beta :\Leftrightarrow \alpha > \beta$ and $\beta = R_n^f \dots R_1^f \alpha$ with $n \geqslant 1$.
3. $\alpha \succ_f^R \beta :\Leftrightarrow \exists n > 0(\alpha \succ_f^{R,n} \beta)$,
4. $\alpha \succ_k^R \beta :\Leftrightarrow \alpha \succ_f^R \beta$ where $f$ is a constant function with value $k$,
5. $\alpha \succeq_k^R \beta :\Leftrightarrow \alpha \succ_k^R \beta$ or $\alpha = \beta$ where $k \geqslant 0$.

Note that $\alpha \succ_k^R \beta$ yields $\alpha[k] \succ_k^R \beta$ and $R_n^f \beta = R_{f(n)}\beta$.

**Definition 6 (Hydra Steps)**

$$\alpha_{f,0} := \alpha,$$
$$\alpha_{f,i+1} := \alpha_{f,i}[1 + f(i)].$$

The Hydra principle ($\mathrm{H}_f$) is the assertion $(\forall\alpha)(\exists i)\alpha_{f,i} = 0$.

Clearly the Hydra principle is closely connected to iterating the operator $Q$. Schematically, we arrive in step $k$ at

$$\alpha_{f,k} = (\dots(\alpha[1 + f(0)])\dots)[1 + f(k - 1)].$$

**Definition 7 (Goodstein Sequences)** Let $m \geqslant 2$.

$$m_{f,0} := m$$
$$m_{f,i+1} := m_{f,i}(1 + f(i) \mapsto 1 + f(i + 1)) - 1.$$

The Goodstein principle ($\mathrm{G}_f$) is the assertion $(\forall m)(\exists i)m_{f,i} = 0$.

In Lemma 7 we prove (following [3]) that Goodstein sequences are intrinsically connected to the operator $P$.

## 2.2 *Sub- and Superprocesses for the Goodstein Principle*

In this paragraph we shall prove that the Goodstein sequences form a subprocess of the Hydra games. Moreover we will see that a Friedman style slowly well orderedness principle with respect to the maximal coefficient is a canonical extension of the standard Hydra game with hydra $\alpha$ for a nondecreasing function $f$ if

$mc(\alpha) \leqslant f(1)$. This follows from the observation

$$mc(\alpha[x]) \leq \max(mc(\alpha), x+1), \tag{1}$$

which is a consequence of Definitions 1.4 and 3.

We investigate first some basic arithmetic of the predecessor operators. Note that a first indication for proving that the Goodstein process is not lasting longer than a corresponding Hydra battle is given in assertion 2 of the next lemma.

**Lemma 1** *Let $\alpha, \beta, \gamma < \varepsilon_0$ and $R \in \{P, Q\}$. Then the following assertions hold:*

1. $\alpha > 0 \Rightarrow \alpha \succcurlyeq_x^Q 1$, *and* $\alpha \succcurlyeq_x^P 0$,
2. $\alpha \succ_x^P \beta \Rightarrow \alpha \succ_x^Q \beta$,
3. $NF(\gamma, \beta)$ *and* $\beta > 0 \Rightarrow R_x^f(\gamma + \beta) = \gamma + R_x^f \beta$,
4. $NF(\gamma, \alpha)$ *and* $\alpha \succ_x^{R,m} \beta \Rightarrow \gamma + \alpha \succ_x^{R,m} \gamma + \beta$,
5. $x \geqslant 1$ *and* $\alpha \succ_x^{R,m} \beta \Rightarrow (\exists n \geqslant m) \left(\omega^\alpha \succ_x^{R,n} \omega^\beta\right)$.

*Proof* Assertions 1 and 2 follow by induction on $\alpha$. For assertion 3 write $\beta =_{\mathrm{NF}} \omega^{\beta_1} + \cdots + \omega^{\beta_m}$ and apply Definition 4. Assertion 4 follows from assertion 3. Assertion 5 follows from assertions 1 and 4 by induction on $\alpha$. □

Now we shall prove the assertion that

$$x \leqslant y, \lambda \in Lim \Rightarrow \lambda[y] \succcurlyeq_z^R \lambda[x] \tag{2}$$

for $z \geqslant 0$ and $R \in \{P, Q\}$. This follows by induction once it is proved for $y = x+1$ and the latter case is dealt with in the next Lemma. Assertion 4 of the next lemma gives a generalization of assertion 2 of Lemma 1.

**Lemma 2** *Let $\alpha, \beta, \gamma < \varepsilon_0$ and $y \geqslant 0$. Then the following assertions hold:*

1. $y > 0$ *and* $\lambda \in Lim \Rightarrow \lambda[x+1] \succcurlyeq_y^Q \lambda[x] + 1$,
2. $\alpha \succcurlyeq_x^Q \beta \succ_x^{P,m} \gamma \Rightarrow (\exists n \geqslant m) \left(\alpha \succ_x^{P,n} \gamma\right)$,
3. $y > 0$ *and* $\lambda \in Lim \Rightarrow \lambda[x+1] \succ_y^P \lambda[x]$,
4. $\alpha > 0$ *and* $y \geqslant x > 0 \Rightarrow \alpha \succcurlyeq_y^Q P_x\alpha + 1$.

*Proof* Note that assertions 1 and 4 imply their strict versions if '+1' is omitted.

Assertion 1 is proved by induction on $\lambda$. Indeed, if we write

$$\lambda =_{\mathrm{NF}} \omega^{\lambda_1} + \cdots + \omega^{\lambda_n},$$

then because of Definition 3 and assertion 4 of Lemma 1 it suffices to prove

$$\omega^{\lambda_n}[x+1] \succ_y^Q \omega^{\lambda_n}[x] + 1.$$

If $\lambda_n \in Lim$, then the induction hypothesis yields $\lambda_n[x+1] \succcurlyeq_y^Q \lambda_n[x]+1$, and therefore

$$\omega^{\lambda_n}[x+1] = \omega^{\lambda_n[x+1]} \succ_y^Q \omega^{\lambda_n[x]+1} \succcurlyeq_y^Q \omega^{\lambda_n}[x]+1,$$

using assertion 5 of Lemma 1. Suppose now $\lambda_n = \alpha+1$. By assertion 1 of Lemma 1 we obtain $\omega^\alpha \succcurlyeq_y^Q 1$, and this yields

$$\omega^{\lambda_n}[x+1] = \omega^\alpha(x+2) = \omega^\alpha(x+1)+\omega^\alpha \succcurlyeq_y^Q \omega^\alpha(x+1)+1 = \omega^{\lambda_n}[x]+1,$$

by assertion 4 of Lemma 1.

Assertion 2 is also proved by induction on $\alpha$. For the non trivial case we suppose $\alpha \neq \beta$. Then $\alpha[x] \succcurlyeq_x^Q \beta \succ_x^{P,m} \gamma$ which implies, by induction hypothesis, the existence of an $n \geqslant m$ such that

$$\alpha[x] \succ_x^{P,n} \gamma.$$

This yields $\alpha \succ_x^{P,k} \gamma$ with $k = n+1$ if $\alpha \notin Lim$ and with $k = n$ if $\alpha \in Lim$ since $P_x\alpha = P_x\alpha[x]$.

Proof of assertion 3. From assertion 1 it follows that $\lambda[x+1] \succcurlyeq_y^Q \lambda[x]+1$. Because of $\lambda[x]+1 \succ_y^P \lambda[x]$ assertion 2 yields $\lambda[x+1] \succ_y^P \lambda[x]$.

Assertion 4 is proved by induction on $\alpha$ using assertion 1. The assertion is trivial for $\alpha \notin Lim$. If $\alpha \in Lim$, then

$$\alpha \succ_y^Q \alpha[y] \succcurlyeq_y^Q \alpha[x] \succcurlyeq_y^Q P_x\alpha[x]+1 = P_x\alpha+1.$$

Here the second inequality follows from assertion 1 by iteration iff $y > x$ (equality holds iff $x = y$), and the last inequality follows by the induction hypothesis. □

The reader may wonder why in the previous lemma the assertions 1 and 4 are stated with a '+1'. The reason is that later in a critical step of the proof of Proposition 2 we need a remainder that is still big enough (expressed in terms of $\omega$-towers, see Corollary 2) to allow for an estimation of the number of needed steps of descents.

**Lemma 3** *Let $\alpha, \beta < \varepsilon_0$, $R \in \{P, Q\}$ and $f, g : \mathbb{N} \to \mathbb{N}$ non decreasing functions. Then the following assertions hold*

1. $\alpha \succ_x^R \beta, x \leqslant y \Rightarrow \alpha \succ_y^R \beta$,
2. $\forall i : g(i) \leqslant f(i), \alpha \succ_g^{R,m} \beta \Rightarrow (\exists n \geqslant m)\left(\alpha \succ_f^{R,n} \beta\right)$.

*Proof* Assertion 1 is proved by induction on $\alpha$ and follows from $\alpha \succ_y^R \alpha[y] \succcurlyeq_y^R \alpha[x] \succcurlyeq_y^R \beta$. The second inequality holds because of Eq. (2) if $\alpha \in Lim$ (equality holds otherwise) and the last one follows from the induction hypothesis.

Assertion 2 is proved by induction on $m$. Assume that $\alpha \succ_g^{R,m} \beta$. Then there exists a $\beta_0$ such that $\alpha \succcurlyeq_g^{R,m-1} \beta_0$ and $\beta = R_{g(m)}\beta_0$. The induction hypothesis yields an $n' \geqslant m-1$ such that $\alpha \succcurlyeq_f^{R,n'} \beta_0$. We have to show that there exists an $n > n'$ such that

$$\beta = R_{f(n)} R_{f(n-1)} \dots R_{f(n'+1)}\beta_0.$$

The assumption yields

$$f(n) \geqslant \dots \geqslant f(n'+1) \geqslant g(m)$$

for all $n \geqslant n' \geqslant m$. Put

$$z_i := f(n'+i).$$

It suffices to show the relation $R_{z_1}\beta_0 \succeq_{z_1}^R \beta$ and the implication $R_{z_i} \dots R_{z_1}\beta_0 \succ_{z_i}^R \beta \Rightarrow R_{z_i+1}\, R_{z_i} \dots R_{z_1}\beta_0 \succeq_{z_i+1}^R \beta$.

Note first that $\beta = R_{g(m)}\beta_0$, hence $\beta_0 \succ_{g(m)}^R \beta$. Since $g(m) \leqslant z_1$ assertion 1 yields $\beta_0 \succ_{z_1}^R \beta$.

Now assume that $R_{z_i} \dots R_{z_1}\beta_0 \succ_{z_i}^R \beta$. Then by assertion 1 we have

$$R_{z_i} \dots R_{z_1}\beta_0 \succ_{z_i+1}^R \beta.$$

This gives

$$R_{z_i+1} R_{z_i} \dots R_{z_1}\beta_0 \succcurlyeq_{z_i+1}^R \beta.$$

□

Now we're able to prove that Goodstein sequences form a subsystem (subsequence) of the Hydra games.

**Corollary 1** *If $f$ is nondecreasing and $\alpha \succ_f^P \beta$, then $\alpha \succ_f^Q \beta$.*

*Proof* The assertion follows essentially from Lemma 2.3. We show the implication

$$\alpha \succ_f^{P,m} \beta \Longrightarrow \exists n \geqslant m, \alpha \succ_f^{Q,n} \beta$$

by induction on $m$. Assume that $\alpha \succ_f^{P,m} \beta$. Then there exists $\beta_0$ such that

$$\alpha \succ_f^{P,m-1} \beta_0$$

and $\beta = P_{f(m)}\beta_0$. The induction hypothesis yields the existence of an $n' \geqslant m-1$ such that $\alpha \succ_f^{Q,n'} \beta_0$. We claim the existence of an $n > n'$ such that

$$Q_{f(n)} \dots Q_{f(n'+1)}\beta_0 = \beta.$$

Let

$$z_i := f(n'+i)$$

It suffices to show the relation $Q_{z_1}\beta_0 \succeq_{z_1} \beta$ and the implication $Q_{z_i} \dots Q_{z_1}\beta_0 \succ_{z_i}^Q \beta \Rightarrow Q_{z_i+1} Q_{z_i} \dots Q_{z_1}\beta_0 \succeq_{z_i+1}^R \beta$.

First we have $\beta_0 \succ_{f(m)}^P \beta$ hence $\beta_0 \succ_{f(m)}^Q \beta$ according to assertion 2 of Lemma 1. Since $z_1 \geqslant f(m)$ we obtain $\beta_0 \succ_{z_1}^Q \beta$ and hence $Q_{z_1}\beta_0 \succcurlyeq_{z_1}^Q \beta$.

Now assume $Q_{z_i} \dots Q_{z_1}\beta_0 \succ_{z_i}^Q \beta$. Assertion 1 of Lemma 3 yields

$$Q_{z_i} \dots Q_{z_1}\beta_0 \succ_{z_i+1}^Q \beta$$

hence $Q_{z_i+1} \dots Q_{z_1}\beta_0 \succcurlyeq_{z_i+1}^Q \beta$. □

We conclude this paragraph with some useful estimates for reductions of $\omega$-towers of an ordinal $\alpha$.

**Corollary 2** *Let $\alpha < \varepsilon_0$. Then the following assertions hold:*

1. $\omega^{\alpha+1} \succ_{x+1}^Q \omega^\alpha 2$,
2. $\omega^{\alpha+1} \succ_{x+1}^Q \omega^\alpha + 1$,
3. $\omega_{h+1}(\alpha+1) \succ_{x+1}^Q \omega_{h+1}(\alpha) + \omega_{h+1}$ *if* $\alpha > 0$,
4. $x > 0 \Rightarrow \omega_{h+1} \succ_{x+1}^P \omega_h$.

*Proof* Assertion 1 is a direct consequence of assertions 1 and 4 of Lemma 1. Indeed, we have

$$\omega^{\alpha+1} \succ_{x+1}^Q \omega^\alpha(x+2) = \omega^\alpha 2 + \omega^\alpha x \succeq_{x+1}^Q \omega^\alpha 2.$$

Assertion 2 follows from assertion 1 by the assertions 1 and 4 of Lemma 1.

Assertion 3 is proved by induction on $h$. First note that an iteration of assertion 5 of Lemma 1 yields

$$\alpha \succ_x^Q \beta \Rightarrow \omega_h(\alpha) \succ_x^Q \omega_h(\beta). \tag{3}$$

Then note that the induction hypothesis implies

$$\omega_h(\alpha+1) \succ_{x+1}^Q \omega_h(\alpha) + 1 \tag{4}$$

by assertion 1 of Lemma 1. Further we have

$$\omega^{\omega_h(\alpha+1)} \succ^Q_{x+1} \omega^{\omega_h(\alpha)+1} \succ^Q_{x+1} \omega^{\omega_h(\alpha)}2 = \omega_{h+1}(\alpha)2 \succ^Q_{x+1} \omega_{h+1}(\alpha) + \omega_{h+1}.$$

The first inequality is obtained by assertion 5 of Lemma 1 applied to (4); the second one by assertion 1 and last one by (3).

Assertion 4 follows from $\omega \succ^P_{x+1} 1$ and assertion 5 of Lemma 1. □

## 2.3 Subrecursive Hierarchies and Counting Functions

Here we recall the standard subrecursive hierarchies which can be used to measure provability strengths with regard to provably recursive functions. Of prime importance is here the Hardy hierarchy $(H_\alpha)_{\alpha<\varepsilon_0}$. We also recall the definition of the slow growing hierarchy $x \mapsto G_x(\alpha)$ which is used for counting purposes.

**Definition 8** Let $f$ be a nondecreasing function, $\alpha, \lambda \leqslant \varepsilon_0$ with $\lambda \in Lim$. Define $G, g, H, h$ as

$$\begin{array}{lll} G_x(0) := 0 & G_x(\alpha+1) := G_x(\alpha)+1 & G_x(\lambda) := G_x(\lambda[x]) \\ g_x(0) := 0 & g_x(\alpha+1) := g_x(\alpha)+1 & g_x(\lambda) := g_x(\lambda[x])+1 \\ H^f_0(x) := x & H^f_{\alpha+1}(x) := H^f_\alpha(x+1) & H^f_\lambda(x) := H^f_{\lambda[f(x)]}(x) \\ h^f_0(x) := x & h^f_{\alpha+1}(x) := h^f_\alpha(x+1) & h^f_\lambda(x) := h^f_{\lambda[f(x)]}(x+1). \end{array}$$

Again we suppress the superscript $f$ in the definitions of $H^f_\xi$ and $h^f_\xi$ for $\xi \leqslant \varepsilon_0$ and $f = id$.

Recall that the Ackermannfunction $A$ is defined as $A(n) := A_n(n)$, where for $k \geqslant 0$ the $(k+1)$th approximation is defined as $A_{k+1}(n) := A_k^{(n+1)}(n)$ (where the upper index denotes iteration) and where $A_0(n) := n + 1$.

Some elementary but crucial properties of $G, g, H, h$ for counting lengths of stepping down processes are provided by the following lemma.

**Lemma 4** *Let* $0 < \alpha < \varepsilon_0$, *then*

1. $\min\{i : \alpha \succ^{P,i}_x 0\} = G_x(\alpha)$,
2. $\min\{i : \alpha \succ^{Q,i}_x 0\} = g_x(\alpha)$,
3. $\min\{i \geqslant x : P^f_{i-1} \dots P^f_x \alpha = 0\} = H^f_\alpha(x)$,
4. $\min\{i \geqslant x : Q^f_{i-1} \dots Q^f_x \alpha = 0\} = h^f_\alpha(x)$.

*Proof* All four assertions are proved by induction on $\alpha$. □

Note that from the previous lemma it follows that $G_x(\alpha) = H^f_\alpha(1)$ and $g_x(\alpha) = h^f_\alpha(1)$ where $f(i) = x$. It also follows that $H^f_\alpha(x)$ and $h^f_\alpha(x)$ are nondecreasing functions.

In the following lemma we prove that the Hydra games and Goodstein sequences don't differ that much, although their definitions imply huge differences in their counting functions (a $P$-step can contain many $Q$-steps). This is formalized in the '+1' in the definition of $h$ compared with $H$ which makes big differences because of recursion, but both functions bound each other. This means that their phase transitions will be closely related.

**Lemma 5** *Let $\alpha, \beta < \varepsilon_0$ and $f$ be a nondecreasing function, then*

1. $NF(\alpha, \beta) \Rightarrow H^f_\alpha\left(H^f_\beta(x)\right) = H^f_{\alpha+\beta}(x)$,
2. $F \in \{H, h\}, \alpha \succcurlyeq^Q_{f(x)} \beta \Rightarrow F^f_\alpha(x) \geqslant F^f_\beta(x)$,
3. $H^f_\alpha(x+1) \geqslant h^f_\alpha(x) \geqslant H^f_\alpha(x)$ *if $f$ is strictly increasing.*
4. $H^f_{\omega^\alpha}(x) \geqslant H^f_\alpha(x)$.
5. $A(n) = H_{\omega^n}(n) \leqslant H_{\omega^\omega}(n)$.

*Proof* All assertions are proved by induction on $\alpha$, for example we prove $H^f_\alpha(x+1) \geqslant h^f_\alpha(x)$. Indeed, we have

$$\begin{aligned} H^f_\alpha(x+1) = H^f_{\alpha[f(x+1)]}(x+1) &\geqslant H^f_{\alpha[f(x)]+1}(x+1) \\ &\geqslant h^f_{\alpha[f(x)]+1}(x) = h^f_{\alpha[f(x)]}(x+1) = h^f_\alpha(x). \end{aligned}$$

The first inequality holds by assertion 2 (by assertion 1 of Lemma 2 the condition is satisfied), the second inequality by the induction hypothesis; equalities hold by definition. For a proof of the last assertion one first proves $A_d(n) = H_{\omega^d}(n)$ for all $n$. □

The last lemma of this section is a useful tool for proving unprovability results. In order to prove provability we use a bound on the complexity of the hydra game with maximal coefficients for a constant function. For this we need a relation between $G$ and $g$. The required properties are contained in the following lemma.

**Lemma 6** *Let $\alpha, \beta < \varepsilon_0$, then*

1. $G_{x+1}(\alpha) \geqslant g_x(\alpha) \geqslant G_x(\alpha)$
2. $\alpha < \beta, mc(\alpha) \leqslant x, \beta \notin Lim \Rightarrow \alpha \leqslant \beta[x+1]$ *and*
   $\alpha < \beta, mc(\alpha) \leqslant x, \beta \in Lim \Rightarrow \alpha < \beta[x+1]$,
3. $\#\{\alpha < \beta : mc(\alpha) \leqslant x\} \leqslant G_{x+1}(\beta)$,
4. $G_x(\alpha) = \alpha(\omega \mapsto x+1)$,
5. $\alpha < \beta, mc(\alpha) \leqslant x \Rightarrow H_\alpha(x) < H_\beta(x)$ *and* $G_x(\alpha) < G_x(\beta)$.

*Proof* Assertion 1 is proved by induction on $x$. Assertion 2 is proved by induction on $\beta$, assertion 3 is proved by induction on $\beta$ using assertion 2. Assertion 4 is proved by induction on $\alpha$. Assertion 5 is proved by induction on $\alpha$. □

## 3 Phase Transitions

In this section we will prove that the phase transition thresholds for the Goodstein sequences, the standard Hydra games and the Friedman-style slowly well-foundedness of $\varepsilon_0$ with regard to the maximal coefficient norm are the same. So, by the previous discussion it is sufficient to prove a sufficiently good lower bound of unprovability for the Goodstein sequences and a sufficiently good upper bound of provability for the Friedman style slowly well-foundedness of $\varepsilon_0$ with regard to maximal coefficients. Proving (un)provability is done by determining whether the step counting function is provably recursive in the Hardy hierarchy. These statements are formalized in the following well known theorem.

**Theorem 1** *Let $T$ denote a standard primitive recursive Kleene predicate for the enumeration of the partial recursive functions. Let $U$ be the corresponding primitive recursive function (producing the output of a terminating computation). Within the language of PA the $T$ predicate is then of complexity $\Sigma_1$. Let $\Phi_e(m) := U(\min\{n : T(e, m, n)\})$. If $\Phi_e$ is provably recursive in the sense that PA $\vdash \forall x \exists y T(e, x, y)$, then there exists an $\alpha < \varepsilon_0$ such that $\Phi_e$ is primitive recursive in and bounded by $h_\alpha$. The function $h_{\varepsilon_0}$ therefore eventually dominates every provably recursive function of PA. Moreover for $\alpha < \varepsilon_0$ the functions $h_\alpha$, $H_\alpha$, $g_\alpha$ and $G_\alpha$ are all provably recursive in PA (and they are eventually dominated by the function $h_{\varepsilon_0}$).*

Let us now investigate appropriate sub- and superprocesses for the Hydra game.

**Definition 9** Define for a number-theoretic function $f$ the assertions $(\mathrm{G'}_f)$, $(\mathrm{H'}_f)$, $(\mathrm{MC}_f)$ as

$$(\mathrm{G'}_f) :\Leftrightarrow (\forall K)(\exists M)\left(\omega_K \succ_f^{P,M} 0\right)$$

$$(\mathrm{H'}_f) :\Leftrightarrow (\forall K)(\exists M)\left(\omega_K \succ_f^{Q,M} 0\right)$$

$$(\mathrm{MC}_f) :\Leftrightarrow (\forall K)(\exists M)(\forall \alpha_0, \ldots, \alpha_M \leqslant \omega_K)$$
$$((\forall i \leqslant M)(mc(\alpha_i) \leqslant K + f(i)) \Rightarrow (\exists i < M)(\alpha_{i+1} \geqslant \alpha_i)).$$

**Corollary 3**

$$PA \vdash (\mathrm{MC}_f) \Rightarrow PA \vdash (\mathrm{H'}_f) \Rightarrow PA \vdash (\mathrm{G'}_f).$$

*Proof* The first implication follows from Eq. (1). The second one follows from Lemmas 5.3 and 4. □

**Lemma 7** 1. $PA \vdash (\mathrm{G}_f) \leftrightarrow (\mathrm{G}'_f)$.
2. $PA \vdash (\mathrm{H}_f) \leftrightarrow (\mathrm{H}'_f)$.

*Proof* The second assertion is immediate since each $\alpha < \varepsilon_0$ is smaller than some $\omega_K$. For a proof of the first assertion we recall from [3] that

$$G_x(P_x\alpha) = P_x(G_x\alpha) \tag{5}$$

holds for $\alpha < \varepsilon_0$. (A proof can be done by induction on $\alpha$.) Moreover we note that if $\alpha = m_{f,i}(f(i) + 1 \mapsto \omega)$, then

$$m_{f,i} = G_{f(i)}(\alpha).$$

Moreover by definition we have

$$\begin{aligned} m_{f,i+1} &= \alpha(\omega \mapsto f(i+1) + 1) - 1 \\ &= P_{f(i+1)}\alpha(\omega \mapsto f(i+1) + 1) \\ &= P_{f(i+1)}G_{f(i+1)}(\alpha). \end{aligned}$$

Assume now that $(\forall m)(\exists i)(m)_{f,i} = 0)$. From this assumption let us prove the assertion $(G'_f)$ by elementary means. Let us assume that $f(0) = 1$ and that

$$e(m) := 2_m.$$

Then

$$PA \vdash (\forall m)(\exists i)(e(m)_{f,i} = 0).$$

Let for given $m \geqslant 2$

$$\alpha(m) := e(m)(2 \mapsto \omega) = \omega_m.$$

By induction on $i$ we show now

$$e(m)_{f,i} = G_{f(i)}\left(P_{f(i)} \dots P_{f(1)}\alpha(m)\right).$$

Indeed, $e(m)_{f,0} = G_{f(0)}\alpha(m)$ holds due to $f(0) = 1$, and for $i > 0$ we have

$$\begin{aligned} e(m)_{f,i} &= P_{f(i)}e(m)_{f,i-1}(f(i-1) + 1 \mapsto f(i) + 1) \\ &= P_{f(i)}G_{f(i-1+1)}\left(P_{f(i-1)} \dots P_{f(1)}\alpha(m)\right) \\ &= G_{f(i)}\left(P_{f(i)} \dots P_{f(1)}\alpha(m)\right). \end{aligned}$$

The second equality holds by induction hypothesis and Lemma 7, the last one by Eq. (5). Therefore

$$\min\{i : e(m)_{f,i} = 0\} = \min\left\{i : \alpha(m) \succ_f^{P,i} 0\right\}.$$

and we are done. The argument is clearly reversible and so from $(G'_f)$ we can also obtain $(G_f)$ by elementary means. □

Note that it suffices to prove termination of the processes under consideration for ordinals of the form $\omega_K$. Also note that because of assertion 3 of Lemma 5 it does not matter if the step counting function is primitive recursive in a function from $(h_\alpha)_{\alpha<\varepsilon_0}$ or from $(H_\alpha)_{\alpha<\varepsilon_0}$.

In the following paragraphs we will recall a well-known basic result, we will provide an improvement and we'll classify the phase transition.

## 3.1 Unprovable Versions

The strategy we use to prove unprovability is to adjust the given ordinal by making a sufficiently big omega tower of it. Iteration will let the step counting function "explode", so that it dominates the function $K \mapsto H_{\omega_K}(1)$.

**Proposition 1** *Let $f = id$, then*

1. *PA* $\nvdash$ $(G_f)$.
2. *PA* $\nvdash$ $(G'_f)$.
3. *PA* $\nvdash$ $(H_f)$.
4. *PA* $\nvdash$ $(H'_f)$.
5. *PA* $\nvdash$ $(MC_f)$.

*Proof* We only have to proof the second assertion. Suppose this assertion would be false, then

$$PA \vdash (\forall K)(\exists M)\left(\omega_{2K+1} \succ_{id}^{P,M} 0\right)$$

hence also

$$PA \vdash (\forall K)(\exists M)\left(\omega_{2K} + \omega \succ_{id}^{P,M} 0\right).$$

Thus the function $k \mapsto \min m : \omega_{2k} + \omega \succ_{id}^{P,m} 0$ is provably recursive in *PA* and by Theorem 1 we find an $\alpha < \varepsilon_0$ such that for all $K$ there is an $M < h_\alpha(K)$ such that

$$\omega_{2K} + \omega \succ_{id}^{P,M} 0.$$

This yields by the definition of the Hardy functions that for all $K$ we would have $H_{\omega_{2K}+\omega}(1) < h_\alpha(K)$. On the other hand, we have by Corollary 2.4

$$\omega_{2K} + \omega \succ_1^P \omega_{2K} \succ_2^P \omega_{2K-1} \succ_3^P \cdots \succ_{K+1}^P \omega_K.$$

This yields $H_{\omega_K}(K+1) \leqslant H_{\omega_{2K}+\omega}(1)$ and this implies that

$$H_{\varepsilon_0}(K) \leq H_{\omega_K}(K+1) < h_\alpha(K)$$

holds for all $K$, yielding a contradiction since $H_{\varepsilon_0}$ eventually dominates $h_\alpha$. □

**Proposition 2** *Let $h \geqslant 0$ and $f(i) := |i|_h$. For a given ordinal $\alpha$ define $\beta := \omega_{h+1}(\alpha) + \omega_{h+1}$. Then there exists an $i \geqslant H_\alpha(1)$ such that $\beta \succ_f^{P,i} \omega_{h+1}(0)$.*

*Proof* We already know that $\min\left\{i : \alpha \succ_{id}^{P,i-1} 0\right\} = H_\alpha(1) =: L$.

By definition we have $f(i) \geqslant 1$ for all $i$. Further we have

$$\begin{aligned}\beta &\succ_1^P \omega_{h+1}(\alpha) + P_1\omega_{h+1}\\ &\vdots\\ &\succ_1^P \omega_{h+1}(\alpha) + \underbrace{P_1 \ldots P_1}_{2_h(2)} \omega_{h+1}\\ &= \omega_{h+1}(\alpha),\end{aligned}$$

since by Lemmas 4.1 and 6.4 $\min\left\{i : \omega_{h+1} \succ_1^{P,i} 0\right\} = G_1(\omega_{h+1}) = 2_h(2)$. So there exists $i_1 \geqslant 2_h(2)$ such that $\beta \succ_1^{P,i_1} \omega_{h+1}(\alpha)$.

Define $\alpha_0 := \alpha, \alpha_k := P_k\alpha_{k-1}$. We now prove by induction that

$$\omega_{h+1}(\alpha_{k-2}) \succ_k^{P,i_k} \omega_{h+1}(\alpha_{k-1}), \qquad \text{for some } i_k \geqslant (k+1)_h(k+1), \tag{6}$$

for all $k = 2, \ldots, L+1$. Since $i_k \geqslant (k+1)_h(k+1)$ we have for all $i > i_k$ that $f(i) \geqslant k+1$. Further we obtain

$$\begin{aligned}\omega_{h+1}(\alpha_{k-1}) &\succcurlyeq_{k+1}^Q \omega_{h+1}(P_k\alpha_{k-1} + 1)\\ &\succ_{k+1}^Q \omega_{h+1}(\alpha_k) + \omega_{h+1}\\ &\succ_{k+1}^P \omega_{h+1}(\alpha_k) + P_{k+1}\omega_{h+1}\\ &\vdots\\ &\succ_{k+1}^P \omega_{h+1}(\alpha_k) + \underbrace{P_{k+1} \ldots P_{k+1}}_{(k+2)_h(k+2)} \omega_{h+1}\\ &= \omega_{h+1}(\alpha_k),\end{aligned}$$

again since

$$\min\left\{i : \omega_{h+1} \succ_{k+1}^{P,i} 0\right\} = G_{k+1}(\omega_{h+1}) = (k+2)_h(k+2).$$

(In the previous calculation the first inequality is obtained by applying assertion 4 of Lemma 2 to assertion 5 of Lemma 1, the second one by assertion 2 of Corollary 2.) This implies by assertion 2 of Lemma 2 that there exists an $i_{k+1} \geqslant (k+2)_h(k+2)$ such that

$$\omega_{h+1}(\alpha_{k-1}) \succ_{k+1}^{P,i_{k+1}} \omega_{h+1}(\alpha_k).$$

Now define the function $g$ as $g(i) = k$ if $i_{k-1} < i \leqslant i_k$ (where $i_0 := 0$) and define $M := \sum_{k=1}^{L} i_k$. Equation (6) yields

$$\beta \succ_g^{P,M} \omega_{h+1}(\alpha_L) = \omega_h.$$

Because $g(i) \leqslant f(i)$ for all $i$ we have by Lemma 3.2 $\beta \succ_f^{P,m} \omega_h$ for some $m \geqslant M > L$. □

To describe the threshold function for the phase transition resulting from the Hydra game it is useful to work with functional inverses of the Hardy functions $H_\alpha$. These are defined as follows:

$$H_\alpha^{-1}(i) = \min\{m > 0 : H_\alpha(m) \geqslant i\}.$$

Then obviously $H_\alpha^{-1}(H_\alpha(i)) = i$ since $H_\alpha$ is strictly increasing. Moreover, for large $\alpha$ the function $H_\alpha^{-1}$ grows very slow and is elementary recursive.

**Lemma 8** *Let $f(i) = |i|_{H_{\varepsilon_0}^{-1}(i)}$ and $m \geqslant 2$. Let $\alpha = \omega_{m+1}(\omega_m) + \omega_{m+1}$ and let $i_0 = H_{\omega_m}(1)$. Then $\alpha \succ_f^{P,i_0} \delta$ for some $\delta \geqslant \omega_m$.*

*Proof* Let $f_m(i) := |i|_m$. For $i \leqslant i_0$ we have

$$H_{\varepsilon_0}^{-1}(i) \leqslant H_{\varepsilon_0}^{-1}(i_0) \leqslant m$$

since $H_{\omega_m}(1) \leqslant H_{\varepsilon_0}(m)$. Thus

$$|i|_{H_{\varepsilon_0}^{-1}(i)} \geqslant f_m(i)$$

for all $i \leqslant i_0$. Then by Proposition 2 there exists an $j_0 \geqslant i_0$ such that

$$\alpha \succ_{f_m}^{P,j_0} \omega_m.$$

Assertion 2 of Lemma 3 implies $\alpha \succ_f^{P,i_0} \delta$ for some $\delta \geqslant \omega_m$ since $f(i) \geqslant f_m(i)$ for $i \leqslant i_0$. □

**Theorem 2 (Phase Transition, Unprovable Version)** *Let*

$$f(i) := |i|_{H_{\varepsilon_0}^{-1}(i)}.$$

*Then the following assertions hold:*

1. $PA \nvdash (\mathrm{G}_f)$.
2. $PA \nvdash (\mathrm{G}'_f)$.
3. $PA \nvdash (\mathrm{H}_f)$.
4. $PA \nvdash (\mathrm{H}'_f)$.
5. $PA \nvdash (\mathrm{MC}_f)$.

*Proof* We need only to prove the first assertion. We prove assertion 1 by contradiction. Assume $PA \vdash (\mathrm{G}_f)$. Let

$$e(m) := 2_{2m+1} + 2_{m+1}.$$

Then

$$PA \vdash (\forall m)(\exists i)(e(m)_{f,i} = 0).$$

Let for given $m \geqslant 2$

$$\alpha(m) := e(m)(2 \mapsto \omega) = \omega_{2m+1} + \omega_{m+1}.$$

Then we see as in the proof of Lemma 7 that

$$e(m)_{f,i} = G_{f(i)}\left(P_{f(i)} \dots P_{f(1)}\alpha(m)\right).$$

This yields by Lemma 8

$$\min\{i : e(m)_{f,i} = 0\} = \min\left\{i : \alpha(m) \succ_f^{P,i} 0\right\} \geqslant H_{\omega_m}(1).$$

Since $m \mapsto H_{\omega_m}(1)$ is not provably recursive in *PA* we obtain a contradiction. □

## 3.2 *Provable Versions*

In assertion 3 of Lemma 6 we proved implicitly a first provable version for the Friedman style assertion $\mathrm{MC}_f$ for constant functions $f$. This observation is used to prove more general provable versions by stating an upper bound $M$ for the lengths of descending sequences. The argument goes roughly as follows: if the function $f$ is nondecreasing, then we have the constant function $f(M)$ as a majorization for $f$. Then we can apply assertion 2 of Lemma 3 and use the formula of assertion 4 of

Lemma 6 to show that if $M$ is chosen sufficiently big and if $f$ does not grow too quickly that then the number of possible descents is less than $M$.

We start with a useful inequality.

**Lemma 9** *Let $n, k > 0$. Then $(2_n)_k(1) \leqslant 2_{n+2(k-1)}$.*

*Proof* This is proved by induction on $k$ by carefully dealing with the exponents involved. □

As an improvement we consider the superlogarithm.

**Proposition 3** *Let $f(i) = \log^*(i) = \min\{d : |i|_d \leqslant 2\}$, then*

1. *PA* $\vdash$ $(\mathrm{G}_f)$.
2. *PA* $\vdash$ $(\mathrm{G}'_f)$.
3. *PA* $\vdash$ $(\mathrm{H}_f)$.
4. *PA* $\vdash$ $(\mathrm{H}'_f)$.
5. *PA* $\vdash$ $(\mathrm{MC}_f)$.

*Proof* We only need to prove the last assertion. Let $K \geqslant 3$ be given. Suppose that

$$\omega_K \geqslant \alpha_n > \cdots > \alpha_0$$

with $mc(\alpha_i) \leqslant K + f(i)$. We show (within *PA*) that

$$n < M(K) := 2_{3 \cdot K}.$$

$M(K)$ as a function of $K$ is provably recursive in *PA*. Assume, for a contradiction that $n \geqslant M(K)$.

Then we have $\alpha_{M(K)} > \cdots > \alpha_0$ with $mc(\alpha_i) \leqslant K + f(i)$. Thus $mc(\alpha_i) \leqslant K + f(M(K))$ for all $i \leqslant M(K)$. Hence

$$\begin{aligned} M(K) &\leqslant \#\{\alpha < \omega_K : mc(\alpha) < K + f(M(K))\} \\ &\leqslant \omega_K(\omega \mapsto K + \log^*(M(K)) + 1) \\ &= \omega_K(\omega \mapsto K + 3 \cdot K + 1) \\ &= (4 \cdot K + 1)_K \\ &< 2_{3 \cdot K} \\ &= M(K), \end{aligned}$$

a contradiction. The last inequality follows from $(4 \cdot K + 1)_K(1) \leqslant (2_K)_K(1) \leqslant 2_{K+2K-2} \leqslant 2_{3 \cdot K}$ which is a consequence of Lemma 9. □

**Theorem 3 (Phase Transition, Provable Version)** *Let $\alpha < \varepsilon_0$ and put*

$$f_\alpha(i) = |i|_{H_\alpha^{-1}(i)}.$$

*Then the following holds*

1. $PA \vdash (\mathrm{G}_{f_\alpha})$.
2. $PA \vdash (\mathrm{G'}_{f_\alpha})$.
3. $PA \vdash (\mathrm{H}_{f_\alpha})$.
4. $PA \vdash (\mathrm{H'}_{f_\alpha})$.
5. $PA \vdash (\mathrm{MC}_{f_\alpha})$.

*Proof* We only prove the last assertion. If $\alpha < \omega$, then $f_\alpha(i) \leqslant \alpha$ for all $i$ and we're done by Lemma 6.3. Now assume $\alpha \geqslant \omega$ and let $\beta := \omega^{\alpha+4} \cdot 2$. Suppose $K$ is given and suppose

$$\omega_K \geqslant \alpha_n > \cdots > \alpha_0$$

with $mc(\alpha_i) \leqslant K + f_\alpha(i)$. Put

$$M(K) := 2_{H_\alpha(H_\beta(K))}.$$

We claim that $n < M(K)$. For, assume otherwise. Then we have $\alpha_{M(K)} > \cdots > \alpha_0$ with $mc(\alpha_i) \leqslant K + f_\alpha(i)$. Then

$$mc(\alpha_i) \leqslant K + f_\alpha(M(K)) \tag{7}$$

for all $i \leqslant M(K)$. This inequality looks obvious but a proof needs a moment's reflection since $f_\alpha$ is not weakly increasing. This could of course be dealt with by replacing $f_\alpha$ by a monotone variant. But a direct argument is available, too. To prove the inequality (7) note the following monotonicity property of $f_\alpha$:

$$f_\alpha(i) = |i|_l \leqslant |H_\alpha(l+1) - 1|_l = f_\alpha(H_\alpha(l+1) - 1)$$

for all natural numbers $l$ and all $i \in \{H_\alpha(l), \ldots, H_\alpha(l+1) - 1\}$. Luckily (7) then follows easily since each $H_\alpha(l+1)$ is much bigger than $H_\alpha(l)$ and moreover since

$$|H_\alpha(l+1) - 1|_l \leqslant |2_{H_\alpha(l+1)}|_{l+1}.$$

We now claim

$$\begin{aligned} M(K) &\leqslant \#\{\alpha < \omega_K : mc(\alpha) < K + f_\alpha(M(K))\} \\ &\leqslant \omega_K(\omega \mapsto K + f_\alpha(M(K)) + 1) \\ &= \left(K + 1 + \left|2_{H_\alpha(H_\beta(K))}\right|_{H_\alpha^{-1}\left(2_{H\alpha\left(H_\beta(K)\right)}\right)}\right)_K (1) \\ &\leqslant \left(K + 1 + \left|2_{H_\alpha(H_\beta(K))}\right|_{H_{\omega^\alpha}(K)}\right)_K (1) \\ &= \left(K + 1 + 2_{H_\alpha(H_\beta(K)) - H_{\omega^\alpha}(K)}\right)_K (1) \end{aligned}$$

$$
\begin{aligned}
&< 2_{H_\alpha(H_\beta(K))} \\
&= M(K),
\end{aligned}
$$

which would yield a contradiction.

The third inequality follows from $H_\alpha^{-1}\left(2_{H_\alpha(H_\beta(K))}\right) \geqslant H_{\omega^\alpha}(K)$, which is equivalent with $2_{H_\alpha(H_\beta(K))} \geqslant H_\alpha(H_{\omega^\alpha}(K))$ which in fact is obvious. The inequality

$$\left(K + 1 + 2_{H_\alpha(H_\beta(K)) - H_{\omega^\alpha}(K)}\right)_K (1) < 2_{H_\alpha(H_\beta(K))}$$

follows by a simple and elementary side calculation (which is left to the reader). □

**Theorem 4** *For every $h \in \mathbb{N}$ let $f_h(i) := |i|_{|i|_h}$. Let $\alpha < \varepsilon_0$. Then*

$$x \mapsto \min\{i \geqslant x : Q_1^f \dots Q_x^f \alpha = 0\}$$

*is elementary recursive.*

*Proof* This is proved likewise as in the previous theorems with $M(K)$ chosen to be $2_{2_{hK}}$. □

## 4 Results Concerning the Fragments of PA

In the last section we consider restricted versions of hydra principles which are related to the fragments $I\Sigma_n$ of Peano arithmetic where the induction scheme is restricted to formulas of quantifier complexity $\Sigma_n$. The corresponding independence results follow basically from the following result $I\Sigma_n \nvdash \forall x \exists y H_{\omega_{n+1}}(x) = y$ which is treated in full detail, e.g., in [5] or [4]. (An adaptation of the methods from [12] to the fragments $I\Sigma_n$ offers of course no problems and is left as an exercise.)

**Definition 10** Define for a function $f$ the assertions $(\mathrm{G}'^n_f)$, $(\mathrm{H}'^n_f)$, $(\mathrm{MC}^n_f)$ as

$$
\begin{aligned}
(\mathrm{G}'^n_f) &:\Leftrightarrow (\forall K)(\exists M)\left(\omega_n(K) \succ_f^{P,M} 0\right) \\
(\mathrm{H}'^n_f) &:\Leftrightarrow (\forall K)(\exists M)\left(\omega_n(K) \succ_f^{Q,M} 0\right) \\
(\mathrm{MC}^n_f) &:\Leftrightarrow (\forall K)(\exists M)(\forall \alpha_0, \dots, \alpha_M \leqslant \omega_n(K)) \\
&\qquad ((\forall i \leqslant M)(mc(\alpha_i) \leqslant K + f(i)) \Rightarrow (\exists i < M)(\alpha_{i+1} \geqslant \alpha_i)).
\end{aligned}
$$

For the principles $(G^n_f)$ and $(H^n_f)$ there exist corresponding combinatorial principles where the base representation of the numbers involved stops at hight $n$ and where the hydras are bounded in hight by $n$ (i.e. they are smaller than $\omega_{n+1}$).

**Corollary 4**

$$\mathrm{I}\Sigma_n \vdash (\mathrm{MC}^n_f) \Rightarrow \mathrm{I}\Sigma_n \vdash (\mathrm{H}'^n_f) \Rightarrow \mathrm{I}\Sigma_n \vdash (\mathrm{G}'^n_f).$$

*Proof* Similarly as before. □

**Proposition 4** *Let $f = id$, then*

1. $\mathrm{I}\Sigma_n \nvdash (\mathrm{G}'^n_f)$,
2. $\mathrm{I}\Sigma_n \nvdash (\mathrm{H}'^n_f)$,
3. $\mathrm{I}\Sigma_n \nvdash (\mathrm{MC}^n_f)$.

*Proof* Similarly as before. □

**Lemma 10** *Assume $n \geqslant 1$*

1. *If $\lambda < \omega_n(m)$ and $\lambda$ is a limit, then $Q_x(\omega_n(m) \cdot \lambda) = \omega_n(m) \cdot (Q_x\lambda)$.*
2. *If $R \in \{P, Q\}$ and $\alpha < \omega_n(m)$ and $\alpha \succ_x^{R,k} \beta$, then there is an $\ell \geqslant k$ such that $\omega_n(m) \cdot \alpha \succ_x^{R,\ell} \omega_n(m) \cdot \beta$.*

*Proof* The first assertion follows by case distinction based on the Cantor normal form of $\lambda$. The second assertion follows by using the first assertion. □

**Proposition 5** *Let $m \geqslant 1$ and $n \geqslant 1$ and $f(i) := \lfloor \sqrt[m]{|i|_{n-1}} \rfloor$. For a given ordinal $\alpha < \omega_n(m)$ define $\beta := \omega_n(m) \cdot \alpha + \omega_n(m)$. Then there exists an $i \geqslant H_\alpha(1)$ and some $\delta > 0$ such that $\beta \succ_f^{P,i} \delta$.*

*Proof* We already know $\min \left\{ i : \alpha \succ_{id}^{P,i} 0 \right\} = H_\alpha(1) =: L$.
By definition we have $f(i) \geqslant 1$ for all $i$. Further we have

$$\begin{aligned} \beta &\succ_1^P \omega_n(m) \cdot \alpha + P_1\omega_n(m) \\ &\vdots \\ &\succ_1^P \omega_n(m) \cdot \alpha + \underbrace{P_1 \dots P_1}_{2_n(m)} \omega_n(m) \\ &= \omega_n(m) \cdot \alpha, \end{aligned}$$

since by Lemmas 4.1 and 6.4 $\min \left\{ i : \omega_n(m) \succ_1^{P,i} 0 \right\} = G_1(\omega_n(m)) = 2_n(m)$. So there exists $i_1 \geqslant 2$ such that $\beta \succ_1^{P,i_1} \omega_n(m) \cdot \alpha$.

Define $\alpha_0 := \alpha, \alpha_k := P_k(\alpha_{k-1})$. We now prove by induction that

$$\omega_n(m) \cdot (\alpha_{k-2}) \succ_k^{P,i_k} \omega_n(m) \cdot (\alpha_{k-1}), \qquad \text{for some } i_k \geqslant (k+1)_n(m), \tag{8}$$

for $k = 2, \dots, L+1$. Because $i_k \geqslant (k+1)_n(m)$ we have for all $i > i_k$ that $f(i) \geqslant k+1$ since $(k+1)_n(m) \geqslant 2_{n-1}((k+1)^m)$ Further we obtain

$$\begin{aligned}
\omega_n(m)\cdot\alpha_{k-1} &\succcurlyeq^{Q}_{k+1} \omega_n(m)\cdot(P_k\alpha_{k-1}+1)\\
&\succ^{Q}_{k+1} \omega_n(m)\cdot\alpha_k + \omega_n(m)\\
&\succ^{P}_{k+1} \omega_n(m)\cdot\alpha_k + P_{k+1}(\omega_n(m))\\
&\vdots\\
&\succ^{P}_{k+1} \omega_n(m)\cdot\alpha_k + \underbrace{P_{k+1}\dots P_{k+1}}_{(k+2)_n(m)}\omega_n(m)\\
&= \omega_n(m)\cdot\alpha_k,
\end{aligned}$$

again since $\min\left\{i : \omega_n(m) \succ^{P,i}_{k+1} 0\right\} = G_{k+1}(\omega_n(m)) = (k+2)_n(m)$. (In the previous calculation the first inequality is obtained by applying assertion 4 of Lemma 2 to assertion 5 of Lemma 1, is obtained by applying assertion 4 of Lemma 2 to assertion 2 of Lemma 10.) This implies by assertion 3 of Lemma 2 that there exists an $i_{k+1} \geqslant (k+2)_n(m)$ such

$$\omega_n(m)\cdot\alpha_{k-1} \succ^{P,i_{k+1}}_{k+1} \omega_n(m)\cdot\alpha_k.$$

Now define the function $g$ as $g(i) = k$ if $i_{k-1} < i \leq i_k$ and define $M := \sum_{k=1}^{L} i_k$. Equation (8) yields

$$\beta \succ^{P,M}_{g} \omega_n\cdot(\alpha_{L-1}) > 0.$$

Because $g(i) \leqslant f(i)$ for all $i$ we have by assertion 2 of Lemma 3 that $\beta \succ^{P,m}_{f} \delta$ for some $m \geqslant M > L$ and some $\delta > 0$. □

**Lemma 11** *Let* $f(i) = \sqrt[H^{-1}_{\omega_n+1}(i)]{|i|_{n-1}}$, $n \geqslant 1$ *and* $m \geqslant 2$. *Let* $\alpha = \omega_n(m)\cdot\omega_n(m)+\omega_n(m)$. *Let* $i_0 := H_{\omega_n(m)}(1)$. *Then* $\alpha \succ^{P,i_0}_{f} \delta$ *for some* $\delta > 0$.

*Proof* Let $f_m(i) := \sqrt[m]{|i|_{n-1}}$. For $i \leqslant i_0$ we have

$$H^{-1}_{\omega_{n+1}}(i) \leqslant H^{-1}_{\omega_{n+1}}(i_0) \leqslant H^{-1}_{\omega_{n+1}}(H_{\omega_n(m+1)}(m)) \leqslant H^{-1}_{\omega_{n+1}}(H_{\omega_{n+1}}(m)) \leqslant m.$$

Thus $\sqrt[H^{-1}_{\omega_n+1}(i)]{|i|_{n-1}} \geqslant f_m(i)$ for all $i \leqslant i_0$. Then by Proposition 5 there exists $j_0 \geqslant i_0$ and some $\delta' > 0$ such $\alpha \succ^{P,j_0}_{f_m} \delta'$. Assertion 2 of Lemma 3 yields $\alpha \succ^{P,i_0}_{f} \delta$ (Just step down $i_0$ steps with $f$ where $f(i) \geqslant f_m(i)$.) □

**Theorem 5 (Phase Transition, Unprovable Version)** *Let*

$$f(i) = \sqrt[H^{-1}_{\omega_n+1}(i)]{|i|_{n-1}}.$$

*Then the following holds:*

1. $I\Sigma_n \nvdash (G'^n_f)$,
2. $I\Sigma_n \nvdash (H'^n_f)$,
3. $I\Sigma_n \nvdash (MC^n_f)$.

*Proof* Similarly as before. □

**Theorem 6 (Phase Transition, Provable Version)** *Let* $\alpha < \omega_{n+1}$ *and put*

$$f_\alpha(i) = \sqrt[H_\alpha^{-1}(i)]{|i|_{n-1}}.$$

*Then the following holds:*

1. $I\Sigma_n \vdash (G'^n_f)$,
2. $I\Sigma_n \vdash (H'^n_f)$,
3. $I\Sigma_n \vdash (MC^n_f)$.

*Proof* Similarly as before. □

FINAL REMARKS: The results of this paper do hold for larger ordinals than $\varepsilon_0$, e.g. for $\Gamma_0$. Let $\lambda$ be a proof-theoretic ordinal for which an assignment of canonical fundamental sequences (with Bachmann property) is given. Assume that $T$ is a natural subsystem of (second order) analysis containing primitive recursive arithmetic. Assume that the proof-theoretic ordinal of $T$ is $\lambda$ and that a profound ordinal analysis of $T$ has been given so that for any $\alpha < \lambda$ we have $T \vdash \forall x \exists y H_\alpha(x) = y$ but $T \nvdash \forall x \exists y H_\lambda(x) = y$. Let $f_\alpha := (i \mapsto G_i(\lambda[H_\alpha^{-1}(i)]))^{-1}$. We expect that the proof given in this paper will then typically generalize to yield

$$\alpha < \lambda \iff T \vdash \mathrm{H}_{f_\alpha} \iff T \vdash \mathrm{G}_{f_\alpha}$$

for the corresponding (parameterized) versions of the Hydra games [1] or Goodstein sequences [13]. (Note that the inverse function of $i \mapsto G_i(\omega_k)$ corresponds to a suitably iterated logarithm so that the general case is in accordance with the results of this paper.) Through this characterization one then will, for example, be able to obtain a new and purely combinatorial characterization of Wainer's first subrecursively inaccessible ordinal $\psi_0\Omega_\omega$ in terms of phase transitions for the Buchholz hydra game [1].

**Acknowledgements** The second author's research has been supported in part by the John Templeton Foundation and by the Flemish Research Organization FWO.

## References

1. W. Buchholz, An independence result for $(\Pi_1^1 - CA) + BI$. Ann. Pure Appl. Log. **33**(2), 131–155 (1987)

2. W. Buchholz, A. Cichon, A. Weiermann, A uniform approach to fundamental sequences and hierarchies. Math. Log. Q. **40**, 273–286 (1994)
3. E.A. Cichon, A short proof of two recently discovered independence results using recursion theoretic methods. Proc. Am. Math. Soc. **87**(4), 704–706 (1983)
4. M. Fairtlough, S.S. Wainer, Hierarchies of provably recursive functions, in *Handbook of Proof Theory, Stud. Logic Found. Math.*, vol. 137, ed. by S.R. Buss (Elsevier, Amsterdam, 1998), pp. 149–207
5. H. Friedman, M. Sheard, Elementary descent recursion and proof theory. Ann. Pure Appl. Logic **71**(1), 1–45 (1995)
6. C.F. Kent, B.R. Hodgson, Extensions of arithmetic for proving termination of computations. J. Symb. Log. **54**, 779–794 (1989)
7. L. Kirby, J. Paris, Accessible independence results for Peano arithmetic. Bull. Lond. Math. Soc. **14**(4), 285–293 (1982)
8. L. Kristiansen, J.C. Schlage-Puchta, A.Weiermann, Streamlined subrecursive degree theory. (English summary) Ann. Pure Appl. Logic **163**(6), 698–716 (2012)
9. M. Loebl, J. Matoušek, On undecidability of the weakened Kruskal theorem, in *Logic and Combinatorics (Arcata, Calif., 1985)* (American Mathematical Society, Providence, 1987), pp. 275–280
10. R. Smith, The consistency strength of some finite forms of the Higman and Kruskal theorems, in *Harvey Friedman's Research on the Foundations of Mathematics* (North-Holland, Amsterdam, 1985), pp. 119–136
11. A. Weiermann, An application of graphical enumeration to *PA*. JSL **68**, 5–16 (2003)
12. A. Weiermann, Classifying the provably total functions of PA. BSL **12**(2), 177–190 (2006)
13. A. Weiermann, G. Wilken, Goodstein sequences for prominent ordinals up to the ordinal of $\Pi^1_1 - CA_0$. Ann. Pure Appl. Logic **164**(12), 1493–1506 (2013)

# Non-deterministic Epsilon Substitution Method for PA and $\mathsf{ID}_1$

**Grigori Mints**

**Abstract** We define a new simplified non-deterministic substitution process for PA and give a simple termination proof. Then the proof is extended to $\mathsf{ID}_1$.

## 1 Introduction

The substitution method introduced for first order arithmetic by D. Hilbert (cf. [4]) employs a formulation where quantifiers are defined in terms of the $\epsilon$-symbol $\epsilon x F[x]$ read "the least $x$ satisfying $F[x]$". Corresponding axioms are *critical formulas*

$$F[t] \to F[\epsilon x F[x]]. \tag{1}$$

The essential part of an arithmetical proof is a finite sequence $E$ of critical formulas. The goal of the epsilon substitution process is to find a *solution*, i.e., $\epsilon$-substitution of numbers $n_1, \ldots, n_k$ for $\epsilon$-terms $e_1, \ldots, e_k$

$$S \equiv (e_1, n_1), \ldots, (e_k, n_k)$$

making true all critical formulas in $E : E \hookrightarrow_S \mathsf{TRUE}$, where $\hookrightarrow_S$ iteratively replaces $e_i$ by $n_i$.

Such a solution provides finitist reduction for proofs of quantifier free sentences and computes numerical instances for provable existential ($\Sigma^0_1$-) formulas. A solution is found by successive updates, that is adding new values $(e, n)$ if the substitution $S$ accumulated by the current stage is not yet a solution. The update process terminates, when a solution is obtained.

The first termination proof for the substitution method for the first order arithmetic PA was given by W. Ackermann [1]. The definition and termination proof was extended to stronger predicative systems in [9, 10]. An extension to the impredicative case of arithmetical inductive definitions $\mathsf{ID}_1$ was outlined in [7] and

G. Mints (✉)
Department of Philosophy, Stanford University, Stanford, CA, USA
e-mail: mints@csli.stanford.edu



R. Kahle, M. Rathjen (eds.), *Gentzen's Centenary*,
DOI 10.1007/978-3-319-10103-3_17

completed in [2]. This formulation and its extension to other subsystems of analysis used complicated notation system for proof-theoretic ordinals. It is not clear how to extend it to subsystems of analysis which do not have a transparent ordinal treatment.

The goal of the present paper is to give a new much simplified formulation of the epsilon substitution method for $\mathsf{ID}_1$ in the hope it allows extension to much stronger systems.

Simplification is achieved due to a modification of the substitution method. The standard reduction step introduced by W. Ackermann not only replaces the default value 0 of the term $\epsilon x F[x]$ in the current substitution $S$ by the "correct" value $n$, but also deletes from $S$ all values of complexity (rank) greater than the complexity of $\epsilon x F$.

Instead we define a *non-deterministic* update process where several values for one and the same $\epsilon$-term are tried in parallel, although each particular update uses only one of these values. This allows to give a short (non-effective) termination proof for $\mathsf{PA}$.

The system $\mathsf{ID}_1$ contains a new predicate $I$ introduced by an inductive definition

$$In \leftrightarrow A[I, n] \tag{2}$$

where $A$ is an arithmetical formula containing $I$ only positively. The corresponding inductive generation free expands $In$ to $A[I, n]$, analyses $A[I, n]$ (causing infinite branching when $A$ begins with $\forall$), expands arising occurrences of $Im$, etc., until true quantifier tree formulas are achieved. Formally this process is reflected in the transfinite induction axiom

$$\forall x(A[F, x] \to F[x]) \to (It \to F[t]). \tag{3}$$

The main new idea in the present extension to $\mathsf{ID}_1$ is the update for the case when this axiom is false under given substitution $S$, i.e.,

$$\forall x(A[F, x] \to F[x]) \hookrightarrow_S \mathsf{TRUE}, \tag{4}$$

but

$$In \hookrightarrow_S \mathsf{TRUE}, \quad F[n] \hookrightarrow_S \mathsf{FALSE}. \tag{5}$$

The update tries to "disprove" $In$, that is to find an infinite path in the inductive generation tree for $In$. The conflict in (5) between the truth-values of $In$ and $F[n]$ is propagated one step further in the inductive generation tree by finding a "preceding" point $m$ in the tree with the same conflict between $Im$ and $F[m]$. The step from $n$ to $m$ can be made locally due to a special feature of the $\epsilon$-language (all quantifiers are already Skolemized) and special normal form of the inductive definition. In the (non-effective) termination proof for $\mathsf{ID}_1$ this results in an infinite path for $In$.

The main new tool in termination proofs is a new notion of a model for $\epsilon$-calculus similar to Henkin's models for second order logic in the sense that not all axioms (in our case, critical formulas) implicit in the principal model are required to be true.

Less significant new feature is the treatment of needed primitive recursive functions: their values are automatically used in computations. This fact is formally reflected in the definition of the standard completion $\bar{S}$ of a substitution $S$.

To simplify presentation we consider a particular but still universal inductively defined set: constructive ordinals.

In Sect. 2 we recall standard notions and elementary results concerning $\epsilon$-calculus, cf. [6, 9]. Then $\epsilon$-models for PA are defined and the update process described. Non-effective termination proof for PA is presented in Sect. 3. In Sect. 4 we reproduce a specially simple normal form for inductive definition of constructive ordinals and define an $\epsilon$-formulation $\mathsf{ID}_1\epsilon(\mathbf{S}_1)$ of our system. Then updates are defined and termination proof for the update process in $\mathsf{ID}_1\epsilon(\mathbf{S}_1)$ is given.

The results of this paper were presented in July 2009 at the Gentzen Centenary Workshop, a part of the Leeds Symposium on Proof Theory and Constructivism, two-week symposium held in the Research Visitors Centre of the School of Mathematics at Leeds. Good working environment during that Symposium allowing ample time for discussions was helpful in refining the results. I especially appreciate discussions with G. Jäger. Continuous communication with W. Buchholz, T. Arai and H. Towsner influenced this paper in many essential respects. The work by two anonymous referees improved presentation and gave a much more distinct form to basic constructions and proofs.

## 2 Non-deterministic Substitution Method: PA$\epsilon$

### 2.1 $\epsilon$-Models for the First Order Arithmetic

The language of the first order arithmetical system PA$\epsilon$ has the constant 0, successor function $S$ and all primitive recursive predicates including equality $=$, as well as Boolean connectives, say $\to, \neg$.

Formulas and $\epsilon$-terms $\epsilon x F[x]$ for variables $x$ and formulas $F[x]$ are defined simultaneously. Quantifiers are defined after that, and these definitions determine a *translation* $F^\epsilon$ of the arithmetical formulas with quantifiers into $\epsilon$-language:

$$(\exists x F[x])^\epsilon := F^\epsilon[\epsilon x F^\epsilon[x]]; \quad (\forall x F[x])^\epsilon := F^\epsilon[\epsilon x \neg F^\epsilon[x]].$$

The terms are $\epsilon$-terms, numerals $S^n 0$ and expressions $S^n t$ where $t$ is a variable or an $\epsilon$-term and $n \geq 0$.

Closed $\epsilon$-term is *canonical* if it does not have proper closed $\epsilon$-subterms.

*Example* The term $\epsilon x(x = \epsilon y(y \neq x))$ is canonical while $\epsilon x(x = \epsilon y(y \neq 5))$ is not canonical.

**Definition 2.1** TRUE(FALSE) denotes the set of all true (false) closed $\epsilon$-free formulas.

**Definition 2.2** $\mathrm{rk}(t) = 0$ if $t$ does not contain $\epsilon$.

For a canonical closed term $t \equiv \epsilon x F[x, t_1[x], \ldots, t_k[x]]$ where $t_1[x], \ldots, t_k[x]$ are all subterms containing only $x$ free,

$$\mathrm{rk}(t) := \max(\mathrm{rk}(t_1[0]), \ldots, \mathrm{rk}(t_k[0])) + 1.$$

For a non-canonical closed term $t[s_1, \ldots, s_n]$ where $s_1, \ldots, s_n$ are all closed proper $\epsilon$-subterms,

$$\mathrm{rk}(t[s_1, \ldots, s_n]) := \max(\mathrm{rk}(s_1), \ldots, \mathrm{rk}(s_k), \mathrm{rk}(t[0, \ldots, 0])).$$

$\mathrm{rk}(F)$ for a closed formula $F$ is the maximum of ranks of closed $\epsilon$-subterms of $F$.

The values of $\epsilon$-terms and formulas in the standard model are defined in a familiar way.

**Definition 2.3 (The Principal $\epsilon$-Model)**

$$\mathcal{P}(\epsilon x F[x]) := \mu n F[n]$$

where $\epsilon x F[x]$ is canonical and $\mu n F[n]$ is the least natural number $n$ satisfying $F[n]$, if such an $n$ exists, 0 otherwise. For a closed non-canonical term we define

$$\mathcal{P}(\epsilon x F[x, \epsilon y B]) := \mathcal{P}(\epsilon x F[x, \mathcal{P}(\epsilon y B)]).$$

if $x$ is not free in $\epsilon y B$. Similarly for formulas:

$$\mathcal{P}(F) \in \{\text{TRUE}, \text{FALSE}\}, \qquad \mathcal{P}(F[\epsilon y B]) := \mathcal{P}(F[\mathcal{P}(\epsilon y B)]).$$

Truth values for closed $\epsilon$-free formulas are already determined by Definition 2.1.

We consider (in analogy with Henkin models for the second order logic) more general models.

Let *Eps* be the set of all canonical $\epsilon$-terms. The predecessor function is defined by $pd(n+1) = n; pd(0) = 0$.

**Definition 2.4 ($\epsilon$-Model)** An $\epsilon$-model is a mapping $\mathcal{M} : Eps \to \mathbb{N}$ such that

$$\text{either } \mathcal{M}(\epsilon x F[x]) = \mu x \mathcal{M}(F[x]) \text{ or } \mathcal{M}(\epsilon x F[x]) = 0$$

and

$$\mathcal{M}(\epsilon x (n = Sx)) = pd(n).$$

Values of closed non-canonical terms and truth values of formulas are computed from inside as for the principal model $\mathcal{P}$, for example:

$$\mathcal{M}(\epsilon x F[x, \epsilon y B[y]]) := \mathcal{M}(\epsilon x F[x, \mathcal{M}(\epsilon y B[y])]).$$

For a closed formula $F$ we write $\mathcal{M} \models F$ if $\mathcal{M}(F) = \top$ (true).

*Critical formulas* are closed formulas

$$F[t] \to F[\epsilon x F[x]] \tag{6}$$

and

$$t \neq 0 \to t = S(\epsilon x(t = Sx)). \tag{7}$$

Define

$$pd(t) := \epsilon x(t = Sx). \tag{8}$$

In fact also formulas

$$F[t] \to \epsilon x F[x] \leq t \tag{9}$$

and

$$pd(St) = t, \quad pd(0) = 0 \tag{10}$$

will be satisfied in our models. Note that always

$$\mathcal{P} \models Cr, \text{ for a critical formula } Cr$$

while for an arbitrary $\mathcal{M}$ this may be false if $\mathcal{M}(\epsilon x F[x]) = 0$.

$E$ denotes a fixed (but arbitrary) finite system of critical formulas. $\mathcal{M}(E)$ means $\mathcal{M}(\&E)$.

Let $\vdash_{p=}$ denote derivability in the quantifier free fragment of arithmetic without induction, that is from classical tautologies and all instances of axioms for equality (also for $\epsilon$-terms), defining equations for primitive recursive predicates and axioms for the successor.

**Lemma 2.5** *If $R[x]$ is $\epsilon$-free, $E \vdash_{p=} R(\epsilon x R[x])$ and $\mathcal{M} \models E$, then $\mathcal{M} \models R[n]$ where $n = \epsilon x R[x]$.*
*In particular if $\mathsf{PA} \vdash R$ for a closed $\epsilon$-free formula $R$, then $R$ is true.*

The idea of this statement goes back to Hilbert.

## 2.2 The Update Process

We assume a fixed finite system $E$ of critical formulas.

**Definition 2.6** A *sequent* is a finite set of *components* of the form

$$(e, n)$$

where $e$ is a canonical $\epsilon$-term, $n$ is a natural number.

$\epsilon$-*substitution* (or simply *substitution*) is a sequent where different components $(e, n)$ have different $e$'s.

We treat an $\epsilon$-substitution as a function from *Eps* to $\mathbb{N}$ and use dom$S$ for $\{e : (e, n) \in S \text{ for some } n\}$.

*Comment.* $(e, n)$ means $e = n$. We don't have in this paper (as opposed to [9], for example), explicit components $(e, ?)$. In a sequent we don't assume $e$'s in different components to be different, so for example

$$(\epsilon xx = 5, 3), (\epsilon x(x = 5), 5)$$

is a legal sequent.

**Definition 2.7 (Default Completion)** $\bar{S}$ for a sequent $S$ means the result of adding to $S$ the default values

$$(\epsilon x(Sn = Sx), n), (\epsilon x(0 = Sx), 0),$$
$$(e, 0) \text{ for all remaining terms } e \notin \text{dom}S.$$

*Comment.* The values for the predecessor function *pd* (cf. (7, 10) reflect a new treatment of primitive recursive functions (compared to [9]). Since $\bar{S}$ is infinite, it is not a substitution. Nevertheless we sometimes treat it as a substitution, for example in Definition 2.10.

For a given substitution $S$ computations under $S$ go by replacing $\epsilon$-terms according to $S$.

**Definition 2.8** If $(e, n) \in S$, then $t[e] \hookrightarrow_S t[n]$ for any term or formula $t$.

We use the same symbol $\hookrightarrow_S$ for multi-step reduction.

$A \hookrightarrow_S$ TRUE(FALSE) means that $A$ reduces to a true (false) $\epsilon$-free closed formula.

**Lemma 2.9** *Every term or formula t has a unique normal (irreducible) form $|t|_S$ with respect to reduction $\hookrightarrow_S$. When S has values for all needed subterms (for example, $S = \bar{U}$ for some substitution U) then $|t|_{\bar{S}}$ is a numeral for a closed term t and $|F|_{\bar{S}}$ is in* TRUE *or* FALSE *for a closed formula F.*

*Proof* Standard. □

**Definition 2.10** Let $l = \mu^+ yF[y]$ for an expression $F[y]$ mean $F[l]$ & $(\forall k < l)\neg F[k]$. Such an $l$ is obviously unique if it exists.

A substitution $S$ is *correct* if for every $(\epsilon x F[x], n) \in S$

$$n = \mu^+ m(F[m]) \hookrightarrow_S \mathsf{TRUE}).$$

A *solution* for a system $E$ of critical formulas is a correct substitution $S$ making $E$ true:

$$E \hookrightarrow_S \mathsf{TRUE}.$$

Note that $\bar{S}$ is used instead of $S$ in checking for a solution.

**Definition 2.11 (The Update Relation)** Let $T$ be a sequent, $S \subseteq T$ a substitution and for some critical formula (6) we have

$$(F[t] \to F[\epsilon x F[x]]) \hookrightarrow_{\bar{S}} \mathsf{FALSE},$$

that is

$$F[t] \hookrightarrow_{\bar{S}} \mathsf{TRUE} \text{ but } F[\epsilon x F[x]] \hookrightarrow_{\bar{S}} \mathsf{FALSE}.$$

Then the corresponding *update* adds to the sequent $T$ the component

$$(\epsilon x |F[x]|_{\bar{S}}, n) \tag{11}$$

where

$$n = \mu^+ m \leq |t|_{\bar{S}}(F[m] \hookrightarrow_{\bar{S}} \mathsf{TRUE}) \tag{12}$$

unless (11) $\in T$.

In this case let $C(S, Cr)$ denote the component (11) for the critical formula (6).

Let $E$ be a (finite or infinite) set of critical formulas.

**Definition 2.12** The non-deterministic update process is the result of generating all possible updates using some fair procedure unless a solution is obtained as a subset of the generated sequent.

In more detail, let $T_0 = \emptyset$ (empty substitution). Fix some enumeration

$$(S_1, Cr_1), \ldots, (S_k, Cr_k), \ldots \tag{13}$$

of all pairs $(S, Cr)$ where $S$ is a correct substitution and $Cr \in E$ is a critical formula such that $Cr \hookrightarrow_{\bar{S}} \mathsf{FALSE}$.

If $T_i$ is already generated and there is a solution $S \subset T_i$ (that is, $E \hookrightarrow_{\bar{S}}$ TRUE), then the process terminates. Otherwise take the first pair $(S_k, Cr_k)$ in (13) for which an update is not yet in $T$, then make the update (11), that is

$$T_{i+1} := T_i \cup \{C(S_k, Cr_k)\}.$$

Let $T = \bigcup_i T_i$.

*Comments.*

1. An alternative would be to take $T_{i+1}$ to be

$$T_i \cup \{C(S, Cr) : \text{correct } S \subset T_i \ \&\ Cr \in E \ \&\ Cr \hookrightarrow_{\bar{S}} \text{FALSE}\}.$$

2. We call this update process non-deterministic to distinguish it from the deterministic process defined by W. Ackermann [1] and extensions of this definition to the second-order case. In our set-up the pool $T$ of possible values of $\epsilon$-terms is formed by adding all "locally needed" values $C(S, Cr)$ in arbitrary order. The enumeration (13) is fixed only for bookkeeping. In this respect our process is similar to a proof-search based on Herbrand Theorem or Gentzen-style calculus (cf., for example, [5]) for classical predicate logic.

**Definition 2.13** We say that $T$ is closed under updates if for any correct $S \subset T$ and any $Cr \in E$ such that $Cr \hookrightarrow_{\bar{S}}$ FALSE we have $C(S, Cr) \in T$.

## 3 Non-effective Termination Proof for PA

Let's use notation $|t|_{\mathcal{M}}$ for an $\epsilon$-model $\mathcal{M}$ similarly to notation $|t|_S$ for a substitution $S$ : $|t|_{\mathcal{M}}$ is the result of replacing all maximal occurrences of closed subterms $s$ in $t$ by $\mathcal{M}(s)$. If $t$ is a closed term, then $|t|_{\mathcal{M}}$ is a natural number.

**Definition 3.1** A *restriction* $\mathcal{M} \downarrow U$ of an $\epsilon$-model $\mathcal{M}$ to a set $U$ of closed terms or formulas is a substitution $S$ consisting of all $\mathcal{M}$-values needed for computation of all expressions in $U$. It is defined inductively.

If $\epsilon x F[x]$ is a closed subterm of some expression in $U$ and $\mathcal{M}(\epsilon x F[x]) \neq 0$ then

$$(\epsilon x |F[x]|_{\mathcal{M}}, \mathcal{M}(\epsilon x F[x])) \in S.$$

If $(\epsilon x G[x], m) \in S$, $\epsilon x F[x]$ is a closed subterm of a formula $G[0] \ \&\ \ldots \ \&\ G[m]$ and $\mathcal{M}(\epsilon x F[x]) \neq 0$, then again $(\epsilon x |F[x]|_{\mathcal{M}}, \mathcal{M}(\epsilon x F[x])) \in S$.

**Lemma 3.2** *For every $\epsilon$-model $\mathcal{M}$ and every set $U$ of expressions $\mathcal{M} \downarrow U$ is a correct substitution that agrees with $\mathcal{M}$: for every component $(\epsilon x F[x], n) \in \mathcal{M} \downarrow U$*

$$\mathcal{M}(\epsilon x F[x]) = n. \tag{14}$$

*Proof* Equality (14) is a part of the definition of $\mathcal{M} \downarrow U$. After this correctness of $\mathcal{M} \downarrow U$ follows from correctness of $\mathcal{M}$. □

For a closed formula $F$ let's write $\mathcal{M} \models F$ for $\mathcal{M}(F) = \top$.

For every sequent $T$ closed under updates let's define a model $\mathcal{M}$ extracted from this sequent.

**Definition 3.3** For a canonical term $\epsilon x F[x]$

$$\mathcal{M}(\epsilon x F[x]) := \begin{cases} n & \textit{if } (\epsilon x F[x], n) \in T \textit{ and } \mu^+ m(\mathcal{M} \models F[m]) = n \\ & \quad \textit{or } \epsilon x F[x] \equiv pd(Sn), \\ 0 & \textit{if } \forall n \neg((\epsilon x F[x], n) \in T \ \& \ M \models F[n]). \end{cases}$$

In fact this is a definition by recursion on $\mathrm{rk}(\epsilon x F[x])$, but $\mathcal{M}$ is not arithmetical even if $T$ is recursively enumerable. When $\mathrm{rk}(\epsilon x F[x]) = 1$ the formula $F[m]$ does not contain $\epsilon$ and its truth value is computable outright. When $\mathrm{rk}(\epsilon x F[x]) > 1$, formulas $F[m]$ have smaller rank, therefore the values $\mathcal{M}(F[m])$ are already defined.

**Lemma 3.4** $\mathcal{M} \models E$.

*Proof* By definition of $\mathcal{M}$ for canonical $\epsilon$-terms $e$ we have

$$\text{If } \mathcal{M}(e) = n \neq 0 \text{ then } (e, n) \in T. \tag{15}$$

$$\text{If } (\epsilon x F[x], n) \in T \text{ and } n = \mu^+ m(\mathcal{M} \models F[m]), \text{ then } n = \mathcal{M}(\epsilon x F[x]). \tag{16}$$

Take a critical formula $Cr \in E$ of the form $F[t] \to F[\epsilon x F[x]]$.

Denote

$$e := \epsilon x F[x], \quad e^* := \mathcal{M}(\epsilon x F[x]), \quad t^* := \mathcal{M}(t).$$

Let

$$S := \mathcal{M} \downarrow Cr.$$

$S \subset T$ by (15). We need to prove $Cr \hookrightarrow_{\bar{S}}$ TRUE, since then $\mathcal{M} \models Cr$ and we are done. If $F[t] \hookrightarrow_{\bar{S}}$ TRUE, $F[\epsilon x F[x]] \hookrightarrow_{\bar{S}}$ FALSE, take the minimal $n \leq t^*$ such that $F[n] \hookrightarrow_{\bar{S}}$ TRUE. Since $S$ contains all needed values of $\mathcal{M}$, in particular $t^* = |t|_{\bar{S}}$, this implies

$$n = \mu^+ m(\mathcal{M}(F[m]) = \top). \tag{17}$$

Since all possible updates are eventually performed, the update to $S$ and $Cr$ adds to $T$ the component $(\epsilon x|F[x]|_{\bar{S}}, n)$. Together with (17) and $\epsilon x|F[x]|_{\bar{S}} \equiv \epsilon x|F[x]|_{\mathcal{M}}$ this implies by (16) $n = \mathcal{M}(\epsilon x F[x])$, and hence $\mathcal{M}(Cr) = \top$ as required.

Critical formulas (7) are true in $\mathcal{M}$ since *pd* has correct value in $\mathcal{M}$. □

**Theorem 3.5** *Non-deterministic update process always terminates in a solving substitution.*

*Proof* Since $\mathcal{M} \models E$ take the restriction $S := \mathcal{M} \downarrow E$. Then $S \subseteq T$ and $E \hookrightarrow_{\bar{S}}$ TRUE. □

## 4 A Substitution Method for $\mathsf{ID}_1(S_1)$

We extend the results and proofs in Sects. 2, 3 from PA to $\mathsf{ID}_1$. To simplify presentation we consider a particular but still universal set. The next subsection coincides with corresponding subsection in [8].

### 4.1 Inductive Definition of Constructive Ordinals

A typical inductively defined set is the set $S_1$ of constructive ordinals introduced by A. Church and S. Kleene (cf. [3, 11]). Every other arithmetically inductively defined set is primitive recursive in $S_1$. Let's describe one particularly simple inductive definition of $S_1$. The author has not found exactly this description in the literature.

**Definition 4.1** $n \in S_1$ iff

$$n = 0 \vee \exists e(n = 2^e \,\&\, e \in S_1) \vee \exists e(n = 3 \cdot 5^e \,\&\, \forall x \exists y(T(e, x, y) \,\&\, U(y) \in S_1))$$

where the last disjunct means that the $e$-th partial recursive function is total and all its values are in $S_1$.

Consider for every $n$ the following primitive recursive tree $\mathcal{T}_n$ whose nodes are finite sequences of natural numbers. Every node is labeled by a natural number.

The root $\emptyset$ is labeled by $n$.

If the node $a$ is labeled by 0, then $a$ is the leaf node: it has no predecessors.

If the node $a$ is labeled by $2^e$, then the immediate predecessor $a * \langle 0 \rangle$ of $a$ is labeled by $e$ and no node $a * \langle k + 1 \rangle$ is in the tree (and similarly below when only one predecessor is labeled).

If the node $a$ is labeled by $3 \cdot 5^e$, then its immediate predecessor at the node $a * \langle x \rangle$ is labeled by $3 \cdot 5^e \cdot 7^{x+1}$.

If the node $a$ is labeled by $3 \cdot 5^e \cdot 7^{x+1}$, then the immediate predecessor $a * \langle 0 \rangle$ is labeled by $3 \cdot 5^e \cdot 7^{x+1} \cdot 11^1$.

If the node $a$ is labeled by $3 \cdot 5^e \cdot 7^{x+1} \cdot 11^{y+1}$ and $T(e, x, y)$ is false, then the immediate predecessor $a * \langle 0 \rangle$ is labeled by $3 \cdot 5^e \cdot 7^{x+1} \cdot 11^{y+2}$.

If the node $a$ is labeled by $3 \cdot 5^e \cdot 7^{x+1} \cdot 11^{y+1}$ and $T(e, x, y)$ is true, then the immediate predecessor $a * \langle 0 \rangle$ is labeled by $U(y)$.

If the node $a$ is labeled by a number $m > 0$ that does not have one of the forms above, then the immediate predecessor $a * \langle 0 \rangle$ of $a$ is labeled by the same number $m$.

**Lemma 4.2** *$n \in S_1$ iff $\mathcal{T}_n$ is well-founded.*

*Proof* 1. If $\mathcal{T}_n$ is well-founded, then $n \in S_1$. This is proved by transfinite induction on $\mathcal{T}_n$. In more detail we prove that
every node with a label $0, 2^e, 3 \cdot 5^e$ is in $S_1$, and
for every node with a label $3 \cdot 5^e \cdot 7^{x+1} \cdot 11^{y+1}$ such that $T(e, x, y)$ is true, $U(y)$ is in $S_1$, and
for every node with a label $3 \cdot 5^e \cdot 7^{x+1} \cdot 11^{y+1}$ such that $T(e, x, y)$ is false, there is an $y' > y$ such that $T(e, x, y')$ is true.
Note that every label has one of these forms since otherwise it generates infinite branch in the tree.
2. If $n \in S_1$, then $\mathcal{T}_n$ is well-founded. Proved similarly by transfinite induction on the inductive definition of $S_1$.

□

**Definition 4.3**

$$p(n, l) := \begin{cases} e & \text{if } n = 2^e, \\ n \cdot 7^{l+1} & \text{if } n = 3 \cdot 5^e, \\ n \cdot 11 & \text{if } n = 3 \cdot 5^e \cdot 7^{x+1}, \\ n \cdot 11 & \text{if } n = 3 \cdot 5^e \cdot 7^{x+1} \cdot 11^{y+1} \ \& \ \neg T(e, x, y), \\ U(y) & \text{if } n = 3 \cdot 5^e \cdot 7^{x+1} \cdot 11^{y+1} \ \& \ T(e, x, y), \\ 0 & \text{otherwise,} \end{cases}$$

$$R(n) :\equiv (\exists e \leq n)(\exists x \leq n)(\exists y \leq n)$$
$$(n = 2^e \vee n = 3 \cdot 5^e \vee n = 3 \cdot 5^e \cdot 7^{x+1} \vee n = 3 \cdot 5^e \cdot 7^{x+1} \cdot 11^{y+1}).$$

Let $WF$ mean "well-founded".

**Lemma 4.4** *$\mathcal{T}_n \in WF \leftrightarrow [n = 0 \vee (Rn \ \& \ \forall l(\mathcal{T}_{p(n,l)} \in WF))]$, and hence relation $\mathcal{T}_n \in WF$ can be given by the inductive definition*

$$In \leftrightarrow (n = 0 \vee (R(n) \ \& \ \forall l Ip(n, l))) \tag{18}$$

*and $n \in \mathbf{S}_1$ is defined inductively by Lemma 4.2.*

*Proof* By cases in the definition of $\mathcal{T}_n$: they are listed by $R(n)$. □

## 4.2 System $\mathsf{ID}_1\epsilon(\mathbf{S}_1)$

Add to the language of $\mathsf{PA}\epsilon$ a new unary predicate $I$ and to the list of axioms a new axiom

$$(A[I,t])^\epsilon \to It \tag{19}$$

and the axiom schema of transfinite induction

$$(\forall x(A[F,x] \to F[x]))^\epsilon \to (It \to F[t]). \tag{20}$$

where $A[I,x] \equiv (x = 0 \vee (R(x) \,\&\, \forall y Ip(x,y)))$ is a formula from (18).

We express the primitive recursive function $p(x,y)$ via the graph $P$ of this function:

$$p(s,t) := \epsilon z P(s,t,z)$$

and include into $\mathsf{ID}_1\epsilon(\mathbf{S}_1)$ the axiom

$$P(s,t,u) \to p(s,t) = u. \tag{21}$$

Let's denote

$$B[I,y,x] :\equiv (x = 0 \vee (R[x] \,\&\, Ip(x,y))),$$
$$e(t) := \epsilon y \neg B[I,y,t]; \qquad e(F,t) := \epsilon y \neg B[F,y,t].$$

Then we have

$$(A[I,x])^\epsilon \equiv B[I,e(x),x],$$

the axiom (19) becomes

$$B[I,e(x),x] \to Ix,$$

and the schema (20) becomes first

$$(A[F,X] \to F[X]) \to (It \to F[t]),$$

where $X := \epsilon x \neg(A[F, x] \rightarrow F[x])$, then

$$(B[F, e(F, X), X] \rightarrow F[X]) \rightarrow (It \rightarrow F[t]).$$

Note that the value $e(n) = k$ is expressed by the formula

$$\neg B[I, k, m] \wedge \forall l < k B[I, l, n]$$

which does not depend on other $\epsilon$-terms except $p(n, k)$, and it is monotonic with respect to $\bar{I}$: if $\neg B[I, k, n]$ computes to TRUE using only given finite number of values of the complement $\bar{I}$ of $I$ (and true values of $p(n, l)$) then by monotonicity the same computation works for any extension of this set of values respecting $\epsilon z P(x, y, z) = p(x, y)$.

**Definition 4.5** A term $\epsilon x A[x]$ is *canonical* if it is closed and does not contain proper closed $\epsilon$-subterms or expressions $It$ with a constant term $t$.

**Definition 4.6** *Components* are expressions of the form

$$(e, n), \bar{I}n,$$

where $e$ is a canonical $\epsilon$-term, $n$, $l$ are natural numbers.

A *sequent* is a set of components.

$\epsilon$*-substitution* or simply substitution is a finite sequent with components of the form

$$(e, n), \bar{I}n,$$

such that the terms $e$ in different components $(e, n)$ are different.

**Definition 4.7** *Computation* under given substitution $S$ (and the symbols $\hookrightarrow_S$, $|F|_S$, etc.) is defined by the same clauses as for PA plus the clauses:

$$\text{If } \bar{I}n \in S \text{ then } In \hookrightarrow_S \text{FALSE}.$$

The *default completion* $\bar{S}$ of the substitution $S$ is the result of adding components

$$(pd(0), 0), \quad (pd(Sn), n), \quad (\epsilon z P(m, n, z), k) \text{ for } k = p(m, n)$$

and $(e, 0)$ for all remaining canonical $\epsilon$-terms not in dom$S$, as well as components $In$ for all $n$ such that $\bar{I}n \notin S$.

For computations with $\bar{S}$ one more clause is added to Definition 2.8:

$$\text{If } In \in \bar{S} \text{ then } In \hookrightarrow_{\bar{S}} \text{TRUE}.$$

Our definition of $\bar{S}$ makes true all critical formulas of the form (7), (21).

Note that the components $\bar{I}n$ and canonical terms $e(n)$ are preserved even when arithmetical $\epsilon$-terms change value.

### 4.3 Updates for $\mathsf{ID}_1\epsilon(S_1)$

As in the case of PA (Sect. 2.2) we assume a fixed system $E$ of critical formulas and are looking for a substitution $S$ solving this system, that is such that

$$Cr \hookrightarrow_{\bar{S}} \mathsf{TRUE}$$

for all critical formulas $Cr \in E$.

Note that below all computations are done for $\bar{S}$, not just $S$. In particular $In \in \bar{S}$ for all $n$ except when already $\bar{I}n \in S$.

Let $T$ be a sequent (called *current sequent* below) which does not contain as a subset any solving substitution and $S \subseteq T$ be a correct substitution. $S$ falsifies some critical formulas in the system $E$ to be solved. We define a corresponding *update*, that is a component or components to be added to the sequent $T$ unless they are there already.

The *update* for the arithmetical critical formula $F[t] \rightarrow F[\epsilon x F[x]]$ and substitution $S$ is defined as before (Definition 2.11): the component $(\epsilon x|F[x]|_{\bar{S}}, n)$ is added with

$$n = (\mu^{+} m \leq |t|_{\bar{S}})(F[m] \hookrightarrow_{\bar{S}} \mathsf{TRUE}).$$

There are no special updates for axioms of the form

$$A[I, t] \rightarrow It.$$

They are in fact implicit in the Definition of $\epsilon$-model for $\mathsf{ID}_1\epsilon(S_1)$, cf. (32) below.

#### 4.3.1 Update for Transfinite Induction Axioms

$Cr$ denotes below the critical formula

$$(\forall x(A[F, x] \rightarrow F[x]))^{\epsilon} \rightarrow (It \rightarrow F[t]), \tag{22}$$

and we assume that

$$Cr \hookrightarrow_{\bar{S}} \mathsf{FALSE}, \tag{23}$$

that is for $n := |t|_{\bar{S}}$

$$(\forall x(A[F, x] \rightarrow F[x])^{\epsilon} \hookrightarrow_{\bar{S}} \mathsf{TRUE}, \tag{24}$$

but

$$In \hookrightarrow_{\bar{S}} \mathsf{TRUE}, \quad F[n] \hookrightarrow_{\bar{S}} \mathsf{FALSE}. \tag{25}$$

Denote $l := |e(F,n)|_{\bar{S}}$.

Recall that intuitively $\bar{I}n \leftrightarrow \exists y \neg B[I,y,n] \leftrightarrow \neg B[I,e(n),n]$.

1. If

$$B[I,l,n] \hookrightarrow_{\bar{S}} \mathsf{FALSE},$$

then the *update* adds to the current sequent $T$ the components

$$\bar{I}n, (e(n),k) \quad (\text{all } k \leq l)$$

and no other update applies to the axiom (22) with the substitution $S$.

2. Assume $B[I,l,n] \hookrightarrow_{\bar{S}} \mathsf{TRUE}$, still with (23), (24), (25) and

$$(A[F,n] \to F[n])^{\epsilon} \hookrightarrow_{\bar{S}} \mathsf{FALSE}.$$

Recall that $In \hookrightarrow_{\bar{S}}$ TRUE means by our default convention that $In \notin \mathrm{dom} S$ and $In \in \bar{S}$.

The *update* adds to the current sequent $T$ the component

$$(\epsilon x \neg |(A[F,x] \to F[x])^{\epsilon}|_{\bar{S}}, n')$$

where $n' := (\mu m \leq n)((A[F,m] \to F[m])^{\epsilon} \hookrightarrow_{\bar{S}} \mathsf{FALSE})$.

*Note.* This has the effect of updating for a virtual critical formula

$$(\forall x(A[F,x] \to F[x]))^{\epsilon} \to (A[F,n] \to F[n])$$

or more precisely

$$\neg(A[F,n] \to F[n]) \to \neg(A[F,X] \to F[X])$$

for $X := \epsilon x \neg(A[F,x] \to F[x])$. In view of (24) this update makes true formula

$$(A[F,n'] \to F[n'])^{\epsilon}.$$

3. Assume that $T$ contains a correct substitution $S$, and for the critical formula (22) denoted below by $Cr$ and for

$$n := |t|_{\bar{S}}, \quad l := |e(F,n)|_{\bar{S}}$$

we have

$$Cr \hookrightarrow_{\bar{S}} \mathsf{FALSE} \quad B[I,l,n] \hookrightarrow_{\bar{S}} \mathsf{TRUE}$$

and

$$(A[F,n] \to F[n])^{\epsilon} \hookrightarrow_{\bar{S}} \mathsf{TRUE}. \tag{26}$$

If $m := p(n,l) = 0$, no update applies to the axiom (22) with the substitution $S$. If $m \neq 0$, note that $F[n] \hookrightarrow_{\bar{S}} \mathsf{FALSE}$ (cf. (25)) and (26) imply

$$A[F,n] \equiv B[F,e(F,n),n] \hookrightarrow_{\bar{S}} \mathsf{FALSE},$$

so we have

$$B[F,l,n] \hookrightarrow_{\bar{S}} \mathsf{FALSE}, \quad B[I,l,n] \hookrightarrow_{\bar{S}} \mathsf{TRUE}. \tag{27}$$

We have $n \neq 0$ since

$$A[F,0] \equiv 0 = 0 \vee (R(0) \,\&\, F[p(0,l)]) \hookrightarrow_{\bar{S}} \mathsf{TRUE}.$$

Recall that

$$B[I,l,n] \equiv (n = 0 \vee (R(n) \,\&\, Ip(n,l))),$$
$$B[F,l,n] \equiv (n = 0 \vee (R(n) \,\&\, F[p(n,l)])).$$

The relations (27) and $n \neq 0$ imply for $m = p(n,l)$ that

$$Im \hookrightarrow_{\bar{S}} \mathsf{TRUE}, \quad F[m] \hookrightarrow_{\bar{S}} \mathsf{FALSE}. \tag{28}$$

The *update* adds to $T$ the components

$$\bar{I}n \text{ and } (e(n),k) \quad (\text{for all } k \leq l), \tag{29}$$

$$\bar{I}m \text{ and } (e(m),k) \quad (\text{for all } k \leq |e(F,m)|_{\bar{S}}). \tag{30}$$

This completes the definition of update.

## 4.4 Termination Proof for $\mathsf{ID}_1$

We extend the definitions of the Sect. 3 from PA to $\mathsf{ID}_1$.

*Eps* denotes now the set of all canonical $\epsilon$-terms in the language of $\mathsf{ID}_1$, and

$$Can := Eps \cup \{In : n \in \mathbb{N}\}.$$

**Definition 4.8 (The Principal $\epsilon$-Model)** The *principal* (standard) model is defined in a familiar way: Let $\mathcal{I}$ be the least fixed point of the inductive definition given by the formula $A$. Define:

$$\mathcal{P}(In) := In; \quad \mathcal{P}(\epsilon x F[x]) = \mu n \mathcal{P}(F[n]).$$

**Definition 4.9 ($\epsilon$-Model)** An $\epsilon$-model is a mapping $\mathcal{M} : Can \to \mathbb{N} \cup \{\top, \bot\}$ such that

$$\mathcal{M}(\epsilon x F[x]) = \mu^{+} x (\mathcal{M}(F[x]) = \top) \text{ or } \mathcal{M}(\epsilon x F[x]) = 0 \tag{31}$$

and

$$\mathcal{M}(In) = \bot \text{ implies } \mathcal{M}(B[I, e(n), n]) = \bot. \tag{32}$$

The definitions for non-canonical expressions are as before (Definition 2.4). For a closed formula $A$ we write $\mathcal{M} \models A$ if $\mathcal{M}(A) = \top$. Note that always

$$\mathcal{P} \models Cr, \text{ for every critical formula } Cr$$

while for an arbitrary $\epsilon$-model $\mathcal{M}$ this may be false both for arithmetical critical formulas and for transfinite induction axioms. However all closed critical formulas of the form (7), (21) and

$$A[I, t] \to It, \quad \text{i.e., } B[I, e(t), t] \to It$$

are true in every $\epsilon$-model.

$\vdash_{p=}$ denotes as before derivability in the quantifier free fragment of arithmetic without induction containing now the predicate symbol $I$ but no special axioms for this symbol.

We have as before:

**Lemma 4.10** *If $R[x]$ is $\epsilon, I$-free, $S$ is solving for all critical formulas in $E$ and $E \vdash_{p=} R(\epsilon x R[x])$,*
*then $\vdash_{p=} R[n]$, in particular $R[n] \in$ TRUE where $n = |\epsilon x R[x]|_S$.*

The update process is defined (with the new notion of update) exactly as in the Definition 2.12.

**Definition 4.11** The non-deterministic update process is the result of generating all possible updates using some fair procedure unless a solution is obtained as a subset of the generated sequent.

**Definition 4.12** A *path* for a sequent $T$ is a sequence

$$n_1, n_2, \ldots \tag{33}$$

(finite or infinite) of positive natural numbers $n_k > 0$ such that there exists a sequence $\{l_k\}$ of natural numbers such that for every $k$ (except possibly the last element of the sequence) either

$$R(n_k) \in \mathsf{FALSE} \;\&\; \forall l(n_{k+l} = n_k) \tag{34}$$

or,

$$\bar{I}n_k, (e(n_k), m) \in T \text{ for all } m \leq l_k \tag{35}$$

and

$$n_{k+1} = p(n_k, l_k), R(n_k) \in \mathsf{TRUE}. \tag{36}$$

A sequence $\{l_k\}$ satisfying (35)–(36) is called a witness for the path (33).

Define

$$\inf(n) :\equiv \text{there is an infinite path for } T \text{ containing } n.$$

*Note 1* Any infinite path (33) for a sequent $T$ is an infinite path in the tree $\mathcal{T}_{n_1}$ (Sect. 4.1), and hence $\inf(n)$ implies $n \notin \mathbf{S}_1$.

**Theorem 4.13** *The update process terminates in a solution after a finite number of steps.*

*Proof* Let $T$ be a sequent closed under updates. Let's define an $\epsilon$-model extracted from $T$.

Define the *rank* $\mathrm{rk}_0$ of $\epsilon$-terms and formulas exactly as before (Definition 2.2). In other words $\mathrm{rk}_0(In) = 0$ and only nesting of variables bound by $\epsilon$ is important.

Define

$$\mathcal{M}(\bar{I}n) = \top :\equiv (\bar{I}n \in T \;\&\; \inf(n)).$$

Note that the computation of $\mathcal{M}(B[I, l, n]) = \bot$ needs only values of $\bar{I}n$-components and values of $p(x, y)$. On the other hand, $\mathcal{M}$ is at best $bool(\Sigma^1_1)$.

Now define for the terms $e(n)$

$$\mathcal{M}(e(n)) = \begin{cases} l & \text{if } l = \mu^+ y((e(n), y) \in T \;\&\; \mathcal{M}(B[I, y, n]) = \bot), \\ 0 & \text{if } \forall y \neg((e(n), y) \in T \;\&\; \mathcal{M}(B[I, y, n]) = \bot). \end{cases}$$

For all remaining canonical $\epsilon$-terms define the values exactly as in Definition 3.3 taking into consideration that the values of all $I$-formulas are already defined. In more detail:

**Definition 4.14** For a canonical term $\epsilon x F[x]$

$$\mathcal{M}(\epsilon x F[x]) := \begin{cases} n & \text{if } \mu^+ m((\epsilon x F[x], m) \in T \;\&\; \mathcal{M}(F[m]) = \top) = n, \\ 0 & \text{if } \forall m \neg((\epsilon x F[x], m) \in T \;\&\; \mathcal{M}(F[m]) = \top). \end{cases}$$

In fact this is a recursion on $\mathrm{rk}_0(\epsilon x F[x])$.

**Lemma 4.15** *1. If $\mathcal{M}(\bar{I}n) = \top$, then $\bar{I}n \in T$;*
*If $\mathcal{M}(\epsilon x F[x]) = n > 0$ for a canonical $\epsilon x F[x]$, then $(\epsilon x F[x], n) \in T$.*
*2. $\mathcal{M}$ is an $\epsilon$-model.*

*Proof* The condition 1 is a part of our definition for $In$ and for canonical $\epsilon$-terms.

Condition (31) is the part of the definition of $\mathcal{M}$.

Let's check the condition (32). Assume $\mathcal{M}(In) = \bot$. Then there is an infinite path (33) with $n_1 = n$ and suitable witness $\{l_k\}$. Note that $n \neq 0$ since $\neg\,\mathrm{inf}(n)$.

If $R(n) \in \mathsf{FALSE}$, then $B[I, e(n), n] \equiv (n = 0 \vee (R(n) \;\&\; Ip(n, e(n))))$ is false for any value of $e(n)$.

Assume now that $R(n) \in \mathsf{TRUE}$ so that (35), (36) hold for $k = 1$. In particular $n_2 = p(n, l_1)$ also belongs to the same infinite path, hence $\mathcal{M}(Ip(n, l_1)) = \bot$, $\bar{I}p(n, l_1) \in T$ by (35) and

$$\mathcal{M}(B[I, l_1, n]) = \mathcal{M}(n = 0 \vee Ip(n, l_1)) = \bot.$$

Together with $(e(n), k) \in T$ for all $k \leq l_1$ this implies $\mathcal{M}(e(n)) \leq l_1$, with $\mathcal{M}(B[I, e(n), n]) = \bot$ as required. □

**Definition 4.16** Extend the Definition 3.1 of restriction to $\mathsf{ID}_1\epsilon(\mathbf{S}_1)$ to include all components $Ip$, $\neg Ip$ needed for computations of $\mathcal{M}$-values for the formulas or terms in question. A *restriction* $\mathcal{M} \downarrow U$ of an $\epsilon$-model $\mathcal{M}$ to a set $U$ of closed terms or formulas is a substitution $S$ consisting of all $\mathcal{M}$-values needed for computation of all expressions in $U$.

By Lemma 4.15.1 we have $S \subseteq T$ for every restriction $S$ of $\mathcal{M}$.

**Lemma 4.17** $\mathcal{M} \models E$.

*Proof* The proof for arithmetical critical formulas is the same as before.

Critical formulas

$$A[I, t] \to It$$

are true by (32).

Consider now a critical formula

$$Cr :\equiv (\forall x(A[F, x] \to F[x]))^{\epsilon} \to (It \to F[t]). \tag{37}$$

Let

$$n := \mathcal{M}(t), \quad l := \mathcal{M}(e(F,n)), \quad m := p(n,l).$$

Denote $Cr[z] := (\forall x(A[F,x] \to F[x]))^{\epsilon} \to (Iz \to F[z])$ and assume

$$\mathcal{M} \models \neg Cr[n], \tag{38}$$

that is

$$\mathcal{M} \models (A[F,X] \to F[X]), \quad \mathcal{M} \models In, \quad \mathcal{M} \models \neg F[n], \tag{39}$$

where $X := \epsilon x \neg(A[F,x] \to F[x])$. We prove that

$$\mathcal{M} \models \neg Cr[m]. \tag{40}$$

Let

$$U := \{e(F,n)\} \cup \{B[I,l,n]\} \cup$$
$$\{(A[F,X] \to F[X]) \to (A[F,n] \to F[n])\} \cup \{A[F,k] \to F[k] : k \leq \mathcal{M}(X)\} \cup \{F[m]\}$$

and take $S := \mathcal{M} \downarrow U$. □

**Lemma 4.18** *Assume $\mathcal{M} \models \neg Cr[n]$. Then the following hold.*

*1. $n \neq 0$.*
*2. $R(n) \in$ TRUE, cf. (36).*
*3. Only case 4.3.1.1 of the update is applicable, and $\bar{I}n \in T$, $\bar{I}m \in T$.*

*Proof* $\mathcal{M} \models A[F,X] \to F[X]$ implies $X = 0$, since otherwise $\mathcal{M} \vdash \neg(A[F,X] \to F[X])$ by correctness. Therefore $\mathcal{M} \models A[F,X] \equiv (X = 0 \vee (R(X) \,\&\, Ip(X,e(X))))$, and hence $\mathcal{M} \models F[X]$, that is $\mathcal{M} \models F[0]$. By (39),

$$n \neq 0.$$

The elements $e(F,n)$ and $B[I,l,n]$ in $U$ ensure that $l = |e(F,n)|_{\bar{S}}$ and (since $Ip(n,l)$ is a subformula of $B[I,n,l]$)

$$B[I,l,n] \hookrightarrow_{\bar{S}} \textsf{FALSE iff } \mathcal{M}(B[I,l,n]) = \bot,$$
$$Ip(n,l) \hookrightarrow_{\bar{S}} \textsf{FALSE iff } \mathcal{M}(Ip(n,l)) = \bot.$$

If update 4.3.1.1 is applicable to $S$, we have $B[I,l,n] \hookrightarrow_{\bar{S}}$ FALSE and (since $R(n) \in$ TRUE) also $Ip(n,l) \equiv Im \hookrightarrow_{\bar{S}}$ FALSE, so $\mathcal{M}(B[I,l,n]) = \bot$, $\mathcal{M}(Im) = \bot$. The update 4.3.1.1 adds to $T$ components $\bar{I}n$ and $(e(n),m)$ for $m \leq l$ (as needed for (35)). By $\mathcal{M}(Im) = \bot$ there is an infinite path for $m = p(n,l)$. Adding $n$ in front of this path and $l_1 = l$ in front of the witness we get the infinite path for $n$ proving $\mathcal{M}(In) = \bot$, a contradiction with (39). Hence update 4.3.1.1 is not applicable.

The next two elements of the list $U$ show that update 4.3.1.2 is not applicable, cf. the proof of Lemma 3.4.

Hence the update 4.3.1.3 is applicable. □

The implication

$$\forall n \exists l(\mathcal{M} \models \neg Cr[n] \to \mathcal{M} \models \neg Cr[p(n,l)])$$

will show by induction that for every $k$ there exists a path $n = n_1, \ldots, n_k$ in $T$ such that $\mathcal{M} \models \neg Cr[n_i]$ and $n_{i+1} = p(n_i, l_i)$ for all suitable $i \leq k$. This in turn implies that there is an infinite path in $T$ beginning with $n$, and hence $\mathcal{M} \models \neg In$ by Lemma 4.18, contradicting (39).

Towards (40) note that the update 4.3.1.3 adds to $T$ (for $m = p(n,l)$) components (29), (30). Moreover, since $F[m] \in U$ we have $\mathcal{M}(F[m]) = \bot$ since $F[m] \hookrightarrow_{\bar{S}} \mathsf{FALSE}$ by (28). As in the case of the update 4.3.1.1 $\mathcal{M}(Im) = \top$ since otherwise inf($m$) implying inf($n$) which contradicts (39). Together with (39) this implies (40) as required. □

**Theorem 4.19** *The non-deterministic update process always terminates in a solution.*

*Proof* Since $\mathcal{M} \models E$, take the restriction of $\mathcal{M}$ that was used in the computation. It is a (finite) solving substitution contained in $\mathcal{M}$, hence in $T$. □

# References


1. W. Ackermann, Zur Widerspruchsfreiheit der Zahlentheorie. Math. Ann. **117**, 162–194 (1940)
2. T. Arai, Epsilon substitution method for ID$_1(\Pi^0_1 \vee \Sigma^0_1)$. Ann. Pure Appl. Logic **121**(2–3), 163–208 (2003)
3. A. Church, S.C. Kleene, Formal definitions in the theory of ordinal numbers. Fundam. Math. **28**, 11–21 (1937)
4. D. Hilbert, P. Bernays, *Grundlagen der Mathematik, Bd.2* (Springer, Berlin, 1970)
5. G. Kreisel, G. Mints, S. Simpson, The use of abstract language in elementary metamathematics. Lect. Notes Math. **253**, 38–131 (1975)
6. G. Mints, Gentzen-type systems and Hilbert's epsilon substitution method. I, in *Logic, Method. and Philos. of Sci. IX*, ed. by D. Prawitz, B. Skyrms, D. Westerstahl (Elsevier, 1994), pp. 91–122
7. G. Mints, *An Approach to An Epsilon-Substitution Method for* ID$_1$. Preprint, no 45 (Institute Mittag-Leffler, MLI, Stockholm, 2001)
8. G. Mints, Non-deterministic epsilon substitution for ID$_1$: effective proof, in *Logic, Construction, Computation*, ed. by U. Berger, H. Diener, P. Schuster, M. Seisenberger (Ontos Verlag, Heusenstamm, 2012), pp. 325–342
9. G. Mints, S. Tupailo, W. Buchholz, Epsilon substitution method for elementary analysis. Arch. Math. Logic **35**, 103–130 (1996)

10. G. Mints, S. Tupailo, Epsilon-substitution method for ramified language and $\Delta^1_1$ comprehension axiom, in *Logic and Foundations of Mathematics*, ed. by A. Cantini, et al. (Kluwer Academic Publishers, Netherlands, 1999), pp. 107–130
11. H. Rogers, *Theory of Recursive Functions and Effective Computability* (McGraw Hill, New York, 1967)

# A Game-Theoretic Computational Interpretation of Proofs in Classical Analysis

**Paulo Oliva and Thomas Powell**

**Abstract** It has been shown by Escardó and the first author that a functional interpretation of proofs in analysis can be given by the product of selection functions, a mode of recursion that has an intuitive reading in terms of the computation of optimal strategies in sequential games. We argue that this result has genuine practical value by interpreting some well-known theorems of mathematics and demonstrating that the product gives these theorems a natural computational interpretation that can be clearly understood in game theoretic terms.

## 1 Introduction

Over the last century, mathematicians and computer scientists have become increasingly interested in understanding the computational content of mathematical proofs. A central feature of modern mathematics is the use of non-constructive methods that allow us to reason about infinitary objects without providing any computational justification. In the 1920s Hilbert's program broadly addressed the task of understanding non-constructive mathematics in computational terms, which led to the development of important proof-theoretic techniques such as cut-elimination, the $\varepsilon$-method and proof interpretations. These were used to obtain significant foundational results such as relative consistency proofs for arithmetic [15] and analysis [27].

In recent decades these metamathematical devices whose roots lie in foundational problems have been employed more directly towards the extraction of programs from non-constructive proofs. This shift of emphasis has its origins in the fundamental work of Kreisel on the "unwinding" of proofs [19, 20], and has now become the focus of a considerable amount of research in logic and computer science.

P. Oliva (✉)
Queen Mary, University of London, School of Electronic Engineering and Computer Science, Godward Square, 10, E1 4FZ, London, England
e-mail: p.oliva@qmul.ac.uk

T. Powell
University of Innsbruck, ICT Gebäude, Technikerstraße 21a, 6020 Innsbruck, Austria



R. Kahle, M. Rathjen (eds.), *Gentzen's Centenary*,
DOI 10.1007/978-3-319-10103-3_18

Among the most powerful tools for extracting constructive information from proofs are *proof interpretations*, which include Freidman's A-translation with realizability [7,13] and Gödel's Dialectica interpretation [15]. The latter in particular is central to the highly successful *proof mining* program (see Kohlenbach [18]), in which the analysis of proofs using the monotone Dialectica interpretation has led to the development of general meta-theorems guaranteeing the extraction of effective uniform bounds from theorems in analysis.

While proof interpretations have been widely applied for both foundational purposes and to obtain numerical information from theorems in mathematics, the *qualitative (computational) behaviour* of their output has received relatively little attention. Indeed, the operational semantics of programs extracted from even relatively simple classical proofs are often very difficult to understand. This is mainly due to two factors:

1. **Higher-order computation**. Even when computing a witness of type $\mathbb{N}$ the constructions involved will work on higher types, usually types 1 and 2. Gödel's primitive recursor itself even for the lowest type is already an object of type 2.
2. **Syntactic nature of proof interpretations and translations**. Extracted programs tend to be hidden beneath a complex layer of syntax that generally accompanies formal translations on proofs.

Moreover, relatively little work has gone into addressing these issues because more often than not proof interpretations are a means to an end—be it a consistency proof or the extraction of a uniform bound—and a qualitative understanding of their output is simply irrelevant.

Nevertheless, the idea of stripping functional interpretations of their syntax and appreciating how they work from a *mathematical* perspective is an interesting one. It has been observed by Gaspar and Kohlenbach [14, 18] that the kind of logical manipulations carried out by the Dialectica interpretation is closely related to the so-called *correspondence principle* between "soft" and "hard" analysis discussed by Tao in [28,29]. In this sense one could potentially view functional interpretations as devices that transform classical proofs into constructive proofs of a "finitized" form of the original theorem, although actually translating their output into what a mathematician would consider a proof is far from straightforward.

In recent work [8, 11, 23] the authors and Escardó have sought to address this problem and better understand programs extracted by the Dialectica interpretation. There it is shown that the Dialectica interpretation of the key combination of classical logic and countable choice can be realised by the *product of selection functions* (as opposed to Spector's bar recursion [27] or modified bar recursion [3, 5, 6]), an intuitive mode of computation that can be characterised as computing optimal strategies in a very general class of sequential games.

Consider, for instance, an $\exists\forall$-theorem $\exists x^X \forall y^Y A(x, y)$, with $A(x, y)$ a decidable predicate, and $x$ and $y$ having types $X$ and $Y$ respectively. We want to think of the type of possible witnesses $X$ as a set of *available moves*, and $Y$ as the set of possible *outcomes* of a game. Note that such game only has one round (and one player!), and the choice of the outcome $y\colon Y$ is thought of as being determined by some external "environment". Such environment would map each move $x$ to an outcome $y$. The predicate $A(x, y)$ is then understood as prescribing what are the good outcomes $y$ given any particular move $x$, i.e. it describes the *goal* of the player in the game by describing which outcomes are good for each choice of move. Hence, the theorem $\exists x^X \forall y^Y A(x, y)$ says that there exists a move $x$ for which all possible outcomes $y$ are considered good, a sort of "winning move".

Now, if the theorem has been proven classically, such a move can be shown to exist without it being explicitly given, or it might be that such $x$ is not even effectively computable in the other parameters of the formula $A(x, y)$. What we should do then is to consider the "constructive" equivalent of the theorem via the negative translation, namely $\neg\neg\exists x^X \forall y^Y A(x, y)$. With the help of the quantifier-free axiom of choice and the Markov principle, this can also be put in the form $\exists\forall$, and in fact that is precisely what the Dialectica interpretation does. In this case we would obtain the (classically) equivalent theorem

$$\exists \varepsilon^{(X \to Y) \to X} \forall p^{X \to Y} A(\varepsilon p, p(\varepsilon p)).$$

Although $x \colon X$ might not be effectively computable, it turns out that the *selection function* $\varepsilon \colon (X \to Y) \to X$ is. Moreover, we can extend our game-theoretic reading and view $p \colon X \to Y$ as a mapping from moves to outcomes. What the selection function $\varepsilon$ does is to pick, for any given such mapping $p$, a move $x = \varepsilon p$ whose corresponding outcome according to $p$, namely $y = px$ is a good outcome for $x$. In other words, if we have access to the "environment" $p \colon X \to Y$ that calculates the outcome of the game for a given move $x$, then we can effectively compute a move whose corresponding outcome is a good one.

Now suppose we are given a countable family of $\exists\forall$-predicates $\exists x \forall y A_n(x, y)$ interpreted by a sequence of selection functions $(\varepsilon_n)$. From $\forall n \exists x \forall y A_n(x, y)$, by countable choice, there exists a sequence $f \colon \mathbb{N} \to X$ satisfying $\forall n, y A_n(fn, y)$. The Dialectica interpretation of $\neg\neg\exists f \forall n, y A_n(fn, y)$ states that for any given functions $q \colon X^{\mathbb{N}} \to Y$ and $\omega \colon X^{\mathbb{N}} \to \mathbb{N}$ there exists a functional $f$ such that

$$A_{\omega f}(f(\omega f), qf).$$

Therefore, thinking of each $A_n(x, y)$ as prescribing the "good" pairs of move-outcome for round $n$, the task of giving a computational interpretation to countable choice corresponds to finding, in $(\varepsilon_n)$, a sequence of moves $f$ which leads to an outcome $y = qf$ that is considered good at round $n = \omega f$.

We will see that the product of the selection functions $(\varepsilon_n)$ calculates such an $f$, and this construction can be viewed as the calculation of an optimal strategy in a sequential game whose "goal" at round $n$—given by the selection function $\varepsilon_n$—is to pick a move which leads to a good outcome according to $A_n$.

The aim of this article is to demonstrate that, *in practice*, program extraction using the product of selection functions and its game-theoretic semantics leads to a much better appreciation of the constructive content of proofs in analysis. We illustrate this using a few well-known classical theorems that have been extensively analysed by proof theorists. In particular we include a detailed analysis of a proof of the Bolzano–Weierstrass theorem, the interpretation of which is by no means trivial, but from which we are nevertheless able to extract a program that can be given a clear description in the language of sequential games.

In the course of the paper our aim is to portray the Dialectica interpretation as an intelligent translation whose output can be read and understood in *mathematical terms*. As such we endeavour to phrase the higher type functionals that arise from the interpretation using a more informal vocabulary. This approach owes a lot to the aforementioned work by Gaspar and Kohlenbach, and it is hoped that our work will complement theirs in forming another small step towards understanding the mathematical significance of proof interpretations.

## 2 Preliminaries

We work in the language of Peano (and Heyting) arithmetic in all finite types $\mathsf{PA}^\omega$ ($\mathsf{HA}^\omega$, respectively). The finite types $\mathbf{T}$ contain a basic type $\mathbb{N}$ and whenever $X, Y \in \mathbf{T}$ then $X \to Y \in \mathbf{T}$, i.e.

$$\mathbf{T} = \{\mathbb{N}, \mathbb{N} \to \mathbb{N}, (\mathbb{N} \to \mathbb{N}) \to \mathbb{N}, \dots\}.$$

Closely related to $\mathsf{PA}^\omega$ is Gödel's system T of primitive recursive functionals of finite type. This quantifier-free calculus is essentially primitive recursive arithmetic $\mathsf{PRA}$ with the schema of recursion extended to all types $X \in \mathbf{T}$, i.e.

$$\mathsf{R}_n(h)(g) \stackrel{X}{=} \begin{cases} g & \text{if } n = 0, \\ h_n(\mathsf{R}_{n-1}(h)(g)) & \text{if } n > 0. \end{cases} \tag{1}$$

For full details of these theories the reader is referred to [1]. Technically by $\mathsf{PA}^\omega$ ($\mathsf{HA}^\omega$) we refer to the *weakly* extensional variants of arithmetic in all finite types, since full extensionality is not sound under the Dialectica interpretation. However, we make no such restriction in the verifying system T.

**Notation** We make informal use of types like $\mathbb{B} \equiv \{0, 1\}$, $\mathbb{Q}$, finite sequences types $X^*$, etc. as elements of these types can be encoded as elements of a suitable type in $\mathbf{T}$. We also make use of the following abbreviations:

- $\mathbf{0}^X$ is the constant zero functional of type $X$.
- $s * t$ is the concatenation of sequences $s\colon X^*$ and $t : X^*$.
- $s * \alpha$ is also used for the concatenation of sequences $s\colon X^*$ and $\alpha : X^{\mathbb{N}}$.
- $s \preceq t$ for "$s$ is a prefix of $t$".
- $\hat{s} \equiv s * \mathbf{0}^{X^{\mathbb{N}}}$ the infinite extension of a finite sequence $s\colon X^*$ with $\mathbf{0}$'s.
- $[\alpha](n) \equiv \langle \alpha 0, \dots, \alpha(n-1) \rangle$ is the initial segment of $\alpha$ of length $n$.

- $\mu n \leq N \, . \, \varphi(n)$ is the bounded search operator that returns the least $n \leq N$ satisfying the decidable predicate $\varphi(n)$ if one exists, or $N$ otherwise.
- Given a functional $f : X \to (Y \to Z)$ and an $x \colon X$, we often write $f_x$ for the functional $f(x)$ of type $Y \to Z$.

## 2.1 The Dialectica Interpretation

We assume that the reader is familiar with Gödel's Dialectica interpretation (details of which are covered in full in [1, 18]), although we recall below a few basic facts to familiarise the reader with our notation and terminology.

The Dialectica interpretation maps formulas $A$ of some specified theory $\mathcal{S}$ to a quantifier-free formula $|A|^x_y$ (whose free variables are $x$, $y$ and the free variables of $A$) definable in some specified quantifier-free system of functionals $\mathsf{F}$. The canonical instance of this mapping is when $\mathcal{S}$ is $\mathsf{HA}^\omega$ and $\mathsf{F}$ is Gödel's system T.

In $|A|^x_y$ we have that $x$ and $y$ stand for (possibly empty) tuples of objects of finite type. We think of $x$ as the *witnessing* variables and $y$ as the *challenge* variables. The intuition is that $A$ is logically equivalent to $\exists x \forall y |A|^x_y$. The translation is formally defined as follows:

**Definition 1 (Gödel's Dialectica Interpretation)** For atomic formulas $P$ we set $|P| := P$, with $x$ and $y$ both empty tuples. Assuming that we have already defined $|A|^x_y$ and $|B|^u_v$, we define

$$\begin{aligned}
|A \wedge B|^{x,v}_{y,w} &:= |A|^x_y \wedge |B|^v_w, \\
|A \vee B|^{x,v,b}_{y,w} &:= (b = 0 \wedge |A|^x_y) \vee (b = 1 \wedge |A|^v_w), \\
|A \to B|^{f,g}_{x,w} &:= |A|^x_{gxw} \to |B|^{fx}_w, \\
|\forall z A(z)|^f_{y,z} &:= |A(z)|^{fz}_y, \\
|\exists z A(z)|^{x,z}_y &:= |A(z)|^x_y.
\end{aligned}$$

We say that $\mathcal{S}$ is (Dialectica) interpreted in $\mathsf{F}$ if whenever $\mathcal{S} \vdash A$ we can construct some $t \in \mathsf{F}$ such that $\mathsf{F} \vdash |A|^t_y$.

In order to interpret *classical* theories, the Dialectica interpretation is typically composed with a negative translation[1] $(\cdot)^N$ to form the so-called ND interpretation. In the remainder of the paper, by the *functional interpretation* of a formula $A$ we specifically mean its ND interpretation, i.e. $|A^N|^x_y$. Hence, a classical theory $\mathcal{T}$ has a functional interpretation in $\mathsf{F}$ if whenever $\mathcal{T} \vdash A$ we can construct some $t \in \mathsf{F}$ satisfying $\mathsf{F} \vdash |A^N|^t_y$.

[1] As in [18] we adopt Kuroda's variant of the negative translation.

In his original paper on the Dialectica interpretation, Gödel proved that Peano arithmetic has a functional interpretation in system T. Later, Spector extended Gödel's result to classical analysis by realizing the Dialectica interpretation of the negative translation of the *axiom of countable choice*

$$\mathsf{AC}_0 \,:\, \forall n \exists x^X A_n(x) \to \exists f^{X^{\mathbb{N}}} \forall n A_n(fn)$$

with a novel form of recursion called *Spector bar recursion* SBR. It was later shown [16, 21] that even *dependent choice*[2]

$$\mathsf{DC} \,:\, \forall s^{X^*} \exists x^X A_s(x) \to \exists f^{X^{\mathbb{N}}} \forall n A_{[f](n)}(fn)$$

could be interpreted using bar recursion. To summarise:

**Theorem 2**

*(a)* $\mathsf{PA}^\omega$ *has a functional interpretation in* T (Gödel [15]).
*(b)* $\mathsf{PA}^\omega + \mathsf{AC}_0$ *and even* $\mathsf{PA}^\omega + \mathsf{DC}$ *have a functional interpretation in* T + SBR (Spector [27], and also [16, 21]).

The main purpose of this paper is to show that these soundness theorems can be reformulated in terms of the product of selection functions, and that this reformulation is better suited towards understanding the behaviour of programs extracted by the Dialectica interpretation.

## 2.2 Outline of Article

We begin in Sect. 3 by introducing the product of selection functions and showing that it can be characterised as an operation that computes optimal strategies in sequential games.

In the main part of the paper we then discuss how the language of selection functions is well suited to capturing the way in which the Dialectica interpretation works, and in particular the product of selection functions directly interprets countable choice.

We then present a short case study (Sect. 6) in which we extract a program from a proof of the Bolzano–Weierstrass theorem via the product of selection functions and demonstrate that our program has a clear game-theoretic semantics.

We conclude by briefly discussing some of the problems we face in gaining a more intuitive understanding of functional interpretations, and outline some potential directions for further research.

[2] In these papers DC is given in a slightly different form to ours.

## 3 Selection Functions

This and the following sections constitute a brief overview of work that is presented in full elsewhere, e.g. the reader is referred to the original paper [9] or a recent survey [11] for a more detailed treatment.

A *selection function* is defined to be any element of type $(X \to R) \to X$ (as in [9] we abbreviate this type as $J_R X$). Closely related to selection functions $\varepsilon \colon J_R X$ is the notion of a *quantifier* $\phi\colon (X \to R) \to R$. A quantifier $\phi$ is *attained* by a selection function $\varepsilon$ if $\phi(p) = p(\varepsilon p)$, for all $p\colon X \to R$. Note that any selection function defines a quantifier, which we denote by $\overline{\varepsilon}(p) = p(\varepsilon p)$. Hence, a quantifier $\phi$ is attainable if $\phi = \overline{\varepsilon}$ for some selection function $\varepsilon$. The intuition is to view $\varepsilon$ as a selector that given a function $p\colon X \to R$ picks a particular element of $x = \varepsilon p$ of $X$ such that $p(x)$ attains $\phi(p)$, as the following examples illustrate.

*Example 3*

(a) The canonical example of a selection function and its associated quantifier is when $R$ forms a set of truth values, e.g. $R = \mathbb{B}$. Hilbert's epsilon term of type $X$, $\varepsilon_X \colon J_{\mathbb{B}} X$ is a selection function which attains the usual existential quantifier $\exists_X$ for predicates over type $X$, since by definition we have

$$\exists x^X\ p(x) \Leftrightarrow p(\varepsilon_X p).$$

(b) By the mean value theorem there exists a selection function $\varepsilon\colon J_{[0,1]}\mathbb{R}$ such that for any continuous function $p\colon [0,1] \to \mathbb{R}$ we have

$$\int_0^1 p(x)dx = p(\varepsilon p).$$

Hence, $\varepsilon\colon ([0,1] \to \mathbb{R}) \to [0,1]$ attains the quantifier $\int_0^1 \colon ([0,1] \to \mathbb{R}) \to \mathbb{R}$.

(c) Assume we are given a position in a game where we have to pick a move in $X$, and a quantifier $\phi$ that given a map $p\colon X \to R$ which assigns to each potential move an outcome in $R$, determines the best possible outcome. A *strategy* for that position can be defined by a selection function $\varepsilon\colon J_R X$ that for each $p$ selects a move $\varepsilon(p)$ that attains the best outcome $\phi(p)$.

The theory of selection functions and quantifiers forms the basis of [9–11]. One of the main achievements of these papers has been to define a product operation on selection functions (along with a corresponding operation on quantifiers which we do not discuss further here). They demonstrate that the product of selection functions is an extremely versatile construction that appears naturally in several different areas of mathematics and computer science, such as fixed point theory (Bekič's lemma), algorithms (backtracking), game theory (backward induction) and, as we also discuss in Sect. 5, proof theory.

In the remainder of the section we define (following [9]) the product of selection functions, and explain in the following section how this procedure can be best understood via the computation of optimal strategies in a certain class of sequential games.

**Definition 4 (Binary Product of Selection Functions [9])** Given a selection function $\varepsilon\colon J_R X$, a family of selection functions $\delta_x\colon J_R Y$ and a predicate $q\colon X \times Y \to R$, let

$$A[x^X] \stackrel{Y}{:=} \delta_x(\lambda y.q(x, y)),$$
$$a \stackrel{X}{:=} \varepsilon(\lambda x.q(x, A[x])).$$

The binary product $\varepsilon \otimes \delta$ of $\varepsilon$ and $\delta$ is another selection function, of type $J_R(X \times Y)$, defined by

$$(\varepsilon \otimes \delta)(q) \stackrel{X\times Y}{:=} \langle a, A[a]\rangle.$$

If $\delta$ is independent of $x$, we call this the *simple* product of selection functions. The general case is then by comparison called the *dependent* product of selection functions.

The binary product constructs a composite selection function on the type $X \times Y$ in the natural way:

*Example 5* Continuing from Example 3 we have:

(a) The product of epsilon operators $\varepsilon_X \otimes \varepsilon_Y$ is an epsilon operator of type $X \times Y$ in the sense that

$$\exists x^X \exists y^Y q(x, y) \Leftrightarrow q((\varepsilon_X \otimes \varepsilon_Y)(q)).$$

(b) For $\varepsilon\colon J_{[0,1]}\mathbb{R}$ as in Example 3 (b), we have

$$\int_0^1 \int_0^1 q(x, y)\, dxdy = q((\varepsilon \otimes \varepsilon)(q))$$

where $q\colon [0, 1]^2 \to \mathbb{R}$ is a given continuous function.

(c) Given strategies $\varepsilon_0$, $\varepsilon_1$ for each round in a two round sequential game with outcome function $q\colon X_0 \times X_1 \to R$, then $(\varepsilon_0 \otimes \varepsilon_1)(q)$ forms a *strategy* for the game which is "compatible" with the local strategies $\varepsilon_0$ and $\varepsilon_1$. This key instance of the product is discussed in more detail in Sect. 4.

As described in [9], we can iterate the binary product of selection functions a finite or an unbounded number of times, where the length of the iteration is dependent on the output of the product in the following sense.

**Definition 6 (Iterated Product of Selection Functions [9])** Suppose we are given a family of selection functions $(\varepsilon_s\colon J_R X)_{s\in X^*}$. The *explicitly controlled unbounded product* of the selection functions $\varepsilon_s$ is defined by the recursion schema

$$\mathsf{EPS}^{\omega}_{s}(\varepsilon) \overset{J_R X^{\mathbb{N}}}{=} \begin{cases} \mathbf{0} & \text{if } \omega(\hat{s}) < |s|, \\ \varepsilon_s \otimes \lambda x.\mathsf{EPS}^{\omega}_{s*x}(\varepsilon) & \text{otherwise} \end{cases} \tag{2}$$

where $s\colon X^*$ and $\omega\colon X^{\mathbb{N}} \to \mathbb{N}$.

The functional $\omega$ acts as a control, terminating the procedure once it has produced a sequence $s$ satisfying $\omega(\hat{s}) < |s|$. The unbounded product is total in any model of bar recursion, which in particular must admit *Spector's condition*:

$$\forall \omega^{X^{\mathbb{N}}\to\mathbb{N}}, \alpha^{X^{\mathbb{N}}} \exists n \left( \omega(\widehat{[\alpha](n)}) < n \right).$$

These include the models of continuous functionals and the majorizable functionals. On the other hand, when $\omega$ is a constant function, say $\omega\alpha = n$, this corresponds to a finite iteration of the binary product and this restricted instance of the product is definable in system T.

By unwinding the definition of the binary product in (2) we obtain an equivalent equation

$$\mathsf{EPS}^{\omega}_{s}(\varepsilon)(q) \overset{X^{\mathbb{N}}}{=} \begin{cases} \mathbf{0} & \text{if } \omega(\hat{s}) < |s|, \\ a_s * \mathsf{EPS}^{\omega}_{s*a_s}(\varepsilon)(q_{a_s}) & \text{otherwise} \end{cases} \tag{3}$$

where $a_s = \varepsilon_s(\lambda x.\overline{\mathsf{EPS}^{\omega}_{s*x}(\varepsilon)(q_x)})$ and $q_x(\alpha) = q(x * \alpha)$. Recall that $\overline{\mathsf{EPS}^{\omega}_{s*x}(\varepsilon)(q_x)}$ abbreviates $q_x(\mathsf{EPS}^{\omega}_{s*x}(\varepsilon)(q_x))$.

For fixed $\omega, \varepsilon$ and $q$ one can think of $\mathsf{EPS}^{\omega}_{s}(\varepsilon)(q)$ as computing an infinite extension to any given finite sequence $s$. The key property of $\mathsf{EPS}$ is that the infinite extension of an initial segment $[\alpha](n)$ of a previous infinite extension $\alpha$ is identical to the original infinite extension. Formally:

**Lemma 7 (cf. [27], Lemma 1)** *Let* $\alpha = \mathsf{EPS}^{\omega}_{s}(\varepsilon)(q)$. *For all* $n$,

$$\alpha = [\alpha](n) * \mathsf{EPS}^{\omega}_{s*[\alpha](n)}(\varepsilon)(q_{[\alpha](n)}). \tag{4}$$

*Proof* By induction on $n$. If $n = 0$, this follows by the definition of $\alpha$. Assume (4) holds for $n$, we wish to show it also holds for $n + 1$. Consider two cases.

(a) If $\omega(s * [\alpha](n) * \mathbf{0}) < |s| + n$, then $\mathsf{EPS}^{\omega}_{s*[\alpha](n)}(\varepsilon)(q_{[\alpha](n)}) = \mathbf{0}^{X^{\mathbb{N}}}$. By induction hypothesis $\alpha = [\alpha](n) * \mathbf{0}$, so that $\alpha(n) = \mathbf{0}^X$. Therefore $\alpha = [\alpha](n+1) * \mathbf{0}$, which, by extensionality, implies

$$\omega(s * [\alpha](n+1) * \mathbf{0}) = \omega(s * [\alpha](n) * \mathbf{0}) < |s| + n < |s| + n + 1.$$

Hence, $\mathsf{EPS}^{\omega}_{s*[\alpha](n+1)}(\varepsilon)(q_{[\alpha](n+1)}) = \mathbf{0}^{X^{\mathbb{N}}}$ so that

$$[\alpha](n+1) * \mathsf{EPS}^{\omega}_{s*[\alpha](n+1)}(\varepsilon)(q_{[\alpha](n+1)}) = [\alpha](n+1) * \mathbf{0} = [\alpha](n) * \mathbf{0} = \alpha.$$

(b) If $\omega(s * [\alpha](n) * \mathbf{0}) \geq |s| + n$, then

$$\alpha \stackrel{\text{(IH)}}{=} [\alpha](n) * \mathsf{EPS}^{\omega}_{s*[\alpha](n)}(\varepsilon)(q_{[\alpha](n)}) \stackrel{(3)}{=} [\alpha](n) * c * \mathsf{EPS}^{\omega}_{s*[\alpha](n)*c}(\varepsilon)(q_{[\alpha](n)*c}),$$

where $c = \varepsilon_{s*[\alpha](n)}(\lambda x.\overline{\mathsf{EPS}^{\omega}_{s*[\alpha](n)*x}(\varepsilon)}(q_{s*[\alpha](n)*x}))$. Hence, $\alpha(n) = c$. Therefore

$$\alpha = [\alpha](n+1) * \mathsf{EPS}^{\omega}_{s*[\alpha](n+1)}(\varepsilon)(q_{[\alpha](n+1)}).$$

□

This lemma is the main building block behind the proof of the following fundamental theorem about EPS.

**Theorem 8 (Main Theorem on EPS)** *Let $q\colon X^{\mathbb{N}} \to R$ and $\omega\colon X^{\mathbb{N}} \to \mathbb{N}$ and $\varepsilon\colon X^* \to J_R X$ be given. Define*

$$\begin{aligned} \alpha &\stackrel{X^{\mathbb{N}}}{=} \mathsf{EPS}^{\omega}_{\langle\rangle}(\varepsilon)(q), \\ p_s(x) &\stackrel{R}{=} \overline{\mathsf{EPS}^{\omega}_{s*x}(\varepsilon)}(q_{s*x}). \end{aligned}$$

*For $n \leq \omega(\alpha)$ we have*

$$\begin{aligned} \alpha(n) &\stackrel{X}{=} \varepsilon_{[\alpha](n)}(p_{[\alpha](n)}), \\ q\alpha &\stackrel{R}{=} \overline{\varepsilon_{[\alpha](n)}}(p_{[\alpha](n)}). \end{aligned} \tag{5}$$

*Proof* Assume $n \leq \omega(\alpha)$. First we argue that

$$(*) \quad n \leq \omega([\alpha](n) * \mathbf{0}).$$

Otherwise, assuming $n > \omega([\alpha](n) * \mathbf{0})$ we would have, by Lemma 7, that $\alpha = [\alpha](n) * \mathbf{0}$. And hence, $n > \omega([\alpha](n) * \mathbf{0}) = \omega(\alpha) \geq n$, which is a contradiction. Hence, we have that

$$\begin{aligned} \alpha(n) &\stackrel{\text{L7}}{=} \mathsf{EPS}^{\omega}_{[\alpha](n)}(\varepsilon)(q_{[\alpha](n)})(0) \\ &\stackrel{(3)+(*)}{=} \varepsilon_{[\alpha](n)}(\lambda x.\overline{\mathsf{EPS}^{\omega}_{[\alpha](n)*x}(\varepsilon)}(q_{[\alpha](n)*x})) \\ &= \varepsilon_{[\alpha](n)}(p_{[\alpha](n)}), \end{aligned}$$

by the definition of $p_s$. For the second equality, we have

$$\begin{aligned} q\alpha &\overset{\text{L7}}{=} q_{[\alpha](n+1)}(\mathsf{EPS}^{\omega}_{[\alpha](n+1)}(\varepsilon)(q_{[\alpha](n+1)})) \\ &= p_{[\alpha](n)}(\alpha(n)) \\ &= \overline{\varepsilon_{[\alpha](n)}}(p_{[\alpha](n)}), \end{aligned}$$

where the last equality uses that $\alpha(n) = \varepsilon_{[\alpha](n)}(p_{[\alpha](n)})$ as previously shown. □

Theorem 8 characterises the product of selection functions as computing a sequence $\alpha$ that forms a kind of sequential equilibrium between the selection functions—expressed by the Eq. (5)—up to a point $\omega\alpha$ parametrised by $\alpha$ itself. The significance of the product is that such equilibria appear naturally in a variety of contexts. In the following we outline perhaps the most illuminating of these contexts, namely the theory of sequential games.

## 4 Sequential Games and Optimal Strategies

One of the most remarkable properties of EPS is that it computes optimal strategies in a certain class of sequential games. The reader is encouraged to consult [11] in conjunction with the relatively concise discussion here.

As in this article we only consider games (in the sense of [11]) where the quantifiers are attainable, we shall incorporate this restriction in the definition of the game itself.

**Definition 9 (Sequential Games of Unbounded Length, [11])** The *type* of a game is given by a pair $(X, R)$ where

- $X$ is the set of possible *moves* at each round.
- $R$ is the set of possible outcomes of the game.

A finite sequence $s\colon X^*$ shall be thought of as a *position* in the game determined by the first $|s|$ moves. An infinite sequence $\alpha\colon X^{\mathbb{N}}$ is called a *play* of the game. An unbounded sequential game of type $(X, R)$ is a triple $(\varepsilon, q, \omega)$ where

- $\varepsilon_s\colon J_R X$ determines the *optimal move* at position $s$.
- $q\colon X^{\mathbb{N}} \to R$ determines, given a play $\alpha\colon X^{\mathbb{N}}$, the outcome of the game.
- $\omega\colon X^{\mathbb{N}} \to \mathbb{N}$ determines the *relevant part* of a play.

The function $q$ is called the *outcome function*, whereas $\omega$ is called the *control function*. Given a play $\alpha$, all moves $\alpha(i)$ for $i \leq \omega\alpha$ are *relevant moves*. In general, a position $s$ is called relevant if $|s| \leq \omega\hat{s}$, i.e. if in a canonical extension of the current position $s$ the current move is considered a relevant move. Games with a finite, fixed number of rounds can be viewed as the special case of an unbounded game in which $\omega$ is constant.

We shall only consider infinite plays which are obtained by some canonical extension of a finite play $s$. Therefore, we think of these as finite games of unbounded length.

The intuition behind Definition 9 is as follows. We think of the selection functions $\varepsilon_s$ as specifying at position $s$ what an optimal move at that point would be if we knew the final outcome corresponding to each of the candidate moves. The selection function takes this mapping $X \to R$ of moves to outcomes and tells us what an optimal move would be in that particular case.

A *strategy* in such game is simply a function next : $X^* \to X$ which determines for each position $s$ what the next move next$(s)$ should be. *To follow a strategy* from position $s$ means to play all following moves according to the strategy, i.e. we obtain a sequence of moves $\alpha(0), \alpha(1), \ldots$ as

$$\alpha(i) = \text{next}(s * [\alpha](i-1)).$$

We call this the *strategic extension of* $s$. The strategic extension of the empty play is called the *strategic play*.

**Definition 10 (Optimal Strategies)** A strategy is said to be *optimal*[3] if the move played at each relevant position $s$ is the one recommended by the selection function $\varepsilon_s$, i.e.

$$\text{next}(s) = \varepsilon_s(\lambda x.q(s * x * \beta)) \tag{6}$$

where $\beta$ is the strategic extension of $s * x$.

The main result of [11] is that the product of selection functions computes optimal strategies:

**Theorem 11 ([11])** *Given a game* $(\varepsilon, q, \omega)$, *the strategy*

$$\text{next}(s) \stackrel{X}{=} \left(\mathsf{EPS}_s^{\omega}(\varepsilon)(q_s)\right)(0) \tag{7}$$

*is optimal, and, moreover,*

$$\alpha \stackrel{X^{\mathbb{N}}}{=} \mathsf{EPS}_s^{\omega}(\varepsilon)(q_s) \tag{8}$$

*is the strategic extension of* $s$, *i.e.* $\alpha(n) = \text{next}(s * [\alpha](n))$.

[3]This is a stronger notion than the one introduced in [11] for the more general case where the quantifiers are not necessarily attainable.

*Proof* We have that

$$\begin{aligned}\alpha(n) &\overset{(8)}{=} \mathsf{EPS}^{\omega}_{s}(\varepsilon)(q_s)(n) \\ &\overset{\text{L7}}{=} \mathsf{EPS}^{\omega}_{s*[\alpha](n)}(\varepsilon)(q_{s*[\alpha](n)})(0) \\ &\overset{(7)}{=} \text{next}(s * [\alpha](n)),\end{aligned}$$

which proves the second claim. Hence, assuming $s$ is a relevant position, i.e. $(*)\ \omega(\hat{s}) \geq |s|$ we Have

$$\begin{aligned}\text{next}(s) &\overset{(7)}{=} \left(\mathsf{EPS}^{\omega}_{s}(\varepsilon)(q_s)\right)(0) \\ &\overset{(3)+(*)}{=} \varepsilon_s(\lambda x.\overline{\mathsf{EPS}^{\omega}_{s*x}(\varepsilon)}(q_{s*x})) \\ &= \varepsilon_s(\lambda x.q(s * x * \beta))\end{aligned}$$

where $\beta = \mathsf{EPS}^{\omega}_{s*x}(\varepsilon)(q_{s*x})$, by the claim just proven above, is the strategic extension of $s * x$. Hence, we have shown (6). □

Therefore in this sense the main Theorem 8 characterises $\mathsf{EPS}$ as a procedure that computes an optimal strategy in the game defined by $(\varepsilon, q, \omega)$. We now show that the product also appears naturally in proof theory, with the advantage that it can be related back to the language of sequential games.

## 5 The Dialectica Interpretation of Classical Proofs

We now show how selection functions and their product are intrinsically connected to the functional interpretation of classical proofs. The key observation is that the language of selection functions elegantly captures the way in which the Dialectica interpretation treats double negations in negative-translated formulas. In particular their product directly interprets the double negation shift, which lies behind the negative translation of the axiom of countable choice.

This means that in many cases the algorithms extracted from classical proofs can be easily phrased in the intuitive language of sequential games. Moreover, though couched in the language of higher type recursive functionals, these games often have a natural informal reading in terms of strategic set-theoretic constructions, making the *mathematical* meaning of the extracted program more perspicuous.

## 5.1 *Interpreting $\Sigma_2$-Theorems*

Suppose are given a $\Sigma_2$-theorem $A \equiv \exists x^X \forall y^Y A_0(x, y)$ where $A_0$ is decidable. The negative translation of $A$ is equivalent[4] to $\neg\neg\exists x \forall y\, A_0(x, y)$, and therefore its functional interpretation is given by

$$|A^N|^{\varepsilon}_{p} = A_0(\varepsilon p, p(\varepsilon p)).$$

In other words, the Dialectica interpretation interprets double negations in front of a $\Sigma_2$ formula with a selection function $\varepsilon\colon J_Y X$. If the predicate $A_0(x, y)$ is thought of as prescribing "good" outcomes $y$ for a particular move $x$ as described in Sect. 1, then $\varepsilon$ implements a strategy that selects a move $x = \varepsilon p$ whose outcome with respect to the mapping $p$ is good.

Thus under the functional interpretation we have the following correspondence:

**$\Sigma_2$ – Theorems $\mapsto$ Selection functions**

The elimination of double negations in an arbitrary negated formula is essentially a (albeit complex) modular iteration of this process, suggesting to us that selection functions and modes of recursion based on selection functions lie behind the functional interpretation of classical proofs in a fundamental way.

There are several ways of characterising the selection function $\varepsilon$ interpreting $A$. For $\Sigma_2$-theorems Kreisel's *no counterexample interpretation* coincides with the functional interpretation and in this sense the constructive interpretation of $A$ is a selection function $\varepsilon$ that refutes an arbitrary "counterexample" functions $p$. We illustrate this in the following example, which demonstrates how selection functions are fundamental to the functional interpretation of pure classical logic.

*Example 12 (Law of Excluded Middle)* Consider the following simple reformulation of the law of excluded middle for $\Sigma_1$-formulas, better known as the *drinkers paradox*:

$$\mathsf{DP} \quad : \quad \exists x^X (\exists y P(y) \to P(x)). \tag{9}$$

Note that DP is intuitionistically equivalent to the $\Sigma_2$ theorem $\exists x \forall y (P(y) \to P(x))$. We ineffectively justify the principle by defining

$$x := \begin{cases} y & \text{for some } y \text{ satisfying } P(y), \\ \mathbf{0}^X & \text{if no such } y \text{ exists.} \end{cases}$$

[4] Assuming stability of atomic formulas.

On the other hand, we can *effectively* justify the principle with the selection function

$$\varepsilon p := \begin{cases} p(\mathbf{0}) \text{ if } P(p(\mathbf{0})), \\ \mathbf{0} \quad\quad \text{if } \neg P(p(\mathbf{0})) \end{cases} \tag{10}$$

that witnesses its functional interpretation:

$$|\mathsf{DP}|^{\varepsilon}_{p} = P(p(\varepsilon p)) \to P(\varepsilon p). \tag{11}$$

The drinkers paradox is essentially the law of excluded middle applied to the $\Sigma_1$-formula $\exists y P(y)$, i.e.

$$\exists b^{\mathbb{B}}(b = 0 \leftrightarrow \exists y P(y))$$

where the boolean $b$ is given by $P(x)$. The mapping $p \colon X \to X$ in the functional interpretation of DP can be seen as a *counterexample function* that attempts to witness $\neg$DP, i.e.

$$\forall x(P(px) \wedge \neg P(x)). \tag{12}$$

The constructive version of the law of excluded middle given by its functional interpretation is the statement that for any $p$ there exists an element $x$ refuting (12), i.e.

$$\forall p \exists x(\neg P(px) \vee P(x)).$$

The selection function $\varepsilon$ above witnesses this statement.

One can alternatively view the selection function $\varepsilon$ interpreting $A$ as an algorithm that produces an arbitrary large *approximation* to the ineffective object $x$ satisfying $\forall y A_0(x, y)$. In fact, when $Y = \mathbb{N}$ the formula $A$ is equivalent to $\exists x^X \forall y \forall i \leq y A_0(x, i)$. Hence, the functional interpretation of $A$ is equivalent to the existence of a selection functions $\varepsilon$ satisfying

$$\forall p \; \forall i \leq p(\varepsilon p) A_0(\varepsilon p, i).$$

In this context $p$ is a function specifying, for each choice of $x$, which values $y$ are required to satisfy $A_0(x, y)$ in some given situation. The selection function $\varepsilon$ should return an "approximation" $x$ for which we have $\forall i \leq p(x) A_0(x, i)$. Hence, $x$ does not need to be a "real" witness to $A$, but only an approximation which is good enough relative to the function $p$. This reading is closer to the notion of a "finitization" of as discussed by Tao in [28], in the sense that we interpret the qualitative statement that there exists some $x$ with the global property $\forall y A_0(x, y)$ by the quantitative statement that there exist approximations $x$ with the local property $\forall i \leq p(x) A_0(x, i)$ for arbitrary $p$.

*Example 13 (Convergence and Metastability)* The functional interpretation of Cauchy convergence

$$\forall k \exists n \forall m \forall i, j \in [n, n+m] \, (\|x_i - x_j\| \leq 2^{-k})$$

(where $i \in [n, n+m]$ is shorthand for $n \leq i \leq n+m$) is a sequence of selection functions $(\varepsilon_k)$ that satisfy

$$\forall k, p \; \forall i, j \in [\varepsilon_k p, \varepsilon_k p + p(\varepsilon_k p)] \, (\|x_i - x_j\| \leq 2^{-k}). \tag{13}$$

In other words, the Cauchy convergence property is equivalent to the existence of a sequence of selection functions $\varepsilon_k$ that compute regions of approximate stability, or *metastability*, of size specified by $p$.

This reformulation of convergence plays a key role in ergodic theory, where one obtains quantitative versions of convergence theorems by extracting explicit bounds on $\varepsilon_k p$ that are highly uniform with respect to $(x_n)$. A simple example is the so-called finite convergence principle discussed in [18, 28], where one can easily show that given $k$ and $p$ a bounded monotone sequence

$$0 \leq x_0 \leq x_1 \leq \cdots \leq 1$$

experiences a period of metastability bounded uniformly by $\tilde{p}(2^k)(0)$ for $\tilde{p}(n) := n + p(n)$. A more involved example of this phenomenon is the quantitative mean ergodic theorem of Avigad et al. in [2].

## 5.2 Interpreting the Axiom of Choice

Classical predicate logic in all finite types $\mathsf{PL}^\omega$ can be extended to encompass most of mathematics through the addition of choice principles. In particular the principle of *finite* choice

$$\mathsf{FC} : \; \forall n \leq N \exists x^X A_n(x) \rightarrow \exists s^{X^*} \forall n \leq N A_n(s_n)$$

is known to be equivalent to induction [24] and therefore we can define full Peano arithmetic (assuming a minimal amount of arithmetic) as $\mathsf{PA}^\omega := \mathsf{PL}^\omega + \mathsf{FC}$, while the further addition of countable choice $\mathsf{AC}_0$ or dependent choice $\mathsf{DC}$ yields a theory sufficient to formalise a large portion of analysis.

Thus a key part of understanding the computational content of classical proofs is to understand the computational interpretation of countable choice principles in the presence of classical logic.

Let us first consider an instance of $\mathsf{AC}_0$ for $\Pi_1$-formulas:

$$\Pi_1\text{-}\mathsf{AC}_0 : \; \forall n \exists x^X \forall y^Y \varphi_n(x, y) \rightarrow \exists f^{\mathbb{N} \rightarrow X} \forall n, y \varphi_n(fn, y),$$

for decidable $\varphi_n$. Its negative translation is equivalent to

$$\forall n \neg\neg \exists x \forall y \varphi_n(x, y) \to \neg\neg \exists f \forall n, y \varphi_n(fn, y),$$

and its Dialectica interpretation is equivalent (using just Markov's principle, which is admitted by the Dialectica interpretation) to the statement[5]

$$\forall \varepsilon, q, \omega \exists f (\forall n, p \varphi_n(\varepsilon_n p, p(\varepsilon_n p)) \to \forall i \leq \omega f \varphi_i(fi, qf)). \tag{14}$$

This constructive interpretation of $\mathsf{AC}_0$ asks for a selection function $F^{\varepsilon} \colon J_{Y \times \mathbb{N}} X^{\mathbb{N}}$ producing an approximation to the sequence $f$, given selection functions $\varepsilon$ interpreting its premise. Such a selection function can be given by

$$F^{\varepsilon}(q, \omega) := \mathsf{EPS}^{\omega}_{\langle\rangle}(\varepsilon)(q).$$

We now prove in detail that the product of selection functions directly realises the functional interpretation of the axiom of choice.

**Theorem 14** *The following hold:*

(a) *The functional interpretation of the schema of finite choice* $\mathsf{FC}$ *is directly witnessed by the finite simple product of selection functions (i.e.* $\omega$ *a constant function).*
(b) *The functional interpretation of the schema of countable choice* $\mathsf{AC}_0$ *is directly witnessed by the (unbounded) simple product of selection functions.*
(c) *The functional interpretation of the schema of dependent choice* $\mathsf{DC}$ *(see Sect. 2.1) is directly witnessed by the dependent product of selection functions.*

*Proof* We prove **(c)**, parts **(a)** and **(b)** both being simple cases of this (proofs which can nevertheless be found in [12] and [8], respectively). Since $A_s(x)$ is equivalent to $\exists \tilde{x}^{\tilde{X}} \forall y |A_s(x)^N|^{\tilde{x}}_y$ it suffices to interpret $\mathsf{DC}$ for $\Sigma_2$-formulas

$$\Sigma_2\text{-}\mathsf{DC} : \ \forall s^{X^*} \exists x^X, \tilde{x}^{\tilde{X}} \forall y |A_s(x)^N|^{\tilde{x}}_y \to \exists f^{X^{\mathbb{N}}} \forall n \exists \tilde{x}^{\tilde{X}} \forall y |A_{[f](n)}(fn)^N|^{\tilde{x}}_y.$$

Moreover, by adding a dummy variable $t$ of type $\tilde{X}^*$ and concatenating the types $X, \tilde{X}$ this follows directly from an instance of $\Pi_1$-$\mathsf{DC}$ i.e.

$$\Pi_1\text{-}\mathsf{DC} : \ \forall s^{X^*}, t^{\tilde{X}^*} \exists x, \tilde{x} \forall y |A_s(x)^N|^{\tilde{x}}_y \to \exists f^{X^{\mathbb{N}}}, \tilde{f}^{\tilde{X}^{\mathbb{N}}} \forall n, y |A_{[f](n)}(fn)^N|^{\tilde{f}n}_y.$$

[5]Note that we are replacing the standard conclusion $\varphi_{\omega f}(f(\omega f), qf)$ with a stronger variant $\forall i \leq \omega f \ \varphi_i(fi, qf)$. This is not essential as one can, given an $\omega$, define $\tilde{\omega}(f) = \mu i \leq \omega(f) \neg \varphi_i(fi, qf)$ so that $\varphi_{\tilde{\omega} f}(f(\tilde{\omega} f), qf)$ implies $\forall i \leq \omega f \ \varphi_i(fi, qf)$. We prefer the version $\forall i \leq \omega f \ \varphi_i(fi, qf)$ since (1) we can directly realise, and (2) it makes the interpretation more intuitive by viewing $\omega f$ as a bound up to which the play $f$ is required to be "optimal".

Therefore it suffices to deal with $\Pi_1$-DC, which in general has a negative translation equivalent to

$$\Pi_1\text{-DC}^N : \forall s^{X^*} \neg\neg\exists x^X \forall y A_s(x, y) \to \neg\neg\exists f \forall n, y A_{[f](n)}(fn, y).$$

The Dialectica interpretation of $\Pi_1$-DC$^N$ is equivalent to

$$|\Pi_1\text{-DC}^N|^{F,p,s}_{\varepsilon,\omega,q} \equiv A_s(\varepsilon_s p, p(\varepsilon_s p)) \to A_{[F](\omega F)}(F(\omega F), qF), \tag{15}$$

omitting, for the sake of readability, the parameters $\varepsilon, \omega$ and $q$ from the functions $F$, $p$ and $s$. In fact, these parameters $(\varepsilon, \omega, q)$ define a sequential game in the sense of Definition 9. Let

$$\begin{aligned} F &= \text{EPS}^{\omega}_{\langle\rangle}(\varepsilon)(q), \\ p_s(x) &= \overline{\text{EPS}^{\omega}_{s*x}(\varepsilon)}(q_{s*x}). \end{aligned}$$

By Theorem 8 we have that $F$ and $p := p_{[F](\omega F)}$ and $s := [F](\omega F)$ are such that $\varepsilon_s p = F(\omega F)$ and $p(\varepsilon_s p) = qF$, and hence, clearly witness $|\text{DC}^N|^{F,p,s}_{\varepsilon,\omega,q}$. □

Theorem 14 proves that under the functional interpretation we have a mapping

**Choice principles ↦ Product of selection functions**

At first glance it may seem strange that an operation that computes optimal strategies in sequential games is related to the axiom of choice is this manner. But if we take a closer look, the game theoretic behaviour of (14) becomes clear. The selection functions $\varepsilon_n$ which realise the premise of (14) can be seen as a collection of strategies each witnessing the $\Sigma_2$-theorems $(A_n)$. The Dialectica interpretation calls for a procedure that takes these pointwise strategies and produces a co-operative selection function $F$ that witnesses $\forall n\, A_n$. Such a procedure is provided naturally by the product of selection functions.

In the following examples we illustrate how the interpretation of theorems that make direct use of the axiom of choice can be given an intuitive game-theoretic constructive interpretation by the product of selection functions.

*Example 15 (Arithmetic Comprehension)* We first give a realizer for the functional interpretation of arithmetic comprehension for $\Sigma^0_1$-formulas, which states that for any $\Sigma_1$-predicate $\varphi$ over $\mathbb{N}$ there exists a set $X$ with

$$\forall n(n \in X \Leftrightarrow \exists y\, \varphi(n, y)).$$

Computing such $X$ is in general not possible. We can, however, try to compute an "approximation" to $X$. For instance, we might ask for an $\tilde{X}$ which only works for a finite number of $n$'s, or an approximation which only checks the existence of $y$'s up to a certain bound (possibly depending on the approximating set $\tilde{X}$). We call these calibrations of the "size" and "depth" of $X$, respectively.

Arithmetic comprehension follows from the formal statement

$$\exists f^{\mathbb{N}\to\mathbb{N}}\forall n(\exists y\ \varphi(n,y)\to\exists k<fn\ \varphi(n,k)),$$

where we define $X := \{n \mid \exists k < fn\ \varphi(n,k)\}$. Again, we cannot (in general) effectively construct $f$ and thus neither $X$, as the above is a direct consequence of countable choice applied the non-constructive statement

$$\forall n\exists x^{\mathbb{N}}(\exists y\ \varphi(n,y)\to\exists k<x\ \varphi(n,k)). \tag{16}$$

But this is just a collection of instances of DP applied to the formulas $P_n(x) := \exists k < x\ \varphi(n,k)$. Therefore defining the sequence of selection functions $(\varepsilon_n)$ by

$$\varepsilon_n p := \begin{cases} p(0) & \text{if } \exists k < p(0)\ \varphi(n,k), \\ 0 & \text{if } \forall k < p(0)\ \neg\varphi(n,k) \end{cases}$$

we have

$$\exists k < p(\varepsilon_n p)\ \varphi(n,k)\to\exists k<\varepsilon_n p\ \varphi(n,k)$$

for any $n$, $p$, and thus by Theorem 8, for any counterexample functionals $\omega$, $q$, setting $F := \mathsf{EPS}^{\omega}_{\langle\rangle}(\varepsilon)(q)$ we have

$$\forall i\leq\omega F(\exists k<qF\ \varphi(i,k)\to\exists k<Fi\ \varphi(i,k)) \tag{17}$$

which is equivalent to the functional interpretation of $\Sigma_1$-CA. So what is the game-theoretic interpretation of our realizer $F$? If we unravel (17), we see that we are essentially constructing a finite set

$$X_F := \{i\leq\omega F \mid \exists k<Fi\ \varphi(i,k)\}$$

that serves as an approximation to $X$ with the property that if $i \leq \omega F$ has a witness for $\varphi$ bounded by $qF$ then we must have $i \in X_F$. In this sense $\omega$ and $q$ can be read as set functions that calibrate the size and depth, respectively, of an approximation to $X$.

The set $X_F$ is constructed as an optimal play in the game $(\varepsilon, q, \omega)$. The job of the selection functions at round $n$ is to decide whether or not to include the number $n$ in the approximation, given that it has already made this decision for $\{0,\ldots,n-1\}$. Its default is to omit $n$ by playing 0, but if the resulting outcome $p_n(0)$ bounds some witness to $n$, it instead adds $n$ and steals this witness as justification.

Therefore in this scenario the product of selection functions forms an intuitive set-theoretic construction, starting with the empty set and strategically adding elements until it reaches the desired approximation. When interpreting a theorem that makes use of arithmetic comprehension as a lemma, we can simply plug in our realizer and impart its game theoretic meaning to better understand the realizer of the main theorem.

Some simple examples of well-known existence theorems that can be given a direct constructive interpretation using this instance of the product can be found in, e.g., Simpson [26], such as the existence of maximal ideals in countable commutative rings or torsion subgroups in countable abelian groups. A more involved consequence of arithmetic comprehension using a more complex game, the Bolzano–Weierstrass theorem, will be discussed in the next section.

*Example 16 (No Injection* $(\mathbb{N} \to \mathbb{N}) \to \mathbb{N}$*)* Following [22] we show that a higher type instance of the product that produces a sequence of *functions* can be used to effectively prove that there is no injection $\Psi \colon (\mathbb{N} \to \mathbb{N}) \to \mathbb{N}$ in any model of functionals in which the unbounded product exists. This time we consider the drinkers paradox applied to the formulas $P_n(f^{\mathbb{N} \to \mathbb{N}}) := (n = \Psi f)$. Defining

$$\varepsilon_n p := \begin{cases} p(\mathbf{0}) \text{ if } n = \Psi(p(\mathbf{0})), \\ \mathbf{0}^{\mathbb{N}^{\mathbb{N}}} \quad \text{if } n \neq \Psi(p(\mathbf{0})), \end{cases}$$

we have

$$\forall n, p(n = \Psi(p(\varepsilon_n p)) \to n = \Psi(\varepsilon_n p)).$$

Let also $qF \overset{\mathbb{N} \to \mathbb{N}}{:=} \lambda k.(F_k(k) + 1)$, i.e. $qF$ is the diagonalisation of $F$, a function that differs from each $F_k$ at point $k$. Finally, let $\omega F = \Psi(qF)$, and define $F := \mathsf{EPS}^{\omega}_{\langle\rangle}(\varepsilon)(q)$. By Theorem 8 and the property of $\varepsilon_n$ above, we obtain

$$\forall n \leq \omega F(n = \Psi(qF) \to n = \Psi(F_n)). \tag{18}$$

Hence, taking $n = \omega F = \Psi(qF)$ in (18) we get

$$\Psi(qF) = \Psi(F_{\Psi(qF)}).$$

But by the definition of $q$ the functions $qF$ and $F_{\Psi(qF)}$ differ at position $\Psi(qF)$. Therefore, we have effectively constructed two functions $\alpha = qF$ and $\beta = F_{\Psi(qF)}$ such that $\Psi(\alpha) = \Psi(\beta)$ but $\alpha \neq \beta$, hence witnessing the non-injectivity of $\Psi$.

How is one supposed to think of the construction above as an optimal strategy in a game? Let us consider the game where at each round $n$ we must play a function $F_n \colon \mathbb{N}^{\mathbb{N}}$. The outcome function $qF$ is the diagonalisation of the sequence of functions (i.e. moves) $F_n$ played at each round $n$. Hence, $qF$ differs from $F_n$, for all $n$. The control function $\omega F$ says that the only relevant move in infinite play $F$ is the move at round $n = \Psi(qF)$. Finally, we must say what the "goal" of the game is. That follows from a definition of what a good strategy is, which in turn can be described by the selection functions $\varepsilon_n$ giving the local strategies at each stage $n$. From the selection functions $\varepsilon_n$ above, it follows that a good strategy behaves as follows: Observe what outcome one obtains by playing a default value $\mathbf{0}$ at round $n$. We will obtain an outcome $r\colon \mathbb{N}^{\mathbb{N}}$ such that either $\Psi(r) = n$ or $\Psi(r) \neq n$. If $\Psi(r) \neq n$, then we know that the current round is not relevant, and hence it is safe to play the default value $\mathbf{0}$. On the other hand, if $\Psi(r) = n$ we have in fact found a value for $\Psi^{-1}$ at point $n$, namely the outcome $r$. So, we play $r$ at round $n$. In this way, we construct a sequence $F_n$ such that either $F_n$ is an inverse of $\Psi$ at point $n$, or $F_n$ has a default value $\mathbf{0}$ but the round $n$ is not relevant in the play $F$ (as $\omega(F) = \Psi(qF) \neq n$). In the play $F$ which one obtains by following the strategy we will have the desired property that the move at the relevant round $n = \Psi(qF)$, i.e. $\alpha = F_n$, and the outcome $\beta = qF$ both map to $n$, but must be distinct since $\beta$ differs from all $F_n$.

## 5.3 The Product Versus Standard Modes of Recursion

A consequence of Theorem 14, and the fact that classical arithmetic and analysis can be formulated as classical logic plus finite and countable choice, respectively, is that the functional interpretation of classical proofs can be given entirely in terms of the product of selection functions. In fact we can reformulate Theorem 2 as follows:

**Theorem 17**

*(a)* $\mathsf{PA}^{\omega}$ *has a functional interpretation in primitive recursive arithmetic plus the finite product of selection functions (see [12] for details).*
*(b)* $\mathsf{PA}^{\omega} + \mathsf{AC}_0$ *and even* $\mathsf{PA}^{\omega} + \mathsf{DC}$ *have a functional interpretation in primitive recursive arithmetic plus the unbounded product of selection functions.*

It is natural then to ask how the product compares to those modes of recursion typically used in the functional interpretation of arithmetic and analysis.

Gödel's primitive recursive functionals of finite type [15] are the computational analogue of induction. In a similar fashion the finite product of selection functions can be seen as a computational analogue of finite choice, which is known to be equivalent to induction [24]. In [12] it is shown that the finite product is in fact equivalent to Gödel's primitive recursors over (a very weak fragment of ) primitive recursive arithmetic, and thus offers an alternative construction of system T.

Countable choice and dependent choice are typically interpreted using Spector's bar recursion [27]. By Theorem 14 (b) and (c) we see that these are also interpreted by the unbounded product, and in [8] it is shown that bar recursion is primitive recursively equivalent to the unbounded product. The whole picture is sketched in Fig. 1.

Of course our point here is that the advantage of using the product as opposed to the other modes of recursion is that it has a highly intuitive semantics, and witnesses extracted using the product often have a clear game-theoretic meaning. This is in

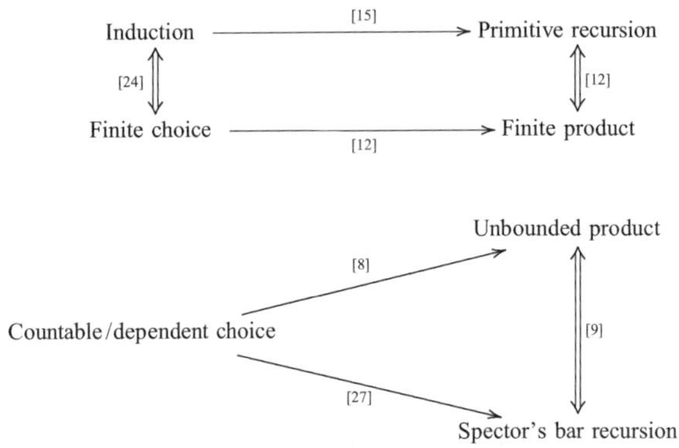


**Fig. 1** Functional interpretation of arithmetic and analysis

stark contrast to other methods, particularly Spector's bar recursion, which are often very difficult to comprehend on a semantic level.

# 6 Interpreting the Bolzano–Weierstrass Theorem

In this section we present a case study in which we formally extract a realizer for the functional interpretation of the Bolzano–Weierstrass theorem using the product of selection functions.

The constructive content of this theorem has been studied before, and in particular a detailed analysis using the Dialectica interpretation and Spector's bar recursion is given in [25].

Our aim here is to show that, even though the Bolzano–Weierstrass theorem is relatively complex from a logical point of view, one can extract from its proof a program whose behaviour can be clearly understood, at least on an informal level, in terms of optimal strategies in sequential games.

As in [25] we analyse a formal proof of the theorem which combines countable choice with *weak König's lemma*—the statement that all infinite binary trees $T$ have an infinite branch:

$$\mathsf{WKL} : \forall n \exists s^{\mathbb{B}^n} T(s) \to \exists \alpha \forall n \, T([\alpha](n))$$

where $T(s)$ is a boolean tree predicate, i.e. a prefix-closed predicate on finite sequences of 0's and 1's. We write $\mathbb{B}^*$ for finite sequences of booleans, and $\mathbb{B}^n$ for sequences of booleans of length $n$. We use the product of selection functions only to interpret the instance of countable choice used in the proof, as this forms the core of the extracted algorithm. Rather than also giving a detailed game-theoretic analysis of $\mathsf{WKL}$ using the product of selection functions (as done in [23]), for simplicity we make use of Howard's well-known realizer of $\mathsf{WKL}$ via binary bar recursion [17] and focus solely on the computational semantics of the main instance of countable choice.

## 6.1 The Bolzano–Weierstrass Theorem

The Bolzano–Weierstrass theorem states that any bounded sequence in $\mathbb{R}^n$ has a convergent subsequence. Here we restrict ourselves to sequences of rational $(x_i)$ in the unit interval $[0, 1]$, as our analysis can be readily generalised. In the language of formal analysis, this instance of the Bolzano–Weierstrass theorem can be written as

$$\mathsf{BW}(x_i) \quad : \quad \exists a^{\mathbb{B}^{\mathbb{N}}}, b^{\mathbb{N}^{\mathbb{N}}} \forall n (bn < b(n+1) \wedge x_{bn} \in I_{[a](n)}),$$

where for a finite sequence of booleans $s$ we define the interval

$$I_s := \left[ \sum_{i=0}^{|s|-1} \frac{s_i}{2^{i+1}}, \sum_{i=0}^{|s|-1} \frac{s_i}{2^{i+1}} + \frac{1}{2^{|s|}} \right]$$

for $|s| > 0$ and $I_{\langle\rangle} := [0, 1]$. We remark that for an infinite sequence $a : \mathbb{B}^{\mathbb{N}}$ we have $I_{[a](n)} \subseteq I_{[a](m)}$ whenever $n \geq m$, and this will be used later.

Intuitively, the infinite boolean sequence $a$ encodes a limit point

$$a^* := \sum_{i=0}^{\infty} \frac{a_i}{2^{i+1}}$$

of the sequence $(x_i)$, and $b$ defines a subsequence converging to this limit point, where $|x_{bn} - a^*| \leq 2^{-n}$ for all $n$. However, a detailed formalisation of $\mathsf{BW}$ would require us to make this intuition precise relative to some appropriate encoding of the real numbers, as in [25].

The functional interpretation of the Bolzano–Weierstrass theorem is given by

$$|\mathsf{BW}(x_i)|_{\psi}^{A,B} = \forall n \leq \psi AB (Bn < B(n+1) \wedge x_{Bn} \in I_{[A](n)}),$$

where, to easy readability, we are omitting the dependency of $A$ and $B$ on $\psi$. Intuitively, the interpreted theorem states that there exist arbitrary large finite approximations $B$ to a convergent subsequence, in the sense that $|x_{Bn} - A^*| \leq 2^{-n}$ for all $n \leq \psi AB$.

## 6.2 A Proof of $\mathsf{BW}(x_i)$

Assume an infinite sequence of rationals $(x_i)_{i \in \mathbb{N}}$ is fixed. Let us prove theorem $\mathsf{BW}(x_i)$, i.e.

$$\exists a^{\mathbb{B}^{\mathbb{N}}}, b^{\mathbb{N}^{\mathbb{N}}} \forall n (bn < b(n+1) \wedge x_{bn} \in I_{[a](n)})$$

directly using WKL and $\mathsf{AC}_0$. We define the predicate $T$ by

$$T(s^{\mathbb{B}^*}, k) \;:=\; |s| < k \wedge \exists i \in (|s|, k] \left(x_i \in I_s\right), \tag{19}$$

where $i \in (|s|, k]$ is shorthand for $|s| < i \leq k$.

**Lemma 18** *By countable choice* $\mathsf{AC}_0$ *there exists a function* $\beta \colon \mathbb{N} \to \mathbb{N}$ *such that*

$$\forall n, s^{\mathbb{B}^n}(\exists k T(s,k) \to T(s, \beta n)). \tag{20}$$

*Proof* By the drinkers paradox we have

$$\forall n, s^{\mathbb{B}^n} \exists l (\exists k T(s,k) \to T(s,l)).$$

By bounded collection over the finite quantifier $\forall s \in \mathbb{B}^n$, and using the fact that $T$ has the monotonicity property $T(s,k) \to T(s,k+l)$ we have

$$\forall n \exists L \forall s^{\mathbb{B}^n}(\exists k T(s,k) \to T(s,L)). \tag{21}$$

Finally, by countable choice we obtain $\beta$ satisfying (20). □

For the rest of the section let $\beta$ be a (classically constructed) function satisfying (20). For now we want to think of $\beta$ as given to us as an oracle. We will then show how one can effectively construct an approximation of $\beta$ which is good enough for the purposes of the proof.

**Corollary 19** *Define* $T^\beta(s) := T(s, \beta(|s|))$. *We have*

*(a)* $\exists k T(s,k) \leftrightarrow T^\beta(s)$, *for all n and* $s^{\mathbb{B}^n}$.
*(b)* $T^\beta(s)$ *is a decidable tree predicate (given the oracle* $\beta$*).*

*Proof (a)* Trivial. *(b)* $T^\beta(s)$ is clearly decidable in the given oracle $\beta$ and the sequence $(x_i)$. It remains to see that it is prefix-closed. Observe that $T(s * t, \beta(|s * t|)) \to T(s, \beta(|s * t|))$, by the definition of $T$ and the fact that $I_{s*t} \subseteq I_s$. Also, by (20), $T(s, \beta(|s * t|)) \to T(s, \beta(|s|))$. Combining the two we have $T^\beta(s * t) \to T^\beta(s)$. □

Hence, given such decidable binary tree predicate $T^\beta(s)$, if we can show that it has branches of arbitrary length then, by weak König's lemma, we have:

**Lemma 20** *The tree* $T^\beta$ *has an infinite branch. That is there exists a sequence* $a \colon \mathbb{B}^{\mathbb{N}}$ *such that*

$$\forall n \underbrace{\left(n < \beta n \wedge \exists i \in (n, \beta n]\left(x_i \in I_{[a](n)}\right)\right)}_{T^\beta([a](n))}. \tag{22}$$

*Proof* This follows from WKL applied to $T^\beta$, once we have shown that the tree $T^\beta(s)$ has branches of arbitrary length $n$. To see that, fix $n$ and let $s$, with $|s| = n$, be the index of the interval $I_s$ which contains $x_{n+1}$. This always exists as for each $n$ the set $(I_s)_{s\colon \mathbb{B}^n}$ covers the unit interval. Then, clearly we have

$$|s| < n + 1 \wedge x_{n+1} \in I_s$$

which implies $T(s, n+1)$. By (20) we obtain $T(s, \beta(|s|)) \equiv T^\beta(s)$. □

For future reference, we define the function $h\colon \mathbb{N} \to \mathbb{B}^*$ as that which, for each $n$, carries out a bounded search for the least $s \in \mathbb{B}^n$ such that $x_{n+1} \in I_s$. This function is primitive recursive in $(x_i)$ by decidability of $x_i \in I_s$, and satisfies $T(hn, n+1)$ for all $n$.

**Theorem 21 (Bolzano–Weierstrass)** *Given a sequence of rationals $(x_i)_{i\in\mathbb{N}}$, there exists $a\colon \mathbb{B}^{\mathbb{N}}$ and $b\colon \mathbb{N}^{\mathbb{N}}$ such that*

$$\forall n (bn < b(n+1) \wedge x_{bn} \in I_{[a](n)}). \tag{23}$$

*Proof* Let $a$ be as in Lemma 20. Define $b$ by

$$\begin{aligned} b0 &:= 0 \\ b(n+1) &:= \mu i \in (bn+1, \beta(bn+1)] \left(x_i \in I_{[a](bn+1)}\right). \end{aligned} \tag{24}$$

By (22), some $i \in (bn+1, \beta(bn+1)]$ satisfying $x_i \in I_{[a](bn+1)}$ always exists. Clearly we have $bn < b(n+1)$. Also $x_{b0} \in I_{\langle\rangle}$ and $x_{b(n+1)} \in I_{[a](bn+1)} \subseteq I_{[a](n+1)}$, since $bn + 1 \geq n + 1$ for all $n$ by induction on $n$. □

## 6.3 Lemma 18 via the Product of Selection Functions

The main ineffective step in the proof above is the existence of the oracle $\beta$ in Lemma 18. To give a computational interpretation of BW we want to be able to produce an approximation to $\beta$ in counterexample functions $\omega, q\colon \mathbb{N}^{\mathbb{N}} \to \mathbb{N}$ for $n$ and $k$ in (20), respectively, i.e.

$$\forall \omega, q \exists \beta \forall n \leq \omega\beta \forall s^{\mathbb{B}^n} (T(s, q\beta) \to T(s, \beta n)). \tag{25}$$

First we need to find selection functions $\delta_n\colon \mathbb{N}^{\mathbb{N}} \to \mathbb{N}$ witnessing the functional interpretation of (21):

$$\exists \delta \forall n, p \forall s^{\mathbb{B}^n} (T(s, p(\delta_n p)) \to T(s, \delta_n p)). \tag{26}$$

Since (21) is just the drinkers paradox combined with bounded collection, appropriate selection functions are constructed in a similar manner to (10), but with some additional primitive recursion.

**Lemma 22** *Let the selection function* $\delta_n \colon (\mathbb{N} \to \mathbb{N}) \to \mathbb{N}$ *be defined as*

$$\delta_n p := p^i(0) \tag{27}$$

*where*

$$i = \mu j \leq 2^n (\forall s^{\mathbb{B}^n} (T(s, p^{j+1}(0)) \to T(s, p^j(0)))).$$

*and* $\mu$ *is the bounded search operator. Then* $\delta$ *witnesses (26).*

*Proof* Note that (26) holds by definition once we show that the bounded search finds some $i \leq 2^n$ satisfying the given predicate. If this were not the case, then we would end up with a sequence $s_0, \ldots, s_{2^n}$ in $\mathbb{B}^n$ such that

$$T(s_i, p^{i+1}(0)) \wedge \neg T_0(s_i, p^i(0)).$$

But this would imply that the $s_i$ form $2^n + 1$ distinct elements of $\mathbb{B}^n$, which is a contradiction. □

**Theorem 23** *Given* $\omega \colon \mathbb{N}^{\mathbb{N}} \to \mathbb{N}$ *and* $q \colon \mathbb{N}^{\mathbb{N}} \to \mathbb{N}$ *let* $\beta \colon \mathbb{N}^{\mathbb{N}}$ *be defined as*

$$\beta := \mathsf{EPS}^{\omega}_{\langle\rangle}(\delta)(q),$$

*with* $\delta$ *as in (27). Then* $\beta$ *witnesses (25).*

*Proof* By Lemma 22, the $\delta_n$ as defined in (27) are such that

$$\forall n, p \forall s^{\mathbb{B}^n} (T(s, p(\delta_n p)) \to T(s, \delta_n p)).$$

For $n \leq \omega\beta$ and $p = p_{[\beta](n)}$, by Theorem 8 we have $\beta n = \delta_n p$ and $q\beta = p(\delta_n p)$, from which we can conclude $\forall s^{\mathbb{B}^n} (T(s, q\beta) \to T(s, \beta n))$. □

## 6.4 A Realizer for $\mathsf{BW}(x_i)$

Finally, we show how the instance of the product of selection functions in Theorem 23, used to interpret the crucial Lemma 18, forms the basis of an algorithm for constructing approximations to BW. We first interpret Lemma 20, making use of Howard's realizer for WKL using a weak, binary form of bar recursion, full details of which can be found in [17].

**Lemma 24** *For any counterexample function* $\varphi\colon \mathbb{B}^{\mathbb{N}} \times \mathbb{N}^{\mathbb{N}} \to \mathbb{N}$ *there exist* $\beta\colon \mathbb{N}^{\mathbb{N}}$ *and* $A\colon \mathbb{B}^{\mathbb{N}}$ *satisfying*

$$\forall n \leq \varphi A\beta \underbrace{\left(n < \beta n \wedge \exists i \in (n, \beta n] \left(x_i \in I_{[A](n)}\right)\right)}_{T([A](n),\beta n)}. \tag{28}$$

*Proof* Assume $\varphi\colon \mathbb{B}^{\mathbb{N}} \times \mathbb{N}^{\mathbb{N}} \to \mathbb{N}$ given. For any given $\gamma\colon \mathbb{N}^{\mathbb{N}}$ let $N^{\gamma}\colon \mathbb{B}^{*} \to \mathbb{N}$ be defined via Howard's binary bar recursion as

$$N^{\gamma}(t) := \begin{cases} 0 & \text{if } \exists s \preceq t\ (\varphi(\hat{s}, \gamma) < |s|), \\ 1 + \max\{N^{\gamma}(t * 0), N^{\gamma}(t * 1)\} & \text{otherwise.} \end{cases}$$

It is quite easy to show that $N^{\gamma}(\langle\rangle)$ is a minimum length for branches $t$ which guarantees that $\varphi(\hat{s}, \gamma) < |s|$ holds for some prefix $s \preceq t$. In other words, for any branch $t$ of length $N^{\gamma}(\langle\rangle)$ we must have $\exists s \preceq t\ (\varphi(\hat{s}, \gamma) < |s|)$. We first construct $\beta$ as in Theorem 23 where we explicitly define counterexamples $\omega$ and $q$ as

$$\omega\gamma := N^{\gamma}(\langle\rangle) \qquad q\gamma := N^{\gamma}(\langle\rangle) + 1$$

to obtain

$$\forall n \leq N^{\beta}(\langle\rangle)\ \forall s^{\mathbb{B}^n} (T(s, N^{\beta}(\langle\rangle) + 1) \to T(s, \beta n)). \tag{29}$$

Let $N = N^{\beta}(\langle\rangle)$ for $\beta$ as just defined. Hence, we have that for any $t$ such that $|t| \geq N$ there is some $s \preceq t$ with $\varphi(\hat{s}, \beta) < |s|$. We then define

$$A := \hat{s}, \qquad \text{where } s = \mu s \preceq h(N)\, (\varphi(\hat{s}, \beta) < |s|)$$

where $h$ is defined after the proof of Lemma 20. Now, by the definition of $h$ we have

$$\underbrace{N < N + 1 \wedge \exists i \in (N, N + 1](x_i \in I_{hN})}_{T(h(N),N+1)}.$$

Also, for $n \leq \varphi(A, \beta) = \varphi(\hat{s}, \beta)$ we must have $n < |s|$ since $\varphi(\hat{s}, \beta) < |s|$, and hence $[A](n) \preceq [A](|s|) = s \preceq h(N)$. Therefore, since $T(h(N), N + 1)$ holds we also have $T([A](n), N + 1)$. Finally, by (29) and the fact that $n \leq N$ we get $T([A](n), \beta n)$, and so we have proved (28). □

We are in a position now to effectively witness an approximation to the Bolzano–Weierstrass theorem, i.e. realise the functional interpretation of (23).

**Theorem 25** *For any counterexample function* $\psi\colon \mathbb{B}^{\mathbb{N}} \times \mathbb{N}^{\mathbb{N}} \to \mathbb{N}$ *there exists* $A\colon \mathbb{B}^{\mathbb{N}}$ *and* $B\colon \mathbb{N}^{\mathbb{N}}$ *satisfying*

$$\underbrace{\forall n \leq \psi AB(Bn < B(n+1) \wedge x_{Bn} \in I_{[A](n)})}_{|\mathsf{BW}(x_i)|_{\psi}^{A,B}}.$$

*Proof* For arbitrary $A$, $\beta$ let $b_{A,\beta}$ be defined as

$$b_{A,\beta}0 := 0, \tag{30}$$

$$b_{A,\beta}(n+1) := \mu i \in (b_{A,\beta}n + 1, \beta(b_{A,\beta}n + 1)] \left(x_i \in I_{[a](b_{A,\beta}n+1)}\right), \text{ else } b_{A,\beta}n + 1.$$

and define $\varphi\colon \mathbb{B}^{\mathbb{N}} \times \mathbb{N}^{\mathbb{N}} \to \mathbb{N}$ by $\varphi(A, \beta) := b_{A,\beta}(\psi(A, b_{A,\beta}))$. Then by Lemma 24 there exists $A\colon \mathbb{B}^{\mathbb{N}}$ and $\beta\colon \mathbb{N}^{\mathbb{N}}$ satisfying (28). Define $B := b_{A,\beta}$. Then clearly $x_{B0} = x_0 \in I_{\langle\rangle}$, and for $n < \psi AB$ we have (now dropping the parameters on $b_{A,\beta}$)

$$Bn + 1 = bn + 1 < b(\psi AB) + 1$$

and hence $Bn + 1 \leq b(\psi AB) = \varphi A\beta$. Therefore by (28) $\exists i \in (Bn{+}1, \beta(Bn{+}1)]$ $(x_i \in I_{[A](Bn+1)})$. But this implies that $x_{B(n+1)} \in I_{[A](Bn+1)} \subseteq I_{[A](n+1)}$. Hence, $x_{Bn} \in I_{[A](n)}$ for all $n \leq \psi AB$, and since $Bi < B(i+1)$ for all $i$ by definition, we're done. □

## 6.5 Understanding the Realizer for $\mathsf{BW}(x_i)$

The main component of our realizer is the construction of a sufficiently good approximation to the function $\beta$, whose job is to determine which intervals $I_s$ are inhabited by some $x_i$ with $i > |s|$. This approximation forms an optimal play in the sequential game given by $(\varepsilon, q, \omega)$, where the role of the selection function $\varepsilon_n$ is fixed: namely to decide which intervals in the partition of size $2^n$ to include in the approximation.

Given that an initial segment $[\beta](n)$ has already been suggested, the strategy implemented by $\varepsilon_n$ is to initially try to include no intervals in $(I_s)_{|s|=n}$ in the approximation by proposing that $\beta(n) = y_0 = 0$ and then testing whether or not $\exists i \in (n, q([\beta](n) * y_0 * \gamma_0)](x_i \in I_s)$ holds for some $s$ of length $n$, where $\gamma_0$ is the optimal continuation of $[\beta](n) * y_0$. If it does, then the proposed value is not good enough, as the selection function has discovered some intervals inhabited by the $x_i$.

It therefore changes its mind and instead proposes $\beta(n) = y_1 = q([\beta](n) * y_0 * \gamma_0)$. It then tests whether or not $\exists i \in (n, q([\beta](n) * y_1 * \gamma_1)](x_i \in I_s)$ holds for some $s$ not already discovered, and if so reflects this new discovery by updating to $\beta(n) = y_2 = q([\beta](n) * y_1 * \gamma_1)$, and so on. Eventually, this process terminates on the $i$th update for some $i \leq 2^n$, and we have

$$\exists i \in (n, q([\beta](n) * y_i * \gamma_i)](x_i \in I_s) \rightarrow \exists i \in (n, y_i](x_i \in I_s)$$

for all $s \in \mathbb{B}^n$, and so $\beta(n) = y_i$ is a sufficiently good approximation for $\beta$ at point $n$. The remaining parameters $\omega$ and $q$ for the game are built using Howard's realizer for WKL to ensure that this approximation to $\beta$ has discovered enough information about the distribution of $(x_i)$ that it has already found an approximation to a convergent subsequence relative to $\psi$ (which in turn takes the form of an intermediate counterexample function $\varphi$).

While a lot more could certainly be said about the behaviour of our realizer, our aim here has been simply to convince the reader that while constructing a realizer for the functional interpretation of the Bolzano–Weierstrass theorem takes a reasonable amount of work, the game theoretic intuition behind the product of selection functions allows us to gain a better understanding of the key operational features of this realizer. We leave a more detailed analysis of the extracted algorithm to future work.

## 7 Further Remarks

In this article we have shown that the language of selection functions and sequential games underlies the Dialectica interpretation of classical proofs in a fundamental way, and we have used the product of selection functions to construct a concise and intuitive computational interpretation of some well-known theorems.

Our motivation has been a more qualitative understanding of functional interpretations, as a response to the fact that formal proof-theoretic methods are becoming increasingly relevant in modern mathematics. We have shown that the product of selection functions is a fundamental construction behind the Dialectica interpretation of classical proofs, and we hope to have convinced the reader that in practice it leads to extracted programs that have an expressive reading in terms of optimal strategies in sequential games.

There is a lot of work to be done towards understanding formal proof-theoretic techniques in mathematical terms, and the question of adapting and refining functional interpretations so that they can be seen as intelligent translations on mathematical proofs as opposed to just syntactic translations on logic sentences forms a very interesting area of research. The authors believe that there are several potentially fruitful avenues for further research.

One is to explore in more detail the link between the Dialectica interpretation and the closely related "correspondence principle" implicitly used in areas like ergodic theory. In particular, the finitary version of theorems discussed by Tao in [28, 29] are strictly speaking related to the *monotone* variant of the Dialectica interpretation, which extracts uniform bounds, or majorants, for realizers of the interpretation. It would be interesting to try to gain an understanding of how the product of selection functions can be used to extract realizers for the monotone interpretation and therefore produce constructive proofs of theorems that can truly be seen as "finitizations" in the sense of Tao.

Another interesting issue is the *efficiency* of the product of selection functions in producing a realizer. For instance, a quick analysis of Example 16 shows that if $\Psi(\lambda n.0) = \Psi(\lambda n.1)$, our program potentially misses this obvious counterexample and eventually produces a much more elaborate one. This highlights the fact that while the product is indeed an intuitive realizer for the axiom of choice, it is far from optimal and refinements of the procedure or even completely different recursion schemata may be more suited to interpreting specific principles, such as the schema of open recursion proposed by Berger in [4] for the realizability interpretation of the minimal bad sequence argument.

A related question is the efficiency of the Dialectica interpretation itself, and the comparison of extracted programs to those obtained using other proof interpretations such as modified realizability. In particular, the modified realizability interpretation of choice has an interesting realizer given by Berardi et al. [3] that was also shown to have a natural game theoretic reading.

We conclude with the remark that our paper belongs to a larger body of recent work on the theory of selection functions and sequential games by the first author and M. Escardó, starting with [9] and surveyed in [11]. Particularly relevant is [23], in which the product of selection functions is used to extract a game theoretic realizer for Ramsey's theorem. This can be seen as an extended case study illustrating the methods employed here.

**Acknowledgements** The first author gratefully acknowledges support of the Royal Society (grant 516002.K501/RH/kk). The second author acknowledges the support of an EPSRC doctoral training grant.

## References

1. J. Avigad, S. Feferman, Gödel's functional ("Dialectica") interpretation, in *Handbook of Proof Theory*, ed. by S.R. Buss. Studies in Logic and the Foundations of Mathematics, vol. 137. (Amsterdam, North Holland, 1998), pp. 337–405
2. J. Avigad, P. Gerhardy, H. Towsner, Local stability of ergodic averages. Trans. Am. Math. Soc. **362**, 261–288 (2010)
3. S. Berardi, M. Bezem, T. Coquand, On the computational content of the axiom of choice. J. Symb. Log. **63**(2), 600–622 (1998)
4. U. Berger, A computational interpretation of open induction. *Proceedings of the Nineteenth Annual IEEE Symposium on Logic in Computer Science*, 2004, ed. by F. Titsworth, pp. 326–334
5. U. Berger, P. Oliva, Modified bar recursion and classical dependent choice. Lect. Notes Log. **20**, 89–107 (2005)
6. U. Berger, P. Oliva, Modified bar recursion. Math. Struct. Comput. Sci. **16**, 163–183 (2006)
7. U. Berger, H. Schwichtenberg, W. Buchholz, Refined program extraction from classical proofs. Ann. Pure Appl. Log. **114**, 3–25 (2002)
8. M.H. Escardó, P. Oliva, Computational interpretations of analysis via products of selection functions, in *Computability in Europe 2010, LNCS 6158*, ed. by F. Ferreira, B. Lowe, E. Mayordomo, L.M. Gomes (Springer, New York, 2010), pp. 141–150

9. M.H. Escardó, P. Oliva, Selection functions, bar recursion, and backward induction. Math. Struct. Comput. Sci. **20**(2), 127–168 (2010)
10. M. Escardó, P. Oliva, Bar recursion and products of selection functions, J. Sym. Log. **80**(1), 1–28 (2015)
11. M.H. Escardó, P. Oliva, Sequential games and optimal strategies. Roy. Soc. Proc. A **467**, 1519–1545 (2011)
12. M.H. Escardó, P. Oliva, T. Powell, System T and the product of selection functions, in *Proceedings of CSL'11*, vol. 12 (LIPIcs, 2011), pp. 233–247
13. H. Friedman, Classically and intuitionistically provably recursive functions, in *Higher Set Theory*, ed. by D. Scott, G. Müller, Lecture Notes in Mathematics vol. 669 (Springer, Berlin, 1978), pp. 21–28
14. J. Gaspar, U. Kohlenbach, On Tao's "finitary" infinite pigeonhole principle. J. Symb. Log. **75**(1), 355–371 (2010)
15. K. Gödel, Über eine bisher noch nicht benützte Erweiterung des finiten Standpunktes. Dialectica **12**, 280–287 (1958)
16. W.A. Howard, Functional interpretation of bar induction by bar recursion. Compos. Math. **20**, 107–124 (1968)
17. W.A. Howard, Ordinal analysis of bar recursion of type zero. Compos. Math. **42**, 105–119 (1981)
18. U. Kohlenbach, *Applied Proof Theory: Proof Interpretations and their Use in Mathematics*. Monographs in Mathematics (Springer, New York, 2008)
19. G. Kreisel, On the interpretation of non-finitist proofs, part I. J. Symb. Log. **16**, 241–267 (1951)
20. G. Kreisel, On the interpretation of non-finitist proofs, part II: Interpretation of number theory. J. Symb. Log. **17**, 43–58 (1952)
21. H. Luckhardt, *Extensional Gödel Functional Interpretation: A Consistency Proof of Classical Analysis*, Lecture Notes in Mathematics. vol. 306 (Springer, Berlin, 1973)
22. P. Oliva, Understanding and using Spector's bar recursive interpretation of classical analysis, *Proceedings of CiE'2006, LNCS 3988*, ed. by A. Beckmann, U. Berger, B.Löwe, J.V. Tucker (Springer, New York, 2006), pp. 423–234
23. P. Oliva, T. Powell, A constructive interpretation of Ramsey's theorem via the product of selection functions. Math. Struct. Comput. Sci. (2014), http://journals.cambridge.org/action/displayAbstract?fromPage=online&aid=9414884&fileId=S0960129513000340. doi:http://dx.doi.org/10.1017/S0960129513000340, 24 p.
24. C. Parsons, On a number theoretic choice schema and its relation to induction. *Intuitionism and Proof Theory: Proceedings of the Summer Conference at Buffalo, N.Y. 1968*, ed. by A. Kino, J. Myhill, R.E. Vesley (Amserdam, North Holland, 1970), pp. 459–473
25. P. Safarik, U. Kohlenbach, On the computational content of the Bolzano-Weierstrass principle. Math. Log. Q. **56**, 508–532 (2010)
26. S.G. Simpson, *Subsystems of Second Order Arithmetic*. Perspectives in Mathematical Logic (Springer, Berlin, 1999)
27. C. Spector, Provably recursive functionals of analysis: a consistency proof of analysis by an extension of principles in current intuitionistic mathematics, in *Recursive Function Theory: Proc. Symposia in Pure Mathematics*, vol. 5, ed. by F.D.E. Dekker (American Mathematical Society, Providence, 1962), pp. 1–27
28. T. Tao, Soft analysis, hard analysis, and the finite convergence principle. Essay, http://terrytao.wordpress.com/2007/05/23/soft-analysis-hard-analysis-and-the-finite-convergence-principle/, 2007
29. T. Tao, The correspondence principle and finitary ergodic theory. Essay, http://terrytao.wordpress.com/2008/08/30/the-correspondence-principle-and-finitary-ergodic-theory/, 2008

# Well-Ordering Principles and Bar Induction

**Michael Rathjen and Pedro Francisco Valencia Vizcaíno**

**Abstract** In this paper we show that the existence of $\omega$-models of bar induction is equivalent to the principle saying that applying the Howard–Bachmann operation to any well-ordering yields again a well-ordering.

## 1 Introduction

This paper will be concerned with a particular $\Pi^1_2$ statement of the form

$$\mathbf{WOP}(f): \qquad \forall X\,[\mathrm{WO}(\mathfrak{X}) \to \mathrm{WO}(f(\mathfrak{X}))] \tag{1}$$

where $f$ is a standard proof-theoretic function from ordinals to ordinals and $\mathrm{WO}(\mathfrak{X})$ stands for '$\mathfrak{X}$ is a well-ordering'. There are by now several examples of functions $f$ familiar from proof theory where the statement $\mathbf{WOP}(f)$ has turned out to be equivalent to one of the theories of reverse mathematics over a weak base theory (usually $\mathbf{RCA}_0$). The first explicit example appears to be due to Girard [7, Theorem 5.4.1] (see also [8]). However, it is also implicit in Schütte's proof of cut elimination for $\omega$-logic [15] and ultimately has its roots in Gentzen's work, namely in his first unpublished consistency proof,[1] where he introduced the notion of a "Reduziervorschrift" [6, p. 102] for a sequent. The latter is a well-founded tree built bottom-up via "Reduktionsschritte", starting with the given sequent and passing up from conclusions to premises until an axiom is reached.

**Theorem 1.1** *Over* $\mathbf{RCA}_0$ *the following are equivalent:*

*(i) Arithmetical comprehension.*
*(ii)* $\forall \mathfrak{X}\,[\mathrm{WO}(\mathfrak{X}) \to \mathrm{WO}(2^{\mathfrak{X}})]$.

[1]The original German version was finally published in 1974 [6]. An earlier English translation appeared in 1969 [5].

M. Rathjen (✉) • P.F. Valencia Vizcaíno
School of Mathematics, University of Leeds, Leeds LS2 JT, England
e-mail: M.Rathjen@leeds.ac.uk



R. Kahle, M. Rathjen (eds.), *Gentzen's Centenary*,
DOI 10.1007/978-3-319-10103-3_19

Another characterization from [7, Theorem 6.4.1], shows that arithmetical comprehension is equivalent to Gentzen's Hauptsatz (cut elimination) for $\omega$-logic. Connecting statements of form (1) to cut elimination theorems for infinitary logics will also be a major tool in this paper.

There are several more recent examples of such equivalences that have been proved by recursion-theoretic as well as proof-theoretic methods. These results give characterizations of the form (1) for the theories $\mathbf{ACA}_0^+$ and $\mathbf{ATR}_0$, respectively, in terms of familiar proof-theoretic functions. $\mathbf{ACA}_0^+$ denotes the theory $\mathbf{ACA}_0$ augmented by an axiom asserting that for any set $X$ the $\omega$-th jump in $X$ exists while $\mathbf{ATR}_0$ asserts the existence of sets constructed by transfinite iterations of arithmetical comprehension. $\alpha \mapsto \varepsilon_\alpha$ denotes the usual $\varepsilon$ function while $\varphi$ stands for the two-place Veblen function familiar from predicative proof theory (cf. [16]). Definitions of the familiar subsystems of reverse mathematics can be found in [17].

**Theorem 1.2 (Afshari and Rathjen [1]; Marcone and Montalbán [9])** *Over* $\mathbf{RCA}_0$ *the following are equivalent:*

*(i)* $\mathbf{ACA}_0^+$.
*(ii)* $\forall \mathfrak{X}\,[\mathrm{WO}(\mathfrak{X}) \to \mathrm{WO}(\varepsilon_{\mathfrak{X}})]$.

**Theorem 1.3 (Friedman [4]; Rathjen and Weiermann [13]; Marcone and Montalbán [9])** *Over* $\mathbf{RCA}_0$ *the following are equivalent:*

*(i)* $\mathbf{ATR}_0$.
*(ii)* $\forall \mathfrak{X}\,[\mathrm{WO}(\mathfrak{X}) \to \mathrm{WO}(\varphi\mathfrak{X}0)]$.

There is often another way of characterizing statements of the form (1) by means of the notion of countable coded $\omega$-model.

**Definition 1.4** Let $T$ be a theory in the language of second order arithmetic, $\mathcal{L}_2$. A *countable coded $\omega$-model of $T$* is a set $W \subseteq \mathbb{N}$, viewed as encoding the $\mathcal{L}_2$-model

$$\mathbb{M} = (\mathbb{N}, \mathcal{S}, \in, +, \cdot, 0, 1, <)$$

with $\mathcal{S} = \{(W)_n \mid n \in \mathbb{N}\}$ such that $\mathbb{M} \models T$ when the second order quantifiers are interpreted as ranging over $\mathcal{S}$ and the first order part is interpreted in the standard way (where $(W)_n = \{m \mid \langle n, m\rangle \in W\}$ with $\langle\,,\rangle$ being some primitive recursive coding function).

If $T$ has only finitely many axioms, it is obvious how to express $\mathbb{M} \models T$ by just translating the second order quantifiers $QX \ldots X \ldots$ in the axioms by $Qx \ldots (W)_x \ldots$. If $T$ has infinitely many axioms, one needs to formalize Tarski's truth definition for $\mathbb{M}$. This definition can be made in $\mathbf{RCA}_0$ as is shown in [17, Definitions II.8.3 and VII.2]. Some more details will be provided in Remark 1.9.

We write $X \in W$ if $\exists n\ X = (W)_n$.

The alternative characterizations alluded to above are as follows:

**Theorem 1.5** *Over* $\mathbf{RCA}_0$ *the following are equivalent:*

*(i)* $\forall \mathfrak{X}\ [\mathrm{WO}(\mathfrak{X}) \rightarrow \mathrm{WO}(\varepsilon_{\mathfrak{X}})]$ *is equivalent to the statement that every set is contained in a countable coded $\omega$-model of* $\mathbf{ACA}$.
*(ii)* $\forall \mathfrak{X}\ [\mathrm{WO}(\mathfrak{X}) \rightarrow \mathrm{WO}(\varphi\mathfrak{X}0)]$ *is equivalent to the statement that every set is contained in a countable coded $\omega$-model of* $\Delta^1_1$-$\mathbf{CA}$ *(or* $\Sigma^1_1$-$\mathbf{DC}$*).*

*Proof* See [12, Corollary 1.8]. □

Whereas Theorem 1.5 has been established independently by recursion-theoretic and proof-theoretic methods, there is also a result that has a very involved proof and so far has only been shown by proof theory. It connects the well-known $\Gamma$-function (cf. [16]) with the existence of countable coded $\omega$-models of $\mathbf{ATR}_0$.

**Theorem 1.6 (Rathjen [12, Theorem 1.4])** *Over* $\mathbf{RCA}_0$ *the following are equivalent:*

*(i)* $\forall \mathfrak{X}\ [\mathrm{WO}(\mathfrak{X}) \rightarrow \mathrm{WO}(\Gamma_{\mathfrak{X}})]$.
*(ii)* *Every set is contained in a countable coded $\omega$-model of* $\mathbf{ATR}_0$.

The tools from proof theory employed in the above theorems involve search trees and Gentzen's cut elimination technique for infinitary logic with ordinal bounds. One could perhaps generalize and say that every cut elimination theorem in ordinal-theoretic proof theory encapsulates a theorem of this type.

The proof-theoretic ordinal functions that figure in the foregoing theorems are all familiar from so-called predicative or meta-predicative proof theory. Thus far a function from genuinely impredicative proof theory is missing. The first such function that comes to mind is of the Bachmann–Howard type. It was conjectured in [14] (Conjecture 7.2) that the pertaining principle (1) would be equivalent to the existence of countable coded $\omega$-models of bar induction, $\mathbf{BI}$. The conjecture is by and large true as will be shown in this paper, however, the relativization of the Bachmann–Howard construction allows for two different approaches, yielding principles of different strength. As it turned out, only the strongest one is equivalent to the existence of $\omega$-models of $\mathbf{BI}$. We now proceed to state the main result of this paper. Unexplained notions will be defined shortly.

**Theorem 1.7** *Over* $\mathbf{RCA}_0$ *the following are equivalent:*

*(i)* $\mathbf{RCA}_0 +$ *Every set $X$ is contained in a countable coded $\omega$-model of* $\mathbf{BI}$.
*(ii)* $\forall \mathfrak{X}\ [\mathrm{WO}(\mathfrak{X}) \rightarrow \mathrm{WO}(\vartheta_{\mathfrak{X}})]$.

Below we shall refer to Theorem 1.7 as the **Main Theorem**.

## 1.1 A Brief Outline of the Paper

Section 1.2 contains a detailed definition of the theory **BI**. Section 2 introduces a relativized version of the Howard–Bachmann ordinal representation system, i.e. given a well-ordering $\mathfrak{X}$, one defines a new well-ordering $\vartheta_{\mathfrak{X}}$ of Howard–Bachmann type which incorporates $\mathfrak{X}$. Section 3 proofs the direction $(i) \Rightarrow (ii)$ of Theorem 1.7. With Sect. 4 the proof of Theorem 1.7 $(ii) \Rightarrow (ii)$ commences. It introduces the crucial notion of a deduction chain for a given set $Q \subseteq \mathbb{N}$. The set of deduction chains forms a tree $\mathcal{D}_Q$. It is shown that from an infinite branch of this tree one can construct a countable coded $\omega$-model of **BI** which contains $Q$. As a consequence, it remains to consider the case when $\mathcal{D}_Q$ does not contain an infinite branch, i.e. when $\mathcal{D}_Q$ is a well-founded tree. Then the Kleene–Brouwer ordering of $\mathcal{D}_Q$, $\mathfrak{X}$, is a well-ordering and, by the well-ordering principle (ii), $\vartheta_{\mathfrak{X}}$ is a well-ordering, too. It will then be revealed that $\mathcal{D}_Q$ can be viewed as a skeleton of a proof $\mathcal{D}^*$ of the empty sequent in an infinitary proof system $T_Q^*$ with Buchholz' $\Omega$-rule. However, with the help of transfinite induction over $\vartheta_{\mathfrak{X}}$ it can be shown that all cuts in $\mathcal{D}^*$ can be removed, yielding a cut-free derivation of the empty sequent. As this cannot be, the final conclusion reached is that $\mathcal{D}_Q$ must contain an infinite branch, whence there is a countable coded $\omega$-model of **BI** containing $Q$, thereby completing the proof of Theorem 1.7 $(ii) \Rightarrow (i)$.

## 1.2 The Theory BI

In this subsection we introduce the theory **BI**. To set the context, we fix some notations. The language of second order arithmetic, $\mathcal{L}_2$, consists of free numerical variables $a, b, c, d, \ldots$, bound numerical variables $x, y, z, \ldots$, free set variables $U, V, W, \ldots$, bound set variables $X, Y, Z, \ldots$, the constant 0, a symbol for each primitive recursive function, and the symbols $=$ and $\in$ for equality in the first sort and the elementhood relation, respectively. The *numerical terms* of $\mathcal{L}_2$ are built up in the usual way; $r, s, t, \ldots$ are syntactic variables for them. *Formulas* are obtained from atomic formulas $s = t$, $s \in U$ and negated atomic formulas $\neg s = t, \neg s \in U$ by closing under $\wedge, \vee$ and quantification $\forall x, \exists x, \forall X, \exists X$ over both sorts; so we stipulate that formulas are in negation normal form.

The classes of $\Pi^1_2$- and $\Sigma^1_n$-*formulae* are defined as usual (with $\Pi^1_0 = \Sigma^1_0 = \bigcup\{\Pi^0_n : n \in \mathbb{N}\}$). $\neg A$ is defined by de Morgan's laws; $A \rightarrow B$ stands for $\neg A \vee B$. All theories in $\mathcal{L}_2$ will be assumed to contain the axioms and rules of classical two sorted predicate calculus, with equality in the first sort. In addition, it will be assumed that they comprise the system $\mathbf{ACA}_0$. $\mathbf{ACA}_0$ contains all axioms of elementary number theory, i.e. the usual axioms for 0, $'$ (successor), the defining equations for the primitive recursive functions, the *induction axiom*

$$\forall X\, [0 \in X \ \wedge \ \forall x(x \in X \rightarrow x' \in X) \rightarrow \forall x(x \in X)],$$

and all instances of *arithmetical comprehension*

$$\exists Z\, \forall x[x \in Z \leftrightarrow F(x)],$$

where $F(a)$ is an *arithmetic formula*, i.e. a formula without set quantifiers.

For a 2-place relation $\prec$ and an arbitrary formula $F(a)$ of $\mathcal{L}_2$ we define

$\mathrm{Prog}(\prec, F) := (\forall x)[\forall y(y \prec x \rightarrow F(y)) \rightarrow F(x)]$ (*progressiveness*)
$\mathbf{TI}(\prec, F) := \mathrm{Prog}(\prec, F) \rightarrow \forall x F(x)$ (*transfinite induction*)
$\mathrm{WF}(\prec) := \forall X\, \mathbf{TI}(\prec, X) :=$
$\forall X(\forall x[\forall y(y \prec x \rightarrow y \in X)) \rightarrow x \in X] \rightarrow \forall x[x \in X])$ (*well-foundedness*).

Let $\mathcal{F}$ be any collection of formulae of $\mathcal{L}_2$. For a 2-place relation $\prec$ we will write $\prec \in \mathcal{F}$, if $\prec$ is defined by a formula $Q(x, y)$ of $\mathcal{F}$ via $x \prec y := Q(x, y)$.

**Definition 1.8** BI denotes the bar induction scheme, i.e. all formulae of the form

$$\mathrm{WF}(\prec) \rightarrow \mathbf{TI}(\prec, F),$$

where $\prec$ is an arithmetical relation (set parameters allowed) and $F$ is an arbitrary formula of $\mathcal{L}_2$.

By **BI** we shall refer to the theory $\mathbf{ACA}_0 + \mathrm{BI}$.

*Remark 1.9* The statement of the main Theorem 1.7 uses the notion of a countable coded $\omega$-model of **BI**. As the stated equivalence is claimed to be provable in $\mathbf{RCA}_0$, a few comments on how this is formalized in this weak base theory are in order. The notion of a countable coded $\omega$-model can be formalized in $\mathbf{RCA}_0$ according to [17, Definition VII.2.1]. Let $\mathbb{M}$ be a countable coded $\omega$-model. Since **BI** is not finitely axiomatizable we have to quantify over all axioms of **BI** to express that $\mathbb{M} \models \mathbf{BI}$. The axioms of **BI** (or rather their Gödel numbers) clearly form a primitive recursive set, $Ax(\mathbf{BI})$. To express $\mathbb{M} \models \phi$ for $\phi \in Ax(\mathbf{BI})$ we use the notion of a *valuation for* $\phi$ from [17, Definition VII.2.1]. A valuation $f$ for $\phi$ is a function from the set of subformulae of $\phi$ into the set $\{0, 1\}$ obeying the usual Tarski truth conditions. Thus we write $\mathbb{M} \models \phi$, if there exists a valuation $f$ for $\phi$ such that $f(\phi) = 1$. Whence $\mathbb{M} \models \mathbf{BI}$ is defined by $\forall \phi \in Ax(\mathbf{BI})\ \mathbb{M} \models \phi$.

## 2 Relativizing the Howard–Bachmann Ordinal

In this section we show how to relativize the construction that leads to the Howard–Bachmann ordinal to an arbitrary countable well-ordering. To begin with, mainly to foster intuitions, we provide a set-theoretic definition working in **ZFC**. This will then be followed by a purely formal definition that can be made in $\mathbf{RCA}_0$.

Throughout this section, we fix a countable well-ordering $\mathfrak{X} = (X, <_X)$ without a maximum element, i.e., an ordered pair $\mathfrak{X} = (X, <_X)$, where $X$ is a set of natural numbers, $<_X$ is a well-ordering relation on $X$, and $\forall v \in X\, \exists u \in X\; v <_X u$. We write $|\mathfrak{X}|$ for $X$.

Firstly, we need some ordinal-theoretic background. Let ON be the class of ordinals. Let AP $:= \{\xi \in \mathrm{ON} : \exists \eta \in \mathrm{ON}[\xi = \omega^\eta]\}$ be the class of additive principal numbers and let E $:= \{\xi \in \mathrm{ON}: \xi = \omega^\xi\}$ be the class of $\varepsilon$-numbers which is enumerated by the function $\lambda\xi.\varepsilon_\xi$.

We write $\alpha =_{NF} \omega^{\alpha_1}+\cdots+\omega^{\alpha_n}$ if $\alpha = \omega^{\alpha_1}+\cdots+\omega^{\alpha_n}$ and $\alpha > \alpha_1 \geq \cdots \alpha_n$. Note that by Cantor's normal form theorem, for every $\alpha \notin \mathrm{E} \cup \{0\}$, there are uniquely determined ordinals $\alpha_1, \ldots, \alpha_n$ such that $\alpha =_{NF} \omega^{\alpha_1} + \cdots + \omega^{\alpha_n}$.

Let $\Omega := \aleph_1$. For $u \in |\mathfrak{X}|$, let $\mathfrak{E}_u$ be the $u^{th}$ $\varepsilon$-number $> \Omega$. Thus, if $u_0$ is the smallest element of $|\mathfrak{X}|$, then $\mathfrak{E}_{u_0}$ is the least $\varepsilon$-number $> \Omega$, and in general, for $u \in |\mathfrak{X}|$ with $u_0 <_X u$, $\mathfrak{E}_u$ is the least $\varepsilon$-number $\rho$ such that $\forall v <_X u\; \mathfrak{E}_v < \rho$.

In what follows we shall only be interested in ordinals below $\sup_{u\in X} \mathfrak{E}_u$. Henceforth, unless indicated otherwise, any ordinal will be assumed to be smaller than that ordinal.

For any such $\alpha$ we define the set $E_\Omega(\alpha)$ which consists of the $\varepsilon$-numbers below $\Omega$ which are needed for the unique representation of $\alpha$ in Cantor normal form recursively as follows:

1. $E_\Omega(0) := E_\Omega(\Omega) := \emptyset$ and $E_\Omega(\mathfrak{E}_u) := \emptyset$ for $u \in |\mathfrak{X}|$.
2. $E_\Omega(\alpha) := \{\alpha\}$, if $\alpha \in E \cap \Omega$.
3. $E_\Omega(\alpha) := E_\Omega(\alpha_1) \cup \cdots \cup E_\Omega(\alpha_n)$ if $\alpha =_{NF} \omega^{\alpha_1} + \cdots + \omega^{\alpha_n}$.

Let $\alpha^* := \max(E_\Omega(\alpha) \cup \{0\})$.

We define sets of ordinals $C_\mathfrak{X}(\alpha,\beta), C^n_\mathfrak{X}(\alpha,\beta)$, and ordinals $\vartheta\alpha$ by main recursion on $\alpha < \sup_{u\in X} \mathfrak{E}_u$ and subsidiary recursion on $n < \omega$ (for $\beta < \Omega$) as follows.

(C0) $\mathfrak{E}_u \in C^n_\mathfrak{X}(\alpha,\beta)$ for all $u \in |\mathfrak{X}|$.
(C1) $\{0,\Omega\} \cup \beta \subseteq C^n_\mathfrak{X}(\alpha,\beta)$.
(C2) $\gamma_1,\ldots,\gamma_n \in C^n_\mathfrak{X}(\alpha,\beta) \wedge \xi =_{NF} \omega^{\gamma_1} + \cdots + \omega^{\gamma_n} \implies \xi \in C^{n+1}_\mathfrak{X}(\alpha,\beta)$.
(C3) $\delta \in C^n_\mathfrak{X}(\alpha,\beta) \cap \alpha \implies \vartheta\delta \in C^{n+1}_\mathfrak{X}(\alpha,\beta)$.
(C4) $C_\mathfrak{X}(\alpha,\beta) := \bigcup\{C^n_\mathfrak{X}(\alpha,\beta): n < \omega\}$.
(C5) $\vartheta\alpha := \min\{\xi < \Omega : C_\mathfrak{X}(\alpha,\xi) \cap \Omega \subseteq \xi \wedge \alpha \in C_\mathfrak{X}(\alpha,\xi)\}$ if there exists an ordinal $\xi < \Omega$ such that $C_\mathfrak{X}(\alpha,\xi) \cap \Omega \subseteq \xi$ and $\alpha \in C_\mathfrak{X}(\alpha,\xi)$. Otherwise $\vartheta\alpha$ will be undefined.
We will shortly see that $\vartheta\alpha$ is always defined (Lemma 2.2).

*Remark 2.1* The definition of $\vartheta$ originated in [10]. An ordinal representation system based on $\vartheta$ was used in [11] to determine the proof-theoretic strength of fragments of Kripke–Platek set theory and in [13] it was used to characterize the strength of Kruskal's theorem.

**Lemma 2.2** *$\vartheta\alpha$ is defined for every $\alpha < \sup_{u\in X} \mathfrak{E}_u$.*

*Proof* Let $\beta_0 := \alpha^* + 1$. Then $\alpha \in C_\mathfrak{X}(\alpha,\beta_0)$ via (C1) and (C2). Since the cardinality of $C_\mathfrak{X}(\alpha,\beta)$ is less than $\Omega$ there exists a $\beta_1 < \Omega$ such that $C_\mathfrak{X}(\alpha,\beta_0) \cap$

$\Omega \subseteq \beta_1$. Similarly there exists for each $\beta_n < \Omega$ (which is constructed recursively) a $\beta_{n+1} < \Omega$ such that $C_{\mathfrak{X}}(\alpha, \beta_n) \cap \Omega \subseteq \beta_{n+1}$. Let $\beta := \sup\{\beta_n : n < \omega\}$. Then $\alpha \in C_{\mathfrak{X}}(\alpha, \beta)$ and $C_{\mathfrak{X}}(\alpha, \beta) \cap \Omega \subseteq \beta < \Omega$. Therefore $\vartheta\alpha \leq \beta < \Omega$. □

**Lemma 2.3**

1. *$\vartheta\alpha \in E$,*
2. *$\alpha \in C_{\mathfrak{X}}(\alpha, \vartheta\alpha)$,*
3. *$\vartheta\alpha = C_{\mathfrak{X}}(\alpha, \vartheta\alpha) \cap \Omega$, and $\vartheta\alpha \notin C_{\mathfrak{X}}(\alpha, \vartheta\alpha)$,*
4. *$\gamma \in C_{\mathfrak{X}}(\alpha, \beta) \iff \gamma^* \in C_{\mathfrak{X}}(\alpha, \beta)$,*
5. *$\alpha^* < \vartheta\alpha$,*
6. *$\vartheta\alpha = \vartheta\beta \implies \alpha = \beta$,*
7. $$\begin{aligned} \vartheta\alpha < \vartheta\beta &\iff (\alpha < \beta \wedge \alpha^* < \vartheta\beta) \vee (\beta < \alpha \wedge \vartheta\alpha \leq \beta^*) \\ &\iff (\alpha < \beta \wedge \alpha^* < \vartheta\beta) \vee \vartheta\alpha \leq \beta^*, \end{aligned}$$
8. *$\beta < \vartheta\alpha \iff \omega^\beta < \vartheta\alpha$.*

*Proof* (1) and (8) basically follow from closure of $\vartheta\alpha$ under (C2).

(2) follows from the definition of $\vartheta\alpha$ taking Lemma 2.2 into account.

For (3), notice that $\vartheta\alpha \subset C_{\mathfrak{X}}(\alpha, \vartheta\alpha)$ is a consequence of clause (C1). Since $C_{\mathfrak{X}}(\alpha, \vartheta\alpha) \cap \Omega \subseteq \vartheta\alpha$ follows from the definition of $\vartheta\alpha$ and Lemma 2.2, we arrive at (3).

(4): If $\gamma^* \in C_{\mathfrak{X}}(\alpha, \beta)$, then $\gamma \in C_{\mathfrak{X}}(\alpha, \beta)$ by (C2). On the other hand, $\gamma \in C^n_{\mathfrak{X}}(\alpha, \beta) \implies \gamma^* \in C^n_{\mathfrak{X}}(\alpha, \beta)$ is easily seen by induction on $n$.

(5): $\alpha^* \in C_{\mathfrak{X}}(\alpha, \vartheta\alpha)$ holds by (4). As $\alpha^* < \Omega$, this implies $\alpha^* < \vartheta\alpha$ by (3).

(6): Suppose, aiming at a contradiction, that $\vartheta\alpha = \vartheta\beta$ and $\alpha < \beta$. Then $C_{\mathfrak{X}}(\alpha, \vartheta\alpha) \subseteq C_{\mathfrak{X}}(\beta, \vartheta\beta)$; hence $\alpha \in C_{\mathfrak{X}}(\beta, \vartheta\beta) \cap \beta$ by (2); thence $\vartheta\alpha = \vartheta\beta \in C_{\mathfrak{X}}(\beta, \vartheta\beta)$, contradicting (3).

(7): Suppose $\alpha < \beta$. Then $\vartheta\alpha < \vartheta\beta$ implies $\alpha^* < \vartheta\beta$ by (5). If $\alpha^* < \vartheta\beta$, then $\alpha \in C_{\mathfrak{X}}(\beta, \vartheta\beta)$; hence $\vartheta\alpha \in C_{\mathfrak{X}}(\beta, \vartheta\beta)$; thus, $\vartheta\alpha < \vartheta\beta$. This shows

$$(a) \quad \alpha < \beta \implies (\vartheta\alpha < \vartheta\beta \iff \alpha^* < \vartheta\beta).$$

By interchanging the roles of $\alpha$ and $\beta$, and employing (6) (to exclude $\vartheta\alpha = \vartheta\beta$), one obtains

$$(b) \quad \beta < \alpha \implies (\vartheta\alpha < \vartheta\beta \iff \vartheta\alpha \leq \beta^*).$$

$(a)$ and $(b)$ yield the first equivalence of (7) and thus the direction "⇒" of the second equivalence. Since $\vartheta\alpha \leq \beta^*$ implies $\vartheta\alpha < \vartheta\beta$ by (5), one also obtains the direction "⇐" of the second equivalence. □

**Definition 2.4** Inductive definition of a set $\mathrm{OT}_{\mathfrak{X}}(\vartheta)$ of ordinals and a natural number $G_\vartheta\alpha$ for $\alpha \in \mathrm{OT}_{\mathfrak{X}}(\vartheta)$.

1. $0, \Omega \in \mathrm{OT}_{\mathfrak{X}}(\vartheta)$, $G_\vartheta 0 := G_\vartheta\Omega := 0$, $\mathfrak{E}_u \in \mathrm{OT}_{\mathfrak{X}}(\vartheta)$ and $G_\vartheta\mathfrak{E}_u = 0$ for all $u \in |\mathfrak{X}|$.

2. If $\alpha =_{NF} \omega^{\alpha_1} + \cdots + \omega^{\alpha_n}$ and $\alpha_1, \ldots, \alpha_n \in \mathrm{OT}_{\mathfrak{X}}(\vartheta)$ then $\alpha \in \mathrm{OT}_{\mathfrak{X}}(\vartheta)$ and $G_\vartheta \alpha := \max\{G_\vartheta \alpha_1, \ldots, G_\vartheta \alpha_n\} + 1$.
3. If $\alpha = \vartheta \alpha_1$ and $\alpha_1 \in \mathrm{OT}_{\mathfrak{X}}(\vartheta)$ then $\alpha \in \mathrm{OT}_{\mathfrak{X}}(\vartheta)$ and $G_\vartheta \alpha := G_\vartheta \alpha_1 + 1$.

Observe that according to Lemma 2.3 (1) and (6) the function $G_\vartheta$ is well-defined. Each ordinal $\alpha \in \mathrm{OT}_{\mathfrak{X}}(\vartheta)$ has a unique normal form using the symbols $0, \Omega, +, \omega, \vartheta$.

**Lemma 2.5** $\mathrm{OT}_{\mathfrak{X}}(\vartheta) = \bigcup\{C_{\mathfrak{X}}(\alpha, 0) : \alpha < \sup_{u \in X} \mathfrak{E}_u\} = C_{\mathfrak{X}}(\sup_{u \in X} \mathfrak{E}_u, 0)$.

*Proof* Obviously $\beta < \sup_{u \in X} \mathfrak{E}_u$ holds for all $\beta \in \mathrm{OT}_{\mathfrak{X}}(\vartheta)$.

$$\beta \in \mathrm{OT}_{\mathfrak{X}}(\vartheta) \Rightarrow \beta \in C_{\mathfrak{X}}(\sup_{u \in X} \mathfrak{E}_u, 0)$$

is then shown by induction on $G_\vartheta \beta$.

The inclusion $C_{\mathfrak{X}}(\sup_{u \in X} \mathfrak{E}_u, 0) \subseteq \mathrm{OT}_{\mathfrak{X}}(\vartheta)$ follows from the fact that $\mathrm{OT}_{\mathfrak{X}}(\vartheta)$ is closed under the clauses (Ci) for $i = 0, 1, 2, 3$. Since $\mathfrak{X}$ is an ordering without a maximal element it is also clear that $\bigcup\{C_{\mathfrak{X}}(\alpha, 0) : \alpha < \sup_{u \in X} \mathfrak{E}_u\} = C_{\mathfrak{X}}(\sup_{u \in X} \mathfrak{E}_u, 0)$. □

If for $\alpha, \beta \in \mathrm{OT}_{\mathfrak{X}}(\vartheta)$ represented in their normal form, we wanted to determine whether $\alpha < \beta$, we could do this by deciding $\alpha_0 < \beta_0$ for ordinals $\alpha_0$ and $\beta_0$ that appear in these representations and, in addition, satisfy $G_\vartheta \alpha_0 + G_\vartheta \beta_0 < G_\vartheta \alpha + G_\vartheta \beta$. This follows from Lemma 2.3 (7) and the recursive procedure for comparing ordinals in Cantor normal form. So we come to see that after a straightforward coding in the natural numbers, we may represent $\langle \mathrm{OT}_{\mathfrak{X}}(\vartheta), < \restriction \mathrm{OT}_{\mathfrak{X}}(\vartheta)\rangle$ via a primitive recursive ordinal notation system. How this ordinal representation system can be directly defined in $\mathbf{RCA}_0$ is spelled out in the next subsection.

## 2.1 *Defining* $\mathbf{OT}_{\mathfrak{X}}(\vartheta)$ *in* $\mathbf{RCA}_0$

We shall provide an explicit primitive recursive definition of $\mathrm{OT}_{\mathfrak{X}}(\vartheta)$ as a term structure in $\mathbf{RCA}_0$. Of course formally, terms or strings of symbols have to be treated as coded by natural numbers since $\mathbf{RCA}_0$ only talks about numbers and sets of numbers. Though, as it is well-known how to do this, we can't be bothered with these niceties.

**Definition 2.6** Given a well-ordering $\mathfrak{X} = (X, <_X)$, i.e., an ordered pair $\mathfrak{X}$ in which $X$ is a set of natural numbers and $<_X$ is a well-ordering relation on $X$, we define, by recursion, a binary relational structure $\vartheta_{\mathfrak{X}} = (|\vartheta_{\mathfrak{X}}|, <)$, and a function $^* : |\vartheta_{\mathfrak{X}}| \to |\vartheta_{\mathfrak{X}}|$, in the following way:

1. $0, \Omega \in |\vartheta_{\mathfrak{X}}|$, and $0^* := 0 =: \Omega^*$.
2. If $\alpha \in |\vartheta_{\mathfrak{X}}|$ and $0 \neq \alpha$ then $0 < \alpha$.

3. For every $u \in X$ there is an element $\mathfrak{E}_u \in |\vartheta_{\mathfrak{X}}|$. Moreover, $(\mathfrak{E}_u)^* := 0$, and $\Omega < \mathfrak{E}_u$. If $u, v \in X$ and $u <_X v$, then $\mathfrak{E}_u < \mathfrak{E}_v$.
4. For every $\alpha \in |\vartheta_{\mathfrak{X}}|$ there is an element $\vartheta\alpha \in |\vartheta_{\mathfrak{X}}|$; and we have $\vartheta\alpha < \Omega$, $\vartheta\alpha < \mathfrak{E}_u$ for every $u \in X$, and $(\vartheta\alpha)^* := \vartheta\alpha$.
5. If $\alpha \in |\vartheta_{\mathfrak{X}}|$ and $\alpha$ is not of the form $\Omega$, $\mathfrak{E}_u$, or $\vartheta\beta$, then $\omega^\alpha \in \vartheta_{\mathfrak{X}}$ and $(\omega^\alpha)^* := \alpha^*$.
6. If $\alpha_1, \ldots, \alpha_n \in |\vartheta_{\mathfrak{X}}|$ and $\alpha_1 \geq \cdots \geq \alpha_n$ with $n \geq 2$, then $\omega^{\alpha_1} + \omega^{\alpha_2} + \cdots + \omega^{\alpha_n} \in |\vartheta_{\mathfrak{X}}|$ and $(\omega^{\alpha_1} + \omega^{\alpha_2} + \cdots + \omega^{\alpha_n})^* := \max\{\alpha_i^* : 1 \leq i \leq n\}$.
7. Let $\alpha = \omega^{\alpha_1} + \cdots + \omega^{\alpha_n} \in |\vartheta_{\mathfrak{X}}|$ and $\beta \in |\vartheta_{\mathfrak{X}}|$, where $\beta$ is of one of the forms $\vartheta\gamma$, $\Omega$, or $\mathfrak{E}_u$.

   (i) If $\alpha_1 < \beta$, then $\omega^{\alpha_1} + \cdots + \omega^{\alpha_n} < \beta$.
   (ii) If $\beta \leq \alpha_1$, then $\beta < \omega^{\alpha_1} + \cdots + \omega^{\alpha_n}$.

8. If $\omega^{\alpha_1} + \cdots + \omega^{\alpha_n}$, $\omega^{\beta_1} + \cdots + \omega^{\beta_m} \in |\vartheta_{\mathfrak{X}}|$ then
   $\omega^{\alpha_1} + \cdots + \omega^{\alpha_n} < \omega^{\beta_1} + \cdots + \omega^{\beta_m}$ iff
   $n < m \wedge \forall i \leq n \; \alpha_i = \beta_i$ or
   $\exists i \leq \min(n, m) \; [(\forall j < i \; \alpha_j = \beta_j) \wedge (\alpha_i < \beta_i)]$.
9. If $\alpha < \beta$ and $\alpha^* < \vartheta\beta$ then $\vartheta\alpha < \vartheta\beta$.
10. If $\vartheta\beta \leq \alpha^*$ then $\vartheta\beta < \vartheta\alpha$.

**Lemma 2.7**

*(i) The set $|\vartheta_{\mathfrak{X}}|$, the relation $<$, and the function $^*$ are primitive recursive in $\mathfrak{X} = (X, <_X)$.*
*(ii) $<$ is a total and linear ordering on $|\vartheta_{\mathfrak{X}}|$.*

*Proof* Straightforward but tedious. □

Of course, $\mathbf{RCA}_0$ does not prove that $<$ is a well-ordering on $|\vartheta_{\mathfrak{X}}|$.

# 3 A Well-Ordering Proof

In this section we work in the background theory

$$\mathbf{RCA}_0 + \forall X \exists Y \; (X \in Y \;\wedge\; Y \text{ is an } \omega\text{-model of } \mathbf{BI})$$

and shall prove the following statement

$$\forall \mathfrak{X} \, (\mathrm{WO}(\mathfrak{X}) \rightarrow \mathrm{WO}(\vartheta_{\mathfrak{X}})) \, ,$$

that is, the part (i) $\Rightarrow$ *(ii)* of the main Theorem 1.7. Some of the proofs are similar to ones in [13, Section 10]. Note that in this theory we can deduce arithmetical comprehension and even arithmetical transfinite recursion owing to [7] and [12], respectively.

Let us fix a well-ordering $\mathfrak{X} = (X, <_X)$, an arbitrary set $Y$ and a countable coded $\omega$-model $\mathfrak{A}$ of **BI** which contains both $\mathfrak{X}$ and $Y$ as elements. In the sequel $\alpha, \beta, \gamma, \delta, \ldots$ are supposed to range over $\vartheta_{\mathfrak{X}}$. $<$ will be used to denote the ordering on $\vartheta_{\mathfrak{X}}$. We are going to work informally in our background theory. A set $U \subseteq \mathbb{N}$ is said to be definable in $\mathfrak{A}$ if $U = \{n \in \mathbb{N} \mid \mathfrak{A} \models A(n)\}$ for some formula $A(x)$ of second order arithmetic which may contain parameters from $\mathfrak{A}$.

**Definition 3.1**

1. $\mathrm{Acc} := \{\alpha < \Omega \mid \mathfrak{A} \models \mathrm{WO}(< \restriction \alpha)\}$,
2. $\mathrm{M} := \{\alpha \colon E_\Omega(\alpha) \subseteq \mathrm{Acc}\}$,
3. $\alpha <_\Omega \beta : \Longleftrightarrow \alpha, \beta \in \mathrm{M} \ \wedge \ \alpha < \beta$.

**Lemma 3.2** $\alpha, \beta \in \mathrm{Acc} \implies \alpha + \omega^\beta \in \mathrm{Acc}$.

*Proof* Familiar from Gentzen's proof in Peano arithmetic. The proof just requires $\mathbf{ACA}_0$. (cf. [16, VIII.§ 21 Lemma 1]). □

**Lemma 3.3** $\mathrm{Acc} = \mathrm{M} \cap \Omega \ (:= \{\alpha \in \mathrm{M} \mid \alpha < \Omega\})$.

*Proof* If $\alpha \in \mathrm{Acc}$, then $E_\Omega(\alpha) \subseteq \mathrm{Acc}$ as well; hence, $\alpha \in \mathrm{M} \cap \Omega$. If $\alpha \in \mathrm{M} \cap \Omega$, then $E_\Omega(\alpha) \subseteq \mathrm{M} \cap \Omega$, so $\alpha \in Acc$ follows from Lemma 3.2. □

**Lemma 3.4** *Let $U$ be $\mathfrak{A}$ definable. Then*

$$\forall \alpha < \Omega \cap \mathrm{M}\, [\forall \beta < \alpha\ \beta \in U \to \alpha \in U] \to \mathrm{Acc} \subseteq U\,.$$

*Proof* This follows readily from the assumption that $\mathfrak{A}$ is a model of **BI**. □

**Definition 3.5** Let $\mathrm{Prog}_\Omega(X)$ stand for

$$(\forall \alpha \in \mathrm{M})[(\forall \beta <_\Omega \alpha)(\beta \in X) \longrightarrow \alpha \in X].$$

Let $\mathrm{Acc}_\Omega := \{\alpha \in \mathrm{M} \colon \vartheta\alpha \in \mathrm{Acc}\}$.

**Lemma 3.6** *If $U$ is $\mathfrak{A}$ definable, then*

$$\mathrm{Prog}_\Omega(U) \to \Omega, \Omega + 1 \in U\,.$$

*Proof* This follows from Lemmas 3.3 and 3.4. □

**Lemma 3.7** $\mathrm{Prog}_\Omega(\mathrm{Acc}_\Omega)$.

*Proof* Assume $\alpha \in \mathrm{M}$ and $(\forall \beta <_\Omega \alpha)(\beta \in \mathrm{Acc}_\Omega)$. We have to show that $\vartheta\alpha \in \mathrm{Acc}$. It suffices to show

$$\beta < \vartheta\alpha \Longrightarrow \beta \in \mathrm{Acc}. \tag{2}$$

We shall employ induction on $G_\vartheta(\beta)$, i.e., the length of (the term that represents) $\beta$. If $\beta \notin E$, then (2) follows easily by the inductive assumption and Lemma 3.2.

Now suppose $\beta = \vartheta\beta_0$. According to Lemma 2.3 it suffices to consider the following two cases:

*Case 1:* $\beta \le \alpha^*$. Since $\alpha \in \mathrm{M}$, we have $\alpha^* \in E_\Omega(\alpha) \subseteq \mathrm{Acc}$; therefore, $\beta \in \mathrm{Acc}$.
*Case 2:* $\beta_0 < \alpha$ and $\beta_0^* < \vartheta\alpha$. As the length of $\beta_0^*$ is less than the length of $\beta$, we get $\beta_0^* \in \mathrm{Acc}$; thus, $E_\Omega(\beta_0) \subseteq \mathrm{Acc}$, therefore $\beta_0 \in \mathrm{M}$. By the assumption at the beginning of the proof, we then get $\beta_0 \in \mathrm{Acc}_\Omega$; hence, $\beta = \vartheta\beta_0 \in \mathrm{Acc}$. □

**Definition 3.8** For every $\mathfrak{A}$ definable set $U$ we define the "Gentzen jump"

$$U^j := \{\gamma \mid \forall\delta\,[\mathrm{M}\cap\delta \subseteq U \rightarrow \mathrm{M}\cap(\delta+\omega^\gamma) \subseteq U]\}.$$

**Lemma 3.9** *Let $U$ be $\mathfrak{A}$ definable.*

*(i)* $\gamma \in U^j \Rightarrow \mathrm{M}\cap\omega^\gamma \subseteq U$.
*(ii)* $\mathrm{Prog}_\Omega(U) \Rightarrow \mathrm{Prog}_\Omega(U^j)$.

*Proof* (i) is obvious. (ii) $\mathrm{M}\cap(\delta+\omega^\gamma) \subseteq U$ is to be proved under the assumptions (a) $\mathrm{Prog}_\Omega(U)$, (b) $\gamma \in \mathrm{M} \wedge \mathrm{M}\cap\gamma \subseteq U^j$ and (c) $\mathrm{M}\cap\delta \subseteq U$. So let $\eta \in \mathrm{M}\cap(\delta+\omega^\gamma)$.

1. $\eta < \delta$: Then $\eta \in U$ is a consequence of (c).
2. $\eta = \delta$: Then $\eta \in U$ follows from (c) and (a).
3. $\delta < \eta < \delta+\omega^\gamma$: Then there exist $\gamma_1, \ldots, \gamma_k < \gamma$ such that $\eta = \delta+\omega^{\gamma_1}+\cdots+\omega^{\gamma_k}$ and $\gamma_1 \ge \cdots \ge \gamma_k$. $\eta \in \mathrm{M}$ implies $\gamma_1, \ldots, \gamma_k \in \mathrm{M}\cap\gamma$. Through applying (b) and (c) we obtain $\mathrm{M}\cap(\delta+\omega^{\gamma_1}) \subseteq U$. By iterating this procedure we eventually arrive at $\delta+\omega^{\gamma_1}+\cdots+\omega^{\gamma_k} \in U$, so $\eta \in U$ holds.

□

**Corollary 3.10** *Let $\mathcal{I}(\delta)$ be the statement that* $\mathrm{Prog}_\Omega(V) \rightarrow \delta \in \mathrm{M} \wedge \delta\cap\mathrm{M} \subseteq V$ *holds for all $\mathfrak{A}$ definable sets $V$. Assume $\mathcal{I}(\delta)$. Let $\delta_0 := \delta$ and $\delta_{n+1} := \omega^{\delta_n}$. Then*

$$\mathcal{I}(\delta_n)$$

*holds for all $n$.*

*Proof* We use induction on $n$. For $n = 0$ this is the assumption. Now suppose $\mathcal{I}(\delta_n)$ holds. Assume $\mathrm{Prog}_\Omega(U)$ for an $\mathfrak{A}$ definable $U$. By Lemma 3.9 we conclude $\mathrm{Prog}_\Omega(U^j)$ and hence $\delta_n \in U^j$ and $\delta_n \cap \mathrm{M} \subseteq U^j$. As clearly $\mathrm{M}\cap 0 \subseteq U$ we get $\omega^{\delta_n}\cap\mathrm{M} \subseteq U$. Since $\mathrm{Prog}_\Omega(U)$ entails $\delta \in \mathrm{M}$ we also have $\delta_{n+1} \in \mathrm{M}$. Thus $\delta_{n+1} \in \mathrm{M} \wedge \delta_{n+1}\cap\mathrm{M} \subseteq U$, showing $\mathcal{I}(\delta_{n+1})$. □

Let $\omega_0(\alpha) := \alpha$ and $\omega_{n+1}(\alpha) := \omega^{\omega_n(\alpha)}$.

**Proposition 3.11** *$\mathcal{I}(\mathfrak{E}_u)$ holds for all $u \in |\mathfrak{X}|$.*

*Proof* Noting that in our background theory $\mathfrak{X}$ is a well-ordering, we can use induction on $\mathfrak{X}$. Note also that $\mathcal{I}(\mathfrak{E}_u)$ is a statement about all definable sets in $\mathfrak{A}$ which is not formalizable in $\mathfrak{A}$ itself. However, in our background theory quantification over all these sets is first order expressible and therefore transfinite induction along $<_X$ is available.

First observe that we have $\mathcal{I}(\Omega + 1)$ by Lemma 3.6. Let $u_0$ be the $<_X$-least element of $|\mathfrak{X}|$. We have $\mathfrak{E}_{u_0} \in \mathrm{M}$ and for every $\eta < \mathfrak{E}_{u_0}$ there exists $n$ such that $\eta < \omega_n(\Omega + 1)$. As a result, using Corollary 3.10, we have

$$\mathrm{Prog}_{\Omega}(U) \rightarrow \mathfrak{E}_{u_0} \cap \mathrm{M} \subseteq U$$

for every $\mathfrak{A}$ definable set $U$.

Now suppose that $u \in |\mathfrak{X}|$ is not the $<_X$-least element and for all $v <_X u$ we have $\mathcal{I}(\mathfrak{E}_v)$. As for every $\delta < \mathfrak{E}_u$ there exists $v <_X u$ and $n$ such that $\delta < \omega_n(\mathfrak{E}_v)$, the inductive assumption together with Corollary 3.10 yields

$$\mathrm{Prog}_{\Omega}(U) \rightarrow \mathfrak{E}_u \cap \mathrm{M} \subseteq U\,.$$

$\mathfrak{E}_u \in \mathrm{M}$ is obvious. □

**Proposition 3.12** *For all $\alpha$, $\mathcal{I}(\alpha)$.*

*Proof* We proceed by the induction on the term complexity of $\alpha$. Clearly, $\mathcal{I}(0)$. By Lemma 3.6 we conclude that $\mathcal{I}(\Omega)$. Proposition 3.11 entails that $\mathcal{I}(\mathfrak{E}_u)$ for all $u \in |\mathfrak{X}|$.

Now let $\alpha = \omega^{\alpha_1} + \cdots + \omega^{\alpha_n}$ be in Cantor normal form. Inductively we have $\mathcal{I}(\alpha_1), \ldots, \mathcal{I}(\alpha_n)$. Assume $\mathrm{Prog}_{\Omega}(U)$. Then $\mathrm{Prog}_{\Omega}(U^j)$ by Lemma 3.9(ii), and hence $\alpha_1 \cap \mathrm{M} \subseteq U^j, \ldots, \alpha_n \cap \mathrm{M} \subseteq U^j$ and $\alpha_1, \ldots, \alpha_n \in \mathrm{M}$. The latter implies $\alpha_1 \in U^j, \ldots, \alpha_n \in U^j$. Using the definition of $U^j$ repeatedly we conclude $\alpha \cap \mathrm{M} \subseteq U$. Moreover, $\alpha \in \mathrm{M}$ since $\alpha_1, \ldots, \alpha_n \in \mathrm{M}$.

Now suppose that $\alpha = \vartheta\beta$. Inductively we have $\mathcal{I}(\beta)$. By Lemma 3.7 we conclude that $\beta \in \mathrm{Acc}_{\Omega}$, and hence $\alpha \in \mathrm{Acc}$. From $\mathrm{Prog}_{\Omega}(U)$ we obtain by Lemma 3.4 that $\xi \in U$ for all $\xi \leq \alpha$. As a result, $\mathcal{I}(\alpha)$. □

**Corollary 3.13** *$\vartheta_{\mathfrak{X}}$ is a well-ordering.*

With the previous Corollary, the proof of Theorem 1.7 (i)⇒(ii) is finally accomplished.

# 4 Deduction Chains

From now on we will be concerned with the part (ii) ⇒ (i) of the main Theorem 1.7. An important tool will be the method of deduction chains. Given a sequent $\Gamma$ and a set $Q \subseteq \mathbb{N}$, deduction chains starting at $\Gamma$ are built by systematically decomposing $\Gamma$ into its subformulas, and adding additionally at the $n$th step the formulas $\neg A_n$ and $\neg \bar{Q}(\bar{n})$, where $(A_n \mid n \in \mathbb{N})$ is an enumeration of the axioms of the theory **BI**, and $\bar{Q}(\bar{n})$ is the atom $\bar{n} \in U_0$ if $n \in Q$ and $\bar{n} \notin U_0$ otherwise. The set of all deduction chains that can be built from the empty sequent with respect to a given set $Q$ forms the tree $\mathcal{D}_Q$. There are two scenarios to be considered.

(i) If there is an infinite deduction chain, i.e. $\mathcal{D}_Q$ is ill-founded, then this readily yields a model of **BI** that contains $Q$.
(ii) If each deduction chain is finite, then this yields a derivation of the empty sequent, $\perp$, in a corresponding infinitary system with an $\omega$-rule. The depth of this derivation is bounded by the order-type $\alpha$ of the Kleene–Brouwer ordering of $\mathcal{D}_Q$. By the well-ordering principle, transfinite induction up to $\mathfrak{E}_{\alpha+1}$ is available, which allows to transform this proof into a cut-free proof of $\perp$ whose depth is less than $\vartheta\mathfrak{E}_{\alpha+1}$.

As the second alternative is impossible, the first yields the desired model.

**Definition 4.1**

1. We let $U_0, U_1, \ldots, U_m, \ldots$ be an enumeration of the free set variables of $\mathcal{L}_2$ and, given a closed term $t$, we write $t^{\mathbb{N}}$ for its numerical value.
2. Henceforth a **sequent** will be a finite list of $\mathcal{L}_2$-formulae *without* free number variables.
3. A sequent $\Gamma$ is **axiomatic** if it satisfies at least one of the following conditions:
   (a) $\Gamma$ contains a true **literal**, i.e., a true formula of either of the forms $R(t_1, \ldots, t_n)$ or $\neg R(t_1, \ldots, t_n)$, where $R$ is a predicate symbol in $\mathcal{L}_2$ for a primitive recursive relation and $t_1, \ldots, t_n$ are closed terms.
   (b) $\Gamma$ contains formulae $s \in U$ and $t \notin U$ for some set variable $U$ and terms $s$, $t$ with $s^{\mathbb{N}} = t^{\mathbb{N}}$.
4. A sequent is **reducible** if it is not axiomatic and contains a formula which is not a literal.

**Definition 4.2** For $Q \subseteq \mathbb{N}$ we define

$$\bar{Q}(n) \Leftrightarrow \begin{cases} \bar{n} \in U_0 & \text{if } n \in Q, \\ \bar{n} \notin U_0 & \text{otherwise.} \end{cases}$$

For some of the following theorems it is convenient to have a finite axiomatization of arithmetical comprehension.

**Lemma 4.3** **ACA**$_0$ *can be axiomatized via a single* $\Pi^1_2$ *sentence* $\forall X C(X)$.

*Proof* [17, Lemma VIII.1.5]. □

**Definition 4.4** In what follows, we fix an enumeration of $A_1$, $A_2$, $A_3, \ldots$ of all the universal closures of instances of ***(BI)***. We also put $A_0 := \forall X\, C(X)$, where the latter is the sentence that axiomatizes arithmetical comprehension.

**Definition 4.5** Let $Q \subseteq \mathbb{N}$. A $Q$**-deduction chain** is a finite string

$$\Gamma_0, \Gamma_1, \ldots, \Gamma_k$$

of sequents $\Gamma_i$ constructed according to the following rules:

1. $\Gamma_0 = \neg\bar{Q}(0), \neg A_0$.
2. $\Gamma_i$ is not axiomatic for $i < k$.
3. If $i < k$ and $\Gamma_i$ is not reducible, then

$$\Gamma_{i+1} = \Gamma_i, \neg\bar{Q}(i+1), \neg A_{i+1}.$$

4. Every reducible $\Gamma_i$ with $i < k$ is of the form

$$\Gamma_i', E, \Gamma_i''$$

where $E$ is not a literal and $\Gamma_i'$ contains only literals. $E$ is said to be the **redex** of $\Gamma_i$.

Let $i < k$ and $\Gamma_i$ be reducible. $\Gamma_{i+1}$ is obtained from $\Gamma_i = \Gamma_i', E, \Gamma_i''$ as follows:

(a) If $E \equiv E_0 \vee E_1$, then

$$\Gamma_{i+1} = \Gamma_i', E_0, E_1, \Gamma_i'', \neg\bar{Q}(i+1), \neg A_{i+1}.$$

(b) If $E \equiv E_0 \wedge E_1$, then

$$\Gamma_{i+1} = \Gamma_i', E_j, \Gamma_i'', \neg\bar{Q}(i+1), \neg A_{i+1}$$

where $j = 0$ or $j = 1$.

(c) If $E \equiv \exists x F(x)$, then

$$\Gamma_{i+1} = \Gamma_i', F(\bar{m}), \Gamma_i'', \neg\bar{Q}(i+1), \neg A_{i+1}, E$$

where $m$ is the first number such that $F(\bar{m})$ does not occur in $\Gamma_0, \ldots, \Gamma_i$.

(d) If $E \equiv \forall x F(x)$, then

$$\Gamma_{i+1} = \Gamma_i', F(\bar{m}), \Gamma_i'', \neg\bar{Q}(i+1), \neg A_{i+1}$$

for some $m$.

(e) If $E \equiv \exists X F(X)$, then

$$\Gamma_{i+1} = \Gamma_i', F(U_m), \Gamma_i'', \neg\bar{Q}(i+1), \neg A_{i+1}, E$$

where $m$ is the first number such that $F(U_m)$ does not occur in $\Gamma_0, \ldots, \Gamma_i$.

(f) If $E \equiv \forall X F(X)$, then

$$\Gamma_{i+1} = \Gamma_i', F(U_m), \Gamma_i'', \neg\bar{Q}(i+1), \neg A_{i+1}$$

where $m$ is the first number such that $U_m$ does not occur in $\Gamma_i$.

The set of $Q$-deduction chains forms a tree $\mathcal{D}_Q$ labeled with strings of sequents.

We will now consider two cases.

**Case I:** $\mathcal{D}_Q$ is not well-founded. Then $\mathcal{D}_Q$ contains an infinite path $\mathbb{P}$. Now define a set $M$ via

$$(M)_i = \{k \mid \bar{k} \notin U_i \text{ occurs in } \mathbb{P}\}.$$

Set $\mathbb{M} = (\mathbb{N}; \{(M)_i \mid i \in \mathbb{N}\}, \in, +, \cdot, 0, 1, <)$.

For a formula $F$, let $F \in \mathbb{P}$ mean that $F$ occurs in $\mathbb{P}$, i.e. $F \in \Gamma$ for some $\Gamma \in \mathbb{P}$.

**Claim:** Under the assignment $U_i \mapsto (M)_i$ we have

$$F \in \mathbb{P} \quad \Rightarrow \quad \mathbb{M} \models \neg F. \tag{3}$$

The Claim will imply that $\mathbb{M}$ is an $\omega$-model of **BI**. Also note that $(M)_0 = Q$, thus $Q$ is in $\mathbb{M}$. The proof of (3) follows by induction on $F$ using Lemma 4.6 below. The upshot of the foregoing is that we can prove Theorem 1.7 under the assumption that $\mathcal{D}_Q$ is ill-founded for all sets $Q \subseteq \mathbb{N}$.

**Lemma 4.6** *Let $Q$ be an arbitrary subset of $\mathbb{N}$ and $\mathcal{D}_Q$ be the corresponding deduction tree. Moreover, suppose $\mathcal{D}_Q$ is not well-founded. Then $\mathcal{D}_Q$ has an infinite path $\mathbb{P}$. $\mathbb{P}$ has the following properties:*

1. *$\mathbb{P}$ does not contain literals which are true in $\mathbb{N}$.*
2. *$\mathbb{P}$ does not contain formulas $s \in U_i$ and $t \notin U_i$ for constant terms $s$ and $t$ such that $s^{\mathbb{N}} = t^{\mathbb{N}}$.*
3. *If $\mathbb{P}$ contains $E_0 \vee E_1$, then $\mathbb{P}$ contains $E_0$ and $E_1$.*
4. *If $\mathbb{P}$ contains $E_0 \wedge E_1$, then $\mathbb{P}$ contains $E_0$ or $E_1$.*
5. *If $\mathbb{P}$ contains $\exists x F(x)$, then $\mathbb{P}$ contains $F(\bar{n})$ for all $n$.*
6. *If $\mathbb{P}$ contains $\forall x F(x)$, then $\mathbb{P}$ contains $F(\bar{n})$ for some $n$.*
7. *If $\mathbb{P}$ contains $\exists X F(X)$, then $\mathbb{P}$ contains $F(U_m)$ for all $m$.*
8. *If $\mathbb{P}$ contains $\forall X F(X)$, then $\mathbb{P}$ contains $F(U_m)$ for some $m$.*
9. *$\mathbb{P}$ contains $\neg C(U_m)$ for all $m$.*
10. *$\mathbb{P}$ contains $\neg \bar{Q}(m)$ for all $m$.*

*Proof* Standard. □

**Corollary 4.7** *If $\mathcal{D}_Q$ is ill-founded, then there exists a countable coded $\omega$-model of* **BI** *which contains $Q$.*

For our purposes it is important that Corollary 4.7 can be proved in $T_0 := \mathbf{RCA}_0 + \forall \mathfrak{X}\,(\mathrm{WO}(\mathfrak{X}) \to \mathrm{WO}(\vartheta_{\mathfrak{X}}))$. To this end we need to show that the semantics of $\omega$-models can be handled in the latter theory, i.e. for every formula $F$ of $\mathcal{L}_2$ there exists a valuation for $F$ in the sense of [17, VII.2.1]. It is easily seen that the principle $\forall \mathfrak{X}\,(\mathrm{WO}(\mathfrak{X}) \to \mathrm{WO}(\vartheta_{\mathfrak{X}}))$ implies

$$\forall \mathfrak{X}\,(\mathrm{WO}(\mathfrak{X}) \to \mathrm{WO}(\varepsilon_{\mathfrak{X}}))$$

(see [1, Definition 2.1]) and thus, by [1, Theorem 4.1], $T_0$ proves that every set is contained in an $\omega$-model of **ACA**. Now take an $\omega$-model containing $\mathcal{D}_Q$ and an infinite branch of $\mathcal{D}_Q$. In this $\omega$-model we find a valuation for every formula by [17, VII.2.2]. And hence Corollary 4.7 holds in the model, but then it also holds in the world at large by absoluteness.

# 5 Proof of the Main Theorem: The Hard Direction Part 2

The remainder of the paper will be devoted to ruling out the possibility that for some $Q$, $\mathcal{D}_Q$ could be a well-founded tree. This is the place where the principle $\forall \mathfrak{X}\,(\mathrm{WO}(\mathfrak{X}) \rightarrow \mathrm{WO}(\vartheta_{\mathfrak{X}}))$ in the guise of cut elimination for an infinitary proof system enters the stage. Aiming at a contradiction, suppose that $\mathcal{D}_Q$ is a well-founded tree. Let $\mathfrak{X}$ be the Kleene–Brouwer ordering on $\mathcal{D}_Q$ (see [17, Definition V.1.2]). Then $\mathfrak{X}$ is a well-ordering. In a nutshell, the idea is that a well-founded $\mathcal{D}_Q$ gives rise to a derivation of the empty sequent (contradiction) in an infinitary proof system.

## 5.1 *Majorization and Fundamental Functions*

In this section we introduce the concepts of majorization and fundamental function. They are needed for carrying through the ordinal analysis of bar induction. More details can be found in [13, Section 4] and [3, I.4] to which we refer for proofs. The missing proofs are actually straightforward consequences of Definition 2.6.

**Definition 5.1** 1. $\alpha \lhd \beta$ means $\alpha < \beta$ and $\vartheta\alpha < \vartheta\beta$.
2. $\alpha \unlhd \beta :\Longleftrightarrow (\alpha \lhd \beta \vee \alpha = \beta)$.

**Lemma 5.2** *1.* $\alpha \lhd \beta \;\wedge\; \beta \lhd \gamma \;\Longrightarrow\; \alpha \lhd \gamma$.
2. $0 < \beta < \varepsilon_0 \;\Longrightarrow\; \alpha \lhd \alpha + \beta$.
*3.* $\alpha < \beta < \Omega \;\Longrightarrow\; \alpha \lhd \beta$.
*4.* $\alpha \lhd \beta \;\Longrightarrow\; \alpha + 1 \unlhd \beta$.
*5.* $\alpha \lhd \beta \;\Longrightarrow\; \vartheta\alpha \lhd \vartheta\beta$.
*6.* $\alpha = \alpha_0 + 1 \Longrightarrow \vartheta\alpha_0 \lhd \vartheta\alpha$.

**Lemma 5.3** $\alpha \lhd \beta,\ \beta < \omega^{\gamma+1} \;\Longrightarrow\; \omega^\gamma + \alpha \lhd \omega^\gamma + \beta$.

**Corollary 5.4** $\omega^\alpha \cdot n \lhd \omega^\alpha \cdot (n+1)$.

**Lemma 5.5** $\alpha \lhd \beta \;\Longrightarrow\; \omega^\alpha \cdot n \lhd \omega^\beta$.

**Definition 5.6** Let $D_\Omega := (\mathrm{OT}_{\mathfrak{X}}(\vartheta) \cap \Omega) \cup \{\Omega\}$. A function $f : D_\Omega \to \mathrm{OT}_{\mathfrak{X}}(\vartheta)$ will be called a *fundamental function* if it is generated by the following clauses:

F1. $Id : D_\Omega \to D_\Omega$ with $Id(\alpha) = \alpha$ is a fundamental function.

F2. If $f$ is a fundamental function, $\gamma \in \mathrm{OT}_{\mathfrak{X}}(\vartheta)$ and $f(\Omega) < \omega^{\gamma+1}$, then $\omega^\gamma + f$ is a fundamental function, where $(\omega^\gamma + f)(\alpha) := \omega^\gamma + f(\alpha)$ for all $\alpha \in D_\Omega$.
F3. If $f$ is a fundamental function, then so is $\omega^f$ with $(\omega^f)(\alpha) := \omega^{f(\alpha)}$ for all $\alpha \in D_\Omega$.

**Lemma 5.7** *Let $f$ be a fundamental function and $\beta \leq \Omega$.*

*(i) If $\alpha < \beta$, then $f(\alpha) < f(\beta)$.*
*(ii) If $\alpha \lhd \beta$, then $f(\alpha) \lhd f(\beta)$.*
*(iii) $(f(\beta))^* \leq \max((f(0))^*, \beta^*)$.*

*Proof* (i) is obvious by induction on the generation of fundamental functions.
(ii) also follows by induction on the generation of fundamental functions, using Lemmas 5.3 and 5.5.
(iii) as well follows by induction on the generation of fundamental functions.

□

**Lemma 5.8** *For every fundamental function $f$ we have $f(\vartheta(f(0))) \lhd f(\Omega)$.*

*Proof* Since $\vartheta(f(0)) < \Omega$, we clearly have $f(\vartheta(f(0))) < f(\Omega)$. Since $0 \lhd \Omega$ and $f$ is a fundamental function, we have $\vartheta(f(0)) < \vartheta(f(\Omega))$ by Lemma 5.7 (ii). Invoking Lemma 5.7 (iii), the latter entails that $(f(\vartheta(f(0))))^* < \vartheta(f(\Omega))$, so that in conjunction with $f(\vartheta(f(0))) < f(\Omega)$ it follows that $\vartheta(f(\vartheta(f(0)))) \lhd \vartheta(f(\Omega))$. As a result, $f(\vartheta(f(0))) \lhd f(\Omega)$. □

## 5.2 The Infinitary Calculus $T_Q^*$

The calculus $T_Q^*$ to be introduced stems from [13, Section 6]. We fix a set $Q \subseteq \mathbb{N}$. Let $\mathcal{L}_2^Q$ be the language of second order arithmetic augmented by a unary predicate $\bar{Q}$. The *formulas* of $T_Q^*$ arise from $\mathcal{L}_2^Q$-formulas by replacing free numerical variables by numerals, i.e. terms of the form $0, 0', 0'', \ldots$ Especially, every formula $A$ of $T_Q^*$ is an $\mathcal{L}_2^Q$-formula. We are going to measure the length of derivations by ordinals. We are going to use the set of ordinals $\mathrm{OT}_{\mathfrak{X}}(\vartheta)$ of Sect. 3.

**Definition 5.9**

1. A formula $B$ is said to be *weak* if it belongs to $\Pi_0^1 \cup \Pi_1^1$.
2. Two closed terms $s$ and $t$ are said to be equivalent if they yield the same value when computed.
3. A formula is called constant if it contains no set variables. The truth or falsity of such a formula is understood with respect to the standard structure of the integers.
4. $\overline{0} := 0, \overline{m+1} := \overline{m}'$.

In the sequent calculus $T_Q^*$ below we shall use the following rules of inference:

$$(\wedge)\quad \vdash \Gamma, A \text{ and } \vdash \Gamma, B \implies \vdash \Gamma, A \wedge B,$$

$$(\vee)\quad \vdash \Gamma, A_i \implies \vdash \Gamma, A_0 \vee A_1 \quad \text{if } i \in \{0,1\},$$

$$(\forall_2)\quad \vdash \Gamma, F(U) \implies \vdash \Gamma, \forall X F(X),$$

$$(\exists_1)\quad \vdash \Gamma, F(t) \implies \vdash \Gamma, \exists x F(x),$$

$$(Cut) \vdash \Gamma, A \text{ and } \vdash \Gamma, \neg A \implies \vdash \Gamma,$$

where in $(\forall_2)$ the free variable $U$ is not to occur in the conclusion.

The most important feature of sequent calculi is cut-elimination. To state this fact concisely, let us introduce a measure of complexity, $gr(A)$, the *grade of a formula* $A$, for $\mathcal{L}_2^Q$-formulae.

**Definition 5.10**

1. $gr(A) = 0$ if $A$ is a prime formula or negated prime formula.
2. $gr(\forall X F(X)) = gr(\exists X F(X)) = \omega$ if $F(U)$ is arithmetic.
3. $gr(A \wedge B) = gr(A \vee B) = max\{gr(A), gr(B)\} + 1$.
4. $gr(\forall x H(x)) = gr(\exists x H(x)) = gr(H(0)) + 1$.
5. $gr(\forall X G(X)) = gr(\exists X G(X)) = gr(G(U)) + 1$, if $G$ is not arithmetic.

**Definition 5.11** Inductive definition of $T_Q^* \overset{\alpha}{\underset{\varrho}{\vdash}} \Gamma$ for $\alpha \in \mathrm{OT}_{\mathfrak{X}}(\vartheta)$ and $\varrho < \omega + \omega$.

1. If $A$ is a true constant prime formula or negated prime formula and $A \in \Gamma$, then $T_Q^* \overset{\alpha}{\underset{\varrho}{\vdash}} \Gamma$.
2. If $n \in Q$ and $t$ is a closed term with value $n$ and $\bar{Q}(t)$ is in $\Gamma$, then $T_Q^* \overset{\alpha}{\underset{\varrho}{\vdash}} \Gamma$.
3. If $n \notin Q$ and $t$ is a closed term with value $n$ and $\neg\bar{Q}(t)$ is in $\Gamma$, then $T_Q^* \overset{\alpha}{\underset{\varrho}{\vdash}} \Gamma$.
4. If $\Gamma$ contains formulas $A(s_1, \ldots, s_n)$ and $\neg A(t_1, \ldots, t_n)$ of grade 0 or $\omega$, where $s_i$ and $t_i$ $(1 \le i \le n)$ are equivalent terms, then $T_Q^* \overset{\alpha}{\underset{\varrho}{\vdash}} \Gamma$.
5. If $T_Q^* \overset{\beta}{\underset{\varrho}{\vdash}} \Gamma_i$ and $\beta \lhd \alpha$ hold for every premiss $\Gamma_i$ of an inference $(\wedge), (\vee), (\exists_1), (\forall_2)$ or $(Cut)$ with a cut formula having grade $< \varrho$, and conclusion $\Gamma$, then $T_Q^* \overset{\alpha}{\underset{\varrho}{\vdash}} \Gamma$.
6. If $T_Q^* \overset{\alpha_0}{\underset{\varrho}{\vdash}} \Gamma, F(U)$ holds for some $\alpha_0 \lhd \alpha$ and a non-arithmetic formula $F(U)$ (i.e., $gr(F(U)) \ge \omega$), then $T_Q^* \overset{\alpha}{\underset{\varrho}{\vdash}} \Gamma, \exists X F(X)$.
7. ($\omega$-rule). If $T_Q^* \overset{\beta}{\underset{\varrho}{\vdash}} \Gamma, A(\overline{m})$ is true for every $m < \omega$, $\forall x A(x) \in \Gamma$, and $\beta \lhd \alpha$, then $T_Q^* \overset{\alpha}{\underset{\varrho}{\vdash}} \Gamma$.
8. ($\Omega$-rule). Let $f$ be a fundamental function satisfying
   (a) $f(\Omega) \unlhd \alpha$,
   (b) $T_Q^* \overset{f(0)}{\underset{\varrho}{\vdash}} \Gamma, \forall X F(X)$, where $\forall X F(X) \in \Pi_1^1$, and
   (c) $T_Q^* \overset{\beta}{\underset{0}{\vdash}} \Xi, \forall X F(X)$ implies $T_Q^* \overset{f(\beta)}{\underset{\varrho}{\vdash}} \Xi, \Gamma$ for every set of weak formulas $\Xi$ and $\beta < \Omega$.

Then $T^*_Q \mid\!\frac{\alpha}{\varrho}\, \Gamma$ holds.

*Remark 5.12* The derivability relation $T^*_Q \mid\!\frac{\alpha}{\varrho}\, \Gamma$ is from [13] and is modelled upon the relation $PB^* \mid\!\frac{\alpha}{n}\, F$ of [3], the main difference being the sequent calculus setting instead of $P$- and $N$-forms and a different assignment of cut-degrees. The allowance for transfinite cut-degrees will enable us to deal with arithmetical comprehension.

*Remark 5.13* If one ruminates on the definition of the derivability predicate $T^*_Q \mid\!\frac{\alpha}{\varrho}\, \Xi$ the question arises whether it is actually a proper inductive definition. The critical point is obviously the condition (c) of the $\Omega$-rule. Note that $T^*_Q \mid\!\frac{\beta}{0}\, \Xi, \forall X F(X)$ occurs negatively in clause (c). However, since $\beta < \Omega$, the pertaining derivation does not contain any applications of the $\Omega$-rule. Thus the definition of $T^*_Q \mid\!\frac{\alpha}{\varrho}\, \Xi$ proceeds via an iterated inductive definition. First one defines a derivability predicate without involvement of the $\Omega$-rule via an ordinary inductive definition, and in a second step defines $T^*_Q \mid\!\frac{\alpha}{\varrho}\, \Gamma$ inductively referring to the first derivability predicate in the $\Omega$-rule.

It will actually be a non-trivial issue how to handle such inductive definitions in a weak background theory.

**Lemma 5.14**

*1.* $T^*_Q \mid\!\frac{\alpha}{\delta}\, \Gamma \;\&\; \Gamma \subseteq \Delta \;\&\; \alpha \trianglelefteq \beta \;\&\; \delta \le \varrho \implies T^*_Q \mid\!\frac{\beta}{\varrho}\, \Delta$,

*2.* $T^*_Q \mid\!\frac{\alpha}{\varrho}\, \Gamma, A \wedge B \implies T^*_Q \mid\!\frac{\alpha}{\varrho}\, \Gamma, A \;\&\; T^*_Q \mid\!\frac{\alpha}{\varrho}\, \Gamma, B$,

*3.* $T^*_Q \mid\!\frac{\alpha}{\varrho}\, \Gamma, A \vee B \implies T^*_Q \mid\!\frac{\alpha}{\varrho}\, \Gamma, A, B$

*4.* $T^*_Q \mid\!\frac{\alpha}{\varrho}\, \Gamma, F(t) \implies T^*_Q \mid\!\frac{\alpha}{\varrho}\, \Gamma, F(s)$ *if $t$ and $s$ are equivalent,*

*5.* $T^*_Q \mid\!\frac{\alpha}{\varrho}\, \Gamma, \forall x F(x) \implies T^*_Q \mid\!\frac{\alpha}{\varrho}\, \Gamma, F(s)$ *for every term $s$.*

*6. If* $T^*_Q \mid\!\frac{\alpha}{\varrho}\, \Gamma, \forall X G(X)$ *and* $gr(G(U)) \ge \omega$, *then* $T^*_Q \mid\!\frac{\alpha}{\varrho}\, \Gamma, G(U)$.

*Proof* Proceed by induction on $\alpha$. These can be carried out straightforwardly. (5) requires (4). As to (6), observe that $\forall X G(X)$ cannot be the main formula of an axiom. □

**Lemma 5.15** $T^*_Q \mid\!\frac{2\cdot\alpha}{0}\, \Gamma, A(s_1, \ldots, s_k), \neg A(t_1, \ldots, t_k)$ *if* $\alpha \ge gr(A(s_1, \ldots, s_k))$ *and $s_i$ and $t_i$ are equivalent terms.*

*Proof* Proceed by induction on $gr(A(s_1, \ldots, s_k))$. Crucially note that if $gr(A(s_1, \ldots, s_k)) = \omega$ then $\Gamma, A(s_1, \ldots, s_k), \neg A(t_1, \ldots, t_k)$ is an axiom according to Definition 5.11 clause (4). □

**Lemma 5.16**

*1.* $T^*_Q \mid\!\frac{2m}{0}\, \neg(0 \in U), (\exists x)[x \in U \wedge \neg(x' \in U)], \overline{m} \in U$,

2. $T^*_Q \mid\!\frac{\omega+5}{0}\, \forall X[0 \in X \wedge \forall x(x \in X \to x' \in X) \to \forall x(x \in X)]$.

*Proof* For (1) use induction on $m$. (2) is an immediate consequence of (1) using Lemma 5.14 (1), the $\omega$-rule, ($\vee$), and ($\forall_2$). □

**Definition 5.17** For formulas $F(U)$ and $A(a)$, $F(A)$ denotes the result of replacing each occurrence of the form $e \in U$ in $F(U)$ by $A(e)$. The expression $F(A)$ is a formula if the bound variables in $A(a)$ are chosen in an appropriate way, in particular, if $F(U)$ and $A(a)$ have no bound variables in common.

**Lemma 5.18** *Suppose $\alpha < \Omega$ and let $\Delta(U) = \{F_1(U), \ldots, F_k(U)\}$ be a set of weak formulas such that $U$ doesn't occur in $\forall X F_i(X)$ $(1 \leq i \leq k)$. For an arbitrary formula $A(a)$ we then have:*

$$T_Q^* \mathop{\vdash}\limits_0^{\alpha} \Delta(U) \implies T_Q^* \mathop{\vdash}\limits_0^{\Omega+\alpha} \Delta(A).$$

*Proof* Proceed by induction on $\alpha$. Suppose $\Delta(U)$ is an axiom. Then either $\Delta(A)$ is an axiom too, or $T_Q^* \mathop{\vdash}\limits_0^{\omega+\omega} \Delta(A)$ can be obtained through use of Lemma 5.15. Therefore $T_Q^* \mathop{\vdash}\limits_0^{\Omega+\alpha} \Delta(A)$ by Lemma 5.14 (1). If $T_Q^* \mathop{\vdash}\limits_0^{\alpha} \Delta(U)$ is the result of an inference, then this inference must be different from $(\exists_2)$, $(Cut)$, and the $(\Omega - rule)$ since $\Delta(U)$ consists of weak formulas, the derivation is cut-free and $\alpha < \Omega$. For the remaining possible inference rules the assertion follows easily from the induction hypothesis. □

**Lemma 5.19** *Let $\Gamma, \forall X F(X)$ be a set of weak formulas. If $T_Q^* \mathop{\vdash}\limits_0^{\alpha} \Gamma, \forall X F(X)$ and $\alpha < \Omega$, then $T_Q^* \mathop{\vdash}\limits_0^{\alpha} \Gamma, F(U)$.*

*Proof* Use induction on $\alpha$. Note that $\forall X F(X)$ cannot be a principal formula of an axiom, since $\exists X \neg F(X)$ does not surface in such a derivation. Also, due to $\alpha < \Omega$, the derivation doesn't involve instances of the $\Omega$-rule. Therefore the proof is straightforward. □

The role of the $\Omega$-rule in our calculus $T_Q^*$ is enshrined in the next lemma.

**Lemma 5.20** $T_Q^* \mathop{\vdash}\limits_0^{\Omega \cdot 2} \exists X F(X), \neg F(A)$ *for every arithmetic formula $F(U)$ and arbitrary formula $A(a)$.*

*Proof* Let $f(\alpha) := \Omega + \alpha$ with $dom(f) := \{\alpha \in OT(\psi) : \alpha \leq \Omega\}$. Then

$$T_Q^* \mathop{\vdash}\limits_0^{f(0)} \forall X \neg F(X), \exists X F(X), \neg F(A) \tag{4}$$

according to Lemma 5.15. For $\alpha < \Omega$ and every set of weak formulas $\Theta$, we have by Lemmas 5.18 and 5.19,

$$T_Q^* \mathop{\vdash}\limits_0^{\alpha} \Theta, \forall X \neg F(X) \implies T_Q^* \mathop{\vdash}\limits_0^{f(\alpha)} \Theta, \neg F(A).$$

Therefore, by Lemma 5.14 (1),

$$T_Q^* \mathop{\vdash}\limits_0^{\alpha} \Theta, \forall X \neg F(X) \implies T_Q^* \mathop{\vdash}\limits_0^{f(\alpha)} \Theta, \exists X F(X), \neg F(A). \tag{5}$$

The assertion now follows from (4) and (5) by the $\Omega$-rule. □

**Corollary 5.21** $T_{\varrho}^{*} \overset{\Omega\cdot 2+1}{\underset{\omega}{\vdash}} \exists X\, \forall y\, (y \in X \leftrightarrow B(y))$ *for every arithmetic formula* $B(a)$.

*Proof* Owing to Lemma 5.20 we have

$$T_{\varrho}^{*} \overset{\Omega\cdot 2}{\underset{0}{\vdash}} \exists X\, \forall y\, (y \in X \leftrightarrow B(y)), \neg\forall y\, (B(y) \leftrightarrow B(y))\,. \tag{6}$$

As Lemma 5.15 yields $T_{\varrho}^{*} \overset{k}{\underset{0}{\vdash}} \forall y\, (B(y) \leftrightarrow B(y))$ for some $k < \omega$, cutting with (6) yields $T_{\varrho}^{*} \overset{\Omega\cdot 2+1}{\underset{\omega}{\vdash}} \exists X\, \forall y\, (y \in X \leftrightarrow B(x))$. □

**Corollary 5.22** *For every arithmetic relation* $\prec$ *(parameters allowed) and arbitrary formula* $A(a)$ *we have* $T_{\varrho}^{*} \overset{\Omega\cdot 2+\omega}{\underset{0}{\vdash}} \forall\vec{X}\, \forall\vec{x}(\mathrm{WF}(\prec) \to \mathrm{TI}(\prec, A))$ *where the quantifiers* $\forall\vec{X}\, \forall\vec{x}$ *bind all free variables in* $\mathrm{WF}(\prec) \to \mathrm{TI}(\prec, A)$.

*Proof* By Lemma 5.20 we have $T_{\varrho}^{*} \overset{\Omega\cdot 2}{\underset{0}{\vdash}} \neg(\mathrm{WF}(\prec))', (\mathrm{TI}(\prec, A))'$ where $'$ denotes any assignment of free numerical variables to numerals. Hence

$$T_{\varrho}^{*} \overset{\Omega\cdot 2+2}{\underset{0}{\vdash}} (\mathrm{WF}(\prec) \to \mathrm{TI}(\prec, A))'$$

by two applications of $(\vee)$. Applying the $\omega$-rule the right number of times followed by the right number of $(\forall_2)$ inferences, one arrives at the desired conclusion. □

## 5.3 The Reduction Procedure for $T_{\varrho}^{*}$

Below we follow [13, Section 7].

**Lemma 5.23** *Let* $C$ *be a formula of grade* $\varrho$. *Suppose* $C$ *is a prime formula or of either form* $\exists X H(X)$, $\exists x G(x)$ *or* $A \vee B$. *Let* $\alpha = \omega^{\alpha_1} + \cdots + \omega^{\alpha_k}$ *with* $\delta \le \omega^{\alpha_k} \le \cdots \le \omega^{\alpha_1}$. *Then we have* $T_{\varrho}^{*} \overset{\alpha}{\underset{\varrho}{\vdash}} \Delta, \neg C \;\&\; T_{\varrho}^{*} \overset{\delta}{\underset{\varrho}{\vdash}} \Gamma, C \implies T_{\varrho}^{*} \overset{\alpha+\delta}{\underset{\varrho}{\vdash}} \Delta, \Gamma$.

*Proof* We proceed by induction on $\delta$.

1. Let $\Gamma, C$ be an axiom. Then there are three cases to consider.

1.1. $\Gamma$ is an axiom. Then so is $\Delta, \Gamma$. Hence $T_{\varrho}^{*} \overset{\alpha+\delta}{\underset{\varrho}{\vdash}} \Delta, \Gamma$.

1.2. $C$ is a true constant prime formula or negated prime formula. A straightforward induction on $\alpha$ then yields $T_{\varrho}^{*} \overset{\alpha}{\underset{\varrho}{\vdash}} \Delta$, and thus $T_{\varrho}^{*} \overset{\alpha+\delta}{\underset{\varrho}{\vdash}} \Delta, \Gamma$ by Lemma 5.14 (1).

1.3. $C \equiv A(s_1, \ldots, s_n)$ and $\Gamma$ contains a formula $\neg A(t_1, \ldots, t_n)$ where $s_i$ and $t_i$ are equivalent terms. From $T_{\varrho}^{*} \overset{\alpha}{\underset{\varrho}{\vdash}} \Delta, \neg A(s_1, \ldots, s_n)$ one receives $T_{\varrho}^{*} \overset{\alpha}{\underset{\varrho}{\vdash}} \Delta, \neg A(t_1, \ldots, t_n)$ by use of Lemma 5.14 (4). Thence $T_{\varrho}^{*} \overset{\alpha+\delta}{\underset{\varrho}{\vdash}} \Delta, \Gamma$ follows by use of Lemma 5.14 (1), since $\neg A(t_1, \ldots, t_n) \in \Gamma$.

2. Suppose $C \equiv A \vee B$ and $T^*_Q \overset{\delta_0}{\underset{Q}{\vdash}} \Gamma, C, A_0$ with $\mathrm{A}_0 \in \{A, B\}$ and $\delta_0 \lhd \delta$. Inductively we get

$$T^*_Q \overset{\alpha+\delta_0}{\underset{Q}{\vdash}} \Delta, \Gamma, A_0 \,. \tag{7}$$

Next use Lemma 5.14 (2) on $T^*_Q \overset{\alpha}{\underset{Q}{\vdash}} \Delta, \neg A \wedge \neg B$ to obtain

$$T^*_Q \overset{\alpha+\delta_0}{\underset{Q}{\vdash}} \Delta, \Gamma, \neg A_0 \,. \tag{8}$$

Whence use a cut on (7) and (8) to get the assertion.

3. Suppose $C \equiv \exists x G(x)$ and $T^*_Q \overset{\delta_0}{\underset{Q}{\vdash}} \Gamma, C, G(t)$ with $\delta_0 \lhd \delta$. Inductively we get

$$T^*_Q \overset{\alpha+\delta_0}{\underset{Q}{\vdash}} \Delta, \Gamma, G(t) \,. \tag{9}$$

By Lemma 5.14 (1) and (5), we also get

$$T^*_Q \overset{\alpha+\delta_0}{\underset{Q}{\vdash}} \Delta, \Gamma, \neg G(t) \,; \tag{10}$$

thus (9) and (10) yield $T^*_Q \overset{\alpha+\delta}{\underset{Q}{\vdash}} \Delta, \Gamma$ by $(Cut)$.

4. Suppose the last inference was $(\exists_2)$ with principal formula $C$. Then $C \equiv \exists X H(X)$ and $T^*_Q \overset{\delta_0}{\underset{Q}{\vdash}} \Gamma, C, H(U)$ for some $\delta_0 \lhd \delta$ and $gr(H(U)) \geq \omega$. Inductively we get

$$T^*_Q \overset{\alpha+\delta_0}{\underset{Q}{\vdash}} \Delta, \Gamma, H(U). \tag{11}$$

By Lemma 5.14 (1) and (6) we also get

$$T^*_Q \overset{\alpha+\delta_0}{\underset{Q}{\vdash}} \Delta, \Gamma, \neg H(U). \tag{12}$$

From (11) and (12) we obtain

$$T^*_Q \overset{\alpha+\delta}{\underset{Q}{\vdash}} \Delta, \Gamma.$$

5. Let $T^*_Q \overset{\delta}{\underset{Q}{\vdash}} \Gamma, C$ be derived by the $\Omega$-rule with fundamental function $f$. Then the assertion follows from the I. H. by the $\Omega$-rule using the fundamental function $\alpha + f$.
6. In the remaining cases the assertion follows from the I. H. used on the premises and by reapplying the same inference. □

**Lemma 5.24** $T^*_Q \overset{\alpha}{\underset{\eta+1}{\vdash}} \Gamma \implies T^*_Q \overset{\omega^\alpha}{\underset{\eta}{\vdash}} \Gamma.$

*Proof* We proceed by induction on $\alpha$. We only treat the crucial case when $T^*_Q \overset{\alpha_0}{\underset{\eta+1}{\vdash}} \Gamma, D$ and $T^*_Q \overset{\alpha_0}{\underset{\eta+1}{\vdash}} \Gamma, \neg D$, where $\alpha_0 \lhd \alpha$, and $gr(D) = \eta$. Inductively this becomes $T^*_Q \overset{\omega^{\alpha_0}}{\underset{\eta}{\vdash}} \Gamma, D$ and $T^*_Q \overset{\omega^{\alpha_0}}{\underset{\eta}{\vdash}} \Gamma, \neg D$. Since $D$ or $\neg D$ must be one of the forms exhibited in Lemma 5.23, we obtain $T^*_Q \overset{\omega^{\alpha_0}+\omega^{\alpha_0}}{\underset{\eta}{\vdash}} \Gamma$ by Lemma 5.23. As $\omega^{\alpha_0} + \omega^{\alpha_0} \lhd \omega^\alpha$, we can use Lemma 5.14 (1) to get the assertion. □

**Theorem 5.25 (Collapsing Theorem)** *Let* $\Gamma$ *be a set of weak formulas. We have*

$$T^*_Q \overset{\alpha}{\underset{\omega}{\vdash}} \Gamma \implies T^*_Q \overset{\vartheta\alpha}{\underset{0}{\vdash}} \Gamma.$$

*Proof* We proceed by induction on $\alpha$. Observe that for $\beta < \delta < \Omega$, we always have $\beta \lhd \delta$.

1. If $\Gamma$ is an axiom, then the assertion is trivial.
2. Let $T^*_Q \overset{\alpha}{\underset{\omega}{\vdash}} \Gamma$ be the result of an inference other than (*Cut*) and $\Omega$-rule. Then we have $T^*_Q \overset{\alpha_0}{\underset{\omega}{\vdash}} \Gamma_i$ with $\alpha_0 \lhd \alpha$ and $\Gamma_i$ being the $i$-th premiss of that inference. $\alpha_0 \lhd \alpha$ implies $\vartheta\alpha_0 \lhd \vartheta\alpha$. Therefore $T^*_Q \overset{\vartheta\alpha_0}{\underset{0}{\vdash}} \Gamma_0$ by the I. H., hence $T^*_Q \overset{\vartheta\alpha}{\underset{0}{\vdash}} \Gamma$ by reapplying the same inference.
3. Suppose $T^*_Q \overset{\alpha}{\underset{\omega}{\vdash}} \Gamma$ results by the $\Omega$-rule with respect to a $\Pi^1_1$-formula $\forall X F(X)$ and a fundamental function $f$. Then $f(\Omega) \unlhd \alpha$ and

$$T^*_Q \overset{f(0)}{\underset{\omega}{\vdash}} \Gamma, \forall X F(X), \tag{13}$$

and, for every set of weak formulas $\Xi$ and $\beta < \Omega$,

$$T^*_Q \overset{\beta}{\underset{0}{\vdash}} \Xi, \forall X F(X) \implies T^*_Q \overset{f(\beta)}{\underset{\omega}{\vdash}} \Xi, \Gamma. \tag{14}$$

The I. H. used on (13) supplies us with $T^*_Q \overset{\vartheta(f(0))}{\underset{0}{\vdash}} \Gamma, \forall X F(X)$. Hence with $\Xi = \Gamma$ we get

$$T^*_Q \overset{f(\vartheta(f(0)))}{\underset{\omega}{\vdash}} \Gamma \tag{15}$$

from (14). Now Lemma 5.8 ensures that $f(\beta) \lhd f(\Omega)$, where $\beta = \vartheta(f(0))$. So using the I. H. on (15), we obtain

$$T^*_Q \overset{\vartheta(f(\beta))}{\underset{0}{\vdash}} \Gamma, \tag{16}$$

thus $T^*_Q \overset{\vartheta\alpha}{\underset{0}{\vdash}} \Gamma$ as $f(\beta) \lhd \alpha$.

4. Suppose $T^*_Q \overset{\alpha_0}{\underset{\omega}{\vdash}} \Gamma, A$ and $T^*_Q \overset{\alpha_0}{\underset{\omega}{\vdash}} \Gamma, \neg A$, where $\alpha_0 \lhd \alpha$ and $gr(A) < \omega$. Inductively we then get $T^*_Q \overset{\vartheta\alpha_0}{\underset{0}{\vdash}} \Gamma, A$ and $T^*_Q \overset{\vartheta\alpha_0}{\underset{0}{\vdash}} \Gamma, \neg A$. Let $gr(A) = n - 1$. Then (Cut) yields

$$T^*_Q \overset{\beta_1}{\underset{n}{\vdash}} \Gamma \tag{17}$$

with $\beta_1 = (\vartheta\alpha_0) + 1$. Applying Lemma 5.24, we get $T^*_Q \overset{\omega^{\beta_1}}{\underset{n-1}{\vdash}} \Gamma$, and by repeating this process we arrive at

$$T^*_Q \overset{\beta_n}{\underset{0}{\vdash}} \Gamma\,,$$

where $\beta_{k+1} := \omega^{\beta_k}$ $(1 \le k < n)$. Since $\vartheta\alpha_0 < \vartheta\alpha$, we have $\beta_n < \vartheta\alpha$; thus, $T^*_Q \overset{\vartheta\alpha}{\underset{0}{\vdash}} \Gamma$. □

## 5.4 Embedding $\mathcal{D}_Q$ into $T^*_Q$

Assuming that $\mathcal{D}_Q$ is well-founded tree, the objective of this section is to embed $\mathcal{D}_Q$ into $T^*_Q$, so as to obtain a contradiction. Let $\mathfrak{X}$ be the Kleene–Brouwer ordering of $\mathcal{D}_Q$. We write $\mathcal{D}_Q \overset{\tau}{\vdash} \Gamma$ if $\Gamma$ is the sequent attached to the node $\tau$ in $\mathcal{D}_Q$.

**Theorem 5.26** $\mathcal{D}_Q \overset{\tau}{\vdash} \Xi \Rightarrow \exists k < \omega\; T^*_Q \overset{\mathfrak{E}_\tau + k}{\underset{\omega}{\vdash}} \Xi$.

*Proof* We proceed by induction on $\tau$, i.e., the Kleene–Brouwer ordering of $\mathcal{D}_Q$.

Suppose $\tau$ is an end-node of $\mathcal{D}_Q$. Then $\Xi$ must be axiomatic and therefore is an axiom of $T^*_Q$, and hence $T^*_Q \overset{\mathfrak{E}_\tau}{\underset{\omega}{\vdash}} \Xi$.

Now assume that $\tau$ is not an end-node of $\mathcal{D}_Q$. Then $\Xi$ is not axiomatic.

If $\Xi$ is not reducible, then there is a node $\tau_0$ immediately above $\tau$ in $\mathcal{D}_Q$ such that $\mathcal{D}_Q \overset{\tau_0}{\vdash} \Xi, \neg\bar{Q}(i), \neg A_i$ for some $i$. Inductively we have

$$T^*_Q \overset{\mathfrak{E}_{\tau_0} + k_0}{\underset{\omega}{\vdash}} \Xi, \neg\bar{Q}(i), \neg A_i$$

for some $k_0 < \omega$. We also have $T^*_Q \overset{0}{\underset{0}{\vdash}} \bar{Q}(i)$ and, using Corollaries 5.21 (if $i = 0$) and 5.22 (if $i > 0$), $T^*_Q \overset{\Omega\cdot 2+\omega}{\underset{\omega}{\vdash}} A_i$. Thus, noting that $\Omega \cdot 2 + \omega \lhd \mathfrak{E}_{\tau_0} + k_0$, and by employing two cuts we arrive at

$$T^*_Q \overset{\mathfrak{E}_{\tau_0} + k_0 + 2}{\underset{\omega+n}{\vdash}} \Xi$$

for some $n < \omega$. By Lemma 5.24 we get $T_Q^* \mid\frac{\omega_n(\mathfrak{E}_{\tau_0}+k_0+2)}{\omega} \Xi$ , and hence $T_Q^* \mid\frac{\mathfrak{E}_\tau}{\omega} \Xi$ since $\omega_n(\mathfrak{E}_{\tau_0} + k_0 + 2) \lhd \mathfrak{E}_\tau$.

Now suppose that $\Xi$ is reducible. $\Xi$ will be of the form

$$\Xi', E, \Xi''$$

where $E$ is not a literal and $\Xi'$ contains only literals.

First assume $E$ to be of the form $\forall x\, F(x)$. Then, for each $m$, there is a node $\tau_m$ immediately above $\tau$ in $\mathcal{D}_Q$ such that

$$\mathcal{D}_Q \models^{\tau_n} \Xi', F(\bar{m}), \Xi'', \neg\bar{Q}(i), \neg A_i$$

for some $i$. Inductively we have

$$T_Q^* \mid\frac{\mathfrak{E}_{\tau m}+k_m}{\omega} \Xi', F(\bar{m}), \Xi'', \neg\bar{Q}(i), \neg A_i$$

for all $m$, where $k_m < \omega$. We also have $T_Q^* \mid\frac{0}{0} \bar{Q}(i)$ and, using Lemma 5.22, $T_Q^* \mid\frac{\Omega\cdot 2+\omega}{0} A_i$ . Thus, noting that $\Omega \cdot 2 + \omega \lhd \mathfrak{E}_{\tau_m} + k_m$, and by employing two cuts there is an $n$ such that

$$T_Q^* \mid\frac{\mathfrak{E}_{\tau m}+k_m+2}{\omega+n} \Xi', F(\bar{m}), \Xi''$$

holds for all $m$. By Lemma 5.24 we get

$$T_Q^* \mid\frac{\omega_n(\mathfrak{E}_{\tau m}+k_m+2)}{\omega} \Xi', F(\bar{m}), \Xi''$$

for all $m$. Whence

$$T_Q^* \mid\frac{\mathfrak{E}_\tau}{\omega} \Xi', F(\bar{m}), \Xi''$$

since $\omega_n(\mathfrak{E}_{\tau_m} + k_m + 2) \lhd \mathfrak{E}_\tau$. A final application of the $\omega$-rule yields

$$T_Q^* \mid\frac{\mathfrak{E}_\tau+1}{\omega} \Xi', \forall x\, F(x), F(\bar{m}), \Xi''$$

i.e., $T_Q^* \mid\frac{\mathfrak{E}_\tau+1}{\omega} \Xi$ .

If $E$ is a redex of another type but not of the form $\exists X B(X)$ with $B(U)$ arithmetic, then one proceeds in a similar way as in the previous case.

Now assume $E$ to be of the form $\exists X\; B(X)$ with $B(U)$ arithmetic. Then there is a node $\tau_0$ immediately above $\tau$ in $\mathcal{D}_Q$ such that

$$\mathcal{D}_Q \models^{\tau_0} \Xi', B(U), \Xi'', \neg\bar{Q}(i), \neg A_i$$

for some $i$ and set variable $U$. Inductively we have

$$T^*_Q \;\Big|\!\frac{\mathfrak{E}_{\tau_0}+k_0}{\omega}\; \Xi', B(U), \Xi'', \neg\bar{Q}(i), \neg A_i$$

for some $k_0 < \omega$. We also have $T^*_Q \;\Big|\!\frac{0}{0}\; \bar{Q}(i)$ and, using Lemma 5.22, $T^*_Q \;\Big|\!\frac{\Omega\cdot 2+\omega}{0}\; A_i$. Thus, noting that $\Omega \cdot 2 + \omega \lhd \mathfrak{E}_{\tau_0} + k_0$, and by employing two cuts there is an $n$ such that

$$T^*_Q \;\Big|\!\frac{\mathfrak{E}_{\tau_0}+k_0+2}{\omega+n}\; \Xi', B(U), \Xi''.$$

By Lemma 5.24 we get

$$T^*_Q \;\Big|\!\frac{\omega_n(\mathfrak{E}_{\tau_0}+k_0+2)}{\omega}\; \Xi', B(U), \Xi''. \tag{18}$$

Lemma 5.20 yields

$$T^*_Q \;\Big|\!\frac{\Omega\cdot 2}{0}\; \exists X B(X), \neg B(U). \tag{19}$$

Cutting $B(U)$ and $\neg B(U)$ out of (18) and (19) we arrive at

$$T^*_Q \;\Big|\!\frac{\omega_n(\mathfrak{E}_{\tau_0}+k_0+2)+1}{\omega}\; \Xi', \exists X B(X), \Xi''.$$

Since $\omega_n(\mathfrak{E}_{\tau_0} + k_0 + 2) + 1 \lhd \mathfrak{E}_\tau$ we get $T^*_Q \;\Big|\!\frac{\mathfrak{E}_\tau}{\omega}\; \Xi', \exists X B(X), \Xi''$, i.e., $T^*_Q \;\Big|\!\frac{\mathfrak{E}_\tau}{\omega}\; \Xi$. □

Below $\emptyset$ stands for the empty sequent and $\tau_0$ denotes the bottom node of $\mathcal{D}_Q$ which is the maximum element of the pertaining Kleene–Brouwer ordering.

**Corollary 5.27** *If $\mathcal{D}_Q$ is well-founded, then $T^*_Q \;\Big|\!\frac{\vartheta(\omega_n(\mathfrak{E}_{\tau_0}+m))}{0}\; \emptyset$ for some $n, m < \omega$.*

*Proof* We have $\mathcal{D}_Q \;\Big|\!\frac{\tau_0}{}\; \neg\bar{Q}(0), \neg A_0$. Thus there is a $k < \omega$ such that

$$T^*_Q \;\Big|\!\frac{\mathfrak{E}_{\tau_0}+k}{\omega}\; \neg\bar{Q}(0), \neg A_0$$

holds by Theorem 5.26. We also have $T^*_Q \;\Big|\!\frac{0}{0}\; \bar{Q}(0)$ and, using Corollary 5.22, $T^*_Q \;\Big|\!\frac{\Omega\cdot 2+\omega}{0}\; A_0$. Thus, noting that $\Omega \cdot 2 + \omega \lhd \mathfrak{E}_{\tau_0} + k$, and by employing two cuts we arrive at

$$T^*_Q \;\Big|\!\frac{\mathfrak{E}_{\tau_0}+k+2}{\omega+n}\; \emptyset$$

for some $n < \omega$. Via Lemma 5.24 we deduce $T_Q^* \overset{\omega_n(\mathfrak{E}_{\tau_0}+k+2)}{\underset{\omega}{\vdash}} \emptyset$, so that by Theorem 5.25 we conclude $T_Q^* \overset{\vartheta(\omega_n(\mathfrak{E}_{\tau_0}+m))}{\underset{0}{\vdash}} \emptyset$ with $m = k + 2$. □

**Corollary 5.28** *$\mathcal{D}_Q$ is not well-founded.*

*Proof* If $\mathcal{D}_Q$ were well-founded, we would have

$$T_Q^* \overset{\vartheta(\omega_n(\mathfrak{E}_{\tau_0}+m))}{\underset{0}{\vdash}} \emptyset \tag{20}$$

for some $n, m < \omega$ by Corollary 5.27. But a straightforward induction on $\alpha < \Omega$ shows that

$$T_Q^* \overset{\alpha}{\underset{0}{\vdash}} \Gamma \;\Rightarrow\; \Gamma \neq \emptyset,$$

yielding that (20) is impossible. □

It remains to show that the result of Corollary 5.28 is provable in $\mathbf{ACA}_0$ from

$$\forall \mathfrak{X}\,(\mathrm{WO}(\mathfrak{X}) \to \mathrm{WO}(\vartheta_{\mathfrak{X}}))\,.$$

Let $\mathbf{S}$ be the theory $\mathbf{ACA}_0$ plus the latter axiom. The main issue is how to formalize the derivability predicate $T_Q^* \overset{\alpha}{\underset{\rho}{\vdash}} \Gamma$ in the background theory $\mathbf{S}$. We elaborated earlier in Remark 5.13 that this seems to require an iterated inductive definition, something apparently not available in $\mathbf{S}$. However, all we need is a fixed point not a proper inductive definition, i.e., to capture the notion of derivability in $T_Q^*$ without the $\Omega$-rule it suffices to find a predicate $\mathcal{D}$ of $\alpha, \rho, \Gamma$ such that

(∗) $\mathcal{D}(\alpha, \rho, \Gamma)$ if and only if $\alpha \in |\vartheta_{\mathfrak{X}}|$, $\rho \leq \omega + \omega$, $\Gamma$ is a sequent, and either $\Gamma$ contains an axiom of $T_Q^*$ or $\Gamma$ is the conclusion of an inference of $T_Q^*$ other than $(\Omega)$ with premisses $(\Gamma_i)_{i \in I}$ such that for every $i \in I$ there exists $\beta_i \lhd \alpha$ with $\mathcal{D}(\beta_i, \rho, \Gamma_i)$, and if the inference is a cut it has rank $< \rho$.

(∗) can be viewed as a fixed-point axiom which together with transfinite induction for $\vartheta_{\mathfrak{X}}$ defines $T_Q^*$-derivability (without $(\Omega)$-rule) implicitly.

How can we find a fixed point as described in (∗)? As it turns out, it follows from [12] that $\mathbf{S}$ proves that every set is contained in a countable coded $\omega$-model of the theory $\mathbf{ATR}_0$. It is also known that $\mathbf{ATR}_0$ proves the $\Sigma^1_1$ axiom of choice, $\Sigma^1_1$-$\mathbf{AC}$ (see [17, Theorem V.8.3]). Moreover, in $\mathbf{ACA}_0 + \Sigma^1_1$-$\mathbf{AC}$ one can prove for every $P$-positive arithmetical formula $A(u, P)$ that there is a $\Sigma^1_1$ formula $F(u)$ such that $\forall x[F(x) \leftrightarrow A(x, F)]$, where $A(x, F)$ arises from $A(x, P)$ by replacing every occurrence of the form $P(t)$ in the first formula by $F(t)$. This is known as the Second Recursion Theorem (see [2, V.2.3]). Arguing in $\mathbf{S}$, we find a countable coded $\omega$ model $\mathfrak{B}$ with $\mathfrak{X} \in \mathfrak{B}$ such that $\mathfrak{B}$ is a model of $\mathbf{ATR}$. As a result, there is a predicate $\mathcal{D}$ definable in $\mathfrak{B}$ that satisfies (∗). As a result, $\mathcal{D}$ is a set in $\mathbf{S}$. To obtain the full derivability relation $T_Q^* \overset{\alpha}{\underset{\rho}{\vdash}} \Gamma$ we have to take the $\Omega$-rule into account. We do this by taking a countable coded $\omega$-model $\mathfrak{C}$ of **ATR** that contains both $\mathfrak{X}$ and $\mathcal{D}$. We then define an appropriate fixed point predicate $\mathcal{D}_\Omega$ using the clauses for defining $T_\varrho^* \mathrel{\substack{\alpha \\ \vdash \\ \rho}} \Gamma$ and $\mathcal{D}$ for the negative occurrences in the $\Omega$-rule.

The upshot is that we can formalize all of this in **S**.

*Remark 5.29* When giving talks about the material of this article, the first author was asked what the proof-theoretic ordinal of the theories that Theorem 1.7 is concerned with might be. He conjectures that it is the ordinal

$$\vartheta(\varphi 2(\Omega + 1))$$

(or $\psi(\varphi 2(\Omega + 1))$ in the representation system based on the $\psi$-function; see [13, Section 3]), i.e. the collapse of the first fixed point of the epsilon function above $\Omega$.

**Acknowledgements** The first author acknowledges support by the EPSRC of the UK through grants EP/G029520/1 and EP/G058024/1.

The authors would also like to thank an anonymous referee for very helpful comments and numerous suggestions. We also thank Anton Setzer for comments on a draft version of this paper.

The results of this article were incorporated in the Ph.D. thesis [18] of the second author.

## References

1. B. Afshari, M. Rathjen, Reverse mathematics and well-ordering principles: A pilot study. Ann. Pure Appl. Logic **160**, 231–237 (2009)
2. J. Barwise, *Admissible Sets and Structures* (Springer, Berlin, 1975)
3. W. Buchholz, K. Schütte, *Proof Theory of Impredicative Subsystems of Analysis* (Bibliopolis, Napoli, 1988)
4. H. Friedman, A. Montalban, A. Weiermann, *Phi Function*, draft (2007)
5. G. Gentzen, in *The Collected Papers of Gerhard Gentzen*. Translated and ed. by M.E. Szabo (North-Holland, Amsterdam, 1969)
6. G. Gentzen, Der erste Widerspruchsfreiheitsbeweis für die klassische Zahlentheorie. Arch. Math. Logik Grundlag. **16**, 97–118 (1974)
7. J.-Y. Girard, *Proof Theory and Logical Complexity*, vol. 1 (Bibliopolis, Napoli, 1987)
8. J.L. Hirst, Reverse mathematics and ordinal exponentiation. Ann. Pure Appl. Logic **66**, 1–18 (1994)
9. A. Marcone, A. Montalbán, The Veblen functions for computability theorists. J. Symb. Logic **76**, 575–602 (2011)
10. M. Rathjen, *Lecture notes: Selected Topics in Proof Theory* (Münster University, Münster, 1989)
11. M. Rathjen, in *Fragments of Kripke-Platek set theory with infinity*. ed. by P. Aczel, J. Simmons, S. Wainer. Proof Theory (Cambridge University Press, Cambridge, 1992), pp. 251–273
12. M. Rathjen, $\omega$-models and well-ordering principles, in ed. by N. Tennant. *Foundational Adventures: Essays in Honor of Harvey M. Friedman* (College Publications, London, 2014), pp. 179–212
13. M. Rathjen, A. Weiermann, Proof–theoretic investigations on Kruskal's theorem. Ann. Pure Appl. Logic **60**, 49–88 (1993)

14. M. Rathjen, A. Weiermann, Reverse mathematics and well-ordering principles, in ed. by S. Cooper, A. Sorbi *Computability in Context: Computation and Logic in the Real World* (Imperial College Press, London, 2011), pp. 351–370
15. K. Schütte, Beweistheoretische Erfassung der unendlichen Induktion in der Zahlentheorie. Math. Ann. **122**, 369–389 (1951)
16. K. Schütte, *Proof Theory* (Springer, Berlin, 1977)
17. S.G. Simpson, *Subsystems of Second Order Arithmetic*, 2nd edn (Cambridge University Press, Cambridge, 2009)
18. P.F. Valencia Vizcaíno, Some uses of cut elimination. PhD Thesis, University of Leeds, 2013

Made in the USA
Monee, IL
13 September 2026

62975068R00315